The Pomegranate

Botany, Production and Uses

The Pomegranate

Botany, Production and Uses

Edited by

Ali Sarkhosh

Department of Horticultural Sciences, University of Florida, Gainesville, Florida, USA

Alimohammad M. Yavari

Pomologist, Private Consultant, Tehran, Iran

Zabihollah Zamani

Department of Horticultural Sciences, University of Tehran, Karaj, Iran

CABI

CABI is a trading name of CAB International

CABI
Nosworthy Way
Wallingford
Oxfordshire OX10 8DE
UK

Tel: +44 (0)1491 832111
Fax: +44 (0)1491 833508
E-mail: info@cabi.org
Website: www.cabi.org

CABI
WeWork
One Lincoln St
24th Floor
Boston, MA 02111
USA

T: +1 (617)682-9015
E-mail: cabi-nao@cabi.org

A catalogue record for this book is available from the British Library, London, UK.

Library of Congress Cataloging-in-Publication Data
Names: Sarkhosh, Ali, editor. | Yavari, Alimohammad, editor. | Zamani, Zabihollah, editor.
Title: The pomegranate : botany, production and uses / edited by Ali Sarkhosh, Alimohammad Yavari, Zabihollah Zamani.
Description: Boston, MA : CAB International, [2021] | Includes bibliographical references and index. | Summary: "Pomegranate cultivation is increasing considerably worldwide for fresh consumption, juice production and medicinal purposes. This is the first comprehensive book by an international team of experts at the fore front of research and provides insights and experiences covering botany, production, processing, health and industrial uses of the fruit"-- Provided by publisher.
Identifiers: LCCN 2020018088 (print) | LCCN 2020018089 (ebook) | ISBN 9781789240764 (hardback) | ISBN 9781789240771 (ebook) | ISBN 9781789240788 (epub)
Subjects: LCSH: Pomegranate.
Classification: LCC QK495.P958 P663 2020 (print) | LCC QK495.P958 (ebook) | DDC 583/.73--dc23
LC record available at https://lccn.loc.gov/2020018088
LC ebook record available at https://lccn.loc.gov/2020018089
References to Internet websites (URLs) were accurate at the time of writing.

ISBN-13: 9781789240764 (hardback)
9781789240771 (ePDF)
9781789240788 (ePub)

Commissioning Editor: Rachael Russell / Rebecca Stubbs
Editorial Assistant: Emma McCann
Production Editor: Marta Patiño

Typeset by Exeter Premedia Services Pvt Ltd, Chennai, India
Printed and bound in the UK by Severn, Gloucester

Contents

Contributors

AbdEl-Fattah Awd, Nahla, Horticultural Research Institute, Agricultural Research Center, Egypt.

Agehara, Shinsuke, Gulf Coast Research and Education Center, University of Florida, Wimauma, FL 33598, USA.

Ahmadpourmir, Hamid, Department of Horticultural Science, Faculty of Agriculture, Ferdowsi University of Mashhad, Mashhad, Iran.

Alizadeh, Mehdi, Horticulture Department, Faculty of Plant Production, Gorgan University of Agricultural Sciences and Natural Resources, Gorgan, Iran.

Asgari, Mahdi, Freelance Agricultural Economist, Tucson, AZ 85719, USA.

Asgary, Sedigheh, Isfahan Cardiovascular Research Center, Isfahan Cardiovascular Research Institute, Isfahan University of Medical Sciences, Isfahan, Iran.

Ayars, James, United States Department of Agriculture (USDA), Agricultural Research Service (ARS), San Joaquin Valley Agricultural Sciences Center, Water Management Research Unit, Parlier, CA, USA.

Azadeh, Zohreh, Medical Plants Research Center, Basic Health Sciences Institute, Shahrekord University of Medical Sciences, Shahrekord, Iran.

Azarmi-Atajan, Farhad, Department of Soil Science and Engineering, College of Agriculture, University of Birjand, Iran.

Babu, Dhinesh, ICAR-National Research Centre on Pomegranate, Solapur-413 255, Maharashtra, India.

Bartual, Julián, Conselleria de Agricultura, Desarrollo Rural, Emergencia Climática y Transición Ecológica. Estació Experimental Agrària d'Elx. 03290 Elx, Alacant, Spain.

Cano-Lamadrid, Maria, Universidad Miguel Hernández, Escuela Politécnica Superior de Orihuela, Departamento Tecnología Agroalimentaria, Grupo Calidad y Seguridad Alimentaria, Ctra. Beniel, km 3.2, 03312-Orihuela, Alicante, Spain.

Carbonell-Barrachina, Ángle A., Universidad Miguel Hernández, Escuela Politécnica Superior de Orihuela, Departamento Tecnología Agroalimentaria, Grupo Calidad y Seguridad Alimentaria, Ctra. Beniel, km 3.2, 03312-Orihuela, Alicante, Spain.

Castillo, Salvador, Department of Food Technology, EPSO, University Miguel Hernández, Ctra. Beniel km 3.2, 03312 Orihuela, Alicante, Spain.

Chater, John M., Department of Botany and Plant Sciences, 900 University Avenue, University of California, Riverside, CA 92521, USA.

Cossio, Ferdinando, Fruit Crops Breeder, Private consultant, Via Scrimiari 59, 37129 Verona, Italy.

Djamali, Morteza, Institut Méditerranéen de Biodiversité et d'Ecologie (IMBE) - UMR CNRS 7263/ IRD 237/Aix-Marseille Université/Avignon Université, Technopôle de l'Environnement Arbois-Méditerranée, Pavillon Villemin BP80, France.

Dodman, Sadegh, Medical Plants Research Center, Basic Health Sciences Institute, Shahrekord University of Medical Sciences, Shahrekord, Iran.

Farzaei, Mohammad Hosein, Pharmaceutical Sciences Research Center, Kermanshah University of Medical Sciences, Kermanshah, Iran.

Ferrara, Giuseppe, Department of Soil, Plant and Food Sciences, University of Bari "Aldo Moro", via G. Amendola 165/A, 70126, Bari, Italy.

Gaikwad, Nilesh N., ICAR-National Research Centre on Pomegranate, NH-65, Solapur-Pune Highway, Kegaon, Solapur-413255, Maharashtra, India.

García-Pastor María E., Department of Food Technology, EPSO, University Miguel Hernández, Ctra. Beniel km 3.2, 03312 Orihuela, Alicante, Spain.

Geetha, G.A., Division of Plant Physiology and Biochemistry, ICAR-Indian Institute of Horticultural Research, Hessaraghatta Lake Post, Bengaluru – 560089, Karnataka, India.

Giancaspro, Angelica, Department of Environmental and Territorial Sciences, University of Bari "Aldo Moro", via G. Amendola 165/A, 70126, Bari, Italy.

Goldansaz, Seyed Hossein, Department of Plant Protection, Faculty of Agriculture, University of Tehran, Karaj 31587, Iran.

Guillén, Fabián, Department of Food Technology, EPSO, University Miguel Hernández, Ctra. Beniel km 3.2, 03312 Orihuela, Alicante, Spain.

Harsur, Mallikarjun M., ICAR-National Research Centre on Pomegranate, Solapur-413 255, Maharashtra, India.

Hassani Mohammad, Department of Horticultural Science, Gorgan University of Agricultural Sciences and Natural Resources, Gorgan, Iran.

Hoseinzadeh-chahkandak, Fatemeh, Social Determinants of Health Research Center, Birjand University of Medical Sciences, Birjand, Iran.

Hou, Lefeng, Research Center of Pomegranate in Zaozhuang, Zaozhuang, China.

Hozeifi, Soroush, Faculty of Medicine, Birjand University of Medical Sciences, Birjand, Iran.

Intrigliolo, Diego S., Spanish National Research Council (CSIC), Center for Applied Biology and Soil Sciences (CEBAS), Campus Universitario Espinardo ed. 25, 30100, Murcia, Spain.

Jasemi, Eghbal, Medicinal Plants Research Center, Institute of Medicinal Plants, ACECR, Karaj, Iran.

Jia, Zhenyu, Department of Botany and Plant Sciences, 900 University Avenue, University of California, Riverside, CA 92521, USA.

Karimi, Hamid R., Department of Horticultural Science, Faculty of Agriculture, Vali-e-Asr University of Rafsanjan, Iran.

Keshvari, Mahtab, Isfahan Cardiovascular Research Center, Isfahan Cardiovascular Research Institute, Isfahan University of Medical Sciences, Isfahan, Iran.

Khayyat, Mehdi, Department of Horticultural Science, College of Agriculture, University of Birjand, Iran.

Li, Jiyu, Institute of Horticultural Research, 40 Nongkenan Road, Anhui Academy of Agricultural Sciences, Hefei 230031, China.

Liu, Chunyan, Institute of Horticultural Research, 40 Nongkenan Road, Anhui Academy of Agricultural Sciences, Hefei 230031, China.

Lorigooini, Zahra, Medical Plants Research Center, Basic Health Sciences Institute, Shahrekord University of Medical Sciences, Shahrekord, Iran.

Maity, Ashish, ICAR-National Research Centre on Pomegranate, Kegaon, Solapur-413255, Maharashtra, India.

Manjunath, Girigowda, University of Horticultural Sciences, Bagalkot, Karnataka, India.

Martínez-Romero, Domingo, Department of Food Technology, EPSO, University Miguel Hernández, Ctra. Beniel km 3.2, 03312 Orihuela, Alicante, Spain.

Massimino Cocuzza, Giuseppe E., Dipartimento di Agricoltura, Alimentazione e Ambiente, Università di Catania, Italy.

Mazzeo, Andrea, Department of Soil, Plant and Food Sciences, University of Bari "Aldo Moro", via G. Amendola 165/A, 70126, Bari, Italy.

Medina Jorge, Department of Food Technology, EPSO, University Miguel Hernández, Ctra. Beniel km 3.2, 03312 Orihuela, Alicante, Spain.

Merhaut, Donald J., Department of Botany and Plant Sciences, 900 University Avenue, University of California, Riverside, CA 92521, USA.

Mirjalili, Mohammad H., Department of Agriculture, Medicinal Plants and Drugs Research Institute, Shahid Beheshti University, G.C. Evin, Tehran, Iran.

Moersfelder, Jeff, National Clonal Germplasm Repository, USDA-ARS, One Shields Avenue, University of California, Davis, CA 95616-8607, USA.

Momtaz, Saeideh, Medicinal Plants Research Center, Institute of Medicinal Plants, ACECR, Karaj, Iran and Toxicology and Diseases Group, the Institute of Pharmaceutical Sciences (TIPS), Tehran University of Medical Sciences, Tehran, Iran.

Mustafayeva, Ziyafet, Institute of Genetics Resources NANA Az1106, Baku city, avenue Azadlig 155 Azerbaijan.

Narzary, Diganta, Department of Botany, Gauhati University, Guwahati 781 014, Assam, India.

Palasciano, Marino, Department of Soil, Plant and Food Sciences, University of Bari "Aldo Moro", via G. Amendola 165/A, 70126, Bari, Italy.

Palou, Lluís, Institut Valencià d'Investigacions Agràries (IVIA), Centre de Tecnologia Postcollita (CTP), 46113 Montcada, Valencia, Spain.

Parashuram Shilpa, ICAR-National Research Centre on Pomegranate, Solapur-413 255, Maharashtra, India.

Patil, Nita B., ICAR-National Research Centre on Pomegranate, NH-65, Solapur-Pune Highway, Kegaon, Solapur-413255, Maharashtra, India.

Patil, Prakash G., ICAR-National Research Centre on Pomegranate, NH-65, Solapur-Pune Highway, Kegaon, Solapur-413255, Maharashtra, India.

Pérez-Gago, María B., Institut Valencià d'Investigacions Agràries (IVIA), Centre de Tecnologia Postcollita (CTP), 46113 Montcada, Valencia, Spain.

Pourbagher-Shahri, Ali Mohammad, Faculty of Medicine, Birjand University of Medical Sciences, Birjand, Iran.

Preece, John E., National Clonal Germplasm Repository, USDA-ARS, One Shields Avenue, University of California, Davis, CA 95616-8607, USA.

Puerto, Herminia, Miguel Hernández University of Elche. Department of Engineering, Ctra. de Beniel km 3.2, 03312 Desamparados, Alicante, Spain.

Qin, Gaihua, Institute of Horticultural Research, 40 Nongkenan Road, Anhui Academy of Agricultural Sciences, Hefei 230031, China.

Rahimi, Roja, Department of Traditional Pharmacy, School of Persian Medicine, Tehran University of Medical Sciences, Tehran, Iran.

Rana, Tikam S., Plant Diversity Systematics & Herbarium Division, CSIR-National Botanical Research Institute, Lucknow, India.

Rezaei, Mehdi, Horticultural Sciences Department, Faculty of Agriculture, Shahrood University of Technology, Shahrood 3619995161, Iran.

Saeedi, Muhammaad A., Citrus Industry Development Manager, National Horticulture Development Organization, Afghanistan.

Saeidi Ghavi, Sara, Institute of Ecology and Nature Conservation Biology, Department of Biology and Preclinical Science, University of Regensburg, Germany.

Sargent, Steven A., Horticultural Sciences Department, University of Florida, Gainesville FL 32611, USA.

Sarkhosh, Ali, Horticultural Sciences Department, University of Florida, Gainesville FL 32611, USA.

Sedaghat, Mahshid, Faculty of Medicine, Birjand University of Medical Sciences, Birjand, Iran.

Selahvarzi, Yahya, Department of Horticultural Science, Faculty of Agriculture, Ferdowsi University of Mashhad, Mashhad, Iran.

Serrano, María, Department of Applied Biology, EPSO, University Miguel Hernández, Ctra. Beniel km 3.2, 03312 Orihuela, Alicante, Spain.

Sharma, Jyotsana, ICAR-National Research Centre on Pomegranate, Solapur-413 255, Maharashtra, India.

Shirazi, Rouhollah, University of Sistan and Baluchestan, Department of Archaeology, Zahedan, Iran & Iranian Center for Archaeological Research (ICAR), Research Center for Cultural Heritage and Tourism (RICHT).

Shirazi, Zohreh, Laboratory of Archaeobotany. World Heritage Site of Shahr-i Sokhta and Regional Museum of Southeast Iran. Culture Heritage, Handicrafts and Tourism of Sistan and Baluchestan, Iran.

Shivashankara, K.S., Division of Plant Physiology and Biochemistry, ICAR-Indian Institute of Horticultural Research, Hessaraghatta Lake Post, Bengaluru – 560089, Karnataka, India.

Shokoohi, Zeinab, Department of Agricultural Economics, School of Agriculture, Shiraz University, Iran.

Rana, Tikam, Plant Diversity Systematics & Herbarium Division, CSIR-National Botanical Research Institute, Lucknow, India.

Singh, Nripendra V., 1ICAR-National Research Centre on Pomegranate, NH-65, Solapur-Pune Highway, Kegaon, Solapur-413255, Maharashtra, India.

Soleymani, Asiye, Department of Chemical Engineering, University of Florida, Gainesville, FL 32603, USA.

Spagnoli, Federica, Italian Institute of Oriental Studies - ISO, Sapienza University of Rome - Archaeological Expedition to Motya, Italy.

Valero, Daniel, Department of Food Technology, EPSO, University Miguel Hernández, Ctra. Beniel km 3.2, 03312 Orihuela, Alicante, Spain.

Vallad, Gary E., Gulf Coast Research and Education Center, University of Florida, Wimauma, FL 33598, USA.

Varasteh, Feryal, Horticulture Department, Faculty of Plant Production, Gorgan University of Agricultural Sciences and Natural Resources, Gorgan, Iran.

Wang Dong, United States Department of Agriculture (USDA), Agricultural Research Service (ARS), San Joaquin Valley Agricultural Sciences Center, Water Management Research Unit, Parlier, CA, USA.

Xavier, KatiaV., Gulf Coast Research and Education Center, University of Florida, Wimauma, FL 33598, USA.

Yavari Alimohammad M., Pomologist, Private Consultant, Tehran, Iran.

Yazdanbakhsh, Nima, Department of Plant Sciences, School of Biology, College of Science, University of Tehran, Tehran, Iran.

Yilmaz, Cenap, Department of Horticulture, Eskisehir Osmangazi University, Eskisehir, Turkey.

Zamani, Zabihollah, Department of Horticulture Science, Faculty of Agriculture, University of Tehran, Karaj 31587, Iran.

Zapata, Pedro J., Department of Food Technology, EPSO, University Miguel Hernández, Ctra. Beniel km 3.2, 03312 Orihuela, Alicante, Spain.

Zarei, Abdolkarim, Department of Plant Production and Genetics (Biotechnology), Jahrom University, 74135-111, Jahrom, Iran.

About Editors

Ali Sarkhosh is currently serving as Assistant Professor and Extension Specialist (fruit crops) in the Horticultural Sciences Department at the University of Florida. Dr. Sarkhosh was born and raised in a farming family, who grow pomegranate and saffron for four generations in Iran. He did his Master thesis and Ph.D. dissertation in pomegranate genetic diversity and breeding under supervision of Professor Zamani (one of the editors in this book) at the University of Tehran in Iran. Dr. Sarkhosh worked as pomegranate expert in Australia for four years where he was responsible for all management aspects of 600 acres of pomegranate production. Toward improving pomegranate production, he published many sources of both technical and referred articles in pomegranate breeding and genetic diversity, postharvest, and production. Dr Sarkhosh has over 10 years of experience with commercial fruit production in U.S., Australia, New Zealand, and Iran. His research with tree fruits has focused on breeding, rootstock and cultivar evaluations, and assessment of cultural practices to optimize yield and production efficiency.

Alimohammad M. Yavari is currently serving as a private pomologist and consultant. Mr. Yavari received his M.S. in horticulture (Pomology) from the University of Urmia, Iran. He worked as research assistant at one of the biggest pomegranate collections and germplasms in the world located in Yazd (a city in central Iran). Mr. Yavari has excellent expertise in all aspects of pomegranate production as well as interacting worldwide with commercial growers and researchers in the field of fruit production, mainly tropical and subtropical fruits. His experiences in pomegranate production are mostly related to harvesting and exporting issues, and training courses for orchardists. He provides public and private sectors with designing projects and writing proposals related to pomegranate production and other tropical and subtropical fruits.

Zabihollah Zamani is currently serving as full Professor in the Horticultural Sciences Department, University of Tehran, Iran. He received his B.S. and M.S. degrees in horticulture science and pomology from the University of Tehran, Iran and his Ph.D. in fruit breeding and biotechnology from University of Sydney, Australia. With over 25 years experience, Dr. Zamani has contributed significant advances in all aspects of pomegranate production including, breeding, biotechnology, physiology, cultural practices, and postharvest. He played an important role in the development of his department in the area of fruit science and biotechnology. Toward improving pomegranate research and development, he published many refereed journal articles, technical reports and extension publications. Dr. Zamani supervised many M.S. and Ph.D. students who did their thesis on fruit crops, specifically on pomegranate. These achievements have been recognized with Dr. Zamani being invited to serve high-profile advisory boards for developing or leading applied research projects in fruit crops including pomegranate.

Preface

The pomegranate, *Punica granatum* L., belongs to the Lythraceae or Punicaceae family. It is one of the oldest known edible fruits and is associated with ancient civilizations of the Middle East.

In many cultures, pomegranate is prominent in numerous myths about different human lives and aspirations. Zoroastrians have planted this tree in their homes as a blessing. In Greek mythology, it was an irrevocable symbol of marriage. In Persian mythology, Esfandiar (a King in ancient Persia) ate a pomegranate and became invincible. In Judaism, the number of pomegranate seeds in a conduit is 613, one for each of the 613 Bible commands. Buddhists consider pomegranate to be one of the three blessed fruits. In Chinese ceramics, pomegranate is associated with fertility, abundance, countless and passionate children, and a blessed future. In the Christian and Bedouin tribes, it is associated with fertility. In Islam, the Qur'an describes a heavenly paradise that contains pomegranates.

From its origin (Persia) in the region now occupied by Iran, Afghanistan, Azerbaijan, etc., the pomegranate spread east to India, China, and also west to the Mediterranean countries like Spain, Morocco, Egypt, Tunisia, and Turkey. The ability of pomegranate trees to adjust to variable climatic conditions is reflected in the wide distribution of the wild forms throughout Eurasia to the Himalayas. It is now widely cultivated in subtropical and tropical areas in many variable climatic conditions in different countries, indicating its flexibility and adaptability to a wide range of climate and biogeography.

Today, in addition to being a fruit, the pomegranate's medicinal properties and application in the food industry have also received the attention of many researchers in various countries, encouraging extensive research. Pomegranate trees can grow naturally in a wide range of climates and adapt to various soil conditions. Trees are sensitive to soils with low drainage and have low growth in these conditions, which also reduces crop quality. The best soil conditions for pomegranate cultivation are deep sand-clay soils. The highest crop growth, yield, and quality can be achieved in areas with hot and long summers. One of the most critical limitations of pomegranate cultivation is its sensitivity to low freezing temperatures. Pomegranate trees can be damaged at temperatures below -10 °C.

There are innumerable amounts of pomegranate cultivars and many germplasm collections have been established in various countries. More than 500 cultivars of pomegranate have been named all over the world, which indicates the plant's genetic diversity. Breeders have accumulated germplasm, which provides a wealth of traits for creating new cultivars and opportunities for the production of pomegranate. The most interesting characteristics of the breeding programmes are bigger fruits, larger arils, soft seeds, higher juice yield, red coloured rind and arils, and higher yield.

The pomegranate and its usage are deeply embedded in human history, and utilization is found in many ancient human cultures as food and a medical remedy. These include its seeds, peels, flowers, and juice. They contain dietary fibre, antioxidants, healthy unsaturated fats, minerals, and other nutrients. Since ancient times, pomegranate has been frequently used as treatments for common ailments in the oldest cultures of the Indus Valley, ancient China, classical Greece, and the Middle East. The chemical composition and pharmacology of pomegranate constituents are of great interest to life scientists in the modern world. Due to the crop's extensive amount of benefits, it has become one of the most favourite fruit among people. Its cultivation has considerably increased worldwide for fresh consumption, juice production, and medicinal purposes, mainly due to the global trend of increased demand. Pomegranate production in the world has expanded more than tenfold over the past twenty years primarily due to the presence of phenolic compounds in pomegranate juice. Those compounds provide relatively high levels of antioxidant activity to human consumers as demonstrated in clinical trials. Much of the research on this fruit has focused on the medicinal and therapeutic properties of pomegranates. However, because of the increased production of this fruit in the past few decades and the expansion of its cultivation in different parts of the world, growers have started to face many problems. Most of these problems can be attributed to the growers' lack of familiarity with the crop's cultural practices.

Pomegranate requires climatic conditions that are conducive to the production of quality fruit. Along with being produced on a good site, soils need to be well-drained, fertile and adequately prepared before planting. Clean water sources are required. The crop is susceptible to many damaging insect pests and severe plant diseases, as well as damage from low winter temperatures, spring frosts and cracking of fruits during ripening. Consequently, these are among the critical areas of on-going research. There has been much research to address these production problems in various parts of the world. Reports have been published, but little exclusive literature is currently available on pomegranate culture worldwide.

Undoubtedly, one of the issues that had made this fruit less popular, especially in Western countries, is the difficulty in separating the edible parts (arils) from the peel. Nowadays, the manufacturing of fully automatic machines has solved the problem of separating arils from the fruit by selling pre-packaged arils in global markets.

Pomegranate: Botany, Production, and Uses, authored by an international team of experts who have been at the forefront of developments in this crop, provide their insights and experiences on pomegranate research. This textbook provides a comprehensive survey of pomegranate growing from a scientific and horticultural perspective covering different issues including botany, production, processing, health, and industrial uses. This book will provide the implementation of scientifically-based horticultural practices that will mitigate production, disease and/or abiotic stresses, and enhance nutrient management, which will increase yield and improve short- and medium-term grower profitability and sustainability. A better understanding of pomegranate fruit and its cultural practices will provide valuable information to a range of principal users, e.g., educators, researchers, students, agriculture extension workers, farming communities, industry personnel, and professionals/practitioners.

Providing information on pomegranate cultural practices on cultivars with horticultural traits such as earliness, high yield, fruit of excellent taste, soft seeds, disease resistance, and low splitting or sunscald rates will increase grower profits and sustainability while increasing crop diversity, which aids in increased crop security. Abiotic disorders such as sunburn and splitting are common issues in all pomegranate production areas around the world, severely affecting fruit production and quality and significantly increasing disease incidences. Pomegranate production in different parts of the world is also occasionally at risk because of the potential for damage to trees, flowers, and fruit from subfreezing temperatures.

The best management strategies that directly address fruit quality and abiotic disorders of existing and new pomegranate cropping systems across the world are also included. In addition to production information, the book also will provide substantial evidence for the beneficial effect of pomegranates. We are confident that this work will be a reference book for a broad spectrum of users, and we look forward to seeing readers' feedback to enhance future editions of this book.

Editors are grateful to all chapter authors who have given their time, without any financial reward, to contribute to this book. We also acknowledge the assistance of people who provided us with pictures and anyone who helped us in this journey and made valuable contributions to the end product.

Ali Sarkhosh
Alimohammad M. Yavari
Zabihollah Zamani

1 Archaeology, History and Symbolism

F. Spagnoli[1]*, R. Shirazi[2], Z. Shirazi[3], S. Saeidi Ghavi Andam[4] and M. Djamali[5]
[1]Italian Institute of Oriental Studies – ISO, Sapienza University of Rome, Rome, Italy; [2]Department of Archaeology, University of Sistan and Baluchestan, Iranian Center for Archaeological Research (ICAR), and Research Center for Cultural Heritage and Tourism (RICHT), Zahedan, Iran; [3]Laboratory of Archaeobotany, World Heritage Site of Shahr-i Sokhta and Regional Museum of Southeast Iran. Cultural Heritage, Tourism and Handicrafts of Sistan and Baluchestan, Zahedan, Iran; [4]Department of Biology and Preclinical Science, Institute of Ecology and Nature Conservation Biology, University of Regensburg, Regensburg, Germany; [5]Institut Méditerranéen de Biodiversité et d'Ecologie (IMBE), Aix-Marseille Université, Avignon Université, Technopôle de l'Environnement Arbois-Méditerranée, Aix-en-Provence, France

1.1 Introduction

According to several botanical reports, a zone including the northern part of the Iranian Plateau, Afghanistan, central Asia, northern India as well as Anatolia is considered as the homeland of the wild ancestors of pomegranate (Fig. 1.1) (Janick, 2005; Hummer *et al.*, 2012). Its later geographical distribution towards the west and the east was a result of cultural interactions and commercial exchanges. In this chapter, first the philological and linguistic evidence of pomegranate in the ancient Near East is presented. Second, since the Iranian territories are considered as the main centre of its origin, the archaeology and history of this fruit have been investigated in this area, and then the study has been extended towards other territories, including Mesopotamia, Egypt, Greece, Cyprus, Levant, Syria and the Iberian Peninsula. Finally, the symbolic values and mythological attestation of pomegranate are studied in various ancient cultures.

1.2 Pomegranate in Ancient Cultures

Pomegranate is a divine gift in the imagery of antiquity. It has so many evocative features: its blossom and flower, the shape of the pome, with a pointed or crowned tip, and its shining red colour have been emblems of power since ancient times; its innumerable ruby-red seeds hint at fecundity; and the regular geometry of the seeds is a replica of the divine order. Pomegranate was depicted and reproduced in ancient art as a major symbol of fertility, abundance, perfection and sanctity (Nigro and Spagnoli, 2018). Its practical function and its importance for human life make pomegranate one of the most symbolic fruits in antiquity as well as in modern times. In ancient cultures pomegranate is always linked with the concept of fertility, even beyond death, and for this reason it often appears in association with deities such as Anahita, Ishtar, Astarte, Era and

*Corresponding author: federica_spagnoli@yahoo.it

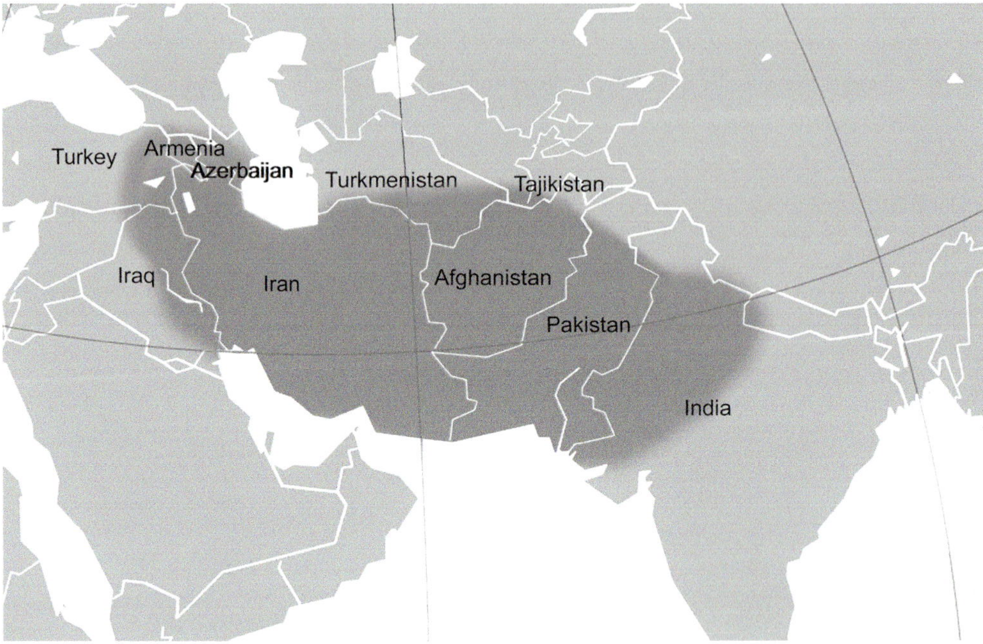

Fig. 1.1. Centre of origin for pomegranate. (Photo: Joseph Postman.)

Demeter. In the Christian culture the notion of Mary's virginity is also related to pomegranate. Its iconographic fortune bears witness to the symbolic transcultural role played by this fruit and the pomegranate tree in ancient Mediterranean, from its farthest oriental origins to modern art and religion (Spagnoli, 2019).

1.3 Ancient Philology

The earliest written evidence for pomegranate comes from Sumerian cuneiform texts. The names for pomegranate in ancient Near Eastern languages are as follows: Sumerian, 'nu-urma'; Akkadian, 'nurmu' (Roth, 1965); Assyrian, 'rummanu' (related to the modern derivations 'rimmon' in Hebrew and 'roumman' in Arabic); and Egyptian (both for the tree and fruit), jhmn/inhmn (Nigro and Spagnoli, 2018). The modern term for pomegranate in Persian is 'anār/انار'. In Middle Persian (Sassanid Pahlavi) the equivalent word for pomegranate is 'nar' or 'narak' and in Sanskrit the term for designation of pomegranate is dalim (MacKenzie, 1971).

1.4 Pomegranate in Iran (Persia)

The earliest archaeobotanical evidence of pomegranate dates back to the 5th millennium BCE. From the Bronze Age, the archaeobotanical data regarding the presence of pomegranate increase considerably. Mesopotamian cuneiform records mentioned pomegranate from the mid-3rd millennium BCE (Zohary *et al.*, 2012). Due to the abundance of wild pomegranate trees in Iran (the central, northern, eastern and north-eastern part of the Iranian Plateau), central Asia (southern Turkmenia) and Afghanistan, it is reasonable to think that the first phase of its cultivation was started in this part of the Near East. The eastern spread of pomegranate cultivation from Iran and central Asia towards north-western China (Xingjian) and then to South-east Asia could be related to the flourishing of trade and commercial contacts alongside the Silk Road.

1.5 Pomegranate in Persian Culture and Literature

In the Iranian religions, pomegranate is considered as a heavenly plant and is counted among the sacred trees in the Zoroastrian religion. Pomegranate's branches and flowers were and still are used during the Zoroastrian rituals. According to the later Zoroastrian texts, it is blessed to use pomegranate's branches in the fire temples. Barsam, a sacred object used during the veneration of fire, was composed of young twigs of pomegranate shrubs considered as a symbol of fecundity and eternity. During the Sassanid period, pomegranate was a sacred fruit and its plentiful grains were a symbol of Anahita's fecundity. In public Iranian folklore, on the first night of winter, called *Yalda*, pomegranate is an essential element that gives health and abundance throughout the year. In the iconographic tradition of the Sassanid period the combination of the pomegranate motif with winged-shaped thistle branches could be interpreted as a divine and fertility symbol. The frequent presence of pomegranate in Persian literature also indicates a popular use of this fruit. In Persian texts, poems and miniatures, pomegranate is a symbol of youth and feminine beauty (Sedaghat, 2017).

1.6 Archaeological Evidence of Pomegranate in Iran

1.6.1 Iron Age

One of the main archaeological corpora related to the iconography of pomegranate in Iran comes from the Iron Age site of Tepe Marlik (Cheragh Ali Tepe), excavated in 1961 by an Iranian mission headed by E.O. Negahban. It is situated in the lower part of the Gohar Rud valley, near Rostam Abad in the province of Gilan (Negahban, 1996). The site dates back to the late 2nd millennium BCE (13th century BCE) and was composed of 53 intact roughly constructed rich tombs made by stones and

Fig. 1.2. Golden necklaces and earrings from Marlik. (Photo: Neda Tehrani and Nima Fakoorzadeh – by courtesy of the National Museum of Iran.)

mortar. Among the archaeological findings, several ornamental items and other objects have been identified related to pomegranate. Several necklaces (Negahban, 1996: Pl. XXVI-163; Pl. XXVII-205/206; Pl. LI-165; Pl. LVI-205/206; Pl. LVIII-210) and earrings (Fig. 1.2) (Negahban, 1996: Pl. XXVIII-363; Pl. LII-170/171) were found in funeral contexts. Besides these objects, four pomegranate cage bells have been discovered in tombs no. 18, 30, 44 and 47 (Negahban, 1996: Pl. CXXXVI-943/944/945/946).

A bronze pomegranate-shaped bell used as clapper was also found in a burial context at Gohar Tepe in Mazandaran. It has been reported from burial AG2IV-20 (Piller *et al.*, 2009) belonging to Iron Age I/II contexts. Another archaeological source for the iconography of pomegranate in Iran is a trove of objects, also known as the Ziwiye Treasure, found at the archaeological site of Ziwiye (dating back from the 9th to 7th centuries BCE), in Kurdistan province. Indeed, in 1948 the French architect A. Godard published a museum catalogue presenting a small group of silver and ceramic artefacts supposedly from a site near Saqqiz in the north-west of Iran (Muscarella, 2013). Among other objects, several pomegranate-shaped decorative elements were found in Ziwiye. One of them is an ivory item in the form of a pomegranate, now kept in the Metropolitan Museum of Art. At the same museum are several golden plaques with winged

Fig. 1.3. Golden earrings and a necklace. (Photo: Neda Tehrani and Nima Fakoorzadeh – by courtesy of the National Museum of Iran.)

creatures approaching stylized trees (dating back to the 8th to 7th centuries BCE) clearly depicting pomegranate motifs.

1.6.2 Achaemenid period

During the Achaemenid period, the iconographic evidence of pomegranate increases considerably. These attestations can be divided into two categories; that is, goldsmithery and rock relief. In the case of goldsmithery, one of the finest objects, actually kept in the National Museum of Iran, is a pair of earrings with two rows of pomegranates encircled in a golden ring. On each of them, a pendent hangs, enfolded by a golden lace (Fig. 1.3, left). This object could be considered as a masterpiece of Achaemenid goldsmithery. An earring in the form of a pomegranate is another example of Achaemenid goldsmithery (Fig. 1.3, middle). Another item is a golden necklace, found recently in the region of Behbahan located in south-east Khuzestan (Razmjou, 2019). Six bulky pomegranates are represented in the lower part of the necklace (Fig. 1.3, right).

Regarding the presence of pomegranate in the rock art of the Achaemenid period, there is a noteworthy rock relief scene from the eastern staircase of Apadana at Persepolis, representing the Achaemenid king, Darius the Great. In his hand, he is holding a flower that easily could be a pomegranate flower with extra blossoms (Figs. 1.4 and 1.5). If it is so, it indicates that pomegranate was an auspicious element not only in religious rituals but also in royal ceremonies (Keshavarzi, 2014). According to historical texts, about one-tenth of the Achaemenid immortal guards carried spears with golden pomegranates on their shafts, and the other nine-tenths of them had spears with silver pomegranates on their shafts (Curtis and Simpson, 2010). This evidence shows sharply the omnipresence of pomegranate in the art and mythology of the Achaemenid period. Moreover, the pollen records from Lake Maharlou near Persepolis also indicate the cultivation of pomegranate in this period.

1.6.3 Pollen evidence of pomegranate cultivation in ancient Persia

Pomegranate (*Punica granatum*) is the most newly discovered cultivated tree in the pollen records of Lake Maharlou basin, some 50 km to the south of Persepolis. Pomegranate has a vast natural distribution from the Balkans to north-western India but is mostly restricted to the Irano-Turanian and Mediterranean floristic regions (Levin, 2006). Despite difficulties in detecting the indigenous species, Reschinger (1966) reported pomegranate natural populations in Iran mainly in the north (Gorgan, Mazandaran and Gilan provinces) as well as northern parts of Azerbaijan, Kurdestan, Qazvin and far east Baluchestan. The documented pomegranate pollen grains were extracted from sediments dating back to the Achaemenid period. The discovery of

Fig. 1.4. A representation of Darius the Great bearing a pomegranate flower with two blossoms. (Photo: Neda Tehrani and Nima Fakoorzadeh – courtesy of the National Museum of Iran.)

P. granatum pollen in the Maharlou record is the first report for southern Iran, which seems too far from the proposed natural stands of the tree in north Iran (Gorgan, Mazandaran and Gilan provinces), north-west Iran (Azerbaijan province), Kurdistan and even in eastern Baluchistan (Reschinger, 1966). Furthermore, the tree is extremely underrepresented in modern pollen rain due to insect- and self-pollination, suggesting that only a few counted pollen grains may indicate large-scale plantations (Morton, 1987). It is worth mentioning that Bottema (1986) identified rare pomegranate pollen in Urmia Lake (north-west Iran). However, due to age uncertainty of that record and the location of the lake in north-west Iran, no solid conclusion can be made about the exact age and possible cultivation of the tree.

Fig. 1.5. Pomegranate flower and blossoms on Persepolis reliefs. (Photo courtesy of Khoobchehr Keshavarzi.)

The domestication practices of pomegranate are thought to have begun in the Transcaucasia-Caspian area and northern Turkey around the late Neolithic period (Levin, 2006; Chandra *et al.*, 2010). Applications of pomegranate have been traced back to the 4th millennium BCE in the ancient Near East and Mediterranean region. Besides having dietary and medicinal properties, pomegranate fruits have been widely used as a fertility symbol to decorate the clothes and jewellery of royal Assyrian women. The fruit also appears in several Assyrian rituals and royal gardens as shown by their rock relief representations (SAA 7, 72, 81). In addition, the pomegranate name appears in PFA 33 from the Persepolis Fortification Archive (PFA[1]). The Elamite administrative texts recorded tree seedling inventories to be planted in five 'paradises' in the Achaemenids heartland (Henkelman, 2013). In this tablet, which is mostly written in Aramaic or Elamite languages, the name for pomegranate (ka-ru-kur) appears with the names of other fruit trees like pear, quince, mulberry, olive, date and apple. In conclusion, the particular pollen dispersal of this species along with the historical records strongly support the hypothesis of fruit tree cultivation and particularly pomegranate cultivation in the Maharlou Lake basin (Tilia and Tilia, 1972–1978; Ward, 2003; Keshavarzi, 2014).

Fig. 1.6. Stucco reliefs representing repeating pomegranates in palmettes from Chal Tarkhan, Rey (left) and Gouriyehr (right). (Photos courtesy of Alimohammad Yavari and Leila Khosrawi, and the National Museum of Iran..)

1.6.4 Sassanid period

In the Sassanid era, pomegranate became a common architectural decorative item (Fig. 1.6). Several motifs of this fruit have been identified on the stucco found in the Sassanid royal complexes including Tell Ḏahab (Cteciphon), Umm az Za'ātir (Ctesiphon), Tell Ma'āridt IV (Kröger, 1982: pl.10/1–2; pl. 21/3–4; pl. 38/3 & 5), the palace of Kish (Kröger, 1982: pl. 81/1–2; pl. 88/4), Chal Tarkhan (Rey) and Gouriyeh (Khosrawi, 2016). It is noteworthy that the motif of pomegranate is very common on some silver vessels of the Sassanid period, too. After the fall of the empire in 651 BCE, traditional decorative motifs of the Sassanid period continued to appear during the Islamic era. Under the Umayyad, the motif of pomegranate was a common architectural decorative element as can be seen in architectural decorations in Syria and Palestine (Baer, 1986). In Kharbat al Mafjar, Hisham's palace, near Jericho, a stucco panel, now kept in the Rockefeller Museum, clearly depicts pomegranate encircled in thistle leaves. It is without any doubt a theme that shows a strong similarity to the stucco frieze of the Sassanid period. The motif of pomegranate has been reported in architectural decoration at Samarra from the Abbasid era (Corsi, 2017). In the Iranian homeland, the motif of pomegranate was found in the archaeological excavations at Darreh Shahr (Seymareh region) dating from the early Islamic era. On a stucco frieze at Darreh Shahr, the motif of pomegranate is depicted in a very natural style (Lakpour, 2011).

1.7 Pomegranate in Bronze Age Ancient Near East and in the Levant

The presence of pomegranate in the Levant, both the original fruit, *Punica protopunica* L., and the domesticates *P. granatum* L. is archaeologically attested from the 4th millennium BCE. The spread of this plant from the Middle to the Near East occurred between the second half of the 4th and the first half of the 3rd millennium BCE, though some specimens reached the Fertile Crescent even before this time (Fateh *et al.*, 2013; Kokaj *et al.*, 2017). This may be connected with Sumerian trade with ancient Iran, Afghanistan, Pakistan and India (Harappa and Mohenjo-daro in the Indus Valley). Long-distance commerce with India was practised by the Sumero-Akkadian city-states of Mesopotamia and Elam (Schmandt-Besserat, 1992), and this route is possibly the one through which the original pomegranate shrub (*P. protopunica* L.) reached Mesopotamia, Anatolia, Syria and Palestine. Sumerians were possibly the protagonists of such diffusion, and they were the originators of the domestication of the pomegranate. Its florid aspect and healthy properties make this fruit suitable for symbolic associations with human fertility, and thus life and death. For this reason, in ancient Mesopotamian art it is often represented with the deities of fertility, fecundity and abundance. The pomegranate tree represents the Tree of Life in Assyrian art (Lurker, 1971; Barnett, 1982; Muthmann, 1982; Moortgat-Correns, 1989).

In ancient Syria and Levant, pomegranate seeds were found in Ebla (Wachter-Sarkady, 1995), Tell es-Sultan/Jericho, Tell el-Jazari/Gezer, Tell el-Hesi (Lipschitz, 1989), Tell es-Sa'idiyeh (Cartwright, 1997) and Arad (Hopf, 1978) in 3rd millennium BCE contexts, showing a capillary distribution at the time of early urbanization. The spread of this species in the Levant increased in the 2nd millennium BCE, and it was found in several rich cities, such as Ebla in the Sacred Area of Ishtar (Matthiae, 1993, Matthiae, 2002), Tell ed-Dab'a Temple III (Bietak, 2009) and in Jericho tombs (1700–1650 BCE; Kenyon, 1960; Hopf, 1969). Over the centuries, pomegranate continued to be a symbolic fruit in the Islamic period; the mosaic of the *diwan* of Qasr Hisham (743 BCE), Jericho, represents the Tree of Life, depicted as a luxuriant pomegranate tree, according to the ancient Mesopotamian iconography. In Late Bronze Age Levant are numerous attestations of pomegranates in jewellery and in other luxury items (Loud, 1948; Nigro, 1994). A symbolic meaning is also evidenced by its presence in cult stands and votives found in Levantine temples.

1.8 Pomegranate in Ancient Egypt

Evidence in Egypt starts from the 2nd millennium BCE – if one excludes a small jar made of breccia stone dating from the Early-Dynastic period (3150–2686 BCE), when renewed relationships with the Levant during the 13th Dynasty favoured the diffusion of the tree. Attestations become significantly more numerous from the 18th Dynasty, thus suggesting that this fruit was imported from Palestine. Furthermore, from Egyptian sources and ethnographic studies we know that ancient Egyptians made wine from the pomegranate (Loret, 1892); the rind was used to treat intestinal diseases and for dyeing leather. The flowers were crushed to make a red dye, which could also be obtained by pressing the peel. The juice from pomegranates was called *schedou*, and the rind was considered to have specific anti-inflammatory properties. Pomegranate juice was used also to trigger the fermentation of wine (Goor, 1967). Pomegranate was an important element in the rich and varied diet of the wealthy classes (Bard, 2015, p. 6).

1.9 Pomegranate in Iron Age Levant

From the Iron Age extensive evidence of pomegranate is found, especially in funerary contexts of the Levant, as its image is found in tombs, decorations, personal ornaments, urns and sarcophagi. This popularity is transversal to cultures and funerary customs, and affects all ethnic groups living in the Levant. The Philistines, one of the Sea Peoples who settled in southern Levant at the beginning of Iron Age, considered the pomegranate as the sacred fruit of Astarte, the goddess of wild nature and fertility. Votive offerings and cult stands, as well as pomegranate-shaped vases and kernoi decorated with these fruits, were found in her cult place (Temple 131) at Tell Qasile (Mazar, 1980a, b). In the Near East and Levant pomegranate has a further symbolic value connected to the kingship (Abram, 2009) and continues to be reproduced on textiles, wood, ivory and precious metals in the form of symbolic ornaments. Already in the 2nd millennium BCE, two small sceptres were found in Tell en-Nami bearing a finial in the shape of a pomegranate. In the 1st millennium BCE similar pomegranate finials were set on the top of bronze or ivory rods and sceptres found in Levantine temples (Tufnell, 1958; Artzy, 1990, 1995). A small ivory pomegranate from the illegal market of antiquities, probably dating to the 8th century BCE, now at the Israel Museum, Jerusalem (Avigad, 1994; Abram, 2009), represents a still-maturing pomegranate fruit with the elongated calyx occupying half its height. As is the case for earlier Late Bronze Age pomegranate finials, its base is slightly hollowed, and probably originally fitted on to a rod. A Paleo-Hebrew inscription is incised on top (Lemaire, 1981, 1984).

Pomegranates are represented on Neo-Assyrian reliefs (Börker-Klähn, 1957–1971) and on Phoenician ivories (Mallowan, 1966). Phoenician glass models of pomegranate were given as funerary offerings in Iron Age Levantine and Egyptian tombs. A key example of how pomegranate could be employed in personal ornaments and robe decoration is offered by the queens' tombs discovered underneath the North-West Palace of Ashurnasirpal II at Nimrud. Diadems, pendants, earrings, beads and engraved representations on ivory boxes are decorated with pomegranates as symbols of fertility (Hussein, 2016). Pomegranate became a common decorative element in the Assyrian and Achaemenid period in furniture, clothes and architecture, as finials or decorations. Typical representations are those on the wall reliefs of the North Palace at Niniveh with King Ashurbanipal under an umbrella with pomegranate-like finials and locks.

1.10 Pomegranate in Greece

During the 2nd millennium BCE, the pomegranate was appreciated as an exotic fruit, rare and refined, by the urban aristocracies of the Near East; it was exchanged as a luxury item and at the same time bore a strong symbolic value both in sacred and funeral spheres. The same symbolism was transmitted to the Greek culture at the beginning of the 1st millennium BCE. In local literary tradition the pomegranate is considered as an allogeneic fruit coming from far territories and mythical places (an exotic fruit of 'Paradise'). As an example, in the luxuriant garden of the palace of Alcinous, king of the Phaeacians, pomegranate is one of the fruit trees – pears, apples, figs and olives – that bears fruit all year round (*Od.* VII, 115). One of the oldest representations of pomegranate in the Greek world is on a pair of gold earrings found in the burial place of the 'Rich Lady of Areiopagus' (Coldstream, 1995), and it continues to appear as a standard element of aristocratic tombs, as part of the grave goods and depicted on funerary vases until the Classical and Hellenistic periods. In Greek mythology the pomegranate acquires a divine value as a fruit of the Underworld in the myth of Demeter and Persephone (Od. XI: 589): as for Persephone, the pomegranate provides nourishment for the journey of the deceased through the afterlife. In this myth, at the end of summer Persephone returns to the Underworld after spending two thirds of the year on earth, exactly when the pomegranate ripens, to be consumed, due to its long shelf life, in autumn and winter, during the death of nature. So the pomegranate symbolizes the nourishment of man in death, but it is at the same time a medium of rebirth, because it sustains and accompanies the new life that awakens in spring, when the pomegranate tree blooms and the cycle starts again. In the myth as in real life, the pomegranate marks the transition from spring/summer to autumn/winter. It has the power to lead man to rebirth (spring) after feeding him during the death (winter). Furthermore, in Greek mythology the pomegranate is also linked to Hera, the wife of Zeus. The golden fruits (in ancient Greek: μῆλη) of the tree that she received from Gaia as a wedding gift had the power to give immortality: the golden pomes were probably pomegranates.

1.11 Pomegranate in Phoenician and Punic Mediterranean

In the Phoenician pantheon, Astarte is the goddess who embodies the power of presiding over natural rhythms and the cycle of seasons, acquiring several aspects of similar goddesses, such as Egyptian Hathor, Greek Hera and, later on, Isis and Demeter. During the 5th century BCE, with the major influence of the Hellenic culture over the Phoenician-Punic world in Sicily, Astarte gathers the chthonic aspects of Demeter in cults, rituals, symbols and iconographies including that of the pomegranate (Ribichini, 2015). It is highly probable that the diffusion of pomegranate in the western Mediterranean is due to Phoenician expansion as indirectly suggested by the Latin name of this fruit. Pomegranate first reached western Sicily (Motya) and North Africa (Utica and Carthage) and then Carthage itself to contribute to its capillary spread over the Balearic Islands, the Iberian Peninsula and Sardinia, from where it was transmitted to the Etruscan and Roman world.

1.12 Carthage and Punic North Africa

> *[...] But the vicinity of Carthage is claimed more particularly as its own by the fruit the name of which is the 'Punic apple'; though by some it is called 'granatum'.*
> Plinius the Elder, *Naturalis Historiae* 13.31[2]

In this renowned part of *Naturalis Historiae*, Plinius the Elder tells us about the provenance and diffusion of *P. granatum* L. from North Africa to other Mediterranean regions, such as Rome. The pomegranate was a very popular fruit in Carthage between the 3rd and the 2nd century BCE (Van Zeist and Bottema, 1982; Lancel, 1992; Van Zeist *et al.*, 2001). Palaeo-botanic analysis revealed that more than half of the seeds found in the area of the harbour of Carthage are *P. granatum* L., thus showing the wide distribution of the fruit, which reached a peak of popularity in North Africa under the Carthaginian rule. The conservation of pomegranates destined for export is the topic of Mago's work. The Carthaginian author, who lived in the 2nd century BCE, was translated into Greek and Latin by Cassius Dionisius, and became a major source for Roman authors such as Plinius the Elder (Plinius the Elder, *Naturalis Historiae* 1.51; 8.84; 10.98). By illustrating Carthaginian agricultural skills, Mago describes the techniques for the preservation of pomegranates to be transported by sea as reported by Latin sources. Mago explains that the best way to ship pomegranates and preserve their fragrance is to sink them into sea water or humid soil (Plinius the Elder, *Naturalis Historiae* 15.20). This may indicate the importance attributed to pomegranate production and trade in Carthage as one of the agricultural industries of the Carthaginian aristocracy, as archaeologically indicated by the great number of seeds of *P. granatum* L. found in the harbour of Carthage, where these fruits underwent the preservation treatments described by Mago.

However, the presence of pomegranate at Carthage goes back to the 7th or even 8th century BCE, when in archaic tombs terracotta replicas of pomegranate were found, like in Levant and in Greece. Pomegranate replicas were also found in funerary sets in the following centuries from the 6th to the 3rd century BCE and also later (Campanella, 2008). Pomegranate representations are very common on Tophet stelas of the 4th and 3rd century BCE, due to the symbology of this fruit in connection with death and rebirth. While in the Archaic period pomegranate seemed to be a prerogative of mercantile aristocracy, since the 5th century BCE it gained great popularity in the carved imagery of the subaltern classes.

1.13 Iberian Peninsula

The presence of pomegranate in the Iberian Peninsula possibly antedates its actual archaeological identification from the 6th century BCE in the coastal region reached by Phoenicians (Mata Parreño *et al.*, 2007; Mata Parreño *et al.*, 2010), who were probably responsible for the introduction of said trees in Andalusia. From that time onwards, the cultivation of pomegranate extended to the whole peninsula not only on the coasts but also in the hinterland. The earliest pomegranate was found in Huelva (Pérez-Jordà *et al.*, 2017) by virtue of the good and stable connections between the Tartessian culture and Phoenicians, who possibly were the source of the plant. In the archaeological records, pomegranate is generally found in funerary contexts. The most ancient attestations are from the necropolis of La Fonteta (Valencia), in aristocratic tombs of the 6th century BCE (Torres Gomariz, 2017). Pomegranate-shaped pottery vessels are in the coeval necropolises of La Bobadilla (Jaén) (Maluquer *et al.*, 1981), and, successively, in tombs of the first half of the 5th century BCE of the necropolis of Tútugi (Granada) (Izquierdo, 1997; Pereira *et al.*, 2004; Mata Parreño *et al.*, 2010), and of the 4th century BCE of Cerro del Santuario (Baza, Granada) (Presedo, 1982; Adroher Aurouz and López Marcos, 1992). In Iberia the presence of pomegranate in archaeological contexts, both funerary and domestic, increases with the strengthening of Carthage military control over the region, as is suggested by the great quantities of seeds of this fruit retrieved in Andalusian harbours, which also shows the importance of the pomegranate trade for the local economy at that time (Almagro-Gorbea *et al.*, 2010; Torres Gomariz, 2017).

1.14 Phoenician and Punic Sardinia

The attestation of pomegranate in Phoenician and Punic Sardinia is mostly connected to the funerary realm. This fruit appears in funerary symbology, as shown by a pomegranate-shaped vase found in a tomb of the Punic necropolis of Olbia (Levi, 1949). From Tharros, in the *Collezione Chessa* now in the National Museum at Sassari, are enlisted some golden pendants in the shape of pomegranates, a common feature also in the homeland, which can be dated from the 6th or 5th century BCE (Crespi, 1868; Moscati and Uberti, 1987). Pomegranate presences increase between the end of the 6th and the 5th century BCE, as the influence of Greek culture affects figurative art, and especially votive choroplastic. From this period and in the following Hellenistic period, the iconography of the enthroned goddess with polos, identified with Demeter, who usually holds in her hands a dove, a torch, a piglet and a pomegranate, becomes very popular (Pesce, 1961; Uberti, 1977). Actually, the circulation and imitation of a certain kind of Greek choroplastic, devoted to Demeter, vehiculated the spread of the iconography of this fruit (Tore, 1989) more than the commercial and cultural relationship between Sardinia and Carthage. Nevertheless, in Sardinia the symbolic value of the pomegranate has survived to the present day: a folk tradition of Sant'Antioco of the Day of the Dead, in the night between 1 and 2 November, is to place a cup of milk and pomegranate outside the front door as offerings for the deceased.

1.15 Pomegranate in Sicily and *Magna Graecia*

Sicily has been considered a fertile and fruitful land since ancient times, as it was described by Aeschylus, the renowned Greek dramatist (τῆς καλλικάρπου Σικελίας λευροὺς γύας: Aesch. *Prom.*, 371). One of the earliest indications of pomegranate in Sicily is in the 8th century BCE, at Motya, the earliest Phoenician colony in Sicily, and it refers to a cult compound devoted to Astarte (Nigro and Spagnoli, 2017). Pomegranates are apparently absent in the 7th and 6th century BCE from Phoenician tombs in Sicily. However, in the following 5th century BCE, with cultural Hellenization, the image of the pomegranate spread, both as part of funerary goods and in iconography. Noticeably, in the 4th and 3th century BCE, majestic pomegranates are depicted on the painted funerary stelas from Lylibaeum (Vento, 2000). The syncretistic process affecting Phoenician Astarte and Greek Demeter (Ribichini, 1995; Spagnoli, 2013) is widespread in Sicily and in *Magna Graecia* (southern Italy), where the eastern Mistress of Animals/Mother Goddess of oriental origins merges with Hera. The pomegranate garlands depicted on the walls, golden and clay pomegranates, and statuettes of Hera holding a pomegranate and a *patera* (plate) in her hands adorned the Patrician hypogea at Paestum (Pontrandolfo and Rouveret, 1992). This symbolism survives across millennia, as testified by the syncretism between Hera and the Holy Mary of the sanctuary of the 'Madonna del Granato' at Capaccio (Salerno, Italy) near Paestum (Puca, 2014). The modern sanctuary was erected upon a Temple of Hera dominating the valley of the Sele river. The cult statue of Holy Mary represents the Virgin holding a sceptre with a pomegranate as finial, replicating the sacred gesture and the attribute of the ancient Greek deity.

Notes

[1] (Abram, 2009) 'PFA is a single ancient artefact integrating Elamite and Aramaic texts in a bureaucratic and archival system ... in the form of sealed, roughly tongue-shaped tablet around a knotted string. It belongs to the branch of regional administration that organized and controlled the intake, storage, and notably the redistribution of locally produced food commodities within the Persepolis economy' (Henkelman, 2013).

[2] [...] *Sed circa Carthaginem punicum malum cognomine sibi vindicat; aliqui granatum appellant.*

References

Abram, M. (2009) The pomegranate: sacred, secular, and sensuous symbol of ancient Israel. *Studia Antiqua* 7, 23–33.

Adroher Aurouz, A. and López Marcos, A. (1992) Reinterpretación cronológica de la necrópolis ibérica del Cerro del Santuario (Baza, Granada). *Florentia Iliberritana* 3, 9–30.

Almagro-Gorbea, M., López Rosendo, M.E., Mederos Martín, A. and Torres Ortiz, M. (2010) Los sarcófagos antropoides de la necrópolis de Cádiz. *Mainake* 32, 357–394.

Artzy, M. (1990) Pomegranate scepters and incense stand with pomegranates found in a priest's grave. *Biblical Archaeology Review* 16(1), 49–51.

Artzy, M. (1995) Nami: a second millennium international maritime trading center in the Mediterranean. In: Gitin, S. (ed.) *Recent Excavations in Israel: A View to the West (Archaeological Institute of America Colloquia & Conference Papers 1)*. Archaeological Institute of America, Dubuque, Iowa, pp. 17–40.

Avigad, N. (1994) The inscribed pomegranate from the 'House of the Lord'. *Israel Museum Journal* 8, 7–16.

Baer, E. (1986) Khirbat al-Mafdjar. In: Bearman, P., Bianquis, T., Bosworth, C.E., van Donzel, E. and Heinrichs, W.P. (eds) *Encyclopaedia of Islam*. Brill, Leiden, Germany, pp. 51–67.
Bard, K.A. (2015) *An Introduction to the Archaeology of Ancient Egypt*. Oxford University Press, Oxford.
Barnett, R.D. (1982) *Ancient Ivories in the Middle East (QEDEM, Monographs of the Institute of Archaeology, the Hebrew University of Jerusalem 14)*. Institute of Archaeology the Hebrew University of Jerusalem, Jerusalem.
Bietak, M. (2009) Near eastern sanctuaries in the eastern Nile delta. *Bulletin d'Archéologie et d'Architecture Libanaises* VI, 209–228.
Börker-Klähn, F. (1957–1971) *Granatapfel: Reallexicon Der Assyriologie und Vorderasiatischen Archäologie II*. de Gruyter, Berlin/New York, pp. 610–630.
Bottema, S. (1986) A late quaternary pollen diagram from lake Urmia (northwestern Iran). *Review of Palaeobotany and Palynology* 47(3–4), 241–261. DOI: 10.1016/0034-6667(86)90039-4.
Campanella, L. (2008) *Il cibo nel mondo fenicio e punico d'Occidente. Un'indagine sulle abitudini alimentari attraverso l'analisi di un deposito urbano di Sulky in Sardegna (Collezione di Studi Fenici 43)*. Serra Editore, Pisa/Rome.
Cartwright, C.R. (1997) Interim report on the archaeobotanical material from the 1996 season of excavations of the Early Bronze Age complex at Tell es-Sa'idiyeh, Jordan. *Palestine Exploration Quarterly* 129, 72–75.
Chandra, R., Babu, K.D., Jadhav, TV. and Teixeira da Silva, J.A. (2010) Origin history and domestication of pomegranate. *Fruit, Vegetable and Cereal Science and Biotechnology* 4(Special Issue 2), 1–6.
Coldstream, J.N. (1995) The rich lady of the Areiopagos and her contemporaries: a tribute in memory of Evelyn Lord Smithson. *Hesperia* 64(4), 391–403. DOI: 10.2307/148498.
Corsi, A.L. (2017) A brief note on early Abbasid stucco decoration: Madinat Al-Far and the first Friday mosque of Isfahan. *Vicino Oriente XXI* 21, 83–95.
Crespi, V. (1868) *Catalogo illustrato della raccolta di antichit sarde del Sig*. Raimondo Chessa, Cagliari, Italy.
Curtis, J. and Simpson, S.J. (2010) *The World of Achaemenid Persia: History, Art and Society in Iran and the Ancient Near East*. I.B. Tauris, London.
Fateh, M.V., Ahmed, S., Ali, M. and Bandyopadhyay, S. (2013) A review on the medicinal importance of pomegranate. *Rajiv Gandhi University of Health Sciences Journal of Pharmaceutical Sciences* 3, 23–25.
Goor, A. (1967) The history of the pomegranate in the Holy Land. *Economic Botany* 21(3), 215–230. DOI: 10.1007/BF02860371.
Henkelman, W.F.M. (2013) Administrative realities: the Persepolis Archives and the archaeology of the Achaemenid heartland. In: Potts, D.T. (ed.) *The Oxford Handbook of Ancient Iran*. Oxford University Press, Oxford, pp. 529–546.
Hopf, M. (1969) Plant remains and early farming in Jericho. In: Ucko, P.J. and Dimbleby, C.W. (eds) *The Domestication and Exploitation of Plants and Animals*. Gerald Duckworth, London, pp. 355–359.
Hopf, M. (1978) Plant remains, strata V-I. In: Amiran, R. (ed.) *Early Arad*. 1. Israel Exploration Society, Jerusalem, pp. 64–81.
Hummer, K.E., Pomper, K.W., Postman, J., Graham, C.J., Stover, E. *et al*. (2012) Emerging fruit crops. In: Badenes, M.L. and Byrne, D.H. (eds) *Fruit Breeding, Handbook of Plant Breeding*. Springer, New York, pp. 97–147.
Hussein, M.M. (2016) *Nimrud. The Queen's Tombs*. Oriental Institute Miscellaneous Publications, Chicago, Illinois.
Izquierdo, I. (1997) Granadas y adormideras en la cultura Ibérica y el contexto del Mediterráneo antiguo. *Pyrenae* 28, 65–98.
Janick, J. (2005) The origins of fruits, fruit growing, and fruit breeding. In: Gilbert, M. and Goldman, I.L. (eds) *Plant Breeding Reviews*. John Wiley & Sons, Inc., Hoboken, New Jersey, pp. 255–320.
Kenyon, K.M. (1960) Excavations at Jericho I. In: *The Tombs Excavated in 1952–1954*. The British School of Archaeology in Jerusalem, London.
Keshavarzi, K. (2014) *A New Approach to Persepolis Based on Avesta and Religious-Ritual Ceremonies*. Behjat Publications, Tehran, Iran.
Khosrawi, L. (2016) Report of excavations at Guriyeh building, Ivan, Ilam Province. unpublished report. Iranian Center for Archaeological Records Archive.
Kokaj, T., Çakalli, A.D. and Ismaili, H. (2017) IBA for rooting influence of some varieties of pomegranate (*Punica granatum* L). *Albanian journal of agricultural sciences* 23, 133–137.

Kröger, J. (1982) *Sassanidischer Stuckdekor*. Verlag Philipp von Zabern, Mainz am Rhein, Germany.
Lakpour, S. (2011) *Archaeological Excavations and Researches of Darreh Shahr (Saymareh)*. Pazineh Publications, Tehran, Iran.
Lancel, S. (1992) *Carthage*. Brépols, Paris.
Lemaire, A. (1981) Une inscription paléo-hébraïque sur grenade en ivoire. *Revue Biblique* 88(2), 236–239.
Lemaire, A. (1984) Probable Head of Priestly Scepter from Solomon's Temple Surfaces in Jerusalem. *Biblical Archaeology Review* 10(1), 24–29.
Levi, D. (1949) Le necropoli puniche di Olbia. *Studi Sardi* 9, 5–120.
Levin, G.M. (2006) *Pomegranate*. Third Millennium Publishing, Tempe, Arizona.
Lipschitz, O. (1989) Plant economy and diet in early Bronze age in Israel: a summary. In: de Miroschedji, P. (ed.) *L'urbanisation de la Palestine l'âge du Bronze Ancien. Bilan et perspectives des recherches actuelles, Actes du colloque d'Emmaüs (20–24 octobre 1986)*. British Archaeological Reports International Series 527, Oxford, pp. 269–277.
Loret, V. (1892) *La flore pharaonique d'après la documentation hiéroglyphique et les spécimens découverts dans les tombes*. E. Leroux, Paris.
Loud, G. (1948) *Megiddo, II. Seasons of 1935–39*, Oriental Institute Publications 62. Oriental Institute, Chicago, Illinois.
Lurker, M. (1971) *Beiträge zu Geschichte, Kultur und Religion des Alten Orients. In Memoriam Eckhard Unger*. Valentin Koerner, Baden-Baden.
MacKenzie, D.N. (1971) *A Concise Pahlavi Dictionary.*. Reprint, Oxford University Press, London, 1986.
Mallowan, M.E.L. (1966) *Nimrud and its Remains*. Vol. 2. Collins, London.
Maluquer, J., Picazo, M. and del Rincón, J. (1981) La necrópolis ibérica de la Bobadilla (Jaén). In: Maluquer, J. and Aubet, M.E. (eds) *Andalucía y Extremadura. Programa de investigaciones protohistóricas dirigido por Juan Maluquer de Motes*. Departamento de Prehistória y Arqueologia, Barcelona, Spain, pp. 1–51.
Mata Parreño, C., Badal García, E., Bonet Rosado, H., Collado Mataix, E., Fabado Alós, F.J. *et al.* (2007) De lo real a l'imaginario. Approcimación a la flora ibérica durante la edad del hierro. *Anales de Arqueología Cordobesa* 18, 93–122.
Mata Parreño, C., Badal García, E., Bonet Rosado, H., Collado Mataix, E., Fabado Alós, F.J. *et al.* (2010) Comida para La eternidad: Saguntum extra. *Papeles del Laboratorio de Arqueología de Valencia* 9,277–286.
Piller, C.K., Mahforouzi, A., Bagherpour, N., Neumann, T. and Ögüt, B. (2009) First preliminary report on the joint Iranian-German excavations at Gohar Tappe, Māzandarān, Iran. *Archäologische Mitteilungenaus Iran und Turan* 41, 1–33.
Matthiae, P. (1993) L'aire sacrée d'Ishtar Ebla: Résultats des fouilles de 1990–1992. *Comptes-Rendus des seances de l'Académie des Inscriptions et Belles-Lettres* 137(3), 613–661.
Matthiae, P. (2002) Tell Mardikh, 1977-1996: Vingt Ans de Fouilles et de Découvertes. La Renaissance d'Ebla Amorrheenne. *Akkadica*101, 1–29.
Mazar, A. (1980a) *Excavations at Tel Qasile Part One. The Philistine Sanctuary: Architecture and Cult Objects*, QEDEM, Monographs of the Institute of Archaeology, the Hebrew University of Jerusalem, 12. Jerusalem.
Mazar, A. (1980b) *Excavations at TEL Qasile Part Two. The Philistine Sanctuary: Various Finds, The Pottery, Conclusions, Appendixes*, QEDEM, Monographs of the Institute of Archaeology, the Hebrew University of Jerusalem, 12. Jerusalem.
Moortgat-Correns, U. (1989) *La Mesopotamia*. Storia Universale dell'Arte, sezione prima: le civilt antiche e primitive, diretta da Sabatino Moscati, Turin, Italy.
Moscati, S. and Uberti, M.L. (1987) *Localia punica. La collezione del Museo Nazionale G.A. Sanna di Sassari (Memorie dell'Accademia Nazionale dei Lincei 29, serie 8°*. Accademia Nazionale dei Lincei, Roma.
Morton, J.F. (1987) *Fruits of Warm Climates*. Creative Resource Systems, Winterville, New York.
Muscarella, O.W. (2013) *Archaeology, artifacts and Antiquities of the ancient near East: sites, cultures, and Provenience*. 62. Brill, Leiden (Original work published 1977) pp. vi–1088.
Muthmann, F. (1982) *Der Granatäpfel: Symbol des Lebens in der alten Welt*. Fribourg, Switzerland.
Negahban, E.O. (1996) *Marlik: The Complete Excavation Report*. The University Museum, University of Pennsylvania Publications, Philadelphia, Pennsylvania.

Nigro, L. (1994) *Ricerche sull'architettura palaziale della Palestina delle et del Bronzo e del Ferro. Contesto archeologico e sviluppo storico (Contributi e Materiali di Archeologia Orientale, 5)*. Università degli Studi di Roma "La Sapienza", Rome.

Nigro, L. and Spagnoli, F. (2017) *Landing on Motya. The earliest Phoenician settlement of the 8th century BCE and the creation of a West Phoenician cultural identity in the excavations of Rome «La Sapienza» University – 2012–2016. Stratigraphy, architecture, and finds*, Quaderni di Archeologia Fenicio-Punica/Colour Monograph 04. Missione Archeologica a Mozia, Rome.

Nigro, L. and Spagnoli, F. (2018) Pomegranate (*Punica granatum* L.) from Motya and its deepest oriental roots. *Vicino Oriente XXII*, 49–90.

Pontrandolfo, A. and Rouveret, A. (1992) *Le tombe dipinte di Paestum*. Franco Cosimo Panini, Modena, Italy.

Presedo, F.J. (1982) *La necrópolis de Baza (Excavaciones Arqueológicas en España 119)*. Ministerio de Cultura. Dirección General de Bellas Artes, Archivos y Bibliotecas, Madrid.

Puca, G. (2014) *La montagna che parla. La Madonna del Granato sul monte Calpazio*. Edizioni il Saggio, Eboli, Italy.

Razmjou, S. (2019) Iran in the Achaemenid period. In: Nokandek, G. and Van Vilsteren, V.T. (eds) *Iran: Cradle of Civilization, Archaeology and History of Iran (on the basis of Iran National Museum Collections)*. Iran National Museum and Drents Museum, Tehran, Iran, pp. 112–130.

Reschinger, K.H. (1966) *Flora Iranica. Punicaceae*. Naturhistorisches Museum Wien, Vienna.

Pereira, L., Chapa, T., Madrigal, A., Uriarte, A. and Mayoral, V. (eds) (2004) *La necrópolis ibérica de Galera (Granada). La colección del Museo Arqueológico Nacional*. Ministerio de Cultura, Subdirección General de Museos Estatales, Madrid.

Pérez-Jordà, G., Peña-Chocarro, L., García Fernández, M. and Vera Rodríguez, J.C. (2017) The beginnings of fruit tree cultivation in the Iberian Peninsula: plant remains from the city of Huelva (southern Spain). *Vegetation History and Archaeobotany* 26(5), 527–538. DOI: 10.1007/s00334-017-0610-6.

Pesce, G. (1961) *Sardegna punica*. Editrice Sarda, Cagliari, Italy.

Ribichini, S. (1995) Flebili dee fenicie. *Rivista di Studi Fenici* 23, 3–35.

Ribichini, S. (2015) Statue greche E culti fenici. In: Giuffre Scibona, C., Mastrocinque, A. and Multari, A. (eds) *Ex pluribus unum. Studi in onore di Giulia Sfameni Gasparro*. Quasar, Rome, pp. 157–167.

Roth, M. (1965) *Chicago Assyrian Dictionary*. University of Chicago Press, Chicago, Illinois.

Schmandt-Besserat, D. (1992) *Before Writing: From Counting to Cuneiform*. University of Texas Press, Austin, Texas.

Sedaghat, N. (2017) An analytical study of sacred and mythological plant motifs during the Sassanid period (rock reliefs stucco and seals). *Seasonal Journal of History* 12(45), 129–148.

Spagnoli, F. (2013) Demetra a Mozia: Evidenze dell'area sacra del Kothon nel V secolo a C. *Vicino Oriente XVII*, 153–165.

Spagnoli, F. (2019) Pomegranate, the divine pome. A brief excursus on the fortune of pomegranate from its deepest oriental roots to the modern Mediterranean culture. *Near Eastern Archaeology Today* 19, 1.

Tilia, G. and Tilia, A.B. (1972–1978) *Studies and Restoration at Persepolis and Other Sites of Fārs*. 16. IsMEO Reports and Memoirs, Libreria Universitaria, Rome.

Tore, G. (1989) Sardinia Antiqua. Saggio di bibliografia fenicio-punica. *Biblioteca Francescana Sarda* 2, 229–427.

Torres Gomariz, O. (2017) La Granada: Usos y significados de una fruta de oriente en occidente. In: Martínez, P. and Sala Sellés, F. (eds) *El Oriente de Occidente. Fenicios y púnicos en el área Ibérica*. VIII edición del Coloquio Internacional del CEFYP en Alicante, Alicante, Spain, pp. 625–640.

Tufnell, O. (1958) *Lachish IV (Tell ed- Duweir). The Bronze Age*. The Welcome Marston Archaeological Research Expedition to the Near East, London.

Uberti, M.L. (1977) Le terrecotte. In: Acquaro, E., Moscati, S. and Uberti, M.L. (eds) *La Collezione Biggio. Antichit puniche a Sant'Antioco*. Consiglio Nazionale delle Ricerche, Rome, pp. 29–36.

Van Zeist, W.V. and Bottema, S. (1982) Vegetational history of the eastern Mediterranean and the near East during the last 20,000 years. In: Bintliff, J.L. and van Zeist, W.V. (eds) *Palaeoclimates, Palaeoenvironments and Human Communities in the Eastern Mediterranean Region in Later Prehistory*, British Archaeological Reports International Series 133. BAR Publishing, Oxford, pp. 277–321.

Van Zeist, W.V., Bottema, S. and van der Veen, M. (2001) *Diet and Vegetation at Ancient Carthage: The Archaeobotanical Evidence*. University of Groningen, Groningen, the Netherlands.

Vento, M. (2000) *Le Stele Dipinte di Lilibeo*. Centro Europeo di Studi Economici e Sociali, Marsala, Italy.

Wachter-Sarkady, C. (1995) Ebla e le condizioni materiali della produzione agricola nell'antico Oriente. In: Matthiae, P., Pinnock, F. and Scandone Matthiae, G. (eds) *Ebla: Alle Origini Della Civilt Urbana*. Electa, Milan, Italy, pp. 242–251.

Ward, C. (2003) Pomegranates in eastern Mediterranean contexts during the late bronze age. *World Archaeology* 34(3), 529–541. DOI: 10.1080/0043824021000026495.

Zohary, D., Hopf, M. and Weiss, E. (2012) *Domestication of Plants in the Old World: The Origin and Spread of Domesticated Plants in South-West Asia*. Oxford University Press, Oxford, pp. 114–115.

2 Taxonomy, Botany and Physiology

D. Narzary[1]*, N. Yazdanbakhsh[2] and T.S. Rana[3]
[1]Department of Botany, Gauhati University, Guwahati, Assam, India; [2]Department of Plant Sciences, School of Biology, University of Tehran, Tehran, Iran; [3]Plant Diversity Systematics & Herbarium Division, CSIR – National Botanical Research Institute, Lucknow, India

2.1 Introduction

The word 'pomegranate' means 'grainy apple' and is the given name for the members of the genus *Punica* L. The genus has been placed in the families *Rosaceae*, *Onagraceae*, *Myrtaceae*, *Lythraceae* and *Punicaceae* at different times by different taxonomists (Rehder, 1949; Rana *et al.*, 2010). At present, the genus is considered to be in the family *Lythraceae* due to its monophyletic relationships with other genera of the family (The Angiosperm Phylogeny Group, 2016; Yan *et al.*, 2019). *Punica* is a distinct entity within the family *Lythraceae* as it shows some unique characteristics not shared by the other members of the family, and therefore, a few taxonomists are still intent on conserving the family name *Punicaceae* for the sake of convenience (Rana *et al.*, 2010; Stepanyan-Gandilyan, 2017). The pomegranate has a long history of usage and cultivation, which started during early civilization in the Mediterranean regions. It is now receiving global attention for its wide adaptability, winter hardiness, ability to grow in marginal soils and striking health benefits in consumption. Pomegranate fruit consumption is increasing globally and the fruit is in high demand, especially in developed countries, due to reported health benefits (Lansky and Newman, 2007; Brodie, 2009; Opara *et al.*, 2009).

Wild pomegranates are as diverse as the cultivated ones, and many cultivars are selected from wild populations. Thus, the pomegranate represents a classical example of domestication of a wild fruit tree (Evreinoff, 1957). Basic botanical information along with different morphological variants that a plant species possesses are the key for germplasm tagging and crop improvement. In addition, the growing conditions and physiological requirements are the points that any grower should know for better crop production and management. The wild relatives of pomegranates that are found growing luxuriantly in the Middle East extending up to the western Himalayan region of India could be the potential genetic stock for further improvement of this economically and medicinally important crop.

2.2 Taxonomy and Systematics of the Genus *Punica*

The genus *Punica* was first described by Linnaeus under the family *Punicaceae*. It has two species, namely, *Punica granatum* L. and *Punica protopunica* Balf. f. The species *P. protopunica*, also known as Socotran pomegranate, is endemic

*Corresponding author: d_narzary@gauhati.ac.in

to the island of Socotra – a part of the Indian Ocean archipelago that lies in the tropical zone (12°30′36″N) (Guarino *et al.*, 1990). The dwarf pomegranate, *P. granatum* var. *nana* Pers., which bears bright scarlet flowers but inedible small fruits, is believed to be of Pacific origin and is often treated as the third species of *Punica* (Melgarejo and Martínez, 1992). Different morphotypes of *Punica* are presented in Fig. 2.1a-e.

The intraspecific classification of *P. granatum* was developed by B.S. Rozanov based on ovary colour, a stable characteristic that is retained when they are grown from seeds (Djavakyants, 2011). Accordingly, there are two subspecies, namely, *P. granatum* subsp. *chlorocarpa* with a green ovary, and *P. granatum* subsp. *porphyrocarpa* with a red ovary. The subsp. *chlorocarpa* includes two botanical varieties – var. *viridicolla* (green collar) and var. *rosaeicolla* (pink collar). Another subsp. *porphyrocarpa* also includes two varieties – var. *rubricolla* (red collar) and var. *cinereicolla* (bluish collar). Subsp. *chlorocarpa* includes all wild forms and some cultivated forms, whereas subsp. *porphyrocarpa* includes only some cultivated forms (Burmistrov, 1993). Most of the widespread cultivated pomegranates are either bluish collar or red collar varieties of the subsp. *porphyrocarpa*. However, the ovary colour of all forms of *P. granatum* distributed worldwide is not yet documented properly.

The present-day cultivated forms of pomegranates are morphologically diverse, which may have resulted due to domestication of wild forms and subsequent artificial selection and cultural practices conducted by people over centuries (Levin, 2006). Therefore, it is difficult to follow a natural system of classification and assign botanical variety to those variants of pomegranates. However, a few workers assigned certain forms of pomegranate to an infraspecific category 'forma' or 'variety', for example, *P. granatum* f. *multiplex*, *P. granatum* f. *albescens*, *P. granatum* f. *flavescens*, *P. granatum* f. *pleniflora*, *P. granatum* f. *legrellei*, *P. granatum* f. *tadshikorum* (brown-gray bark), *P. granatum* f. *turcomanica* (ash-grey bark), *P. granatum* f. *nigro violacea* (black-purple fruits), *P. granatum* var. *parvifolia* (small-leaf variety), *P. granatum* var. *pyrocarpa* (pear-like fruits), *P. granatum* var. *macrocarpa* (large fruits), *P. granatum* var. *nana* f. *flore pleno*, *P. granatum* var. *nana* f. *flore pleno minore*, *P. granatum* var. *nana* f. *racemosa*, *P. granatum* var. *nana* f. *legrellei*, etc. (Rehder, 1949; Levin, 2006).

2.2.1 Phylogeny

The genus *Punica* is monophyletic to the other genera of the family *Lythraceae* according to DNA-based phylogenetic studies (Conti *et al.*, 1997; Narzary *et al.*, 2016; Yan *et al.*, 2019). In a phylogenetic study of genera in the family *Lythraceae* using chloroplast and nuclear ribosomal gene sequences, *Punica* consistently appeared as a sister to *Pemphis* with high bootstrap supports in parsimony and likelihood analyses (Huang *et al.*, 2002). Conversely, internal transcribed spacer (ITS)-sequence analysis revealed a sister relationship of *Punica* to *Woodfordia* (Narzary *et al.*, 2016). Xylem anatomy suggests *P. protopunica* as the ancestor of the genus (Shilkina, 1973).

Punica, being tropical in origin (Burmistrov, 1993), has subsequently adapted and naturalized to the subtropical areas of the Mediterranean region extending up to the western Himalayas. The wild pomegranates, which are mainly propagated through seeds, are less diverse than the cultivated forms. Pomegranate seeds have a high tocopherol content, a natural auto-antimutagen that might have contributed to the persistence of *Punica* as a relict genus (Burmistrov, 1993). Pomegranate being a self- as well as a cross-pollinated plant might have enabled the adaptive trait combinations to cope with diverse climatic conditions through the process of natural selection. On the other hand, the absence of tocopherol in the vegetative organs might have contributed to somatic mutations, thus creating variations among the pomegranates in the course of evolution. The frost-resistant bluish collar variety (var. *cinereicolla*) is considered to be evolved from the less frost-resistant red collar variety (var. *rubricolla*) within the subsp. *porphyrocarpa*. Frost sensitivity varies with cultivars, and the soft-seeded cultivars are generally less frost-hardy than the hard-seeded ones (Glozer and Ferguson, 2011).

Levin (2006) opined that *Punica* evolved to have a xerophilic and cryophilic habit by the process of convergent evolution, and the advanced species *P. granatum* acquired traits such as thorns, xylopodium, vegetation mobility, etc. in addition to the specialized fruit character that the genus acquired at a very early stage of its evolution. Although a wide diversity is observed

Fig. 2.1. Different types of pomegranate: (a) Socotran pomegranate (*Punica protopunica*); (b) wild pomegranate (*P. granatum*); (c) dwarf pomegranate (*P. granatum* var. *nana*); (d) commercial pomegranate (*P. granatum* cv. Bhagwa); (e) ornamental pomegranate (*P. granatum* var. *pleniflora*). (Photos: (a) David Eickhoff; (b, d) Teixeira da Silva *et al.*, 2013 (Elsevier); (c) Diganta Narzary; and (e) Ali Sarkhosh.)

among pomegranates at the infraspecific levels, its further evolution beyond the species rank might have been slowed down due to its higher-level specialization, which occurred at a very fast pace of initial evolution.

2.2.2 Distinguishing features of the genus *Punica*

The genus *Punica* is close to *Lythraceae* in pollen morphology (Patel *et al.*, 1984; Graham *et al.*, 1990), and wood anatomy, especially the occurrence of intraxylary phloem (Bridgwater and Baas, 1978; Graham *et al.*, 1993). However, some distinct features make *Punica* unique in the order Myrtales. The salient features of *Punica* are the ovules with thick, multilayered outer integument and unicellular archesporium (Huang *et al.*, 2002), the union of the ovary with the receptacle of the thalamus, the fruit with leathery pericarp, and the pulpy seeds with edible sarcotesta and rolled seed cotyledons (Watson and Dallwitz, 1992).

2.2.3 Centre of origin and distribution of pomegranates

Pomegranate has been used and domesticated by humans since time immemorial. There are various speculative views on its origin and domestication. According to Vavilov's Centre of Origin, pomegranate belongs to the central Middle East, which includes the interior of Asia Minor, the Transcaucasus, Iran and the highlands of Turkmenistan. Wild pomegranates grow today in central Asia from Iran and Turkmenistan to northern India. de Candolle determined Persia (Iran) and its surroundings as its origin (Goor and Lieberman, 1956). None the less, *Punica* appears to be a relic of a much wider Mediterranean distribution of clear Gondwanan origin (Burmistrov, 1993), and the same is supported by the tropical origin of its ancestor *P. protopunica*. The species *P. granatum* might have evolved under natural selection and domestication, with the latter extending its range well beyond its native area.

The cultivation of pomegranate began in prehistoric times and it is estimated that its domestication began some time in the Neolithic era (Levin, 2006a; Still, 2006), initially in the Transcaucasia-Caspian region and northern Turkey (Zohary and Spiegel-Roy, 1975; Harlan, 1992). Evidence for the use of pomegranates in the Middle East is dated at over 5000 years ago. Pomegranate artefacts and relics dating to 3000 BCE onwards have been found in Egypt, Israel, Armenia and Mesopotamia (Goor and Lieberman, 1956; Still, 2006; Stepanyan, 2007). Carbonized fragments of pomegranate rinds dating from the early Bronze Age have been found in Jericho and Arad, Israel, in Nimrod, Lebanon, and in Egypt (Still, 2006), as well as in Armenia (Stepanyan, 2007). The domestication process took place independently in various regions and not only in the Mediterranean region. Pomegranate was probably introduced to China from central Asia or Europe during the Han dynasty (207 BCE) and then naturalized in west China (Chandra *et al.*, 2010). It is now widely grown in many parts of China and Indonesia (Morton, 1987).

It is speculated that pomegranate travelled the Indian peninsula from Iran in about the 1st century CE and was reported growing in Indonesia in 1416. The Greeks and the successor empires distributed the pomegranate all over Europe. This tree appeared in Spain around 800 BCE thanks to the Moors. Spanish sailors and missionaries took the pomegranate from the Mediterranean region to the New World soon after Cortez conquered Mexico in 1521. In 1769, Spanish missionaries introduced pomegranate trees into California, and they now grow from the southern USA to Chile and Argentina, probably reaching the highest quality in the arid regions of California, Arizona and northern Mexico (Goor and Lieberman, 1956; Morton, 1987).

Naturally growing (wild) pomegranates are reported from the countries bordering the Mediterranean regions including Iran, Afghanistan, India and Pakistan (Nasir and Ali, 1972; Muradoglu *et al.*, 2006). In India, wild pomegranate trees were reported in the warm valleys and hill slopes of Himachal Pradesh, Uttarakhand, and Jammu and Kashmir. The pomegranate represents a classic example of domestication of a wild fruit tree (Evreinoff, 1953). Introduction into culture did not occur at any one centre, but was a parallel process extending over the whole wild range. As with central Asia, other countries within the natural range, such as Afghanistan and Iran, have their own local selections. Despite the intensive selective pressure due to domestication, it is claimed that the probable progenitor has a very similar appearance to the domesticated form, with the latter having larger seeds and fruits, non-dehiscent fruits and seeds, and different seed or fruit colour (Harlan, 1992; Teixeira da Silva *et al.*, 2013).

The ability of pomegranate trees to adjust to variable climatic conditions is reflected in the wide distribution of the wild forms throughout Eurasia to the Himalayas (Levin, 2006). It is now widely cultivated in subtropical and tropical areas in many variable climatic conditions in different countries, such as Algeria, Armenia, Azerbaijan, Afghanistan, China, France, India, Iran, Israel, Spain, Turkey, Tunisia, Morocco, Greece, Japan, Cyprus and Egypt, indicating its flexibility and adaptability to a wide range of climatic and biogeographic conditions. Commercial orchards of pomegranate trees are now grown in the Mediterranean basin (North Africa, Egypt, Israel, Syria, Lebanon, Turkey, Greece, Cyprus, Italy, France, Spain, Portugal) and Asia (Iran, Iraq, India, China, Afghanistan, Bangladesh, Myanmar, Vietnam, Thailand), and in the former Soviet Republics (Kazakhstan, Turkmenistan, Tajikistan, Kirghistan, Armenia and Georgia). In the New World, pomegranates are grown in the USA and Chile. New orchards are now established in South Africa, Australia, Argentina and Brazil (Holland *et al.*, 2009).

As an ancient and widespread fruit, pomegranate cultivars were spread throughout different regions and continents. Thus, some cultivar names have considerable synonymy, in which the same basic genotype is known by different names in different regions. Synonymy is further encouraged by the fact that husk and aril colour can vary markedly when grown in different regions. However, on several occasions these names provide a clue to the origin of the cultivar. For instance, the 'Kabul' or 'Kandhari' cultivar in the Indian collections hints at their possible origin in the Kabul and Qandahar regions of Afghanistan. The same is true for 'Afghansky', 'Washingtonsky', 'Iran 29–3' and 'Kalifornijsky' cultivars of the Turkmenistan collection (Levin, 1994). Furthermore, in Chinese the pomegranate is referred to as 'An Shi Liu', meaning 'the fruit of Kabul', which reflects its Afghanistan origin (Holland *et al.*, 2009).

A number of characteristics vary between pomegranate genotypes (Stover and Mercure, 2007). Due to the ~13% outcrossing that occurs in pomegranate (Jalikop and Kumar, 1990), seedlings are not 'true to type', resulting in a plant-to-plant variation, which may explain the morphological changes that have occurred during domestication. Pomegranate selections are made on the basis of rind and aril colour, fruit size, sugar and acid contents, resistance to biotic and abiotic stresses, yield, keeping quality, seed hardiness, etc. (Harlan, 1992; Holland *et al.*, 2009).

2.2.4 Pomegranate germplasm collection centres

More than 500 cultivars of pomegranate have been named all over the world, indicating its genetic diversity, from which only 50 are grown commercially (Teixeira da Silva *et al.*, 2013). Since this practice has drastically reduced the genetic diversity of modern cultivars, it is extremely important to conserve the gene pool of wild forms and lesser-known cultivars to maintain the genetic base for future crop improvement programmes (Rana *et al.*, 2007). International groups focusing on collection and conservation of pomegranate need to evaluate the germplasm from around the world for sustainable development and posterity (Mars, 2000; Fadavi *et al.*, 2006; Still, 2006; Zamani *et al.*, 2007). Local pomegranate germplasm collections have been established in several Mediterranean countries, including Spain, Morocco, Tunisia, Greece, Turkey and Egypt (Mars, 2000). In order to prevent duplication and accurately assess the variation that exists in these collections, each accession needs to be characterized not only in terms of its morphological variation, but should also include a genome-wide survey of the genetic diversity (Still, 2006).

Germplasm collections from the wild, semi-wild and cultivated forms from different geographical regions are important to develop a good repository for future breeding and improvement of pomegranates. The largest collection, containing 1117 accessions, is in the Garrygala Research Station in Turkmenistan (Levin, 1996; Still, 2006; Holland *et al.*, 2009). This collection contains specimens collected from a geographical region that is a part of the central Asian region considered as the origin of pomegranate, as well as Transcaucasia and other countries such as Spain, the USA, Iran, Tajikistan and India. A collection of 370 accessions in the Nikita Botanical Gardens in Ukraine includes accessions from central Asia, Transcaucasia, Iran, Afghanistan, Spain, Italy and the USA (Yezhov *et al.*, 2005). Further important centres include the Iranian collections in Tehran, Saveh, Yazd and Markazi (Fadavi *et al.*, 2006; Zamani *et al.*, 2007), the Indian collection and China's collection. More than 700 accessions have been reported from the Yazd pomegranate collection in Iran (Zamani *et al.*, 2007; Nemati *et al.*, 2012).

Since wild pomegranates have been reported to grow on the slopes of the Himalayas and most of the reported evergreen pomegranate cultivars originate from India (Singh *et al.*, 2006), the Indian collections might be interesting. The presence of wide genetic diversity among the wild and cultivated Indian pomegranates has already been reported (Ranade *et al.*, 2009; Narzary *et al.*, 2010a, b). Three Indian collections of pomegranate have been reported (Mars, 2000), one being the highly diverse collection of 90 accessions in the National Bureau of Plant Genetic Resources Regional Station, Phagli, Shimla (Rana *et al.*, 2007). There are several reports available which indicate that Chinese pomegranate germplasm also possesses significant diversity (Feng *et al.*, 2006; Yang *et al.*, 2007).

The US National Clonal Germplasm Repository in Davis, CA has almost 200 pomegranate accessions including accessions from all over the world (Stover and Mercure, 2007). Smaller collections are held in Alata Horticultural Research Institute, Çukurova University Adana and Izmir in Turkey (Özgüven *et al.*, 1997; Özgüven and Yilmaz, 2000; Holland *et al.*, 2009), south-east regions of Armenia (Stepanyan, 2007), Tunisia (Mars and Marrakchi, 1999; Mars, 2000; Hasnaoui *et al.*, 2010), the Newe Ya'ar Research Center in Israel and Thailand (Holland *et al.*, 2009).

Some of the popular cultivars grown in different biogeographic regions are: *Wonderful*, *Spanish Ruby*, *Early Wonderful*, *Granada* and *Early Foothill* in the USA; *Mollar de Elche* and *Valenciana* in Spain; *Achik-Dona*, *Kazake-Anar*, *Kai-Achik-Anar* and *Kzyl-Anar* in Uzbekistan; *Bala-Miursel*, *Guilosha Pink*, *Kaim-Nar*, *Krmyzy-Kabukh*, *Nazik-Kabukh* and *Shakh-Nar* in Azerbaijan; *Devedisi*, *Kadi*, *Lefon*, *Misk*, *Zivzik*, *Cekirdeksiz* and *Hicaz* in Turkey; *Kandhari* and *Bedana* in Afghanistan; *Ganesh*, *Bhagua*, *Mridula*, *Bedana*, *Jalore*, *Jyothi*, *Sindhuri* and *Arakhta* in India; and *Rabab*, *Malase Saveh*, *Yousef-Khani*, *Malase Yazdi*, *Daneh-Siah*, *Shisheh-Kap*, *Bajestani*, *Naderi* and *Ghojagh* in Iran (Mohseni, 2009; Varasteh *et al.*, 2009).

2.3 Botany

2.3.1 Habitat

Pomegranate is primarily subtropical and naturally adapted to regions with cool winters and hot summers, except the Socotran pomegranate, which is tropical and endemic to the Socotra Island and found in moist woodland at an altitude of 300–1200 m (Miller, 2004). Pomegranate grows well in areas where winter temperature is not less than −12°C, with long, hot summers, and a warm dry even autumn (Djavakyants, 2011). Naturally growing pomegranates can be found at altitudes ranging from 300–1800 m above sea level on rocky mountain slopes, on riverbank sands, and even on gravels and alkaline soils, but generally not in marshy place (Thakur *et al.*, 2010; Djavakyants, 2011). Pomegranate grows on different soils ranging from sand and shingle to heavy clay, with the exception of highly saline or very calcareous, alkaline soils. It prefers slightly acidic soil (pH 5.5–6.5) and can tolerate up to pH 7.5. Even though pomegranates have good drought resistance, they will grow well in soils of high moisture content (Burmistrov, 1993).

2.3.2 Habit

The Socotran pomegranate (*P. protopunica*) usually grows as a small tree attaining a height of 2.5–4.5 m. However, it may become prostrate when growing in some upland, plateau regions (Miller, 2004). The dwarf pomegranate (*P. nana*) does not exceed a height of 1.5 m (Holland *et al.*, 2009) and is widely used as an ornamental plant due to its miniature habit (Stover and Mercure, 2007).

The common pomegranate (*P. granatum*) is a shrub or small tree of 3–10 m in height. A fully grown wild tree may vary in appearance from drooping to erect with a curved trunk, heavily branched, more or less spiny, and may live longer than 100 years (Morton, 1987). It usually develops only one stem or trunk in the beginning when grown from a seed or vegetative cutting, but naturally tends to develop multiple trunks, owing to the suckers produced from the base of the plant, and thus has a bushy or shrubby appearance in natural conditions (Levin, 2006a).

Wild pomegranates are usually deciduous (Levin, 2006), and cultivated forms from temperate regions are also deciduous, while those grown in tropical regions may be evergreen or deciduous depending on the cultivar type. Wild Indian pomegranates lose their leaves in winter, whereas leaf-shed is an adaptation to a drier and more temperate climate (Burmistov, 1993). Deciduous types are usually erect and produce many suckers as compared with the evergreen types (Jagtap *et al.*, 1992).

2.3.3 Stem

The stem of a pomegranate tree when newly grown is smooth, covered by a red-brown bark that later becomes grey, and is often quadrangular at young stage. Branches are stiff, angular and often spiny. Young branches, developed from the vegetative growth of the recent year, are numerous and thin. The colour of the bark in young branches depends on the variety. In some varieties, bark colour varies from pink to purple, while in others it is light green with pink/purple spots or stripes. Young branches sometimes have thorns at their tips that are already visible in the axils in the young bloom. Branches may grow as short, medium or long shoots, and usually the short- and medium-length shoots may develop flower buds.

Upon maturation, the pink colour of the bark starts to disappear, the bark becomes light grey in the second year and darkens as the branches matures (Goor and Lieberman, 1956). In old trees, the bark tends to split, and detaches from the trunk in certain cases. The wood colour is light yellow. The stem and bark contain alkaloids (Holland *et al.*, 2009; Teixeira da Silva *et al.*, 2013; Smith, 2014; Erkan and Dogan, 2018). A wood-like underground stem called a xylopodium performing the function of vegetative reproduction and an accumulating organ is often found in pomegranate, and is important for survival under extreme conditions (Levin, 2006).

2.3.4 Stem anatomy

Young stems are tetragonal, with four wings that are soon lost. The bark is about 6 cm across, dingy-grey and finely fissured (Burmistrov, 1993). Scattered secretory cells are found in the cortex and pith (Fig. 2.2). It possesses cork cambium which is initially deep-seated. Stem nodes are unilacunar. Primary vascular tissues appear in a cylinder, without separate bundles, and are bicollateral. The phloem is internal, and cortical or medullary bundles are absent. Secondary thickening develops from a conventional cambial ring. Primary medullary rays are narrow. The wood is diffuse-porous. The vessels are very to moderately small, solitary, radially paired and appear in radial multiples. The vessels have vestured pits and simple end-walls, but lack spiral thickening. The axial xylem has libriform fibres, including septate fibres, but lacks fibre tracheids and spiral thickening. The parenchyma is scantily paratracheal (or absent), and the secondary phloem is not stratified. There is no 'included' phloem seen, and the wood is not storied.

The phloem of stem contains calcium oxalate druses (Lersten and Horner, 2005), and the accumulation of druses in the phloem of the pomegranate stem possibly provides protection from bark-boring insects.

2.3.5 Leaf

Pomegranate leaves are deciduous, and are arranged in alternate, opposite or subopposite positions, often crowded on short lateral shoots. The leaf is simple with short petiole, and is exstipulate, glossy, entire, glabrous, eglandular, non-sheathing, pinnately veined, cross-venulate, lanceolate, oblanceolate, oblong or sometimes obcordate. In *P. protopunica*, leaves are circular or elliptic with an incision at the top or are obtuse (Levin, 2006). The size

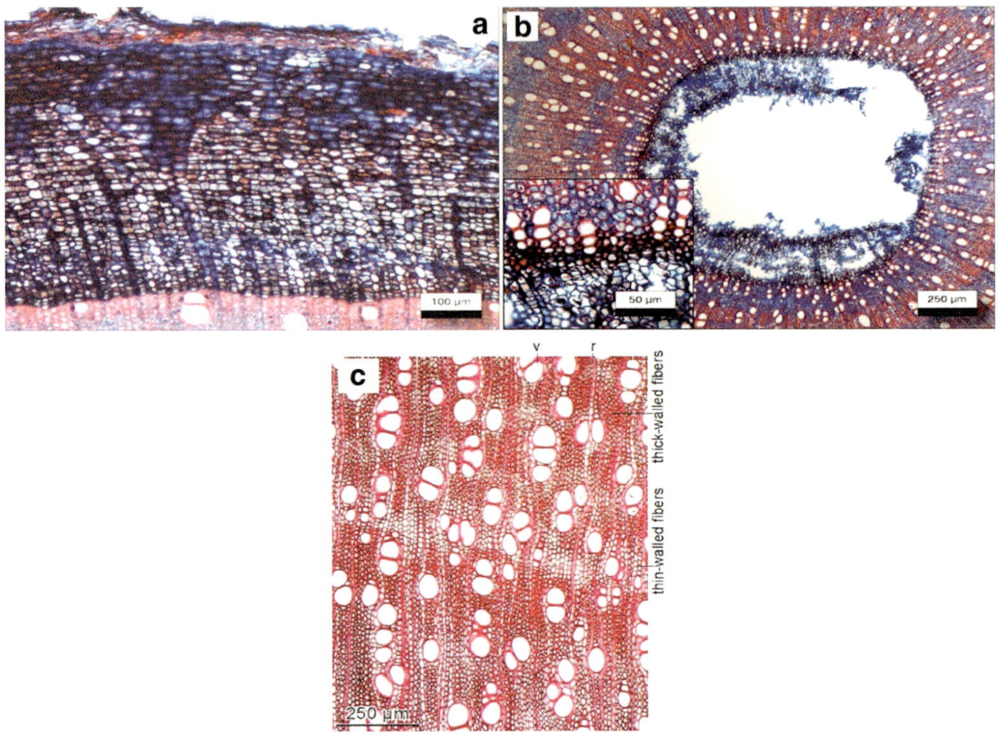

Fig. 2.2. Stem anatomy of pomegranate (a) Bark with some dilated rays, sclereids scattered or irregularly dispersed, crystal druses present, cell content in parenchyma cells, phloem uniform, phellem homogeneous and distinct in polarized light. (b) Wood with diffuse-porous xylem, vessels arranged in short radial multiples, fibres thin to very thick-walled, more than 20 rays per millimetre. Pith not visible in polarized light, pith shape square, unlignified cells present, cell content present (dark staining substances), prismatic crystals present, crystal druses present, vascular bundles not distinct, tracheary elements of metaxylem in distinct radial rows. (c) Enlarged portion of (b) showing vessels (v), rays (r), and fibres (thin- and thick-walled). (Photos: (a, b) Crivellaro and Schweingruber (2013) (Springer); (c) Schweingruber *et al.* (2011) (Springer-Verlag).)

of the leaves varies from 2–11 cm in length and 1–3 cm in width. Upon maturation, the reddish young leaves turn bright green, with the upper face being darker, while the petiole maintains its reddish colour (Holland *et al.*, 2009; Fawole and Opara, 2013a).

Pomegranate is heterophyllous, possessing two types of leaves: obcordate (5 cm in length and 2 cm in width) and lanceolate (11 cm in length and 3 cm in width). Both the types bear the same anatomical features but differ in the developmental aspects leading to their final shape. However, they are distinguishable according to the dormant buds on the basis of shape and especially the apex (Rajaei and Yazdanpanah, 2015).

2.3.6 Leaf anatomy

The leaf lamina is dorsiventral, and the epidermis has crystal idioblasts containing large, solitary crystals on the boundary between palisade and spongy mesophyll, with secretory substances (Holland *et al.*, 2009). The stomata are mainly confined to one surface (abaxial) and are anomocytic. The leaves have minor veins without phloem transfer cells and possess druses in the midrib. Crystals do not occur in the phloem or xylem of the midvein, but in the surrounding midrib parenchyma cells. Early in leaf development small prismatics occur in midveins. Consequently, a greater number of druses appear. Prismatics originate in the subpalisade layer parallel with the vascular bundles, and expand in opposite directions, many of them appearing finally as massive crystalline pillars extending from upper to lower epidermis (Lersten and Horner, 2005).

Pomegranate has an apical foliar nectary gland. Each leaf bears a single apical nectary consisting of a closely packed mass of densely staining cells, surrounded by a layer of larger cells that are more vacuolated. The epidermis at the apex of the nectary bulges to form a bump over a small chamber. The leaf apex has no stomata or other specialized pores for nectar to escape from, although separation of epidermal cells and conspicuous ruptures may occur. The midvein, together with two to four lateral bundles, converge on the nectary, and while the xylem stops before reaching the nectary, the phloem curves about halfway around the nectariferous tissue. Nectar droplets are composed of roughly equal amounts of fructose, glucose and sucrose (Turner and Lersten, 1983).

2.3.7 Buds

Buds of pomegranates are small (0.2 cm), brownish-green and turnip-shaped (Burmistrov, 1993). Levin (2006) reported larger-sized buds located in the top part of brachiblasts (spurs) and smaller-sized axile buds of vegetative sprouts. However, Rajaei and Yazdanpanah (2015) reported two types of ecodormant scaly buds in cv. Rabbab-e-Neyriz – narrow buds with sharp scales, and large bud with oval-shaped scales. Each bud was reported to bear four scales, a pair of transition leaves and two types of primordial leaves. Narrow buds were reported to develop only the vegetative shoots, whereas large buds can develop either reproductive or vegetative shoots.

2.3.8 Inflorescence

The inflorescence is a dichasial cyme bearing two to seven flowers and is placed terminally or axillary. Flowers are primarily borne subterminally, on short lateral branches older than 1 year, although some cultivars flower on spurs. Flowers can appear as solitary, in pairs or with a few in clusters. The solitary flowers mostly appear on spurs along the branches, and the clusters are terminal (Singh *et al.*, 1978; Josan *et al.*, 1979b; Pal *et al.*, 2014).

2.3.9 Flower and flower types

The pomegranate flowers are almost sessile (*P. protopunica* with pedicels), actinomorphic, bisexual, heterostylous and have a brightly coloured hypanthium. Nectaries are located between the stamens and the ovary base (Fahan, 1976). Heterostyly is common in pomegranates and thus the flowers may be of hermaphrodite, staminate or intermediate type, which is urcerate, campanulate and tubular in shape (Fig. 2.3). The hermaphrodite flowers in pomegranate may be either pin- or thrum-type (Babu, 2010). The ratio of different flowers varies among cultivars and season to season (Martinez *et al.*, 2000).

Hermaphrodite flowers are perfect flowers, having well-formed female (stigma, style, ovary) and male (filaments and anthers) parts and have been referred to as 'fertile,' 'vase-shaped' and 'bisexual' flowers. Because the hermaphroditic flowers are the type that set fruit, they are commonly referred to as 'female' flowers, albeit with some inaccuracy (Wetzstein *et al.*, 2011b). They are long-styled, have urceolate (pitcher-like) calyx and a larger, well-developed ovary (Fig. 2.4a–f). Their stigma is at the anthers' height or emerging above them, thus allowing self-pollination as well as pollination by insects. The intermediate type of flowers bears a shorter style and tubular calyx (Babu, 2010), opens 7–8 days later than the hermaphrodite flowers and leads to underdeveloped fruits (Burmistrov, 1993).

Staminate flowers are smaller, with a campanulate (bell-shaped) calyx, have poorly developed or no pistil with atrophied ovaries (Fig. 2.4g–l). Thus, their role is more accurately depicted as functionally male flowers (i.e. flowers are not strictly male), but with degenerated female parts (Wetzstein *et al.*, 2011b). Male flowers typically drop and fail to set fruit (Shulman *et al.*, 1984; Holland *et al.*, 2009).

2.3.10 Calyx

The calyx is tubular, persistent, fleshy and valvate. It consists of five to eight sepals, fused in

Fig. 2.3. Types of flowers in pomegranate tree. Bell-shaped male flower (left), and vase-shaped bisexual flower (right). (Photos: Ali Sarkhosh.)

their base, which form a red vase shape. Upon fruit set, the sepals will not drop and stay as an integral part of the fruit as it matures to generate a fruit crowned with a prominent calyx (Babu, 2010).

2.3.11 Corolla

The corolla is polypetalous, imbricate, with five to eight petals, clawed to sessile; colour usually ranges from pink-orange to orange-red, although other petal colours like white, creamy or variegated are also found in ornamental types. Petals are obovate, slightly wrinkled and inserted between the shorter calyx lobes. The number of petals usually equals the numbers of sepals, except in the ornamental 'Double Fower' cultivar, which bears an unusually high petal number due to modification of stamens into petals (Holland *et al.*, 2009). Furthermore, after opening of a flower, petals persist for about 6–7 days in the case of ornamental cultivars (Babu, 2010) compared with only 1–3 days in the case of other pomegranates.

2.3.12 Androecium

Stamens are numerous (100–300 or more), attached to the androphore and are arranged in five or six circles (three or four circles in *P. protopunica*) (Levin, 2006). The filaments are free, filiform, about 1 cm long and red in colour, while the anthers are dorsifixed, versatile, bilocular, elliptical in shape, yellowish in colour, tetrasporangiate and dehisce via longitudinal slits (Fig. 2.5a–e). The endothecium develops fibrous thickenings. The pollen grains are numerous and shed as single grains.

Pollen grains of pomegranate are aperturate, colporate and two-celled. Erdtman (1971) first studied the pollen morphology under light microscope, and reported that pollen grains were prolate (26 × 18 μm) in equatorial view with three colpi. Subsequently, a pollen study on Iranian pomegranate cultivars under the scanning electron microscope (SEM) reported prolate (elliptical) pollen morphology having differences in exine pattern with foveolate ornamentation (Varasteh and Arzani, 2009). Yang *et al.* (2015)

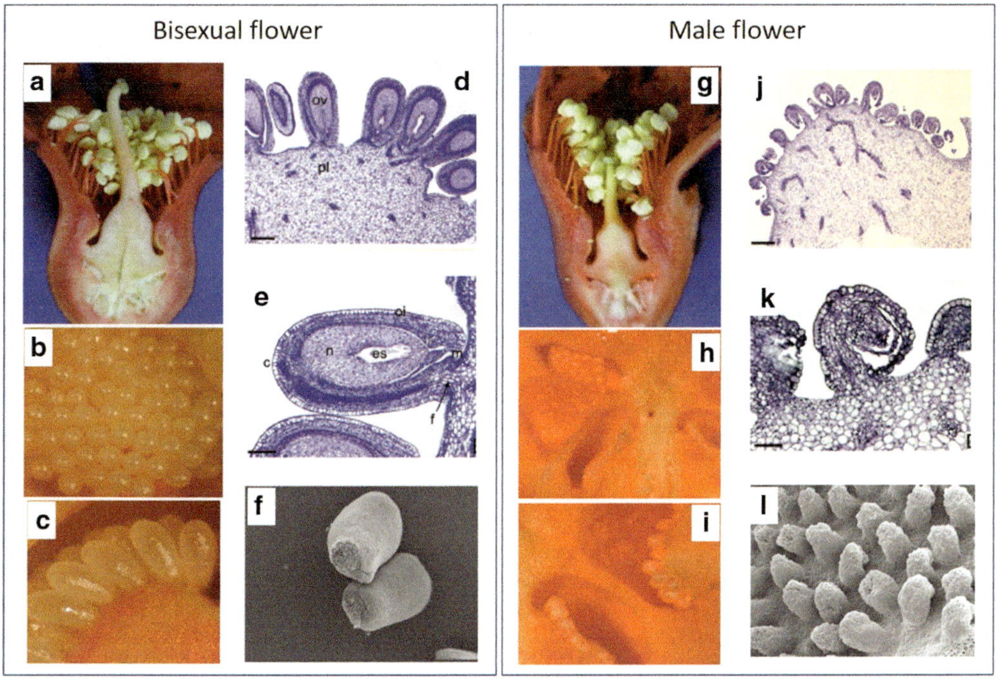

Fig. 2.4. Comparison of ovule development between bisexual and male flowers. Bisexual flower (a–f): (a) flower with a long pistil; (b, c) well developed numerous ovules; (d, e) (enlarged) – sectional view of ovules; (f) functional ovules with smooth outer surface and micropylar opening near funicule scar viewed under scanning electron microscope. ov – ovule, pl – placenta, f – funiculus, oi – outer integument, m – micropyle, n – nucellus, c – chalaza, es – embryo sac. Male flower (g–l): (g) flower with short pistil; (h, i) ovule degeneration; (j, k) (enlarged) – sectional view of degenerate ovules; (l) underdeveloped and collapsed ovules viewed under scanning electron microscope. Scale: d, f, j, l = 100 µm; e = 50 µm; k = 25 µm. (Photos: Wetzstein *et al*., 2011b, American Society for Horticultural Science.)

gave a more detailed account of pollen morphology of Chinese pomegranate cultivars where pollen grains were reported to be tricolpi monads that were rounded or triangular in polar view, and prolate (P/E <2) or perprolate spheroid in equatorial view. Further, the polar axis varied from 22.75–31.50 µm in length, and the equatorial axis (E) varied from 12.75–17.75 µm. The membrane form of the aperture hole was with or without evagination, and exine ornamentation was psilate verrucate to micromesh verrucate to scabrate verrucate.

2.3.13 Gynoecium

The gynoecium is syncarpous and the number of carpels (polysperm cavities) is usually eight, but varies among cultivars. The carpels are superimposed in two whorls, usually five on top and three underneath forming a syncarpic ovary (Babu, 2010). The ovary is multilocular, inferior with axile placentation (in *P. protopunica*) or axile and parietal (in *P. granatum*). Differential growth of carpels in *P. granatum* superposes them in two or three layers with axile (the lower) or ostensibly parietal (the upper) placentation, and a peculiar type of fruit is formed, which is termed as a 'balusta', a very distinctive character of *Punica* (Nath and Randhawa, 1959).

The stigma is discoid, capitates, papillate and covered with copious exudates (Fig. 2.5f–k). The style is thin with a stylopodium at the base (*P. protopunica* is without a stylopodium) and lengths may vary from flower to flower. Ovules are numerous, 20–50 per locule, thick, elliptical,

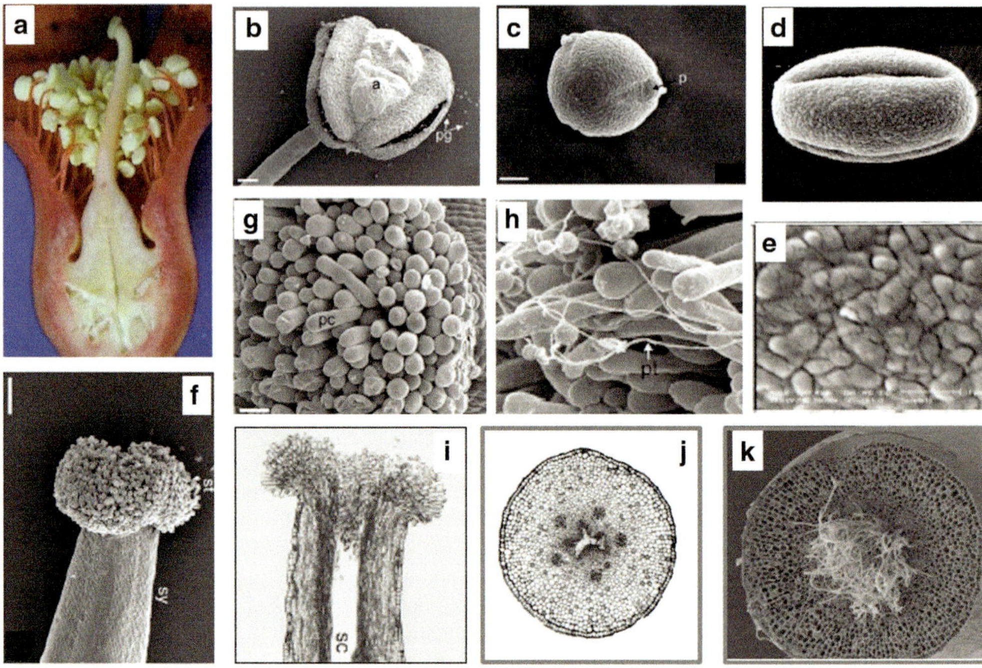

Fig. 2.5. Male reproductive structure and pollen germination events in pomegranates: (a) Longitudinal section view of a bisexual flower showing a centrally located carpel surrounded by numerous stamens; (b) stamen with dehiscing pollens; (c) pollen under light microscope; (d) pollen shape (3000×) under scanning electron microscope; (e) exine pattern of a pollen (20,000×) under scanning electron microscope; (f) stigma and style; (g) disc-shaped stigma with elongated papillary cells; (h) germinating pollens over stigmatic papillae; (i) longitudinal fracture of stigma and style; (j) cross-section of the style of an unpollinated flower showing vascular strands and stylar canal, (k) cross-section of the style of a pollinated flower showing massive growth of pollen tubes within the stylar canal. a – anther, pg – pollen grains, p - pore, pc – papillary cells, pt – pollen tube, st – stigma, sy – style, sc – stylar canal. Scale: (b, f, i) 200 μm; (c) 5 μm; (g, h) 50 μm; (j, k) 100 μm. (Photos: (a–c, f–k) Wetzstein *et al.*, 2011b (American Society for Horticultural Science); (d, e) Varasteh and Arzani, 2009 (Springer).)

anatropous, bitegmic and crassinucellate; the outer integument is multilayered contributing to the micropyle, and the archesporium is unicellular.

2.3.14 Pollination and pollinators

Both self- and cross-pollination occurs in pomegranate (Gammie and Patwardhan, 1929; Nath and Randhawa, 1959; Nalawadi *et al.*, 1973; Morton, 1987; Jalikop and Kumar, 1990; Karale *et al.*, 1993). Cross-pollination in pomegranate is due to protogyny. Insects, mainly black ants (*Camponotus* sp.), honey bees (*Apis* sp.) and lemon butterfly (*Papilio demoleus* Linn.) or hummingbirds are reported as principal pollinators (Nath and Randhawa, 1959; Morton, 1987; Melgarejo *et al.*, 2000b). Furthermore, beetles belonging to the genera *Cetonia* and *Trichodes* are reported to affect both cross- and self-pollination in pomegranate (McGregor, 1976). Wind pollination (anemophily) is also reported to occur, though infrequently, as indicated by the quite low concentration of pollen grains in the atmosphere (Morton, 1987).

2.3.15 Pollen germination and fertilization

Pollen receipt on the stigmatic surface initiates a series of pollination events leading to fertilization (Wetzstein *et al.*, 2011b). After pollination, pollen tubes reach the base of stigmas within 24 h of pollen germination. Pollen growth through the style is within a central stylar canal that can accommodate the growth of hundreds of pollen tubes. Pollen tubes reach the ovules, and sperm cells enter via the micropyle and fertilization occurs. The fertility of pollen grains produced by hermaphrodite and male flowers differs by 6–20% and 14–28%, respectively. Both size and fertility of pollen grains vary with the cultivar and season (Morton, 1987). Each aril (seed) is the result of an independent fertilization event.

2.3.16 Fruit

The fruit of pomegranate is a non-climacteric and indehiscent berry with two or three layers of locules and a leathery rind representing the hypanthium. The fruit is globose or somewhat flattened at the top, 5–12 cm in diameter and weighing 200–650 g, sometimes more. It is crowned by the thick tubular prominent calyx, which is maintained to maturity and forms a distinctive feature of the pomegranate fruit. Depending on the variety, the apex of the crown is almost closed to widely open. Typically, the fruits ripen 5–8 months after flowering, depending on the variety and prevailing temperature (Morton, 1987; Kader, 2006; Holland *et al.*, 2009).

The fruit is coriaceous and woody (Holland *et al.*, 2009). It has a spongy mesocarp (albedo) and is divided by a horizontal diaphragm and vertical septal membranes made up of inedible papery tissue into several chambers that are organized in a non-symmetrical way. The lower part of the fruit usually contains two or three chambers, while the upper part has six to nine chambers. Each chamber is filled by many seeds, which are crowded on thick, spongy placentae.

2.3.17 Rind

Pomegranate fruit has a smooth pericarp also known as rind or husk, and is connected to the tree with a short stalk. The rind (peel) is tough, thick and pliable. It comprises three parts: (i) the exocarp, which provides a cuticle layer and fibrous mat, (ii) the mesocarp (also known as the albedo) is the spongy tissue, and (iii) the endocarp or inner fruit wall. At some parts it is protuberant (verrucous) where placenta or the arils get attached. Sepal membranes are the extensions of the endocarp in the form of papery tissue that further compartmentalizes groups of arils (Stover and Mercure, 2007).

The peel colour of the fruit varies from green, pink, reddish, to dark red, with some cultivars having an external black colour. There are some exceptional cultivars with black skin from the early fruit development that remains black until ripening time (Babu, 2010). The peel thickness varies from 1.5–4.24 mm depending on genotype, and cultural practices (Erkan and Dogan, 2018). Anatomically the rind (pericarp) bears a cuticle layer on the outer surface of pomegranate fruits followed by epidermal cells organized by one layer immediately underneath the cuticle (exocarp). The epidermis forms the outermost cell layer on the structure of pomegranate fruit peel. Lens-shaped lenticels, which function as stomata, are found on the peel of the pomegranate fruit and are evenly distributed on the epidermis. Parenchyma cells in the peel of the pomegranate are in the isodiametric shape, and vascular bundles and sclerenchyma cells are located between parenchyma cells. Protective sclerenchyma cells are densely distributed in the peel of fruit (Yazici *et al.*, 2011).

2.3.18 Seeds and arils

The seeds are non-endospermic, prismatic in shape, with pulpy testa and woody tegmen. Groups of seeds are compartmentalized by the white septal membrane but do not attach to it. Sarcotesta (the outer layer of testa) is well developed in *P. granatum*, whereas it is less developed in the case of *P. protopunica*. The sarcotesta in pomegranate cannot be called a true 'aril' although the latter is frequently used in the literature. It develops entirely from the outer epidermal cells of the seed and elongates to a very large extent in a radial direction. A turgor pressure developed by the sap of these cells preserves their characteristic external shape. The

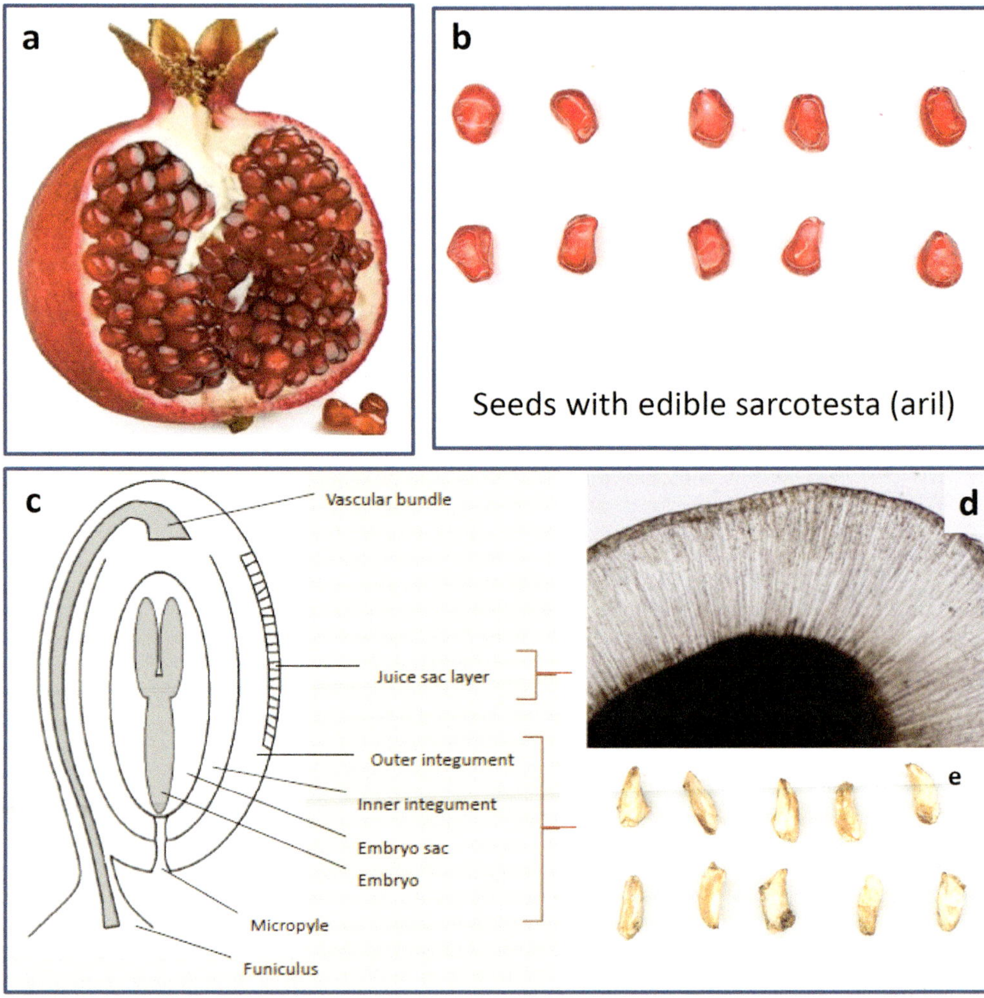

Fig. 2.6. Fruit and seeds of pomegranate: (a) fruit with persistent calyx and numerous seeds inside; (b) seeds with edible arils; (c) diagrammatic view of seed and aril; (d) aril section showing a single layer of giant tubular cells and seed (black); (e) seeds separated from edible arils. (Photos: (a, c) Porter and Wetzstein, 2014; (b, e) Ali Sarkhosh; (d) Wetzstein *et al.*, 2015 (Acta Horticulturae).)

sarcotesta (aril) consists of a single layer of giant tubular cells that radiate outward from the surface of a centrally located seed (Cui *et al.*, 2004; Wetzstein *et al.*, 2011a) (Fig. 2.6). Juice cells are tightly packed, polygonal in cross-section and have a thick-walled outer cell surface that is covered by a cuticle.

The colour of the edible juicy arils can range from virtually colourless to deep red, depending on the variety. The arils account for about 55–60% of the total fruit weight, containing 80% juice and 20% seeds (Erkan and Kader, 2011), although these also depend on the cultivar and growing conditions. The size and taste of arils and hardness of seeds vary among different cultivars. Seed softness is attributed to its content of sclerenchyma tissue. Varieties known as seedless contain seeds that are soft. In general, the wild genotypes have lower aril ratios as well as higher seed hardness and toughness (Al-Said *et al.*, 2009). The number of locules and arils per fruit varies, but may be as high as 1300 (Stover and Mercure, 2007).

No correlation exists between the colours of the fruit's outer skin and the arils, which could be similar or different depending on the variety or cultivar. The external skin colour neither indicates the extent of ripening of the fruit, nor its readiness for consumption. The fruit can attain its final colour long before arils are fully ripened. Cultivars are categorized as sweet, sweet/sour or sour depending on acidity levels (Levin, 2006; Holland *et al.*, 2009; Teixeira da Silva *et al.*, 2013; Erkan and Dogan, 2018). The sugar content of wild pomegranate is comparable with that of the cultivated types, but the sourer taste in the former is due to higher (sometimes double) acid content (Burmistrov, 1993).

2.3.19 Embryo

The embryo in pomegranate is well differentiated (oily), large, spirally rolled, achlorophyllous and straight, has two cotyledons and germinate phanerocotylar (Watson and Dallwitz, 1992).

2.3.20 Root

The root system of pomegranate is flexible and roots are seen growing vertically and horizontally in matured trees. The vertically growing roots may reach a depth of 170–180 cm after 4–6 years, while the surface roots may extend beyond the canopy limits, which often helps in anchoring to rocky mountain slopes and soils in natural habitats (Burmistrov, 1993); Rathore *et al.*, 2013). In contrast, the root distribution in pomegranates propagated by stem cutting is shallow, often penetrating less than 60 cm and rarely above 90 cm into the soil, although it also depends on the soil and environmental conditions and cultural practices (Hiwale *et al.*, 2011). Rathore *et al.* (2013) reported higher feeder root densities post rainy season compared with spring.

2.3.21 Cytology

The basic chromosome number in *Punica* is x = 7, and the diploid chromosome numbers recorded in *P. protopunica* and *P. granatum* are 2n = 14 and $2n = 16$, respectively (Sheidai and Noormohammadi, 2005). Pomegranates bear small chromosomes varying significantly in size. Different pomegranate cultivars exhibit differences in chromosome size and morphology, and the occurrence of $n = 8$ in the case of *P. granatum* is a factor of advancement from an evolutionary point of view (Levin, 2006a). The diploid chromosome of *P. granatum* ($2n = 16$) is reported to carry 1.4 pg of DNA containing 1412 M base pairs (Ohri, 2002; Bennett and Leitch, 2005).

Among pomegranate cultivars diploid chromosome numbers 2n = 16 and 18 have been reported. Nath and Randhawa (1959) reported $2n = 16$ in Indian cultivars 'Dholka', 'Ganesh', 'Kandhari', 'Muskat White' and 'Patiala', but $2n = 18$ in ornamental cultivar 'Double Flower'. Accordingly, Raman *et al.* (1971) reported $2n = 18$ in 'Vellodu' and some Kashmiri varieties with one or two quadrivalent associations at meiosis stage. A tetraploid plant with $2n = 32$ was obtained from the cultivar GB-1 ($2n = 16$) by air-layering (IBPGR, 1986) where the flowers and fruit of the plant exceeded the size of the original form, while the pollen sterility reached 85.4% as compared with only 7.4% in diploid plants (Das and Sur, 1968). A tetraploid pomegranate obtained *in vitro* from the dwarf variety 'Nana' had shorter roots, wider but shorter leaves. It grew and flowered normally in pots and produced more pollen grains per anther but the viability of pollen was significantly less than that of the diploid form (Shao *et al.*, 2003).

Iranian pomegranate cultivars have mostly bivalent chromosomes along with a low number of quadrivalents (Sheidai and Noormohammadi, 2005). In addition, a significant variation in meiotic characteristics such as chiasma frequency, chromosome pairing and segregation was detected among cultivars, which was attributed to their genomic differences. Furthermore, some of the cultivars possess B-chromosomes (also called accessory or supernumerary chromosomes), which can significantly modulate the meiotic characteristics, while some cultivars could produce unreduced gametes. The presence of up to five B-chromosomes that were smaller than A-chromosomes, not paired among themselves or with A-chromosomes, was reported in 23 out of 50 studied Iranian cultivars (Sheidai, 2007). A significant increase or decrease in the number of chiasmata was also reported in case of cultivars containing B-chromosomes with a change in genetic recombination. The differences in karyotypes among the pomegranates can be useful for resolving issues related to their evolution, phylogeny and intraspecific variations (Holland *et al.*, 2009).

2.4 Juvenility, Phenology and Reproductive Biology

2.4.1 Juvenile period

Pomegranate has a characteristically short juvenile period of 1–2 years, which is independent of seedling or cutting propagation (Babu, 2010). When grown from seeds, a small proportion of pomegranate seedlings, even in dwarf cultivar 'Nana', develop flowers in their first year of growth but will bear fruits in their second year (Terakami *et al.*, 2007; Holland *et al.*, 2009). Although the first-year fruits are usually smaller, the fruit colour characteristics of young plants are similar to those of mature trees. Most seedlings will flower and bear fruit from their second or third year of growth (Glozer and Ferguson, 2011). In pomegranates, the time required for young plants to get established from seeds (juvenile) to flower is similar to that of those established from cuttings of mature plants (Holland *et al.*, 2009).

2.4.2 Phenology

The pomegranate tree shows different phenological characteristics through its vegetative cycle in response to changing temperature. Melgarejo *et al.* (1997) reported 13 phenological stages, namely bud in winter dormancy (A), bud swelling (B), red tip (C), leaf sprouting and development (D, D2, D3, D4), flower bud sprouting and development (E, E2, E3), open flower (F), petals fall (G), fruit setting (H), young fruit (I), fruit growth (J), second bud sprouting (K), fruit ripening (L) and leaf fall (M), during the annual cycle of a common pomegranate variety 'Mollar de Elche' grown in Spain (Table 2.1.). However, the phenological pattern of pomegranates may vary from cultivar to cultivar, and also based on phytogeographical conditions. For example, the time required for completion of flower bud development in Indian cultivars is between 20 and 27 days (Nalawadi *et al.*, 1973; Josan *et al.*, 1979b) in contrast to 17 days (E+E2 + E3) in the case of the Spanish variety (Melgarejo *et al.*, 1997).

2.4.3 Reproductive biology

Naturally growing pomegranates reproduce and propagate by seed in the wild (Burmistrov, 1993), as the sexual reproduction in pomegranate is robust and mature seeds are viable and germinate easily. Pomegranate trees generally produce multiple suckers that sprout from the stem either underground or aboveground and traditionally suckers were used for propagation. Pomegranates are very easy to root from cuttings, and this is the major method for pomegranate propagation in horticulture. Orchard establishment can be done by directly planting the cuttings in the soil or by planting potted nursery trees. The latter method sometimes is preferred because it assures a better uniformity and successful establishing of the trees. To achieve the desirable traits of the mother plant, vegetative propagation is recommended as the plants raised from seeds may differ from their parents due to cross-pollination.

2.4.4 Flowering habit

The flowering habit of pomegranate is influenced by the climatic condition of the region where it is grown. In a tropical climate, pomegranates flower almost throughout the year, but in the subtropics, flowering occurs over a shorter period of the year in late spring (Stover and Mercure, 2007). In the northern hemisphere, flowering occurs in April–May but may continue throughout the summer, particularly in the case of young trees. Although such flowers are fertile, in Mediterranean climatic conditions the fruits of later flowers do not mature properly as the plants enter the cooler season and the dormancy period. Flowering occurs about 1 month after bud break on newly developed branches of the same year, mostly on spurs or short branches (Holland *et al.*, 2009).

Table 2.1. Phenological stages of pomegranate (reproduced from Melgarejo *et al.*, 1997).

Growth stage	Duration days	Development	Fleckinger code	BBCH code
Bud in winter dormancy	61	Bud greyish brown, completely closed, deeply linked to the twig and sharply pointed at its tip	A	00
Bud swelling	11	Bud swells, becomes paler and rounder in shape	B	01
Red tip	6	Bud opens; new shoot comes up with spear shaped and red tip	C	09
Sprouting of first leaves	6	First leaves appear; leaves furled and bright red with a pale midrib green	D	10
Leaf separation	4	New leaves separate	D2	10
Leaf growth	12	Leaves grow in length and width; colour changes from bright red to light green	D3	11
Lengthening of internodes	119	Rapid shoot growth; internodes lengthen	D4	31
Appearance of the flower buds	3	Greenish flower buds appear on shoots, buds become red after a few days; sepals visible and close together	E	51
Swollen calyx	11	Buds increase in size, become pear-shaped; differences between male and hermaphrodite flowers become apparent in the shape and the colour of the calyx; terminal branches bud together with several flowers usually abscises	E2	55
Opening of calyx	3	Sepals open, folded red petals appear inside; petals unfold and the pistil anthers become visible at later stage	E3	59
Open flower	6	Calyx and petals open totally, petals unfold over the sepals. Anthers of the stamen change to deep yellow when the pollen is ripe. Pollination takes place	F	61
Petals fall	2	Petals wither and fall; the calyx turns colour from red to orange-red; stamens bend towards the longitudinal axis of the flower and the anthers become greyish-yellow. The terminal part of the style withers	G	67

Continued

Table 2.1. Continued

Growth stage	Duration days	Development	Fleckinger code	BBCH code
Fruit setting	10	The fertilized ovary grows in size and the base of the calyx swells; the stamens wither and the fruit slam changes from orange-red to greenish brown	H	69
Young fruit	17	The fruit increases in size rapidly and the colour turns from greenish brown to green	I	71
Fruit growth	90	The fruit enlarges almost to its final size through cell enlargement; the sepals form a crown, the dry stamen being inside	J	73
Second bud sprouting	45	Resumption of shoot growth on the tree	K	39
Fruit ripening	35	The fleshy seeds change from white to pinkish-red or red; the skin of the fruit changes from green to greenish yellow, and finally to brownish-yellow with reddish patches (according to cultivar)	L	81, 85
Leaf fall	57	Leaves turn yellowish, and fall; and when complete, winter dormancy starts	M	93

Punica protopunica flowers and fruits appear from December to January and throughout the summer, and it bears smaller and less palatable fruits than *P. granatum* (Rana *et al.*, 2010).

Some pomegranate cultivars flower three times a year (Shulman *et al.*, 1984; Mars, 2000). Pomegranate flowers in spring in north India, but in central and south India, it flowers almost throughout the year. In evergreen cultivars grown in subtropical central and western India, there are three distinct flowering seasons, namely *ambe bahar* (January–February flowering), *mrig bahar* (June–July flowering) and *hasth bahar* (September–October flowering). The flowering is more profuse and yield is higher in *ambe bahar* compared with other flowering seasons (Ranpise *et al.*, 2014). Spring flowering gives fruits in summer. In evergreen cultivars, floral buds of spring flush are borne on mature wood of the 1-year-old shoot, whereas the flowers that appear during July–August are borne on the current year's growth. In deciduous cultivars, the flowers are borne on the current season's growth between July and August as well as some flower buds during June on spurs developed on older shoots.

Varieties vary with respect to their flowering season. Under Delhi conditions, 'Dholka', 'Kandhari', 'Muscat White' and 'Patiala' flower only once a year; 'G.B.1' and 'Japanese Dwarf' flower twice and 'Double Flower' blooms three times a year (Nath and Randhawa, 1959). In 'Japanese Dwarf', vegetative growth and flower initiation takes place simultaneously.

In subtropical climates of the northern hemisphere, flowering occurs from the end of March till the middle of May (Singh *et al.*, 1978). In the Central Valley of California, pomegranate blooms from early May to November, with most flowering from mid-May to early June (Stover and Mercure, 2007).

Wild pomegranate flowers from the middle of April to the end of May in the temperate climate of Himachal Pradesh, Jammu and Kashmir, and Uttarakhand, and two off-season blooms of much less intensity also appear during

July and November (Parmar and Kaushal, 1982; Rana *et al.*, 2003).

2.4.5 Sex ratio

There are three conditions of sex generally present in pomegranate, namely, male, hermaphrodite (bisexual, sometimes said to be female) and intermediate (Nath and Randhawa, 1959; Nalawadi *et al.*, 1973; Singh *et al.*, 1978; Josan *et al.*, 1979a; Parmar and Kaushal, 1982; Shulman *et al.*, 1984). Flower development can follow a path of either male or bisexual flowering. Since only bisexual flowers are capable of producing fruit, higher ratios of female:male flower types will promote higher production. Female flowers are vase-shaped and fertile with a normal ovary capable of developing fruits. The male flowers are bell-shaped and drop without fruit set (Wetzstein *et al.*, 2011b). The intermediate flowers are tubular in shape with a short style sometimes developing a fertile ovary leading to fruit set (Goor and Lieberman, 1956; Nalawadi *et al.*, 1973; Assaf *et al.*, 1991). If fruit set takes place in such flowers they may drop before maturity, even if some fruits that reach maturity become misshaped. Therefore, cultivars with a higher vase-to-bell shape ratio will have a higher fruit yield potential.

The number of vase-shaped flowers determines the fruit set capacity and shows a positive correlation with the bearing capacity (El Sese, 1988; Chaudhari and Desai, 1993). The total number of flowers and the ratio of bisexual to male flowers also vary with season, plant age, position within the plant and the environment. The percentage of vase-shaped flowers was recorded as 43–66% among the Israeli cultivars (Assaf *et al.*, 1991), 78–86% in the Turkish 'Hicaznar' cultivar (Gozlekci and Kaynak, 2000) and 53–80% among the local Indian cultivars (Nalawadi *et al.*, 1973).

The ability to manipulate the relative ratio of flower types to the environmental conditions can be very advantageous. Due to the high costs associated with developing female flowers, repression of such flowers under poor environmental conditions may help conserve limited resources that would not permit high fruit/seed yield. This repression is proposed as a mechanism to optimize the allocation of resources to male and female function (Bertin, 1982). Furthermore, pollen spread and cross-pollination increase when the number of male flowers is greater, and thus can be a way to spread genes (Herlihy and Eckert, 2002; Tanurdzic and Banks, 2004). However, it is not clear how seasonal changes and genotypic differences affect sex expression in pomegranates (Wetzstein *et al.*, 2015).

2.4.6 Andromonoecy and heterostyly

Pomegranate flowers exhibit heterostyly. Three kinds of flowers with different pistil lengths are found in pomegranate. The pistil is pin type (longer or equal to the length of stamens) in hermaphrodite flowers, and thrum type (shorter than stamens) in intermediate flowers, but rudimentary in staminate flowers (Babu, 2010).

In pomegranates, andromonoecy, where both bisexual and male flowers are found on the same plant, is common (Wetzstein *et al.*, 2011b). Bisexual flowers contain a discoid stigma covered with copious exudate, elongated stigmatic papillae and a single elongated style through which pollen tubes grow towards the ovules (Fig. 2.5). In contrast, male flowers possess short pistils devoid of a well-developed stylar canal and ovules (Fig. 2.4g–l), and thus fruit set is unsuccessful.

2.4.7 Petaloidy and self-sterility

Ornamental pomegranate cultivars have a higher number of petals due to petaloidy wherein the stamens are modified into petals. Petaloidy leads to functional male sterility and flowers that are incapable of setting fruits (Babu, 2010).

2.4.8 Anthesis and pollen viability

Anthesis in pomegranate flowers is completed within 3–5 h (Josan *et al.*, 1979b). As an early indicator of floral bud unfolding, some cracks appear at the apex of the bud. Consequently, the sepals separate gradually so that the crumpled scarlet petals become exposed and start to bulge out. The bulging petals force the sepals to grow from an incurved to straight form and finally reach a slightly outcurved form. The corolla opens within 3–4 h upon full bloom and stretches from its inflated and crumpled stage (Babu, 2010).

As a vital phenomenon, pollen viability decides fruit set. While pollen germination in pomegranate is reported to be strongly influenced by temperature, both flower types produce pollen grains with a similar size of ~20 µm (Evreinoff, 1957). Maximum pollen germination is recorded at 25–35°C (greater than 74%), which reduces significantly at 15°C (58%) and 5°C (10%). With respect to fertility, however, pollens produced from hermaphrodite flowers outcompete those produced by male flowers (Nath and Randhawa, 1959). The size and fertility of pollen grains vary with cultivar and season (Morton, 1987).

2.4.9 Dehiscence

The time of dehiscence of anther varies in different cultivars, and no general sequence was found at the time of anthesis. Generally, it begins with the opening of the flowers, but sometimes is delayed by 2–3 h. In most varieties, dehiscence starts after the flowers are in full bloom. In 'Patiala' cultivar, dehiscence starts prior to anthesis, whereas in 'Muskat White', it starts afterwards. Dehiscence is also affected by temperature and atmospheric humidity (Nath and Randhawa, 1959; Josan *et al.*, 1979a; Mir *et al.*, 2012; Fawole and Opara, 2013b).

2.4.10 Impact of pollination on fruit set

As the pomegranate is both self- and cross-pollinated, its pollination may be affected by cultural and environmental factors. Cross-pollinations result in higher fruit set in pomegranates than self-pollination (Nath and Randhawa, 1959; Levin, 1978; Jayesh and Kumar, 2004; Derin and Eti, 2001; Mir *et al.*, 2012). Therefore, co-cultivation of at least two different cultivars in an orchard is recommended in order to encourage cross-pollination and higher fruit set (Derin and Eti, 2001; Vazifeshenas *et al.*, 2015).

2.4.11 Stigma receptivity

Pomegranate flowers are receptive to and capable of setting fruit before petals are fully open. Reproductive function reduces with flower age, which can be due to stigma degradation, a decline in production of stigmatic exudate and loss of ovule longevity. Thus, the time a flower is functional for may be an important determinant of reproductive success (Steinacher and Wagner, 2010).

In pomegranate, the stigma attains maturity 1 day prior to anthesis. The receptivity is maintained until the second day after anthesis but gradually decreases throughout the next day and abruptly becomes non-receptive afterwards (Nath and Randhawa, 1959). Other studies have reported stigma receptivity lasting for 5 days (Josan *et al.*, 1979a) or 2–3 days (Melgarejo *et al.*, 2000a).

2.4.12 Flower vigour

The position of pomegranate flowers and their type is reported to affect their size characteristics as well as ovule development (Wetzstein *et al.*, 2013). Single flowers, as well as those positioned terminally within a cluster, are shown to be significantly larger than the lateral flowers. Furthermore, more flowers with poor ovule development were recorded among lateral flowers, which may negatively affect fruiting and yield.

Pomegranate fruit size is highly correlated with the number of arils per fruit. This implies that crop strategies directed towards increasing aril number would lead to larger fruits. Each pomegranate flower contains a remarkable number of ovules, which may exceed 3000 in some flowers. Large flowers consistently exhibit high numbers of ovules (2032–2090 per flower), whereas small flowers not only possess fewer ovules, but also present greater variability in their number (<1950 per flower), with some having no well-developed ovule (Wetzstein *et al.*, 2011b). Apart from the factors related to ovules, other pistil-related characteristics, such as those differentiating bisexual and male flowers, may also affect flower vigour. Typically, bisexual flowers produce ample amounts of stigmatic and stylar exudate for pollen capture, hydration and tube growth within the stylar canal. In male flowers, however, the stigmatic exudate production is limited. In this respect, although pollen grains can germinate in male flowers, pollen tubes lack directional growth and extension into the style. Considering vigour in female flowers spanning from large flowers with high vigour to intermediate types with repressed pistil development, stigmatic secretory development and the ability of styles to support a high number of pollen tubes may limit reproductive capability, especially in weak female flowers.

Adequate pollination of highly receptive flowers promotes high fertilization rates and high fruit set. Subsequent embryo development, filling of juice cells, and the development of colour and ingredients leads to the production of large, high-quality fruits (Wetzstein *et al.*, 2015). Fruit size and weight are highly correlated with the aril number and mass. Limitations in determination of fruit size may reflect reproductive issues such as poor ovule development, insufficient pollination, inadequate fertilization or poor flower vigour (Wetzstein *et al.*, 2011b). While the plants may live for a long time, finally their vigour declines and the plant becomes non-productive (Teixeira da Silva *et al.*, 2013).

2.5 Developmental Biology

Research performed on pomegranate developmental biology is inadequate, and there is ample scope for research on pomegranate development to understand the detailed mechanisms of bud, leaf, flower and fruit development, as well as the genetic regulatory pathways governing these processes. Below we review the current knowledge on pomegranate development.

2.5.1 Embryo development and seed germination

The embryo sac development is *Polygonum*-type because polar nuclei fuse together prior to fertilization. Three antipodal cells (uninucleate) are non-proliferating and synergid cells are elongated. Endosperm development is of nuclear type. In *P. granatum*, the ovule is anatropous, bitegmic and crassinucellate with a long funicle (Johri *et al.*, 1992). The outer integument forms the micropyle. The archesporial cell produces a parietal cell, and the parietal tissue is three-layered. The megaspore mother cell undergoes meiotic divisions, leading to a linear type of embryo sac. The polar nuclei fuse in the centre of the embryo sac, and the antipodals are ephemeral. The seeds are exalbuminous, and the embryo is straight with contorted cotyledons. The outer epidermis of the testa comprises a palisade of very long, thin-walled, columnar pulpy cells (sarcotesta), and a mesophyll of many layers of sclerotic cells and inner layers of radially elongated cells (mesotesta), and an inner epidermis of longitudinal tracheas with annular-spiral lignified thickenings (Johri *et al.*, 1992).

Anatomical studies performed on pomegranate seeds revealed that except in the thickening of mesotesta cell walls, there are not many differences in the structure and the radial length of the mesotesta and the sarcotesta of the soft and hard seeds of pomegranate fruit (Pujari and Rane, 2015). While in soft seeds, thick-walled sclerotic cells appear only near the tegmen, in the hard seeds all mesotesta cells, with the exception of a few near the sarcotesta, possess thickened walls (Pujari and Rane, 2015).

Seeds of pomegranate have non-deep physiological dormancy (Shalimu *et al.*, 2015). These seeds go through a series of temperature-driven changes in their capacities for physiological responses to various factors between dormancy and non-dormancy. Comparing seed germination biology of four Chinese pomegranate cultivars, the authors concluded warm moist followed by cold moist stratification were the best treatments for breaking dormancy.

2.5.2 Bud development

The buds on the secondary growth are lateral and appear on the axils of the leaves. The terminal buds at the very tip may produce a thorn, grow into a flower or clusters of flowers, or simply fall off. Having no real terminal buds, growth

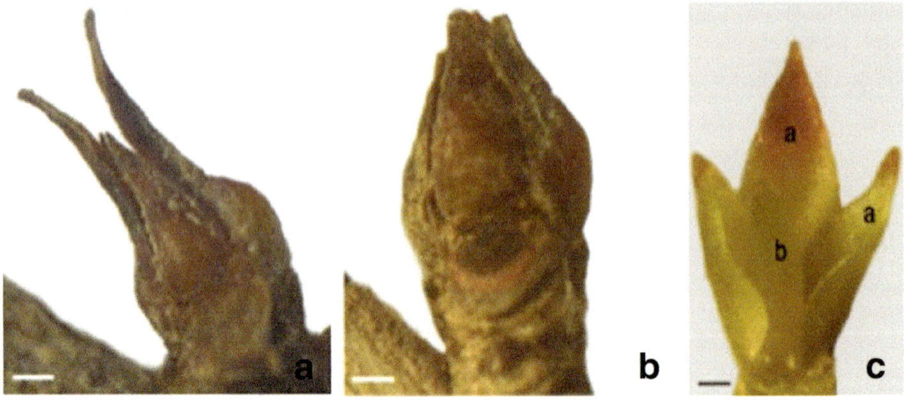

Fig. 2.7. Pomegranate dormant bud morphology: (A) narrow bud with sharp scale; (B) large bud with oval-shaped scale; (C) two kinds of leaves in a sprouting bud (a, leaf with cuspidate apex; b, leaf with acute apex). Scale bar: 1 mm. (Photos: Rajaei and Yazdanpanah, 2015 (Elsevier).)

occurs through the activity of the lateral buds (Ashton, 2006).

Two types of dormant buds (Fig. 2.7); namely, (i) narrow bud with sharp scales, and (ii) large bud with oval-shaped scales, each with four sturdy brown scales, were reported in an Iranian pomegranate cultivar (Rajaei and Yazdanpanah, 2015). While narrow buds developed only into vegetative shoots, large buds not only developed into vegetative shoots, but also went through a reproductive phase. Buds could immediately enter the reproductive phase giving rise either to a floral primordium (a simple flower bud) or a single bud containing two or more flower buds (a compound flower bud). These buds could also enter the reproductive phase after a vegetative period and give rise either to a shoot with floral primordia (a mixed flower bud) or a single bud containing a mixed bud and four to eight flower buds (a compound flower bud).

Dormant buds bear two types of scales, one pair of transition leaves, and two types of normal leaves. During bud breaking, dormant buds first become round, then swollen and the scales gradually open. Swollen buds elongate and become spear-shaped (Rajaei and Yazdanpanah, 2015).

2.5.3 Leaf development

Transition leaves of dormant buds are generally bright green in colour, have a very wide-ovate shape and appear in a decussate arrangement. However, they may differ in terms of size, number and especially apex morphology. Two types of leaf have been reported in dormant buds (Rajaei and Yazdanpanah, 2015): (i) obcordate shaped having a cleft in the apex consisting of two pairs, appear as outermost leaves, varying from 1.3 mm in length and 0.6 mm in width, with cuspidate apex; and (ii) lanceolate or oblanceolate shaped, which appear in higher numbers, varying from 1 mm in length and 0.5 mm in width, with an acute apex (Fig. 2.8).

After bud breaking, leaves become red-greenish, develop rapidly and change to green by maturity. Mature obcordate leaves are about 5 cm in length and 2 cm in width with cuneate base and short petiole, leathery, shining and entire, sometimes undulated with brochidodromous venation. Mature lanceolate leaves are about 11 cm in length and 3 cm in width, mucronate at apex and similar to obcordate leaf type in the other characteristics cited above.

To know the histological changes during leaf development, Rajaei and Yazdanpanah (2015) studied leaf development in an Iranian pomegranate at six stages. The first three stages were enclosed in the buds, and the next three stages were after emergence of the leaves from the bud.

During the first ontogenic stage of development, which occurred at the beginning of January, leaves were enclosed in dormant

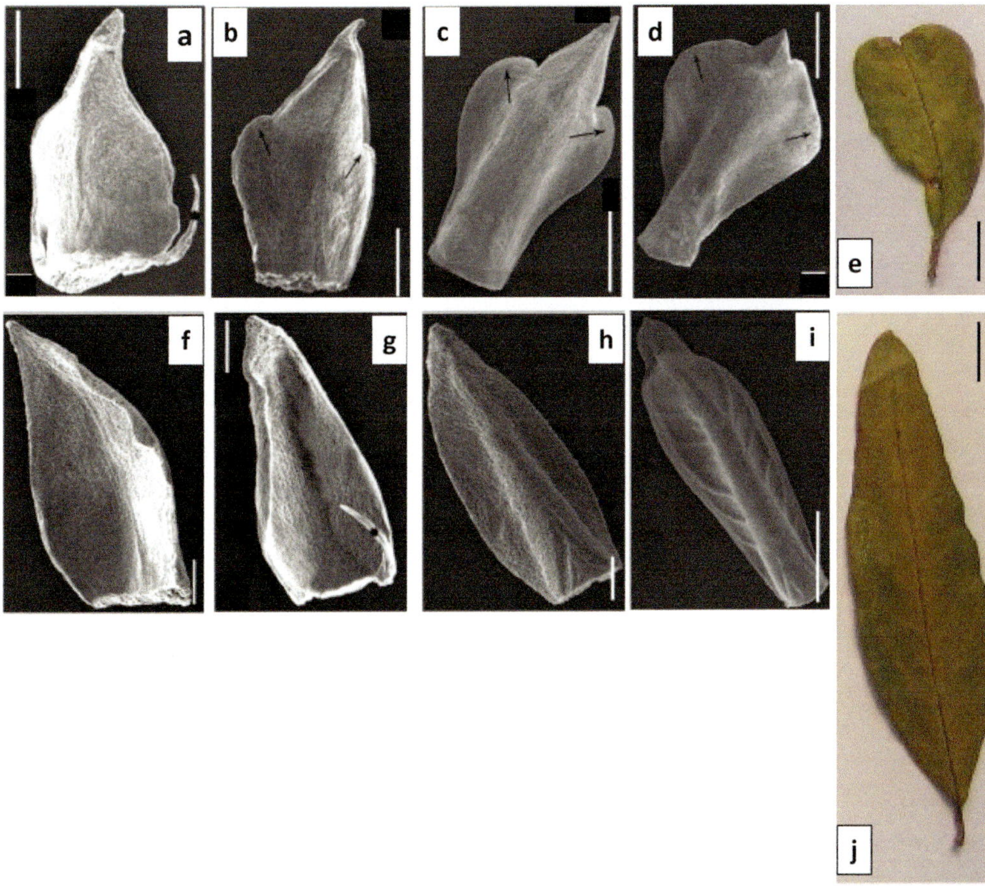

Fig. 2.8. Developmental stages of two types of leaves in pomegranate: (a–e) development of obcordate leaf with mucronate apex; (a) leaf from dormant bud; (b–d) transitional stages; (e) a matured leaf. (f–j) development of oblanceolate leaf with acute apex: (f) leaf from dormant bud; (g–i) transitional stages; (j) a matured leaf). Scale: (a, b) 500 μm; (f, g, h) 200 μm; (c, d, i) 1 mm; (e, j) 1 cm. (Photos: (a–d, f–i) Rajaei and Yazdanpanah, 2015 (Elsevier); (e, j) Stepanyan-Gandilyan, 2017.)

buds. At this stage, epidermis (both upper and lower) had a thin cuticle and granular phenolics. Mesophyll cells were thin-walled, arranged compactly, actively proliferating and uniform. Vascular tissues were differentiated in the median vein, but the lateral veins were still not formed.

During the second stage, in which leaves were enclosed within the swollen buds, epidermal cells were compact, and were round or tetragonal in appearance. Ground system cells were enlarged compared with the previous stage, and few lateral veins could be detected. Consequently, as the leaves were emerging from the bud (third stage), intercellular spaces appeared within the ground tissue and more divisions were obvious in the phloem cells of the vascular system.

In the fourth stage, epidermal cells were elongated vertically, cuticle was thicker in the red leaves compared with the enclosed leaves and several rows of cells beneath the upper epidermis were differentiated into collenchyma. Palisade parenchyma was distinguishable in the mesophyll, while the vascular system in these red leaves was characterized by bicollateral bundles and more differentiated lateral veins. Leaves appeared red-greenish in the fifth stage with the main modifications occurring in their ground tissue and the formation of angular collenchyma.

Reaching the sixth stage, green mature leaves were covered with a very thick cuticle. Upper stomatal cells were larger than the lower ones, stomata were flat, and parenchymal cells were characterized by abundant intercellular spaces, existence of druse crystals and granular phenolic compounds. Mesophyll was composed of a single row of tightly packed palisade cells, several rows of loosely packed spongy cells and a single row of palisade-like cells elongated vertically. Numerous chloroplasts, some of which formed idioblasts with solitary prismatic crystals, were detectable in palisade and spongy cells. Several unicentrally arranged druses were detectable in phloem parenchyma and bicollateral veins were narrower towards leaf margins.

2.5.4 Flower bud development

The reproductive cycle begins with the commitment of vegetative buds to floral organogenesis. Flower (reproductive) buds are located close to the apical meristem and distinguishable from the dormant bud stage (Fig. 2.9). Upon floral evocation, six sepal primordia are detectable within the spear-shaped buds at the initial stage (Fig. 2.10). Consequently, numerous small papillae laterally cover the sepals (Rajaei and Yazdanpanah, 2015). Based on the weight (size) of flower buds, 12 developmental stages have been proposed for flower buds, ranging from pinhead to ready-to-crack stages (Fig. 2.11). Furthermore, a gradual increase in flower bud weight has been detected throughout these stages (Babu, 2010).

Under tropical conditions, flower bud development occurs at varied times. The time from the initiation of flower bud growth to anthesis may range from 14 to 28 days depending on the variety and climatic conditions (Babu, 2010). In a tropical climate of the northern hemisphere, flowering occurs during the last week of March and the second week of May, with quite frequent occurrence of several distinct flushes on the same tree (Singh *et al.*, 1978; Fouad *et al.*, 1979). Flower bud development in Indian cultivars completes within 20 to 27 days (Nalawadi *et al.*, 1973). Among 24 pomegranate genotypes grown under a semi-arid climate, the time required for bud development varied between 27 days (for 'Japanese Dwarf') and 14 days (for 'Patiala') (Meena *et al.*, 2011). In semi-arid tropics of western India, flower bud development in 'Ganesh' took 20.4 days, which was higher than 'Bhagawa' (19.3 days), 'Mridula' (17.5 days) and 'Arakta' (16.2 days) (Babu *et al.*, 2009).

Not only the bud development, but also the bud dormancy is shown to vary with cultivar. Bud dormancy (defined as the time between abscission of more than 50% of the leaves, and the appearance of vegetative buds or more than 10% new growth) among 24 genotypes grown under a semi-arid climate ranged from 56 days in 'Sur Sukkar' to 66 days in 'Kazak Anar' and 'Kali Shirin' and 67 days in 'Gul-e-Shah' (Meena *et al.*, 2011).

Fig. 2.9. Shoot apex of pomegranate: (a) vegetative shoot; (b) vegetative apex visualized using scanning electron microscope. Apex subtended by two leaf primordia. Scars denote removed microscopic leaves; (c) reproductive shoot; (d) reproductive apex is showing a central floral apex with sepal primordia under scanning electron microscope. Precocious axillary buds have formed. (Photos: Porter and Wetzstein, 2014.)

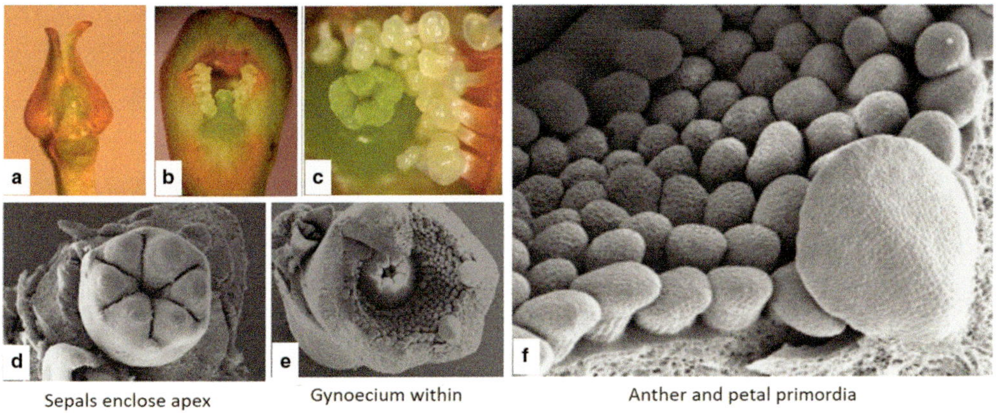

Fig. 2.10. Floral organogenesis: (a) flower bud; (b) longitudinal section view of (a); (c) transverse section view of (a); (d) scanning electron microscope view of (a) after removal of bud scales and leaves; (e) scanning electron microscope view of (a) after removal of sepals; (f) enlarged view of a portion of (e). (Photos: Porter and Wetzstein, 2014.)

Fig. 2.11. Flower bud development stages. (Photos: Alimohammad Yavari.)

2.5.5 Flower development

Control of flowering is known as an important adaptive trait and is influenced by environmental and genetic cues. As a critical step towards reproductive growth, vegetative apices will be induced to become reproductive and undergo floral organogenesis. In general, the time taken from initiation to flowering is affected by growth habitat and climatic range of the species (Reddy, 2011). Floral initiation in many temperate fruit crops occurs one or two seasons prior to the flower emergence time. In this respect, flowering initiates in summer or autumn, prior to winter dormancy. In the case of tropical and warm temperature species, however, flowering initiates just prior to anthesis, flowers appear on both old wood and new growth, and flowering varies with cultivar (Reddy, 2011).

It is not clear when pomegranate floral induction occurs. According to the microscopic dissections, meristems in dormant buds of 'Wonderful' are vegetative, and flower organogenesis is not evident prior to growth resumption in spring (Wetzstein *et al.*, 2015). In early spring, however, a floral meristem with primordia developed along the margins of a central reproductive apex could be detected and so is considered as a precocious axillary bud in early stages of floral development.

It is reported that pomegranate flower undergoes ten developmental stages. In the early bloom stage, the flower resembles a small pear, and is either greenish at the basal part and reddish at the apex, or is entirely dark red. Upon maturation, the sepal becomes orange-red to deep red, depending on the variety, while petals remain orange-red or pink, and rarely turn white. A good correlation is reported between the colour of the sepals and the final colour of the fruit skin where cultivars with a darker-red flower will usually have a deep-red fruit skin colour (Nalawadi *et al.*, 1973; Feng *et al.*, 1998; Wang, 2003; Babu, 2010).

The flowering duration (the number of days spanning the appearance of flower bud to full bloom) also varies with cultivar. Among 24 genotypes grown under a semi-arid climate, flowering duration ranged from 16 days (for 'Sur Sukkar') to 39 days (for 'Dholka') (Meena *et al.*, 2011). Modified cultural practices, including the application of chemical treatments to promote flowering may be considered as ways to improve flower production (Chaudhari and Desai, 1993; Reddy, 2011).

2.5.6 Male generative sphere

Deposition of viable pollens on to a functional pistil is required for fruit development. Pollen receipt on the surface of the stigma initiates pollination events, which lead to fertilization. Pollen tubes grow through a central stylar canal (Fig. 2.5k) and reach the base of stigma within 24 h after pollination. The stylar canal can accommodate hundreds of pollen tubes forming clusters of chord-like structures (Wetzstein *et al.*, 2011b). Upon successful guidance in the pistil, the pollen tube reaches the ovules, and sperm cells enter the micropyle ultimately leading to the fertilization events. In the case of male flowers, the pollen germination and tube growth lack directionality and fail to penetrate the stigma, and therefore fertilization is unsuccessful (Wetzstein *et al.*, 2011b).

2.5.7 Female generative sphere

In contrast to the large number of stamens, pomegranate flowers contain a single pistil. This elongated pistil is terminated by a discoid stigma, and possesses an inferior ovary (Fig. 2.5). The ovary, however, contains numerous ovules. The stigma is covered with elongated papillae, which are coated with abundant exudate (Wetzstein *et al.*, 2011b). Stigmatic exudate is believed to participate in directing pollen tubes towards and into the stigma through directional gradients of water potential and guiding their chemotropic growth (Kim *et al.*, 2003; Higashiyama and Takeuchi, 2015).

Successful fertilization and fruit set is only possible in well-developed pistil as in the case of hermaphrodite flowers. Stigmatic papillae of male flowers hold little exudate but can support pollen germination. However, pollen tubes are rarely observed in styles. Ovules in male flowers are somehow rudimentary and exhibit various stages of degeneration (Wetzstein *et al.*, 2011b). The intermediate flowers also exhibit various degrees of ovary degeneration (Goor and Lieberman, 1956; Nalawadi *et al.*, 1973; Assaf *et al.*, 1991). Their styles may equal the length of the long-styled flowers or are as short as the short-styled ones. Occasionally, long-styled intermediate flowers become fertilized, but the fruits often drop early, and if matured, will be misshaped and defective (Hodgson, 1917; Nath and Randhawa, 1959).

2.5.8 Metaxenia

The effect of foreign pollen on maternal/ovular tissue, referred to as 'metaxenia', has been reported in pomegranates (Levin, 2006). Xenia effects due to cross-pollination in pomegranate have already been demonstrated by some workers (Derin and Eti, 2001; Karimi and Mirdehghan, 2015; Vazifeshenas *et al.*, 2015; Gharaghani *et al.*, 2017). The pollen source influences fruit properties such as fruit shape, size, weight, rind colour and thickness, as well as seed characteristics including seed size, hardness, and aril colour and weight. Occasional occurrence of several arils with different colours within a single pomegranate fruit is due to metaxenia (Holland *et al.*, 2009).

2.5.9 Fruit development

Fruiting in pomegranate plants may start in the second year, but the substantial bearing of fruits appears after 3–5 years' maturity in plants. Flowering and fruit set last about 1 month (Holland *et al.*, 2009). Aril development completes within 80 days from fruit set, and is associated with a progressive increase in total sugars, reducing sugars and anthocyanin pigments. This period is accompanied by a significant reduction in total phenolics, ascorbic acid and acidity, followed by a steady state. Following fruit set, the sepals' skin colour changes progressively from orange-red to green. The colour changes again in later stages of fruit maturation to reach the final characteristic colour of the ripe fruit. The external fruit colour, which varies between yellow, green and pink, changes to fully red, deep pink or deep purple, depending on the variety and the ripening stage. The black pomegranate, one of the exceptional cultivars, acquires its black skin very early in development and keeps it until ripening. The thickness of the skin (leathery pericarp) varies among cultivars (Holland *et al.*, 2009; Babu, 2010).

The pomegranate fruit growth pattern, from fruit set to maturity, has been characterized as a simple sigmoidal curve (Ben-Arie *et al.*, 1984; Varasteh *et al.*, 2008), with periods of fast fruit growth rate set apart from periods of slow growth (Kumar and Purohit, 1989). The initial rapid fruit growth occurs due to cell divisions and is characterized by growing kernel tissue and increasing testa hardness (Shulman *et al.*, 1984). Consequently, a slowdown in fruit growth occurs (Gozlekci and Kaynak, 2000). While the kernel ceases growing, the aril progressively grows as the fruit proceeds to its final size through cell enlargement (Ben-Arie *et al.*, 1984; Shulman *et al.*, 1984; Melgarejo *et al.*, 1997). This growth pattern varies among cultivars. The 'Mule's Head' cultivar follows a simple sigmoid curve, but in 'Wonderful' cultivar, the growth pattern is reported to be linear (Shulman *et al.*, 1984) or with a slow growth phase detected as fruit diameter reaches 52.5 mm (Ben-Arie *et al.*, 1984). The average fruit weight and volume of the 'Malas-e-Torsh-e-Saveh' cultivar grown in Iran increase rapidly until 45 days after fruit set and continue more slowly until harvest time (Varasteh *et al.*, 2008). Furthermore, fruit growth measurement of the Omani cultivars grown in the Al-Jabal Al-Akhadar area (Al-Yahyai *et al.*, 2009) and 'Wonderful' cultivar grown in Australia (Weerakkody *et al.*, 2010) revealed a linear fruit growth pattern, although the measurements started several weeks after fruit set.

2.5.10 Aril development

Arils are the juice-containing, multifaceted structures in mature fruit. Each aril is produced from an ovule through an independent fertilization event within the ovary. For fruit development to occur, embryo growth and development must proceed after fertilization. Each aril possesses a centrally located seed that nests the developing embryo. The layers of the seedcoat whose structure determines seed hardness are derived from the integuments that are the protective layers of the ovule. The juice-containing outer region of arils is composed of giant tubular cells that can reach up to 2 mm long. These cells elongate rapidly with significant amounts of aril growth occurring late in the season. During the last 2 months preceding harvest, aril increases 40–50% in mass. Embryo abortion and poor aril growth can lead to unfruitfulness and/or poor fruit sizing. Strong aril development, filling of juice cells and colour/ingredient formation are prerequisites to high-quality crop production (Wetzstein *et al.*, 2015).

2.5.11 Fruit ripening

Fruit ripening involves several changes in fruit characteristics from flowering and fruit set to maturity and senescence. These changes include physical, structural, biochemical and physiological changes, reflecting differences in fruit appearance during maturation and ripening among cultivars (Ben-Arie *et al.*, 1984; Shulman *et al.*, 1984; Al-Maiman and Ahmad, 2002; Holland *et al.*, 2009; Shwartz *et al.*, 2009).

The outer skin colour does not indicate fruit ripening or its readiness for consumption since fruits can attain their final colour long before the arils are fully ripened. The time taken from flowering to maturity and senescence varies among different genotype/cultivars, growing locations, climatic conditions and seasons (Shulman *et al.*, 1984; Gil *et al.*, 1995; Holland *et al.*, 2009; Fawole and Opara, 2013b). The most pronounced difference in ripening time among cultivars is not derived from the differences in flowering time but rather from the time taken from anthesis to ripening. Fruits ripen within 5–8 months or 135–180 days after anthesis, depending on the cultivar and climate (Morton, 1987; Kader, 2006; Holland *et al.*, 2009). In Israel, fruits matured more rapidly in the hot valley region than in the coastal plain (Shulman *et al.*, 1984).

Pomegranate fruits are non-climacteric and should be harvested only after reaching the fully mature stage to ensure the best eating quality (Ben-Arie *et al.*, 1984; Shulman *et al.*, 1984; Erkan and Kader, 2011). Although the calyx opening, distinctive peel colour and a metallic sound on tapping are recommended as indicators for pomegranate ripening in certain cultivars (Mir *et al.*, 2012), they are not reliable and cannot be applied in all cultivars. A scientific maturity index for pomegranate cultivars is yet to be established (Fawole and Opara, 2013b). In general, maturity indices for the fruits are cultivar dependent, depending on skin and aril colour, acid level (titrable acids, TA), total soluble solids (TSS) content and TSS/TA ratio (Al-Maiman and Ahmad, 2002; Erkan and Kader, 2011). In 'Wonderful', the harvesting is preferred when the soluble sugar varied from 15–17%, and the acids are lower than 1.85%. In Bejestan pomegranates of Iran, the optimum harvesting time was found suitable when the soluble solids content reached 17.5% (Sherafatian, 1994).

2.6 Physiology of Pomegranate

2.6.1 Growth physiology

Pomegranate is a C_3 plant, and the tree requires a long, hot and dry season for producing a good yield of high-quality fruit (Holland *et al.*, 2009). Sugars are transported as oligosaccharides + sucrose, or as sugar alcohols + oligosaccharides + sucrose.

Although pomegranates have been adapted to a wide range of climatic conditions, Mediterranean-like climates provide optimal growth conditions. This includes high exposure to sunlight, mild winters and long, hot, dry summers without rain during the last stages of fruit development. Under such conditions, the fruit will develop to its best size and optimal colour and sugar accumulation (Holland *et al.*, 2009). Chilling is required for breaking bud dormancy in deciduous pomegranates grown in temperate climates (Soloklui *et al.*, 2017). However, pomegranate is sensitive to frost; sweet pomegranates being more sensitive than the sour ones. Annual shoots are damaged at 14–15°C subzero, perennial branches at 16–17°C subzero and the aerial parts of the shrubs at 18–19°C subzero temperatures (Djavakyants, 2011).

Pomegranates flower at an average daily temperature of 20°C and pollen germinates at 12°C, but 20–25°C is optimum (Djavakyants, 2011). Fruit develops well in the arid and semi-arid regions of the subtropical climates, and a temperature of 38°C with dry climate during fruit development produces the best-quality fruits (Glozer and Ferguson, 2011). Areas with high relative humidity or rain are totally unsuitable for cultivation, as fruits produced under such conditions tend to taste less sweet and are prone to cracking. For ornamental gardening, pomegranate can be grown in all climatic conditions throughout the globe except the polar regions.

Pomegranate can adapt itself to different types of soil, but is sensitive to low drainage. The best fruits are produced on deep, water-absorbing and fertile loamy soil with good drainage (Glozer and Ferguson, 2011). Light to sandy soils can also be used for pomegranate cultivation if irrigated well.

Light affects pomegranate bearing and fruit quality. Therefore, summer pruning is sometimes needed to remove suckers and water sprouts. Winter pruning is mostly used when there is a need to induce new growth, eliminate broken or intrusive branches, and/or control the tree height (Holland *et al.*, 2009).

Pomegranate is considered as a drought-resistant crop since it enjoys the heat and thrives well in arid and semi-arid areas, even under desert conditions (Aseri *et al.*, 2008; Rodríguez *et al.*, 2012). Grown in such areas, however, the crop requires regular irrigation during the dry season in order to reach optimal yield and fruit quality (Sulochanamma *et al.*, 2005). Deficit irrigation during fruit ripening has adverse effects on fruit size and total yield (Rodríguez *et al.*, 2012). As irrigation timing further affects ripening time, irrigation is done strategically in order to direct the time of fruit yield in Indian evergreen pomegranates (Sonawane and Desai, 1989).

Pomegranate can be irrigated by alternative water sources, mainly recycled water and saline water. With salts increasing in recycled water, usage of recycled water is strongly connected to salinity (Raviv *et al.*, 1998). Although the mode and magnitude of the plant response is cultivar dependent, the crop is considered as moderately tolerant to salinity (Maas, 1993; Naeini *et al.*, 2005, 2006; Okhovatian-Ardakani *et al.*, 2010; Borochov-Neori *et al.*, 2013). Thus the crop is amenable to irrigation with saline water, although it requires constant irrigation to leach the salt and prevent the detrimental effects of increasing salinity. In response to irrigation with saline water, pomegranate tissues accumulate sodium, chloride and potassium, and the concentrations of these ions increases in parallel to the salt concentration within the irrigation water (Naeini *et al.*, 2006). Pomegranate plants can tolerate up to 40 mM NaCl in irrigation water, above which, growth parameters such as the length of the main stem, length and number of internodes, as well as leaf surface area are severely affected (Naeini *et al.*, 2006). Irrigation with saline water also results in higher vegetative growth (Holland *et al.*, 2009).

In pomegranate roots, the Casparian strip is reported (Figs. 2.12 and 2.13) to form earlier in the exodermis than in the endodermis (Tuladhar and Nii, 2014). The endodermal Casparian strip blocks the free apoplastic movement of several ions and heavy metals, as well as fluorescent dyes, and a similar function is considered for the Casparian strip formed at exodermis (Chen *et al.*, 2011; Tuladhar and Nii, 2014). The suberin and lignin accumulation in root endodermal cells and cell layers newly formed beyond the endodermis is completely different from that in endodermis-like cells. The accumulation of lignin in addition to suberization or together with suberin in the same cell layer distinguishes these cells from the periderm observed in other roots. Suberin deposition in the outermost tissues of pomegranate roots may contribute to its ability to tolerate and adapt to various soil environments (Tuladhar and Nii, 2014).

2.6.2 Physical changes during fruit maturation

In pomegranates, the fruit volume, fruit weight, total aril weight and total aril number are highly correlated (Wetzstein *et al.*, 2011a, Wetzstein *et al.*, 2015). The fruit weight in 'Bhagwa' in South Africa increased from 107 g, 54 days after the full bloom of the flowers to 322 g when harvested 139 days after full bloom (Fawole and Opara, 2013a). During this period, the fruit size changed linearly with the diameter increasing from 60–84 mm and its length changing from 54–75 mm. Fruits of the 'Wonderful' grown in Condobolin, Australia also presented a linear increase in fruit mass, reaching the maximum value of 675 g per fruit 14 weeks after the fruit set where fruit diameter increased faster than the length (Weerakkody *et al.*, 2010). These physical changes that may even occur after the optimum harvest stage are attributed to the fruit growth and are presumably caused by the expansion of the cells due to the uptake of water and nutrients (Shwartz *et al.*, 2009). Among four Israeli pomegranates, fruits that ripened in early summer and during the winter were significantly smaller than those that ripened at the end of summer and autumn (Borochov-Neori *et al.*, 2011). In the case of the Iranian 'Malas Yazdi' cultivar, the peel (with mean dry weight of 22.33 g) contributed more to the fruit mass early in the season, but from the middle to the end of the season, the arils (with mean dry weight of 35.03g at the end of the season) performed the dominant role

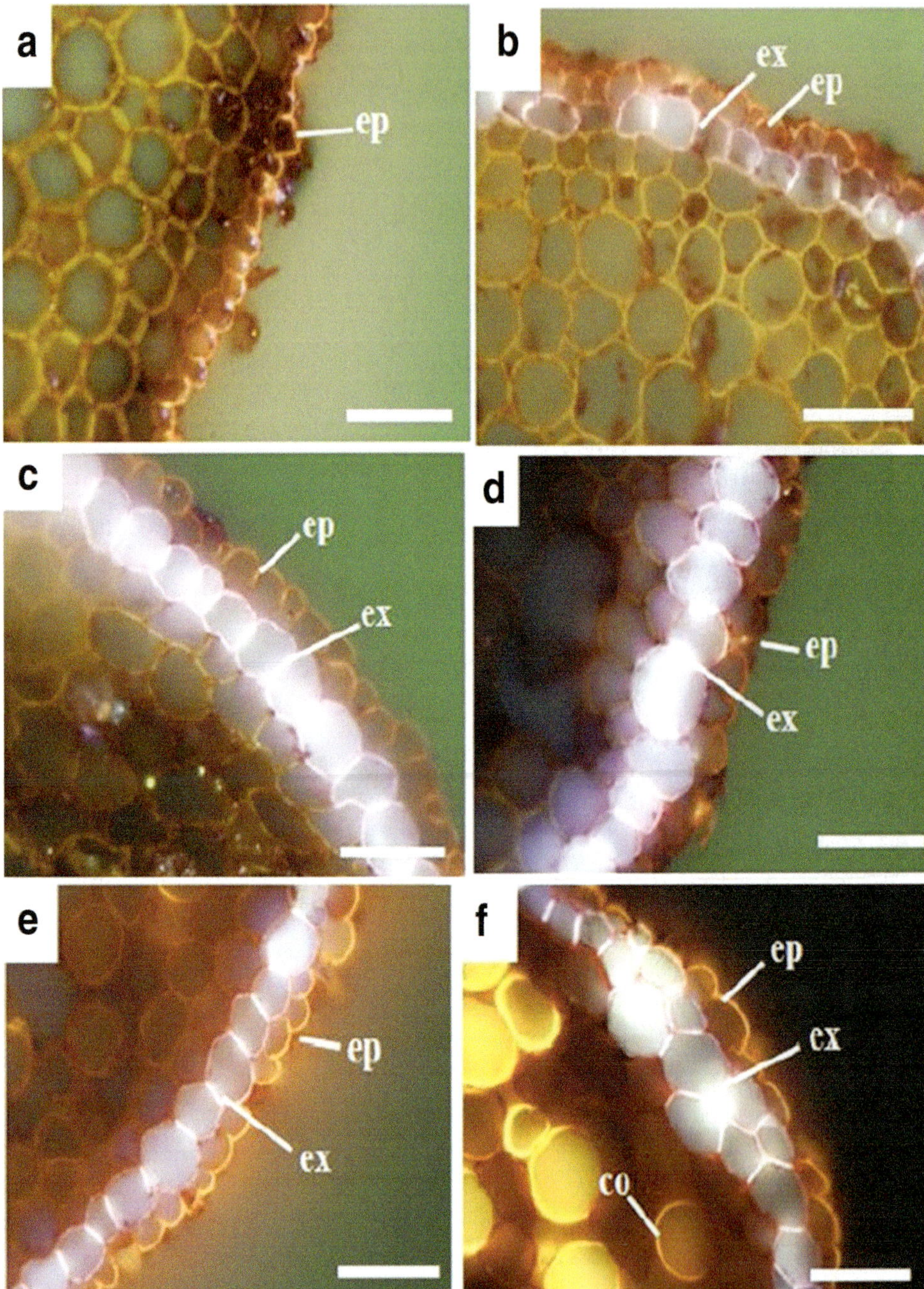

Fig. 2.12. Pomegranate root exodermis development at a different distance from the root tip (panel a: 10 mm, b: 20 mm, c: 60 mm, d: 80 mm, e: 110 mm, f: 140 mm). Root samples stained with berberine hemisulfate-aniline blue-safranin O. Lignin-encrusted rhizodermis turned reddish orange under the fluorescent microscope, and suberized areas in exodermis appeared milky white. Casparian strip formed in one or more layers at the exodermis. ep, rhizodermis/epidermis; ex, exodermis; co, lignified cortical cell. Scale = 50 μm. (Photos: Tuladhar and Nii, 2014.)

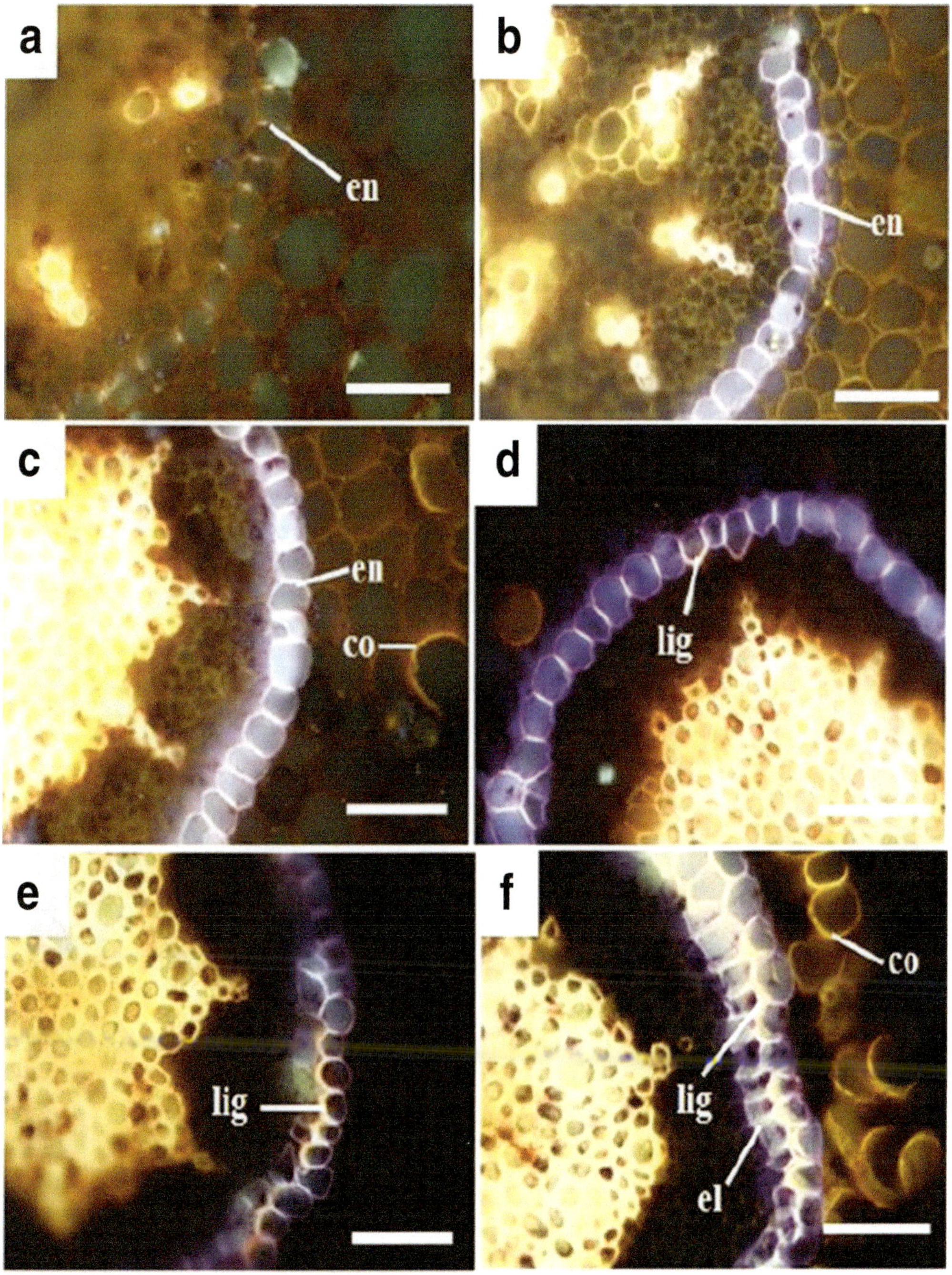

Fig. 2.13. Pomegranate root endodermis development at a different distance from the root tip (panel a: 20 mm, b: 60 mm, c: 80 mm, d: 110 mm, e: 130 mm, f: 150 mm). Root samples stained with berberine hemisulfate-aniline blue-safranin O. Lignin-encrusted cortical cells and few endodermal cells appear reddish orange. co, lignified cortical cell; en, endodermis; el, endodermis-like cells; lig, lignin accumulation. Scale = 25 μm. (Photos: Tuladhar and Nii, 2014.)

(Mirdehghan and Rahemi, 2007). Fruit arils constituted about 50% of the fruit weight in 'Mule's Head' and 'Wonderful' cultivars in Israel (Shulman *et al.*, 1984), and 57–66% in Spanish 'Mollar' (Sánchez *et al.*, 1996), while in the South African-grown 'Bhagwa' they comprised less than 50% of the semi-ripe fruit weight (until 110 days after full bloom), and increased to 58% of the fruit weight at full ripe stage (Fawole and Opara, 2013a). Aril weight increases 40–50% during the last 2 months prior to harvest (Wetzstein *et al.*, 2015).

Average fruit juice yield for 'Wonderful' cultivar grown in Condobolin, Australia amounted to 37% of the fruit weight (Weerakkody *et al.*, 2010) and ranged from 57–67% in the case of local pomegranate cultivars in Oman (Al-Said *et al.*, 2009). While the fruit juice yield for 'Mule's Head' and 'Wonderful' cultivars was less than 25% during late immature stages, it increased to 35–40% and 40–45% at harvest time (Shulman *et al.*, 1984). A wider range, but lower juice content of 18–40% measured for 'Wonderful' grown in Israel was attributed to the differences in climatic conditions (Shulman *et al.*, 1984). Between the immature and full-ripe stages, the juice content of the 'Bhagwa' grown in South Africa increased from 29 to 54% (Fawole and Opara, 2013a). Furthermore, early summer fruits of Israel-grown cultivars had the highest juice content compared with late summer, autumn and midwinter ones at the mature stage (Borochov-Neori *et al.*, 2011).

The colour of the fruit skin, as well as the aril may change during the development and ripening of pomegranate fruits (Al-Said *et al.*, 2009; Holland *et al.*, 2009). In case of the Spanish 'Mollar de Elche' cultivar, the fleshy arils change from white to pinkish-red while the fruit peel changes from green to greenish yellow, and finally to brownish yellow with reddish patches (Melgarejo *et al.*, 1997), whereas among different accessions of 'Wonderful' red pigmentation increased significantly in fruit parts during ripening (Ben-Arie *et al.*, 1984; Shulman *et al.*, 1984; Shwartz *et al.*, 2009).

The colour of pomegranate juice is influenced by several pre- and postharvest factors. Fruits grown on the coastal plains developed more intense colour than those grown in warmer valleys in Israel (Shulman *et al.*, 1984). Another report revealed an inverse correlation between the intensity of the red colour of the arils with the cumulated amount of heat during fruit development and ripening using a Chroma meter (Borochov-Neori *et al.*, 2009). With the colour being expressed in CIE Color Space, the a* value (redness) for fruit aril increased while the L* value (luminosity, ranging from 0 for pure black to 100 for white) declined during the ripening period of a mid-season cultivar in parallel to a gradual change in aril colour from white to pink. Similar results were also reported for 'Ruby' grown in South Africa (Fawole and Opara, 2013c). Colour parameters detected from 'Codpa' grown in Chile also presented the highest L* and h° values at green maturity stage (Labbé *et al.*, 2010).

2.6.3 Biochemical changes during fruit maturation

As the pomegranate fruit matures on the tree, a reduction in the titrable acidity and parallel increase in total soluble sugars (TSS), pH and colour intensity is observed (Kader, 2006). TSS increased significantly during three major fruit developmental stages in 'Rabbab-e-Fars' cultivar (Zarei *et al.*, 2011). During the development of 'Bhagwa' fruits grown in South Africa, TSS increased about 150% between 54 days after full bloom and the harvest time at 165 days after full bloom (Fawole and Opara, 2013a). Similarly, the TSS increased from 10.30°Brix in immature fruit at 20 days after fruit set to 19.56°Brix in fully ripe fruit at 140 days after fruit set (Zarei *et al.*, 2011), and from 13% in 40-day-old for 'Ganesh' fruit grown in India to 16.3% on day 140 (Kulkarni and Aradhya, 2005). The lowest TSS for 'Ruby' grown in South Africa was detected at an early growth stage (54 days after full bloom) and increased significantly by harvest time on day 139 (Fawole and Opara, 2013c). In contrast, the TSS remained almost constant between green unripe fruit and full-ripe stage in 'Taifi' with a slight increase from 16.4 to 16.9% (Al-Maiman and Ahmad, 2002) corroborating the results reported in 'Wonderful' grown in California, USA (Kader *et al.*, 1984), and remained constant throughout the ripening process in the Spanish clones 'ME5', 'ME17' and

'MO6' (Legua *et al.*, 2000) as well as the sweet 'Mollar' cultivar (Gil *et al.*, 1995).

The juice of fully mature fruits contains 12–16% sugars, mainly consisting of glucose and fructose (Sánchez *et al.*, 1996; Al-Maiman and Ahmad, 2002; Shwartz *et al.*, 2009; Fawole and Opara, 2013c, Fawole and Opara, 2013a), with glucose being the predominant sugar (Legua *et al.*, 2000). The concentration of glucose and fructose in South African-grown 'Bhagwa' and 'Ruby' increased significantly during fruit maturation (Fawole and Opara, 2013a), as is reported for '121–2' and '101–2' accessions of the 'Wonderful' pomegranate (Shwartz *et al.*, 2009). The rapid increase of the total sugar content in South African-grown 'Bhagwa' detected between immature and early half-ripe stages (54–110 days after full bloom) has been attributed to intense fruit expansion during the maturity stages (Fawole and Opara, 2013a). In 'Ganesh', the amount of reducing sugars remained unchanged during the first 80 days after full bloom but significantly increased when fruit became fully ripe at day 140 (Kulkarni and Aradhya, 2005).

As fruit maturation proceeds, the titrable acidity (TA) of pomegranate juice declines with different rates among cultivars and growing regions. In South African-grown 'Ruby' and 'Bhagwa', the TA decreased from 0.39 to 0.31% and from 0.62 to 0.38%, respectively, during day 54 after full bloom and either day 139 ('Ruby') or day 165 ('Bhagwa'), (Fawole and Opara, 2013a, Fawole and Opara, 2013c). For 'Malas-e-Torsh-e-Saveh' cultivar grown in Iran, the TA increased early after fruit set and declined later on throughout the growing season (Varasteh *et al.*, 2008). Similarly, a decrease in TA has been reported during fruit development of 'Wonderful', 'Taifi', 'Codpa' and 'Ganesh' cultivars (Ben-Arie *et al.*, 1984; Gil *et al.*, 1995; Al-Maiman and Ahmad, 2002; Kulkarni and Aradhya, 2005; Labbé *et al.*, 2010).

The pH value has an inverse correlation with TA and has been reported to increase with maturity in 'Taifi' (Al-Maiman and Ahmad, 2002), has no significant changes at different maturity stages of 'Mollar' (Sánchez *et al.*, 1996) and increases with fruit ripening in 'Rabbab-e-Fars' (Zarei *et al.*, 2011).

The composition and concentration of organic acids can also affect the perception of sweetness and sourness (Jalikop, 2007). Pomegranate fruit juice has been reported to contain several organic acids, often with citric acid being the major one that accounts for titrable acidity (Melgarejo *et al.*, 2000b; Poyrazoğlu *et al.*, 2002; Schwartz *et al.*, 2009Shwartz *et al.*, 2009). The concentrations of other organic acids including tartaric, malic and oxalic acids vary between different cultivars. Among Turkish pomegranate varieties, citric, malic and oxalic acids were the major organic acids (Poyrazoğlu *et al.*, 2002).

The amount and composition of different organic acids varied among 'ME5', 'ME17' and 'MO6' clones, with malic acid as the most predominant one, followed by citric acid (Legua *et al.*, 2000). During fruit development, the dominant organic acids in 'Ruby' were tartaric, citric and malic acid (Fawole and Opara, 2013c). In '101–2' and '121–2' accessions of 'Wonderful' grown under the same agro-climatic conditions, citric acid was predominant in '101–2' but least abundant in '121–2' compared with other organic acids (Shwartz *et al.*, 2009). Furthermore, oxalic and succinic acids were only found in accession '121–2,' and the concentrations decreased significantly with advancing maturity (Shwartz *et al.*, 2009).

In the 'Ruby' cultivar, tartaric acid was the most abundant organic acid during the immature stage, and its concentration decreased as development proceeded, whereas citric and malic acids became unquantifiable at advanced maturity stages (Fawole and Opara, 2013c). Several studies have shown strong correlations between organic acids and titrable acidity in mature pomegranate fruit (Shwartz *et al.*, 2009; Hasnaoui *et al.*, 2011; Mena *et al.*, 2011).

Ascorbic acid (vitamin C) is another important acid found in pomegranate fruit whose concentration tends to decline during fruit maturation in 'Ganesh' (Kulkarni and Aradhya, 2005), 'Taifi' (Al-Maiman and Ahmad, 2002) and 'Rabbab-e-Fars' (Zarei *et al.*, 2011). On the contrary, other reports indicate increasing ascorbic acid levels during fruit development in two 'Wonderful' accessions grown in Israel (Shwartz *et al.*, 2009) as well as 'Bhagwa' grown in South Africa (Fawole and Opara, 2013a).

One of the desirable changes that occur during fruit maturation and ripening is the loss of astringency, which is primarily due to the decline in the amount of phenolic compounds

(Al-Maiman and Ahmad, 2002; Kulkarni and Aradhya, 2005). In 'Rabbab-e-Fars', the lowest total phenolic concentration in fruit (786.20 mg/100 g) was detected at commercial harvest (Zarei *et al.*, 2011). For 'Wonderful', about 50% decline in total phenolic content was detected in the first growing season, while during the second season, an initial increase was followed by rapid declining levels (Weerakkody *et al.*, 2010). An increasing amount of phenolic compounds was detected for 'Malas Yazdi' cultivar at early maturity stage, which declined thereafter as maturation proceeded (Mirdehghan and Rahemi, 2007).

Total phenolic content might also be affected by the seasons. A study performed on Israeli-grown pomegranate cultivars showed that midwinter-ripened fruits had the highest level of total phenolic compounds, compared with early summer-, late summer- and autumn-ripened fruits (Borochov-Neori *et al.*, 2011). In Chilean-grown Codpa cultivar, the highest total phenolic concentration was found in fruit juice at the green maturity stage, which declined significantly as maturation proceeded (Labbé *et al.*, 2010).

Although juice containing very high concentrations of phenolic compounds could be less desirable due to high astringency (Kader, 2006), several reports have demonstrated that a significant reduction in phenolic compounds in pomegranate coincides with a sharp decline in juice antioxidant capacity during fruit development (Borochov-Neori *et al.*, 2009; Shwartz *et al.*, 2009; Labbé *et al.*, 2010; Weerakkody *et al.*, 2010; Borochov-Neori *et al.*, 2011). Common phenolic compounds in pomegranate juice include ellagic acid derivatives and hydrolysable tannins (Gil *et al.*, 2000; Shwartz *et al.*, 2009; Fischer *et al.*, 2011).

The presence of anthocyanins gives the pomegranate juice its red colour (Shulman *et al.*, 1984). It has been shown that most cultivars contain the same anthocyanin compounds, irrespective of the growing region, whose relative abundance differs among cultivars (Gil *et al.*, 1995; Alighourchi *et al.*, 2008; Fawole and Opara, 2013a, Fawole and Opara, 2013c). The total amount of anthocyanins stays almost constant during the first stages of fruit ripening followed by a significant increase in later stages (Sánchez *et al.*, 1996; Kulkarni and Aradhya, 2005; Shwartz *et al.*, 2009; Zarei *et al.*, 2011). Winter-ripened fruits have higher anthocyanin content than those harvested in early summer, late summer or autumn (Borochov-Neori *et al.*, 2011).

Six anthocyanins, the 3-glucosides and 3, 5-diglucosides of pelargonidin, cyanidin and delphinidin constitute the anthocyanin profile of pomegranate juice (Gil *et al.*, 1995; Hernández *et al.*, 1999; Borochov-Neori *et al.*, 2011). The same anthocyanins exist in all cultivars but present in different abundancies depending on cultivar, maturation stage and the geographical source of the fruit, with delphinidin 3,5-diglucoside being the dominant pigment in early ripening stages and the monoglucoside derivatives of cyanidin 3-glucoside and delphinidin 3-glucoside increasing in the later stages (Gil *et al.*, 1995; Hernández *et al.*, 1999).

Limited information is available regarding changes in mineral nutrients during pomegranate fruit maturation. Potassium (K^+) is the most abundant mineral nutrient in the fruit of Taifi cultivar, followed by Na^+ and Ca^{2+} (Al-Maiman and Ahmad, 2002). The highest concentration of K^+ was detected in unripe arils, followed by the juice obtained from fully ripe fruits. With advancing maturity, the concentration of P, Na and Ca in arils increases significantly, while the concentration of Mg, Na, and Ca in juice decreases. Similarly, K is reported to be the most abundant element in 'Hicaznar' in all fruit parts (Gozlekci *et al.*, 2011). The profile of the nutrient concentration in fruit peel of 'Hicaznar' changed during fruit development, with P having the highest concentration at unripe maturity stage, Mg, Mn, Zn and Cu at an immature stage, and Ca and Fe at fully mature stage (Gozlekci *et al.*, 2011).

Studying the changes in macro- and micronutrients in fruit arils and the peel of 'Malas Yazdi' cultivar showed that macronutrients accumulated in parallel to fruit development (from days 10–140 after full bloom) while the micronutrients presented declining values (Mirdehghan and Rahemi, 2007).

2.6.4 Physiological changes during fruit maturation

Fruit respiration in 'Wonderful' presented a gradual decline during and after the first month upon fruit set (Ben-Arie *et al.*, 1984). CO_2 output in young fruits rose initially for a day and thereafter declined gradually. With advancing maturity, CO_2 production became progressively less pronounced. Only a trace amount of ethylene production was detected throughout all maturity stages, even when fruits were sealed for 4 h in respiration jars, but exogenous ethylene induced 30 and 100% higher respiration rates in 1- and 5-month-old fruits, respectively. Due to relatively low respiration rates and the low amount of ethylene evolved during fruit development and ripening, the pomegranate fruit is classified as non-climacteric (Lee *et al.*, 1974). In parallel, when comparing 'Mule's Head' with 'Wonderful', fruit respiration rate was found to decline with advancing maturity, low CO_2 evolution was detected during the mature and ripening period (with 'Mule's Head' having slightly higher levels), and both cultivars had no measurable ethylene production (Shulman *et al.*, 1984). Exposure to ethylene may immediately increase fruit respiration rate and ethylene production, however, the effect of ethylene treatment on respiration rapidly declines. The pomegranate fruit will not mature after harvest and should be harvested only when fully mature. Once the fruit is harvested, it keeps respiring at a relatively low rate, which further decreases with time. Storage at low temperature can keep the respiration rate low. Therefore, storing pomegranate fruits in 0 and 4.5°C temparatures (depending on cultivar) at 85% relative humidity can keep them well for several months after harvesting (Mukerjee, 1958; Kader *et al.*, 1984; Prasad *et al.*, 2010).

2.6.5 Fruit splitting

Fruits of most known pomegranate cultivars eventually split when they over ripen. The fruits of some cultivars tend to split at much earlier stages of fruit development or at higher frequencies than others (Tabatabaei and Sarkhosh, 2006), whereas some are resistant to splitting (Trapaidze and Abuladze, 1998). The sudden change in moisture content due to heavy rain or irrigation during the dry season induces splitting of matured fruits, but the extent of fruit splitting can be reduced significantly by regular irrigation, particularly drip irrigation (Prasad *et al.*, 2003; Pal *et al.*, 2014). It is known that following the end of the dry season, rainfall on mature pomegranate fruits can induce rapid fruit splitting. A few reports indicate that spraying with gibberellic acid (GA_3) at 150 ppm or benzyl adenine (BA) at 40 ppm could significantly reduce pomegranate fruit splitting (Sepahi, 1986; Mohamed, 2004; Yilmaz and Özgüven, 2006). Other studies indicate that application of boron may reduce fruit splitting (Singh *et al.*, 2003; Sharma and Belsare, 2011; Khalil and Aly, 2013; Sahu *et al.*, 2013).

2.7 Conclusions

The pomegranate fruit is morphologically unique in the plant kingdom. The pomegranate has drawn wide attention globally as it thrives well in diverse biogeographic conditions and has tremendous utility in food and medicines. There is scant information and knowledge available on the genetics, physiology and biology of pomegranate, particularly the reproductive bud development, pathways of two morphologically distinct leaf development, induction of flowering, flower development, fruit development, root physiology and stress responses of this plant. Information in these fields is significantly important for further improvement of pomegranate crops for quality fruits and higher yields. Furthermore, research on the molecular aspects and genetic control of organogenesis in pomegranate is required to improve this economically and medicinally important crop.

References

Al-Maiman, S.A. and Ahmad, D. (2002) Changes in physical and chemical properties during pomegranate (*Punica granatum* L.) fruit maturation. *Food Chemistry* 76(4), 437–441. DOI: 10.1016/S0308-8146(01)00301-6.

Al-Said, F.A., Opara, L.U. and Al-Yahyai, R.A. (2009) Physico-chemical and textural quality attributes of pomegranate cultivars (*Punica granatum* L.) grown in the Sultanate of Oman. *Journal of Food Engineering* 90(1), 129–134. DOI: 10.1016/j.jfoodeng.2008.06.012.

Al-Yahyai, R., Al-Said, F. and Opara, L. (2009) Fruit growth characteristics of four pomegranate cultivars from Northern Oman. *Fruits* 64(6), 335–341. DOI: 10.1051/fruits/2009029.

Alighourchi, H., Barzegar, M. and Abbasi, S. (2008) Anthocyanins characterization of 15 Iranian pomegranate (*Punica granatum* L.) varieties and their variation after cold storage and pasteurization. *European Food Research and Technology* 227(3), 881–887. DOI: 10.1007/s00217-007-0799-1.

Aseri, G.K., Jain, N., Panwar, J., Rao, A.V. and Meghwal, P.R. (2008) Biofertilizers improve plant growth, fruit yield, nutrition, metabolism and rhizosphere enzyme activities of pomegranate (*Punica granatum* L.) in Indian Thar Desert. *Scientia Horticulturae* 117(2), 130–135. DOI: 10.1016/j.scienta.2008.03.014.

Ashton, R. (2006) Meet the pomegranate. In: Ashton, R., Baer, B. and Silverstein, S. (eds) *The Incredible Pomegranate Plant & Fruit*. 6. Third Millennium Publishing, Tempe, Arizona, pp. 3–8. Available at: http://3mpub.com

Assaf, R., Bar-Ya'akov, I., Dagan, M., Fahima, M. and Hatib, K. (1991) Pomegranate floral biology and trials to increase productivity. *Alon Hanotea* 45, 461–471.

Babu, K.D. (2010) Floral biology of pomegranate (*Punica granatum* L.). *Fruit, Vegetable and Cereal Science and Biotechnology* 4(2), 45–50.

Babu, K.D., Chandra, R., Jadhav, V.T. and Sharma, J. (2009) Blossom biology of pomegranate cv. 'Bhagawa' under semiarid tropics of western India.. *Abstracts of 2nd International Symposium on Pomegranate and Minor Including Mediterranean Fruits*, University of Agricultural Sciences, Dharwad, India, 23–27 June, pp. 88–89.

Ben-Arie, R., Segal, N. and Guelfat-Reich, S. (1984) The maturation and ripening of the Wonderful pomegranate. *Journal of the American Society for Horticultural Science* 117, 100–104.

Bennett, M.D. and Leitch, I.J. (2005) Nuclear DNA amounts in angiosperms: progress, problems and prospects. *Annals of Botany* 95(1), 45–90. DOI: 10.1093/aob/mci003.

Bertin, R.I. (1982) The evolution and maintenance of andromonoecy. *Evolutionary Theory* 6, 25–32.

Borochov-Neori, H., Judeinstein, S., Tripler, E., Harari, M., Greenberg, A. *et al.* (2009) Seasonal and cultivar variations in antioxidant and sensory quality of pomegranate (*Punica granatum* L.) fruit. *Journal of Food Composition and Analysis* 22(3), 189–195. DOI: 10.1016/j.jfca.2008.10.011.

Borochov-Neori, H., Judeinstein, S., Harari, M., Bar-Ya'akov, I., Patil, B.S. *et al.* (2011) Climate effects on anthocyanin accumulation and composition in the pomegranate (*Punica granatum* L.) fruit arils. *Journal of Agricultural and Food Chemistry* 59(10), 5325–5334. DOI: 10.1021/jf2003688.

Borochov-Neori, H., Lazarovitch, N., Judeinstein, S., Patil, B.S. and Holland, D. (2013) Climate and salinity effects on color and health promoting properties in the pomegranate (*Punica granatum* L.) fruit arils). *ACS Symposium Series, Tropical and Subtropical Fruits: Flavors, Color, and Health Benefits* 1129, 43–61.

Bridgwater, S.D. and Baas, P. (1978) Wood anatomy of the Punicaceae. *IAWA Bulletin* 1, 3–6.

Brodie, L. (2009) Pomegranate production in South Africa. *South African Fruit Journal* 8, 30–35.

Burmistrov, L.A. (1993) Pomegranate culture in central Asia. WANATCA (West Australian Nut and Tree Crop Association), Yearbook 17. Available at: https://www.growables.org/information/TropicalFruit/PomWANATCA.htm (accessed accessed 17 July 2020).

Chandra, R., Babu, D.K., Jadhav, V.T. and Teixeira da Silva, J.A. (2010) Origin, history and domestication of pomegranate. *Fruit, Vegetables and Cereal Science and Biotechnology* 4(2), 1–6.

Chaudhari, S.M. and Desai, U.T. (1993) Effect of plant growth regulators on flower sex in pomegranate. *Indian Journal of Agricultural Sciences* 63, 34–35.

Chen, T., Cai, X., Wu, X., Karahara, I., Schreiber, L. *et al.* (2011) Casparian strip development and its potential function in salt tolerance. *Plant Signaling & Behavior* 6(10), 1499–1502. DOI: 10.4161/psb.6.10.17054.

Conti, E., Litt, A., Wilson, P.G., Graham, S.A., Briggs, B.G. *et al.* (1997) Interfamilial relationships in Myrtales: molecular phylogeny and patterns of morphological evolution. *Systematic Botany* 22(4), 629–647. DOI: 10.2307/2419432.

Crivellaro, A. and Schweingruber, F.H. (2013) *Atlas of Wood, Bark and Pith Anatomy of Eastern Mediterranean Trees and Shrubs with a Special Focus on Cyprus*. Springer, Berlin, Heidelberg, Germany.

Cui, S.-mao., Sasada, Y., Sato, H. and Nii, N. (2004) Cell structure and sugar and acid contents in the arils of developing pomegranate fruit. *Engei Gakkai zasshi* 73(3), 241–243.

Das, P.K. and Sur, S.C. (1968) Tetraploidy in pomegranate (*Punica granatum* L). *Technology Bihar* 5, 8–126.

Derin, K. and Eti, S. (2001) Determination of pollen quality, quantity and effect of cross-pollination on the fruit set and quality in the pomegranate. *Turkish Journal of Agriculture and Forestry* 25(3), 169–173.

Djavakyants, M.Y. (2011) Pomegranate distribution and diversity in Uzbekistan. In: Turdieva, M.K., Kayimov, A.K., Baymetov, K.I., Mustafina, F.U. and Butkov, E.A. (eds) *Conservation and Sustainable Use of Biodiversity of Fruit Crops and Wild Fruit Species – The Proceedings of the International Scientific and Practical Conference, 23–26 August 2011*. Tashkent, Uzbekistan, pp. 84–86.

El Sese, A.M. (1988) Physiological studies on flowering and fruiting habits of some pomegranate cultivars under Assiut conditions. *Assiut Journal of Agricultural Sciences* 19, 320–336.

Erdtman, G. (1971) *Pollen Morphology and Plant Taxonomy: Angiosperms*. Hafner Publishing Company, New York, pp. 10–18.

Erkan, M. and Dogan, A. (2018) Pomegranate/Roma – *Punica granatum*. In: Rodrigues, S., de Oliveira Silva, E. and de Brito, E.S. (eds) *Exotic Fruits Reference Guide*. Academic Press, London, pp. 355–361.

Erkan, M. and Kader, A.A. (2011) Pomegranate (*Punica granatum* L.). In: Yahia, E.M. (ed.) *Postharvest Biology and Technology of Tropical and Subtropical Fruits*. Woodhead Publishing, Cambridge, UK, pp. 287–311.

Evreinoff, V.A. (1953) Pomological studies of the pomegranate. *Annales de l'Ecole nationale superieure agronomique* 1, 141–154.

Evreinoff, V.A. (1957) A contribution to the study of the pomegranate. *Journal d'Agriculture Tropicale et de Botanique Appliquée* 4, 124–138.

Fadavi, A., Barzegar, M. and Hossein Azizi, M. (2006) Determination of fatty acids and total lipid content in oilseed of 25 pomegranates varieties grown in Iran. *Journal of Food Composition and Analysis* 19(6–7), 676–680. DOI: 10.1016/j.jfca.2004.09.002.

Fahan, A. (1976) *The Leaf, the Flower, the Seed. Plant Anatomy*. Hakkibutz Hameuhad Publ House Ltd, Jerusalem, pp. 171–212.

Fawole, O.A. and Opara, U.L. (2013a) Changes in physical properties, chemical and elemental composition and antioxidant capacity of pomegranate (cv. ruby) fruit at five maturity stages. *Scientia Horticulturae* 150, 37–46. DOI: 10.1016/j.scienta.2012.10.026.

Fawole, O.A. and Opara, U.L. (2013b) Developmental changes in maturity indices of pomegranate fruit: a descriptive review. *Scientia Horticulturae* 159, 152–161. DOI: 10.1016/j.scienta.2013.05.016.

Fawole, O.A. and Opara, U.L. (2013c) Effects of maturity status on biochemical content, polyphenol composition and antioxidant capacity of pomegranate fruit arils (cv. 'Bhagwa'). *South African Journal of Botany* 85, 23–31. DOI: 10.1016/j.sajb.2012.11.010.

Feng, Y.Z., Chen, D.J., Song, M.T., Zhao, Y.L. and Li, Z.H. (1998) Assessment and utilization of pomegranate varieties resources. *Journal of Fruit Sciences* 15, 370–373.

Feng, Y.Z., Song, M.T. and Han, D.B. (2006) The general status of pomegranate germplasm resources in China. *China Fruits* 4, 57–58.

Fischer, U.A., Carle, R. and Kammerer, D.R. (2011) Identification and quantification of phenolic compounds from pomegranate (*Punica granatum* L.) peel, mesocarp, aril and differently produced juices by HPLC-DAD-ESI/MS(n). *Food Chemistry* 127(2), 807–821. DOI: 10.1016/j.foodchem.2010.12.156.

Fouad, M.M., Barkat, M.R. and El-yazal, S.A. (1979) Bud burst activity, flowering and fruit set of 'Grenouillere' and 'Manfaloti' pomegranate cultivars under Giza conditions. Research Bulletin of Faculty of Agriculture and Animal Science . Ain Shams University, Cairo.

Gammie, G.A. and Patwardhan, G.B. (1929) Field garden and orchard crops of the Bombay presidency. *Bombay Agriculture Department Bulletin* 30, 249.

Gharaghani, A., Ghasemi Soloklui, A.A., Oraguzie, N. and Zare, D. (2017) Pollen source influences fruit quality, aril properties, and seed characteristics in pomegranate. *International Journal of Fruit Science* 17(3), 333–348. DOI: 10.1080/15538362.2017.1318733.

Gil, M.I., García-Viguera, C., Artés, F. and Tomás-Barberán, F.A. (1995) Changes in pomegranate juice pigmentation during ripening. *Journal of the Science of Food and Agriculture* 68(1), 77–81. DOI: 10.1002/jsfa.2740680113.

Gil, M.I., Tomás-Barberán, F.A., Hess-Pierce, B., Holcroft, D.M. and Kader, A.A. (2000) Antioxidant activity of pomegranate juice and its relationship with phenolic composition and processing. *Journal of Agricultural and Food Chemistry* 48(10), 4581–4589. DOI: 10.1021/jf000404a.

Glozer, K. and Ferguson, L. (2011) Pomegranate production in Afghanistan. UC-Davis College of Agricultural & Environmental Sciences. Available at: https://ucanr.edu/sites/Pomegranates/files/164500.pdf (accessed 26 November 2018).

Goor, A. and Lieberman, J. (1956) *The Pomegranate*. State of Israel, Ministry of Agriculture, Agricultural Publication Section, Tel Aviv, Israel, pp. 5–57.

Gozlekci, S. and Kaynak, L. (2000) Physical and chemical changes during fruit development and flowering in pomegranate (*Punica granatum* L.) cultivar Hicaznar grown in Antalya region. *Options Méditerranéennes Série A, Séminaires Méditerranéens*, 79–85.

Gozlekci, S.G., Ercisli, S., Okturen, F. and Sonmez, S. (2011) Physico-chemical characteristics at three development stages in pomegranate cv. "Hicaznar". *Notulae Botanicae Horti Agrobotanici Cluj-Napoca* 39(1), 241–245. DOI: 10.15835/nbha3918985.

Graham, A., Graham, S.A., Nowicke, J.W., Patel, V. and Lee, S. (1990) Palynology and systematics of the Lythraceae. III. Genera *Physocalymma* through *Woodfordia*, *Addenda*, and conclusions. *American Journal of Botany* 77(2), 159–177. DOI: 10.1002/j.1537-2197.1990.tb13543.x.

Graham, S.A., Crisci, J.V. and Hoch, P.C. (1993) Cladistic analysis of the Lythraceae *sensu lato* based on morphological characters. *Botanical Journal of the Linnean Society* 113(1), 1–33. DOI: 10.1111/j.1095-8339.1993.tb00326.x.

Guarino, L., Miller, T., Baazara, M. and Obadi, N. (1990) Socotra: the island of bliss revisited. *Diversity* 6(3–4), 28–31.

Harlan, J.R. (1992) *Crops and Man*, 2nd edn. 53711. American Society of Agronomy, Madison, Wisconsin.

Hasnaoui, N., Mars, M., Chibani, J. and Trifi, M. (2010) Molecular polymorphisms in Tunisian pomegranate (*Punica granatum* L.) as revealed by RAPD fingerprints. *Diversity* 2(1), 107–114. DOI: 10.3390/d2010107.

Hasnaoui, N., Mars, M., Ghaffari, S., Trifi, M., Melgarejo, P. *et al.* (2011) Seed and juice characterization of pomegranate fruits grown in Tunisia: comparison between sour and sweet cultivars revealed interesting properties for prospective industrial applications. *Industrial Crops and Products* 33(2), 374–381. DOI: 10.1016/j.indcrop.2010.11.006.

Herlihy, C.R. and Eckert, C.G. (2002) Genetic cost of reproductive assurance in a self-fertilizing plant. *Nature* 416(6878), 320–323. DOI: 10.1038/416320a.

Hernández, F., Melgarejo, P., Tomás-Barberán, F.A. and Artés, F. (1999) Evolution of juice anthocyanins during ripening of new selected pomegranate (*Punica granatum*) clones. *European Food Research and Technology* 210(1), 39–42. DOI: 10.1007/s002170050529.

Higashiyama, T. and Takeuchi, H. (2015) The mechanism and key molecules involved in pollen tube guidance. *Annual Review of Plant Biology* 66(1), 393–413. DOI: 10.1146/annurev-arplant-043014-115635.

Hiwale, S.S., More, T.A. and Bagle, B.G. (2011) Root distribution pattern in pomegranate 'Ganesh' (*Punica granatum* L). *Acta Hortulturae* 890, 323–326.

Hodgson, R.W. (1917) The pomegranate. *California Agricultural Experiment Station Bulletin* 276, 163–192.

Holland, D., Hatib, K. and Bar-Ya'akov, I. (2009) Pomegranate: botany, horticulture, breeding. *Horticultural Review* 35, 127–191.

Huang, Ye-lin., Shi, Su-hua., Huang, Y. and Shi, S. (2002) Phylogenetics of Lythraceae *sensu lato*: a preliminary analysis based on chloroplast *rbc* L gene, *psa A – ycf 3* spacer, and nuclear rDNA internal transcribed spacer (ITS) sequences. *International Journal of Plant Sciences* 163(2), 215–225. DOI: 10.1086/338392.

IBPGR (1986) *Punica granatum* (pomegranate). In: *Genetic Resources of Tropical and Sub-tropical Fruits and Nuts (Excluding Musa)*. International Board for Plant Genetic Resources, Rome, pp. 97–100.

Jagtap, D.B., Desai, U.T. and Masalkar, S.D. (1992) Assessment of pomegranate germplasm for vegetative and fruit characters. *Annals of Arid Zone* 31(3), 217–219.

Jalikop, S.H. (2007) Linked dominant alleles or inter-locus interaction results in a major shift in pomegranate fruit acidity of 'Ganesh' × 'Kabul Yellow'. *Euphytica: Netherlands Journal of Plant Breeding* 158(1–2), 201–207. DOI: 10.1007/s10681-007-9443-1.

Jalikop, S.H. and Kumar, P.S. (1990) Use of a gene marker to study the mode of pollination in pomegranate (*Punica granatum* L.). *The Journal of Horticultural Science* 65(2), 221–223. DOI: 10.1080/00221589.1990.11516050.

Jayesh, K.C. and Kumar, R. (2004) Crossability in pomegranate (*Punica granatum* L). *The Indian Journal of Horticulture* 61(3), 209–210.

Johri, B.M., Ambegoakar, K.B. and Srivastava, P.S. (1992) *Comparative Embryology of Angiosperms 1, 2*. Springer, Berlin.

Josan, J.S., Jawanda, J.S. and Uppal, D.K. (1979a) Studies on the floral biology of pomegranate. I. Sprouting of vegetative buds, flower bud development, flowering habit, time and duration of flowering and floral morphology. *Punjab Horticultural Journal* 2, 59–65.

Josan, J.S., Jawanda, J.S. and Uppal, D.K. (1979b) Studies on the floral biology of pomegranate III. Mode of pollination, fruit development and fruit cracking. *Punjab Horticultural Journal*4, 134–138.

Kader, A.A. (2006) Postharvest biology and technology of pomegranates. In: Seeram, N.P., Schulman, R.N. and Heber, D. (eds) *Pomegranates: Ancient Roots to Modern Medicine*. CRC Press Taylor & Francis Group, Boca Raton, Florida, pp. 211–220.

Kader, A.A., Chordas, A. and Elyate, S. (1984) Responses of pomegranate to ethylene treatment and storage temperature. *Hilgardia*38(7), 14–15.

Karale, A.R., Supe, V.S., Kaulgud, S.N. and Kale, P.N. (1993) Pollination and fruit set studies in pomegranate. *Journal of Maharashtra Agricultural Universities* 18(3), 364–366.

Karimi, H.R. and Mirdehghan, S.H. (2015) Effects of self, open, and supplementary pollination on growth pattern and characteristics of pomegranate fruit. *International Journal of Fruit Science* 15(4), 382–391. DOI: 10.1080/15538362.2015.1009974.

Khalil, A. and Aly, S.H. (2013) Cracking and fruit quality of pomegranate (*Punica granatum* L.) as affected by pre-harvest sprays of some growth regulators and mineral nutrients. *Journal of Horticultural Science & Ornamental Plants* 5, 71–76.

Kim, S., Mollet, J.-C., Dong, J., Zhang, K., Park, S.-Y. *et al.* (2003) Chemocyanin, a small basic protein from the lily stigma, induces pollen tube chemotropism. *Proceedings of the National Academy of Sciences of the United States of America* 100(26), 16125–16130. DOI: 10.1073/pnas.2533800100.

Kulkarni, A.P. and Aradhya, S.M. (2005) Chemical changes and antioxidant activity in pomegranate arils during fruit development. *Food Chemistry* 93(2), 319–324. DOI: 10.1016/j.foodchem.2004.09.029.

Kumar, B.P. and Purohit, A.G. (1989) Studies on fruit growth and development in pomegranate. *Journal of Maharashtra Agriculture University* 14, 187–189.

Labbé, M., Pena, A. and Saenz, C. (2010) Antioxidant capacity and phenolic composition of juices from pomegranates stored in refrigeration. *International Conference on Food Innovation*, Valencia, Spain, pp. 25–29.

Lansky, E.P. and Newman, R.A. (2007) *Punica granatum* (pomegranate) and its potential for prevention and treatment of inflammation and cancer. *Journal of Ethnopharmacology* 109(2), 177–206. DOI: 10.1016/j.jep.2006.09.006.

Lee, S.W., Kim, K.S. and Kim, S.D. (1974) Studies on changes in the composition of the pomegranate fruit during maturation changes in sugars, organic acids, amino acids, and the respiration rate. *Journal of Korean Society of Horticultural Science* 15, 57–68.

Legua, P., Melgarejo, P., Martinez, M. and Hernandez, F. (2000) Evolution of sugars and organic acid content in three pomegranate cultivars (*Punica granatum* L.). *Options Méditerranéennes Série A, Séminaires Méditerranéens* 42, 93–97.

Lersten, N.R. and Horner, H.T. (2005) Development of the calcium oxalate crystal macropattern in pomegranate (*Punica granatum*, Punicaceae). *American Journal of Botany* 92(12), 1935–1941. DOI: 10.3732/ajb.92.12.1935.

Levin, G.M. (1978) The floral biology of pomegranate (*Punica granatum* L.) in Southwest Turkmenistan. *Turkm SSR Ylymlar Akad Habar Biol Ylymlaryn* 5, 31–38.

Levin, G.M. (1994) Pomegranate (*Punica granatum*) plant genetic resources in Turkmenistan. *Plant Genetic Resources Newsletter* 97, 31–36.

Levin, G.M. (1996) Pomegranate (*Punica granatum* L.) collection research in Turkmenistan. *Plant Genetic Resources Newsletter* 106, 1020–1362.

Levin, G.M. (2006) *Pomegranate*. Third Millennium Publishing, Tempe, Arizona.

Levin, G.M. (2006a) *Pomegranate Roads: A Soviet Botanist's Exile from Eden*, 1st edn. Floreant Press, Forestville, California, pp. 15–183.

Maas, E.V. (1993) Testing crops for salinity tolerance. In: Maranville, J.W., Baligar, B.V., Duncan, R.R. and Yohe, J.M. (eds) *Proceedings of Workshop on Adaptation of Plants to Soil Stresses, 1–4 August 1993, INTSORMIL, Pub. No. 94-2*. University of Nebraska, Lincoln, Nebraska, pp. 234–247.

Mars, M. (2000) Pomegranate plant material: genetic resources and breeding, a review. *Options Méditerranéennes Série A, Séminaires Méditerranéens* 42, 55–62.

Mars, M. and Marrakchi, M. (1999) Diversity of pomegranate (*Punica granatum* L.) germplasm in Tunisia. *Genetic Resources and Crop Evolution* 46(5), 461–467. DOI: 10.1023/A:1008774221687.

Martinez, J.J., Melgarejo, P. and Martinez, F. (2000) Study of the floral morphology of the pomegranate clones: PTO8, CRO1 and ME14. In: Melgarejo-Moreno, P., Martinez-Nicolas, J.J. and Martinez-Tome, J. (eds) *Production, Processing and Marketing of Pomegranate in the Mediterranean Region: Advances in Research and Technology*. CIHEAM-IAMZ, Options Méditerranéennes: Série A 42, Zaragoza, Spain, pp. 105–113.

McGregor, S.E. (1976) Tree fruits & nuts and exotic tree fruits & nuts. *Insect Pollination of Cultivated Crop Plants*. Available at: https://www.ars.usda.gov/ARSUserFiles/20220500/OnlinePollinationHandbook.pdf (accessed 26 November 2018).

Meena, K.K., Singh, R., Pareeka, S. and Kashyap, P. (2011) Evaluation of pomegranate (*Punica granatum* L.) genotypes for morphological and flowering characteristics under semiarid climate. *Acta Horticulturae* 890, 233–238.

Melgarejo, P., Legua, P., Martinez, M. and Martinez, J.J. (2000a) Contribution to a better knowledge of the quality of pomegranate pollen (*Punica granatum* L. *Options Méditerranéennes Série A, Séminaires Méditerranéens* 42, 115–121.

Melgarejo, P., Salazar, D.M. and Artés, F. (2000b) Organic acids and sugars composition of harvested pomegranate fruits. *European Food Research and Technology* 211(3), 185–190. DOI: 10.1007/s002170050021.

Melgarejo, P. and Martínez, R. (1992) *El Granado*. Mundi-Prensa, Madrid.

Melgarejo, P., Martinez-Valero, R., Guillamón, J.M., Amoros, A. and Miro, M. (1997) Phenological stages of the pomegranate tree (*Punka granatum* L.). *The Annals of applied biology* 130(1), 135–140. DOI: 10.1111/j.1744-7348.1997.tb05789.x.

Mena, P., García-Viguera, C., Navarro-Rico, J., Moreno, D.A., Bartual, J. *et al.* (2011) Phytochemical characterisation for industrial use of pomegranate (*Punica granatum* L.) cultivars grown in Spain. *Journal of the Science of Food and Agriculture* 91(10), 1893–1906. DOI: 10.1002/jsfa.4411.

Miller, A. (2004) *Punica protopunica*. The IUCN Red List of Threatened Species 2004 e.T30404A9544416. Available at: http://dx.doi.org/10.2305/IUCN.UK.2004.RLTS.T30404A9544416.en (accessed 24 May 2019).

Mir, M.M., Umar, I., Mir, S.A., Rehman, M.U., Rather, G.H. *et al.* (2012) Quality evaluation of pomegranate crop – a review. *International Journal of Agriculture and Biology* 14, 658–667.

Mirdehghan, S.H. and Rahemi, M. (2007) Seasonal changes of mineral nutrients and phenolics in pomegranate (*Punica granatum* L.) fruit. *Scientia Horticulturae* 111(2), 120–127. DOI: 10.1016/j.scienta.2006.10.001.

Mohamed, A.K.A. (2004) Effect of gibberellic acid (GA3) and benzyladinine (Ba) on splitting and quality of Manfalouty fruits. *Assiut Journal of Agricultural Sciences* 35(3), 11–21.

Mohseni, A. (2009) The situation of pomegranate orchards in Iran. *Acta Horticulturae* 818, 35–42. DOI: 10.17660/ActaHortic.2009.818.3.

Morton, J. (1987) Pomegranate. In: Morton, J.F. (ed.) *Fruits of Warm Climates*. Miami, Florida, pp. 352–355.

Mukerjee, P.K. (1958) Storage of pomegranates (*Punica granatum* L.). *Scientific Culture* 24, 94.

Muradoglu, F., Balta, M.F. and Ozrenk, K. (2006) Pomegranate (*Punica granatum* L.) genetic resources from Hakkari, Turkey. *Research Journal of Agriculture and Biological Sciences* 2(6), 520–525.

Naeini, M.R., Khoshgoftarmanesh, A.H., Lessani, H. and Fallahi, E. (2005) Effects of sodium chloride-induced salinity on mineral nutrients and soluble sugars in three commercial cultivars of pomegranate. *Journal of Plant Nutrition* 27(8), 1319–1326. DOI: 10.1081/PLN-200025832.

Naeini, M.R., Khoshgoftarmanesh, A.H. and Fallahi, E. (2006) Partitioning of chlorine, sodium, and potassium and shoot growth of three pomegranate cultivars under different levels of salinity. *Journal of Plant Nutrition* 29(10), 1835–1843. DOI: 10.1080/01904160600899352.

Nalawadi, U.G., Farooqui, A.A., Dasappa, M.A., Gubbaiah, N.R., Sulikeri, G.S. *et al.* (1973) Studies on the floral biology of pomegranate (*Punica granatum* L.). *Mysore Journal of Agricultural Sciences* 7(2), 213–225.

Narzary, D., Rana, T.S. and Ranade, S.A. (2010a) Genetic diversity in inter-simple sequence repeat profiles across natural populations of Indian pomegranate (*Punica granatum* L.). *Plant Biology* 12(5), 806–813. DOI: 10.1111/j.1438-8677.2009.00273.x.

Narzary, D., Rana, T.S. and Ranade, S.A. (2010b) Molecular analyses of genetic diversity in pomegranates. *Fruit, Vegetable and Cereal Science and Biotechnology* 4(2), 126–143.

Narzary, D., Ranade, S.A., Divakar, P.K. and Rana, T.S. (2016) Molecular differentiation and phylogenetic relationship of the genus *Punica* (Punicaceae) with other taxa of the order Myrtales. *Rheedea* 26(1), 37–51.

Nasir, E. and Ali, S.I. (1972) *Flora of West Pakistan*. 501. Fakhri Pring Press, Karachi.

Nath, N. and Randhawa, G.S. (1959) Studies on cytology of pomegranate (*Punica granatum* L.). *Indian Journal of Horticulture* 16, 210–215.

Nemati, Z., Tehranifar, A., Farsi, M., Mirshamsikakhki, A., Nemati, H. *et al.* (2012) Evaluation of genetic diversity of Iranian pomegranate cultivars using fruit morphological characteristics and AFLP markers. *Notulae Botanicae Horti Agrobotanici Cluj-Napoca* 40(1), 261–268. DOI: 10.15835/nbha4017369.

Özgüven, A.I. and Yilmaz, C. (2000) Pomegranate growing in Turkey. *Options Méditerranéennes Série A, Séminaires Méditerranéens* 42, 41–48.

Özgüven, A.I., Tatli, H., Coskun, M., Daskan, Y. and Kuden, A.B. (1997) Fruit characteristics of some Mediterranean and Aegean pomegranate varieties under ecological conditions of Adana, Turkey. *Acta Horticulturae* 441, 345–349.

Ohri, D. (2002) Genome size variation in some tropical hardwoods. *Biologia Plantarum* 45(3), 455–457. DOI: 10.1023/A:1016290222360.

Okhovatian-Ardakani, A.R., Mehrabanian, M., Dehghani, F. and Akbarzadeh, A. (2010) Salt tolerance evaluation and relative comparison in cuttings of different pomegranate cultivars. *Plant, Soil and Environment* 56(4), 176–185. DOI: 10.17221/158/2009-PSE.

Opara, L.U., Al-Ani, M.R. and Al-Shuaibi, Y.S. (2009) Physico-chemical properties, vitamin C content, and antimicrobial properties of pomegranate fruit (*Punica granatum* L.). *Food and Bioprocess Technology* 2(3), 315–321. DOI: 10.1007/s11947-008-0095-5.

Pal, R.K., Babu, K.D., Singh, N.V., Maity, A. and Gaikwad, N. (2014) Pomegranate research in India – status and future challenges. *Progressive Horticulture* 46(2), 184–201.

Parmar, C. and Kaushal, M.K. (1982) *Wild Fruits of Sub-Himalayan Region*. Kalyani Publishers, New Delhi, India, pp. 74–77.

Patel, V.C., Skvarla, J.J. and Raven, P.H. (1984) Pollen characters in relation to the delimitation of Myrtales. *Annals of the Missouri Botanical Garden* 71(3), 858–969. DOI: 10.2307/2399170.

Porter, J. and Wetzstein, H.Y. (2014) The biology of pomegranates: all about flowers, fruit and arils. Florida Pomegranate Association, University of Georgia/Purdue University, 10 October 2014.. Available at: https://crec.ifas.ufl.edu/extension/pomegranates/pdfs/Porter_2014.pdf (accessed 2 July 2019).

Poyrazoğlu, E., , Gokmen, V. and Artik, N. (2002) Organic acids and phenolic compounds in pomegranates (*Punica granatum* L.) grown in Turkey. *Journal of Food Composition and Analysis* 15(5), 567–575. DOI: 10.1016/S0889-1575(02)91071-9.

Prasad, R.N., Bankar, G.J. and Vashishtha, B.B. (2003) Effect of drip irrigation on growth, yield and quality of pomegranate in arid region. *Indian Journal of Horticulture* 60, 140–142.

Prasad, R.N., Chandra, R. and Teixeira da Silva, J.A. (2010) Postharvest handling and processing of pomegranate. *Fruit, Vegetables and Cereal Science and Biotechnology* 4(2), 88–95.

Pujari, K.H. and Rane, D.A. (2015) Concept of seed hardness in pomegranate – I) Anatomical studies in soft and hard seeds of 'Muskat' pomegranate. *Acta Horticulturae* 1089,97–104. DOI: 10.17660/ActaHortic.2015.1089.11.

Rajaei, H. and Yazdanpanah, P. (2015) Buds and leaves in pomegranate (*Punica granatum* L.): phenology in relation to structure and development. *Flora – Morphology, Distribution, Functional Ecology of Plants* 214, 61–69. DOI: 10.1016/j.flora.2015.05.002.

Raman, V.S., Manimekalai, G. Sreerangaswamy, S.R, Rangaswami, S.R. (1971) Chromosome behaviour at meiosis in *Punica granatum* L. *Cytologia* 36(3), 400–404. DOI: 10.1508/cytologia.36.400.

Rana, T.S., Datt, B. and Rao, R.R. (2003) *Flora of Tons Valley, Garhwal Himalaya (Uttaranchal)*. Bishen Singh Mahendra Pal Singh, Dehradun, India, p. 166.

Rana, J.C., Pradheep, K. and Verma, V.D. (2007) Naturally occurring wild relatives of temperate fruits in Western Himalayan region of India: an analysis. *Biodiversity and Conservation* 16(14), 3963–3991. DOI: 10.1007/s10531-007-9201-7.

Rana, T.S., Narzary, D. and Ranade, S.A. (2010) Systematics and taxonomic disposition of the genus *Punica* L. *Fruit, Vegetable and Cereal Science and Biotechnology* 4(2), 19–25.

Ranade, S.A., Rana, T.S. and Narzary, D. (2009) Spar profiles and genetic diversity amongst pomegranate (*Punica granatum* L.) genotypes. *Physiology and Molecular Biology of Plants* 15(1), 61–70. DOI: 10.1007/s12298-009-0006-x.

Ranpise, S.A., Sankaran, M., Babu, K.D., Prakash, J. and Hiwale, S.S. (2014) Pomegranate. In: Ghose, S.N. (ed.) *Tropical and Sub-tropical Fruit Crops: Crop Improvement and Varietal Wealth.*JAYA Publishing House, New Delhi, pp. 549–579.

Rathore, R.S., Kaushik, R.A. and Chandra, A. (2013) Determination of root distribution in pomegranate by root excavation technique. *Agricultural Science Digest* 33(2), 142–145.

Raviv, M., Krasnovsky, A., Medina, S. and Reuveni, R. (1998) Assessment of various control strategies for recirculation of greenhouse effluents under semi-arid conditions. *The Journal of Horticultural Science and Biotechnology* 73(4), 485–491. DOI: 10.1080/14620316.1998.11511003.

Reddy, Y.N. (2011) Certain new approaches to the production problems of pomegranate. *Acta Horticulturae* 890, 287–293.

Rehder, A. (1949) *Bibliography of Cultivated Trees and Shrubs Hardy in the Cooler Temperate Regions of the Northern Hemisphere*. Arnold Arboretum of Harvard University, Jamaica Plain, Massachusetts, pp. 484–485.

Rodríguez, P., Mellisho, C.D., Conejero, W., Cruz, Z.N., Ortuño, M.F. *et al.* (2012) Plant water relations of leaves of pomegranate trees under different irrigation conditions. *Environmental and Experimental Botany* 77, 19–24. DOI: 10.1016/j.envexpbot.2011.08.018.

Sahu, P., Sharma, N. and Sharma, D.P. (2013) Effect of in-situ moisture conservation, forchlorfenuron and boron on growth, fruit cracking and yield of pomegranate cv. Kandhari under rainfed conditions of Himachal Pradesh. *Indian Journal of Horticulture* 70(4), 501–505.

Sánchez, R., Marín, J.G. and Artés, F. (1996) Quality changes in pomegranates during ripening and cold storage. *Zeitschrift für Lebensmittel-Untersuchung und -Forschung* 202(6), 481–485. DOI: 10.1007/BF01197269.

Schweingruber, F.H., Borner, A. and Schulze, E.-D. (2011) *Atlas of Stem Anatomy in Herbs, Shrubs and Trees (Vol. 1)*. Springer-Verlag, Berlin, Heidelberg, Germany.

Sepahi, A. (1986) GA3 concentration for controlling fruit cracking in pomegranates. *Iran Agricultural Research* 5, 93–99.

Shalimu, D., Li, K., Baskin, C.C., Baskin, J.M. and Liu, Y. (2015) Seed germination biology of four pomegranate (*Punica granatum*) cultivars from Xinjiang, China. *HortScience* 50(6), 826–829. DOI: 10.21273/HORTSCI.50.6.826.

Shao, J., Chen, C. and Deng, X. (2003) In vitro induction of tetraploid in pomegranate (*Punica granatum*). *Plant Cell, Tissue and Organ Culture* 75(3), 241–246. DOI: 10.1023/A:1025871810813.

Sharma, N. and Belsare, C. (2011) Effect of plant bio-regulators and nutrients on fruit cracking and quality in pomegranate (*Punica granatum* L.) 'G-137' in Himachal Pradesh. *Acta Horticulturae* 890,347–352. DOI: 10.17660/ActaHortic.2011.890.48.

Sheidai, M. (2007) B-chromosome variability in pomegranate (*Punica granatum* L. cultivars). *Caryologia* 60, 251–256.

Sheidai, M. and Noormohammadi, Z. (2005) Chromosome pairing and unreduced gamete formation in nineteen pomegranate (*Punica granatum* L.) cultivars. *Cytologia* 70(3), 257–265. DOI: 10.1508/cytologia.70.257.

Sherafatian, D. (1994) The effect of harvesting date and temperature during storage on keeping quality of pomegranate. *Seed and Plant* 10, 25–31.

Shilkina, I.A. (1973) On the xylem anatomy of the genus *Punica* L. *Botanicheskii Zhurnal* 58, 1628–1630.

Shulman, Y., Fainberstein, L. and Lavee, S. (1984) Pomegranate fruit development and maturation. *Journal of Horticultural Science* 59(2), 265–274. DOI: 10.1080/00221589.1984.11515196.

Shwartz, E., Glazer, I., Bar-Ya'akov, I., Matityahu, I., Bar-Ilan, I. *et al.* (2009) Changes in chemical constituents during the maturation and ripening of two commercially important pomegranate accessions. *Food Chemistry* 115(3), 965–973. DOI: 10.1016/j.foodchem.2009.01.036.

Singh, R.P., Kar, P.L. and Dhuria, H.S. (1978) Studies on behaviour of flowering and sex expression in some pomegranate cultivars. *Plant Science* 10, 29–31.

Singh, D.B., Sarma, B.D. and Bhargava, R. (2003) Effect of boron and GA3 to control fruit cracking in pomegranate (*Punica granatum*). *Current Agronomy* 27, 125–127.

Singh, D.B., Samadia, D.K. and Kingsly, A.R.P. (2006) Conservation, characterization and evaluation of pomegranate germplasm under arid ecosystem of India. In: *1st International Symposium on Pomegranate and Minor Mediterranean Fruits, Abstracts Contributed Papers*. 16–19. Adana, Turkey.

Smith, R.E. (2014) *Pomegranate: Botany, Postharvest Treatment, Biochemical Composition and Health Effects*. Nova Science Publishers Inc, New York.

Soloklui, A.A.G., Gharaghani, A., Oraguzie, N., Eshghi, S. and Vazifeshenas, M. (2017) Chilling and heat requirements of 20 Iranian pomegranate cultivars and their correlations with geographical and climatic parameters, as well as tree and fruit characteristics. *HortScience: A Publication of the American Society for Horticultural Science* 52(4), 560–565. DOI: 10.21273/HORTSCI11614-16.

Sonawane, P.C. and Desai, P.C. (1989) Performance of staggered cropping in pomegranate. *Journal of Maharashtra Agricultural University* 14, 341–342.

Steinacher, G. and Wagner, J. (2010) Flower longevity and duration of pistil receptivity in high mountain plants. *Flora – Morphology, Distribution, Functional Ecology of Plants* 205(6), 376–387. DOI: 10.1016/j.flora.2009.12.012.

Stepanyan, N. (2007) Armenian wild pomegranate: a rare and relic fruit. *Bioversity International Bioversity Newsletter for Europe* 6.

Stepanyan-Gandilyan, N.P. (2017) Taxonomic revision of the family *Punicaceae*. *Novitates Systematicae Plantarum Vascularium* 48, 110–117.

Still, D.W. (2006) Pomegranates: a botanical perspective. In: Schulman, R.N., Seeram, N.P. and Heber, D. (eds) *Pomegranates, Ancient Roots to Modern Medicine*. Taylor & Francis Group, Boca Raton, Florida, p. 199.

Stover, E. and Mercure, E.W. (2007) The pomegranate: a new look at the fruit of paradise. *HortScience: A Publication of the American Society for Horticultural Science* 42(5), 1088–1092. DOI: 10.21273/HORTSCI.42.5.1088.

Sulochanamma, B.N., Yellamanda Reddy, T., Subbi Reddy, G., Reddy, T.Y. and Reddy, G.S. (2005) Effect of basin and drip irrigation on growth, yield and water use efficiency in pomegranate cv. Ganesh. *Acta Horticulturae* 696,277–279. DOI: 10.17660/ActaHortic.2005.696.48.

Tabatabaei, S.Z. and Sarkhosh, A. (2006) Analysis and comparison of salinity tolerance among 10 Iranian commercial pomegranate cultivars. *ISHS 1st International Symposium Pomegranate and Minor Mediterranean Fruits, Abstracts contributed papers, 16–19 October*, Adana, Turkey.

Tanurdzic, M. and Banks, J.A. (2004) Sex-determining mechanisms in land plants. *The Plant Cell* 16 Suppl, S61–S71. DOI: 10.1105/tpc.016667.

Teixeira da Silva, J.A., Rana, T.S., Narzary, D., Verma, N., Meshram, D.T. *et al*. (2013) Pomegranate biology and biotechnology: a review. *Scientia Horticulturae* 160, 85–107. DOI: 10.1016/j.scienta.2013.05.017.

Terakami, S., Matsuta, N., Yamamoto, T., Sugaya, S., Gemma, H. *et al*. (2007) Agrobacterium-mediated transformation of the dwarf pomegranate (*Punica granatum* L. var. nana). *Plant Cell Reports* 26(8), 1243–1251. DOI: 10.1007/s00299-007-0347-2.

Thakur, N.S., Bhat, M.M., Rana, N. and Joshi, V.K. (2010) Standardization of pre-treatments for the preparation of dried arils from wild pomegranate. *Journal of food science and technology* 47(6), 620–625. DOI: 10.1007/s13197-010-0091-4.

The Angiosperm Phylogeny Group (2016) An update of the angiosperm phylogeny group classification for the orders and families of flowering plants: ApG IV. *Botanical Journal of the Linnean Society* 181(1), 1–20. DOI: 10.1111/boj.12385.

Trapaidze, T.G. and Abuladze, L.S. (1998) Pomegranate cultivars resistant to cracking. *Subtropicheskie Kul'tury* 2, 95–97.

Tuladhar, A. and Nii, N. (2014) An anatomical study of developmental changes in maturing root tissues of pomegranate (Punica granatum L.) and formation of a unique type of periderm. *Plant Root* 8, 33–41.

Turner, G.W. and Lersten, N.R. (1983) Apical foliar nectary of pomegranate (*Punica granatum*: Punicaceae). *American Journal of Botany* 70(4), 475–480. DOI: 10.1002/j.1537-2197.1983.tb07875.x.

Varasteh, F., Arzani, K., Zamani, Z. and Tabatabaei, S.Z. (2008) Physico-chemical seasonal changes of pomegranate (*Punica granatum* L.) fruit 'malas-e-torsh-e-saveh' in Iran. *Acta Horticulturae* 769, 255–258. DOI: 10.17660/ActaHortic.2008.769.36.

Varasteh, F., Arzani, K., Zamani, Z. and Mohseni, A. (2009) Evaluation of the most important fruit characteristics of some commercial pomegranate (*Punica granatum* L.) cultivars grown in Iran. *Acta Horticulturae* 818, 103–108. DOI: 10.17660/ActaHortic.2009.818.13.

Varasteh, F. and Arzani, K. (2009) Classification of some Iranian pomegranate (*Punica granatum*) cultivars by pollen morphology using scanning electron microscopy. *Horticulture, Environment, and Biotechnology* 50(1), 24–30.

Vazifeshenas, M.R., Tehranifar, A., Davarnejad, G.H. and Nemati, H. (2015) Self and cross-pollination affect fruit quality of Iranian pomegranate 'Malas-e-Yazdi'. *Advances in Environmental Biology* 9(2), 1299–1301.

Wang, H.X. (2003) The characteristics of Mudanhua pomegranate variety and its cultural techniques. *South China Fruits* 32, 49–50.

Watson, L. and Dallwitz, M.J. (1992) The families of flowering plants: descriptions, illustrations, identification, and information retrieval. Available at: http://www.delta-intkey.com/angio/www/punicace.htm (accessed 8 July 2018).

Weerakkody, P., Jobling, J., Infante, M.M.V. and Rogers, G. (2010) The effect of maturity, sunburn and the application of sunscreens on the internal and external qualities of pomegranate fruit grown in Australia. *Scientia Horticulturae* 124(1), 57–61. DOI: 10.1016/j.scienta.2009.12.003.

Wetzstein, H.Y., Zhang, Z., Ravid, N. and Wetzstein, M.E. (2011a) Characterization of attributes related to fruit size in pomegranate. *HortScience* 46(6), 908–912. DOI: 10.21273/HORTSCI.46.6.908.

Wetzstein, H.Y., Ravid, N., Wilkins, E. and Martinelli, A.P. (2011b) A morphological and histological characterization of bisexual and male flower types in pomegranate. *Journal of the American Society for Horticultural Science* 136(2), 83–92. DOI: 10.21273/JASHS.136.2.83.

Wetzstein, H.Y., Yi, W., Porter, J.A. and Ravid, N. (2013) Flower position and size impact ovule number per flower, fruit set, and fruit size in pomegranate. *Journal of the American Society for Horticultural Science* 138(3), 159–166. DOI: 10.21273/JASHS.138.3.159.

Wetzstein, H.Y., Porter, J.A. and Ravid, N. (2015) Reproductive biology of pomegranate from flowering to fruit development. *Acta Horticulturae* 1089,21–28. DOI: 10.17660/ActaHortic.2015.1089.1.

Yan, M., Zhao, X., Zhou, J., Huo, Y., Ding, Y. *et al*. (2019) The complete chloroplast genomes of *Punica granatum* and a comparison with other species in Lythraceae. *International Journal of Molecular Sciences* 20(12), 2886. DOI: 10.3390/ijms20122886.

Yang, R.P., Long, W.H., Zhang, H., Xu, B. and W.X, L. (2007) RAPD analysis of 25 *Punica granatum* germplasm resources collected in Yunnan province. *Journal of Fruit Science* 24, 226–229.

Yang, S., Yuan, Z., Yin, Y., Zhao, X., Feng, L. *et al*. (2015) Pollen morphology of pomegranate (*Punica granatum* L.) from different eco-geographical populations in China. In: Yuan, Z., Wilkis, E. and Wang, D. (eds) *Proceedings of the Third International Symposium on Pomegranate and Minor Mediterranean Fruits*. Acta Horticulturae 1089, pp. 269–278.

Yazici, K., Cevik, M.S. and Kaynak, L. (2011) Anatomy of pomegranate (*Punica granatum* L. 'Hicaznar') fruit exocarp. *Acta Horticulturae* 890, 215–220.

Yezhov, V.N., Smykov, A.V., Smykov, V.K., Khokhlov, S.Y., Zaurov, D.E. *et al*. (2005) Genetic resources of temperate and subtropical fruit and nut species at the Nikita botanical gardens. *HortScience* 40(1), 5–9. DOI: 10.21273/HORTSCI.40.1.5.

Yilmaz, M. and Özgüven, A.I. (2006) The effect of some plant nutrients, gibberellic acid and pinolene treatments on the yield, fruit quality and cracking in pomegranate. *Acta Horticulturae* 818, 205–212.

Zamani, Z., Sarkhosh, A., Fatahi, R. and Ebadi, A. (2007) Genetic relationships among pomegranate genotypes studied by fruit characteristics and RAPD markers. *The Journal of Horticultural Science and Biotechnology* 82(1), 11–18. DOI: 10.1080/14620316.2007.11512192.

Zarei, M., Azizi, M. and Bashir-Sadr, Z. (2011) Evaluation of physicochemical characteristics of pomegranate (*Punica granatum* L.) fruit during ripening. *Fruits* 66, 121–129.

Zohary, D. and Spiegel-Roy, P. (1975) Beginnings of fruit growing in the old world. *Science* 187(4174), 319–327. DOI: 10.1126/science.187.4174.319.

3 Production and Growing Regions

G. Ferrara[1], Y. Selahvarzi[2*], H. Ahmadpourmir[2], A. Mazzeo[1] and A. Giancaspro[3]
[1]Department of Soil, Plant and Food Sciences, University of Bari 'Aldo Moro', Bari, Italy; [2]Department of Horticultural Science, Ferdowsi University of Mashhad, Mashhad, Iran; [3]Department of Environmental and Territorial Sciences, University of Bari 'Aldo Moro', Bari, Italy

3.1 Introduction

Today, pomegranate is cultivated throughout the world in Mediterranean, subtropical and tropical climatic conditions. Orchards can be found in the Middle East and Caucasus region, North and tropical Africa, the Indian subcontinent, central Asia, the drier parts of Southeast Asia, the Mediterranean Basin, North and South America, and Australia. In recent years, pomegranate fruit has become more common in the commercial markets of Europe and Western countries. In many areas of the world, in particular in Asia, pomegranate is also known and used for different medicinal purposes.

Official Food and Agriculture Organization of the United Nations (FAO) data for pomegranate are lacking and world pomegranate production is estimated to be around 5 million metric tonnes (t) (Jaime *et al.*, 2013) with increasing production year after year. The main world production is located in the Middle East and India. In the off-season period of the northern hemisphere (from late February to early May) the southern hemisphere supplies <50,000 t to the northern hemisphere market. In the following text, pomegranate growth and production statistics from different countries and regions are presented and discussed. Climatic conditions of the growing areas are also presented, with chilling hours (November–February in the northern hemisphere; June–September in the southern hemisphere) ranging from 0 (tropical areas) to values up to 2000 in areas with long and cold winter periods. Data indicate very wide conditions for pomegranate cultivation in the world.

3.2 Iran

According to the literature, Iran is known to be the native area of pomegranate (Holland *et al.*, 2009), from where this species spread throughout the world. Pomegranate is deeply integrated with Persian culture and images of this fruit have been carved on the Persepolis relics, which date back to ancient times of the Persian Empire. Iran is one of the biggest producing countries in the world in close competition with India and China (Ebrahimi, 2015). Pomegranate is the fifth fruit among the top-produced horticultural Iranian products after grape, olive, apple, orange and date (Ministry of Agriculture Jihad, 2016). The reason for the extensiveness of pomegranate production lies in the unique pedo-climatic and geographical characteristics of the country. Iran is located at 25° N to 40° N, and one-fourth of the country is covered by desert lands

*Corresponding author: selahvarzi@um.ac.ir

with an arid and subtropical climate, which is suitable for cultivation of pomegranate trees. Pomegranate, pistachio, date palm and fig are among the few fruit crop species that can grow and thrive, becoming noticeable cultivations in desert areas of the country. Pomegranates are nowadays grown all over the country except for the Hamadan province because of the very cold winters. In the region of Fars (Shiraz), an important area for pomegranate production in Iran, annual mean daily temperature ranges between a minimum of 8.7°C and a maximum of 25.1°C, with an average of 16.8°C; the annual total precipitation is 316 mm (https://en.climate-data.org/) with 1573 chilling hours (SoDa Service, 2019).

Pomegranate is used as fresh fruit or processed into jam, syrup and concentrate, and is used as flavour for ice cream and local foods. 'Fesenjan' is one of the ancient and most favourite foods in Persian culture, of which pomegranate sauce is the main ingredient. Pomegranate fruits and arils are used for decorative purposes on different occasions and in festivals in many cities, similar to what happened in ancient times. Pomegranate peel is still used for dyeing wool and clothes in some towns and villages throughout the country.

Production of pomegranate in Iran is almost one-fourth of the whole world production, with most of the orchards irrigated and only a few hectares under rain-fed management. Production is growing, due to the increasing demand of the local market and it almost doubled between 2010 and 2015 (from 555,000 to 1,086,000 t) (Fig. 3.1.). However, only 1.5% of the total domestic pomegranates are exported (Ministry of Agriculture Jihad, 2016).eading pomegranate importers include Azerbaijan, Japan, Armenia, Canada, Singapore, Germany, Hungary, Greece, South Korea, the UK, Malaysia and Denmark (Mohseni, 2009). To meet the increasing demand from importing countries, in later years (from 2008 onwards) there has been an increase in pomegranate orchards up to 90,000 ha in 2018.

Fars is the main Iranian province where pomegranate is cultivated (Kohansal and Rahimi, 2013) with a production of about 273,000 t from 20,445 ha, followed by Yazd province (99,239 t on 6439 ha), and Semnan (47,052 t on 4037 ha). Esfahan, Qom, Southern Khorasan, Lorestan,

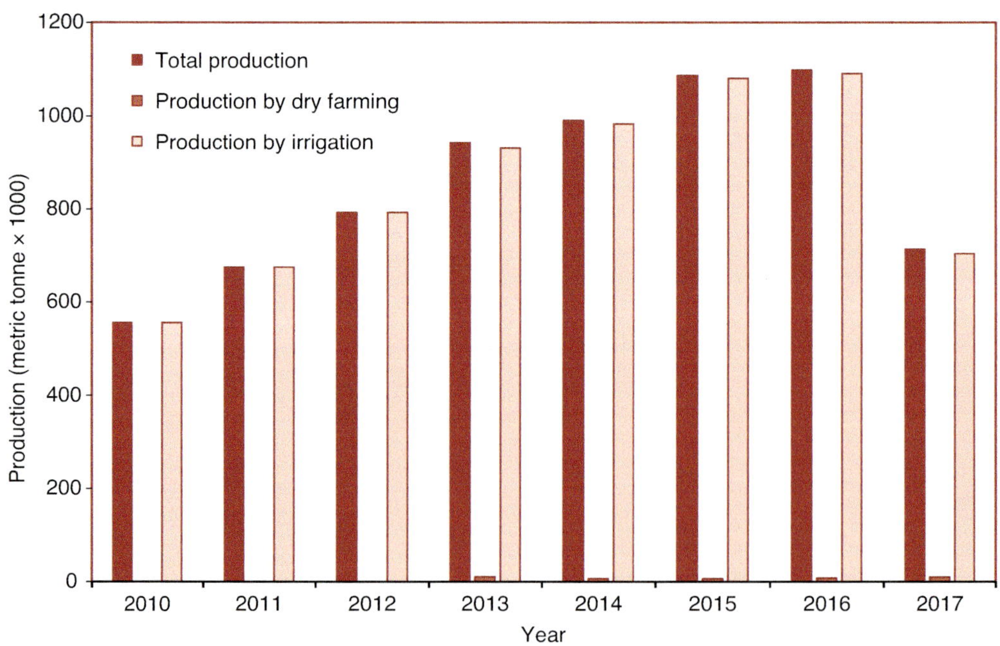

Fig. 3.1. Pomegranate production in Iran between 2010 and 2017. (Source: Ministry of Agriculture Jihad, 2016.)

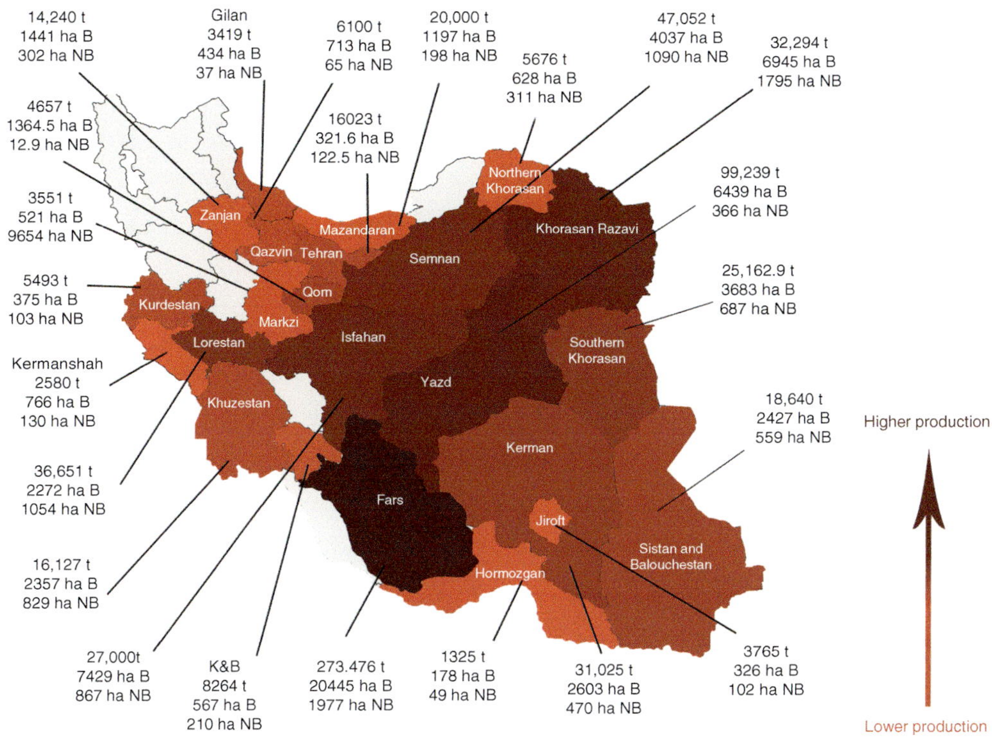

Fig. 3.2. Pomegranate production in provinces of Iran (B, bearing; NB, non-bearing orchards). (Source: Ministry of Agriculture Jihad, 2016.)

Khuzestan, Sistan and Baluchistan, Tehran and Kerman are other provinces with production above 10,000 t/year (Fig. 3.2).

Although there are many cultivars and genotypes of pomegranate in Iran (more than 760 accessions in the Iran National Pomegranate Collection), the most important commercial Iranian pomegranate cultivars are: 'Malas Saveh', 'Robab-e-Neyriz', 'Shishekap Ferdows', 'Naderi Badrood', 'Khazar', 'Aalak', 'Naderi Kashan', 'Bajestani' and 'Malas Yazdi'.

3.3 India

Pomegranate is among the most cultivated fruit species in India; production has risen tremendously in recent years so that India has become the world's leading pomegranate-producing country followed by Iran (https://numerical.co.in). Pomegranate was probably first spread from Iran to the Indian peninsula more than 2000 years ago (Holland *et al.*, 2009) or much earlier. Nowadays, pomegranate is cultivated in most states of the country, and there are at least 14 repositories for pomegranate accessions located in different areas (Holland and Bar-Ya'akov, 2008). More common in India is the dry–wet tropical climate, significantly drier than the tropical monsoon climate, it prevails through the interior peninsular territories.

India has surpassed Iran in terms of both cultivated area and production in the past few years. According to recent data (https://numerical.co.in), nearly 2.8 million t of pomegranates on 220,000 ha were produced in 2018 (Fig. 3.3, Fig. 3.4), because of the considerable increase in planted orchards in the past few years. Despite the massive production, Indian pomegranates are mainly consumed in local markets, and only 7% of total production is exported. Major importers are: the United Arab Emirates (UAE), the Netherlands, the UK, Russia, Saudi Arabia and Bangladesh. The

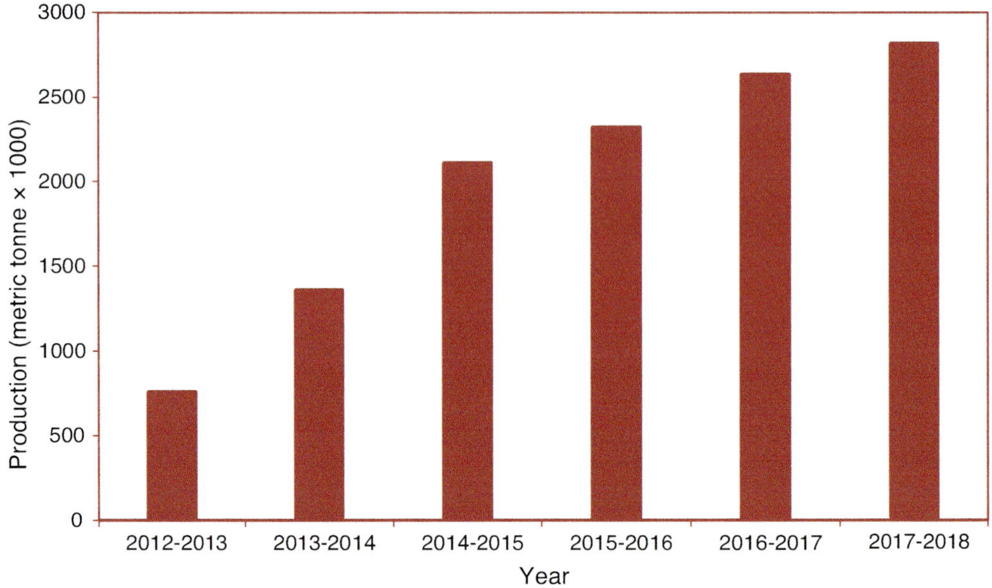

Fig. 3.3. Pomegranate production in India between 2012 and 2018. (Source: Horticulture Statistics Division, Department of Agriculture, Cooperation and Farmers Welfare.)

export period mainly falls in January–February (Holland *et al.*, 2009).

In India, in the geographical areas where pomegranate is mainly produced (Solapur), annual mean daily temperature ranges between a minimum of 20.5°C and a maximum of 33.7°C, with an average of 27.1°C; the annual total precipitation is 713 mm (https://en.climate-data.org/) with no chilling hours because of the tropical climate (SoDa Service, 2019).

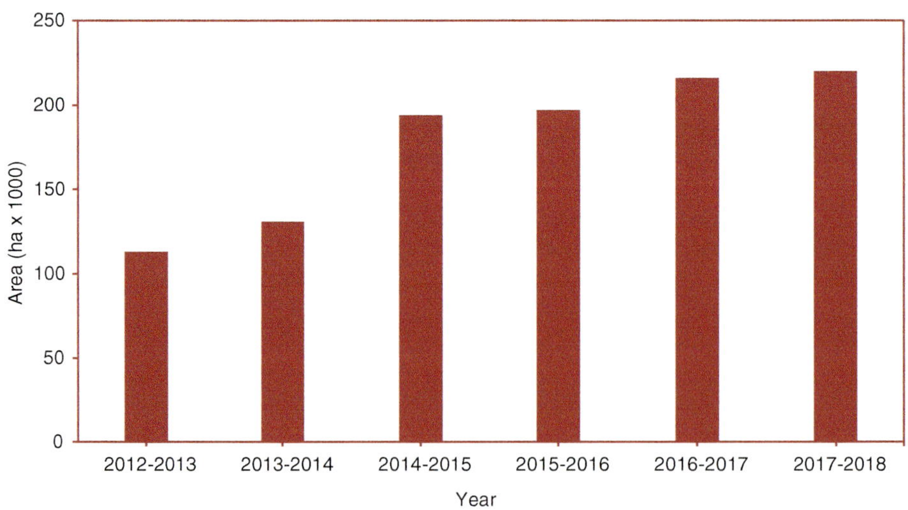

Fig. 3.4. Pomegranate orchards in India. (Source: Horticulture Statistics Division, Department of Agriculture, Cooperation and Farmers Welfare.)

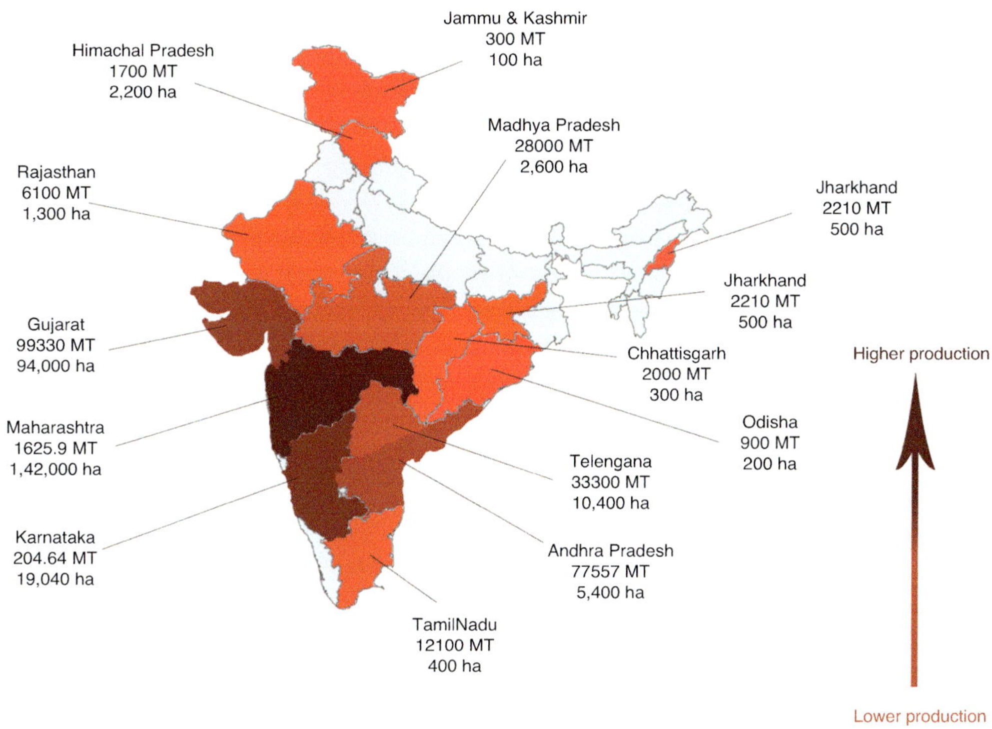

Fig. 3.5. Pomegranate production regions of India. (Source: Horticulture Statistics Division, Department of Agriculture, Cooperation and Farmers Welfare.)

Maharashtra state (districts of Solapur, Sangli, Nasik, Ahmednagar, Pune, Dhule, Aurangabad, Satara, Osmanabad and Latur) is the leading producer of pomegranate followed by Karnataka (districts of Bijapur, Belgaum and Bagalkot), Andhra Pradesh, Gujarat and Tamil Nadu (Fig. 3.5).

'Bhagwa' is the most common cultivar with a sweet taste, red arils and skin, and soft seeds. Other popular cultivars are 'Ganesh' with yellowish-pink seeds and arils, soft seeds and sweet taste, 'Mridula' with dark red arils and skin, 'Arakta' with soft seeds and dark red arils, and 'Muskat Red' with red skin and pink arils. Despite the small fruit size of 'Ganesh' and 'Mridula', proper fruit thinning can increase their size up to 350 g (Holland *et al.*, 2009). The cultivar Ganesh is also grown to produce juice (Aulakh, 2004; Singh, 2004), however the light red colour of the peel, fruit cracking, thin rind (prone to physical damages) and low yield/tree are all negative aspects limiting the export rate (Holland *et al.*, 2009) and it is mainly used for domestic consumption. The cultivars Mridula and Bahgwa originate from crosses between Ganesh and Gul Shah red cultivars. Pomegranate cultivars in the different Indian states are shown in Table 3.1. 'Mridula', 'Bhagwa' and 'Ganesh' are grown as evergreen cultivars because of the climatic conditions in the states of Maharashtra and Gujarat. In Karnataka state, peak harvest season falls from February to the end of March. The major harvest in Andhra Pradesh is from mid-April to the end of May.

Pomegranates from India have regular importers, with most of the production being exported to the nearby traditional markets like the Middle East and India's neighbours. Pomegranate fruits are usually exported in January–February to Europe, mainly to the Netherlands, the UK and some eastern European countries.

Table 3.1. Pomegranate cultivars in the different Indian states (Source: Holland *et al.*, 2009).

State	Cultivar
Maharashtra	Bhagwa, Alandi, Phule Arakta, Mridula, Ganesh, Karadi, Muskat
Karnataka	Madhugiri, Bassein Seedless, Jyothi, Madhugiri, Paper Shell
Gujarat	Dholka, Muskat Red, Kandhari, Ganesh
Himachal Pradesh	Kandhari
Andhra Pradesh	Ganesh, Paper Shell, Muskat Red, Spanish Ruby
Rajasthan	Jodhpuri Red, Jodhpuri White, Jalore Seedless
Haryana	Chawla, Nabha, Country Large Red, Ganesh, Muskat Red, Paper Shell
Tamil Nadu	Velludu, Kabul Red, Yercaud 1
West Bengal	Ruby, Mridula, Ganesh

3.4 China

Pomegranate has been cultivated in China for more than 2000 years (Jing *et al.*, 2012). China is considered among the biggest pomegranate producers in the world. However, despite the massive pomegranate production in many areas of the country, at the present time pomegranate industry plays a minor role in the national economy. Pomegranate in China is grown under old-fashioned and unsuitable conditions in many producing areas, where cultivation methods resemble traditional gardening arts rather than efficient modern agriculture. Moreover, pomegranate cultivation in China has to face many other problems, such as a wide number of homonymies, lack of efficient strategies for the prevention of diseases and pests, lack of optimization of fruit storage conditions and marketing mechanisms, lack of adoption of effective traceability tools, and poor scientific attention and research funding (Cao *et al.*, 2015). In China, in the geographical region of Sichuan (important centre of Huili), annual mean daily temperature ranges between a minimum of 14.8°C and a maximum of 27.1°C, with an average of 14.8°C; the annual total precipitation is 831 mm (https://en.climate-data.org/) with 1320 chilling hours (SoDa Service, 2019).

There are no official reports for production amount or cultivation areas; however, in 2012 the total production was estimated at 1,600,000 t over an area of 110,000 ha (Sarig and Galili, 2012). Since pomegranate can adapt to a variety of climatic and soil conditions, it grows in many regions of China, especially in the provinces of Sichuan, Chongqing, Shandong, Shaanxi, Anhui and Henan (Wang *et al.*, 2010; Sarig and Galili, 2012). Other provinces where pomegranate is produced on a smaller scale include Xinjiang, Hebei, Guangdong, Anhui, Ningxia and Yunnan (Fig. 3.6). In addition to domestic production, the fruit is also imported from Egypt and other countries.

In China, there are many climatic differences from one region to another depending on their topographical features. Southern regions have a tropical climate where summer lasts all year and vegetation is lush and always green. In the north-eastern region of the country, summer is short and cool, and winter is hard. In the south-west, winter is mild and summer is cool. In the Tibetan area the climate is cold and dry. However, most of the Chinese territory falls into the temperate zone where the four seasons are very distinct periods of high temperatures, with consistent rainfall, alternating with periods of low temperatures with low rainfall.

There are over 200 different cultivars of pomegranate grown in China (Feng *et al.*, 2006), which vary from province to province in terms of weight, colour, skin thickness, seed hardness, sugar content and other components affecting the taste (Sarig and Galili, 2012). There are no reliable data on top grown cultivars. However, some of the best-known cultivars, along with their geographical origin, are presented in Table 3.2 (Lihua *et al.*, 2013). 'Taishan Dahongshiliu' is a well-known and widely cultivated cultivar, first discovered and named in 1984 in a home garden in the Taishan Mountains of Shandong province (Shi, 1991).

Fig. 3.6. Main pomegranate production regions in China. (Source: Sarig and Galili, 2012.)

This variety is high yielding with large, soft-seeded fruits, which are resistant to cracking (Mars, 2000). Despite the vast diversity of cultivars, Zhang *et al.* (2012) reported that many cultivars are named by local farmers, thus the same landrace might be given different names in different regions (synonymies), or some different landraces might share the same name (homonymies). This situation urges a rapid and efficient solution for unambiguous identification of pomegranate cultivars in China in order to create an accurate varietal list.

3.5 Spain

Pomegranate was first introduced in Spain by the Phoenicians in 600 BCE. Now, Spain is the biggest pomegranate producer and exporter in Europe (Bartual *et al.*, 2015), with a yield of 18.5 t/ha (Chandra *et al.*, 2010). In Spain, the region of Murcia is where pomegranate is mostly cultivated. In this region, in particular, in Alicante province, annual mean daily temperature ranges between a minimum of 12.1°C and a maximum

Table 3.2. Most important pomegranate cultivars in different regions of China (Source: Zhao *et al.*, 2013).

State/region	Cultivar
Shandong	Bingtangshiliu, Yihongyihao, Mapicao, Lushiliu, Taishan Dahongshiliu
Shanxi	Bairixue, Moshiliu, Mozishiliu, Yushiliu, Jingpitian
Yunnan	Hongpibaizi, Qingpibaizi, Lvpisuan, Huopao, Nuoshiliu
Hennan	Tianhongmi, Heyinruanzi, Mudanhonghua, Suanhongpishiliu, Yueyuehong
Anhui	Dabenzi, Huohulu, Manaozi, Shuifenpi, Yushizi
Xinjiang	Suanshiliu, Tianshiliu, Hongpishiliu
Sichuan	Shuijingshiliu, Qingpiruanzi, Huilihongpi

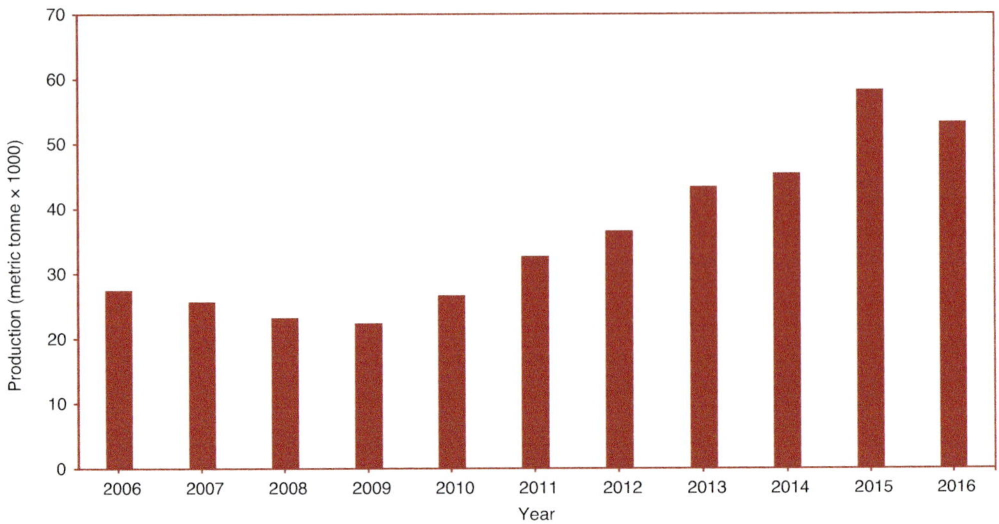

Fig. 3.7. Pomegranate production in Spain from 2006 to 2016. (Source: MAPA, 2017.)

of 24.1°C, with an average of 18.1°C; the annual total precipitation is 293 mm (https://en.climate-data.org/) with 540 chilling hours (SoDa Service, 2019). The main destination countries for Spanish export are Germany, the UK, Italy, Russia and France. According to official reports, the overall production of pomegranate in Spain, aside from a drop between 2006 and 2009, has almost always been on the rise since 2009, with up to 53,187 t of pomegranates harvested in 2016 (Fig. 3.7). The increase in pomegranate production is mainly due to the establishment of new orchards, following the increasing demand of consumers for a fruit with scientifically assessed healthy properties. The total cultivated area, for both bearing and non-bearing orchards, is shown in Fig. 3.8.

Spain is located in the Iberian Peninsula of Europe at approximately 36° N to 43.5° N. There are different climates in the country: hot

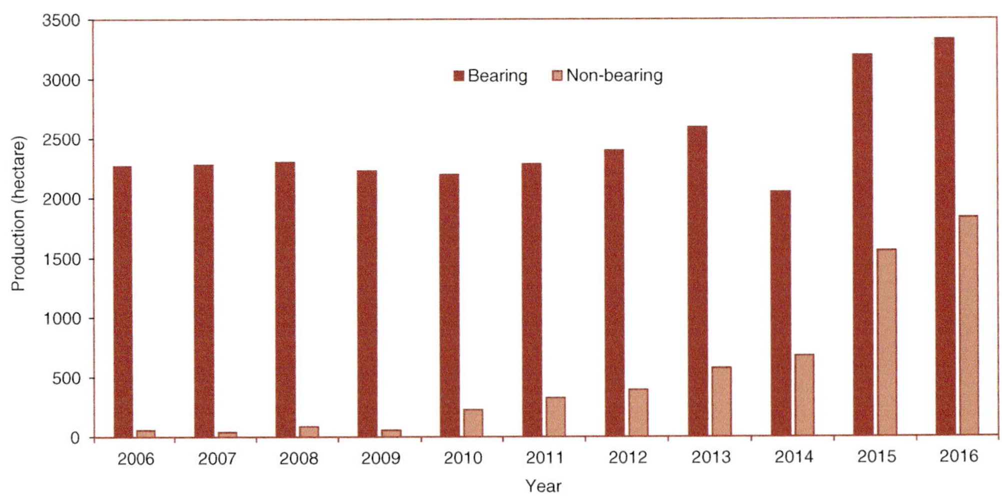

Fig. 3.8. Production area of pomegranate in Spain from 2006 to 2016, for bearing and non-bearing orchards. (Source: MAPA, 2017.)

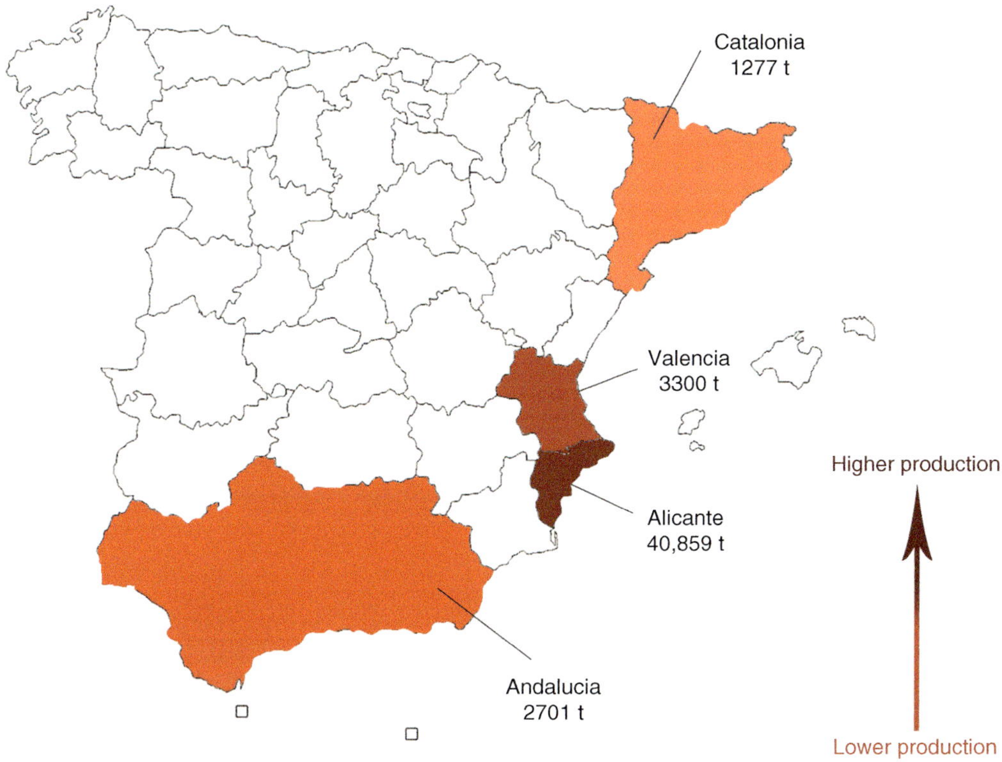

Fig. 3.9. Pomegranate cultivation in provinces of Spain. (Source: Bartual *et al.*, 2015.)

Mediterranean summer is the dominant climate in the south, whereas oceanic and warm summers are the dominant climates in the north of the country. Pomegranate production in Spain is concentrated in Valencia and Andalusia, and mostly in Alicante province, in particular in the regions of Elche, Alberta and Crevillente (Calin-Sanchez *et al.*, 2010, Fig. 3.9). Alicante is responsible for 84% of the pomegranate production in the area, especially in the Bajo Segura and Bajo Vinalopo regions (MAPA, 2011).

Since 1980 there has been a high increase in the cultivated area, reaching the historical peak of 3300 ha in 2000 and >3300 ha in more recent years (Bartual *et al.*, 2015). Also, pomegranate imports have notably increased since 2002 (Melgarejo *et al.*, 2012), reflecting the market growth for this fruit, which is becoming more and more appreciated by consumers.

At least 40 different cultivars have been reported to be cultivated in Spain, each with specific features (Holland *et al.*, 2009). Two main varietal groups of commercial interest can be identified: the 'Valencianas' and the 'Mollars'. Cultivars within the 'Valencianas' group are characterized by early harvesting period (between August and September), sweet flavour and soft seeds. On the other hand, the varietal group called 'Mollars' includes the most important cultivars, which are characterized by excellent quality, soft seeds and late harvesting time (between mid-September and mid-November) (Stover and Mercure, 2007; Melgarejo *et al.*, 2015). Beside 'Mollar' and 'Valenciana', the more recently introduced cultivar 'Wonderful' is one of the most cultivated and well-known pomegranate cultivars for juice production (Melgarejo *et al.*, 2012). Other cultivars include 'Agria de Albatera', 'Agria de Blanca', 'Agridulce de Ojos', 'Albar de Bianca' and 'Borde de Albatera', named according to flavour or skin/aril colour (Holland *et al.*, 2009).

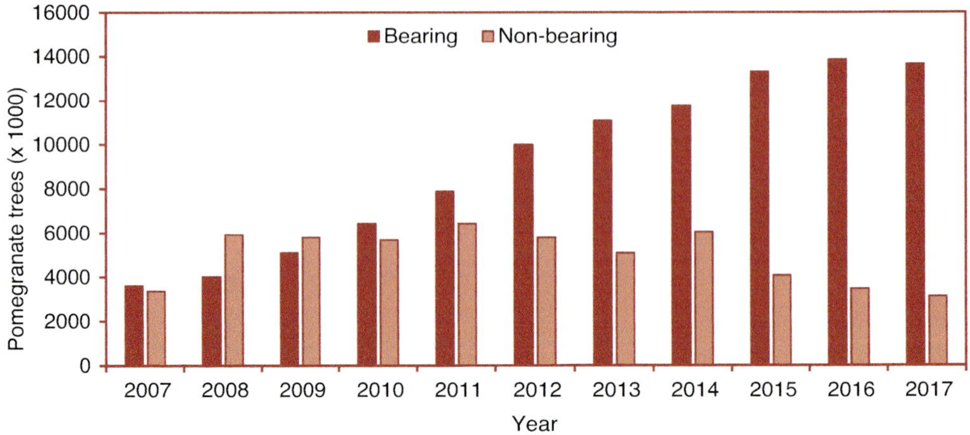

Fig. 3.10. Pomegranate trees in Turkey from 2007 to 2017. (Source: Ministry of Food, Agriculture and Livestock of Turkey.)

3.6 Turkey

Turkey is one of the centres of origin of pomegranate, which was initially domesticated in the northern Transcaucasian–Caspian region (Zohary and Spiegel-Roy, 1975; Harlan, 1992), and then spread all over the country including both mountainous and coastal areas (Ozguven and Yilmaz, 2000). Due to its adaptability to all kinds of soils and high tolerance to drought and salinity, pomegranate is one of the major cultivated fruits in the country, in particular in the southern regions (Özgüven *et al*., 2015). In Turkey, the region of Antalya is the main region for pomegranate cultivation; here, annual mean daily temperature ranges between a minimum of 13.3°C and a maximum of 24.0°C, with an average of 18.6°C; the annual total precipitation is 1009 mm (https://en.climate-data.org/) with 886 chilling hours (SoDa Service, 2019). In recent years, production of pomegranate has rapidly increased as a result of new orchards, use of popular cultivars and public awakening as to the positive effects of this fruit on human health (Ozcani and Unaldi, 2007). Figure 3.10 depicts the number of cultivated pomegranate trees in 2007–2017 in Turkey. However, fruits produced in traditional orchards are not fully appropriate for the modern market. Many old orchards have been decommissioned, and only a few local cultivars are propagated in commercial nurseries and planted in new orchards.

Pomegranate production in Turkey has been increasing in the past two decades. Annual pomegranate production in Turkey was about 445,750 t in 2015, and production has rapidly increased from year to year. In 2016–2017, 465,200 t of pomegranates were harvested, which represents a 4.1% growth compared with 2015–2016 (Turkish Statistical Institute, 2018, www.turkstat.gov.tr). Figure 3.11 pictures the pomegranate production trend from 2007 to 2016. Along with the increase in production, fresh pomegranate exports also increased (from 3591 t in 2000 to 151,174 t in 2015) and they were primarily directed to countries like Greece, the Netherlands, Germany, Russia, Ukraine and Bulgaria (Akcaoz *et al*., 2009). The favourite cultivar of importing countries is 'Hicaznar' (Özgüven *et al*., 2015).

Located at 42° 06′ — 35° 51′, Turkey has eight different climates, among which Mediterranean and cold semi-arid climates in the west and south-west of the country are the best suited for pomegranate growing. In Turkey, pomegranates are harvested between September and December, with the peak of harvest falling between the beginning of September and mid-November (Salgado, 2017). Today pomegranates are cultivated in 53 of the 81 cities of the country, prevalently in regions facing the Mediterranean Sea (Ozcani and Unaldi, 2007) (Fig. 3.12). Pomegranate orchards in Turkey are mostly located in the Mediterranean, Aegean, south-east

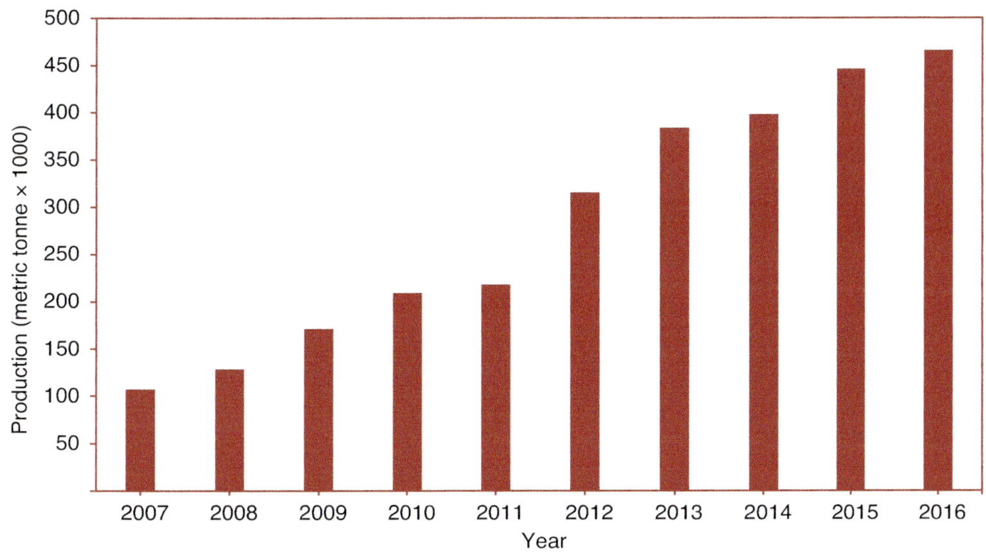

Fig. 3.11. Pomegranate production of Turkey from 2007 to 2016. (Source: Ministry of Food, Agriculture and Livestock of Turkey.)

Anatolia and Marmara regions, with 59.5%, 25.6%, 10.5%, and 2% of the total production, respectively (Kurt and Sahin, 2013). South-eastern regions of the country do not have suitable conditions for commercial production, because of the rainy and snowy climate, which is too cold in winter (Ozguven and Yilmaz, 2000).

The Mediterranean region has a rich pomegranate genetic diversity useful for breeding programmes. However, in recent years, pomegranate resources have suffered deep genetic erosion due to vulnerability to biotic and abiotic stresses and loss of agricultural land due to intensive urbanization of the country (Caliskan and Bayazit, 2012). In Turkey, pomegranates are used as either fresh fruit and juice or to produce value-added products such as syrup, vinegar, citric acid, dye, medicine and concentrated solutions (Ozkan, 2003, Ozkan, 2003). Processed pomegranates are becoming

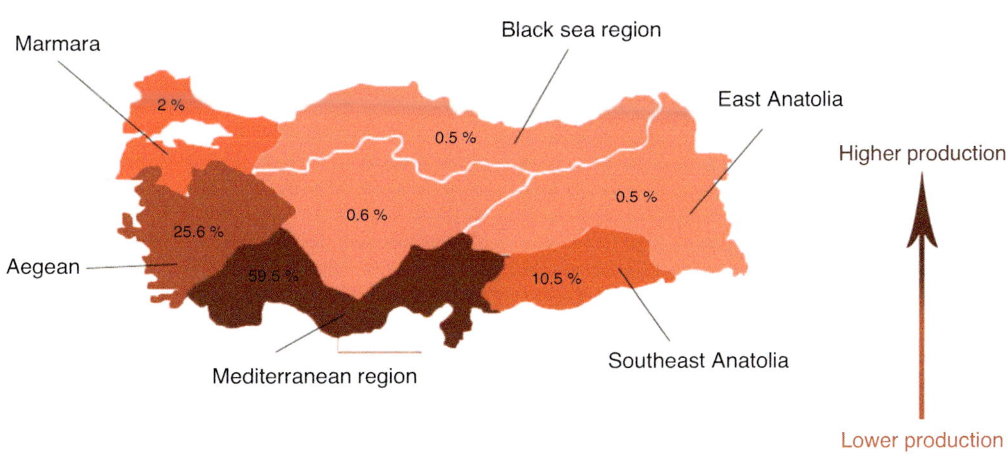

Fig. 3.12. Pomegranate-producing regions in Turkey. (Source: Kurt and Sahin, 2013.)

more popular and their production is reported to have increased by 150% in the past few years (Akdag, 2009; Yücel, 2010; Ministry of Economy, 2013).

The suitable ecological conditions of Turkey allow a wide number of pomegranate cultivars to grow all over the country, varying in terms of taste, acid level, seed hardness, skin colour, etc. Local landraces are numerous and very useful for some of their characteristics such as resistance to diseases, pests, cold, drought, and other biotic and abiotic stresses. Most of the Turkish cultivars are sweet-sour, and red coloured (Janic, 2009). 'Cekirdksiz', 'Ernar', 'Hatay', 'Hicaznar', 'Fellahyemez', 'Izmir 1', 'Izmir 1264', 'Izmir 1265', 'Janarnar', 'Katrbas' and 'Lefan' are the most frequently grown commercial cultivars (Ozguven *et al.*, 2006). 'Hicaznar', which is characterized by sweet-sour taste and hard seeds, is the most common cultivar (Holland *et al.*, 2009) followed by 'Lefan', 'Janarnar' and 'Izmir'.

3.7 Afghanistan

Agriculture represents the second main source of Afghanistan income (Surgul, 2016). However, it is still managed by old-fashioned techniques resulting in low-quality fruits and quantity yields. Fruit trees cover 9% of total agricultural production, and 2% of the cultivated fruit species is represented by pomegranates (Fitrat and Verma, 2014). In the area of Kandahar, where pomegranate is a very important crop, annual mean daily temperature ranges between a minimum of 10.2°C and a maximum of 27.5°C, with an average of 18.8°C; the annual total precipitation is 176 mm (https://en.climate-data.org/) with 1346 chilling hours (SoDa Service, 2019). Pomegranate in the Chinese language means 'fruit of Kabul', which points to Afghanistan as the region of origin of this fruit (Holland *et al.*, 2009). In Afghanistan the climate is generally arid continental, with cold and relatively rainy winters (with a rainy peak in spring), and hot, sunny summers.

An FAO survey in 2003 estimated the total pomegranate production area of Afghanistan to be about 2500 ha, which is now claimed to have almost quadrupled according to official reports. Current pomegranate production is now growing, whereas between 1996 and 2003 production rate fell to less than half because of both unstable political conditions and drought seasons, which caused a strong decrease in pomegranate cultivation (Fig. 3.13).

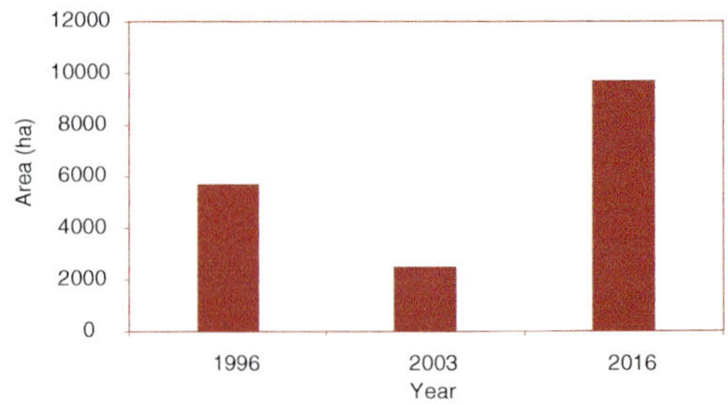

Fig. 3.13. Afghanistan pomegranate cultivation area. (Source: FAO survey 2003, Afghanistan Ministry of Agriculture, Irrigation and Livestock.)

According to the Agriculture Ministry of Afghanistan, total pomegranate production was 99,871 t in 2016, cultivated on 9721 ha. Recent data report production of 150,000 t for 2018 with an increasing production year after year (Edgardo Giordani, unpublished data). Kandahar, with 72,100 t of fruits, ranks first in the top-producing provinces, followed by Kapisa, Urozgan and Farah with 7200, 5590 and 4480 t, respectively. Balkh with 3174 t and Helmand with 1986 t are other important provinces for pomegranate production in Afghanistan (Fig. 3.14). Production amount and planted area of provinces are shown in Table 3.3. Other provinces where pomegranate is cultivated are Samangan, Nemroz, Wardak, Ghazni and Paktika (Finetto, 2011). The province of Kandahar was responsible for about 72% of the pomegranate production and half of the cropping area in 2016. Pomegranate has always been a very important crop in Afghanistan, ranking fifth after grape, almond, apricot and apple in terms of cultivation area and production amount, as it fits very well the environmental conditions of the country, which are the best for fruit development and ripening. Pomegranate harvesting season in Afghanistan starts from summer and lasts until autumn (Surgul, 2016). The southern areas of Afghanistan are subtropical zones with arid and semi-arid climates providing the best conditions for pomegranates to thrive. Afghanistan produces over 48 of the

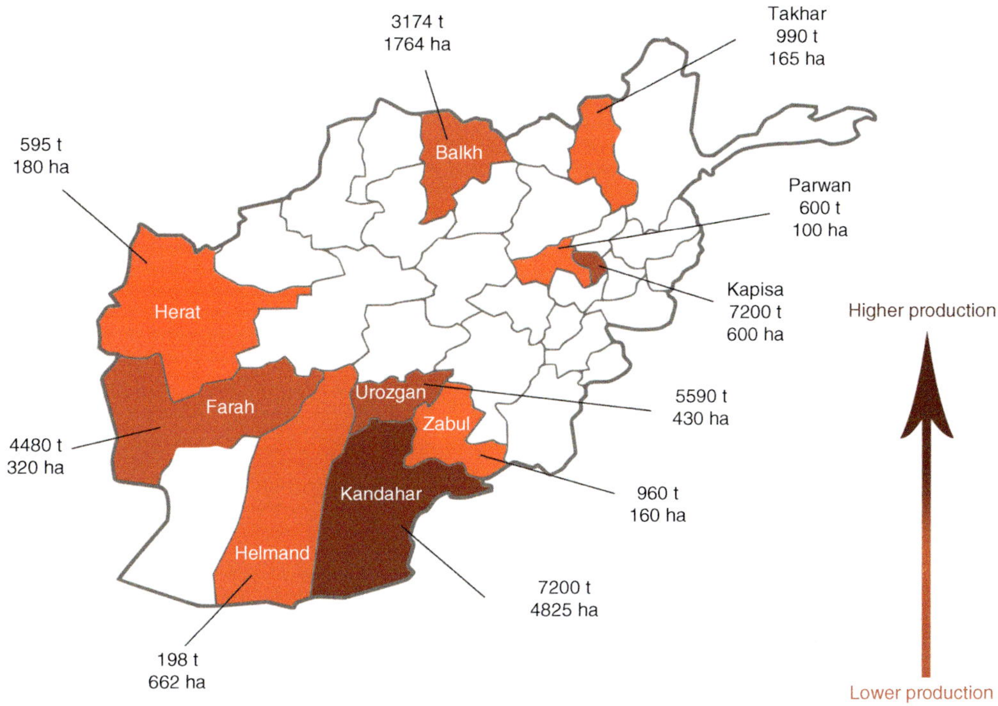

Fig. 3.14. Top pomegranate-producing provinces of Afghanistan with more than 500 t of production in 2016. (Source: Agricultural Ministry of Afghanistan.)

world's best cultivars; in fact, Surgul (2016) named it 'the country of pomegranate fruit'. There are different pomegranate cultivars grown in the various provinces of Afghanistan. Nangarhar is responsible for the production of 79 cultivars, and Kandahar for 54 (Fitrat and Verma, 2014). Most cultivars are sold in local markets, whereas only 5% of pomegranate production is destined for export to countries such as Pakistan, India, the UEA, Saudi Arabia and the USA (Finetto, 2011). Out of 48 commercially grown cultivars in Afghanistan, 20 are sweet tasting, 17 sweet-sour and 11 sour (Fitrat and Verma, 2014). Finetto (2011) lists 'Kandahari' and 'Bedana' as two of the most important cultivars of the country.

3.8 Egypt

In ancient Egypt, the pomegranate tree was considered a symbol of life-giving forces of fertility, and the fruits were used in traditional medicine (Shaheen *et al.*, 2016). The most popular pomegranate cultivars in Egypt are 'Baladi' and 'Wonderful'. The cultivar 'Baladi' is dark red inside, and lighter red on the skin. 'Wonderful' is grown extensively in particular in the newly reclaimed areas. Pomegranate is mostly cultivated in upper Egypt, especially in Assiut. In this area, annual mean daily temperature ranges between a minimum of 15.1°C and a maximum of 30.2°C, with an average of 22.6°C; the annual total precipitation is almost insignificant with 2–6 mm (https://en.climate-data.org/) and 289 chilling hours (SoDa Service, 2019). The harvesting season starts from September and reaches its peak by the end of the year. In Egypt there are three types of climate: the Mediterranean climate of the northern coast, desert climate in the central and southern inland areas, and the climate of the Red Sea coast, which is also desert but a little milder. Although the environmental conditions of some areas of Egypt are

Table 3.3. Pomegranate production in different provinces of Afghanistan in 2016 (Source: Agricultural Ministry of Afghanistan).

Province	Production (t)	Area (ha)
Total	99,871	9721
Kapisa	7200	600
Parwan	600	100
Nangarhar	450	30
Laghman	70	14
Baghlan	36	12
Paktika	480	80
Khost	100	20
Nooristan	300	50
Badakhshan	360	89
Takhar	990	165
Balkh	3174	1764
Ghor	0	170
Daykundy	400	50
Urozgan	5590	430
Zabul	960	160
Kandahar	72,100	4825
Helmand	1986	662
Herat	595	180
Farah	4480	320

suitable for pomegranate cultivation, fruits present problems of low quality caused by cracking, sunburn, lack of internal colouring, and some fungal and pest infestation. Egyptian pomegranates are mainly exported to Canada and Russia, apart from local consumption.

3.9 United States

Pomegranate was first introduced in the USA by Spanish settlers in 1769 (Arena *et al.*, 2000). The USA is known as one of the most important pomegranate-producing countries. However, there is little official information on the production amount. According to a census in 2017, pomegranate was cultivated in 13,308 ha in 2012, which represented a 34% growth compared with 2005 when there were 9921 ha on 599 pomegranate farms.

Most pomegranate cultivation is located in California, with 13,041 ha, followed by Texas, with only 93 ha. The climate of California differs from one location to another depending on the latitude. It is Mediterranean on the southern coast, and temperate along the northern coast. Fresno and Kern Counties are the main pomegranate production areas in California. In the county of Fresno, annual mean daily temperature ranges between a minimum of 9.3°C and a maximum of 24.6°C, with an average of 16.9°C; the annual total precipitation is 272 mm (https://en.climate-data.org/) with 819 chilling hours (SoDa Service, 2019). Pomegranate-producing states of the USA are presented in Fig. 3.15. In 2012, about 90% of pomegranates were produced in California with 282,817 t and a yield of 26 t/ha (California County Agricultural Commissioners' Reports) with a value of US$115.4 million (California Department of Food and Agriculture, 2018). California exports its pomegranates to Japan, South Korea and Australia, but the largest importer is Canada. On average, exports amount to 1–2 million boxes (www.fresh plaza.com).

There is a relatively limited number of pomegranate cultivars in the USA. 'Wonderful' is the most important and cultivated one (Holland and Bar-Ya'akov, 2008), which was introduced in Florida in 1896 (Levin, 2006). The fruit is large with red arils, sweet-sour taste and semi-hard seeds. The external appearance of the fruit is very appealing with a glossy red colour. The several Israeli landraces of 'Wonderful' are either 'Wonderful' seedlings (most likely) or sports. However, the American 'Wonderful' is genetically distinguishable from any of the Israeli 'Wonderful' landraces by using molecular markers (Giancaspro *et al.*, 2017). The American 'Wonderful' fruit is much harder and less prone to mechanical aril extraction than the Israeli landraces, but these differences could reflect variations in growth and climatic conditions. 'Wonderful' landraces are also grown in Western Europe and Chile and in countries where new orchards have been established (Stover and Mercure, 2007). Many new commercial cultivars were introduced in the USA during the 20th century, but none of them could be a replacement for 'Wonderful'

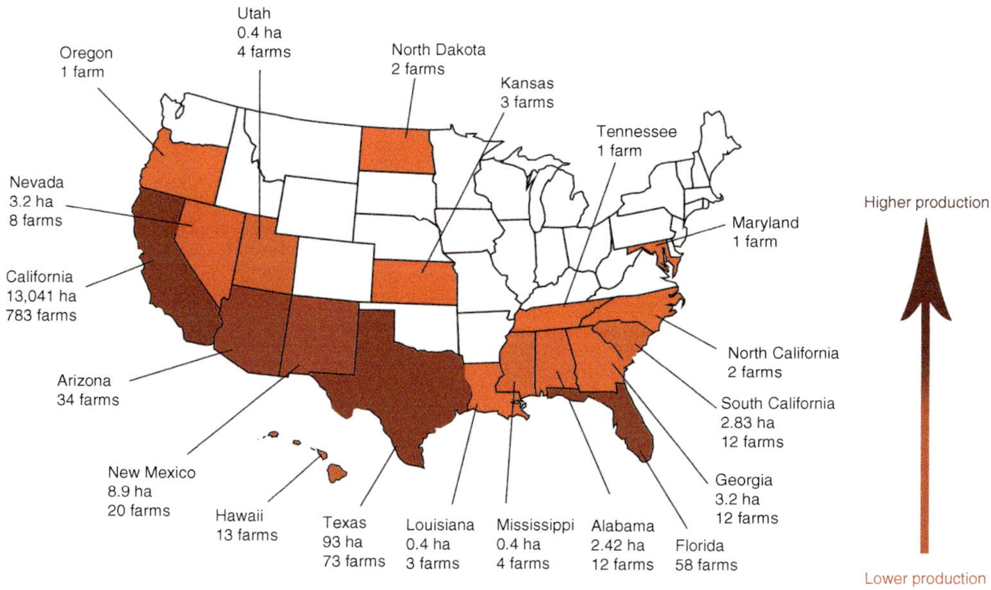

Fig. 3.15. States producing pomegranate in the USA. (Source: NASS, 2014.)

for the production of juices (Holland and Bar-Ya'akov, 2008). Among these newly released cultivars there are: 'Early Wonderful', which ripens 2 weeks earlier than 'Wonderful'; 'Granada', with dark red peel that ripens 1 month earlier than 'Wonderful'; 'Eversweet', with soft and sweet pale pink arils; 'Ambrosia', with very large fruits (Holland *et al.*, 2009; www.bayflora.com/pomegranates.html), 'Eve', with red large late-ripening fruits; and 'Sharp Velvet' and 'Sweet', with dark burgundy skin and arils, along with compact forms of trees (Holland *et al.*, 2009).

Other cultivars grown in California to a smaller scale are 'Balegal', 'Cloud', 'Fleishman', 'Crab', 'Francis', 'Green Globe', 'Home', 'King', 'Phoenicia' and 'Utah Sweet' (California Rare Fruit Growers, 1997). Also, several ornamental pomegranate cultivars, such as 'California Sunset', are sold in the USA. At least two originated in Japan and include the 'Double Flower' cultivars: 'Nochi Shibori' and 'Toyosho', according to the Davis repository list (USDA, 2007). As well as California, pomegranate breeding has been recently initiated in Florida in order to increase the number of pomegranate cultivars available for cultivation in the USA (Holland and Bar-Ya'akov, 2008).

3.10 Azerbaijan

Pomegranate is one of the three most important fruit crops in the Republic of Azerbaijan, along with apple and grape, and is grown for both domestic consumption and export. Azerbaijan is the origin place of several plant species because of its exceptional climate and edaphic factors, in addition to its very favourable geographical position. Located at the boundary between the temperate and subtropical belt, Azerbaijan has nine types of climates (Khalilov *et al.*, 2015). Given such a variety of climatic conditions, Azerbaijan enjoys a vibrant vegetation cover (Hajiyeva *et al.*, 2018) and a huge variety of pomegranate accessions. The area that accounts for the highest pomegranate production is Goychay, known also for a pomegranate festival. In this region, the annual mean daily temperature ranges between a minimum of 9.5°C and a maximum of 19.8°C, with an average of 14.7°C; the annual total precipitation is 449 mm (https://en.climate-data.org/) with 1904 chilling hours (SoDa Service, 2019). According to the Azerbaijan State Statistical Committee, a total of 156,797 t of pomegranates grown on 22,627 ha was

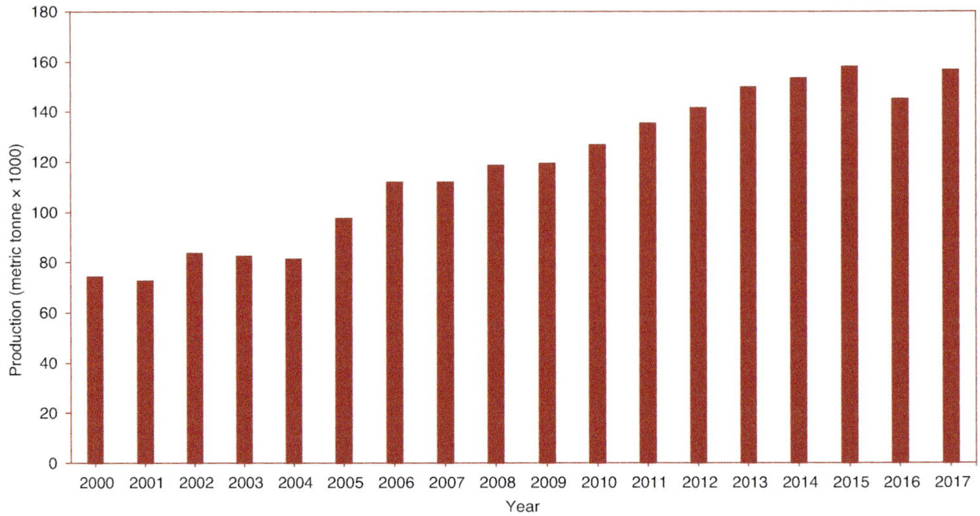

Fig. 3.16. Increase in pomegranate production from 2000 to 2017 in Azerbaijan. (Source: .)

recorded in 2017, which showed an 8% increase compared with 2016. The peak of production dates to 2015 with 158,101 t, followed by a relatively significant drop to 145,104 t in 2016, despite the expansion of the areas of cultivation. Azerbaijan is divided into ten economic regions, each containing several cities. Except for a very few cases, pomegranate production in all the regions and cities has always been on the increase in the past decade. Figure 3.16 shows the production trend over the past 18 years and Fig. 3.17. depicts the areas dedicated to pomegranate cultivation.

The majority of orchards are located in the Aran region, the largest of the 10 economic districts, with a surface of 23,375 km^2. This region consists of 18 cities located either at or under sea level, where the summers are hot, and winters are mild (Fig. 3.18; Nazarov, 2011). Some of these cities are Goychay, Ujar, Kurdmair, Hajigabul, Sabirabad, Saalty, Agdash, Agjabedi, Bilasuvar, Salyan, Goranboy and Shamkir,

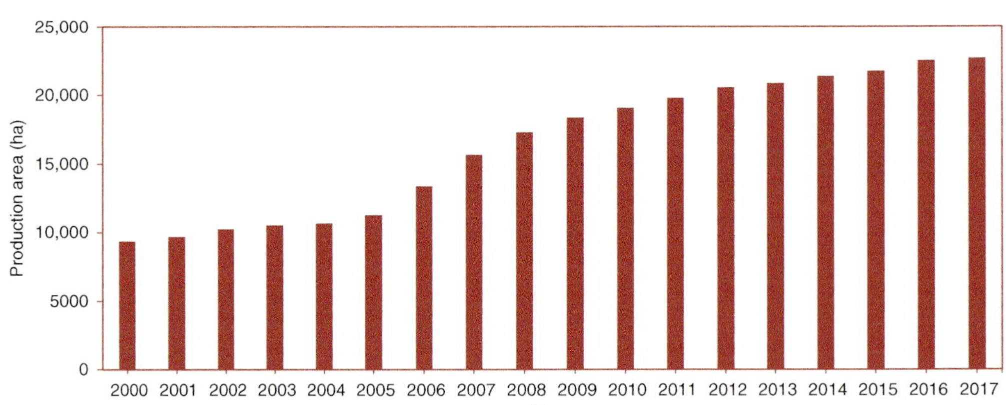

Fig. 3.17. Areas (ha) for pomegranate cultivation in Azerbaijan from 2000 to 2017. (Source: The State Statistical Committee of the Republic of Azerbaijan, 2017.)

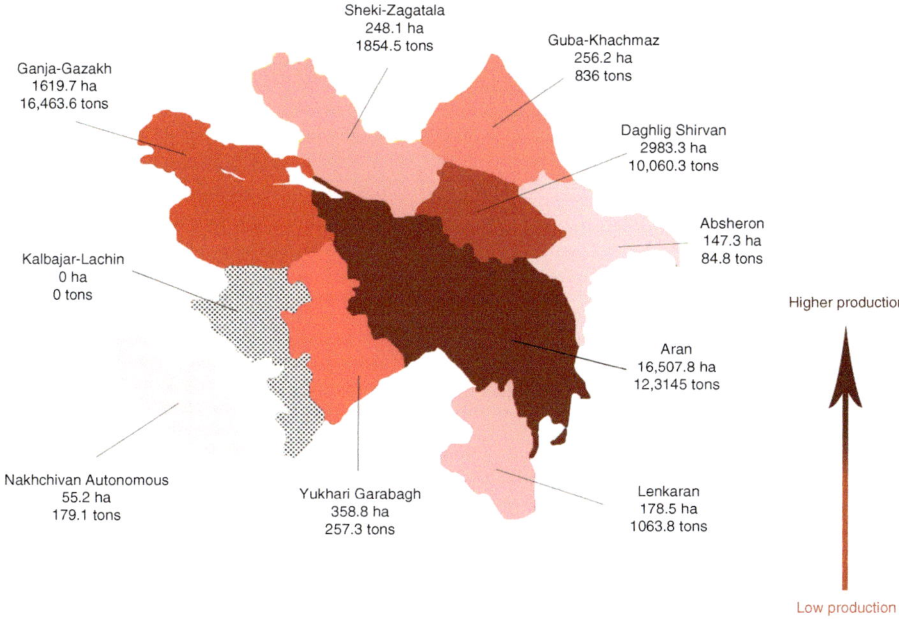

Fig. 3.18. Main provinces in Azerbaijan dedicated to pomegranate cultivation. (Source: R98.)

which possess a total area dedicated to pomegranate cultivation larger than >400 ha. The reason for such large pomegranate production lies in the excellent soil composition and the unique microclimate. Goychay district is the main pomegranate production area, with 67 different cultivars and 30% of the total pomegranate harvest, followed by Ganja-Gazakh and Daghlig Shirvan. The Aran region accounts for 78% of the total production and 72% of the orchards. The economic district of Ganja-Gazakh consists of 11 cities, nine of which are involved in pomegranate production, ranking second with 16,463 t (10% of the total domestic production). The best-producing city is Shakmir, with 7930 t and 455 ha of orchards.

The third greatest pomegranate producer region in Azerbaijan is Daghlig Shirvan, which accounts for 6.5% of the total domestic production while owning 13% of the pomegranate orchards. All the remaining regions play a minor role in the pomegranate industry of the country (Fig. 3.18.). Kalbajar-Lachin is the only region with no pomegranate production. The average production in the whole country is 7.4 t/ha.

Cultivars 'Gulovsha', 'Vesel', 'Shandi', 'Shirin', 'Girmizigabig', 'Vilash' and 'Bala Mursal' are the most widespread in Azerbaijan (Hajiyeva *et al.*, 2018). Vilash is the most interesting among these cultivars in terms of cold storage and resistance to damage during transport. However, for the further development of pomegranate production, attempts to introduce new cultivars and to establish modern orchards are being taken (www.freshplaza.com).

Azerbaijani pomegranates are widely exported to Russia, Georgia and Ukraine, whereas export to European countries is very low. The export revenues for 2017 amounted to US$10 million.

3.11 Tunisia

Pomegranate was introduced to Tunisia at the time of the Phoenicians as in Spain. Tunisia is located at 34°00′ N, 9°00′ E and enjoys a warm desert climate in almost 85% of the country. The climate is Mediterranean, with mild, rainy winters and hot, sunny summers on the northern coast, while it is semi-desert or desert in inland areas. Pomegranate is cultivated throughout Tunisia except in the regions that

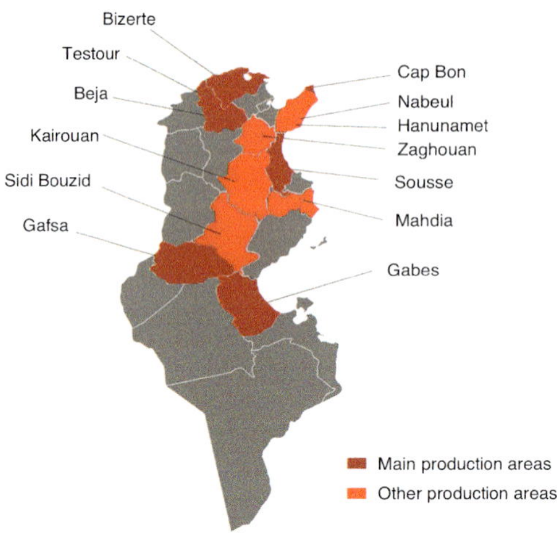

Fig. 3.19. Pomegranate-producing regions and cities in Tunisia. (Source: Mansour *et al.*, 2011.)

can face possible frost damage to trees (Mansour *et al.*, 2011). In Tunisia, the area surrounding Gabes is very important for pomegranate production and here the annual mean daily temperature ranges between a minimum of 14.2°C and a maximum of 24.6°C, with an average of 19.3°C; the annual total precipitation is 177 mm (https://en.climate-data.org/) with 330 chilling hours (SoDa Service, 2019). Main production is achieved in the oases of Gabes and Gafsa in southern Tunisia, Cap Bon in the north, Kairouan, Sidi Bouzid and Mahdia in the centre, and Sousse and Bizerte on the coast. Gabes is the first producing region with 30,000 t of fruits over an area of about 3000 ha, including more than 1.5 million pomegranate trees, with a total of 5 million pomegranate trees all over the country. This region accounts for about 37% of total domestic production. Here, pomegranates are grown with date palms, which help pomegranate trees to grow much better by providing shade that reduces air dryness and prevents tree damage (Hafez Hamid, 2016, personal communication). A pomegranate festival is held annually in the Qatana region, located in the south-east of Tunisia and 500 km away from Gabes, which attracts a considerable mass of visitors with the attempt to exhibit different uses of pomegranates and show cultural aspects of this fruit. Top-producing cities and regions of Tunisia are illustrated in Fig. 3.19.

There are no official estimates for pomegranate production; however, unofficial statements indicate a reduced production until a few years ago. Production was estimated to be 74,000 t in 2012, with an increase of 2.7% compared with 2011. 2010 was the least productive year, with 67,000 t. From 2008 to 2010, about 6% of the production was exported, whereas this amount dropped to 1.7% in 2011 and 1.9% in 2012

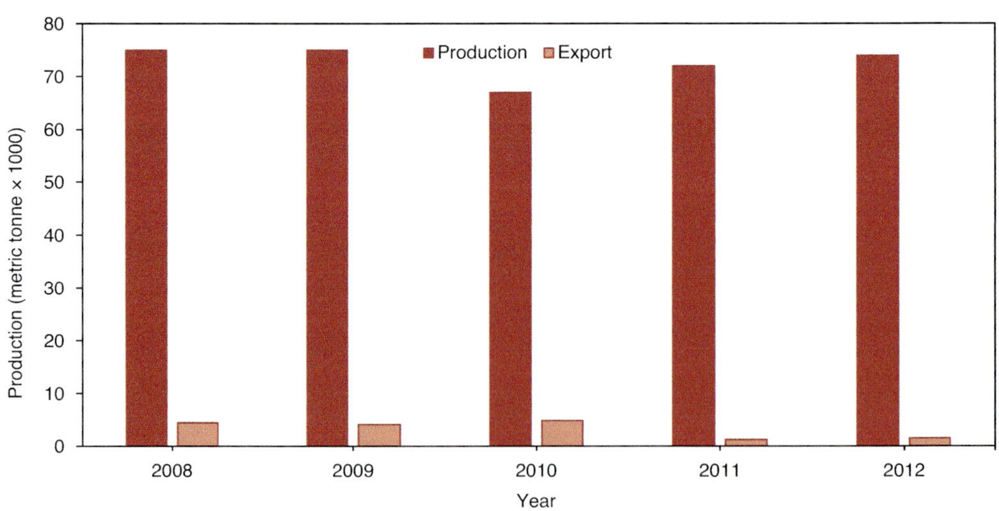

Fig. 3.20. Pomegranate production and export in Tunisia from 2008 to 2012. (Source: Freshplaza, 2013.)

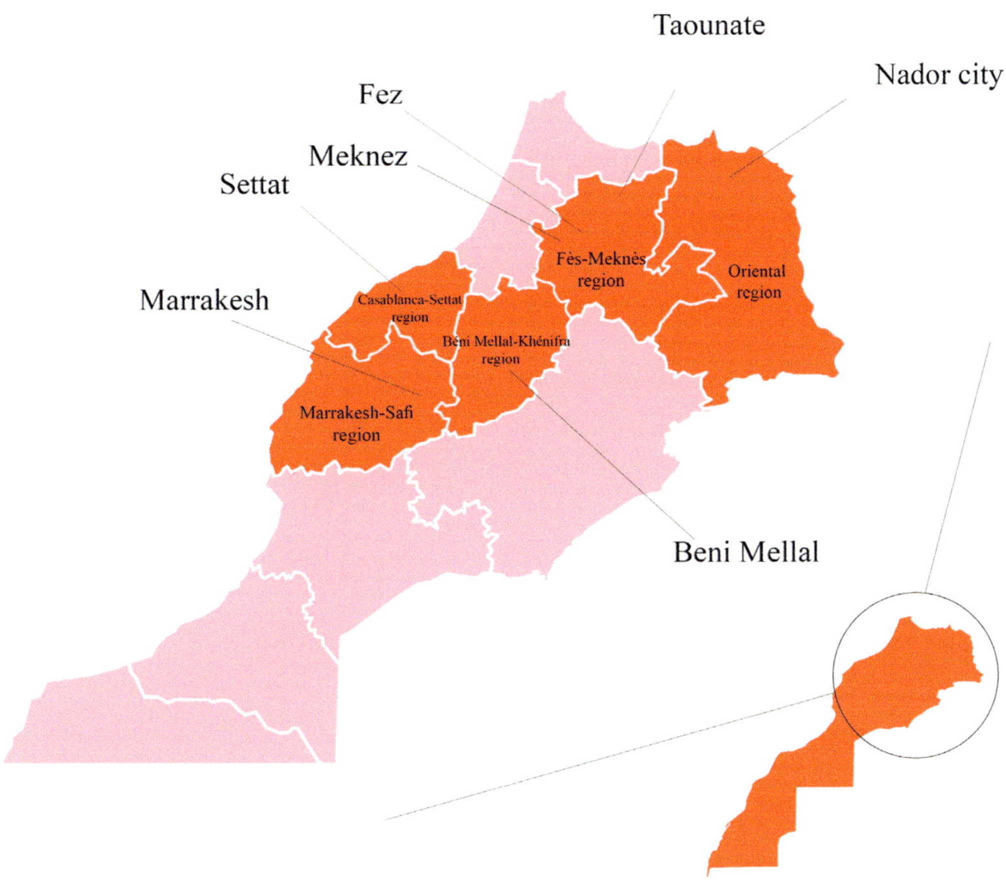

Fig. 3.21. Main pomegranate production provinces in Morocco. (Source: Walali *et al*., 2003; Oukabli, 2004.)

(Fig. 3.20). Importing countries are Libya, with 44% of the exported share; France with 20%; Italy with 8%; Kuwait with 8%; Qatar with 8%; Algeria with 5%; the UAE with 4%; Saudi Arabia with 2%; and Sweden with 1%. The estimated total area of cultivation was 5000 ha in 2012 (Mars and Marrakchi, 1999), which was more than 2.5 times the production area compared with 1980.

Tunisian pomegranates are harvested in a period of 4 months ranging from September to late December, depending on the variety; the majority is usually sold in local markets (Mars and Marrakchi, 1999). Tunisian citizens consume pomegranates as fresh fruit or process them into juice, jam and syrup. Medicinal use (Zouaoui and Zouaoui Skandrani, 1992) is another traditional way of consuming the fruit along with use for making cakes, wine, drinks and fragrances (Aviram and Dornfeld, 2001). There are about 15 different cultivars of pomegranate spread across the country. Almost all the pomegranate cultivars are sweet-tasting: 'Gabsi', 'Tounsi', 'Zehri', 'Chefli', 'Mezzi', 'Jebali', 'Garoussi', 'Kalaii', 'Zaghouani', 'Andalousi' and 'Bellahi' (Mars and Marrakchi, 1999). Other minor cultivars are 'Beyounsi', 'Florepleno', 'Panache', 'Gabsi Khadouri', 'Garoussi Sahel' and 'Nabil' (Holland and Bar-Ya'akov, 2008).

3.12 Morocco

Pomegranate cultivation has noticeably increased in the past few years in Morocco, and traditional plantings have been improved towards a more commercial cultivation (Haddioui and Valero, 2012). However, official statistics are not yet available. Use of drip irrigation systems has greatly contributed to the increase in pomegranate cultivation and improvement of fruit quality (Regional Directorate of Agriculture). At the same time, market demand has also risen following consumers' awareness of its benefits for human health (Martínez *et al.*, 2006; Çam *et al.*, 2009). However, despite the increasing demand for and production of this fruit and its important role among other crops, there is still little research and funded scientific investigation (Legua *et al.*, 2012a).

The total pomegranate planted area in Morocco accounts for 5823 ha, with a production of 64,656 t of fruits, showing a 25% increase in area and 11% increase in production amount compared with the 5 years before (4625 ha, 58,000 t) (Hmid *et al.*, 2018). Recent data for pomegranate cultivation report an area of 5000 ha and production of 58,000 t (Freshplaza, 2018) Most of the pomegranate fruits are consumed in local markets and only a minor part (0.5%) is destined for export. The fruits are either consumed fresh or processed into juice (Haddioui and Valero, 2012). Morocco, located at 31.7917° N and 7.0926° W, overlooks the Mediterranean Sea in the north and the Atlantic Ocean in the west. Important pomegranate orchards are located in the area of Beni Mellal, where the annual mean daily temperature ranges between a minimum of 10.3°C and a maximum of 26.3°C, with an average of 18.3°C; the annual total precipitation is 493 mm (https://en.climate-data.org/) with 971 chilling hours (SoDa Service, 2019). Suitable climates are the main reasons for the cultivation of this species in Morocco. The northern regions of the country have warm Mediterranean climate, while warm-desert climate lies in the southern regions. The main pomegranate production areas are concentrated in the northern regions of the country (Haddioui and Valero, 2012). Beni Mellal, Marrakesh, Settat, Nador, Fez, Taounate, Meknes and some southern oases are the most important producing provinces in which pomegranate represents a major economic product (Walali *et al.*, 2003; Oukabli, 2004; Fig. 3.21.). Tadla (a town in Beni Mellal province) has annual production of 29,094 t on a cultivated surface of 1400 ha, showing a yield average of 20 t/ha, which is almost twice the overall yield of the whole country (Maghres News, 2010). In Morocco, pomegranate harvest season falls in November until the end of February (Tridge, 2018, www.tridge.com).

Table 3.4. Pomegranate cultivars (and cultivation regions) of Morocco (Source: Haddioui and Valero, 2012).

Sweet varieties	Sour varieties
Sefri (Beni Mellal)	Wonderful
Kharaji (Bzou)	Negro
Mesri (Meknes)	Monstruoso
Laroussi (Fez)	
Zheri	

Despite the lack of proper classification, there are several cultivars, some of which share the same name (homonymies). These cultivars are named due to the shape of the fruit, colour of the skin or growing zone (Haddioui and Valero, 2012). Moroccan pomegranate cultivars are usually classified into two groups: sweet and sour pomegranates. Sweet cultivars are consumed fresh, while sour ones are usually processed for juices (Haddioui and Valero, 2012). Table 3.4 shows the two groups of cultivars along with their cultivation regions. Some of the cultivars grown in Morocco are 'Sefri', 'Ruby', 'Rouge Marrakech', 'Jaune Marrakech', 'Ounk Hman', 'Bouaâdime' (Legua *et al.*, 2012b), 'Jaibi', 'Kharazi' (Maghres News, 2010), 'Gjeigi', 'Grenade Jaune', 'Gordo de Javita', 'Djeibali', 'Dwarf Ever Green' and 'Onuk Hmam' (Holland and Bar-Ya'akov, 2008).

3.13 Italy

Pomegranate is not among the main fruit crops in Italy and in 2009 only 9 ha of orchards were under cultivation. Compared with olive, table grape, sweet cherry, peach, apple and pear, which are important crops of the country, pomegranate production is still very low. However, in the past few years, the growing interest of consumers in this healthy fruit and the opportunity to diversify the offer of fruits on the market has stimulated the establishment of pomegranate orchards. For 2019, statistical data report a productive area of 1033 ha (1234, including non-bearing areas) with a total harvested production of 13,956 t of fruits (National Institute of Statistics, 2018, http://dati.istat.it/). This product is used for the local market and because of the still low production, exported only in small amounts. Therefore, the local Italian market is supplied with pomegranates imported from countries like Iran, Turkey, Israel and Spain, and from the southern hemisphere in winter–spring. However, in recent years there has been a growing demand for domestic production and attention for this fruit, given the increased consumer awareness of its healthy properties and nutraceutical virtues (Cristofori *et al*., 2011; Ferrara *et al*., 2011, Ferrara *et al*., 2014). Apart from the statistical data reported above, the area dedicated to pomegranate orchards is increasing year after year, in particular in regions of southern Italy.

Italy is located south of the temperate belt of the boreal hemisphere and enjoys a wide variety of climates, ranging from temperate continental in the north to warm Mediterranean in the centre, and warm dry Mediterranean in the south. The southern and insular regions are the ideal areas for the growth of pomegranate trees as they enjoy long, hot and dry summers, and mild, rainy winters. Main producing regions are Sicilia, Puglia, Veneto and Lazio, with 371, 258, 210 and 61 ha, respectively (National Institute of Statistics, 2018). In Italy, an important area of pomegranate cultivation is in the Puglia region, in particular in Taranto province (Palagiano). In this area the annual mean daily temperature ranges between a minimum of 11.9°C and a maximum of 20.5°C, with an average of 16.2°C; the annual total precipitation is 445 mm (https://en.climate-data.org/) with 636 chilling hours (SoDa Service, 2019). All the producing regions are shown in Fig. 3.22, along with their production. Puglia region is becoming the leading region for pomegranate cultivation and processing both in terms of increased areas

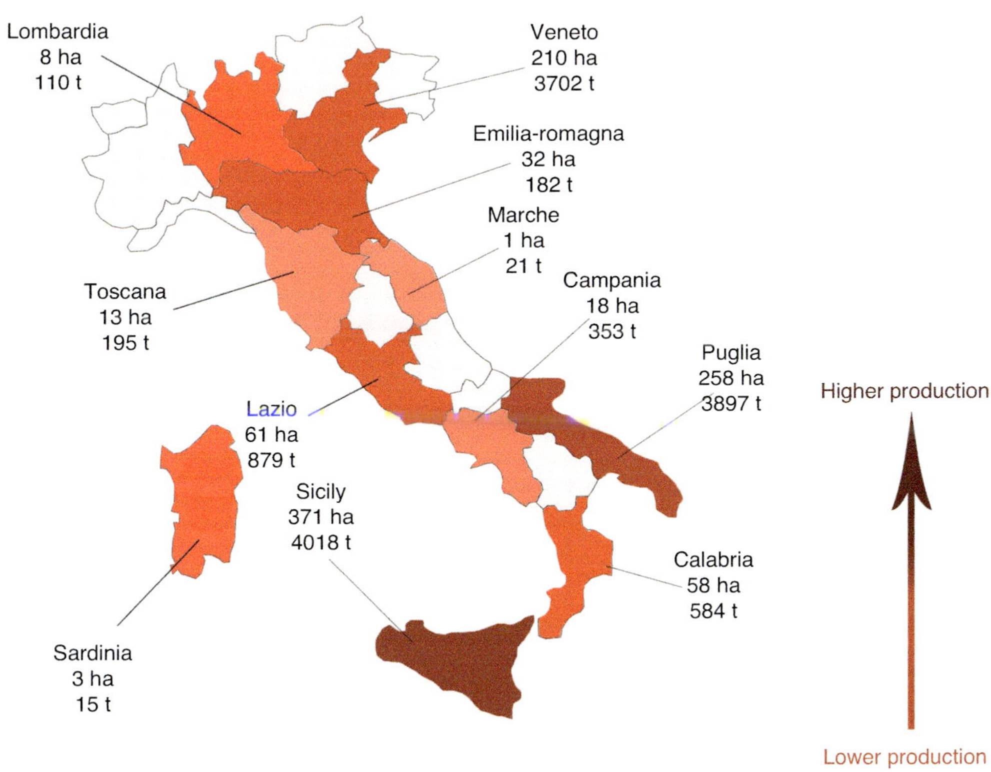

Fig. 3.22. Pomegranate area (ha) and production (t) in regions of Italy. (Source: Istat, 2020.)

of cultivation and presence of modern processing facilities for producing juices, ready-to-eat arils and other products (Giuseppe Ferrara, unpublished data).

The pomegranate harvesting period in Italy falls between September and December; however, with plantations of early ripening cultivars, along with the utilization of new cooling storage facilities and micro-perforated films, growers can supply the market at almost any time of the year, counting on the 4–5-month shelf-life period (Giuseppe Ferrara, unpublished data).

Pomegranate diffusion in Italy, and in particular in Puglia, dates back to the 3rd to 4th century BCE as indicated by the famous tomb of the pomegranate at Egnazia. Moreover, pottery representing pomegranate fruits in tombs (4th to 2nd century BCE) or floor mosaics in Roman villas testify to the diffusion and importance of this fruit tree in Puglia before and during the Roman Empire (Giancaspro *et al.*, 2017).

There are various pomegranate landraces/cultivars but many of them with a very limited presence (Ferrara *et al.*, 2014). The most important cultivar before the establishment of modern pomegranate orchards was 'Dente di Cavallo', sweet tasting with a yellowish-pink skin and pinky arils with hard seeds, which has different landraces in various regions of the country similarly to 'Wonderful' in the USA and Israel. In the new modern orchards, the most commonly grown pomegranates are 'Wonderful', 'Wonderful One', 'Akko', 'Shany', 'Emek' and 'Hicaz', whereas local cultivars such as 'Dente di Cavallo', 'Neirana', 'Acido' and many others are cultivated as only a few trees in small orchards. Pomegranate fruits are consumed fresh and as juice; recent uses include marmalade, ice creams, liquors, cakes, cookies, etc. In order to offer new cultivars, a breeding programme was started at the University of Bari 'Aldo Moro' in 2018 using either international or local cultivars.

3.14 Greece

Greece is a country in the Mediterranean basin where pomegranate is one of the oldest cultivated fruit species (Tsagkarakis, 2012). Greek culture is deeply influenced by pomegranate fruit, where it is referred to as a symbol of life and prosperity, bringing fertility, welfare and happiness (Hoza and Plisiotis, 2010). This fruit is also used in different celebrations like weddings and New Year ceremonies. Moreover, some special cultivars, bearing beautiful flowers, are used as ornamental trees. Until recently, pomegranates have been used exclusively as fresh fruits or for making home-made derived products (Hoza and Plisiotis, 2010). The pomegranate industry has been rising only since scientific reports of its beneficial health effects began to be published (Thomidis, 2014). Greece is located at 35–41.5° N and 19.5–26.4° E, facing the Mediterranean Sea in the south. There are nine different climates in the country, among which the warm Mediterranean is characteristic of most regions of the country and all the Aegean islands, providing suitable conditions for pomegranate growing. In Greece, pomegranate orchards can be found in the southern part of the country (Peloponnese, plain of Skala) where the annual mean daily temperature ranges between a minimum of 14.0°C and a maximum of 22.0°C, with an average of 17.9°C; the annual total precipitation is 608 mm (https://en.climate-data.org/) with 400 chilling hours (SoDa Service, 2019).

According to the Greek Ministry of Agricultural Development and Food (2018), total pomegranate cultivation increased from 202 ha in 1994 to 8093 ha in 2011; however, domestic production does not meet the market demand, which has to be supplied by imported pomegranates from Turkey and Spain (Hoza and Plisiotis, 2010). The Aegean islands, Crete, Peloponnese, central Greece and Macedonia regions are the major producing areas in the country (Pontikis, 1987). Since 2007, new areas have been dedicated to pomegranate cultivation, including land in Thessaloniki, Kavala, Rodopi and Serres (Tsagkarakis, 2012, Fig. 3.23). Depending on the cultivar, maturation and harvesting time might differ; however, Greek pomegranates are generally harvested between September and October.

There is a very large number of local cultivars grown across the country; however, in recent years 'Wonderful' has replaced most of them, becoming the most cultivated all over the country (Drogoudi *et al.*, 2012). Some of the other local cultivars are 'Glykies Pometi', 'Karavelos', 'Dill', 'Rhododendron', 'Chiprodija',

Fig. 3.23. Pomegranate production provinces of Greece. (Source: Tsagkarakis, 2012.)

'Coarse Grass' ('Kambadiko'), 'Vassilios', 'Fur', 'Xhorodrodia' ('Agrokabátika'), 'Ermioni', 'Politika', 'Glykia', 'Patras', 11010, 11015 and 11041 (Stamatia, 2017). The majority of these cultivars have red skin colour, sweet taste and medium–hard seeds.

3.15 Israel

Pomegranate was introduced for the first time in Israel from Iran and has been grown in this country for thousands of years. Due to the decorative value of pomegranate in Israel, the external appearance of this fruit is the main factor for selection; hence, colour and size are prioritized over the edible characteristics. Red skin, nice crown and sweet pink-red arils with a pleasant aroma and soft seeds are appreciated traits for a favourable pomegranate cultivar (Blumenfeld *et al.*, 2000). The coincidence of pomegranate ripening period and the beginning of Jewish New Year has led to a wide use of this fruit for decorative and blessings purposes in related ceremonies, along with providing edible juice and fruit (Rymon, 2015). The publication of a revolutionary paper on the positive effects of pomegranate juice on human health by Aviram *et al.* (2000) stimulated a wide expansion of pomegranate orchards in response to a considerable increase of market demand (Rymon, 2015). According to the Israeli Department of Information, 66,500 t of pomegranates were produced in Israel in 2017, representing a minor increase in production compared with 2016 (Fig. 3.24). The total cultivated area dedicated to commercial pomegranates is about 3000 ha (2500 bearing). In Israel, in the region of Negev, with semi-arid climate, the annual mean daily temperature ranges between a minimum of

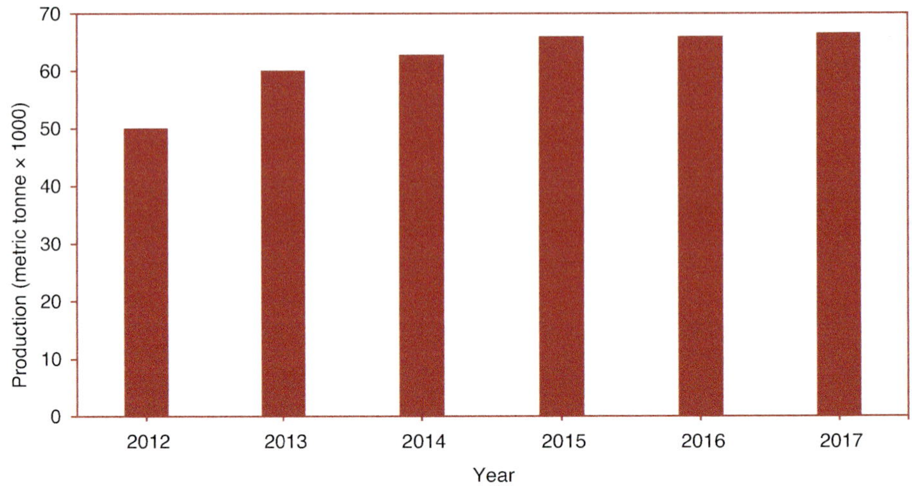

Fig. 3.24. Pomegranate production in Israel between 2012 and 2017. (Source: Central Bureau of Statistics of Israel, 2018.)

13.1°C and a maximum of 25.5°C, with an average of 19.3°C; the annual total precipitation is 229 mm (https://en.climate-data.org/) with 386 chilling hours (SoDa Service, 2019).

Export of pomegranates from Israel has significantly increased in the past few years, tripling between 2010 and 2015. In 2013, half of total production was exported, while 20% was used for industrial processing, and the remaining part was sold in the local market for fresh fruit consumption and ready-to-drink juices (Rymon, 2015). Export destination countries include Western and Central Europe (mainly the European Union and Switzerland, 56% of the total exports), Eastern European countries (Russia and Ukraine, 40%) and a small percentage to Canada, Jordan, Hong Kong and Sri Lanka (www.israelagri.com). More than half of the production areas are located in southern Israel and the Negev region, both located in the warm-desert climatic belt. Another 25% of the production comes from Moshavot (cooperative villages) in the Shomron region (the area around Zichron Yaakov and the Sharon district) in the northern West Bank. The remaining 19% of orchards are located in the north of the country (Galilee, Golan Heights and Hula Valley) with warm Mediterranean climate (www.israelagri.com). Pomegranate-ripening season in Israel starts from August for early-ripening cultivars such as 'Shany', and from October for other cultivars such as 'Wonderful'.

'Wonderful' is the most popular and commonly grown cultivar in Israel, which is said to have been imported from the USA (Atsmon, 1956). This cultivar, characterized by the hardness of its seeds (Rymon, 2015) represents two-thirds of the total production. Most Israeli citizens originating from Western European countries prefer sweet-sour cultivars, among which 'Wonderful' is the most popular, whereas citizens originating from Middle East countries usually prefer non-acidic soft-seeded cultivars such as 'Malisi' (Holland *et al.*, 2009). Among early-ripening cultivars are 'Mule Head', 'Shany' and 'Akko' (Rymon, 2015), which are sweet-sour and soft seeded. Other common cultivars grown in Israel, all with a skin colour from light red to deep red, are the following: 'Asmar', 'Ras-el-Bghil' (medium hardness), 'Red Lufani' (nearly seedless), 'Emek' (soft-seeded), 'Bent Elbash' and 'Herskovitz' (hard seeds), which has a sour taste and red skin.

3.16 Portugal

In Portugal, pomegranate is cultivated in the mountain areas of the southern region; for example, in Algarve it grows as a wild shrub at the border of orchards. The climate is cool and rainy in the north, and gradually becomes warmer and sunnier in the south of the country; in the far south, the Algarve region has a dry and sunny microclimate. In the region of Algarve, the annual mean daily temperature ranges between a minimum of 12.7°C and a maximum of 21.8°C, with an average of 17.2°C; the annual total precipitation is 501 mm (https://en.climate-data.org/) with 241 chilling hours (SoDa Service, 2019). The most popular cultivar in Portugal is 'Assaria', and it is used to bud some wild accessions (Miguel *et al.*, 2004). In addition, other local cultivars are diffused, but they are less appreciated for fresh consumption because of acidity, small arils and high skin/arils ratio, and are used for liquor preparation sometimes mixed with arils of 'Assaria'. In recent years there has been a growing interest in Portugal in pomegranate cultivation and new orchards have been established with international cultivars such as 'Wonderful', 'Akko' and 'Shany'.

3.17 Albania

In Albania, wild pomegranate was domesticated during the Bronze Age, from the 3rd to the 2nd millennium BCE. Albania is one of the Mediterranean countries where the wild ancestors of cultivated pomegranate can be found, due to its favourable climatic and soil characteristics. The climate is Mediterranean on the coast with mild, rainy winters and hot, sunny summers, while it becomes slightly more continental inland, but it is cold only in mountain areas. Due to the very favourable Mediterranean and continental climate, pomegranate grows easily in this country, showing a substantial genetic variability, in every area of the coast and even in colder areas, up to altitudes of 600 m. In the local language, the fruit is called 'shege'; in fact, being a symbol of beauty and prosperity, many women in Albania are named Shege.

Pomegranates in Albania are mainly diffused in the regions of Shkoder, Lezha, Durres and Tirana down to the south of the country in Tepelen and Gjirokastra. Wild pomegranates grow mainly on the slopes around Shkoder and Lezha in the north of the country, whereas several examples of organic cultivation can be found in many parts of the coastal area of Saranda. In the area of Shkoder, annual mean daily temperature ranges between a minimum of 10.7°C and a maximum of 19.7°C, with an average of 15.1°C; the annual total precipitation is 1783.0 mm (https://en.climate-data.org/) with 1640 chilling hours (SoDa Service, 2019). In Albania, pomegranate is grown either for fresh fruits or for processing (juices); pomegranates are also used for the pharmaceutical industry and as an ornamental tree especially in dry areas, steep lands and stony soils without irrigation.

In the past few years, many attempts have been made to valorize the economic benefits of this fruit, especially for export, in order to contribute to the economic development of remote areas in Albania. However, at the moment, Albania is not competitive for pomegranate export; a lot of work still needs to be done by local producers to ensure the necessary quality of fruits, in order to conform to the current consumer preferences and meet the European Union import standards (Christie, 2007). There are various local cultivars grown in the different areas of the country such 'Devedishe', 'Tivarash' and 'Majhoshe' grown in Shkoder, Tirana and Gjirokastra districts. The colour of the fruit ranges from green-yellow and sweet taste ('Tivarash') to red and sweet-sour taste ('Majhoshe') (Koka, 2015).

3.18 Southern Hemisphere

In the southern hemisphere, pomegranate season starts from late February and continues until May. Considering the world production of pomegranate, <50,000 t comes from the southern hemisphere, where the crop is relatively new to cultivation. The countries of the southern hemisphere supply the northern hemisphere market from late February to early May. Less than 5% of total pomegranate production is grown in the southern hemisphere, and it is concentrated mainly in Peru, Chile and South Africa (Fig. 3.25). Although volumes are growing rapidly from 2012 to 2018 (PPECB, 2020),

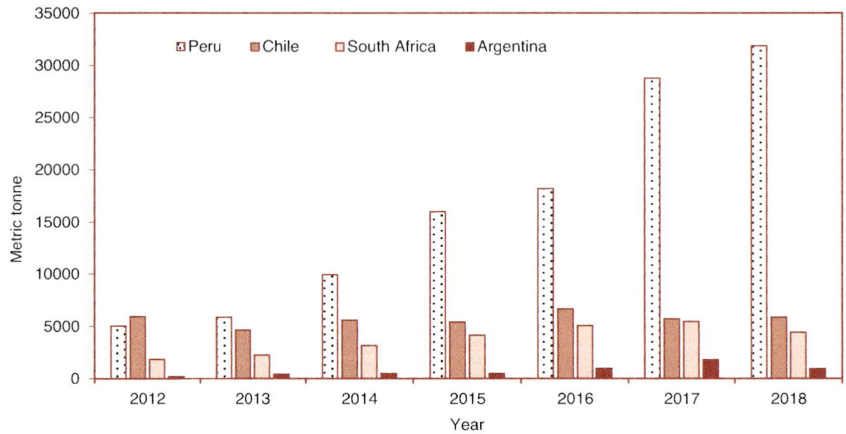

Fig. 3.25. Growth in exports per southern hemisphere country (total tonnes) from 2012 to 2018. (Source: PPECB, https://ppecb.com/)

there is still a long way to go before the southern hemisphere can satisfy the demand levels coming from the northern hemisphere during the winter time.

Most of the imported pomegranates from countries such as Peru and South Africa go through the Netherlands for successive packaging. From here, pomegranates are distributed all over Europe, with Germany as the most important importer together with the UK. The export of the southern hemisphere is mainly directed to Europe and Russia (60%), the USA and Canada (16%), and the UK (10%), followed in small amounts by the rest of the world (Middle East, 6%; Far East (FE) and Asia, 5%; Africa, 2%; and Latin America, 1%) (PPECB, 2020, https://ppecb.com/) (Fig. 3.26).

3.19 Peru

Peru is the main pomegranate producer in the southern hemisphere. Production for national consumption or export is concentrated in the regions of Arequipa, Ica, Lima and Ancash. For example, Santa Cruz, a district located in the province of Palpa in Ica, has a tropical climate that is optimal for the cultivation of pomegranate, as it accelerates the fruit-ripening process allowing an earlier harvest (up to 1 month) (www.cbi.eu). In the region of Arequipa, the annual mean daily temperature ranges between a minimum of 6.5°C and a maximum of 22.5°C, with an average of 14.5°C; the annual total precipitation is 75 mm (https://en.climate-data.org/) with 853 chilling hours (SoDa Service, 2019).

According to ProGranada's producer association, there are around 2000 ha of orchards dedicated to the export of fresh pomegranates and arils, producing an annual crop of >16,000 t. Most of the Peruvian pomegranates are exported to the European market; in particular, the Netherlands, which is the leading importer with a 43% share, followed by Russia and the UK, with 15 and 10%, respectively. The remaining 32% of the market is divided between the USA (7%), Canada (4%), Belgium (4%), the UAE (3%), Saudi Arabia (3%), Hong Kong (3%) and other countries (8%). Importing countries from Peru are depicted in Fig. 3.27.

In the past few years, there has been a significant increase in the production and export of pomegranates in Peru together with other fruits such as table grape. Exported amounts have risen from 340 t in 2006 to 9842 t in 2014, a very impressive increase in a few years. Pomegranate harvesting in Peru is from February till May. Different cultivars of pomegranate are grown in Peru; both early ripening 'Emek', 'Akko' and 'Shany', and late ripening

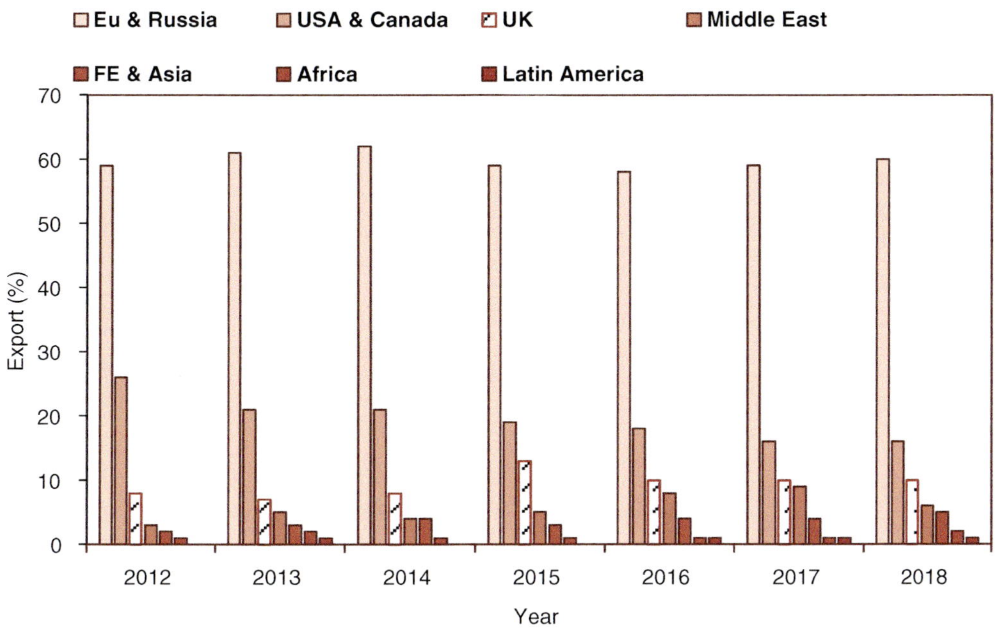

Fig. 3.26. Destination markets (%) for southern hemisphere exports (from 2012 to 2018). (Source: PPECB, https://ppecb.com/)

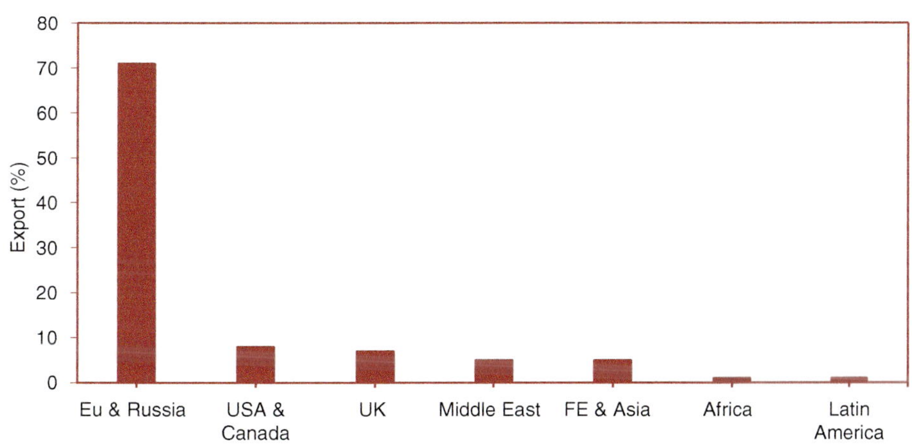

Fig. 3.27. Peru pomegranate exports per market (2018). (Source: Simfruit, https://www.simfruit.cl/.)

such as 'Wonderful'. 'Wonderful' is the most widely planted cultivar, mainly because it has an excellent shelf-life, lasting 60–80 days after harvesting in controlled atmosphere packaging. The early-ripening cultivars ('Akko', 'Emek' and 'Shany') are not grown on such a large scale due to their shorter shelf-life (around 25–30 days).

3.20 South Africa

South Africa, with its Mediterranean climate, is optimal to allow pomegranate production (Levin, 1995; Shwartz *et al.*, 2009). Temperatures can be around 30°C in summer (December–February) and 18°C

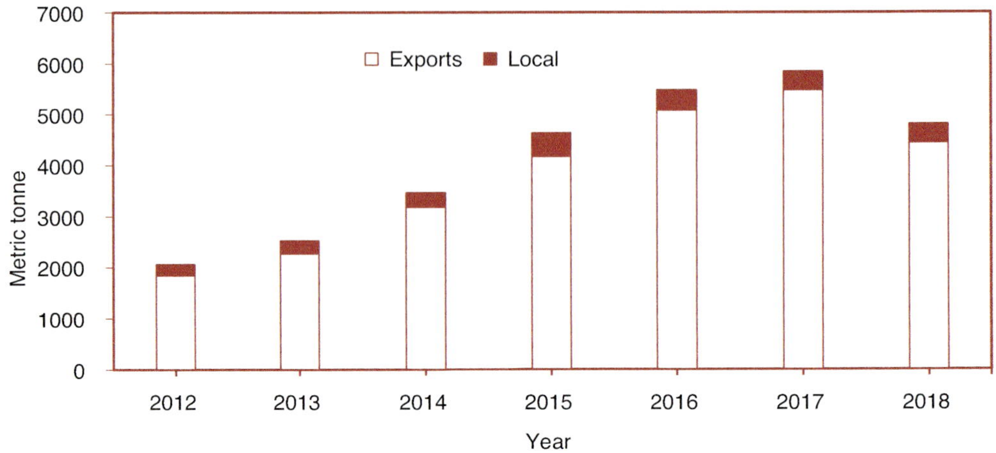

Fig. 3.28. Export and local market trends of South Africa's pomegranates. (Source: POMASA, https://www.sapomegranate.co.za/)

in winter (July–August), and the rains fall mainly between June and August. Winters are prevalently cold and dry. In the South African regions that account for the highest pomegranate production (Western Cape) such as the districts of Bergrivier, Wellington, Malmesbury and Bonnievale, the annual mean daily temperature ranges between a minimum of 10.6°C and a maximum of 22.3°C, with an average of 16.4°C; the annual total precipitation is 802 mm (https://en.climate-data.org/) with 641 chilling hours (SoDa Service, 2019). Under these conditions, the fruit can develop to its best size and optimal colour and sugar accumulation, without splitting. Almost 80% of pomegranate produced locally is exported to international markets. South African pomegranate exports increased by 193.8% from 2012 to 2017 with a decrease in 2018 (Fig. 3.28). Most of South Africa's exports are directed to the European Union and the Middle East. In particular, the European Union accounts for 56% of all exports, while the Middle East accounts for 16%. South Africa usually provides the European market at the end of March, when it is the only supplier of pomegranates to the northern hemisphere (Farmer's Weekly, 2019).

On a total area of approximately 1000 ha, the most popular cultivars in South Africa are 'Wonderful' (68%), 'Hershkowitz' (14%) and 'Acco' (11%) with only a few hectares for the other cultivars. 'Herschkovitz' and 'Akko' are harvested from the end of February to the end of March. 'Wonderful' is harvested from early April to early May. Some Indian cultivars such as 'Bagwa' and 'Ganesh' are also grown in South Africa, but on a much smaller scale. Almost 50% of South African pomegranate orchards are 7–10 years old (Freshplaza, 2018).

The main area for growing pomegranates in South Africa is in the Western Cape (80%) and Northern Cape, with 932 ha of commercially grown pomegranates in the country in 2018 (Fig. 3.29). The Pomegranate Growers' Association of South Africa was established in 2009, and it is now one of the largest pomegranate projects in the southern hemisphere, which supply pomegranates across the globe including to Canada, the Middle East, the Far East and the rest of Africa (Phaleng and Lubinga, 2018).

3.21 Chile

Pomegranate was first introduced to America by Spanish settlers in 1769, and is now cultivated in many parts of Chile. Chile lies between 17° and 56° latitudes south of the Equator, at the western side of the continent as a long and narrow coastal country. The central regions of the country are characterized by a Mediterranean

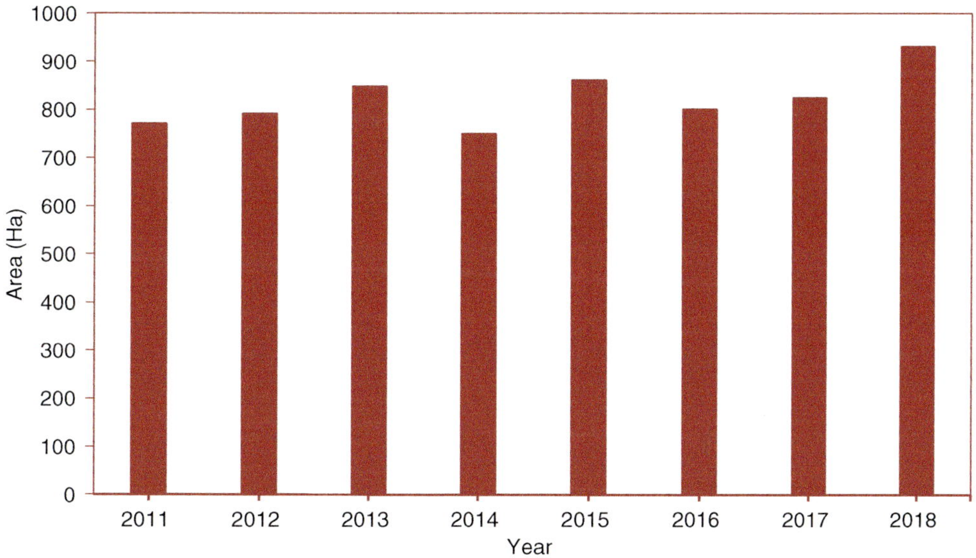

Fig. 3.29. Total pomegranate planted area (ha) in South Africa from 2012 to 2018. (Source: POMASA, https://www.sapomegranate.co.za/)

climate with hot and dry summers and cold, wet winters (www.britannica.com/science/Mediterranean-climate). Pomegranate is grown in the valleys towards the Andes chain, from 300 to 700 m above sea level, and is traditionally harvested from mid-March until early May. In Chile, pomegranate is cultivated in the region of Coquimbo where the annual mean daily temperature ranges between a minimum of 10.9°C and a maximum of 20.6°C, and an average of 15.7°C; the annual total precipitation is 127 mm (https://en.climate-data.org/) with 375 chilling hours (SoDa Service, 2019).

In the past decade, the surface area dedicated to pomegranate cultivation in Chile has increased significantly from 250 ha in 2007 to 1150 ha in 2015 (ODEPA, 2015), mainly due to the rise of its popularity owing to the nutraceutical properties of its phenolic compounds, including phenolic acids, flavonoids (anthocyanins), tannins (proanthocyanidins) and hydrolyzable tannins (ellagitannins and gallotannins) (Li *et al.*, 2015; Mphahlele *et al.*, 2016). The best conditions for growing pomegranates in Chile are in the interior of the transversal valleys (Andes foothills) of the northern cultivation zones: especially the Atacama and Coquimbo Regions, which therefore constitute 83% of the country's pomegranate production area. These zones present long, dry and hot summers, high solar irradiance, slightly colder winters and nights than coastal regions, and dry spring and autumn, which all favour the production of high-quality fresh pomegranates (Franck, 2012).

The principal importers of pomegranates from Chile are the countries of the northern hemisphere, prevalently the USA and Russia (Fig. 3.30). According to the Association of Fruit Exporters of Chile (ASOEX), the pomegranate export season in 2018 ended with 5580 t exported, 2% less than in the previous season. The main export market was represented by North America, with 54.7% of total exports, followed by Europe with 42.7% (97.4% of total exports). The Far East ranked third with 1.3% of total exports (i.e. 73 t), Latin America ranked fourth with only 0.8% and the Middle East got 0.6% of total exports (SimfruitSimfruit, 2020, www.simfruit.cl).

Among the most popular pomegranate cultivars grown in Chile are: 'Ambrosia', 'Foothill', 'Smith', 'Angel Red', 'Early Wonderful', 'Wonderful', 'Sripanya', 'Ruby', 'Rosy', 'Shir', 'Kessari', 'Akko', 'Hershkovitz', 'Manfalouti' and 'Gabsi Grenades' (Tridge, 2018).

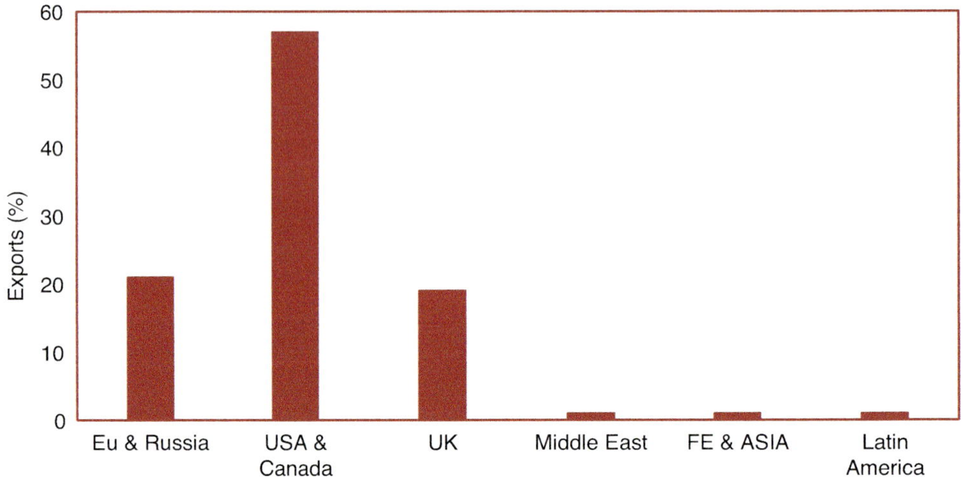

Fig. 3.30. Chile pomegranate exports per market in 2018. (Source: Simfruit, https://www.simfruit.cl/)

3.22 Australia

Pomegranate is relatively new to Australia; in fact, according to an unofficial report there are only <600 ha of orchards dedicated to pomegranate cultivation in this country, which can hardly meet the local market demand. In Australia, pomegranate is cultivated in the territory of Victoria, where the annual mean daily temperature ranges between a minimum of 9.8°C and a maximum of 19.8°C, with an average of 14.8°C; the annual total precipitation is 666 mm (https://en.climate-data.org/) with 814 chilling hours (SoDa Service, 2019). Pomegranate trees have long been grown in Australia, but they were not much destined for commerce; rather, fruits were consumed by local people, sold in local markets or used as ornamental trees. Australia has been commercially growing pomegranates for approximately 15 years, and this commerce is operated by only two big companies: 'PomLife' in Victoria and 'Pomegranates Australia' in South Australia. The reason for the rarity of pomegranate production in Australia lies in some challenges such as: the long distance between markets and orchards, lack of cool storage and packaging facilities, and the 'dieback' disease causing yellowing and falling of the leaves and decline of the tree (Davidson, 2014). Eccles (2009) also mentions lack of good cultivars and low market demand as other reasons. Despite all the cultivation challenges and the low production level, pomegranate should be considered a big potential crop for the country, taking into consideration the increasing awareness of its health benefits and consequently consumers' demand (Eccles, 2009). Lately, in Australia, there is a higher demand for pomegranate-derived concentrated juice rather than for fresh fruits. Juice extracted from the arils is traditionally used to produce wine and grenadine, whereas the remaining waste is employed in the cosmetic and pharmaceutical industries (Davidson, 2014).

Pomegranate trees find the optimal climate in the Mediterranean and tropical regions. Subtropical regions are not generally suitable for pomegranate growing, since the harvest time coincides with rainy season, leading to fruit splitting and fungal diseases; however, there are some Indian cultivars, such as 'Ganesh' and 'Bhagwa', that are partially tolerant to such conditions and can be cultivated as evergreen (Eccles, 2009). Pomegranate orchards in Australia are located in south-east (South Australia, Victoria and New South Wales) and south-west (Perth region) of the country with subtropical and Mediterranean climates. The best Australian production was around 4000 t in 2013; since then it has dropped as low as half, but is now increasing again. The pomegranate market is small, but has been growing considerably over the past 2–3 years.

Pomegranate orchards in the south-east of the country are placed near the Murray-Darling basin, which well provides the moisture for those groves; on the other side, commercial production in Western Australia demands a more consistent irrigation. Regions with regular rain during autumn and summer causing fruit cracking are not suitable for pomegranate production. Geraldton and Albany are other western provinces tending to commercially produce pomegranates. Condobolin and Dareton, located in new south Wales, along with Sunraysia, Shepparton and Goulburn valley regions in Victoria are also considered to be the host of the first large-scale pomegranate plantings between 2007 and 2010 (Davidson, 2014).

Australian pomegranates are commonly harvested between March and April (Eccles, 2009). Local markets are usually satisfied by importing from the USA (mostly 'Wonderful' fruits) (Eccles, 2009). Australia encounters a wide range of different pomegranate cultivars differing in terms of the acid level of the arils and hardness of the seeds, which have been lately tested in some research centres in order to find the most suitable for Australian conditions.

References

Akcaoz, H., Ozcatalbas, O. and Kizilay, H. (2009) Analysis of energy use for pomegranate production in Turkey. *Journal of Food, Agriculture & Environment* 7, 475–480.

Akdag, E. (2009) Türkiye meyve suyu v.b. ürünler sanayi raporu (report on Turkish fruit juice industry). *Meyve Suyu Endüstrisi Derneği (Turkish Fruit Juice Industry Association – MEYED).*

Arena, E., Fallico, B. and Maccarone, E. (2000) Influence of carotenoids and pulps on the color modification of blood orange juice. *Journal of Food Science* 65(3), 458–460.

Atsmon, J. (1956) The pomegranate. State of Israel, Ministry of Agriculture, Agricultural Publications Section, Tel Aviv.

Aulakh, P.S. (2004) Evaluation of pomegranate cultivars grown in the lower Shivalik's of Punjab. *Haryana Journal of Horticultural Sciences,* 33, 81–82.

Aviram, M. and Dornfeld, L. (2001) Pomegranate juice consumption inhibits serum angiotensin converting enzyme activity and reduces systolic blood pressure. *Atherosclerosis* 158(1), 195–198. DOI: 10.1016/S0021-9150(01)00412-9.

Aviram, M., Dornfeld, L., Rosenblat, M., Volkova, N., Kaplan, M. *et al.* (2000) Pomegranate juice consumption reduces oxidative stress, atherogenic modifications to LDL, and platelet aggregation: studies in humans and in atherosclerotic apolipoprotein E-deficient mice. *The American Journal of Clinical Nutrition* 71(5), 1062–1076. DOI: 10.1093/ajcn/71.5.1062.

Bartual, J., Fernandez-Zamudio, M.A. and De-Miguel, M.D. (2015) Situation of the production, research and economics of the pomegranate industry in Spain. *Acta Horticulturae* 1089, 345–349.

Blumenfeld, A., Shaya, F. and Hillel, R. (2000) Cultivation of pomegranate. *Options Méditerranéennes Ser. A* 42pp., 143–147.

California Department of Food and Agriculture (2018) California agricultural production statistics. Available at: www.cdfa.ca.gov/Statistics/ (accessed 14 June 2018).

California Rare Fruit Growers (1997) Pomegranate fruit facts. Available at: www.crfg.org/pubs/ff/ pomegranates.html (accessed 5 July 2018).

Calin-Sanchez, A., Martinez, J.J., Vazquez-Araujo, L., Burlo, F., Melgarejo, P. *et al.* (2010) Volatile composition and sensory quality of Spanish pomegranates (*Punica granatum* L.). *Journal of the Science of Food and Agriculture* 91, 586–592.

Caliskan, O. and Bayazit, S. (2012) Phytochemical and antioxidant attributes of autochthonous Turkish pomegranates. *Scientia Horticulturae* 147, 81–88.

Çam, M., Hışıl, Y. and Durmaz, G. (2009) Classification of eight pomegranate juices based on antioxidant capacity measured by four methods. *Food Chemistry* 112(3), 721–726. DOI: 10.1016/j.foodchem.2008.06.009.

Cao, S.Y., Haoxian, L., Juan, N., Pinli, Y., Hong-Hua, T. *et al.* (2015) An overview of cultivation, scientific research, and industrialization of Chinese pomegranate. *Acta Horticulturae* 1089, 369–373.

Central Bureau of Statistics of Israel (2018) Homepage. Available at: www.cbs.gov.il (accessed 4 August 2018).
Chandra, R., Jadhav, V.T. and Sharma, J. (2010) Global scenario of pomegranate (*Punica granatum* L.) culture with special reference to India. *Fruit, Vegetable and Cereal Science and Biotechnology* 4, 7–18.
Christie, S. (2007) Pomegranates. In: *Albanian Export Opportunities to Europe and the Region*. OTF Group, Belmont, Massachusetts, p. 21.
Cristofori, V., Caruso, D., Latini, G., Dell'Agli, M., Cammilli, C. *et al.* (2011) Fruit quality of Italian pomegranate (*Punica granatum* L.) autochthonous varieties. *European Food Research and Technology* 232, 397–403.
Davidson, D. (2014) The current pomegranate situation in Australia. *RIRDC Publication No 14/* 087, 1–19.
Drogoudi, P., Vassilakakis, M., Thomidis, T.H., Navrozidis, E. and Pantelidis, G. (2012) Handbook on cultivation of pomegranate.. NAGREF, Naoussa, Greece.
Ebrahimi, M.S. (2015) Production and supply of pomegranate in Iran. *Ekonomika APK* 7, 121–125.
Eccles, J. (2009) An R&D strategy for the Australian pomegranate industry. *Rural Research and Development Corporation Publication* 9, 1–23.
Farmer's Weekly (2019) A look at South Africa's pomegranate production. Available at: https://www.farmersweekly.co.za/crops/field-crops/project-pomegranate-production/ (accessed 29 July 2019).
Feng, Y.Z., Song, M.T. and Han, D.B. (2006) The general status of pomegranate germplasm resources in China. *China Fruits* 4, 57–58.
Ferrara, G., Cavoski, I., Pacifico, A., Tedone, L. and Mondelli, D. (2011) Morpho-pomological and chemical characterization of pomegranate (*Punica granatum* L.) genotypes in Apulia region, southeastern Italy. *Scientia Horticulturae* 130, 599–606.
Ferrara, G., Giancaspro, A., Mazzeo, A., Giove, S.L., Matarrese, A.M.S. *et al.* (2014) Characterization of pomegranate (*Punica granatum* L.) genotypes collected in Puglia region, southeastern Italy. *Scientia Horticulturae* 178, 70–78.
Finetto, G.A. (2011) Pomegranate industry in Afghanistan: opportunities and constraints. *Acta Horticulturae* 890, 45.
Fitrat, K. and Verma, M.K. (2014) Potential and challenges of fruit production in Afghanistan. *ICAR – Indian Agricultural Research Institute*, 1–27.
Franck, N. (2012) The cultivation of pomegramate cv 'Wonderful' in Chile. *Options Méditerranéennes. Série A, Séminaires Méditerranéens* 103, 97–99.
Freshplaza (2013) Tunisia: pomegranates – an exotic fruit enjoyed by both Europe and Arab countries. Available at: https://www.freshplaza.com/article/113581/Tunisia-Pomegranates-an-exotic-fruit-enjoyed-by-both-Europe-and-Arab-countries%20%20%20%202013/ (accessed 12 June 2020).
Freshplaza (2018) California pomegranates available soon. Available at: https://www.freshplaza.com/article/2199073/california-pomegranates-available-soon/ (accessed 28 October 2018).
Giancaspro, A., Mazzeo, A., Giove, L.S., Zito, D., Marcotuli, I. *et al.* (2017) Exploiting DNA-based molecular tools to assess genetic diversity in pomegranate (*Punica granatum* L.) selections and cultivars. *Fruits* 72(5), 292–305. DOI: 10.17660/th2017/72.5.5.
Haddioui, A. (2012) La culture Du grenadier (Punica granatum L.) Au Maroc. In: *II International Symposium on the Pomegranate*. 103. CIHEAM/Universidad Miguel Hernández, Zaragoza, pp. 79–81.
Hajiyeva, S.V., Akparov, Z.I., Hasanov, N.A., Mustafayeva, Z.P., Hajiyev, E.S. *et al.* (2018) Issr analysis of variability of cultivated form and varieties of pomegranate (*Punica granatum* L.) from Azerbaijan. *Russian Journal of Genetics* 54(2), 188–197. DOI: 10.1134/S1022795418020072.
Harlan, J.R. (1992) *Crops and Man*, 2nd edn. American Society of Agronomy, Madison, Wisconsin, p. 284.
Hmid, I., Hanine, H., Elothmani, D. and Oukabli, A. (2018) The physico-chemical characteristics of Morrocan pomegranate and evaluation of the antioxidant activity for their juices. *Journal of the Saudi Society of Agricultural Sciences* 17(3), 302–309. DOI: 10.1016/j.jssas.2016.06.002.
Holland, D. and Bar-Ya'akov, I. (2008) The pomegranate: new interest in an ancient fruit. *Chronica Horticulturae* 48, 11–15.
Holland, D., Hatib, K. and Bar-Ya'akov, I. (2009) Pomegranate: botany, horticulture, breeding. *Horticultural Reviews* 35, 127–191.
Hoza, D. and Plisiotis, N. (2010) Pomegranate, a fruit growing species of major interest in Greece. *Journal of Horticulture, Forestry and Biotechnology* 14, 260–261.
Janic, J. (2009) *Horticultural Reviews*. 35. Purdue University, West Lafayette, Indiana, pp. 1–552.

Jing, P., Ye, T., Shi, H., Sheng, Y., Slavin, M. *et al.* (2012) Antioxidant properties and phytochemical composition of China-grown pomegranate seeds. *Food Chemistry* 132(3), 1457–1464. DOI: 10.1016/j.foodchem.2011.12.002.

Khalilov, H., Shalbuzov, N. and Huseyn, R. (2015) *Country report: Azerbaijan*. Research Institute of Agricultural Economics, Azerbaijan.

Kohansal, M.R. and Rahimi, M. (2013) Investigating factors marketing pomegranate in Fars. *International Journal of Agronomy and Plant Production* 4, 2759–2763.

Koka, T. (2015) Collection and documentation of pomegranate germplasm in Albania. *Acta Horticulturae* 1089, 375–378. DOI: 10.17660/ActaHortic.2015.1089.50.

Kurt, H. and Sahin, G. (2013) A study of agricultural geography: pomegranate (*Punica granatum* L.) cultivation in Turkey. *Marmara Geographical Review* 27, 551–574.

Legua, P., Melgarejo, P., Abdelmajid, H., Martínez, J.J., Martínez, R., Ilham, H. *et al.* (2012a) Total phenols and antioxidant capacity in 10 Moroccan pomegranate varieties. *Journal of Food Science* 77(1), C115–C120. DOI: 10.1111/j.1750-3841.2011.02516.x.

Legua, P., Melgarejo, P., Martínez, J.J., Martínez, R. and Hernández, F. (2012b) Evaluation of Spanish pomegranate juices: organic acids, sugars, and anthocyanins. *International Journal of Food Properties* 15(3), 481–494. DOI: 10.1080/10942912.2010.491931.

Levin, G.M. (1995) Genofund of pomegranate in Turkmenistan (to the 60th anniversary of its creation). *Problems of Desert Development C/C of Problemy Osvoeniia Pustyn*, 84–89.

Levin, G.M. (2006) *Pomegranate Roads: a Soviet Botanist's Exile from Eden*, 1st. Floreant Press, Forestville, California, pp. 15–183.

Li, X., Wasila, H., Liu, L., Yuan, T., Gao, Z. *et al.* (2015) Physicochemical characteristics, polyphenol compositions and antioxidant potential of pomegranate juices from 10 Chinese cultivars and the environmental factors analysis. *Food Chemistry* 175, 575–584. DOI: 10.1016/j.foodchem.2014.12.003.

Lihua, Z., Mingyabg, L., Guangze, C., Tiianchun, P. and Chenghai, S. (2013) Assessment of the genetic diversity and genetic relationships of pomegranate (*Punica granatum* L.) in China using RAMP markers. *Scientia Horticulturae* 151, 63–67.

Maghres News (2010) The children of Abdullah leadership of the Bani Amir East Province of Faqih bin Saleh. Available at: https://www.maghress.com/alittihad/118932 (accessed 28 July 2018).

Mansour, E., Haddad, M., Abid, M., Bachar, K. and Ferchichi, A. (2011) Selection of pomegranate (*Punica granatum* L.) in south-eastern Tunisia. *African Journal of Biotechnology* 10, 9352–9361.

MAPA (2017) Statistical yearbook 2016. Available at: https://www.mapa.gob.es/es/estadistica/temas/publicaciones/anuario-de-estadistica/2017/default.aspx?parte=3&capitulo=13&grupo=9&seccion=14 (accessed 15 May 2020).

Mars, M. (2000) Pomegranate plant material: genetic resources and breeding, a review. *Options Mediterraneennes Serie A, Seminaires Mediterraneens* 42, 55–62.

Mars, M. and Marrakchi, M. (1999) Diversity of pomegranate (*Punica granatum* L.) germplasm in Tunisia. *Genetic Resources and Crop Evolution* 46(5), 461–467. DOI: 10.1023/A:1008774221687.

Martínez, J.J., Melgarejo, P., Hernández, F., Salazar, D.M. and Martínez, R. (2006) Seed characterisation of five new pomegranate (*Punica granatum* L.) varieties. *Scientia Horticulturae* 110(3), 241–246. DOI: 10.1016/j.scienta.2006.07.018.

Melgarejo, P., Martínez, J.J., Hernandez, F., Legua, P., Melgarejo-Sanchez, P. *et al.* (2012) The pomegranate tree in the world: its problems and uses. II International Symposium on the pomegranate. *Options Méditerranéennes. Séries A: Mediterranean Seminars*, 11–26.

Melgarejo, P., Martinez, J.J., Hernandez, F., Legua, P., Martinez, R. *et al.* (2015) The pomegranate tree in the world: new cultivars and uses. *Acta Horticulturae* 1089, 327–332.

Miguel, G., Fontes, C., Antunes, D., Neves, A. and Martins, D. (2004) Anthocyanin concentration of "Assaria" pomegranate fruits during different cold storage conditions. *Journal of Biomedicine and Biotechnology* 2004(5), 338–342. DOI: 10.1155/S1110724304403076.

Ministry of Agriculture Jihad (2016) Statistics of agricultural productions. Available at: https://www.maj.ir/Index.aspx?page_=form&lang=1&PageID=11583&tempname=amar&sub=65&methodName=ShowModuleContent (accessed 10 August 2019).

Ministry of Economy (2013) Fruit Juice and Concentrates. Republic of Turkey - Ministry of Economy. Available at: www.economy.gov.tr (accessed 23 October 2020).

Mohseni, A. (2009) The situation of pomegranate orchards in Iran. *Acta Horticulturae* 818, 35–42. DOI: 10.17660/ActaHortic.2009.818.3.

Mphahlele, R.R., Fawole, O.A., Makunga, N.P. and Opara, U.L. (2016) Effect of drying on the bioactive compounds, antioxidant, antibacterial and antityrosinase activities of pomegranate peel. *BMC Complementary and Alternative Medicine* 16, 143. DOI: 10.1186/s12906-016-1132-y.

NASS (2014) Agricultural statistics 2014. *U.S. Government Printing Office*.

National Institute of Statistics (2018) Agriculture. AKA, ISTAT. Available at: https://www.istat.it/en/agriculture?data-and-indicators (accessed 8 July 2018).

Nazarov, E. (2011) Azerbaijan country report. Available at: www.adrc.asia/countryreport/AZE/2011/FY2011A_AZE_CR.pdf (accessed 25 September 2018).

ODEPA (2015) Panorama de la AGRICULTURA, Chilean agriculture overview. Available at: https://www.odepa.gob.cl (accessed 5 October 2020).

Oukabli, A. (2004) Transfert de technologie en agriculture: Le Grenadier, des variétés performantes pour La culture. *Bulletin mensuel d'information et de liaison du PNTTA* 123.

Ozcani, E. and Unaldi, U.E. (2007) Ecology of pomegranate and its economics in Turkey. *International Symposium on Geography, Environment and Culture in the Mediterranean Region*, 1–11.

Ozguven, A.I. and Yilmaz, C. (2000) Pomegranate growing in Turkey. *Options Mediterraneennes Serie A, Seminaires Mediterraneens* 42, 41–48.

Ozguven, A.I., Yilmaz, C., Rehber, Y. and Yilmaz, M. (2006) The adaptation of different pomegranate varieties in the North Cyprus ecological conditions. *Acta Horticulturae* 818, 461–464.

Özgüven, A.I., Gültekin, U., Gözlekçi, S., Yilmaz, I., Yilmaz, C. *et al.* (2015) A review of the economics and the marketing of the pomegranate industry in Turkey. *Acta Horticulturae* 1089, 221–228.

Ozkan, Y. (2003) Determination of pomological characteristics of Niksar district pomegranates (*Punica granatum* L.) of the Tokat province. *Acta Horticulturae* 598, 199–203.

Phaleng, L. and Lubinga, M. (2018) South African fruit trade flow. 30. National Agricultural Marketing Council (NAMC), Pretoria, South Africa.

Pontikis, K. (1987) *Fruit Crops*. Agricultural University of Athens, Athens.

PPECB (2020) South Africa's official Perishable produce export certification agency. Available at: https://ppecb.com (accessed 15 May 2020).

Rymon, D. (2015) On the economics and marketing of pomegranate in Israel. *Acta Horticulturae* 1089, 189–196.

Salgado, M.O. (2017) Pomegranate, South Africa September. MBA. Available at: https://www.sapex.co.za/pdf/Sapex-Pomegranate-Symposium/2.pdf (accessed 18 June 2018).

Sarig, Y. and Galili, A. (2012) The pomegranate industry in China – current status and future challenges. *Options Méditerranéennes. Série A, Séminaires Méditerranéens* 103, 261–264.

Shaheen, S.A., Ali, A.A. and El-Bolok, T.K. (2016) Improving pomegranate fruit quality by using some practices. *Egyptian journal of horticulture* 43(2), 259–276.

Shi, C.D. (1991) 'Taishan Dahongshiliou', a top quality pomegranate cultivar. *China Fruits* 4(27)(in Chinese, English abstract).

Shwartz, E., Glazer, I., Bar-Ya'akov, I., Matityahu, I., Bar-Ilan, I. *et al.* (2009) Changes in chemical constituents during the maturation and ripening of two commercially important pomegranate accessions. *Journal of Agricultural and Food Chemistry*115, 965–973.

Simfruit (2020) Portal Oficial de la Fruta Chilena de Exportación de ASOEX (Exportadores de Frutas de Chile A.G.). Available at: https://www.simfruit.cl/ (accessed 17 May 2019).

Singh, D.B. (2004) Screening of pomegranate (*Punica granatum*) cultivars for arid ecosystem. *The Indian Journal of Horticulture* 74, 604–606.

SoDa Service (2019) MERRA-2 meteorological Re-analysis. Available at: http://soda-pro.com/web-services/meteo-data/merra (accessed 3 July 2019).

Stamatia, R. (2017) Pomegranate: bioactive ingredients and health benefits. *Technological Educational Institute of Crete Nutrition and Nutrition Department*, 1–57(in Greek).

Stover, E. and Mercure, E.W. (2007) The pomegranate: a new look at the fruit of paradise. *HortScience* 42(5), 1088–1092. DOI: 10.21273/HORTSCI.42.5.1088.

Surgul (2016) Profitability of pomegranate (*Punica granatum*) cultivation in Kandahar province of Afghanistan. Available at: http://krishikosh.egranth.ac.in/handle/1/93692 (accessed 13 March 2018).

Thomidis, T. (2014) Fruit rots of pomegranate (cv. Wonderful) in Greece. *Australasian Plant Pathology* 43, 583–588.

Tridge (2018) Pomegranate. Available at: https://www.tridge.com/intelligences/pomegranate/variety (accessed 19 April 2018).

Tsagkarakis, A.E. (2012) First record of *Siphoninus phillyreae* on pomegranate in Greece. *Entomologia Hellenica* 21(1), 39–43. DOI: 10.12681/eh.11516.

Turkish Statistical Institute (2018) Crop production statistics. Available at: www.turkstat.gov.tr/PreTablo.do?alt_id=1001 (accessed 7 June 2018).

U.S. Department of Agriculture, ARS, The National Clonal Germplasm Repository (NCGR) at Davis (2007) Repository inventory of available accessions for Punica granatum. Available at: www.ars.usda.gov/Main/docs.htm?docid¼12856

Walali, I.D., Skiredj, A. and Elattir, H. (2003) L'amandier, l'olivier, le figuier, le grenadier. Bulletin Mensuel d'Information et de Liaison du PNTTA. Transfert de Technologie en Agriculture, Ministère de l'Agriculture et du Développement rural.

Wang, C.Y., Qu, H.Y., Dai, X.C., Shi, L.L., Liu, Y.J. *et al.* (2010) Pharmacological effects, safety and application of pomegranate (*Punica granatum* L.). *World Sci. Tech/Mod. Traditional Chinese Med. Mater. Med.12961968 (in Chinese with English abstract)*.

Yücel, H. (2010) Fruit juice and concentrates. *IGEME, Export Promotion Center of Turkey*.

Zhang, Y.P., Tan, H.H., Cao, S.Y., Wang, X.C., Yang, G. *et al.* (2012) A novel strategy for identification of 47 pomegranate (*Punica granatum*) cultivars using RAPD markers. *Genetics and Molecular Research* 11(3), 3032–3041. DOI: 10.4238/2012.May.30.1.

Zhao, L., Li, M., Cai, G., Pan, T. and Shan, C. (2013) Assessment of the genetic diversity and genetic relationships of pomegranate (*Punica granatum* L.) in China using ramp markers. *Scientia horticulturae* 151, 63–67.

Zohary, D. and Spiegel-Roy, P. (1975) Beginnings of fruit growing in the old world. *Science* 187(4174), 319–327. DOI: 10.1126/science.187.4174.319.

Zouaoui, M. and Zouaoui Skandrani, F. (1992) Grenade. In: *La santé par les produits de la nature*. Editions Jugurtha International, Tunis, pp. 147–154.

4 Biodiversity, Germplasm Resources and Breeding Methods

A. Zarei[1]*, Z. Zamani[2]* and A. Sarkhosh[3]
[1]*Department of Plant Production and Genetics (Biotechnology), Jahrom University, Jahrom, Iran;* [2]*Department of Horticulture Science, University of Tehran, Karaj, Iran;* [3]*Horticultural Sciences Department, University of Florida, Gainesville, Florida, USA*

4.1 Introduction

Pomegranate (*Punica granatum* L.), a fruit-bearing deciduous shrub or small tree, is a fruit species that has been cultivated from ancient times and is one of the oldest edible fruits. Due to its nutritional and pharmacological properties, over the past few decades this fruit has become more popular across the world. Therefore, the number of research studies on different aspects of this fruit crop is increasing considerably; a fact that is obvious from the number of documents published during recent years. Biodiversity studies and screening of the existing accessions in different pomegranate-producing regions were among the research activities that started at early stages and contributed to the majority of pomegranate studies in the leading pomegranate producer countries.

4.2 Germplasm Resources and Biodiversity

Historical evidence shows that Persia (the ancient name for Iran, an area that spanned from Asia Minor to India) and some surrounding areas are the first locations of pomegranate cultivation. According to de Candolle, Iran and its surrounding areas are the central origins of pomegranate (Goor and Liberman, 1956). Vavilov believed that this fruit tree originated in the Near East (Chandra *et al.*, 2010a). Levin (1994) reported that pomegranate is native to Iran and gradually transferred west to Asia Minor and Mediterranean Basin countries and east to India. Goor and Liberman (1956) stated that south-west Asia is determined as the centre of origin of pomegranate. Wild forms of pomegranate trees have been reported in Iran, Afghanistan, Pakistan, Turkmenistan and north India (Narzary *et al.*, 2009, 2010). According to Rechinger (1969), the natural distribution of pomegranate trees in Iran is on the northern coasts of Iran and also some southern domains of the Alborz mountains range and west forests towards the Balouchistan.

Based on the reports of most researchers as well as the distribution of wild pomegranates in the world, it seems likely that this fruit tree originated from the Iranian Plateau and specifically somewhere around the southern parts of the Caspian Sea. The vast distribution of natural pomegranate forests on the shore of the Caspian Sea in the north of Iran, and also in plain forests like the forests of the Zagros mountains in Lorestan, Kurdistan, Charmahal Bakhtiyari, Fars,

*Corresponding author: zzamani@ut.ac.ir; zarei@jahromu.ac.ir

Baluchistan and the southern domains of the Alborz mountain range in the west and east of Manjil and the Lushan valley confirmed this matter. Therefore, Persia is the centre of origin and centre of diversity of this fruit tree and contains the richest pomegranate gene pool in the world, which can be utilized in breeding programmes.

It is estimated that pomegranate cultivation started in prehistoric times and its domestication began in the Neolithic era in the regions of Transcaucasia and the Caspian shores (Harlan, 1992; Levin, 2006; Still, 2006). Edible pomegranates were cultivated and used in Persia by 3000 BCE (Stover and Mercure, 2007). Numerous relics and artefacts of pomegranate have been found that are estimated to have been created more than 5000 years ago (Goor and Liberman, 1956; Still, 2006). The epigraphs of pomegranate shapes exist in Persepolis, representing Iranian's cognition of pomegranate in the age-old period. In Iran, this fruit tree is called Anar or Nar, and several ancient towns and villages have their names from this fruit.

Pomegranate probably was introduced from central Asia to China along the Silk Road about 2000 years ago (Yuan *et al.*, 2007). The introduction of pomegranate to the western regions, including North Africa and Europe took place through the Mediterranean region. This fruit was transferred to the south of Spain by the Muslims soon after the occupation of this country (711 CE), and Spanish sailors and missionaries introduced the pomegranate from the Mediterranean region to the New World including Mexico and California in the 1760s (Goor and Liberman, 1956). Today this fruit species is grown on four continents (in the north and south hemisphere) in different climate conditions mainly due to its tolerance for various unfavourable conditions; a fact that is reflected by a wide distribution of natural populations and wild forms of this fruit species throughout Eurasia to the Himalayas.

The main cultivation areas are throughout the Middle East and Caucasus region, central Asia, India, northern Africa and some parts of South-east Asia. California and Arizona are among the leading regions of pomegranate cultivation in North America. Due to the recent awareness of its various beneficial compounds, pomegranate is becoming more popular in the European countries and western hemisphere, and demand for this fruit is increasing in the commercial markets of these regions.

Nearly all of the pomegranates varieties that are now cultivated in Iran, Afghanistan and the Mediterranean regions are local accessions that have been selected by natives and propagated vegetatively (Levin, 1995; Mars and Marrakchi, 1999; Chandra *et al.*, 2010a). Levin (2006) defined three mega centres and five macro centres for the origin and diversity of pomegranate. The natural habitat of this fruit species, the eastern part of the Middle East region, which includes Iran and Afghanistan, was considered as the primary centre of *Punica granatum*. Mediterranean basin countries and eastern Asia were considered as the secondary and tertiary mega centres for this fruit species, respectively. These two centres were formed during the process of introduction of pomegranate westwards and eastwards from the primary mega centre. In addition to the three mega centres, Levin (2006) also suggested five macro centres for this fruit tree, which include the Middle East, eastern Asia, the Mediterranean, South Africa and America. Pomegranate accessions are highly non-uniform in different centres. According to Chandra *et al.* (2010a), the principal conglomerates of pomegranate varieties were developed through domestication. Selections have been made in the proximity of the natural habitats and then transferred to the surrounding regions through vegetative or sexual propagation, and these introductions formed the basis of local varieties and subsequently their further distribution. Therefore, local planted materials are the primary gene pools for this plant. Wild forms still exist in Iran, Turkmenistan, Dagestan and Transcaucasia and may be considered as the secondary gene pool. Therefore, Turkmenistan and the north-east of Iran are considered among the richest areas for the secondary gene pool of pomegranate (Still, 2006). Hybridization between local and wild forms is probably still taking place in these regions. The relative wild -type species (*Punica protopunica*), which is native to Socotra Island, may be the tertiary gene pool. There are no obvious botanical differences between cultivated and wild pomegranate accessions except for *P. protoponica* (Holland *et al.*, 2009). However, molecular study on wild and relative types of pomegranate has not been carried out to the extent to enable the determination of phylogenic relationships among them. The fruits of the wild pomegranates are small with thicker rinds and extremely high acidity compared with cultivated types (Bist *et al.*, 1994). Pomegranates of some regions are well known for their special features. According to Levin (2006), the large soft-seed pomegranate varieties grow in Kandahar, Afghanistan. Accordingly, various frost hardy varieties grow in Dashnabad, Uzbekistan.

Although there are many phenotypically different pomegranate accessions throughout the world, only a few have gained commercial acceptance and are cultivated in each region. Therefore, due to the selection of superior and elite accessions for breeding purposes as well as orchard establishment, most of the existing genetic diversity was not included in the commercial cultivation of this fruit species. This selective process might have resulted in genetic erosion and bottleneck, which might confine subsequent breeding programmes in the future. On the other hand, during severe selection and breeding programmes, the genetic variability of primary populations might have been reduced to the level of diversity of our domesticated or hybrid varieties, which are only a small fraction of the previous diversity. Nowadays, it is well established that agriculture sustainability is highly dependent on the diversification of the genetic background of species. Broadening the genetic bases is highly important not only for the preservation of crop production in the cultivated regions but also for expanding the crop production from the borders of arable regions. Fortunately, the importance of genetic diversity conservation for crop production and improvement is recognized by researchers, and they are actively exploring different approaches to retain plant diversity; the germplasm conservation action is one of these. However, conservation and evaluation of wild accessions of pomegranate have not received sufficient attention and should be considered more seriously in the germplasm conservation strategies.

4.2.1 Germplasm collections

At present, more than 500 pomegranate cultivated varieties (cultivars) are recognized around the world. However, only a few (<50) are commercially cultivated (Chandra *et al.*, 2010a). Therefore, only small portions of the total genetic diversity of the pomegranate are actively utilized worldwide. Collecting and conservation of visually different pomegranate accessions have been started as one of the preliminary activities in different pomegranate producer countries with the purpose of characterization, selection, introduction and utilization in the breeding programmes. There are several pomegranate collections in different pomegranate producer regions.

Until recently, the establishment of pomegranate orchards in Iran has been traditionally done using cuttings collected from superior and productive genotypes. Meanwhile, different cultivars were planted together in the same orchard, and hence, homogeneous orchards are scarce. Although, from the modern horticultural activities point of view, homogeneous orchards are preferred, such heterogenous orchards are, to some extent, a guarantee for the conservation of different plant genetic resources as an *in situ* collection. This type of orchard establishment may also result in higher productivity of trees due to the occurrence of cross-pollination of different genotypes in vicinity. This also may lead to maintenance of an acceptable amount of fruit production in the case of unfavourable conditions such as frost, drought, pest and disease stresses. Moreover, Iran, as the main centre of origin and diversity of this fruit species, is the home of several *ex situ* pomegranate collections. The pomegranate collection of Varamin (a city located in the south-east of Tehran, Iran) is the first *ex situ* pomegranate collection of Iran, and contains about 300 pomegranate accessions. Later, the establishment of the Iran National Pomegranate Collection started in 1987. After a comprehensive survey in different provinces of Iran, 762 cultivars and genotypes that were visually distinct were collected from each region and a comprehensive pomegranate germplasm collection was established in Yazd city located in the centre of Iran (Fig. 4.1) (Behzadi Shahrbabaki, 1997). By adding 20 new specimens, the numbers of pomegranate genotypes reached 782 samples in recent years. This collection consisted of four replicates of each genotype that are planted in an augmented design in five blocks with five control cultivars from Yazd province including 'Malas-e-Yazdi', 'Shirine-Shahvar','Goroch-Shahvar', 'Zaghe-Yazdi' and

Fig. 4.1. Pomegranate germplasm collection in Yazd (central Iran). (Photos: Abdolkarim Zarei and Alimohammad Yavari.)

'Tough-Gardan'. Although some regions of Iran have no delegate in this collection (such as Hamedan province), one of the outstanding properties of this collection is that the samples resident to this collection are not restricted to the commercial and marketable cultivars.

The collected accessions represent a relatively good diversity of Iranian pomegranates, hence will be useful as materials for breeding programmes with various purposes. However, the lack of wild-type pomegranates, as well as the scarcity of international varieties are among the main drawbacks of this collection. Saveh (Markazi province) is the home of another pomegranate collection in Iran that includes 540 cultivars; some of these are the same as the Yazd collection. Several smaller collections have been established in other cities in the format of agricultural research stations as well as universities, such as a collection at the horticultural research station of the University of Tehran in Karaj city (Zamani *et al.*, 2007; Zarei *et al.*, 2009).

Afghanistan is also considered as a major centre of pomegranate diversification. Traditional cultivation of pomegranate began in ancient times in this country as one of the provinces of Persia; hence, there are excellent quality pomegranate landraces in different regions of this country including Kandahar and Farah provinces. In addition, scattered populations of wild pomegranates are reported in different regions of this country including Kandahar, Baikh, Farah, Kapisa, Samangan, Nagharhar and Heart (Saeedi *et al.*, 2012). Similar to Iranian pomegranate orchards, Afghanistan pomegranate orchards are also highly heterogeneous and each can serve as an *in situ* collection.

The Perennial Horticulture Development Project (PHDP) is a programme started by the European Commission–Europe Aid Programme in 2006 with the aim of supporting horticultural crops in Afghanistan. After a survey of different areas of the country during the years 2006–2008 by local experts, fruit growers and nurserymen, an *in situ* national collection of pomegranate was established in Afghanistan. More than 950 accessions were registered and labelled. According to the marketability, some outstanding accessions were included in this project. Afterward, the *ex situ* Pomegranate National Collection of Afghanistan was founded as duplicates in Kandahar and Nangarhar provinces in 2009. This collection included 59 Afghanistan pomegranate genotypes from *in situ* collections along with 20 introduced cultivars from other countries. Six replicated trees from each clone are conserved in each collection. Samples resident to this collection are evaluated phenotypically, and descriptors of their phenological data are listed. This collection is considered as the official repository for pomegranate in Afghanistan and is the main source of plant materials for the establishment of mother stock nurseries that provide traced cuttings for the propagation of this fruit species (Saeedi *et al.*, 2012).

The world's most extensive pomegranate collection was established in south-western Turkmenistan at the Garrygala (previously known as Kara-Kala) experimental station for plant genetic resources. This collection was launched by the famed Russian botanist Gregory Levin, who collected 1117 pomegranate accessions from 37 countries in four continents from the 1960s through the fall of the Soviet Union. These accessions are maintained at Garrygala in field genebanks. The Garrygala collection includes pomegranate samples with economically valuable attributes that might be utilized in breeding programmes (Holland *et al.*, 2009). Besides the cultivated samples, this collection also includes some wild materials. Fruit with large arils, high juice content, thin peel, high content of bioactive compounds such as vitamin C and tolerance to frost, sunburn and pests and diseases as well as long postharvest shelf-life are among the prominent traits reported in the Garrygala collection. To facilitate the management and better utilization of diversity in the vast collections, core collections are established for different plant species. Accordingly, the pomegranate accessions in the Garrygala research collection have been explored and selected based on the desirable qualitative and quantitative attributes of fruits and a core collection with 10% of the main collection size was established from the best specimens. To retain the highest genetic diversity in the core collection as much as possible, Vavilov's law of homologous series has been regarded in the establishment of the core collection. Due to better management, this collection will be suitable for breeding work on pomegranate. The Turkmenistan mountains also contain various bushes of wild pomegranates, which grow along river banks, or as creeping shrubs on rocky slopes or clinging to steep hillsides (Robertson, 2009).

More than 100 accessions from the Turkmenistan pomegranate collection were transferred to the USA and planted in an orchard in a Germplasm Repository at the Wolfskill Ranch (Robertson, 2009).

In India, some of the wild pomegranate germplasm from the western Himalayas, along with exotic samples, were collected at the National Field Gene Bank of pomegranate (India). This collection was established in 2007 and included 59 exotic samples, 10 cultivars, 57 genetic materials and 61 wild types in a total of 187 germplasm accessions (Chandra *et al.*, 2010b). Moreover, in this country, seven agricultural universities and six Indian Council of Agricultural Research (ICAR) institutes maintain pomegranate germplasms in their repositories (Table 4.1).

The pomegranate collection of European Minor Fruit Tree Species with 116 pomegranate accessions located in Italy and Spain, the collection of Vavilov Research Institute of Plant Industry in St. Petersburg, Russia with 800 accessions and Nikita Botanical Garden in Yalta, Ukraine with 370 accessions are among the largest European collections of this fruit species, with multitudinous samples (Frison and Serwinski, 1995; Yezhov *et al.*, 2005; Seeram *et al.*, 2006). A pomegranate collection with 58 specimens was created in Miguel Hernandez University, containing pomegranate samples from different south-eastern Spain localities at Alicante, Spain (Martinez-Nicolas *et al.*, 2016). Aegean Agricultural Research Institute in Izmir, Turkey also protects 158 pomegranate samples as *ex situ* germplasm (Frison and Serwinski, 1995). It is reported that more than 100 pomegranate accessions have been collected in the agriculture research organization of Israel, including both local (including semi-wild accessions) and introduced samples from other regions including China, India, central Asia, Turkey, Spain and other Mediterranean basin countries as well as the USA (Ophir *et al.*, 2014).

According to Mars and Marrakchi (1999), two pomegranate collections containing 63 samples have been established in Tunisia. These highly divergent collections include 20 local landraces from different regions of Tunisia. Reports from China also indicate the existence of diverse pomegranate resources in this country. It is reported that 238 pomegranate cultivars are growing in different regions of China and a germplasm collection is

Table 4.1. Germplasm collections of pomegranate worldwide, according to available information.

Country	Centre	Location	Number of samples	Reference
Turkmenistan	Turkmenian Experimental Station of Plant Genetic Resources	Garrygala	1117	Levin, 2006
Russia	N.I. Vavilov Research Institute of Plant Industry	St. Petersburg	800	Frison and Serwinski, 1995
Iran	National Iranian Pomegranate Collection	Yazd	786	Behzadi Shahrbabaki, 1997
Iran	Agricultural research stations of Saveh	Saveh	540	Behzadi Shahrbabaki, 1997
Azerbaijan	Unknown	Unknown	200–300	Levin, 1995
Tajikistan	Unknown	Unknown	200–300	Levin, 1995
Ukraine	Unknown	Unknown	200–300	Levin, 1995
Uzbekistan	Schroeder Uzbek Research Institute	Tashkent	200–300	Levin, 1995
Afghanistan	Pomegranate National Collection of Afghanistan	Kandahar and Nangarhar	79	Saeedi *et al.*, 2012
Ukraine	Nikita Botanical Gardens	Yalta, Crimea	370	Yezhov *et al.*, 2005
USA	US National Clonal Germplasm Repository	Davis, CA	≈200	Stover and Mercure, 2007; USDA, 2007
USA	Fruit and Nut Germplasm	Davis, CA	59	Seeram *et al.*, 2006
USA	Germplasm Repository	Wolfskill Ranch	>100	Robertson, 2009
Turkey	Plant Genetic Resources Department, Aegean Agricultural Research Institute	Izmir	158	Frison and Serwinski, 1995
China	Different provinces	Unknown	238	Feng *et al.*, 2006
China	Yunnan	Unknown	>25	Yang *et al.*, 2007
India	National Bureau of Plant Genetic Resources Regional Station	Phagli, Shimla	90	Rana *et al.*, 2007
India	Three collections (unknown locations)	Unknown	>30	Gulick and Van Sloten, 1984
India	Central Institute of Arid Horticulture	Rajasthan	190	Chandra *et al.*, 2010a

Continued

Table 4.1. Continued

Country	Centre	Location	Number of samples	Reference
India	National Research Centre of pomegranate	Maharashtra	187	Chandra *et al.*, 2010a
India	National Bureau of Plant Genetic Resources	New Delhi	170	Chandra *et al.*, 2010a
India	Mahatma Phule Krishi Vidyapeeth	Maharashtra	61	Chandra *et al.*, 2010a
India	Sardarkrishinagar Dantiwada Agricultural University	Gujarat	52	Chandra *et al.*, 2010a
India	Central Arid Zone Research Institute	Rajasthan	34	Chandra *et al.*, 2010a
India	Acharaya NG Ranga Agricultural University	Andhra Pradesh	29	Chandra *et al.*, 2010a
India	Tamil Nadu Agricultural University	Tamil Nadu	24	Chandra *et al.*, 2010a
India	Indian Institute of Temperate Horticultural Research	Karnataka	20	Chandra *et al.*, 2010a
Israel	Newe Ya'ar Research Center, Agricultural Research Organization	Yizre'el Valley, Haifa	>100	Bar-Ya'akov *et al.*, 2003, Bar-Ya'akov *et al.*, 2007 ; Ophir *et al.*, 2014
Thailand	Five locations in Chiang Mai, one in Bangkok	Chiang Mai and Bangkok	29	Thongtham, 1986
Tunisia	Two collections	Gabes and South Tunisia	63	Mars and Marrakchi, 1999
Turkey	Alata Horticultural Research Institute	Erdemli	>180	Onur, 1983; Onur and Kaska, 1985; Holland *et al.*, 2009
Turkey	Cukurova University	Adana	33	Ozguven *et al.*, 1997, Ozguven *et al.*, 2000
Turkey	Aegean University	Izmir	158	Frison and Serwinski, 1995
Spain	Escuela Politécnica Superior de Orihuela-Miguel Hernandez University	Alicante	59	Martinez-Nicolas *et al.*, 2016
Spain	Valencian Agricultural Research Institute (IVIA)	Elche	35	Bartual *et al.*, 2012
European Minor Fruit Tree Species, EC Project GENRES 29	Eleven locations	Spain, Italy	116	Seeram *et al.*, 2006

Continued

Table 4.1. Continued

Country	Centre	Location	Number of samples	Reference
Egypt	Unknown	Upper Egypt, Lower Egypt and north Sinai	13	Hassan and Abd-El Gawad, 2013
Morocco	Pomegranate Collection of Ahl Souss	Beni Mellal and Meknès	Unknown	Ajal *et al.*, 2015

established in Yunnan (Feng *et al.*, 2006; Yang *et al.*, 2007).

Moreover, collection, evaluation, characterization and *ex situ* conservation of pomegranate germplasms are in progress in different countries including Italy, France, Ukraine, Portugal, Hungary, Germany, Greece, Cyprus, Albania, Tajikistan, Morocco, Egypt and Jordan. However, there is insufficient information about conserved genotypes in most of the mentioned countries. For *P. protopunica*, the wild relative of pomegranate, seeds from plants growing in Socotra Island in Yemen were collected and a germplasm collection was established (Guarino *et al.*, 1990).

Although there are several pomegranate collections in different regions, it is estimated that some of the accessions might be identical owing to being introduced from the same source. For instance, a significant proportion of the accessions in the Davis collection of the USA, the Newe Ya'ar collection of Israel, the Garrygala collection of Turkmenistan, as well as Tunisia, Thailand and Indian collections were introduced from other countries. In the Iranian collections, although there are no foreign cultivars, there are several duplicates, and most of the genotypes in the Saveh collection are the same as the Yazd collection. Molecular and morphological evaluations of some of the existing germplasms indicate that there are genotypes with the same name but different characteristics (homonyms) or identical genotypes with different names (synonyms) in some collections. Therefore, more accurate characterization of collections, for example by molecular markers, will help to better manage the existing germplasms by eliminating redundancy and replacing them with new different samples as well as by the establishment of the core collections. In addition, *ex situ* conservation of wild types of pomegranate should be also considered in countries that have wild forests of this fruit species, such as Iran and Afghanistan.

4.3 Cytological Study

The reported chromosome numbers of pomegranate varies in different studies (n = 8, 9; $2n$ = 16, 18) (Smith, 1976). Nath and Randhawa (1959) reported $2n$ = 16 for six Indian cultivars, and $2n$ = 18 for the ornamental cv. 'Double Flower'. It is reported that the diploid chromosome ($2\,n$ = 16) of pomegranate contains 1.4 pg (=1412 Mbp) of DNA (Ohri, 2002; Bennett and Leitch, 2005). However, the results of flow cytometry analysis estimated that the genome size of pomegranate would be 356.98 Mb (Qin *et al.*, 2017). The chromosome numbers in 'Vellodu' and 'Kashmiri' cvs. were reported to be $2n$ = 18 with one or two quadrivalent associations at meiosis (Raman *et al.*, 1963). It is reported that a tetraploid pomegranate with $2n$ = 32 was obtained from cv. 'GB-1' ($2n$ = 16) by air-layering (IBPGR, 1986). A spontaneous tetraploid clone ($2n$ = 32) was reported in India, with flowers and fruit exceeding the size of the original form, and the sterility of its pollen was 84.4% as compared with 7.4% in a normal diploid genotype (Das and Sur, 1968). However, most of the records indicated that *P. granatum* is a diploid species possessing 16 chromosomes (n = 8) (Gill *et al.*, 1981; Ue *et al.*, 1992; Sheidai *et al.*, 2005). *Punica protopunica* species has a diploid genome ($2n$ = 14) and a haploid number of chromosomes of n = 7. Therefore, from an evolutionary point of view based on the lower number of chromosomes as well as xylem anatomy this species is considered as the progenitor of the *Punica* genus (Shilkina, 1973).

According to different cytological studies, most of the pomegranate's chromosomes are small. However, in spite of their small size, the chromosomes are suitable for cytogenetic studies. Sheidai *et al.* (2005) studied the meiotic behaviours considering ploidy level, chiasma frequency and chromosome association in 22 Iranian genotypes. Among the eight bivalents that usually occur during metaphase I of meiosis in the nucleus of this plant, seven chromosomes have equal sizes but one pair was much bigger than the others. According to karyotype analysis of some Egyptian pomegranates, the chromosome lengths varied from 1.067 μm to 2.265 μm in different genotypes (Hassan and Abd-El Gawad, 2013). Considering the centromer position, chromosomes 1, 2, 3, 4 and 5 are median centromeric while chromosomes 6, 7 and 8 are median to sub-median centromeric. Significant differences were reported between chiasma frequency, chromosomal pairing and their segregation in different pomegranate genotypes, reflecting their genomic differences (Sheidai *et al.*, 2005). It is noteworthy that even though pomegranate is a diploid species, and it is expected only to create bivalents in metaphase I of meiosis, observation of some quadrivalents were reported in a number of studied genotypes ('Berit', 'Malase-Peyvandi', 'Malase-Toghi', 'Malas-Gardanboland', 'Bejestoni-Torsh', 'Dadashi-Peivandie', 'Shirin-Poostnazok', 'Anbari-Danehghermez', 'Shahvar-Torsh', 'Goltorsh-Mamooli Taft', 'Atabaki Jahrom', 'Malas-Aghdaei' and 'Bejestoni-Poostkoloft') (Sheidai *et al.*, 2005, Sheidai *et al.*, 2012). The occurrence of quadrivalents might be attributed to the translocation between two pairs of chromosomes. Heterozygote translocations can create new linkage groups and induce new genetic changes in their place of occurrence in the chromosomes. Because of the relatively large sizes of reported quadrivalents, it is surmised that the largest chromosome pairs in the genome were involved. In all the studied genotypes, univalents were observed at a low frequency that ranged from 0.4 to 2.4. The small sizes of pomegranate chromosomes and hence, the early termination of chiasmata in them, may be the mainspring of the usual occurrence of univalents (Sheidai *et al.*, 2005). These authors also observed the formation of laggard chromosomes in their studied genotypes and attributed this phenomenon to the regular occurrence of univalents due to the small size of chromosomes. Chiasma frequency and distribution is controlled genetically (Rees and Jones, 1977). Therefore, significant differences in frequency and distribution of chiasma as well as rod and ring bivalents among genotypes in the same location may reflect their genetic differences.

Some abnormal chromosomal segregation has also also reported in pomegranate and this phenomenon has been ascribed to chromosome stickiness and formation of laggard chromosomes and micronuclei (Sheidai *et al.*, 2005). Chromosome stickiness and their tardy segregation were observed from early stages of prophase to the end of meiosis. Genetic and environmental factors, as well as their interaction, have been stated as the main reasons for chromosome stickiness in various plant species, and the same may be true in pomegranate genotypes (Nirmala and Rao, 1996). Spindle apparatus abnormality is another interesting feature in most Iranian pomegranate genotypes, which subsequently results in the formation of multipolar cells in telophase I and II, which would be normally bipolar (Sheidai *et al.*, 2005). Owing to the role of spindle apparatus in the chromosome of balancing and alignment during metaphase, any irregularity in its formation may have resulted in the random subgrouping. Several reasons have been cited for abnormality of the spindle apparatus such as environmental effects, duality of the nucleus in foreign cytoplasm and disharmonious gene interaction (Nirmala and Rao, 1996). Vorsa and Bingham (1979) reported the giant pollen grains as a sign of 2n pollen production. In some Iranian pomegranate genotypes, such as 'Agha-Mohamad-Ali' and 'Malase-Peyvandi', pollen grains with significantly larger sizes than normal samples were observed, which presumably formed from the occurrence of multipolar cells in these genotypes. Also several unusual tetrads were formed that may have resulted from such abnormal cells (Sheidai *et al.*, 2005). Grouping of pomegranate genotypes based on their cytological attributes was in accordance with the geographical distribution of samples, hence, cytological attributes were an indication of genotypes' geographical distribution.

Moreover, high frequencies of B-chromosomes have been observed in pomegranate (Sheidai, 2007; Sheidai *et al.*, 2012). In fact, among 50 evaluated pomegranate genotypes, 23 samples possessed from zero to five B-chromosomes (Sheidai *et al.*, 2012). B-chromosomes are supernumerary chromosomes and smaller than standard chromosomes (also known as A-chromosomes). In low frequency they usually confer some benefits to the plants possessing them, but when they are present in high numbers, act inversely and decrease plant growth and vigour. These types of chromosomes are found in all eukaryotic phyla and are thought to stand for a specific type of selfish DNA (Houben, 2017). B-chromosomes show high polymorphism and are able to influence the chiasma frequency, distribution and interaction of chromosomes either directly or indirectly by the effect on meiosis regulator genes located on the A-chromosomes (Camacho *et al.*, 2000). B-chromosomes in pomegranate genotypes are much smaller than A-chromosomes and presumably are not paired with their standard counterpart. Although these chromosomes can be transfered to the daughter cells during cell division, very often they are lagged and excluded from daughter cells. Like other plants, this is an inhibitory system for accumulation of B-chromosomes that may be unwholesome to them. Effects of B-chromosomes are somehow variable in pomegranate cultivars; in some genotypes they cause a significant reduction in total chiasmas while in others they increase chiasmas (Sheidai *et al.*, 2012). These variable responses to the B-chromosomes may be due to differences in the genomic background of the genotypes. The presence of B-chromosomes increases the intercalary chiasmata significantly in some pomegranate genotypes. An increase in frequency of intercalary chiasma might increase genetic variation through the inclusion of the genes located in the intercalary positions of the chromosome arms in the genetic recombination process. This phenomenon increases the mean value of quadrivalents, which subsequently causes a higher amount of heterozygote translocation and finally results in greater genetic diversity in the resulting gametes. A decrease in the number of univalents and an increase in the number of ring bivalents may improve the adequate segregation of chromosomes in anaphase, which subsequently results in higher fertility of pollen in the B-chromosome-containing cultivars. In addition, it is reported that cultivars from different geographic regions of Iran have different numbers of B-chromosomes (Sheidai *et al.*, 2012). However, cultivars from provinces that have no diversity in B-chromosomes (Tehran and Semnan) have lower genetic diversity and did not separate in a plot derived from principal component analysis. Therefore, these types of chromosomes may contribute to the diversity of pomegranate genotypes.

High positive and negative correlations have been reported between some of the pomegranate cytological characteristics and climate conditions. Total chiasma frequency had a significant negative correlation with the mean number of rod bivalents and univalents. A significant positive correlation was reported between annual rainfall and the mean number of rod bivalents. Although elevation did not have any significant correlation with cytological characteristics, longitude showed a significant positive correlation with the number of ring bivalents and negative correlation with the number of univalents and intercalary chismata (Sheidai *et al.*, 2012). There was a significant positive correlation between pollen fertility and total chiasmata. Therefore, these differences may be used in planning cultivar selection and hybridization. However, even though significant correlations were recorded between some of the cytological characteristics and different climate conditions, no significant affinity was observed among the cultivars collected from the same geographic origin in the depicted cluster based on their cytological characteristics (Sheidai *et al.*, 2012).

4.4 Unreduced Pollen Grains

Formation of unreduced pollen grains has also been reported in some Iranian pomegranate genotypes (Sheidai *et al.*, 2012). Abnormalities during either megasporogenesis (embryo sac formation) or microsporogenesis (pollen formation) may result in unreduced diploid gametes, which finally produce the sporophyte rather than the gametophyte chromosome numbers. These types of gametes are able to produce offspring with higher chromosome sets than normal ones – the process known as sexual polyploidization (Villeux, 1985). Tripolar cell formation due to failure in anaphase I is considered as the main reason underlying the production of the 2n gametes in different pomegranate genotypes. Interaction between the genome and environmental factors is considered as the main reason for chromosome stickiness in different plant species (Baptista-Giacomelli *et al.*, 2000); this may hold true for pomegranate genotypes.

Spindle abnormalities is one of the main causes of unreduced gamete (2n) production, which can result in polyploid and aneuploidy gametes. Abnormal spindles may be produced as a result of the duality of the nucleus in the foreign cytoplasm and environmental factors (Nirmala and Rao, 1996). Large-sized grains are an indicator for production of 2n pollen (Vorsa and Bingham, 1979). It is reported that different pomegranate cultivars varied greatly based on their grain sizes, and the giant grains are considered as 2n pollen (Sheidai *et al.*, 2012). Based on the difference in the cytological attributes of pomegranate genotypes in the same region, it could be concluded that cytogenetic features along with geographic regions contribute greatly to the genetic variability of pomegranate genotypes. These findings emphasize the need for comprehensive exploration of different pomegranate producer areas and increasing the genotype sampling in each geographic region as much as possible to establish a thorough germplasm collection as genetic stock for future breeding programmes and hybridization purposes.

4.5 Morphological Studies on Pomegranate

Evaluation of morphological attributes among the existing genetic resources is one of the preliminary steps of each breeding programme (Fig. 4.2). Evaluation, description and classification of plant germplasms can facilitate the process of identifying the individuals with appropriate traits for utilization as parents in hybridization programmes. Due to the widespread distribution of pomegranate across the world, assessment of existing genotypes using different marker systems is one of the research aspects of the pomegranate that is widely investigated. Morphological diversity of different pomegranate accessions has been evaluated in different pomegranate producer regions (Zamani, 1990; Mars and Marrakchi, 1999; Sarkhosh *et al.*, 2005, 2006; ; Martínez *et al.*, 2006; Zamani *et al.*, 2010; Ferrara *et al.*, 2011, 2014; Karimi and Mirdehghan, 2013; Zarei *et al.*, 2013; Khadivi-Khub *et al.*, 2015 ; Martinez-Nicolas *et al.*, 2016; Zarei, 2017; Dandachi *et al.*, 2017; Khadivi *et al.*, 2018). Besides descriptive statistics, estimation of simple correlations among traits, clustering and factor analysis were among the main analyses that have been conducted in different studies. Nearly all the reports indicate

Fig. 4.2. Morphological attributes among some pomegranate accessions. (Photo: Doron Holland.)

that pomegranate genotypes exhibit relatively high pomological diversity and there are wide ranges of different fruit characteristics (both physical and biochemical attributes) among existing accessions. Fruit-related characteristics have high potential for discrimination among different pomegranate accessions. However, it is stated that tree-related characteristics including leaf and flower attributes have lower discrimination power for pomegranate genotypes (Martinez-Nicolas *et al.*, 2016). Fruit taste has a prominent role in the clustering of evaluated accessions (Zamani, 1990; Sarkhosh *et al.*, 2005; Zamani *et al.*, 2010; Zarei, 2017). This could be attributed to the differences in titratable acidity (TA) between sweet and sour genotypes rather than total soluble solids (TSS). In some cases, sour genotypes have higher TSS content compared with the sweet ones. On the other hand, the amount of TA (not TSS) determines the final taste of the pomegranate fruit. Beside the taste, fruit and aril colour, male and hermaphrodite flowers, petal width as well as seed hardness (where both soft and hard seed genotypes were included in the investigation) have been found to be important in pomegranate grouping (Zamani *et al.*, 2007; Zarei *et al.*, 2013; Martinez-Nicolas *et al.*, 2016; Dandachi *et al.*, 2017). High variation among TA, seed hardness as well as fruit and aril colour in different pomegranate accessions makes these traits very important in pomegranate classification analysis.

A positive correlation has been reported between seed hardness and TA, and colour-related characteristics including fruit and aril colour, as well as anthocyanin content (Khadivi-Khub *et al.*, 2015; Zarei, 2017). Antioxidant activity has also shown positive correlations with the amount of ascorbic acid and gallic acid in arils (Sarkhosh *et al.*, 2008). Woody portion percentage has been negatively correlated with seed length (Martínez *et al.*, 2006; Zarei *et al.*, 2013; Zarei, 2017). Some desirable characteristics including fruit and aril size, juice, TSS, vitamin C and aril colour have been reported to be favourably associated with each other (Desai *et al.*, 1994). No strong correlation has been reported between colour of fruit skin and the arils (Jalikop, 2010, Jalikop, 2011). In addition, seed hardness has been negatively correlated with aril juices (Elhem *et al.*, 2011). It is documented that the number of arils has a linear relationship to fruit

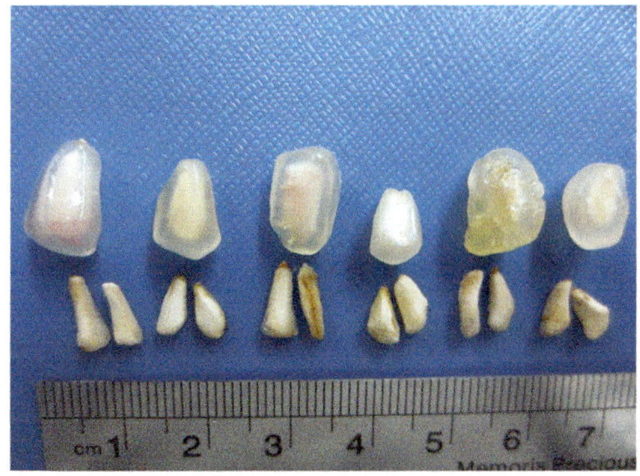

Fig. 4.3. Seeds and arils of different soft- and hard-seeded pomegranate genotypes. Name of genotypes from left to right are as follows: 'Bihasteh-Ravar' (soft seed), 'Malase-Esfahani' (hard seed), 'Bihasteh-Najafabad' (soft seed), 'Torshe-Zabol' (hard seed), 'Bihasteh-Sangan' (soft seed) and 'Malase-Shirine-Saveh' (hard seed). (Photo: Abdolkarim Zarei.)

weight and there is no relationship between fruit weight and juice volume (Kumari *et al.*, 2017). Hard-seeded varieties have significantly higher fruit weight and size than semi-soft or soft-seeded varieties (Holland *et al.*, 2009; Kumari *et al.*, 2017), but soft-seeded accessions have relatively bigger and juicier arils than hard-seeded ones (Fig. 4.3) (Zarei *et al.*, 2013). In addition, there is a high positive correlation between seed softness and length of the seed (Fig. 4.4) (Zarei *et al.*, 2013). Although it has been reported that larger fruits do not necessarily contain larger arils, the number of arils per fruit has been strongly correlated with fruit size (Hazel *et al.*, 2011). A comprehensive morphological evaluation revealed that in some cases the name of the cultivar describing a trait of the fruit does not reflect the real characteristic of the fruit. For instance, in the evaluation of Iranian soft-seed germplasm Sarkhosh *et al.* (2008) identified that some of the genotypes harbouring the soft seed name (bidaneh in Farsi) were not real soft-seed varieties. They attributed their observations to the environmental effects on this characteristic or more probably to mislabelling during germplasm establishment. Some discrepant results were reported regarding morphological characteristics from different countries. Although a high negative correlation was reported between seed hardness and aril and seed length

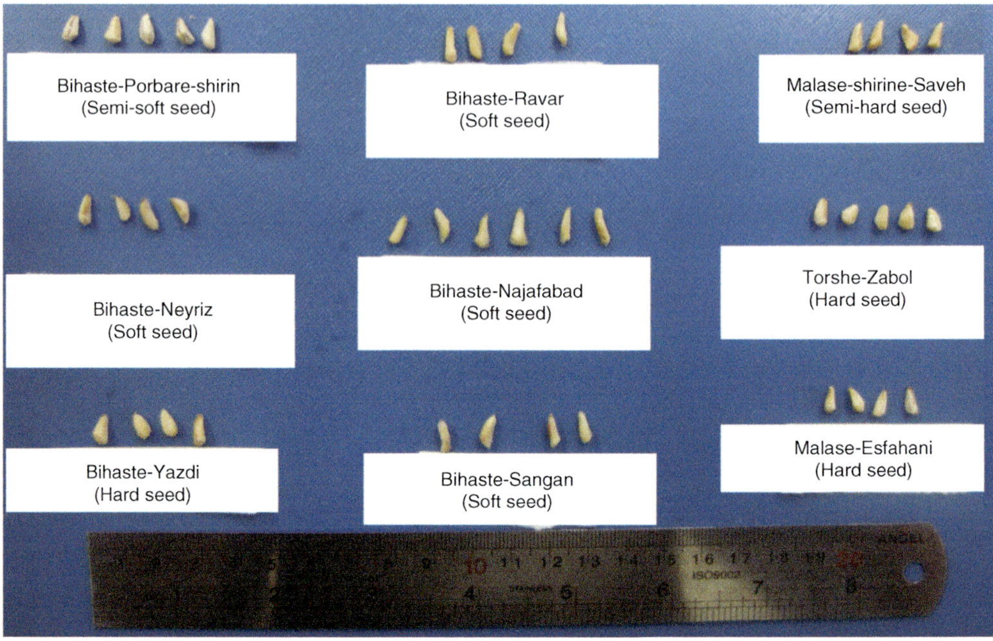

Fig. 4.4. Seed variability in hard- and soft-seeded pomegranate genotypes. (Photo: Abdolkarim Zarei.)

in Iranian accessions (Zarei *et al.*, 2013; Zarei, 2017), there was no significant correlation between these two characteristics in the Chinese genotypes (Lu *et al.*, 2006). Moreover, reports on Iranian pomegranate accessions indicated a higher woody portion percentage compared with Spanish genotypes (Martínez *et al.*, 2006; Zarei *et al.*, 2013). According to Jalikop (2010) fruit size, sugar, acidity, colour of peel and aril as well as ripening may be influenced by the environment. Therefore, great care should be taken when using pomological, morphological or physiological traits. In fact, traits such as colour of peel and aril, fruit size, time of ripening, TSS and acidity are consequences of various factors and strongly influenced by the environment.

4.6 Molecular Markers and Their Application on Pomegranate

With the advent of molecular markers, germplasm characterization and identification of genotypes became easier than with morphological characterization alone. Molecular markers analysis is one of the most comprehensive aspects of pomegranate studies. Different molecular marker systems have been exploited to evaluate genetic basis and relatedness of different pomegranate accessions including random amplification of polymorphic DNA (RAPD), inter-simple sequence repeats (ISSRs), amplified fragment length polymorphism (AFLP), inter-retrotransposon amplified polymorphism (IRAP), directed amplification of minisatellite DNA (DAMP), simple sequence repeats (SSRs) and single nucleotide polymorphism (SNP) (Table 4.2).

RAPD molecular markers are among the most prevalent marker types used successfully for evaluation of genetic diversity and germplasm management of this fruit species, respectively. Results of the first report of RAPD marker application on 28 local and commercial Iranian pomegranate genotypes indicated low genetic diversity among the germplasm and clonal propagation of pomegranate was stated as the main reason behind this (Talebi Bedaf *et al.*, 2003). After this, RAPD molecular markers have been extensively used for evaluation of different pomegranate accessions (Sarkhosh *et al.*, 2006; Sheidai, 2007; Zamani *et al.*, 2007, Zamani

Table 4.2. Different molecular markers studies in pomegranate and summary of their results. The reports are presented chronologically

Type of molecular markers and number of primers	Place of sampling	Number of samples	Number of polymorphic and monomorphic fragments	The main objective of study	The main findings	Reference
RAPD; 16 primers	Iran	24 genotypes	Total of 178 fragments produced, 102 were polymorphic	Evaluation of genetic diversity	High genetic diversity among studied genotypes; usefulness of RAPD for genetic diversity assessment in pomegranate	Sarkhosh *et al*., 2006
AFLP: 10 primer combinations	Iran	11 genotypes	187 bands produced, 70 bands were polymorphic	Genetic diversity evaluation	Low polymorphism between genotypes but different banding pattern between genotypes with the same name as well as in different geographic regions	Rahimi *et al*., 2006
DAMP and RAPD; 5 DAMP and 21 RAPD	Western Himalayas of India	49 wild genotypes	RAPD: 445 bands, 421 polymorphic; DAMP: 144 bands, 140 polymorphic;	Evaluation of genetic diversity and comparison of RAPD and DAMP markers	DAMP markers were more suitable for genetic diversity studies in pomegranate; wild pomegranate resources had high potential for utilization in breeding programmes	Narzary *et al*., 2009

Continued

Table 4.2. Continued

Type of molecular markers and number of primers	Place of sampling	Number of samples	Number of polymorphic and monomorphic fragments	The main objective of study	The main findings	Reference
RFLP; Synthetic oligonucleotide primers for amplification of 18S-28S rDNA intergenic spacer	Spain	10 specimens	18S-28S rDNA: a 750bp that digestion with HhaI yielded a total of 12 fragments from which eight were polymorphic	Evaluation of genetic diversity and comparison of morphological and molecular markers	Low correlation was recorded between morphological and genetic traits of pomegranate	Melgarejo *et al.*, 2009
RAPD; 14 primers out of 29 screened primers	Yazd collection, Iran	21 soft-seed genotypes	146 bands; 43 polymorphic	Evaluation of genetic structure of soft-seed genotypes and comparison with morphological attributes	Data from fruit characteristics and RAPD markers were complementary for genetic discrimination in soft-seed pomegranate accessions	Sarkhosh *et al.*, 2009
RAPD (26 primers) and morphological attributes	Iran	39 samples; two parents along with their 37 F1 progenies	From 325 fragments, 70 were polymorphic	Evaluation of RAPD markers for identification of hybrid progenies	Most of the progenies had similar banding pattern to their maternal parent	Zamani *et al.*, 2010
SSR: designing of 11 primer pairs	Tunisia	27 accessions	25 alleles (1–4 alleles per locus)	Development of new SSR markers for pomegranate	11 new SSR loci were reported	Hasnaoui *et al.*, 2010

Continued

Table 4.2. Continued

Type of molecular markers and number of primers	Place of sampling	Number of samples	Number of polymorphic and monomorphic fragments	The main objective of study	The main findings	Reference
Chloroplast SSR; 16 out of 25 loci were polymorphic	Iran	51 cultivars	39 alleles	Genetic diversity assessment	Chloroplast microsatellite markers were useful tools to study the genetic diversity of pomegranate genotypes; significant differences among groups of genotypes were according to cultivation types and geographical regions	Norouzi *et al.*, 2012
SSR: 59 EST-SSR were designed; 18 loci tested from which 12 were polymorphic	China; east–central	42 accessions	45 alleles (2–5 alleles per locus on average 2.8 alleles per locus)	EST SSR development and genetic diversity evaluation	Distinctive genetic background of the accessions from east–central China	Jian *et al.*, 2012
SRAP: 13 primer combinations	Iran, five regions	63 wild, and ornamental pomegranate genotypes	250 bands, 133 polymorphic	Genetic diversity assessment	SRAP markers could be powerful tools and an effective marker system for determining the genetic diversity and population genetic structure of the pomegranate	Soleimani *et al.*, 2012

Continued

Table 4.2. Continued

Type of molecular markers and number of primers	Place of sampling	Number of samples	Number of polymorphic and monomorphic fragments	The main objective of study	The main findings	Reference
AFLP: 26 primer combinations	Iran	two parents and their 62 F1 progenies	216 markers; 114 markers were scored in 'Malas Danesiyah Esfahani' and 130 in 'Bihaste Danesefid Ravar'	Construction of genetic linkage map	A consensus map constructed using 67 AFLP markers covered 334.5 cM and eight marker groups	Sarkhosh *et al*., 2012
RAMP; 14 pairs of primers	Seven provinces of China	46 pomegranate genotypes	127 bands, 113 polymorphic	Evaluation of genetic diversity and comparison of molecular markers and morphological traits	RAMP detected high level of genetic diversity in Chinese pomegranates; molecular classification was not consistent with the morphological/ agronomical classification or geographic origin	Zhao *et al*., 2013

Continued

Table 4.2. Continued

Type of molecular markers and number of primers	Place of sampling	Number of samples	Number of polymorphic and monomorphic fragments	The main objective of study	The main findings	Reference
Different molecular markers including 10 RAPD; 4.8S rRNA, ITS-1 and ITS-2 gene regions and morpho-physiological	India	13 samples	RAPD: 119 fragments 92.44% polymorphism	Evaluation of genetic diversity and barcode system efficiency	4.8S rRNA gene region was found to be highly conserved (99.34%) followed by ITS-1 (96.58%) and ITS-2 (89.21%); ITS rDNA region suggested as a reliable indicator of phylogenetic interrelationships, especially ITS-2 as probable DNA barcode at higher levels	Singh *et al.*, 2013
Two chloroplastic regions, petA–psaJ, trnC–trnD and four DNA barcodes (trnH–psbA, ITS, rbcL, matK)	Iran	18 genotypes	psbE–petL in petA–psaJ region revealed 1300 nucleotides with 4.29% genetic diversity among genotypes	Evaluation of intra-species relationship in pomegranate	Two regions, psbE–petL and trnH–psbA, were more suitable for determining intra-species relationships of pomegranate	Hajiahmadi *et al.*, 2013

Continued

Table 4.2. Continued

Type of molecular markers and number of primers	Place of sampling	Number of samples	Number of polymorphic and monomorphic fragments	The main objective of study	The main findings	Reference
18 RAPD primers and 26 morphology attributes	Iran, northern part	50 samples: including 37 wild and 12 cultivated genotypes and one miniature pomegranate	229 fragments 174 polymorphic	Evaluation of genetic diversity in wild pomegranates of Iran and comparison with cultivated ones	High molecular and morphological variability indicated that this germplasm includes rich and valuable plant materials for pomegranate breeding; poor correlation between RAPD and morphological traits	Zamani *et al*., 2013
SNP: 480 SNPs	Israel; Agricultural Research Organization (ARO) germplasm including international accessions	105 accessions	480 SNP markers	SNP and SSR discovery	Only 10.7% of the SNPs showed minor allele frequencies indicating pomegranate genome is highly diverse; China and Iran, composed of mainly unknown country origin pomegranate	Ophir *et al*., 2014

Continued

Table 4.2. Continued

Type of molecular markers and number of primers	Place of sampling	Number of samples	Number of polymorphic and monomorphic fragments	The main objective of study	The main findings	Reference
SSR: 171 loci detected, out of which 167 were polymorphic	India; germplasm collection of Indian Institute of Horticultural Research, Bengaluru	12 genotypes	951 alleles (1–14 alleles per locus)	Mining and characterization of SSRs from pomegranate by pyrosequencing	Shotgun 454 sequencing provided a cost-effective and efficient method for development of SSR markers in pomegranate; 108 markers have PIC values more than 0.5 and are useful for genetic diversity studies	Ravishankar *et al.*, 2015
SNP: 1092 and 7 quantitative traits	Israel	Two cultivars and their 76 F2 progenies	-	Construction of genetic map of pomegranate based on transcript markers enriched with QTLs for fruit quality traits	A novel genetic map of pomegranate enriched with quantitative trait loci (QTLs) for seven traits was constructed	Harel-Beja *et al.*, 2015
SSR: 16 loci	Iran, Fars province	50 genotypes from five populations	47	Genetic diversity and population structure assessment	A relatively high level of genetic admixture was found among accessions from different regions, suggesting that there is a high level of genetic exchange between individual genotypes	Zarei and Sahraroo, 2018

et al., 2010, Zamani *et al.*, 2013; Sarkhosh *et al.*, 2009; Zarei *et al.*, 2009; Adivapp *et al.*, 2010; Narzary *et al.*, 2010; Ercisli *et al.*, 2011; Noormohammadi *et al.*, 2012; Zhang *et al.*, 2012).

RAPD molecular markers also have been used successfully for other purposes. For example, Akbar *et al.* (2008) examined RAPD marker efficiency to examine pollination type of pomegranate cv. 'Malase-Tourshe-Saveh' and reported that offspring banding patterns were the same as their parents and attributed their results to the self-pollinating nature of this fruit tree. However, this is not a definite result and to confirm it the use of controlled pollination as well as more accurate molecular markers such as SSRs are needed. In addition, Zamani *et al.* (2010) used RAPD markers to evaluate offspring derived from different pollination systems of pomegranate cv. 'Malase-Tourshe-Saveh' and observed that most of the offspring had a similar banding pattern to their maternal plant and reported that cytoplasmic inheritance might play an important role in pomegranate crossing. Moreover, RAPD markers were used to compare the drought stress tolerance of 'Ganesh' and 'Nana' cultivars as well as their progenies with various degrees of water stress tolerance (Adivapp *et al.*, 2010). However, the authors did not observe any distinct banding pattern between the parents or the bulk of the offspring (Adivapp *et al.*, 2010). The RAPD molecular system has proven its usefulness for evaluation of genetic structure among Iranian soft-seed pomegranates (Sarkhosh *et al.*, 2009) as well as for the wild types of this fruit species (Ercisli *et al.*, 2011; Zamani *et al.*, 2013). Kanwar *et al.* (2010a) used RAPD markers to compare *in vitro* culture-raised plants from cotyledonary callus with their wild types and observed 26% dissimilarity between these plants; the authors attributed these genetic differences to the somaclonal variation under *in vitro* conditions.

Other molecular markers including ISSRs (Ghobadi *et al.*, 2006; Narzary *et al.*, 2010; Talebi Bedaf *et al.*, 2011; Noormohammadi *et al.*, 2012; Hajiyeva *et al.*, 2018), direct amplification of minisatellite DNA (DAMD) (Narzary *et al.*, 2009), single primer amplification reaction (SPAR) (Soleimani *et al.*, 2012), inverse sequence tagged repeat (ISTRs), sequence-related amplified polymorphism (SRAP) (Amar and El-Zayat, 2017), random amplified microsatellite polymorphism (RAMP) (Zhao *et al.*, 2013) and AFLP (Rahimi *et al.*, 2006; Yuan *et al.*, 2007; Sarkhosh *et al.*, 2011; Ajal *et al.*, 2015) also have been used successfully to evaluate genetic diversity among different pomegranate genotypes. In addition, AFLP markers have also been used to study the geographic and host-associated population variations of carob moth in Iranian pomegranate genotypes (Mozaffarian *et al.*, 2008).

Different combinations of markers also have been used in parallel to investigate their efficiency and to find their association. Narzary *et al.* (2010) evaluated the efficiency of three types of DNA-based molecular markers for assessing genetic diversity among wild pomegranate samples and identified that DAMD markers are more powerful than ISSR and RAPD. Melgarejo *et al.*, (2009) used 18S–28S rDNA intergenic spacer-RFLP to evaluate ten pomegranate accessions and reported that this method is efficient for cultivar identification and all of the accessions were differentiated according to their genetic profiles. However, a low correlation was observed between morphological attributes and genetic profiles. Genetic relationships among pomegranates of Taiwan and their relatedness in seedlings and cuttings introduced from China, the USA and Singapore was evaluated using internal transcribed spacer (ITS) sequences of rDNA, ISSR and RAPD markers (Wang *et al.*, 2007). The authors reported that 4.8S rRNA and ITS sequences showed no differences among eight samples of five lines. In addition, they observed that ISSR clustering was in accordance with the origin of the individual pomegranate lines and reported that the (GA)n dinucleotide motif sequence presented a major part of the pomegranate genome through ISSR analysis (Wang *et al.*, 2007). Different phylogenetic studies have been carried out to distinguish *Punica* from other genera and using different plant barcoding systems including rbcL sequence (Conti *et al.*, 1993, 1996, 1997), ITS sequences of nuclear rDNA (Shi *et al.*, 2000), the rbcL gene, the psaA-ycf3 spacer and the nuclear ITS regions including 4.8S ribosomal gene (Huang *et al.*, 2002). *Punica granatum* was previously classified in the Punicaceae family and was considered as the only genus in this family. However, recent phylogenic studies placed this fruit species in the Lythraceae (Teixeira da Silva *et al.*, 2013). *Punica protopunica*, a close relative of *P. granatum*, is more similar to a Lythraceae taxon in some characteristics. However, there is still no clear resolution of the taxonomic status of the genus *Punica* and the currently accepted taxonomy is according to that of the APG-III, whereby *Punica* is considered as a genus included in family Lythraceae (Angiosperm Phylogeny Group, 2009). Different barcode systems including 4.8S rRNA, ITS1 and ITS2 have been tested for evaluation of genetic variability in pomegranate genotypes, and the ITS regions, especially ITS2, were suggested as the best reliable barcode systems for phylogenetic evaluation of pomegranate (Singh *et al.*, 2013). In addition, results obtained from evaluation of two chloroplastic regions (petA–psaJ and trnC–trnD) and four nuclear DNA regions (trnH–psbA, ITS, rbcL, matK) as the plant barcode systems for evaluation of genetic variability in pomegranate genotypes indicated that psbE–petL and trnH–psbA were more suitable for determining intra-species relationships of pomegranate (Hajiahmadi *et al.*, 2013).

Molecular marker systems also have been used successfully to create the linkage map of pomegranate. The first linkage map of pomegranate was constructed using 62 F_1 offspring from a cross between a commercial pomegranate ('Malas Danesiyah Esfahani') and a soft-seed genotype ('Bihaste Danesefid Ravar') using AFLP markers (Sarkhosh *et al.*, 2012). These authors obtained 216 useful markers from 26 primer combinations that covered 434.3 centiMorgans (cM) of the genome in the 'Malas Danesiyah Esfahani' which included seven major groups and five minor groups, with a mean map distance between adjacent markers of 6.5 cM. In 'Bihaste Danesefid Ravar' seven major linkage groups and three minor groups were constructed with a mean map distance between adjacent markers of 4.21 cM, which covered 344.2 cM of the genome. A consensus map constructed using 67 AFLP markers covered 334.5 cM and eight marker groups. Another genetic map was also constructed by quantitative trait loci (QTLs) of seven traits and 1092 NP markers using a population of 76 F_2 progeny derived from the crossing of 'Nana' and 'Black' pomegranates (Harel-Beja *et al.*, 2015). This map covers 1141 cM with an average of 1.17 cM between markers over 11 linkage groups. Twenty-five QTLs were reported for plant height as well as fruit characteristics including TSS, seed hardness, fruit and aril weight, fruit perimeter and aril colour.

Microsatellites or SSRs have become the markers of choice for genetic fingerprinting studies due to their high levels of allelic variation, high reproducibility, their co-dominant nature and high information content, which allows them to deliver more information per unit assay than other marker systems (Madhou *et al.*, 2013). This type of molecular marker has been shown to be an abundant source of polymorphism in eukaryotic genomes and could be a useful tool for genetic diversity studies especially in species such as pomegranate that do not show a high level of diversity using other marker systems. Koohi-Dehkordi *et al.* (2007) made the first attempt to isolate and identify SSR markers from pomegranate. These authors isolated five microsatellites from Iranian pomegranate cv. 'Malase-Yazdi'. However, only one of these five loci was polymorphic among different pomegranate accessions (Zarei *et al.*, 2009). During the past decade a set of SSR markers has been developed for *P. granatum* in different pomegranate producer regions and used for understanding the genetic identification and structure of this genus (Hasnaoui *et al.*, 2010; Currò *et al.*, 2010; Pirseyedi *et al.*, 2010; Soriano *et al.*, 2011; Kazemialamuti *et al.*, 2012; Ravishankar *et al.*, 2015). These SSRs have been successfully used for evaluation of genetic diversity in Tunisia and Italy (Hasnaoui *et al.*, 2012), Iran (Pirseyedi *et al.*, 2010; Basaki *et al.*, 2011; Basaki *et al.*, 2013; Norouzi *et al.*, 2012; Noormohammadi *et al.*, 2012; Zarei and Sahraroo, 2018), Israel and Italy (Ferrara *et al.*, 2014), Tunisia (Jbir *et al.*, 2012), Pakistan (Nafees *et al.*, 2015), Kurdistan (Sanjar, 2015), as well as in Italian, Spanish and Turkish pomegranate genotypes (Currò *et al.*, 2010). Results of SSR analysis in different pomegranate genotypes indicated that the average number of alleles per locus was about two or three, which is similar to other self-pollinated plants such as peach (Aranzana *et al.*, 2002; Bouhadida *et al.*, 2007) and much lower than cross-pollinating plants such as walnut (Pollegioni *et al.*, 2011;

Ebrahimi *et al.*, 2016; Ebrahimi *et al.*, 2017) and apricot (Wang *et al.*, 2014). In addition, the efficiency of universal chloroplastic microsatellite loci was proven in 52 Iranian pomegranate genotypes (Norouzi *et al.*, 2012). Several studies have also been conducted to estimate the association between SSR marker loci and phenotypic traits, and some loci were reported to have a strong association with seed softness, aril size, fruit weight, TA and bacterial blight (Basaki *et al.*, 2011; Basaki *et al.*, 2013; Singh *et al.*, 2015).

Recent progress in *in silico* approaches will enable researchers to retrieve expressed sequence tag (EST) sequences from the National Center for Biotechnology Information (NCBI) and develop EST-SSRs or gene-based SSRs. This method for the development of EST-SSR markers is more convenient and they can be isolated with higher efficiency and at a lower expense than genomic sequence-derived SSRs (Wang *et al.*, 2012). In addition, microsatellites that are present in the coding regions are more conserved and possess high reproducibility and higher interspecific/intraspecific transferability than SSRs in non-coding regions (Haq *et al.*, 2014). In recent years, different transcriptomic studies have been carried out on *P. granatum* and some EST sequences from different tissues of this fruit tree have been deposited in the public databases. EST sequences of pomegranate were explored and a set of EST-SSRs was reported in pomegranate (Jian *et al.*, 2012; Zarei and Ebrahimi, 2017). Results of data mining revealed that dinucleotide repeats were the most abundant sequences in pomegranate followed by trinucleotide and tetranucleotide repeats, respectively. Among the dinucleotide repeats, AG/GA/CT/TC were reported as the most frequent ones. Thanks to the comprehensive transcriptomics data on pomegranate, several thousands of SNP markers have been identified in this species (Ophir *et al.*, 2014).

Generally, results of different molecular markers studies of pomegranate germplasms have indicated lower genetic differentiation among commercial genotypes, but when the wild accessions were included in the study relatively higher genetic differentiations were reported. In addition, most researchers observed some cases of homonymous and synonymous genotypes in their investigations. In the majority of the studies, which used pomological features in combination with molecular ones, relatively low correlations were reported between these two markers. However, where tree characteristics were included in the morphological studies, relatively higher correlation was reported. In addition, in some cases grouping of accessions in the cluster analyses were in accordance with the geographic regions of sampling. Therefore, it is probable that the geographic region somehow will affect the genome rearrangement.

All the authors of reported morphological studies on pomegranate cultivars have a broad consensus on the extensive diversity in pomegranate accessions. However, in the case of molecular markers, there are conflicting reports about genetic diversity in this fruit species. While most of the reports represented the existence of a high level of diversity, a few authors had the opposite opinion and reported relatively low genetic diversity among the pomegranate genotypes (Talebi Bedaf *et al.*, 2003; Ghobadi *et al.*, 2006; Rahimi *et al.*, 2006; Hasnaoui *et al.*, 2012). The similarity coefficients that were recorded between evaluated accessions in the latter two studies were in the range of 0.29–0.93, which implies relatively moderate level of genetic diversity. Such statements about pomegranate diversity could be due to the selection of unsuitable plant materials, but also may be attributed to the misinterpreting of the results. Even these low coefficients (0.29) are the same similarity coefficients that were observed among some different species. Moreover, results from all of the morphological studies have indicated high differentiation between evaluated germplasms. However, the phenotypical differences should result from genetic differences. Pomegranate is a vegetatively propagated species and most of the cultivars are recognized to be mostly self-pollinated. Hence, it is reasonable to expect a lower degree of genetic differentiation among the specimens of this type of species than those tree species that are cross-pollinated and seedling-propagated such as walnut. Therefore, by considering all reports it could be concluded that plant material sampling, as well as unsuitable primer types may explain these results. It is probable that diversity in cultivated pomegranates has not been greatly expanded due to their method of propagation, but this is not true of the wild types (as shown in some research), which are not affected by human disturbance.

4.7 Marker-Assisted Selection

Different studies have also been conducted on pomegranate with the aim of evaluating the marker–trait association and identifying informative markers. The results of such experiments would be very important for the utilization of molecular markers in breeding programmes and marker-assisted selection at the early stages of breeding programmes. The results of evaluating the association between RAPD markers and fruit attributes in the progenies of pomegranate cv. 'Malase-Tourshe-Saveh' revealed significant linkage of seed softness with two markers, and each of the fruit juice TSS/TA and peel thickness traits with six molecular markers (Zamani *et al.*, 2010). Singh *et al.* (2015) used 44 SSR loci and fruit characteristics on 88 pomegranate accessions of wild-type and cultivated varieties and identified four linked markers for fruit weight, TA and bacterial blight severity. They sequenced the linked markers and observed that the bacterial linked marker is a hypothetical protein (GB no. FN677540) of *Prunus persica*. SSR loci associated with calyx height, fruit diameter, calyx shape, fruit size, seed softness and aril size have also been reported in pomegranate (Basaki *et al.*, 2011; Basaki *et al.*, 2013). In addition, as was stated before, two linkage maps have been constructed in *P. granatum* using AFLP (Sarkhosh *et al.*, 2012) and SNP (Harel-Beja *et al.*, 2015) molecular markers. Although high associations between several QTLs and SNP markers were reported in pomegranate, lack of genome sequences (Harel-Beja *et al.*, 2015) as well as phenotypic data (Sarkhosh *et al.*, 2012) postponed the utilization of these linkage maps for early selection in pomegranate breeding programmes. Therefore, the availability of high-resolution genome sequences would enable researchers to anchor these data on the genomic sequences and highly facilitate the identification and characterization of respective genes.

4.8 Genome Size and Genomics

The genome of pomegranate cv. 'Dabenzi' has been sequenced at 100X coverage using Illumina paired-end reads of libraries with insert sizes of 170 bp to 40 kb (Qin *et al.*, 2017). The final genome size assembled in this method is reported to be 328.38 Mb (about 92% of that estimated using the flow cytometry method). Also the chromosome numbers are reported to be 2n = 2x = 18 for this cultivar. According to homology-based analysis, 154.3 Mb of repetitive sequences (46.1% of the pomegranate assembly) were identified. Based on the analysis of repetitive elements it was found that retrotransposons comprise 40.5% of the total genome sequences. Among different types of retrotransposons, long terminal repeats (LTRs) comprised 18.9% and non-autonomous LTRs comprised 19.8% of the pomegranate genome, respectively. The *Gypsy* type comprises 9.8% and the *Copia* type 4.8% of pomegranate's LTR retrotransposons. Moreover, it was estimated that *Copia* elements have a recent or ongoing wave of retrotransposition. According to transcriptomics analysis of different tissues, 29,229 protein-coding gene models were annotated with the average gene length of 2574.61 bp. Clustering of 29,229 annotated genes based on the sequence similarity yielded 13,640 gene families and 6230 orphan genes that could not be clustered with any genes from similar plants. Gene ontology (GO) annotation and KEGG pathway enrichment analysis for orphan genes revealed that the majority of these genes are involved in the secondary metabolite and phenylpropanoid pathways. Polyphenolic compounds, which are produced from the phenylpropanoid pathway, are the main secondary metabolites in pomegranate and are responsible for the health beneficial properties of this fruit. Therefore, some of these orphan genes might be responsible for the biosynthesis of these secondary compounds and could ultimately have contributed to the high adaptation of this fruit species. However, to complete the available data about pomegranate genome sequences, it is necessary to retrieve and compare sequences from various cultivars and use different functional genomics tools to validate the existing sequences and assign a function to each section of the genome.

Another report about the genomics of pomegranate is presented by Luo *et al.* (2020) conducted on 'Tunisia' cultivar (probably originated in Tunisia but cultivated in China), which is a soft-seed pomegranate. The main purpose of this work was dissecting the genetic divergence between soft-seeded ('Tunisia') and hard-seeded

('Taishanhong' and 'Dabenzi') cultivars. The Pacific Biosciences (PacBio) Sequel platform was used for the single molecule real-time (SMRT) sequencing and a total of 20.94 Gb PacBio long reads were used, resulting in a 320.31 Mb assembly, being close to the estimated size based on flow cytometry. In this work the chromosome numbers for 'Tunisia' were reported to be 2n = 2x = 16.

About 33,600 coding genes were predicted in the 'Tunisia' genome, with an average coding sequence length of 2300 bp. Approximately 70% (23,357) of the genes were supported by transcriptome profiling using Illumina and SMRT-based RNA-seq data. Approximately 51% of the 'Tunisia' assembly sequences were identified as repetitive elements, including retrotransposons and DNA transposons, showing a higher proportion of repeats in the 'Tunisia' genome than in 'Dabenzi' and 'Taishanhong'.

A genetic diversity evaluation of 26 pomegranate varieties (by sequence coverage of 97.91%), with various geographical origins and differences in seed hardness trait divided them into three main clusters, one for the hard-seeded ones and the other two for soft and semi-soft-seeded ones. Sucrose transport protein expression evaluated by seed transcriptome sequencing revealed that two genes (*SUC8-like* and *SUC6*) are important for controlling seed development and may be related to differences in seed hardness between varieties. They concluded that seed hardness genes differ from those related to cold tolerance, implying that it is possible to breed new pomegranate cultivars that are both freezing tolerant and soft-seeded using artificial hybridizations or genetic manipulations.

According to the Azerbaijan Institute of Botany (2018), the nucleus genome of 'Azerbaijan Guloyshe' pomegranate has been sequenced and results were deposited at NCBI. However, no more details are provided in the literature.

Yuan *et al.* (2018) reported a 274-Mb high-quality draft pomegranate genome sequence, covering approximately 81.5% of the estimated 336 Mb genome, containing 30,903 genes.

Based on phylogenomic analysis they concluded that this fruit tree belongs to the Lythraceae family rather than the monogeneric Punicaceae family. Also, they proved by comparative analyses that pomegranate and *Eucalyptus grandis* shared the paleotetraploidy event. Based on integrated genomic and transcriptomic analyses, they provided insights into the molecular mechanisms underlying the biosynthesis of ellagitannin-based compounds, the colour formation in both peels and arils during pomegranate fruit development, and the unique ovule development processes that are the main characteristics of pomegranate.

Chloroplast (cp) genome of three different pomegranate cultivars has been sequenced by Yan *et al.* (2019) using a genome skimming approach. The cp genomes displayed the typical quadripartite structure of angiosperms, and their length ranged from 156,638 to 156,639 bp, encoding 113 unique genes with 17 duplicated in the inverted regions. They checked the sequence diversity and concluded that it was extremely low with no informative sites, suggesting that pomegranate cp genome sequences may not be suitable for investigating the genetic diversity of pomegranate genotypes. Further, they analysed the codon usage pattern and identified the potential RNA editing sites. Comparative cp genome analysis with other species within Lythraceae revealed that the gene content and organization were highly conserved.

Based on the complete chloroplast genomes of the order Myrtales that were previously released, and phylogenetic analysis, *P. granatum* formed a single clade with other species from Lythraceae with a high supporting value.

4.9 Breeding Criteria

Generally, breeding objectives in fruit trees can be divided into two sections: the rootstock and the scion. Although pomegranates are propagated by stem cutting and grafting is not practised, using grafted pomegranate trees might gain importance in the future for specific purposes. Input traits and output traits are two classes of traits with commercial significance (Mou and Scorza, 2011). Input traits are those that are related to the sustainability of yield and productivity and are highly important for producers, growers and handlers. These traits included tree yield, tolerance to unfavourable conditions, as well as disease and pest resistance. The output traits are mostly focused on the fruit quality and its nutritional properties, which are of primary importance to consumers, though the producers participate in their benefits as well. So far, most of the objectives in pomegranate breeding have been dictated by indigenous markets and included improving fruit appearance, organoleptic quality, seed softness, juice content and tree yield. However, as a prerequisite for inclusion in international markets, as well as consumer acceptance, criteria such as colour of peel and arils, fruit size and shape, easily separable arils, volume and taste of juice, appropriate ripening time, as well as good postharvest quality and shelf-life and fruit handling should be considered. In addition, pomegranate is gaining more attention and becoming a desired fruit across the world for its pharmacological properties, hence, increasing the nutritional contents as well as their beneficial bioactive compounds should be considered in the breeding programmes. Moreover, breeding against physiological disorders including fruit cracking, sunburn, internal break down of arils or aril browning as well as tolerance to unfavourable conditions such as water deficiency, salinity, frost and hot, dry conditions are among the input traits that are receiving more attention. Region-specific breeding objectives such as resistance to the pomegranate fruit moth in Iran as well as bacterial blight in India need to be considered by breeders.

Attractive colour of fruit skin and arils are among the most important sensory characteristics of pomegranate. Red and dark red colours are desirable for fresh consumption and export purposes in most countries. In the majority of commercial pomegranate cultivars, red rind and aril colour, sweet-sour taste and medium fruit size (250–350 g) are alike. 'Malase-Tourshe-Saveh', 'Rabab', 'Shishe-Kepe-Ferdous', 'Malase-Yazdi' and 'Bejestani' are among the well-known pomegranate cultivars in Iran with the aforementioned characteristics. Taste is an important sensory attribute of pomegranate, which highly affects consumer acceptance. Pomegranates are divided into three main groups based on their taste: sweet (low acidic), sweet-sour (intermediate acidic) and sour (highly acidic). The sweet-sour taste, with a range of sweeter to sour tastes, is preferred by consumers to the other two types in many countries and should be considered in the breeding programmes.

Fruit size in pomegranate varies greatly (from <100 g to >600 g) depending on cultivars and growing conditions. However, medium-sized fruits with the average weight of 250–350 g are more desirable and should be considered for commercial cultivars. Thick peeled fruits may be preferred due to their higher tolerance to cracking in the orchard and damage during transportation. Increasing the storage and shelf-life of fruit is also desirable, and can be attained through breeding for cultivars with fruits to be conserved in lower temperatures during the storage period. Tree yield is an important input trait that should be considered in the breeding programmes. Extending the ripening season to produce very early up to very late ripening pomegranate cultivars is one of the main objectives in the countries that do not have such genotypes. Types of consumption, fresh or processed, also affect the breeding programmes. According to Feng *et al.* (2006) pomegranate cultivars in China are classified into four groups according to their consumption types: fresh consumption, processing, decorative, and both decorative and consumption purposes. For example, colour of aril juice as well as seed softness would be of great importance for canned pomegranate arils. In addition, highly productive trees with thornless strong branches are preferred.

4.9.1 Other breeding objectives related to scion characters

In scion breeding, breeders should also consider seed softness, fruit cracking, pest and disease resistance, freezing and frost tolerance or cold hardiness, juice content and sunburn tolerance. Seed softness is a desirable trait with high commercial, nutritional and health benefit importance. There are several soft-seed cultivars in pomegranate producer regions. At least 21 soft-seed accessions have been collected in Iran's national pomegranate collection in Yazd city (Sarkhosh *et al.*, 2008, 2011). Soft- and semi-soft-seeded cultivars showed more common fruit attributes such as yellow colour for fruit and aril, sweet taste or low acidity, low fruit yield per tree, low fruit weight, high aril juice and large arils than the hard-seeded cultivars. Therefore, these types of cultivars were suggested to be used for breeding soft-seeded cultivars.

Resistance to pests and diseases are among the factors that should be considered in pomegranate breeding projects. Pomegranate is attacked by about 45 species of insects and fruit is most vulnerable to the attack of the pest (Mir *et al.*, 2012). Pomegranate production in Iran is severely hampered by pomegranate fruit moth (carob moth), *Ectomyelois ceratoniae* (Lep: Pyralidae). It is the main pest of Iran's pomegranate orchards and also a major constraint for other pomegranate producing countries including Turkey, Tunisia, Iraq and Algeria (Azqandi *et al.*, 2015; Demirel, 2016). The pest larva highly reduces the fruit quality by entering the fruit from inside the crown leading to subsequent fungus contamination of the damaged parts (Fig. 4.5). It is estimated that about 30% of annual pomegranate production in Iran is damaged by this pest (Azqandi *et al.*, 2015). Due to the egg placement habit of this moth on the pomegranate fruit stamens inside the crown as well as the usual hanging of pomegranate fruit on the branches, chemical control of this pest is very hard and most methods of controlling this pest have failed. However, gathering and destruction of leftover infected fruits as well as proper fertilization can reduce the damage for the upcoming season (Olyaie Torshiz *et al.*, 2017). Fruit sensitivity to the pomegranate fruit moth also differs among various genotypes and black-skinned fruits or those containing thick peel are less infested by the pest. High contents of polyphenolic compounds in the skin of black-skinned fruits and thick and rough fruit peel, which can hamper the larva's movement inside the fruit, may be among the affecting factors for lower infestation of these types of fruits. In addition, because of the closed nature of the fruit crown in some genotypes, the fruit borer cannot oviposit inside the crown; hence these types of genotypes are relatively more resistant to the pest. Therefore, regarding carob moth resistance, a closed fruit crown could be considered as a suitable trait for transferring to new commercial cultivars. Reduced damage by carob moth was obtained by rendering male moths infertile through gamma irradiation (120 and 160 Gray) and subsequent release of the infertile males into the orchards to reduce hatching (Zolfagharieh *et al.*, 2008). The fruit borer (*Virachola isocrates*) is considered as the most important pest of pomegranate in India, which causes two-fifths of the damage to the fruit (Kakar *et al.*, 1987). *Aphis punicae* Passerini is also among the insects that highly affect the leaves, twigs and blossom of pomegranate mainly during April to early July. Gene transferring technology can assist breeders to produce insect-resistant pomegranate through introducing pest-resistance genes such as those obtained from *Bacillus thuringiensis*, the *cry* gene, to the commercially susceptible pomegranate cultivars.

Sunburn of fruit skin is another environmentally caused limiting factor resulting in serious economic losses to pomegranate producers all around the world. Sunburn peel damage appears in the form of large brown to black spots on the fruit skin, which decreases marketability as well as juiciness of the arils under the affected region (Fig. 4.6) (Hosein-Beigi *et al.*, 2019). Some of the pomegranate cultivars are sensitive to intense sunlight. Shading and spraying with kaolin are among the horticultural practices that have been employed to combat this disorder (Melgarejo *et al.*, 2004; Parashar and Ansari, 2012; Ghorbani *et al.*, 2015; Olyaie Torshiz *et al.*, 2017). However, having a high-density canopy can reduce the direct exposure of fruit to sunlight. Fruit cracking or splitting is another pomegranate disorder adversely affecting pomegranate production. According to available records, sunburn and cracking damages may account for losses of up to 40–50% of the total harvest in pomegranate production areas (Yazici and Ercişli, 2017). This disorder not only reduces the marketability and consumer acceptance, but also provides easy entry of insects, fungi, yeasts and bacteria to the fruit and makes them more susceptible to environmental stresses. Imbalanced irrigation, high fluctuation in day and night temperatures, low organic soils, mineral deficiency, rain at harvest time, air humidity as well as rind moisture and flexibility are among the factors that will have a profound effect on pomegranate fruit cracking. Cracking and sunburn in pomegranate fruits may emerge separately. However, it is reported that the ratio of cracking increases in fruits that are subjected to sunburn (Yazici and Ercişli, 2017; Hosein-Beigi *et al.*, 2019). This is due to the effects of sunburn, which reduces the fruit skin moisture ultimately resulting in lower pliability of the skin and subsequent cracking. The likelihood of fruit cracking increases as the ripening goes on

Fig. 4.5. Carob moth larva (a), adult pest (b), infestation in the pomegranate flower (c), their effects on the fruit of pomegranate cv. Bajestani (d–f). (Photos: Abdolkarim Zarei.)

Fig. 4.6. Cracking and sunburn as two main physiological disorders in pomegranate cv. 'Malase-Torshe-Saveh'. (Photos: Abdolkarim Zarei.)

(with the progress of cold seasons); therefore, early ripening cultivars are less susceptible to this disorder than late ripening ones. In addition, genotypes containing thicker fruit peel are relatively more tolerant than those with thinner fruit skin. Therefore, breeding early ripening or thicker peel cultivars can be exploited to reduced fruit cracking.

Bacterial blight disease caused by *Xanthomonas axonopodis* pv. *punicae* is considered as one of the main constraints to pomegranate production in India, with 60–90% incidence in some regions (Hosamani *et al.*, 2016). The wild pomegranates in the northern parts of Iran are also affected by this disorder. The disease causes brown to black spots on the fruit, stem and leaves followed by subsequent defoliation and plant death in severe cases. Therefore, there have been several attempts to produce bacterial blight-resistant cultivars in humid regions (Jalikop *et al.*, 2006; Swetha, 2011; Madhvi, 2015; Hosamani, 2012; Hosamani *et al.*, 2016).

4.9.2 Criteria for rootstock improvement

Due to the effects of global environment changes in different parts of the world, the need for breeding cultivars tolerant to unfavourable conditions such as frost, heat, humidity, salinity and drought has gained more attention during the past decade in pomegranate-producing regions. In addition, the recent popularity of this fruit across the world has increased the demand for the cultivation of this fruit in new regions, which may not have the ideal conditions for pomegranate production. Therefore, pomegranate is grown on a wide range of climates and soils. These reasons indicate the need for breeding new cultivars with specific characteristics for each condition. Pomegranate rootstocks need to be bred for tolerance to water deficiency, salty, poor and calcareous soils, winter frost, noxious microorganisms (including bacterial blight), pests and diseases as well as wood borers. Controlling tree architecture would be of great importance for establishment of modern orchards with mechanized horticultural practices. Dwarf rootstocks might be desirable for easy management and mechanical harvesting.

From the perspective of plant architecture, sucker-free genotypes are preferred. Similar to other fruit species, the use of dwarf rootstocks is desirable for the establishment of compact orchards for easy handling of horticultural practices as well as yield improvement. Increasing tree tolerance to unfavourable conditions such as water deficiency, salty soils and water, hot, dry climate, frost hardiness, pest and disease resistance should be considered as region-specific breeding objectives.

4.10 Rootstock Studies

Although pomegranate is considered as moderately tolerant to salinity (Allen *et al.*, 1998), water scarcity and salinity are becoming severe problems in many pomegranate-growing regions. Therefore, several studies have been conducted for identification of salt-tolerant varieties that could be used as rootstock as well as the parent for hybridization. It is reported that a rooted cutting of 'Malas-e-Shirin' cultivar was tolerant to up to 40 mM of sodium chloride (Hasanpour *et al.*, 2014). Increase in salt concentration up to 30 mM increased leaf dry and fresh weight in 'Shishe-Kab' while decreasing these factors in the 'Rabab' pomegranate (Hasanpour, 2012). Although the higher concentration of salt (up to 60 mM) adversely affected both cultivars, 'Shishe-Kab' had higher K in the aerial parts of the plant and showed higher tolerance to salty conditions than 'Rabab' (Hasanpour, 2012). Salt stress up to 8.5 dS/m had no adverse effects on the yield of 'Shishe-Kab' pomegranate (Tavousi *et al.*, 2016) but water deficiency significantly reduced quantitative and qualitative characteristics in this cultivar. Therefore, the authors considered this cultivar as a salt-tolerant and drought-sensitive pomegranate. Salt tolerance of various pomegranate cultivars has been evaluated and two cultivars of 'Malas-e-Yazdi' and 'Tab-o-Larz' and two cultivars of 'Gabri' and 'Khafr-e-Jahrom' were reported as the most salinity-tolerant and salinity-sensitive cultivars, respectively (Okhovatian *et al.*, 2010). It is reported that salt-tolerant rootstock can significantly increase the salinity tolerance of sensitive scion cultivars either through restriction of the uptake or transport of Na and Cl ions from the root to the shoot or maintaining sufficient levels of K in the scion (Karimi and Hassanpour, 2017). These authors used two salt-resistant pomegranate cultivars of 'Tab-o-Larz' and 'Malas-e-Yazdi' as the rootstock for 'Gabri' (a salt-sensitive cultivar) and reported that Na and Cl concentrations decreased in the scion. There was a difference between the two rootstocks in terms of higher salt tolerance in the scion and it was obtained by using the 'Tab-o-Larz' rootstock (Karimi and Hassanpour, 2017).

Different combinations of the scion–rootstock were investigated to evaluate their effects on tree size, vigour and yield of pomegranate (Vazifeshenas *et al.*, 2009). It was reported that grafted plants had a higher yield than the own-rooted ones, and grafting led to lower vigour and tree size compared with the own-rooted ones (Vazifeshenas *et al.*, 2009). 'Torsh-Mamoly-Zabol' was reported as a rootstock that can be used for reducing tree size and height, as well as the fruit sunburn disorder; therefore might be a useful rootstock for the establishment of high-density orchards. Pomegranate cv. 'Golnar-Farsi' was also suggested to be used as rootstock for increasing fruit yield and decreasing sucker production (Vazifeshenas *et al.*, 2009).

Pomegranate drought stress tolerance has also been the subject of many research studies (Hasanpour, 2012; Hassani Moghadam *et al.*, 2014; Ebtadaei and Shekafande, 2016; Pourghayoumi *et al.*, 2017). Hassani Moghadam *et al.* (2014) evaluated the drought tolerance of six Iranian pomegranate cultivars at three water deficiency levels including

80%, 60% and 40% field capacity. High variation was reported among different cultivars and 'Rabab' and 'Malase-Yazd' were reported as the most tolerant while 'Nadery Badroud' was considered as the most sensitive cultivar to drought stress. Accordingly, 'Rabab' pomegranate showed higher tolerance to drought stress compared with 'Shishe Kep' (Ebtadaei and Shekafande, 2016). Results of another screening of some famous Iranian pomegranate cultivars against drought stress also indicated that 'Rabab' as well as 'Ghojagh' have good tolerance to water stress (Pourghayoumi *et al.*, 2017). A lower level of lipid peroxidation under drought stress and marked reduction in malondialdehyde concentration after rewatering were reported in these two cultivars. The up-regulation of cytosolic glutathione reductase and glutation peroxidase was associated with drought stress tolerance of these two cultivars. The high drought tolerance of 'Ghojagh' was attributed to efficient osmotic adjustment and for 'Rabab' higher antioxidant capacity and efficient reactive oxygen species (ROS) scavenging was suggested as the main factor for increasing drought tolerance (Pourghayoumi *et al.*, 2017). These authors reported that 'Malase-Yazdi' was the most sensitive cultivar to drought stress and cytosolic glutathione reductase was completely suppressed under severe water deficiency (Pourghayoumi *et al.*, 2017). Most of the investigations indicated that 'Rabab' is a drought hardy cultivar and may be considered for cultivation in regions that are facing water shortage as well as in breeding programmes aiming to increase drought tolerance.

Cold hardiness is another limiting factor restricting pomegranate cultivation. Pomegranate has considerable soil and climatic adaptations and withstands frosty conditions but will not survive long below −15°C (Ghasemi Soloklui *et al.*, 2012). In Iran pomegranate is traditionally grown on the margins of desert, where the winter temperature may plunge down to −20°C, causing serious damage to the tree. In fact, pomegranate cultivated in such regions (like central and northeastern parts of Iran) may encounter freezing damage almost every 10–15 years. The last pervasive winter freeze in the Iranian pomegranate orchards occurred in 2007, when −21°C temperatures lasted for 3 days and destroyed more than 8000 ha of pomegranate orchards in Saveh, Isfahan, Ghom, Karaj, Yazd, Khorasan and other pomegranate producer provinces. Moreover, freezing tolerance has been one of the main objectives of breeding programmes in Turkmenistan and Russia (Levin, 1979, Levin, 2006; Jalikop, 2011). Despite the high economic importance of freezing tolerance for pomegranate producers, there are few studies about monitoring and selection of pomegranate geneotypes for this desirable trait. Ghasemi Soloklui *et al.* (2012) studied winter freezing tolerance of seven Iranian commercial pomegranate cultivars at three different stages (autumn, midwinter and late winter) and reported that acclimation and deacclimation occurred at different times for each cultivar and this factor played a key role in freezing tolerance of different cultivars, especially in autumn and late winter. 'Poost Sefid Bafgh' was reported as having high cold tolerance early in autumn, but susceptible to the freezing that occurs during winter. 'Naderi', 'Yusef Khani', 'Malas Saveh' and 'Rabab' were reported as the highest midwinter cold hardy cultivars while 'Poost Sefid Bafgh' and 'Shishe Kap' were susceptible at this stage. The authors concluded that although an increase in proline content was observed during the period of cold hardening, there was a stronger correlation between LT50 and soluble carbohydrates compared with proline, particularly in the early stages of acclimation (Ghasemi Soloklui *et al.*, 2012). Freezing sensitivity is a cultivar-dependent factor, and it is stated that soft-seed cultivars are generally less hardy than hard-seed ones (Jalikop, 2011). Therefore, wider screening of existing genotypes against cold stress may result in the identification of freezing-tolerant genotypes for utilization in breeding programmes. In addition, modern breeding strategies could assist the breeders in breeding cold hardy cultivars. For instance, *in vitro* selection and mutagenesis have good capacity for this purpose. Finally, genetic transformation approaches have high potential to produce freezing-tolerant pomegranates through up-regulation of cold-related transcription factors such as CBF and DREB1, or through introducing antifreeze proteins.

4.11 Classical Breeding

4.11.1 Breeding programmes

Different breeding strategies have been used or have the capacity to be used for pomegranate improvement. As pomegranate accessions have a wide range of morphological diversity, most of the desirable traits that a plant breeder may look for are likely present in the existing germplasms. Therefore, selection and subsequent clonal propagation of genotypes with desirable characteristics are considered as one of the widely used strategies for improvement of this fruit species. In fact, most of the commercially cultivated cultivars grown in different pomegranate producer regions in various countries are the results of human selection. In addition, breeding new cultivars may be achieved both through classical approaches such as hybridization of appropriate genotypes followed by screening and selection among the hybrid progenies and/or backcrossing with the recipient parent, and or through modern breeding strategies including mutagenesis, *in vitro* selection, as well as genetic transformation.

4.11.2 Selection

Selection is the first and preliminary breeding strategy that has been used for the improvement of pomegranate in different regions. The majority of cultivated pomegranate cultivars are the result of human selections from naturally occurring genotypes. In nearly all of the pomegranate producer regions, most of the pomegranate cultivars were selected based on the demands of local consumers. The preferences for attributes were not the same in all pomegranate regions. Therefore, most of the cultivars that are extensively cultivated reflect the local priorities of each country or region. For example, the majority of the Iranian commercial cultivars are characterized by red colour of arils and rind, sweet-sour taste with an average weight of 250 g, while the traditional Indian and Spanish cultivars are characterized by soft seeds and low-acid taste (Raina, 2013). Recent awareness of the beneficial properties of pomegranate fruit has increased the world demand for this fruit species and hence elevated the economic importance of pomegranate for export. However, international markets constantly affect fruit selection criteria, which will have an increasing role in pomegranate breeding. In addition, in recent years, global warming has influenced the majority of the world regions, hence, synchronizing breeding programmes with new climate conditions would be of great importance for retaining the current production and increasing it to keep pace with the increasing trend of market demands. Climate change might necessitate farmers using plants that are more tolerant to abiotic stresses such as winter freezes, water deficiency, salty water and soils, high pH, heavy metals, as well as humidity.

4.11.3 Hybridization studies and progress

Breeding through hybridization enables the breeders to transfer desirable traits from one genotype to another and is one of the main approaches that can be used for the production of new cultivars. Pomegranate crossing studies have been started in different pomegranate producer regions with the aim of breeding new cultivars.

Breeding through crossing in pomegranate is a little easier than breeding of most other fruit crops. Pomegranate has relatively large flowers making it convenient for emasculation, pollen collection and hybridization (Fig. 4.7). In addition, removing of sepals during emasculation can serve as a good indicator for hybridized flowers and there is no need for bagging and caging. Moreover, the breeder is able to establish a large hybrid population as a hybridized fruit can produce abundant numbers of seeds with relatively high germination rate. The shorter juvenile phase of the tree compared with other tree crops also facilitates the evaluation of fruit characteristics in the hybrid progeny.

Like other plant species, hybridization in pomegranate involves the crossing of appropriate genotypes, creating large hybrid populations and evaluating their characteristics for identification of promising or superior hybrids. After performing necessary tests including uniformity, distinctness and stability, the superior

Fig. 4.7. Pollination method in pomegranate tree: (a) selection of appropriate flower bud for emasculation; (b) removal of calyx cup; (c) removal of stamens; (d) emasculated flower; (e) stored pollen grains in vial; (f) pollen application with camel's hair brush. (Photos: Donald Sangam.)

hybrids may be regarded as new cultivars by releasing for public cultivation. However, in most cases, the promising hybrids could be used as the plant materials for subsequent breeding programmes including backcross especially in the cases where wild types are involved in the crossing (a situation that usually occurs in the disease resistance programmes). The process of progeny selection (for desirable characters) after each backcross cycle of selected progenies with the recipient parent is repeated several times to eliminate the undesirable traits of the wild parent. Once suitable plants are obtained, they should be proliferated by vegetative propagation methods including cutting, layering and or micropropagation. Before release phase, the superior selections should be subjected to new cultivars tests, which include replicated trials (over times and over places) for evaluation of their stability, distinctness and uniformity, all of which are required for releasing new plant cultivars.

Various attempts have been made to breed new pomegranate varieties in recent decades. Pomegranate trees usually have three flowering peaks. This characteristic sometimes makes natural hybridization impossible. Therefore, pollen conservation might be needed for controlled pollination. Pollen viability of different pomegranate cultivars (as tested by acetocarmine) has been evaluated at different storage temperatures (−196°C, −20°C, 4°C and room temperature) (Raina, 2013). The maximum pollen viability (82.0%) in 'Ganesh' was obtained at −20°C followed by −196°C (69.5%), 4°C (56.1%) and room temperature (22.8%) for a 9-week storage period. The same trend was reported for pollen germination (on 10% sucrose medium) and the germination rate of 'Ganesh' pollen was 54%, 36%, 24% and 9% at −20°C, −196°C, 4°C and room temperature, respectively (Raina, 2013). Results from 50 Iranian pomegranate cultivars indicated that pomegranate pollen fertility varies between 83.0 and 99.3% (Sheidai *et al.*, 2012).

This fruit species is capable of both open and self-pollination and has been described as self-pollinated, cross-pollinated, highly cross-pollinated or often cross-pollinated. Fruit set was reported to be 79 and 43.3% for intact open and self-pollinated flowers and 26.4 and 66.2% for the same after emasculation (Karale *et al.*, 1993). It is reported that removing of the panicle bag after hybridization substantially increases fruit set percentage (Singh *et al.*, 1980). These authors also expressed that due to the small stigmatic surface, high fruit set (3.85%) resulted from the first deposited pollens and there is hardly any chance of contamination with foreign pollen. Hybridization studies revealed that some economic characteristics including fruit peel weight, fruit juice acidity, fruit weight, aril to fruit ratio, tree yield and number of fruits per tree showed high heritability and high genetic advance (Manohar *et al.*, 1981).

It has been reported that 'Sverkhrannil', a newly bred cultivar at Turkmenistan Academy of Agricultural Sciences, Garrygala (Turkmenistan), is very early ripening and has small seeds (Levin, 1990a). However, it seems that the most intensive breeding programmes with the objective of pomegranate improvement are underway in India. In this country, pomegranate cv. 'Ganesh' was extensively used as parent in combination with other cultivars including 'Bedana' and 'Jyoti' (characterized by brownish skin, medium-size fruit, sweet taste and soft seeds), 'Kabul' (characterized by yellow-red skin, large fruit, sweet taste and hard seeds), 'Nana' and 'Kabul Yellow' (Nageswari *et al.*, 1999).

'Ruby' is a hybrid cultivar characterized by the sweet taste, low tannin, dark red and non-sticky arils with soft seeds (Pareek, 1996). A number of 2900 pomegranate hybrids of different crosses including F_2 progenies were developed by Prasanna Kumar (1998) and promising hybrids containing the high fruit quality of 'Ganesh' and '18 Kabul' and the deep red colour of 'Gulsha Rose Pink' have been observed. Seven Indian cultivars ('Ganesh', 'Daru', 'Kabul Yellow', 'Amilidana', 'Ruby', 'Nana' and 'Double Flower') were crossed to investigate the effects of different cross combinations as well as season on fruit set (Jayesh and Kumar, 2004). Parent type affected the fruit set and the highest fruit set was observed in the crosses involving 'Ruby', 'Ganesh' and 'Daru', while the lowest fruit set was recorded in the selfing of 'Kabul Yellow' and the crosses that this cultivar was involved in, which was attributed to the poor cross- and self-compatibility behaviour of this cultivar (Jayesh and Kumar, 2004). Selfing resulted in the maximum fruit set and all crosses produced the maximum fruit set in winter months probably due to the low temperature and high humidity conditions, which subsequently increased the percentage of viable pollens (Jayesh and Kumar, 2004). In order to produce high-yielding and soft-seeded fruits with blood red aril colour, 69 crosses were performed among six soft-seed and three hard-seed pomegranate cultivars (Kumar, 2012). Fruit set ranged from 37.60 to 80.95%, while fruit retention varied from 8.14 to 29.60% among different cross combinations. A relatively high percentage of seed germination (from 74.12 to 94.53%) was reported for all crosses and 21,968 seedlings were raised for further evaluation (Kumar, 2012).

Jalikop (2003) performed a series of hybridization investigations between 'Ganesh' and a recessive rosette mutant clone of 'Kabul Yellow' to reveal the inheritance of the rosette phenotype that previously was observed in the 'Kabul Yellow'. 'Ganesh' is a soft-seed evergreen seedling selection from 'Alandi' with poor fruit quality that is principally grown in India, while 'Kabul Yellow' is a cultivar with no pomological significance (Jalikop, 2003). According to the results of this investigation, the F_2 progeny of 'Ganesh' × rosette mutant 'Kabul Yellow' segregated into recombinant forming rosette (0.12) and normal types. Therefore, the rosette phenotype was attributed to a recessive mutant gene (ascribed as *rg* symbol) in 'Kabul Yellow'. The author observed that seeds of F_2 and BC_1 had very low seed fertility and concluded that F_1 possibly carried a heavy load of lethal genes. In addition, a tendency to regress to normal state was observed in some of the rosette progeny (nine progeny out of 31) and attributed these findings to the role of cryptic transposable elements (Jalikop, 2003). 'Amlidana' is a hybrid obtained from a cross between 'Ganesh' × 'Nana' that is characterized by its more acidic anardana (the processed dry form of arils) and higher fruit yield per tree than their parents (Jalikop *et al.*, 2000).

It is reported that two main Indian export pomegranate cultivars, namely 'Bahgwa' and 'Mridula', which are characterized by their soft seeds, red skin colour and evergreen trees, are selected from offspring resulted from crossing of 'Ganesh' and a Russian cultivar 'Gul Shah Red' (Holland *et al.*, 2009).

A pomegranate breeding project in Turkey began in 1997 with the objective of producing pomegranate hybrids with red rind, dark red aril and soft seeds in Aegean Agricultural Research (Onur *et al.*, 1999; Küçük, 2007). Reciprocal crosses as well as selfing were performed among five standard cultivars: İzmir 16, İzmir 23, İzmir 1445, İzmir 1465 and İzmir 1513. Some of the progeny were planted and 38 superior offspring were selected for further evaluation (Küçük, 2007). Hybridization studies were also performed in order to breed sunburn- and cracking-resistant cultivars in Turkey (Yazici and Ercişli, 2017). The results of the evaluation of 69 selected offspring derived from selfing, open pollination and crossing of 'Hicaznar' (a late ripening cultivar) with 'Fellahyemez', 'Ernar', 'Seedless IV' and 'SeedlessVI' indicated a wider range of ripening time than their parents, which subsequently influenced the cracking and sunburn. The authors stated that early ripening individuals were more tolerant to fruit cracking and sunburn while the late season genotypes were the most susceptible to these disorders (Yazici and Ercişli, 2017).

Pomegranate breeding in Spain began in 2002 through crossing of selected genotypes at Valencian Agricultural Research Institute (IVIA) in Elche (Bartual *et al.*, 2012). The main objectives of this breeding project are stated as: sweet, dark red aril and fruit peel, extending the ripening season as well as resistance to cracking and sunburn (Bartual *et al.*, 2012). Hybridization studies in Spain were carried out using local cultivar 'Mollar de Elche' (the most known Spanish pomegranate, characterized by soft seeds, sweet taste and yellowish pink fruit) with international 'Wonderful' cultivar with the objectives of breeding new varieties containing darker fruit skin and higher antioxidant content, while keeping the soft seeds (Bartual *et al.*, 2015). Preliminary results indicated a wide range of variability in soluble solids (12.9–18.1), TA (0.29–1.91% citric acid), pH (2.7–4.1), juice content (17.1–39.4%), and skin and juice colour among F_1 seedlings (Bartual *et al.*, 2015).

The objectives of pomegranate breeding in Israel were established based on the demands of European markets including appealing skin and aril colour (particularly bright red colour) and seed softness (Holland *et al.*, 2009). The majority of breeding programmes were based on seedling selection from existing cultivars. Extending the ripening season to produce very early and very late ripening pomegranate cultivars was one of the main objectives in this region. 'Emek' is one of the successful outcomes of this programme. 'Emek' is an early ripening (mid to late August in Israel conditions) high-quality fruit with red colour of aril and fruit peel, sweet or sweet-sour taste, medium-sized fruit and soft seeds that was bred in Newe Ya'ar Research Center of Israel (Holland *et al.*, 2014). The tree is reported to be self-fertile and self-pollinated, medium sized with vigorous growth and good productivity. This new cultivar was selected from screening of ~800 seedlings originated from open pollination of P.G.128–29, a cultivar in the Israel Plant Gene Bank in the Newe Ya'ar pomegranate collection (Holland *et al.*, 2014). This cultivar was characterized by pink red skin, bright red arils, soft seeds and sweet taste with heavier fruit and earlier ripening time than 'Shani-Yonay' (Bar-Ya'akov *et al.*, 2007). 'Kamel', a selection from Newe Ya'ar, which resembles 'Wonderful' with regard to fruit quality and tree architecture, is one of the newly bred cultivars, which ripens a month earlier than 'Wonderful'. Cross-based breeding programmes also initiated in this region to breed very early ripening cultivars that are tolerant to the negative effects of heat on the anthocyanin content of rind and aril. There are hybrid populations derived from 'Wonderful' and an evergreen cultivar from India and their F_2 selfed progenies were used to study the inheritance of anthocyanin as well as the evergreen phenotype (Bar-Ya'akov *et al.*, 2007).

Recently pomegranate breeding in Sri Lanka has begun with the purpose of increasing tree yield as well as producing high-quality fruits with red colour of peel and arils and soft seeds with three local cultivars 'Daya', 'Nayana' and 'Nimali' as well as a promising accession, 'Kalpitiya Red', using the full diallele method (Kumari *et al.*, 2017). The authors observed wider ranges for most of the studied attributes in the hybrid plants than their parents as well as positive heterosis for tree yield, number and weight of fruits, 100 arils weight, TSS, juice volume and pH over the mid and superior parents.

Nahla *et al.* (2014) crossed pomegranate cv. 'Manfaloty' from Egypt with five other cultivars ('Eversweet', 'Sweet', 'Aca', 'Wonderful' and 'Blackberry') and evaluated physical and chemical parameters in the fruits of these crosses. They reported that fruit set ranged from 0 to 100% in different combinations. The authors also reported that the source of pollen grains significantly affected qualitative and quantitative fruit characteristics and suggested that 'Sweet' cultivar is the appropriate pollinator for improving physical and chemical attributes of 'Manfaloty' cv. Xenia. Effects were also reported in Iranian and Indian pomegranate cultivars (Gharaghani *et al.*, 2017; Singh *et al.*, 2017). It is reported that the pollen source not only affects fruit set and its characteristics, but also influences seed germination (Singh *et al.*, 2017). In fact, it seems that seeds derived from some of the crosses are sterile. Therefore, it could be concluded that some of the pomegranate genotypes are cross-incompatible and identification of the proper pollinators in pomegranate orchards is of great importance.

As mentioned previously, freezing injury is one of the restricting factors for pomegranate cultivation in temperate regions. One of the main objectives of pomegranate breeding programmes in such regions is obtaining freezing-tolerant cultivars. Efforts for breeding freezing-tolerant cultivars were made in Turkmenistan (Levin, 1979) and hardy hybrid plants were obtained by successive hybridization (Levin, 2006).

Most of the breeding programmes in China have been focused on fruit quality. These include good-quality fruit 'Dui Hong 1', 'Duo Qing 11' and 'Duo Bai 2' (Liang and Cheng, 1991), 'Zaoxuan 018' and 'Zaoxuan 027' (Wang *et al.*, 2006), soft-seeded 'Hongmanaozi' (Wang *et al.*, 2006) and early ripening 'Yushiliu 4' (Zhao *et al.*, 2006).

From the scant breeding programmes underway in Iran, a breeding programme started at the Department of Horticultural Science, University of Tehran can be mentioned. The purpose of this programme was to obtain soft-seed genotypes by obtaining progeny from selfing, open pollination and crossing a commercial cultivar ('Malase-Tourshe-Saveh') with a soft-seeded genotype ('Bihaste-Ravare') (Zamani *et al.*, 2010). However, the number of hybridization studies in Iran as the main centre of origin and centre of diversity of pomegranate is very low, which may be due to the extensive variations available in the existing cultivars and genotypes that have been selected and cultivated in different geographic regions in this country during the long pomegranate cultivation history.

4.11.4 Inheritance of traits

There are few reports about pomegranate hybridization with the aim of uncovering the heritability of desirable attributes. The role of maternal inheritance was emphasized in pomegranate crosses by Zamani *et al.* (2010). The reciprocal crossing of 'Daru' and 'Ganesh' in India resulted in progeny with pink coloured arils, high acidity and hard seeds (Jalikop *et al.*, 2005). Moreover, it was found that white colour of aril was recessive to the pink colour, low acidity was always recessive to high acidity and soft seed was recessive to hard seed (Jalikop *et al.*, 2005).

In a breeding programme with the aim of breeding bacterial nodal blight (*Xanthomonas campestris* pv. *parthenii*) resistant pomegranate cultivars by using 'Daru' cultivar, it was found that resistance to this disease is controlled by recessive genes (Jalikop *et al.*, 2005). Ataseven Isik (2006) studied the inheritance of some fruit characteristics including colour of peel and aril, taste and seed softness using different combinations of crosses between 'Fellahyemez', sweet-sour 'Hicaznar' and sweet 'Ernar', and reported that when a sweet-sour and the sweet cultivars were hybridized, 40% of progeny had a sweet taste, while 90% of progeny resulting from two sweet tasting parents were sweet-tasting themselves. Karale and Desai (2000) investigated the heterosis of some pomegranate fruit attributes in individual hybrids compared with midparental values and reported that the highest heterosis was recorded for the percentage of juice weight followed by aril weight, fruit weight, fruit perimeter and rind thickness. The authors reported that none of the crosses recorded significant heterosis over their superior parent in a favourable direction for the evaluated attributes.

It is reported that seed softness is a multigenic characteristic (Jalikop *et al.*, 2005; Jalikop, 2010; Lu *et al.*, 2006; Zamani *et al.*, 2010) and many pomegranate cultivars are heterogeneous for this trait (Glozer and Ferguson, 2008). According to their inheritance pattern, hard seeds and red and pink aril colour are dominant over soft seeds and white arils (Jalikop *et al.*, 2005). The results of a Spanish breeding programme also indicated that hard seeds and red aril colour traits are dominant over soft seeds and pale aril colour (Bartual *et al.*, 2015). It seems that colour of fruit and flowers as well as vegetative parts of the plant are monogenic traits, and red and pink colours are dominant to yellow. A single dominant gene is considered as the cause of the red colour of the fruit peel, flower bud, leaf border and petiole base in 'Ganesh' cultivar, while the recessive form of this gene is responsible for the yellow colour of the aforementioned parts in the 'Kabul Yellow' cultivar (Jalikop and Kumar, 1990). However, according to Jalikop (2011) fruit peel colour is a polygenic trait and is affected by sunlight. Strebkova (1974) observed that mottled pink petals, double petals and bright petal colour were dominant. The flower type is controlled monogenically and the ornamental flower type called double flower (in which the stamens are changed into the petals) is dominant over the normal flower type (Strebkova, 1974). According to Jalikop (2007) 'Double Flower' types are able to produce fruit and it is suggested that the cultivated fruit type pomegranates with normal flowers took part in the development of this pomegranate flower type. Pollination of 'Double Flower' flowers by pollen of 'Ganesh' (a sweet cultivar) resulted in the production of hard seed and highly acidic F_0 fruits (Jalikop, 2007). Therefore, it is suggested that occurrence of a dominant mutation in a homeotic locus (Df) of a sour genotype may have been responsible for stamens' conversion into petals. So the 'Double Flower' is considered as a heterozygous (Dfdf) specimen that has been propagated clonally (Jalikop and Kumar, 2009).

According to Jalikop (2011) large fruit size is dominant to small fruit size and red/pink aril colour is dominant to light pink/white and is affected by temperature. It is also reported that high acidity is dominant over sweet or low acidity of fruit and the gene symbol SS has been assigned to this locus (Jalikop, 2007). The author stated that a major gene determines high acidity while some modifying genes with small effects to some extent cause the variations that are within the sweet or sour genotypes. Because some progeny derived from two sweet pomegranate cultivars ('Kabul Yellow' and 'Ganesh') were highly acidic, this was attributed to the linked dominant alleles or epistatic effects of vicinity loci, which is similar to the overdominance effect (pseudo-overdominance).

It is reported that in pomegranate a single recessive gene controls rosette growth habit (Jalikop, 2011). In addition, based on the occurrence of the rosette growth habit in the F_2 progenies of 'Ganesh' and 'Kabul Yellow', it was concluded that a recessive mutant locus in 'Kabul Yellow' is responsible for the rosette growth nature and the gene symbol of *rg* was assigned to this locus (Jalikop, 2003). Moreover, fruit weight and perimeter, TSS, colour and weight of aril, plant height, as well as seed hardness were reported to be inherited quantitatively and some of their QTLs were mapped on the pomegranate genetic map (Harel-Beja *et al.*, 2015).

4.12 Breeding Through Biotechnology

In addition to classical breeding programmes, new breeding methods are also opening new avenues in pomegranate breeding programmes. These methods include gamma irradiation (Kerkadze, 1987), chemical mutagenesis (Shao *et al.*, 2003; Matuskovic and Micudova, 2006), tetraploid induction (Shao *et al.*, 2003) and genetic transformation (Terakami *et al.*, 2007).

4.12.1 Mutation breeding

Mutation breeding is a technique that uses different mutagen agents to induce sudden heritable changes in the genetic materials (and in turn in the resultant characteristics) that are not derived from recombination or genetic segregation (Van Harten, 1998). Mutation in genetic material is considered as a natural phenomenon that occurs spontaneously in nature and plays an important role in expanding the genetic diversity of different organisms and subsequent evolution of different species. However, some artificial mutation agents were discovered that are able to accelerate this phenomenon. These agents are called mutagens and are grouped into two broad categories; namely physical and chemical mutagens (Mba *et al.*, 2010). Cosmic, gamma and X-rays are among the most widely used physical mutagens, and alkylating agents including ethyl methanesulphonate (EMS), diethyl sulfate (DES), ethyleneimine (EI), methyl nitrozourethane (MNU), ethyl nitrosourea (ENU) and azides are among the most widely used chemical mutagens. Although it is documented that frequency and types of mutations are direct results of dosage and rate of exposure or administration of the mutagen, chemical mutagens are preferably used for induction of point mutations, while physical agents induce gross lesions such as chromosome abbreviations or rearrangements (Kunzang *et al.*, 2017). However, because the majority of the mutants are recessive and have little or no use, success in mutation breeding is highly dependent on the creation of a large mutated population to find a desirable type, which is the main reason for lack of popularity of this method among fruit breeders.

4.12.2 Gamma irradiation

Mutagenesis through irradiation is one of the approaches that have been used successfully for many plant species including pomegranate with the aim of producing new cultivars as well as expanding the existing diversity. Seeds and cuttings of pomegranate were subjected to the 1–40 kR of gamma radiations and new variants with a wider range in fruit yield, size and quality were reported (Akhundzadeh *et al.*, 1977). Based on their observations, Akhundzadeh *et al.* (1977) reported that sweet-tasting cultivars were more sensitive to gamma radiation than sour or sweet-sour cultivars. Gamma radiation of seeds, cuttings and pollen of pomegranate at 5–10 kR resulted in a high level of variability (Akhundzadeh, 1981). It is reported that pomegranate cultivar 'Karabakh' resulted from gamma-irradiated plants (Kerkadze, 1987). Good quality of fruit and juice as well as good keeping quality were obtained in the seedlings irradiated by 10–20 kR gamma rays (Levin, 1990b). From the second day after culture, gamma rays 250 R/h to a total of 4–64 KR were used to induce mutation in the leaf explants of *P. granatum* var. 'Nana' (Omura *et al.*, 1987). High doses (64 kR) of gamma radiation severely inhibited callus and bud formation and reduced numbers of shoots on each explant. Regenerated plants were significantly different from normal ones with regard to leaf shape and growth habit as well as pollen viability (19.7% pollen grains were sterile). The frequency of variations was very high in the leaf shape. Narrow leaves, thread-like, round, lanceolate, slender and curly leaves were some of the irregularities and the two latter were the most frequent. Growth habit variations were reported, such as plantlets with upright habits containing more leaves as well as more dwarfed plants. The dwarf phenotype was reported from irradiated pomegranate plants (Mohan Jain, 2010; Ulukapi and Gul Nasircilar, 2015). In order to obtain bacterial blight-resistant variants, seeds of 'Kesar' cultivar were treated with various gamma ray concentrations and significantly different morphological characteristics were reported in the resulting plants, which included different plant height, number of branches, canopy spread and number of bacterial blight lesions on leaves (Sangamesh, 2014). However, after 12 days of bacteria inoculation on 1-year-old plants, moderately resistant variants were observed at gamma ray doses of 10, 15, 20 and 25 Gy. It is also reported that 'Taihanghong', 'Hongmanaozi' and 'Hongyushizi' are among the new varieties that were obtained from natural mutation (bud sports) of three Chinese cultivars (Zhao, 2007).

4.12.3 Chemical mutagenesis

Chemical mutagenesis has also been used for inducing mutation in pomegranate. EMS has become the mutagenic agent of choice, due to its effectiveness, ease of handling as well as detoxification through hydrolysis for disposal of nitrose compounds. Therefore, EMS has been used extensively for many plant species to induce mutation. Cotyledon explants of pomegranate were subjected to different concentrations of EMS (0.1–0.9%) for 1, 2, 3 and 3.5 h. It is reported that the lethal dose (LD50) of EMS for pomegranate was observed at 0.3% for 1 h (Sangamesh, 2014). The optimum dose of EMS for the maximum percentage of pomegranate survival was reported to be 0.1–0.2%. Usage of different chemical mutagens such as natrium azid for inducing mutation in pomegranate have reported (Matuskovic and Micudova, 2006). Promising mutants of pomegranate were also obtained with N,N-dimethyl-N-nitrosurea treatments (Levin, 1990b).

4.12.4 Colchicine treatment

Polyploidization is a technique usually used in different horticultural crops with the purpose of expanding vegetative as well as reproductive organs. This technique is widely used in ornamental plants for obtaining new plant architecture, colour and bigger flowers, as well as in medicinal plants for increasing plant mass. However, in fruit species this technique could be used for production of polyploid plants, which could serve as parents in breeding programmes with the purpose of obtaining triploid progenies. Tetraploid plants usually have larger plant organs than diploid ones, which may lead to the evolution of variants with larger fruit and arils with improved quality. Tetraploid pomegranate was obtained under *in vitro* conditions by shoot node explants (Shao *et al.*, 2003). For this purpose, 10 mg/l colchicine was used in the solid MS (Murashige and Skoog, 1962) medium supplemented with 0.1 mg/l 1-naphthaleneacetic acid (NAA), 1.0 mg/l benzyladenine (BA) and 400 mg/l activated charcoal. It is reported that after a 30-day period, 20% of regenerated plants were tetraploid in this medium. The tetraploid plants were differentiated from diploid plants by shorter roots as well as wider and shorter leaves. These plants grew normally, but their flowers were shorter and bulkier (with larger diameter) than the flowers of normal plants. Tetraploid flowers had higher numbers of pollen grains per anther with lower viability than diploid flowers. In addition, it is reported that shoots cultured on the same medium supplemented with 5000 mg/h colchicine for 114 h did not induce tetraploid plants, while three morphological mutants that were characterized by their narrow leaves were detected with the same concentration for 96 h. Further investigations revealed that these mutants were mixoploids that produced diploid and tetraploid plants when subcultured (Shao *et al.*, 2003). These authors concluded that since tetraploid pomegranate plants were semi-fertile, they could serve as parents in hybridization programmes for crossing diploid and tetraploid plants to produce seedless triploids (Shao *et al.*, 2003). However, as the edible part of pomegranate is fleshy tissue growing from the seed integument surface cells, seedless plants are not preferred in this species, and tetraploid plants might have the potential for different horticultural purposes other than seedless plants. Meanwhile, for triploid production in pomegranate, endosperm culture could be an alternative strategy. As pomegranate trees are easily propagated by cutting and layering as well as micropropagation methods, there is no robust barrier to multiplication of triploid plants.

4.12.5 Tissue culture studies

Pomegranate has been subjected to several tissue culture studies over the past few years. Several protocols have been developed for direct *in vitro* propagation of pomegranate through axillary shoot proliferation from different explants including nodal segments (Zhang and Stolz, 1991; Naik *et al.*, 1999, 2000; Murkute *et al.*, 2004; Kanwar and Rachna Kashyap, 2004), cotyledonary node (Sharon and Sinha, 2000) and shoot tips (Murkute *et al.*, 2004). In addition, indirect plant regeneration has been reported from different explants including leaf (Omura *et al.*, 1987; Murkute *et al.*, 2002), cotyledon (Murkute *et al.*, 2002; Deepika and Kanwar, 2008; Kanwar *et al.*, 2010b) and anther (Moriguchi *et al.*, 1987). According to Kalalbandi *et al.* (2014) among different substances that were used for surface sterilization of explants (NaOCl, $HgCl_2$ and Na methiolate) 0.1% mercuric chloride for 10 min resulted in the maximum survival (90.58%) and the minimum microbial contamination (9.52%).

Due to the high polyphenol content of different pomegranate organs, culture browning is one of the impediments in pomegranate tissue culture studies (Murkute *et al.*, 2003). Different approaches have been used to overcome this problem. Use of absorbents (such as activated charcoal, polyvinylpyrrolidone (PVP)), addition of antioxidant to the media or soaking explants in antioxidant solution prior to culture establishment, as well as frequent subculture, are among the most common procedures that have been used successfully to overcome culture browning for establishment of pomegranate tissue culture (Mahishni *et al.*, 1991; Al-Wasel, 1999; Naik *et al.*, 2000; Murkute *et al.*, 2004; Singh and Khawale, 2006; Singh *et al.*, 2007; Agamy *et al.*, 2009) . However, specific approaches have also been successfully tested in pomegranate including rapid subculturing on the first and third days after inoculation and reducing explant size (Murkute *et al.*, 2004; Chaugule *et al.*, 2005; Singh and Patel, 2016) as well as sealing the cut edges of the explant with sterile wax to reduce phenolic compound exudation (Singh *et al.*, 2007), which resulted in a higher rate of plant regeneration.

Different media have been used for tissue culture optimization of two pomegranate cultivars ('Manfalouty' and 'Nab El-Gamal') such as MS (1962), Nitsch & Nitsch (N&N, Nitsch and Nitsch, 1969) and woody plant medium (WPM), among which WPM resulted in the better survival of explants, plantlet height and number of leaves per shoot (Agamy *et al.*, 2009). Among different plant growth regulators used for proliferation, BA (1.0 mg/l) and kinetin resulted in the highest growth (six shoots per explant) and the lowest (three shoots per explants), respectively. The same trend was reported for the number of leaves per plantlet. For rooting medium, 0.25 mg/l NAA and 0.25 mg/l indole-3-butyric acid (IBA) induced the highest value of rooting for 'Manfalouty' and 'Nab El-Gamal' cultivars, respectively, while half-strength WPM resulted in longer roots than full-strength WPM for both cultivars.

Some of the studies used *P. granatum* var. '*nana*' for optimization of tissue culture media. Bonyanpour and Khosh-Khui (2013) analysed different concentrations of BA and NAA in MS medium for callus induction in a local genotype of dwarf pomegranate. It was reported that the highest callus induction was attained in the medium supplemented with 1.0 mg/l BA and 0.2–0.4 mg/l NAA after 40 days, while the highest callus growth was reported from 1.0 mg/l BA and 1.0 mg/l NAA. According to these authors, medium supplemented with 5 mg/l BA and 0.1 mg/l NAA resulted in the highest number of shoots (seven shoots per explant). However, the highest shoot proliferation was attained in the WPM containing 5 mg/l kinetin. In order to induce root formation on the shoot explants, WPM supplemented with 0.2 mg/l IBA resulted in the maximum root percentage (100%) and root growth (2.06 cm and two roots in each explant). Different juvenile explants of 'Kandhari Kabuli' have been tested for callus induction in MS medium supplemented with different concentrations of NAA (8–13 μM) and BA (9–18 μM) (Deepika and Kanwar, 2010). The maximum percentage of callus induction and regeneration rate was reported from cotyledon, hypocotyl, internode and leaf explants, respectively. Half-strength MS medium supplemented with 500 mg/l activated charcoal was shown to be the best condition for root induction of 'Kandhari Kabuli' cultivar.

According to Kalalbandi *et al.* (2014) shoot tip was the best explant for culture establishment and low microbial contamination in 'Bhagawa' and the highest number of shoots was obtained in the MS medium containing 2.0 mg/l 6-benzylaminopurine (BAP) while the best rooting condition resulted from half-strength MS medium supplemented with 8 mg/l NAA. Patil *et al.* (2011) also used MS and WPM supplemented with 0.2–2 mg/l BAP, 0.1–1 mg/l NAA, 0.5–2.5 mg/l $AgNO_3$ and 30 mg/l adenine sulfate for establishment of tissue culture in nodal explants of 'Bhagava' cultivar and reported that MS medium resulted in the higher plantlet survival, and the highest rate of proliferation (10–15 shoots per explant) was obtained in MS medium containing 1.8 mg/l BAP, 0.9 mg/l NAA, 1 mg/l silver nitrate and 30 mg/l adenine sulfate. Equal rooting responses were reported in both media supplemented with 0.5 mg/l NAA and 0.5 mg/l IBA; however, IBA resulted in thicker roots.

In another report, results of MS and WPM compared in two Iranian pomegranate cultivars ('Malase Saveh' and 'Yousef Khani') revealed that WPM (supplemented with 4.7–9.2 μM kinetin) was more efficient for micropropagation of these cultivars, while half-strength WPM medium containing 4.4 μM NAA was more effective for rooting (Valizadehkaji *et al.*, 2013).

4.12.6 Somatic embryogenesis

Somatic embryogenesis could provide appropriate plant materials for clonal multiplication, genetic transformation and the production of synthetic seeds. Under *in vitro* condition, somatic embryos might be obtained directly from explants cultured on the medium or indirectly after callus induction from explants. Both direct and indirect somatic embryogenesis have been reported from pomegranate.

Induction of somatic embryogenesis has been reported in pomegranate from various explants including cotyledon, hypocotyl, stem and leaf segments of 3-weeks-old seedlings (Jaidka and Mehra, 1986), petals (Nataraja and Neelambika, 1996), immature and mature zygotic embryos from unripened and ripened fruits (Bhansali and Raj Bhansali, 1990; Kanwar *et al.*, 2010a), shoot tip (Helaly *et al.*, 2014) and hypocotyl (Bharose *et al.*, 2014). MS medium was the most frequently used for somatic embryogenesis in pomegranate. It is reported that MS medium supplemented with sucrose, casein hydrolysate, L-glutamine and myo-inositol can induce embryos from somatic tissues of pomegranate (Bhansali and Raj Bhansali, 1990). According to Bharose *et al.* (2014) the greatest amount of embryogenic callus was obtained from hypocotyl explants on MS medium supplemented with 0.5 mg/l 2, 4-D and 0.5 mg/l kinetin. Helaly *et al.* (2014) used shoot tip explants of four pomegranate genotypes for induction of embryonic callus on the MS medium supplemented with different concentrations of 2,4-D. The authors were able to produce embryonic calli within 6 weeks in MS medium supplemented with 1 mg/l 2,4-D and reported that embryonic callus induction levels were significantly different among pomegranate genotypes. However, embryogenesis was also successful in other basal media such as B5 and RBM-I (revised basal medium I, the name assigned by the authors for a medium used for pomegranate embryogenesis) (Bhansali and Raj Bhansali, 1990; Sinha and Sharon, 1997).

Different auxins including IBA, NAA, indole-3-acetic acid (IAA), and 2,4-D singly or in combination with Kin and BAP were reported to be effective plant growth regulators for embryogenesis induction in pomegranate (Bhansali and Raj Bhansali, 1990; Nataraja and Neelambika, 1996; Sinha and Sharon, 1997; Kanwar *et al.*, 2010a). Surprisingly, application of 2,4-D, which is considered as an effective hormonal treatment for inducing somatic embryos in most plant species (Rai *et al.*, 2010), is not usually the best auxin for *P. granatum*. Beside the concentrations, the ratio of plant growth regulators is also an important determinant of a culture's destination. Generally, high ratios of auxin/cytokinin are essential for the embryo induction stage, while reducing or eliminating the auxin promotes embryo development. Coconut water has also been reported as an essential ingredient of embryogenesis induction media for pomegranate (Steward *et al.*, 1964). In addition, several studies showed the best response in dark conditions (Bhansali and Raj Bhansali, 1990; Kanwar *et al.*, 2010a). An increase in the concentration of sucrose (6%) and a decrease in the media strength (for example using half-strength MS medium) as well as adding activated charcoal (0.3%) in the medium is required to give rise to mature somatic embryos in the plantlets (Bhansali and Raj Bhansali, 1990; Nataraja and Neelambika, 1996). Low germination rates (e.g. 4.45% in a report of Kanwar *et al.*, 2010a) were reported as the major limiting factor for somatic embryogenesis induction in pomegranate. In fact, for commercial application of somatic embryos, the germination rate should be up to 80–85% (Rout *et al.*, 2006). Therefore, more investigations are needed in this plant species to improve the germination rate of somatic embryos to plantlet.

4.12.7 Somaclonal variation and *in vitro* selection

Tissue culture conditions can induce higher mutation rates in the genetic materials compared with normal conditions: the phenomenon that is called somaclonal variation (Larkin and Scowcroft, 1981). It is reported that the frequency of somaclonal variation in the whole plant can be 10,000 times higher than spontaneous mutation (Teixeira da Silva *et al.*, 2013). These types of variations can arise from chemicals used under long-term *in vitro* conditions and the bare nature of explants in the tissue culture that have no protective tissues and are more exposed to the chemicals used. In fact, the tissue culture conditions provide the appropriate situation for developing a mutated cell into the whole plant. Such variations can expand the existing genetic diversity and serve as a potential valuable source for breeding programmes. However, in spite of many claims regarding the potential applications of such variations in plant breeding, there is no report about major crop species that have resulted from these variants.

Tissue culture techniques can also be used for selection of plant materials that are tolerant to specific stresses, the phenomenon that is known as '*in vitro* selection'. In fact, by using different lethal and sublethal concentrations of a stressful agent under *in vitro* conditions containing dividing cells, the researchers are able to select the mutated cells (the so-called escape) that are more tolerant to that unfavourable condition and could be of high importance in breeding programmes. So far this aspect of tissue culture techniques has been widely used for breeding objectives, especially in selection for stress tolerance. From the scant *in vitro* selection studies in pomegranate, we can mention the attempt for producing bacterial resistance in this fruit species. To do this, pomegranate cultures were screened for resistance to *Xanthomonas axonopodis* pv. *punicae*, which is responsible for bacterial blight disease, and *Ceratocystis fimbriata*, which is responsible for pomegranate wilt disease (Madhvi, 2015). For this purpose small pieces of calli from pomegranate cv. 'Kandhari Kabuli' were cultured on callus proliferation medium supplemented with 5–40% of bacterial culture filtrate and 5–50% fungal culture filtrate. Selection against bacterial disease showed that about 13% calli survival was observed at the 25% level of bacterial culture filtrate and no survival was detected at the 30% level. In the case of pomegranate wilt, 40% fungal culture filtrate resulted in about 19% calli survival and no survival was detected at higher concentrations.

4.12.8 Anther culture and haploid production

Haploid production through *in vitro* culture of the anther (androgenesis) or ovule (gynogenesis) is one of the most valuable outcomes of tissue culture technique and can be used to shorten the period that is required for breeding programmes, especially in fruit crop species. The haploid plants can be converted into homozygous diploid plants that are excellent materials for different purposes including crossing as well as the study of inheritance of different attributes. There are a few reports of unprofitable attempts with the aim of producing haploid plants through anther culture of pomegranate. One of the reports is about dwarf pomegranate cv. 'Issaizakuro'. Its anthers were cultured at the uninucleate to binucleate stage on the MS, Miller (M) (Miller, 1965) and N&N media in dark conditions (Moriguchi *et al.*, 1987). Among three tested basal media, M and MS were superior to N&N and 20% of anther cultures produced calli after 30 days on these two media supplemented with 5 μM BAP and NAA in dark conditions. Although the authors were successfully able to produce shoots from these calli after 3 weeks of transferring the explants on to half-strength MS medium supplemented with 0.5 μM NAA and 2.0 μM BAP, all of the regenerated shoots had the diploid chromosome number ($2n = 18$) and no haploid was detected. Supplementary histological studies revealed that calli were originated from the somatic cells of anther wall (tapetal cells) and not from microspores resulting from microsporogenesis. In addition, shoot regeneration capacity from anther-derived calli has been found to be very low, as out of 2391 cultures only 10 were regenerated to shoots with low proliferation (one or two per explant). Low-frequency shoot regeneration capacity of anther-derived calli may indicate that haploid production may be lethal in pomegranate and it could be concluded that some of the anther-derived calli were haploid in this investigation and were not able to convert to a viable plantlet. According to Naik and Chand (2011), Mascarenhas *et al.* (1988) tried to produce haploid plants from anther culture of 'Ganesh' and 'Muskat' pomegranate cultivars. These authors reported that anther culture on the N&N medium supplemented with 0.5 mg/l kinetin, 0.5 mg/l IAA and 0.2 mg/l BA resulted in compact, nodular calli with embryoid-like structures. However, only root and leaves were regenerated on Nitsch medium containing 0.5–1.0 mg/l IAA and 0.1–4.3 mg/l BA, respectively.

4.12.9 Genetic transformation

With the advent of new biotechnology approaches, the plant breeding cycle has been highly accelerated. Genetic transformation is one of the procedures that has led to significant progress in plant breeding during recent decades. Genetic transformation studies in pomegranate are far fewer in number than on model plants or staple crops. However, there are several reports on pomegranate transformation using different systems, which mostly were carried out with the purpose of optimization of transferring methods. Some of the authors indicated that pomegranate is a difficult-to-transform tree (Valizadeh Kaji *et al.*, 2014; Valizadeh Kaji and Abbasifar, 2017), while some others encountered minor problems with pomegranate transformation (Terakami *et al.*, 2007; Helaly *et al.*, 2014). Most of the transformation studies in pomegranate used *Agrobacterium tumefaciens* as the transformation vehicle. Transformation using *Agrobacterium* offers several advantages over direct gene transfer methods. This method reduces the copy number of the transgene, potentially leading to fewer problems with transgene co-suppression and instability.

Terakami *et al.* (2007) reported the first investigation on transgenic pomegranate. These authors subjected leaf segments of dwarf pomegranate (*P. granatum* var. '*Nana*') to the *Agrobacterium tumefaciens* strain EHA105 harbouring a binary vector pBin19-*sgfp* with green fluorescent protein (*gfp*) and neomycin phosphotransferase (*nptII*) genes as reporter and selectable markers, respectively. Putative transformed shoot segments were selected on the MS medium containing 0.5 μM NAA, 4.0 μM BA, 50.0 mg/l kanamycin, 10 mg/l meropenem and 0.3% gellan gum after 6–8 months. A relatively high transformation rate (at least one transformant per adventitious shoot) was reported for dwarf pomegranate, and the inheritance of the transgene was confirmed in the T_1 generation. In addition to high transformation rate, transformed dwarf pomegranate plants were able to produce fruit in a short period (within 3 months) after transferring to the pot. Therefore, based on these two characteristics, dwarf pomegranate was suggested as a good model plant that can be exploited for transformation studies in different fruit tree species (Terakami *et al.*, 2007).

Agrobacterium tumefaciens strain EHA 105 harbouring the plasmid pBinBt1 (containing *Cry1A(b)* and *nptII* genes) was also used to investigate the efficiency of different explants (somatic embryo, shoot bud and cotyledon) as well as regeneration methods (callus cultures using embryonic explants, direct regeneration of shoot buds and cell suspension of cotyledon explants) for transferring *Cry1A(b)* gene into pomegranate 'Kandhari Kabuli' (Verma *et al.*, 2014). Among the three different examined explants and regeneration systems, cotyledonary calli on cell suspension methods had the highest transformation frequency (13.54%) followed by embryonic explants via indirect regeneration (8.7%) and shoot bud explants through direct regeneration (3.33%). However, the plating efficiency of transgenic cells derived from cotyledon explants in the cell suspension, which plays an important role in the regeneration of transformants, was very low. Among three different regeneration methods, regeneration through callus-derived embryos was reported as the best system (23.33% regeneration) for producing transgenic tissues. These authors reported that although indirect regeneration and transformation (embryo and cotyledon) resulted in a better response than direct nodal explants (without passing from the callus phase), owing to the polyploidy and mixploid cells that can result in the production of chimeric shoots, this procedure is not a reliable method. This phenomenon causes the stage of transgenic tissue selection to become very difficult and expensive. Therefore, regeneration through nodal explants, where no callus phase is required, is reported as the best system for producing true to type plants. Due to the opportunity it allows to produce a transformed plant with single-cell origin, cell culture was reported as an alternative protocol for producing genetic transformation in pomegranate (Verma *et al.*, 2014). With these types of transformed plants, it is possible to minimize the chance of obtaining chimeric transgenics. The presence of the *Cry1Ab* gene was confirmed at DNA level of all putative transgenic shoots (Verma *et al.*, 2014). Therefore, these systems were proposed to have the potential to be used for producing pest-resistant pomegranate. Helaly *et al.* (2014) used *A. tumefaciens* strain LBA4404 harbouring the binary vector pBI-121 for transformation of embryonic callus of two pomegranate genotypes with high ('Manfalouty') and low ('Araby') regeneration capacity. The authors reported that transformants had a lower concentration of H_2O_2 and the higher specific activities of catalase (CAT) and superoxide dismutase (SOD) than control plants and suggested that this protocol can be used for production of stress-tolerant pomegranate variants (Helaly *et al.*, 2014).

A reliable transformation method for inoculation of *in vitro*-derived shoot explants of 'Yousef Khani' pomegranate was reported using *A. tumefaciens* strain LBA4404 harbouring the binary vector pBI121 containing *nptII* and *ß-glucuronidase (gus)* genes (Valizadeh Kaji *et al.*, 2014). GUS staining was positive for 32 shoots of 59 newly proliferated ones on WPM selection medium. The putative transgenic shoots were cut and subcultured twice in the selection media for final *gus* analysis and six putative transformed plantlets were obtained after 3 months of selection. Confirmation of *gus* and *nptII* genes as well as stable integration of T-DNA was performed using PCR and Southern blot analysis, respectively. Although the transformation efficiency in this study was relatively low (1.6%), the authors reported that *in vitro* proliferated shoots are excellent explants for producing transgenic pomegranate (Valizadeh Kaji *et al.*, 2014). In addition, high frequency of escapes and chimeric shoots were reported when using this method.

Agrobacterium tumefaciens strain EHA105 harbouring pBin19 binary vector, containing the *nptII* and *GFP* genes was used for optimization of callus transformation in pomegranate cv. 'Rabab' (Valizadeh Kaji and Abbasifar, 2017). In a preliminary study, the authors optimized the callus induction and regeneration method and reported that callus induction was the highest from the internodal stem explants on WPM containing 12 μM BA and 8 μM NAA while the highest shoot regeneration (69.33%) and number of shoots per piece of callus (7.16%) were attained on the same medium supplemented with 12 μM BA and 2 μM NAA (Valizadeh Kaji and Abbasifar, 2017).

As well as the studies carried out with the purpose of transformation optimization in pomegranate, there are also some reports with the objectives of obtaining new lines with resistance to bacterial blight disease (Hosamani *et al.*, 2016). Plant ferredoxin-like protein (PFLP) was used in the pCAMBIA 2301 binary vector under CaMV35S promoter and *nptII* selection marker for transformation of leaf, node, petal and cotyledon calli of cv. 'Bhagwa'. One leaf-derived and two cotyledonary-derived plants showed the presence of the *PFLP* gene using the PCR method. However, no information was provided about the expression of the *PFLP* gene in the transformed plants, or the integration incidents (i.e. using Southern blot analysis), or the degree of bacterial blight resistance in the transgenic plants under greenhouse or field conditions. In addition, similar to other reports, the reported transformation frequency was very low.

Several efforts have been made to produce bacterial blight-resistant pomegranate plants in south India, where this disease is becoming a serious problem (Nungshilepden, 2009; Swetha, 2011; Hosamani, 2012). All these reports used 'Bhagwa' cultivar and *A. tumefaciens* LBA 4404 strain harbouring pCAMBIA construct containing CaMV35S promoter driving antimicrobial peptide (*AMP*) gene and *nptII* as a selectable marker. Although the main objective of all these investigations was stated as the production of bacterial blight-resistant varieties, the majority of their investigations were dedicated to the optimization of transformation procedures, especially for evaluating the different explants' efficiency, and no analyses of resistance capacity of produced transformants were reported. Among the different examined explants, including leaf, petal, node and cotyledon, callus obtained from cotyledonary explants showed the highest transformation and regeneration response (Nungshilepden, 2009; Swetha, 2011; Hosamani, 2012).

Although pomegranate does not seem to be a recalcitrant plant species, some tissue culture-associated problems (such as exudation of phenolics and browning of media and explants, microbial contamination and recalcitrant *in vitro* tissues) have caused it to be considered as a difficult plant for tissue culture propagation (Naik and Chand, 2011; Hosamani *et al.*, 2016). Therefore, transformation methods, enabling circumvention of the tissue culture establishment phase problems, are always preferred by researchers. Pollen grain-mediated transformation is one of these methods, which bypasses the requisite for *in vitro* culture. Briefly, with this method, the foreign gene (in the form of naked DNA produced by sonication or through an *Agrobacterium* vector system) is transferred into pollen grains, and then stigmas are pollinated with transformed pollen grains (Yang *et al.*, 2016). Digestion of plasmid by nucleases is one of the impediments to pollen transformation that could be minimized by sonication (Wang *et al.*, 2001). Yang *et al.* (2016) studied the effects of different sonic-related parameters, including intensity, processing duration and treatment times, on the transformation of pomegranate pollen grains using a plasmid DNA harbouring *gfp* gene. Although all of the sonic-related factors were significantly important for transformation of pollen grains of pomegranate, the ultrasonic intensity was the most important followed by processing duration and treatment time. The sonication intensity of 150W, treatment duration of 5s, for a total of seven times and each time with an interval of 10s, was reported as the best treatment (Yang *et al.*, 2016).

Pomegranate is a rich source of polyphenolic compounds including hydrolyzable tannins (predominantly punicalagins). Therefore, increasing the contents of such beneficial compounds is highly appreciated for pharmaceutical purposes. Due to their high growth rate, hairy root induction is one of the most effective methods and has been used routinely for increasing the plant-derived metabolites under *in vitro* conditions. Ono *et al.* (2012) evaluated the efficiency of three strains (MSU440, 15834 and A4) of *Agrobacterium rhizogenes* harbouring a binary vector containing a yellow fluorescent protein (YFP) as reporter gene, for transformation of different explants (radicle, cotyledon and leaf) of pomegranate cv. 'Wonderful'. Both bacterial strain and explants significantly affected the transformation efficiency. *Agrobacterium rhizogenes* strain MSU440 was the most effective strain for obtaining transgenic hairy roots and cotyledon explants resulted in the highest transgenic roots. The authors reported that pomegranate hairy root culture is a good system for expressing heterologous genes.

Most of the transformation reports on pomegranate used *A. tumifaciens* as the vehicle, but examined different explants as well as transformation systems with the aim of optimizing the transformation procedure. *Agrobacterium tumefaciens* strain LBA4404 has been used in the majority of pomegranate transformation studies. This could be attributed to the easy accessibility and controlling of this strain. It is reported that the elimination of LBA4404 from plant tissues is relatively easy at low concentrations of antibiotics (Maheswaran *et al.*, 1992). Although low frequency of transformation rate using this strain is reported in other fruit species, the effects of cultivar and developmental stage of the explant can not be ignored (Petri and Burgos, 2005).

Although some efforts have been dedicated to obtaining bacterial blight- and pest-resistant pomegranates, no definitive analyses of resistance were performed on the transformants. Methods to introduce genes, either from pomegranates or other organisms, into existing pomegranate cultivars are now well established and permit the targeted modifications of existing pomegranate cultivars. This may provide the means to reduce losses due to disease and pesticide usage in classic cultivars without changing their good attributes. Frost tolerance, seed softness, low suckering and resistance to the fruit borers, especially *Spectrobates ceratonia* (as the most destructive pomegranate pest in Iran) are some of the proposed subjects that should be considered in future breeding programmes.

4.12.10 Transcriptomics studies

The advent of new next-generation sequencing (NGS) tools could be considered as a milestone in biology-related studies, which introduced a remarkable change in the acquired data and finally our perception of different features of living organisms. In fact, new technologies have enabled researchers to shift from gene-by-gene studies of different phenomena towards studying whole living systems as a complete unit. This creates a big picture of different phenomena in a way that is more similar to the natural situation of living cells, which include the interaction of different components. These improvements have resulted in the appearance of various high-throughput methods in biology including different omics studies, such as comparative genomics, transcriptomics, proteomics and metabolomics, which serve as powerful tools for revealing the genetic basis of different phenomena in living organisms.

Transcriptomics studies have gained attention in recent years. In order to investigate the biosynthesis of natural products and compounds of pomegranate that affect pigmentation, flavour and nutritional value, Ono *et al.* (2011) explored the transcriptomics landscape of pomegranate fruit peel cv. 'Wonderful' using RNA-seq by Illumina genome analyser platform.

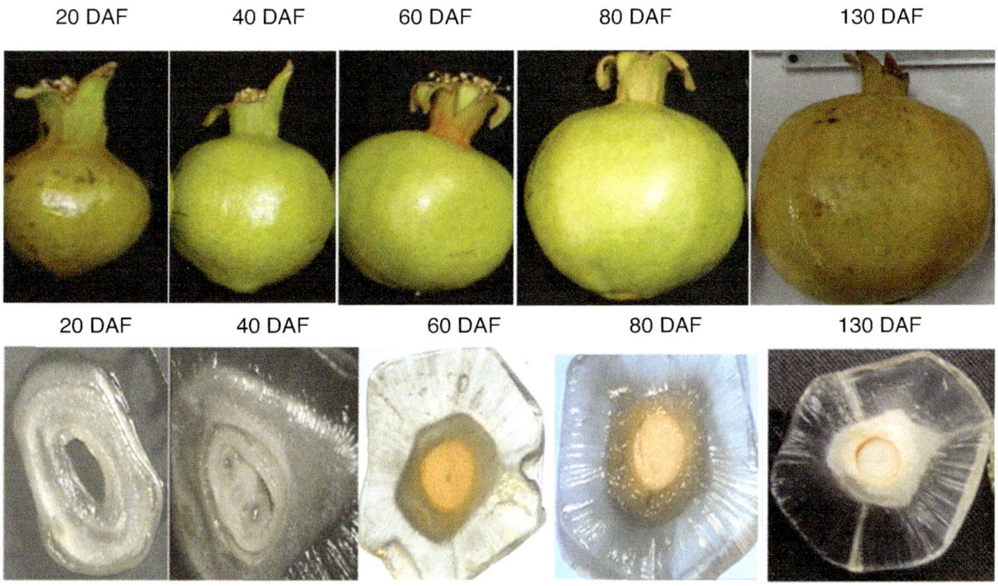

Fig. 4.8. Arils and fruits of pomegranate 'Bihaste-Ravar', a soft-seeded pomegranate accession at different developmental stages of fruit ripening. DAF, days after full bloom. (Photos: Abdolkarim Zarei.)

After analysing over 100 million raw sequence reads, 9839 transcriptome assemblies with the size of >200 bp were constructed and some candidate genes for hydrolyzable tannis, flavonoid, anthocyanin, terpenoid and fatty acid biosynthesis or regulation were identified. In addition, these authors identified 115 EST-SSRs from peel transcripts, designed specific primers for 77 SSR markers and reported that the NGS approach is an economical and effective method for gene identification and molecular markers in non-model plants (Ono *et al*., 2011).

Due to its importance as an organoleptic trait in different pomegranate producer regions, the seed softness characteristic has been the subject of various transcriptomics studies (Kacar *et al*., 2016; Zarei *et al*., 2016a, Zarei *et al*., 2016b; Dong *et al*., 2018; Luo *et al*., 2018b). Zarei *et al*. (2016b) examined the expression level of some of the cell wall-related biosynthesis genes in the integuments of soft- and hard-seeded pomegranate cultivars during different developmental stages of fruit development (Fig. 4.8) and reported that different monolignol ratios and lignin polymerized structures might have a more important role than lignin content in the development of seed softness/hardness characteristics. Xue *et al*. (2017) reported that lignin- and cellulose-related genes were highly expressed in the hard-seed cultivar, while programmed cell death and flavonoid-related genes were slightly higher in the soft-seed cultivar. The authors also found that in addition to lignin, biosynthesis of cellulose, hemicellulose and callose might be related to the testa formation in pomegranate. It is reported that in addition to the MYB and NAC transcription factors, which affect the lignin expression genes in several plants, WRKY transcription factor is also involved in the development of lignin in pomegranate (Xue *et al*., 2017). Recent results of microRNA and mRNA expression profiles of soft- and hard-seeded pomegranate cultivars imply that several microRNA are involved in seed softness development, which could regulate the enzymes, transcription factors and proteins involved in seed harness (Luo *et al*., 2018b). A predicted microRNA (novel-mir367) was found to regulate coniferylaldehyde dehydrogenase (which is involved in the final step of biosynthesis of guaiacyl and syringyl monolognols), and pectate lyase (which is a depolymerizing enzyme that degrades plant cell walls), and three microRNA was found to down-regulate the UDP-glucuronate decarboxylase (which is required for xylan formation) and may alter cell wall structure by increasing the cellulose (Luo *et al.*, 2018b).

According to reports, NAC, WRKY and MYC are among the transcription factors that are related to seed hardness in pomegranate (Xue *et al.*, 2017). Among the transcription factors regulating microRNAs, mdm-miR164e directly down-regulated NAC1, while mdm-miR172 up-regulated WRKY, MYC and AP2 to block seed hardness (Luo *et al.*, 2018a). According to these authors, some other microRNAs also were differentially expressed in the soft- and hard-seeded cultivars that were involved in the regulation of inorganic phosphate transporter (lack of energy limits phosphate deposition in seed vacuoles and soft seeds) and down-regulation of homeobox-leucine zipper proteins (which may limit seed hardness) as well as a CLIP-associating protein, which prevents xyloglucan formation and has a subsequent effect on the cellulose structure. Therefore, seed softness is a complex biological process regulated by the miRNA–mRNA network in pomegranate.

Laccases along with peroxidases are two key lignin biosynthesis genes that are responsible for lignin polymerization. With the aim of shedding light on the seed softness characteristic in pomegranate, Dong *et al.* (2018) cloned, characterized and screened the expression of laccase genes from pomegranate (*PgLAC*). Sequence characterization of *PgLAC* indicated that this gene was most closely related to the *LAC5* orthologue identified in *Eucalyptus grandis* (EgLAC5). *Eucalyptus grandis* is one of the most studied members of the Myrtaceae family, which is in the same order (Myrtales) as pomegranate, is reported to be close in sequence with pomegranate and is used as a reference for assembling pomegranate short reads. The expression of *PgLAC* increased from 20 to 80 days after flowering in the pomegranate fruit, which is the key stage of seedcoat hardening. The expression of this gene decreased from this stage to ripening. In addition, *PgLAC* showed higher expression in the cultivars with high lignin content. According to Dong *et al.* (2018) *PgLAC* may be a good candidate gene for reducing the lignin content and subsequently the seed hardness of pomegranate. However, the expression of *PgLAC* was detected in different tissues including leaf, petal and stem, and its down-regulation may cause a profound effect on the tree structure. Hence, the practical application of this gene for reducing seed hardness in pomegranate is highly dependent on the availability of a seedcoat-specific promoter for driving the transgene.

The RNAseq approach has also been used to shade light on the molecular basis of female sterility in pomegranate, which can result in the formation of two flower types in this species (Chen *et al.*, 2017). Cessation in ovule development after the formation of the inner integument primordium is the main reason for the development of female sterility in pomegranate and genes influencing the ovule development may be crucial factors in this phenomenon. *INNER OUTER* and *AINTEGUMENTA* homologous genes were two regulators of integument development, which were down-regulated in the female flowers at critical stages of ovule development (Chen *et al.*, 2017). Upstream regulators of these two genes, including *AGAMOUS*-like and *SPOROCYTELESS* homologue genes were also differentially expressed in the female sterile and fertile flowers at critical stages of ovule development. In addition, reports indicated that ethylene response signal genes such as ethylene-resistant (*ETR*) and ethylene-responsive factors (*ERF1/2*) might be involved in female sterility of pomegranate (Chen *et al.*, 2017). This interpretation was supported by an increase in fertile flowers after ethephon spraying.

High-throughput sequencing of the RNA pool from young seedlings to mature fruit has been used to identify microRNAs (miRNA) in pomegranate, and both conserved and pomegranate-specific miRNAs were identified (Saminathan *et al.*, 2016). Results indicated that miR157 was the most frequent family of miRNAs, followed by miR156, miR166 and miR168, respectively. In addition, 24 nucleotide miRNAs comprised more than 50% of the small RNA tags followed by 21 nucleotides (21.86%) and 23 nucleotides (8.95%). The expression pattern of predominant and novel miRNAs revealed that miR156, miR156a, miR159a, miR159b and miR319b were up-regulated during the final stages of fruit development and the authors concluded that miRN156, which is highly expressed at the final stages of fruit development, may be involved in the regulation of anthocyanin biosynthesis by reducing SPL transcription factor. Ten novel miRNAs were reported in pomegranate that targeted different transcription factors and hormone-related regulators. According to the KEGG pathway, these miRNAs are involved in ascorbate and linolenic acid, starch and sucrose metabolism, RNA transport, plant hormone signalling pathways and the circadian cycle (Saminathan *et al.*, 2016).

Due to its importance as a source of health beneficial substances as well as consumer acceptance, the anthocyanin biosynthesis pathway has also been the subject of various transcriptomics studies in pomegranate (Ben-Simhon *et al.*, 2011, 2015; Zhao *et al.*, 2015; Rouholamin *et al.*, 2015; Luo *et al.*, 2018a). Several anthocyanin biosynthesis genes have been identified in pomegranate and have been successfully cloned and characterized. According to available records, it is suggested that WD40 protein may have a prominent role in the biosynthesis of anthocyanin during fruit development (Ben-Simhon *et al.*, 2011). In addition, it was found that the expression of WD40, along with MYB (*PgAn1*) and bHLH (*PgAn2*) transcription factors, is required to regulate downstream regulatory genes of the anthocyanin biosynthesis pathway (Ben-Simhon *et al.*, 2011). Rouholamin *et al.* (2015) detected the highest and the lowest expression of bHLH (*AN1*) transcripts in the white/green-skinned fruits and bright red/black-skinned fruits, respectively. In addition, green and bright red fruit skins had the highest and the lowest expression of MYB (*AN2*) transcription factor (Rouholamin *et al.*, 2015). Moreover, the highest and the lowest expression levels of *DFR* gene were recorded in black and white fruit skins and this finding was attributed to the major effect of WD40 transcription factor on this structural gene (Rouholamin *et al.*, 2015). Lack of expression of anthocyanidin synthase (*ANS*) gene in the peel of white colour pomegranate was suggested as the main factor responsible for the white phenotype in pomegranate skin and *PgANS* was proposed as the key gene involved in anthocyanin biosynthesis (Zhao *et al.*, 2015). Gene expression analysis of anthocyanidin synthase (*ANS*) (also called leucoanthocyanidin dioxygenase; *LDOX*) in a 'white' pomegranate, which lacked anthocyanin in all tissues of the plant including flowers, fruits and leaves, also revealed that none of the tissues of this accession was able to synthesize transcription of this gene (Ben-Simhon *et al.*, 2015). The presence of an SNP and an insertion in the *PgLDOX* in all 'white' accessions was observed by molecular analysis. Further investigations revealed that the SNP is a synonymous and is not completely linked with the 'white' phenotype, while the insertion is completely linked with this phenotype and segregated as a recessive single gene trait in the segregating populations (Ben-Simhon *et al.*, 2015).

RNA-sequencing technology has also been used in pomegranate to study the genes that are involved in the fruit peel colour development, and several candidate genes corresponding to the general reaction and glycolsylation of the pigments and compounds have been identified from pomegranate peel (Luo *et al.*, 2018a).

4.12.11 Proteomics studies

There are few reports on protein contents of pomegranate fruit, and some of them are limited to specific proteins. Two lipid transfer proteins from fruit juice were purified by SDS-PAGE and then were characterized by mass spectrometry (Zoccatelli *et al.*, 2007). Trends in total protein content as well as some enzymatic activities were screened in various pomegranate genotypes during different stages of fruit development (Zarei *et al.*, 2016a). Total water-soluble non-storage proteins (Yang *et al.*, 2012) and class III chitinase (Yang *et al.*, 2011) were isolated from pomegranate seeds and it was reported that out of 120 protein spots that were resolved with two-dimensional electrophoresis, seven abundant spots with low molecular mass were identified by liquid chromatography–tandem mass spectrometry. Globulins, albumins, glutelin and prolamin were reported as the main components of seed storage proteins in pomegranate (Elfalleh *et al.*, 2010; Elfalleh *et al.*, 2012). Pomegranate aril proteome using a gel free shotgun proteomics approach revealed that out of 1488 proteins only six were unique to pomegranate species (Capriotti *et al.*, 2013). In addition, these authors reported that due to the long phylogenetic distance with model plants, many proteins had been assigned only one peptide in homology search analysis.

Seed hardness has been the subject of two proteomics studies in pomegranate (Cao *et al.*,

2015; Niu *et al.*, 2018). Two-dimensional electrophoresis gels have been used to compare the proteom profile of 'Zhongnonghong' (soft seed) and 'Sanbai' (hard seed) pomegranates (Cao *et al.*, 2015). Out of 892 protein spots, 76 were differentially expressed. The soft-seeded cultivar possessed 14 up-regulated and 10 down-regulated proteins, some of which were characterized by MALDI-TOF-TOF MS, of which three proteins were hypothesized to play a role in seed softness and the two remaining were involved in seed protection from adverse stress during fruit development.

Niu *et al.* (2018) also compared the protein differences of hard- and soft-seeded pomegranates and identified 1940 proteins. These authors reported that most of the 399 differentially expressed proteins in the soft-seed and hard-seed genotypes were involved in post-transcriptional modification and carbohydrate metabolism. The results of transcript expression of 14 proteins involved in cell wall biosynthesis revealed that lignin-related differentially expressed proteins had lower expression, while cellulose biosynthesis-related proteins had higher expression at both mRNA and protein levels in the soft-seed genotype than in the hard-seed genotype at an early stage of fruit development (60 days after flowering). In addition, cell wall degradation-related proteins showed higher expression level in the soft-seed genotype at both studied stages (60 and 120 days after flowering). Collectively the authors suggested that lignin and cellulose play opposing roles in cell wall formation in pomegranate seeds. The same conclusion was reached in some of the previous studies that had the aim of revealing the basis of seed softness (Zarei *et al.*, 2016a; Saminathan *et al.*, 2016; Xue *et al.*, 2017).

In addition, the isobaric tags for the relative and absolute quantitation (iTRAQ) proteome-level analysis method have been used in pomegranate to identify proteins related to the pomegranate fruit peel colours in Tunisian pomegranates (Luo *et al.*, 2018a). The authors analysed and compared two red-peel and white-peel pomegranate cultivars at two different stages of fruit colouring and ripening and found 27 differentially abundant proteins. These authors observed that many proteins had differential abundance between the two fruit development stages, and there were more differential proteins at the second stage. The authors also combined the proteomics and transcriptomics data and reported that anthocyanin, stilbenoid, diarylheptanoid, gingerol, flavonoid and phenylpropanoid are among the main biosynthesis pathways that contribute to the final colour of pomegranate fruit peel. They finally concluded that a complex transcriptional and translational network regulates the colour formation of pomegranate fruit peel.

4.13 Conclusions and Future Prospects

Punica granatum is a highly diverse plant species with a high level of adaptation capacity that is already cultivated from east to west on five continents, and the cultivation area is rising in different regions. There are several germplasm collections of this fruit species around the world but there is a need for a more comprehensive evaluation of the existing collections and their enrichment. In addition, there is very limited information about some of the existing germplasms, and their capacity is not exploited for the genetic improvement of this fruit species. Specifically, the wild accessions need to be considered more seriously in the regions that have rich materials. Even though there are few reports about the cytological characteristics of pomegranate, differences have been reported between pomegranate varieties in meiosis and other features of cell division, which are an indication of genetic diversity in this fruit tree. Molecular marker studies are ample and different markers have been successfully exploited to assess the genetic basis of various cultivated and wild types of this fruit across different regions. In fact, molecular marker analysis is one of the most comprehensively covered aspects of pomegranate studies. Many studies have also been conducted to optimize tissue culture condition and, despite the many remaining challenges, tremendous progress has been made in this area. Most pomegranate cultivars are selected by exploiting natural seedling variations (chance seedlings) and some by deliberate application of breeding techniques. Clone selections from the existing cultivars may also have been another source for expansion of the cultivated materials. Mutation and polyploidy work in pomegranate has resulted in a small number of cultivars. Although hybridization studies are increasing and progress has been made in the areas of inheritance patterns of some important traits as well as the introducion of new varieties, to keep pace with the increasing demand for this fruit tree across the world, more investigations are needed to breed new cultivars with special characteristics. Most of the genetic transformation studies in pomegranate were conducted with the objective of protocol optimization, and practical application of this approach could significantly affect pomegranate breeding programmes. The majority of transcriptomics and proteomics studies are restricted to a few important fruit traits such as seed softness and colour-related attributes. However, other aspects of this fruit tree, including tolerance to different biotic and abiotic stresses, should also be considered in different omics studies. With the advent of NGS technology, it has become more feasible to obtain a large amount of sequence data in plant species. So far, some pomegranate genomics studies have been carried out and there is a need for more exploitation of these data to reveal the pomegranate genome features including genic regions, the number of genes, their positions as well as functions, characterization of regulatory elements and subsequent application of such data in molecular breeding of pomegranate. Therefore, greater progress is needed in the fields of genomics, gene mapping and functional genomics of pomegranate, which have lagged behind those in other fruit crops such as apple, grape and peach. This progress is bound to continue as current techniques are refined and new ones are developed so the required cost is reduced to be applicable in fruit species like pomegranate.

References

Adivapp, N., Jalikop, S.H., Ananthanarayanan, T.V., Srinivas Rao, N.K. and Laxman, R.H. (2010) Use of molecular markers in crop improvement of *Punica granatum* L. for water stress tolerance. *Acta Horticulturae* 865, 69–72. DOI: 10.17660/ActaHortic.2010.865.7.

Agamy, E.S.Z., Mostafa, R.A.A., Shaaban, M.M. and El-Mahdy, M.T. (2009) *In vitro* propagation of Manfalouty and Nab El-gamal pomegranate cultivars. *Research Journal of Agriculture and Biological Sciences* 5(6), 1169–1174.

Ajal, E.A., Jbir, R., Legua, P., Martínez, J.J., Martinez, R. *et al.* (2015) Genetic diversity of Moroccan pomegranate (*Punica granatum* L.) cultivars using AFLP markers. *Australian Journal of Crop Science* 9(1), 22–29.

Akbar, S., Mirlohi, A. and Seyed Tabatabaei, B. (2008) Study on pollination type of pomegranate by RAPD molecular markers. In: *Abstract Book of Fifth National Congress of Biotechnology*. 208. Tehran, Iran.

Akhundzadeh, I.M. (1981) Radiation mutagenesis in subtropical crops. *1-ya-Vses-konf-po-prikl-radiobiol: -Teor-prikl-aspekty-radiats-biol-tekhnol,-10-12-noyab,-1981-Tez-dokl*, 50–51.

Akhundzadeh, I.M., Fedorova, E.E., Mamedov, G.M. and Iskenderova, Z.D. (1977) Study of the cytogenetic characteristics of pomegranate. *Ispol'-z-biofiz-metodov-v-genet-selektsion-eksperimente*, 8–9.

Al-Wasel, A.S.A. (1999) *In vitro* clonal propagation of 'Al-Belehi' pomegranate (*Punica granatum* L). *Journal of King Saudi University* 11, 3–14.

Allen, R.G., Pereira, L.S., Raes, D. and Smith, M. (1998) Crop evapotranspiration: guidelines for computing crop water requirements. In: *FAO Irrigation and Drainage 56*. UN-FAO, Rome.

Amar, M.H., El-Zayat, M.A.S. and Egyptian Deserts Gene Bank, Desert Research Center, Egypt., (2017) Utilization of ISTR, ISSR and SRAP molecular markers to reveal and classify Egyptian pomegranates (*Punica granatum* L.). *Plant Omics* 10(05), 237–246. DOI: 10.21475/poj.10.05.17.pne794.

Angiosperm Phylogeny Group (2009) An update of the Angiosperm Phylogeny Group classification for the orders and families of flowering plants: ApG III. *Botanical Journal of the Linnean Society* 161(2), 105–121. DOI: 10.1111/j.1095-8339.2009.00996.x.

Aranzana, M.J., García-Mas, J., Carbó, J. and Arús, P. (2002) Development and variability analysis of microsatellite markers in peach. *Plant Breeding* 121(1), 87–92. DOI: 10.1046/j.1439-0523.2002.00656.x.

Ataseven Isik, E. (2006) Similarities between pomegranate parents and crosses regarding some fruit characteristics. In: *ISHS 1st International Symposium on Pomegranate and Minor Mediterranean Fruits, Abstracts contributed papers, 16–19 October*. ISHS, Adana, Turkey, p. 67.

Azerbaijan Institute of Botany (2018) The pomegranate genome first read by Azerbaijani scientists. Available at: https://botany.az/en/news/1350 (accessed 26 November 2019).

Azqandi, S.R., Kazazi, M. and Abdul Ahadi, F. (2015) Ectomyelois ceratoniae Zeller (Lep., Pyralidae) and its control procedures in Iran.. *Journal of Applied Environmental and Biological Sciences* 5(12s), 743–747.

Baptista-Giacomelli, F.R., Pagliarini, M.S. and Almeida, J.L. de. (2000) Meiotic behavior in several Brazilian oat cultivars (*Avena sativa* L.). *Cytologia* 65(4), 371–378. DOI: 10.1508/cytologia.65.371.

Bar-Ya'akov, I., Hatib, K., Abed Elhadi, F. and Holland, D. (2003) Pomegranate cultivars in Israel: past and present. *Alon Hanotea* 57, 125–129.

Bar-Ya'akov, I., Trainin, T., Hefetz, H., Hatib, K. and Holland, D. (2007) Improving pomegranate cultivars in Israel. In: *ISHS 1st International Symposium Pomegranate and Minor Mediterranean Fruits, Abstracts contributed papers, 16–19 October*. ISHS, Adana, Turkey, p. 2.

Bartual, J., Valdés, G., Andreu, J., Lozoya, A., Garcia, J. *et al.* (2012) Pomegranate improvement through clonal selection and hybridization in Elche. *Options Méditerranéennes* 103, 71–74.

Bartual, J., Palou, L. and Pérez-Gago, M.B. (2015) Characterization of fruit traits from 'Mollar de Elche' pomegranate progenies. *Acta Horticulturae* 1106, 25–30. DOI: 10.17660/ActaHortic.2015.1106.5.

Basaki, T., Choukan, R., Nekouei, S.M.K., Mardim, M., Majidim, E. *et al.* (2011) Association analysis for morphological traits in pomegranate (*Punica geranatum* L.) using microsatellite markers.. *Middle-East Journal of Scientific Research* 9, 410–417.

Basaki, T., Nejat, M.A., Nejad, R.J., Faraji, S. and Keykhaei, F. (2013) Identification of simple sequences repeat (SSR) informative markers associated with important traits in pomegranate (*Punica granatum* L). *International Journal of Agronomy and Plant Production* 4, 575–583.

Behzadi Shahrbabaki, H. (1997) Genetic diversity of pomegranate genotypes in Iran. Agricultural Education Publication, Karaj, Iran.

Ben-Simhon, Z., Judeinstein, S., Nadler-Hassar, T., Trainin, T., Bar-Ya'akov, I. *et al.* (2011) A pomegranate (*Punica granatum* L.) WD40-repeat gene is a functional homologue of *Arabidopsis* TTG1 and is involved in the regulation of anthocyanin biosynthesis during pomegranate fruit development. *Planta* 234(5), 865–881 DOI: 10.1007/s00425-011-1438-4.

Ben-Simhon, Z., Judeinstein, S., Trainin, T., Harel-Beja, R., Bar-Ya'akov, I. *et al.* (2015) A 'white' anthocyanin-less pomegranate (*Punica granatum* L.) caused by an insertion in the coding region of the leucoanthocyanidin dioxygenase (LDOX; ANS) gene. *PLoS ONE* 10(11), e0142777. DOI: 10.1371/journal.pone.0142777.

Bennett, M.D. and Leitch, I.J. (2005) Nuclear DNA amounts in angiosperms: progress, problems and prospects. *Annals of Botany* 95(1), 45–90. DOI: 10.1093/aob/mci003.

Bhansali, R.R.A.J. and Raj Bhansali, R. (1990) Somatic embryogenesis and regeneration of in plantlets in pomegranate. *Annals of Botany* 66(3), 249–253. DOI: 10.1093/oxfordjournals.aob.a088022.

Bharose, A.A., Khandagale, D.J. and Damse, D.N. (2014) Regeneration media standardization of pomegranate (*Punica granatum* L.) cv. Bhagava from shoot tip, cotyledonary node, hypocotyls and leaves. *Vegetos* 27(2), 338–342.

Bist, H.S., Srivastava, R. and Sharma, G. (1994) Variation in some promising selections of wild pomegranate (*Punica granatum* L.). *Horticultural Journal* 7(1), 67–70.

Bonyanpour, A. and Khosh-Khui, M. (2013) Callus induction and plant regeneration in *Punica granatum* L. 'Nana' from leaf explants. *Journal of Central European Agriculture* 14(3), 75–83. DOI: 10.5513/JCEA01/14.3.1285.

Bouhadida, M., Casas, A.M., Moreno, M.A. and Gogorcena, Y. (2007) Molecular characterization of Miraflores peach variety and relatives using SSRs. *Scientia Horticulturae* 111(2), 140–145. DOI: 10.1016/j.scienta.2006.10.018.

Camacho, J.P.M., Sharbel, T.F. and Beukeboom, L.W. (2000) B-chromosome evolution. *Philosophical Transactions of the Royal Society of London. Series B: Biological Sciences* 355(1394), 163–178. DOI: 10.1098/rstb.2000.0556.

Cao, S.., Niu, J., Cao, D., Li, H.., Xue, H. *et al.* (2015) Comparative proteomics analysis of pomegranate seeds on fruit maturation period (*Punica granatum* L.). *Journal of Integrative Agriculture* 14(12), 2558–2564.

Capriotti, A.L., Caruso, G., Cavaliere, C., Foglia, P., Piovesana, S. *et al.* (2013) Proteome investigation of the non-model plant pomegranate (*Punica granatum* L.). *Analytical and Bioanalytical Chemistry* 405(29), 9301–9309. DOI: 10.1007/s00216-013-7382-3.

Chandra, R., Dhinesh Babu, K., Jadhav, T.V. and Teixeira da Silva, J.A. (2010a) Origin, history and domestication of pomegranate. *Fruit, Vegetable and Cereal Science and Biotechnology* 4(Special Issue 2), 1–6.

Chandra, R., Jadhav, V.T. and Sharma, J. (2010b) Global scenario of pomegranate (*Punica granatum* L.) culture with special reference to India. *Fruit, Vegetable and Cereal Science and Biotechnology* 4(Special Issue 2), 7–18.

Chaugule, R.R., More, T.A., Kamble, A.B. and Karale, A.R. (2005) Studies on micropropagation in pomegranate (Punica granatum L.). *Recent Trends in Horticultural Biotechnology (Vol I) and II-ICAE National Symposium on Biotechnological Interventions for Improvement of Horticultural Crops Issues and Strategies*, Vellanikkar, India, pp. 195–199.

Chen, L., Zhang, J., Li, H., Niu, J., Xue, H. *et al.* (2017) Transcriptomic analysis reveals candidate genes for female sterility in pomegranate flowers. *Frontiers in Plant Science* 8, 1430. DOI: 10.3389/fpls.2017.01430.

Conti, E., Fischbach, A. and Sytsma, K.J. (1993) Tribal relationships in Onagraceae: implications from rbcL sequence data. *Annals of the Missouri Botanical Garden* 80(3), 672–684. DOI: 10.2307/2399853.

Conti, E., Litt, A. and Sytsma, K.J. (1996) Circumscription of Myrtales and their relationships to other rosids: evidence from *rbc* L sequence data. *American Journal of Botany* 83(2), 221–233. DOI: 10.1002/j.1537-2197.1996.tb12700.x.

Conti, E., Litt, A., Wilson, P.G., Graham, S.A., Briggs, B.G. *et al.* (1997) Interfamilial relationships in Myrtales: molecular phylogeny and patterns of morphological evolution. *Systematic Botany* 22(4), 629–647. DOI: 10.2307/2419432.

Currò, S., Caruso, M., Distefano, G., Gentile, A. and La Malfa, S. (2010) New microsatellite loci for pomegranate, *Punica granatum* (Lythraceae). *American Journal of Botany* 97(7), e58–e60. DOI: 10.3732/ajb.1000143.

Dandachi, F., Hamadeh, B., Youssef, H., Chahine, H. and Chalak, L. (2017) Diversity assessment of the Lebanese germplasm of pomegranate (*Punica granatum* L.) by morphological and chemical traits. *Annals of Agricultural Sciences* 62(1), 89–98. DOI: 10.1016/j.aoas.2017.05.004.

Das, P.K. and Sur, S.C. (1968) Tetraploidy in pomegranate (*Punica granatum* L). *Technology Bihar* 5, 8–126.

Deepika, R. and Kanwar, K. (2008) Efficient *in vitro* shoot multiplication and root induction enhanced by rejuvenation of microshoots in *Punica granatum* cv. Kandhari Kabuli. *National Seminar on Physiological and Biotechnological Approaches to Improve Plant Productivity*, Hisar, India, 15–18 March, p. 24.

Deepika, R. and Kanwar, K. (2010) *In vitro* regeneration of *Punica granatum* L. plants from different juvenile explants. *Journal of Fruit and Ornamental Plant Research* 18(1), 5–22.

Demirel, N. (2016) A study on occurrence and population trends of the carob moth, *Ectomyelois ceratoniae* (Zeller) (Lepidoptera: Pyralidae) in pomegranate orchards by using pheromone traps. *Entomology and Applied Science Letters* 3(1), 26–31.

Desai, U.T., Masalkar, S.D. and Choudhari, S.M. (1994) Association of fruit characters in pomegranate. *Annals of Arid Zone* 33(2), 157–158.

Dong, L., Xiong, F., Liu, N., Wang, Q. and Zhang, S. (2018) Molecular cloning and expression analysis of *PgLAC* in pomegranate. *American Journal of Molecular Biology* 8, 145–154.

Ebrahimi, A., Zarei, A., Lawson, S., Woeste, K.E. and Smulders, M.J.M. (2016) Genetic diversity and genetic structure of Persian walnut (*Juglans regia*) accessions from 14 European, African, and Asian countries using SSR markers. *Tree Genetics & Genomes* 12(6), 114. DOI: 10.1007/s11295-016-1075-y.

Ebrahimi, A., Zarei, A., Zamani Fardadonbeh, M. and Lawson, S. (2017) Evaluation of genetic variability among ‘Early Mature’ *Juglans regia* using microsatellite markers and morphological traits. *PeerJ* 5(2), e3834. DOI: 10.7717/peerj.3834.

Ebtadaei, M. and Shekafande, A. (2016) Morpho-physiological changes of two cultivars of pomegranate ‘Rabab’ and ‘Shisheh Gap’ under water stress conditions. *Iranian Journal of Horticultural Science and Technology* 17(2), 209–220.

Elfalleh, W., Nasri, N., Sarraï, N., Guasmi, F., Triki, T. *et al.* (2010) Storage protein contents and morphological characters of some Tunisian pomegranate (*Punica granatum* L.) cultivars. *Acta Botanica Gallica* 157(3), 401–409. DOI: 10.1080/12538078.2010.10516217.

Elfalleh, W., Hannachi, H., Guetat, A., Tlili, N., Guasmi, F. *et al.* (2012) Storage protein and amino acid contents of Tunisian and Chinese pomegranate (*Punica granatum* L.) cultivars. *Genetic Resources and Crop Evolution* 59(6), 999–1014. DOI: 10.1007/s10722-011-9739-9.

Elhem, M., Abdennaceur, B., Mansour, H., Mabrouka, A., Khouloud, B. *et al.* (2011) Selection of pomegranate (*Punica granatum* L.) in south-eastern Tunisia. *African Journal of Biotechnology* 10(46), 9352–9361. DOI: 10.5897/AJB10.1959.

Ercisli, S., Gadze, J., Agar, G., Yildirim, N. and Hizarci, Y. (2011) Genetic relationships among wild pomegranate (*Punica granatum*) genotypes from Coruh Valley in Turkey. *Genetics and Molecular Research* 10(1), 459–464. DOI: 10.4238/vol10-1gmr1155.

Feng, Y.Z., Song, M.T. and Han, D.B. (2006) The general status of pomegranate germplasm resources in China. *China Fruits* 4, 57–58.

Ferrara, G., Cavoski, I., Pacifico, A., Tedone, L. and Mondelli, D. (2011) Morpho-pomological and chemical characterization of pomegranate (*Punica granatum* L.) genotypes in Apulia region, Southeastern Italy. *Scientia Horticulturae* 130(3), 599–606. DOI: 10.1016/j.scienta.2011.08.016.

Ferrara, G., Giancaspro, A., Mazzeo, A., Giove, S.L., Matarrese, A.M.S. *et al.* (2014) Characterization of pomegranate (*Punica granatum* L.) genotypes collected in Puglia region, Southeastern Italy. *Scientia Horticulturae* 178, 70–78. DOI: 10.1016/j.scienta.2014.08.007.

Frison, E.A. and Serwinski, J. (1995) Directory of European institutions holding crop genetic resources collections, vol. 1, Holdings. In: *Plant Genetic Resources Institute*, 4th edn. Available at: www.ecpgr.cgiar.org/publications/directories/direct94.htm

Gharaghani, A., Ghasemi Soloklui, A.A., Oraguzie, N. and Zare, D. (2017) Pollen source influences fruit quality, aril properties, and seed characteristics in pomegranate. *International Journal of Fruit Science* 17(3), 333–348. DOI: 10.1080/15538362.2017.1318733.

Ghasemi Soloklui, A.A., Ershadi, A. and Fallahi, E. (2012) Evaluation of cold hardiness in seven Iranian commercial pomegranate (*Punica granatum* L.) cultivars. *HortScience* 47(12), 1821–1825. DOI: 10.21273/HORTSCI.47.12.1821.

Ghobadi, S., Khoshkhoy, M. and Seyed-Tabatabaei, B. (2006) Phylogenetic relationships among some Iranian pomegranate accessions revealed by inter-simple sequence repeat (ISSR) markers. *Journal of Horticultural Science and Technology* 6, 11–20.

Ghorbani, M., Dabbagh, G.R., Yousefi, S., Khademi, S. and Taki, M. (2015) The effect of application of different kinds of covers on the sunburn and internal qualities of pomegranate in Iran. *Biological Forum* 7(1), 64–68.

Gill, B., Bir, S.S. and Bedi, Y.S. (1981) Cytological studies on woody Euphorbiaceae from north and central India. *New Botanist* 8, 35–44.

Glozer, K. and Ferguson, L. (2008) *Pomegranate production in Afghanistan*. UC Davis, California, p. 39.

Goor, A. and Liberman, J. (1956) The pomegranate. In: Atsmon, J. (ed.) *The Pomegranate. Bulletin of the Ministry of Agriculture*. State of Israel, Ministry of Agriculture, Tel Aviv, pp. 5–57.

Guarino, L., Miller, T., Baazara, M. and Obadi, N. (1990) Socotra: the island of bliss revisited. *Diversity* 6, 28–31.

Gulick, P. and Van Sloten, D.H. (1984) *Directory of Germplasm Collections. 6-1 Tropical and Subtropical Fruits and Tree Nuts*. IBPGR, Rome.

Hajiahmadi, Z., Talebi, M. and Sayed-Tabatabaei, B.E. (2013) Studying genetic variability of pomegranate (*Punica granatum* L.) based on chloroplast DNA and barcode genes. *Molecular Biotechnology* 55(3), 249–259. DOI: 10.1007/s12033-013-9676-2.

Hajiyeva, S.V., Akparov, Z.I., Hasanov, N.A., Mustafayeva, Z.P., Hajiyev, E.S. *et al.* (2018) Issr analysis of variability of cultivated form and varieties of pomegranate (*Punica granatum* L.) from Azerbaijan. *Russian Journal of Genetics* 54(2), 188–197. DOI: 10.1134/S1022795418020072.

Haq, S.U., Jain, R., Sharma, M., Kachhwaha, S. and Kothari, S.L. (2014) Identification and characterization of microsatellites in expressed sequence tags and their cross transferability in different plants. *International Journal of Genomics* 2014(3), 1–12Article ID 863948. DOI: 10.1155/2014/863948.

Harel-Beja, R., Sherman, A., Rubinstein, M., Eshed, R., Bar-Ya'akov, I. *et al.* (2015) A novel genetic map of pomegranate based on transcript markers enriched with QTLs for fruit quality traits. *Tree Genetics & Genomes* 11(5). DOI: 10.1007/s11295-015-0936-0.

Harlan, J.R. (1992) *Crops and Man*, 2nd ed. Agronomy and Crop Science Society of America, Madison, Wisconsin.

Hasanpour, Z. (2012) Response of two pomegranate cultivars on the drought and salt stress. MSc thesis in Horticultural Science. Valie Asr University, Rafsenjan, Iran.

Hasanpour, Z., Karimi, H.R. and Mirdehghan, S.H. (2014) Effects of salinity and water stress on eco-physiological parameters and micronutrients concentration of pomegranate (*Punuca granatum* L.). *Journal of Plant Nutrition* 38, 1–13.

Hasnaoui, N., Buonamici, A., Sebastiani, F., Mars, M., Trifi, M. *et al*. (2010) Development and characterization of SSR markers for pomegranate (*Punica granatum* L.) using an enriched library. *Conservation Genetics Resources* 2(S1), 283–285. DOI: 10.1007/s12686-010-9191-8.

Hasnaoui, N., Buonamici, A., Sebastiani, F., Mars, M., Zhang, D. *et al*. (2012) Molecular genetic diversity of *Punica granatum* L. (pomegranate) as revealed by microsatellite DNA markers (SSR). *Gene* 493(1), 105–112. DOI: 10.1016/j.gene.2011.11.012.

Hassan, N.A. and Abd-El Gawad, M.H. (2013) Morphological karyotype analysis of eleven pomegranate cultivars. *American-Eurasian Journal of Agriculture & Environmental Science* 13(11), 1562–1567.

Hassani Moghadam, E., Esna-Ashari, M. and Rezainejad, A. (2014) Effect of drought stress on some physiological characteristics in six commercial Iranian pomegranate (*Punica granatum* L cultivars). *Plant Production Technology* 15(1), 1–11.

Hazel, Y., Zibin, Z., Nadav, R. and Michael, E.W. (2011) Characterization of attributes related to fruit size in pomegranate. *HortScience* 46(6), 908–912.

Helaly, M.N., El-Hosieny, H., Tobias, N., Alsudays, I., Abdelaziz Omar, S. *et al*. (2014) *In vitro* studies on regeneration and transformation of some pomegranate genotypes. *Australian Journal of Crop Science* 8(2), 307–316.

Holland, D., Bar-Ya'akov, I. and Hatib, K. (2014) 'Emek', a red and very early-ripening new pomegranate cultivar. *HortScience* 49(7), 968–970. DOI: 10.21273/HORTSCI.49.7.968.

Holland, D., Hatib, K. and Bar-Ya'akov, I. (2009) Pomegranate: botany, horticulture, breeding. *Horticultural reviews* 35, 127–191.

Hosamani, A.M. (2012) Development of transgenic pomegranate with disease resistant gene for control of bacterial blight. MSc Thesis. University of Agricultural Sciences, Bengaluru, India.

Hosamani, A., Mohandas, S., Manjula, V., Kerure, P. and Geetha, M.S. (2016) *Pflp* gene transformation in pomegranate (*Punica granatum* L.) resistance to bacterial blight disease (*Xanthomonas axonopodis* pv. *punicae*). *Electronic Journal of Plant Breeding* 7(4), 864–870. DOI: 10.5958/0975-928X.2016.00117.4.

Hosein-Beigi, M., Zarei, A., Rostaminia, M. and Erfani-Moghadam, J. (2019) Positive effects of foliar application of Ca, B and GA3 on the qualitative and quantitative traits of pomegranate (*Punica granatum* L.) cv. 'Malase-Torshe-Saveh'. *Scientia Horticulturae* 254, 40–47. DOI: 10.1016/j.scienta.2019.04.081.

Houben, A. (2017) B chromosomes – a matter of chromosome drive. *Frontiers in Plant Science* 08(618), 210. DOI: 10.3389/fpls.2017.00210.

Huang, Y. Shi, S., Huang, Y. and Shi, S. (2002) Phylogenetics of Lythraceae *sensu lato*: a preliminary analysis based on chloroplast *rbc* L gene, *psa A – ycf 3* spacer, and nuclear rDNA internal transcribed spacer (ITS) sequences. *International Journal of Plant Sciences* 163(2), 215–225. DOI: 10.1086/338392.

IBPGR (1986) *Punica granatum* (pomegranate). In: *Genetic Resources of Tropical, Sub-Tropical Fruits and Nuts (Excluding Musa)*. International Board for Plant Genetic Resources, Rome, pp. 97–100.

Jaidka, K. and Mehra, P.N. (1986) Morphogenesis in *Punica granatum* (pomegranate). *Canadian Journal of Botany* 64(8), 1644–1653. DOI: 10.1139/b86-220.

Jalikop, S.H. (2003) Rosetted siblings in F2 of a cross in pomegranate (*Punica granatum* L.) can be useful model for resetting investigations. *Euphytica* 131(3), 333–342. DOI: 10.1023/A:1024007429088.

Jalikop, S.H. (2007) Linked dominant alleles or inter-locus interaction results in a major shift in pomegranate fruit acidity of 'Ganesh' × 'Kabul Yellow'. *Euphytica* 158(1-2), 201–207. DOI: 10.1007/s10681-007-9443-1.

Jalikop, S.H. (2010) Pomegranate breeding. *Fruit, Vegetable and Cereal Science and Biotechnology* 4(02), 26–34.

Jalikop, S.H. (2011) Breeding of pomegranate and annonaceous fruits. *Acta Horticulturae* 890, 191–197. DOI: 10.17660/ActaHortic.2011.890.25.

Jalikop, S.H. and Kumar, P.S. (1990) Use of a gene marker to study the mode of pollination in pomegranate *(Punica granatum L.)*. *Journal of Horticultural Science* 65(2), 221–223. DOI: 10.1080/00221589.1990.11516050.

Jalikop, S.H. and Kumar, P.S. (2009) 'Double Flower' pomegranate originated from a hardseeded acidic pomegranate by spontaneous dominant gene 'df' mutation. *International Symposium on Pomegranate and Minor including Mediterranean Fruits*, Dhrawad, India, 23–27 June.

Jalikop, S.H., Tiwari, R.B. and Kumar, P.S. (2000) 'Amlidana': a new pomegranate hybrid. *Indian Horticulture* 47, 22–23.

Jalikop, S.H., Rawal, R.D. and Kumar, R. (2005) Exploitation of sub-temperate pomegranate Daru in breeding tropical varieties. *Acta Horticulturae* 696, 107–112. DOI: 10.17660/ActaHortic.2005.696.18.

Jalikop, S.H., Kumar, P.S., Rawal, R.D. and Ravindra, K. (2006) Breeding pomegranate for fruit attributes and resistance to bacterial blight. *Indian Journal of Horticulture* 63, 352–358.

Jayesh, K.C. and Kumar, R. (2004) Crossability in pomegranate (*Punica granatum* L.). *Indian Journal of Horticulture* 61(3), 209–210.

Jbir, R., Zehdi, S., Hasnaoui, N., Ben Dhiaf, A., Mars, M. *et al.* (2012) Microsatellite polymorphism in Tunisian pomegranates (*Punica granatum* L.): cultivar genotyping and identification. *Biochemical Systematics and Ecology* 44, 27–35.

Jian, Z.-H., Liu, X.-S., Hu, J.-B., Chen, Y.-H. and Feng, J.-C. (2012) Mining microsatellite markers from public expressed sequence tag sequences for genetic diversity analysis in pomegranate. *Journal of Genetics* 91(3), 353–358. DOI: 10.1007/s12041-012-0185-z.

Kacar, Y.A., Akgol, M., Simsek, O. and Donmez, D. (2016) Determination of genes involved in lignification of pomegranate seeds by transcriptome sequencing. *Journal of Biotechnology & Biomaterials* 06(08), 54(Suppl.). DOI: 10.4172/2155-952X.C1.068.

Kakar, K.L., Dogra, G.S. and Nath, A. (1987) Incidence and control of pomegranate fruit borers, *Virachola isocrates* Fb. and *Deudorix epijarbas* Moore. *Indian Journal of Agricultural Science* 57, 749–752.

Kalalbandi, B.M., Waskar, D.P., Khandare, V.S. and Gorad, D.S. (2014) Micropropagation studies on pomegranate var. *Bhagwa. Indian Journal of Horticulture* 71(4), 564–566.

Kanwar, K. and Rachna Kashyap, A. (2004) In vitro propagation of wild pomegranate (*Punica granatum* L.). *National Seminar on IPR of Horticultural Crops*, Solan, India, 12–13 October.

Kanwar, K., Joseph, J. and Deepika, R. (2010a) Comparison of *in vitro* regeneration pathways in *Punica granatum* L. *Plant Cell, Tissue and Organ Culture* 100(2), 199–207. DOI: 10.1007/s11240-009-9637-4.

Kanwar, K., Thankur, K., Verma, V. and Sharma, R.K. (2010b) Genetic variability of *in vitro* raised plants of *Punica granatum* L. by RAPDs. *Fruit, Vegetable and Cereal Science and Biotechnology*. *Special issue* 4(2), 144–147.

Karale, A.R., Supe, V.S., Kaulgud, S.N. and Kale, P.N. (1993) Pollination and fruit set studies in pomegranate. *Journal of Maharashtra Agriculture University* 18, 364–366.

Karale, A.R. and Desai, U.T. (2000) Study of heterosis for fruit characters in inter cultivar crosses of pomegranate (*Punica granatum* L.). *Indian Journal of Genetics and Plant Breeding* 60, 191–196.

Karimi, H.R. and Hassanpour, N. (2017) Effects of salinity, rootstock and position of sampling on macro nutrient concentration of pomegranate cv. Gabri. *Journal of Plant Nutrition* 40(16), 2269–2278.

Karimi, H.R. and Mirdehghan, S.H. (2013) Correlation between the morphological characters of pomegranate (*Punica granatum* L.) traits and their implications for breeding. *Turkish Journal of Botany* 37, 355–362.

Kazemialamuti, M., Zeinalabedini, M., Mousavi Derazmahalleh, M., Roodbar Shojaie, T., Poor Irandoost, H. *et al.* (2012) Extensive genetic diversity in Iranian pomegranate (*Punica granatum* L.) germplasm revealed by microsatellite markers. *Scientia Horticulturae* 146, 104–114.

Kerkadze, I.G. (1987) Radiation mutagenesis in subtropical crops. *Radiatsionnyi Mutagenez i ego rol' v Evoliutsii i Selektsi*, 231–254.

Khadivi, A., Ayenehkar, D., Kazemi, M. and Khaleghi, A. (2018) Phenotypic and pomological characterization of a pomegranate (*Punica granatum* L.) germplasm collection and identification of the promising selections. *Scientia Horticulturae* 238, 234–245. DOI: 10.1016/j.scienta.2018.04.062.

Khadivi-Khub, A., Kameli, M., Moshfeghi, N. and Ebrahimi, A. (2015) Phenotypic characterization and relatedness among some Iranian pomegranate (*Punica granatum* L.) accessions. *Trees* 29(3), 893–901. DOI: 10.1007/s00468-015-1172-9.

Koohi-Dehkordi, M., Sayed-Tabatabaei, B.E., Yamchi, A. and Danesh-Shahraki, A. (2007) Microsatellite markers in pomegranate. *Acta Horticulturae* 760, 179–184. DOI: 10.17660/ActaHortic.2007.760.23.

Küçük, E. (2007) Improvement of new pomegranate (*Punica granatum* L.) cultivars by cross breeding in Aegean Region. Project in Aegean Agricultural Research Institute, İzmir, Turkey. Available at: http://agris.fao.org/agris-search/search.do?recordID=TR2010000701

Kumar, R. (2012) Studies on hybridization in pomgranate (*Punica granatum* L.). MSc thesis in Horticultural Science, College of Horticulture, Dr. Yashwant Singh Parmar University of Horticulture and Forestry, Nauni. India.

Kumari, S.A.S.M., Weerawarna, S.B.A., Fahim, M., Randunu, R.P.D., Ikram, M. *et al.* (2017) Development of pomegranate hybrids. *Annals of Sri Lanka Department of Agriculture* 19, 36–59.

Kunzang, L., Deep Ji, B., Kiran, K., Shivendu, P. and Singh, S. (2017) Mutation studies in fruit crops: a review. *International Journal of Current Microbiology and Applied Sciences* 6(12), 3620–3633.

Larkin, P.J. and Scowcroft, W.R. (1981) Somaclonal variation — a novel source of variability from cell cultures for plant improvement. *Theoretical and Applied Genetics* 60(4), 197–214. DOI: 10.1007/BF02342540.

Levin, G.M. (1979) Pomegranate breeding in the USSR and the role of the Vavilov Institute of plant industry (Vir) collection in providing initial material. *Subtropicheskie Kul'tury* 5, 75–79.

Levin, G.M. (1990a) Breeding pomegranate. *Refrativnyl Zhumal* 10, 33–99.

Levin, G.M. (1990b) Induced mutagenesis in pomegranate. In: *Dostizheniya-nauki-v-praktiku: Kratkie-tezisy-dokladov-k-predstoyashchei-nauchnoi-konferent-sii:-Puti-uskoreniya-selektsionnogo-protestenii*, pp. 126–128.

Levin, G.M. (1994) Pomegranate (*Punica granatum* L.) plant genetic resource in Turkmenistan. *Plant Genetic Resource Newsletter* 97, 31–36.

Levin, G.M. (1995) Genofund of pomegranate in Turkmenistan (to the 60th anniversary of its creation). *Problems of Desert Development* 3, 84–89.

Levin, G.M. (2006) *Pomegranate Roads: A Soviet Botanist's Exile from Eden*. Floreant Press, Forestville, California, p. 183.

Liang, C.C. and Cheng, Y.D. (1991) Selection of elite pomegranate. *Journal of Fruit Science* 8, 59–60.

Lu, L.J., Gong, X.M. and Zhu, L.W. (2006) Study on seed hardness of pomegranate cultivars in China. *Journal of Anhui Agricultural University* 33(3), 356–359.

Luo, X., Cao, D., Li, H., Zhao, D., Xue, H. *et al.* (2018a) Complementary iTRAQ-based proteomic and RNA sequencing-based transcriptomic analyses reveal a complex network regulating pomegranate (*Punica granatum* L.) fruit peel colour. *Scientific Reports* 8(1), 12362. DOI: 10.1038/s41598-018-30088-3.

Luo, X., Cao, D., Zhang, J., Chen, L., Xia, X. *et al.* (2018b) Integrated microRNA and mRNA expression profiling reveals a complex network regulating pomegranate (*Punica granatum* L.) seed hardness. *Scientific Reports* 8(1), 9292. DOI: 10.1038/s41598-018-27664-y.

Luo, X., Li, H., Wu, Z., Yao, W., Zhao, P. *et al.* (2020) The pomegranate (*Punica granatum* L.) draft genome dissects genetic divergence between soft- and hard-seeded cultivars. *Plant Biotechnology Journal* 18(4), 955–968. DOI: 10.1111/pbi.13260.

Madhou, M., Normand, F., Bahorun, T. and Hormaza, J.I. (2013) Fingerprinting and analysis of genetic diversity of Litchi (Litchi chinensis Sonn.) accessions from different germplasm collections using microsatellite markers. *Tree Genetics & Genomes* 9(2), 387–396. DOI: 10.1007/s11295-012-0560-1.

Madhvi, S. (2015) *In vitro* selection of *Punica granatum* L. cv. Kandhari Kabuli against bacterial blight and pomegranate wilt. PhD Thesis. College of Horticulture, University of Horticulture and Forestry, Solan, India.

Maheswaran, G., Welander, M., Hutchinson, J.F., Graham, M.W. and Richards, D. (1992) Transformation of apple rootstock M26 with *Agrobacterium tumefaciens*. *Journal of Plant Physiology* 139(5), 560–568. DOI: 10.1016/S0176-1617(11)80370-6.

Mahishni, D.M., Muralikrishna, A., Shivashankar, G. and Kulkarni, R.S. (1991) Shoot tip culture method for rapid clonal propagation of pomegranate (*Punica granatum* L.). *Horticulture New Technologies and Applications. Proceedings of the International Seminar on New Frontiers in Horticulture*, Bangalore, India, 25–28 November, pp. 215–217.

Manohar, M.S., Tikka, S.B.S. and Nathu, L. (1981) Phenotypic variation and its heritable components in some biometric characters in pomegranate (*Punica granatum* L.). *The Indian Journal of Horticulture* 38, 187–190.

Mars, M. and Marrakchi, M. (1999) Diversity of pomegranate (*Punica granatum* L.) germplasm in Tunisia. *Genetic Resources and Crop Evolution* 46(5), 461–467. DOI: 10.1023/A:1008774221687.

Martinez-Nicolas, J.J., Melgarejo, P., Legua, P., Garcia-Sanchez, F. and Hernández, F. (2016) Genetic diversity of pomegranate germplasm collection from Spain determined by fruit, seed, leaf and flower characteristics. *PeerJ* 4(1), e2214. DOI: 10.7717/peerj.2214.

Martínez, J.J., Melgarejo, P., Hernández, F., Salazar, D.M. and Martínez, R. (2006) Seed characterisation of five new pomegranate (*Punica granatum* L.) varieties. *Scientia Horticulturae* 110(3), 241–246. DOI: 10.1016/j.scienta.2006.07.018.

Mascarenhas, A.F., Nair, S., Iyer, R.S. and Gupta, P.K. (1988) Genetic improvement of fruit crops through tissue culture. *Genetic Manipulation in Crops, Proceedings of International Symposium in Genetic*

Manipulation in Crops; The Third International Symposium on Haploidy; The First International Symposium on Somatic Cell Genetics in Crops, Beijing, p. 41.

Matuskovic, J. and Micudova, O. (2006) Practices with chemical mutagen (natriumazid) on growth habit Punica granatum L. In: *ISHS, 1st International Symposium Pomegranate and Minor Mediterranean Fruits, Abstracts contributed papers*. 41. ISHS, Adana, Turkey, pp. 16–19.

Mba, C., Afza, R., Bado, S. and Anthony, P. (2010) Induced mutagenesis in plants using physical and chemical agents. In: Davey, M.R. (ed.) *Plant Cell Culture: Essential Methods*. John Wiley and Sons, Ltd, Chichester, UK, pp. 111–130.

Melgarejo, P., Martínez, J.J., Hernández, F., Martínez-Font, R., Barrows, P. *et al.* (2004) Kaolin treatment to reduce pomegranate sunburn. *Scientia Horticulturae* 100(1–4), 349–353. DOI: 10.1016/j.scienta.2003.09.006.

Melgarejo, P., Martínez, J.J., Hernández, F., Martínez, R., Legua, P. *et al.* (2009) Cultivar identification using 18S–28S rDNA intergenic spacer-RFLP in pomegranate (*Punica granatum* L.). *Scientia Horticulturae* 120(4), 500–503. DOI: 10.1016/j.scienta.2008.12.013.

Miller, C.O. (1965) Evidence for the natural occurrence of zeatin and derivatives: compounds from maize which promote cell division. *Proceedings of the National Academy of Sciences* 54(4), 1052–1058. DOI: 10.1073/pnas.54.4.1052.

Mir, M.M., Umar, I., Mir, S.A., Rehman, M.U., Rather, G.H. *et al.* (2012) Quality evaluation of pomegranate crop – a review. *International Journal of Agriculture & Biology* 14(4), 658–667.

Mohan Jain, S. (2010) Mutagenesis in crop improvement under the climate change. *Romanian Biotechnological Letters* 15(2), 88–106.

Moriguchi, T., Omura, M., Matsuta, N. and Kozaki, I. (1987) *In vitro* adventitious shoot formation from anthers of pomegranate. *Horticultural Science* 22, 947–948.

Mou, B. and Scorza, R. (2011) *Transgenic Horticultural Crops: Challenges and Opportunities*. CRC Press, Boca Raton, Florida, p. 347.

Mozaffarian, F., Mardi, M., Sarafrazi, A. and Ganbalani, G.N. (2008) Assessment of geographic and host-associated population variations of the carob moth, *Ectomyelois ceratoniae*, on pomegranate, fig, pistachio and walnut, using AFLP markers. *Journal of insect science* 8(6), 1–9. DOI: 10.1673/031.008.0601.

Murashige, T. and Skoog, F. (1962) A revised medium for rapid growth and BIO assays with tobacco tissue cultures. *Physiologia Plantarum* 15(3), 473–497. DOI: 10.1111/j.1399-3054.1962.tb08052.x.

Murkute, A.A., Patil, S., Patil, B.N. and Kumari, M. (2002) Micropropagation in pomegranate, callus induction and differentiation. *South Indian Horticulture* 50(1,3), 49–55.

Murkute, A.A., Patil, S. and Mayakumari. (2003) Exudation and browning in tissue culture of pomegranate. *Agricultural Science Digest* 23, 29–31.

Murkute, A.A., Patil, S. and Singh, S.K. (2004) *In vitro* regeneration in pomegranate cv. Ganesh from mature trees. *The Indian Journal of Horticulture* 61(3), 206–208.

Nafees, M., Jaskani, M.J., Ahmed, S. and Awan, F.S. (2015) Morpho-molecular characterization and phylogenetic relationship in pomegranate germplasm of Pakistan. *Pakistan Journal of Agricultural Science* 52(1), 97–106.

Nageswari, K., Manivannan, K., Thamburaj, S. and Balakrishnamoorthy, G. (1999) A note on the performance of hybrid progenies of pomegranate. *South Indian Horticulture* 47, 139–140.

Nahla, A., El-Taweel, A.A. and Aly, A.A. (2014) Studies on cross pollination between Manfaloty pomegranate and some evaluated import cultivars. *British Journal of Applied Science & Technology* 4(25), 3701–3715. DOI: 10.9734/BJAST/2014/10518.

Naik, S.K. and Chand, P.K. (2011) Tissue culture-mediated biotechnological intervention in pomegranate: a review. *Plant Cell Reports* 30(5), 707–721. DOI: 10.1007/s00299-010-0969-7.

Naik, S.K., Pattnaik, S. and Chand, P.K. (1999) *In vitro* propagation of pomegranate (*Punica granatum* L. cv. Ganesh) through axillary shoot proliferation from nodal segments of mature tree. *Scientia Horticulturae* 79(3–4), 175–183. DOI: 10.1016/S0304-4238(98)00218-0.

Naik, S.K., Pattnaik, S. and Chand, P.K. (2000) High frequency axillary shoot proliferation and plant regeneration from cotyledonary nodes of pomegranate (*Punica granatum* L.). *Scientia Horticulturae* 85(4), 261–270. DOI: 10.1016/S0304-4238(99)00149-1.

Narzary, D., Mahar, K.S., Rana, T.S. and Ranade, S.A. (2009) Analysis of genetic diversity among wild pomegranates in Western Himalayas, using PCR methods. *Scientia Horticulturae* 121(2), 237–242. DOI: 10.1016/j.scienta.2009.01.035.

Narzary, D., Rana, T.S. and Ranade, S.A. (2010) Genetic diversity in inter-simple sequence repeat profiles across natural populations of Indian pomegranate (*Punica granatum* L.). *Plant Biology* 12(5), 806–813. DOI: 10.1111/j.1438-8677.2009.00273.x.

Nataraja, K. and Neelambika, G.K. (1996) Somatic embryogenesis and plantlet formation from petal cultures of pomegranate (*Punica granatum* L.). *Indian Journal of Experimental Biology* 34(7), 719–721.

Nath, N. and Randhawa, G.S. (1959) Studies on cytology of pomegranate (*Punica granatum* L.). *Indian Journal of Horticulture* 16, 210–214.

Nirmala, A. and Rao, P.N. (1996) Genesis of chromosome numerical mosaism in higher plants. *The Nucleus* 39, 151–174.

Nitsch, J.P. and Nitsch, C. (1969) Haploid plants from pollen grains. *Science* 163(3862), 85–87. DOI: 10.1126/science.163.3862.85.

Niu, J., Cao, D., Li, H., Xue, H., Chen, L. *et al.* (2018) Quantitative proteomics of pomegranate varieties with contrasting seed hardness during seed development stages. *Tree Genetics & Genomes* 14(1), 1–12. DOI: 10.1007/s11295-018-1229-1.

Noormohammadi, Z., Fasihee, A., Homaee-Rashidpoor, S., Sheidai, M., Ghasemzadeh Baraki, S. *et al.* (2012) Genetic variation among Iranian pomegranates (*Punica granatum* L.) using RAPD, ISSR and SSR markers. *Australian Journal of Crop Breeding* 6(2), 268–274.

Norouzi, M., Talebi, M. and Sayed-Tabatabaei, B.-E. (2012) Chloroplast microsatellite diversity and population genetic structure of Iranian pomegranate (*Punica granatum* L.) genotypes. *Scientia Horticulturae* 137, 114–120. DOI: 10.1016/j.scienta.2012.01.034.

Nungshilepden (2009) Transformation of pomegranate (*Punica granatum* L.) by amp (antimicrobial peptide) gene to confer resistance against bacterial blight of pomegranate. MSc Thesis. University of Agricultural Sciences, GKVK, Bangalore, India.

Ohri, D. (2002) Genome size variation in some tropical hardwoods. *Biologia Plantarum* 45(3), 455–457. DOI: 10.1023/A:1016290222360.

Okhovatian, A.R., Mehrabanian, M., Dehghani, F. and Akbarzadeh, A. (2010) Salt tolerance evaluation and relative comparison in cuttings of different pomegranate cultivars. *Journal of Plant Soil Environment* 4, 176–184.

Olyaie Torshiz, A., Goldansaz, S.H., Motesharezadeh, B., Asgari-Sarcheshmeh, M.A. and Zarei, A. (2017) Effect of organic and biological fertilizers on pomegranate trees: yield, cracking, sun burning and infestation to pomegranate fruit moth *Ectomyelois ceratoniae* (Lepidoptera: Pyralidae). *Journal of Crop Protection* 6(3), 327–340.

Omura, M., Matsuta, N., Moriguchi, T., Kozaki, I. and Sanada, T. (1987) Establishment of tissue culture method in dwarf pomegranate (*Punica granatum* L.) and application for the induction of variants. *Bulletin of the Fruit Tree Research Station* 14, 17–144.

Ono, N.N., Britton, M.T., Fass, J.N., Nicolet, C.M., Lin, D. *et al.* (2011) Exploring the transcriptome landscape of pomegranate fruit peel for natural product biosynthetic gene and SSR marker discovery. *Journal of Integrative Plant Biology* 53(10), 800–813. DOI: 10.1111/j.1744-7909.2011.01073.x.

Ono, N.N., Bandaranayake, P.C.G. and Tian, L. (2012) Establishment of pomegranate (*Punica granatum*) hairy root cultures for genetic interrogation of the hydrolyzable tannin biosynthetic pathway. *Planta* 236(3), 931–941. DOI: 10.1007/s00425-012-1706-y.

Onur, C. (1983) *Selection of pomegranate cultivars from Mediterranean region. Master's Thesis, Research and Training Center of Horticulture Crops, Ministry of Agriculture, Forestry and.* Village Affairs, Erdemli, Turkey.

Onur, C., Tibet, H. and Isik, E.A. (1999) *Cultivar breeding of pomegranate (Punica granatum L.) by hybridization*. Turkish National Horticultural Congress, Ankara, Turkey, pp. 58–61.

Onur, C. and Kaska, N. (1985) Selection of pomegranate (*Punica granatum* L.) from Mediterranean region of Turkey. *Doga Bilim Dergisi, D2 TarM ve Ormanclk* 9, 25–33.

Ophir, R., Sherman, A., Rubinstein, M., Eshed, R., Sharabi Schwager, M. *et al.* (2014) Single-nucleotide polymorphism markers from de-novo assembly of the pomegranate transcriptome reveal germplasm genetic diversity. *PLoS ONE* 9(2), e88998. DOI: 10.1371/journal.pone.0088998.

Ozguven, A.I., Tatli, H., Coskun, M. and Daskan, Y. (1997) Fruit characteristics of some Mediterranean and Aegean pomegranate varieties under ecological conditions of Adana, Turkey. *Acta Horticulturae* 441, 345–349.

Ozguven, A.I. and Yilmaz, C. (2000) Pomegranate growing in Turkey. *Options Méditerranéennes Série A, Séminaires Méditerranéens* 42, 41–48.

Parashar, A. and Ansari, A. (2012) A therapy to protect pomegranate (*Punica granatum* L) from sunburn. *International Journal of Comprehensive Pharmacy* 3, 1–3.

Pareek, O.P. (1996) Indian jujube and pomegranate. In: Paroda, R.S. and Chadha, K.L. (eds) *50 years of Crop Science Research in India*. ICAR, New Delhi, pp. 557–573.

Patil, V.M., Dhande, G.A., Thigale, D.M. and Rajput, J.C. (2011) Micropropagation of pomegranate (*Punica granatum* L.) 'Bhagava' cultivar from nodal explant. *African Journal of Biotechnology* 10, 18130–18136.

Petri, C. and Burgos, L. (2005) Transformation of fruit trees. useful breeding tool or continued future prospect? *Transgenic Research* 14(1), 15–26. DOI: 10.1007/s11248-004-2770-2.

Pirseyedi, S.M., Valizadehghan, S., Mardi, M., Ghaffari, M.R., Mahmoodi, P. *et al.* (2010) Isolation and characterization of novel microsatellite markers in pomegranate (*Punica granatum* L.). *International Journal of Molecular Sciences* 11(5), 2010–2016. DOI: 10.3390/ijms11052010.

Pollegioni, P., Woeste, K., Olimpieri, I., Marandola, D., Cannata, F. *et al.* (2011) Long-term human impacts on genetic structure of Italian walnut inferred by SSR markers. *Tree Genetics & Genomes* 7(4), 707–723. DOI: 10.1007/s11295-011-0368-4.

Pourghayoumi, M., Rahemi, M., Bakhshi, D., Aalami, A. and Kamgar-Haghighi, A.A. (2017) Responses of pomegranate cultivars to severe water stress and recovery: changes on antioxidant enzyme activities, gene expression patterns and water stress responsive metabolites. *Physiology and Molecular Biology of Plants* 23(2), 321–330. DOI: 10.1007/s12298-017-0435-x.

Prasanna Kumar, B. (1998) Pomegranate. In: Chatttopadhyay, T.K. (ed.) *A Textbook on Pomology*. 3. Kalayani Publishers, Ludhiana, India, pp. 166–188.

Qin, G., Xu, C., Ming, R., Tang, H., Guyot, R. *et al.* (2017) The pomegranate (*Punica granatum* L.) genome and the genomics of punicalagin biosynthesis. *Plant Journal* 91, 1108–1128.

Rahimi, T., Sayed-Tabatabaei, B.E., Sharif Nabi, B. and Ghobadi, C. (2006) Genetic relationship of some of the Iranian pomegranate (*Punica granatum* L.) genotypes by AFLP markers. *Iranian Journal of Agricultural Science* 36(6), 1373–1379.

Rai, M.K., Asthana, P., Jaiswal, V.S. Jaiswal, U., Singh, S.K, Jaiswal, U. (2010) Biotechnological advances in guava (*Psidium guajava* L.): recent developments and prospects for further research. *Trees* 24(1), 1–12. DOI: 10.1007/s00468-009-0384-2.

Raina, D. (2013) Assessment of genetic divergence and hybridization studies in pomegranate germplasm.. PhD Thesis in Horticulture. Department of Fruit Science, College of Agriculture, Punjab Agricultural University.

Raman, V.S., Kesavan, P.C., Manimekalai, G., Alikhan, W.M. and Rangaswami, S.R. (1963) Cytological studies in some tropical fruit plants – banana, Annona, guava and pomegranate. *South Indian Horticulture* 11, 27–33.

Rana, J.C., Pradheep, K. and Verma, V.D. (2007) Naturally occurring wild relatives of temperate fruits in Western Himalayan region of India: an analysis. *Biodiversity and Conservation* 16(14), 3963–3991. Available at: www.springerlink.com/content/f7kv824161506520/ DOI: 10.1007/s10531-007-9201-7.

Ravishankar, K.V., Chaturvedi, K., Puttaraju, N., Gupta, S. and Pamu, S. (2015) Mining and characterization of SSRs from pomegranate (*Punica granatum* L.) by pyrosequencing. *Plant Breeding* 134(2), 247–254. DOI: 10.1111/pbr.12238.

Rechinger, K.H. (1969) *Flora Iranica, no 66*. Akademische Druck-und Verlagsanstalt, Graz, Austria.

Rees, H. and Jones, R.N. (1977) *Chromosome Genetics (Genetics, Principles and Perspectives)*. Edward Arnold, London, p. 151.

Robertson, L. (2009) Pomegranate roads. *Redwood Empire Chapter Newsletter*, 4.

Rouholamin, S., Zahedi, B., Nazarian-Firouzabadi, F. and Saei, A. (2015) Expression analysis of anthocyanin biosynthesis key regulatory genes involved in pomegranate (*Punica granatum* L.). *Scientia Horticulturae* 186, 84–88. DOI: 10.1016/j.scienta.2015.02.017.

Rout, G.R., Mohapatra, A. and Jain, S.M. (2006) Tissue culture of ornamental pot plant: a critical review on present scenario and future prospects. *Biotechnology Advances* 24(6), 531–560. DOI: 10.1016/j.biotechadv.2006.05.001.

Saeedi, A.M., Mohammad, G., Samadi, G.R., Abdiani, S. and Giordani, E. (2012) The pomegranate national collection of Afghanistan. In: Melgarejo, P. and Valero, D. (eds) *Second International Symposium on the Pomegranate*. CIHEAM/Universidad Miguel Hernández, Zaragoza, Spain, pp. 57–60.

Saminathan, T., Bodunrin, A., Singh, N.V., Devarajan, R., Nimmakayala, P. *et al.* (2016) Genome-wide identification of microRNAs in pomegranate (*Punica granatum* L.) by high-throughput sequencing. *BMC Plant Biology* 16(1), 1–16. DOI: 10.1186/s12870-016-0807-3.

Sangamesh, K. (2014) *Micropropagation and mutation studies in pomegranate (Punica granatum L.). MSc Thesis in Crop Improvement and Biotechnology*. University of Horticultural Sciences, Bagalkot, India.

Sanjar, D.Y. (2015) Application of microsatellite SSR markers in a number of pomegranate (*Punica granatum* L.) cultivars in Kurdistan region/Duhok Province. *International Journal of Chemical and Biomedical Science* 1(3), 117–122.

Sarkhosh, A., Zamani, Z., Fatahi, R. and Ebadi, A. (2005) Analysis of some quantitative and qualitative traits in pomegranate genotypes. *Journal of Science and Technology in Agriculture and Natural Resources* 4, 147–159.

Sarkhosh, A., Zamani, Z., Fatahi, R. and Ebadi, A. (2006) RAPD markers reveal polymorphism among some Iranian pomegranate (*Punica granatum* L.) genotypes. *Scientia Horticulturae* 111(1), 24–29. DOI: 10.1016/j.scienta.2006.07.033.

Sarkhosh, A., Zamani, Z., Fatahi, R., Ebadi, A. and Ranjbar, H. (2008) Evaluation of Iranian soft-seed pomegranate accessions by using simple and multivariate analyses. *Tree and Forestry Science and Biotechnology* 2(1), 18–25.

Sarkhosh, A., Zamani, Z., Fatahi, R. and Ranjbar, H. (2009) Evaluation of genetic diversity among Iranian soft-seed pomegranate accessions by fruit characteristics and RAPD markers. *Scientia Horticulturae* 121(3), 313–319. DOI: 10.1016/j.scienta.2009.02.024.

Sarkhosh, A., Zamani, Z., Fatahi, R., Hassani, M.E., Wiedow, C. *et al.* (2011) Genetic diversity of Iranian soft-seed pomegranate genotypes as revealed by fluorescent-AFLP markers. *Physiology and Molecular Biology of Plants* 17(3), 305–311. DOI: 10.1007/s12298-011-0070-x.

Sarkhosh, A., Zamani, Z., Fatahi, R., Wiedow, C., Chagné, D. *et al.* (2012) A pomegranate (*Punica granatum* L.) linkage map based on AFLP markers. *The Journal of Horticultural Science and Biotechnology* 87(1), 1–6.

Seeram, N.P., Schulman, R.N. and Heber, D. (eds) (2006) *Pomegranates: Ancient Roots to Modern Medicine*. CRC Press, Taylor & Francis Group, Boca Raton, Florida, p. 262p..

Shao, J., Chen, C. and Deng, X. (2003) *In vitro* induction of tetraploid in pomegranate (*Punica granatum*). *Plant Cell, Tissue and Organ Culture* 75(3), 241–246. DOI: 10.1023/A:1025871810813.

Sharon, M. and Sinha, S. (2000) Plant regeneration from cotyledonary node of *Punica granatum* L. *Indian Journal of Plant Physiology* 5(4), 344–348.

Sheidai, M. (2007) B-chromosome variability in pomegranate (*Punica granatum* L.) cultivars. *Caryologia* 60(3), 251–256. DOI: 10.1080/00087114.2007.10797944.

Sheidai, M., Kolahizadeh, S., Noormohammadi, Z., Azani, N. and Nikoo, M. (2012) Correlation between geography and cytogenetic diversity in pomegranate (*Punica granatum* L.) cultivars in Iran. *Acta Botanica Brasilica* 26(4), 953–965. DOI: 10.1590/S0102-33062012000400025.

Sheidai, M., Mahmood, K. and Nasre-Esfahani, S. (2005) Cytogenetical study of some Iranian pomegranate (*Punica granatum* L) cultivars, Caryologia. *International Journal of Cytology, Cytosystematics and Cytogenetics* 58(2), 132–139.

Shi, S., Huang, Y., Tan, F., He, X. and Boufford, D.E. (2000) Phylogenetic analysis of the Sonneratiaceae and its relationship to Lythraceae based on its sequences of nrDNA. *Journal of Plant Research* 113(3), 253–258. DOI: 10.1007/PL00013926.

Shilkina, I.A. (1973) On the xylem anatomy of the genus *Punica* L. *Botany* 58, 1628–1630.

Singh, R.P., Kar, P.L. and Dhuria, H.S. (1980) Floral biology studies in some pomegranate cultivars. *Haryana Journal of Horticulture Science* 9, 7–11.

Singh, N.V., Singh, S.K. and Patel, V.B. (2007) *In vitro* axillary shoot proliferation and clonal propagation of 'G 137' pomegranate (*Punica granatum*). *Indian Journal of Agricultural Sciences* 77, 505–508.

Singh, S.K., Meghwal, P.R., Pathak, R., Gautam, R. and Kumar, S. (2013) Genetic diversity in *Punica granatum* revealed by nuclear rRNA, internal transcribed spacer and RAPD polymorphism. *National Academy Science Letters* 36(2), 115–124. DOI: 10.1007/s40009-013-0120-8.

Singh, N.V., Abburi, V.L., Ramajayam, D., Kumar, R., Chandra, R. *et al.* (2015) Genetic diversity and association mapping of bacterial blight and other horticulturally important traits with microsatellite markers in pomegranate from India. *Molecular Genetics and Genomics* 290(4), 1393–1402. DOI: 10.1007/s00438-015-1003-0.

Singh, D., Sangma, D. and Kumar, K. (2017) Hybridization studies on pomegranate (*Punica granatum* L) cultivars and wild germplasm accessions. *International Journal of Farm Sciences* 7(2), 119–126.

Singh, S.K. and Khawale, R.N. (2006) Plantlet regeneration from the nodal segments of pomegranate (*Punica granatum* L.) cv. Jyoti. In: Kumar, A. and Roy, S. (eds) *Plant Biotechnology and its Applications in Tissue Culture*. I.K. International Publishing House Pvt. Ltd, New Delhi, pp. 105–113.

Singh, P. and Patel, R.M. (2016) Factors affecting *in vitro* degree of browning and culture establishment of pomegranate. *African Journal of Plant Science* 10(2), 43–49.
Sinha, S. and Sharon, M. (1997) Somatic embryogenesis and plant regeneration from roots of *Punica granatum* L. *Horticultural Science* 32, 545–546.
Smith, P.M. (1976) Minor crops. In: Simmonds, N.W. (ed.) *Evolution of Crop Plants*. Longman, London, pp. 301–324.
Soleimani, M.H., Talebi, M. and Sayed-Tabatabaei, B.E. (2012) Use of SRAP markers to assess genetic diversity and population structure of wild, cultivated, and ornamental pomegranates (*Punica granatum* L.) in different regions of Iran. *Plant Systematics and Evolution* 298(6), 1141–1149. DOI: 10.1007/s00606-012-0626-4.
Soriano, J.M., Zuriaga, E., Rubio, P., Llácer, G., Infante, R. *et al.* (2011) Development and characterization of microsatellite markers in pomegranate (*Punica granatum* L.). *Molecular Breeding* 27(1), 119–128. DOI: 10.1007/s11032-010-9511-4.
Steward, F.C., Mapes, M.O., Kent, A.E. and Holsten, R.D. (1964) Growth and development of cultured plant cells. *Science* 143(3601), 20–27. DOI: 10.1126/science.143.3601.20.
Still, D.W. (2006) Pomegranate: a botanical perspective. In: Seeram, N.P., Schulman, R.N. and Heber, D. (eds) *Pomegranates: Ancient Roots to Modern Medicine*. CRC Press, Taylor & Francis Group, Boca Raton, Florida, pp. 199–210.
Stover, E. and Mercure, E.W. (2007) The pomegranate: a new look at the fruit of paradise. *HortScience* 42(5), 1088–1092. DOI: 10.21273/HORTSCI.42.5.1088.
Strebkova, A.D. (1974) The double-petalled mottled-pink pomegranate (*Punica granatum* var Legrellei Vanh.) and its origin. *Ma'-ruzalar-AzSSR-Elmar-Akad* 30(3), 84–87.
Swetha, H.G. (2011) Development of transgenic pomegranate cv. Bhagwa for bacterial blight resistance. MSc thesis. University of Agricultural Sciences, GKVK, Bengaluru, India.
Talebi Bedaf, M., Sharifi Neia, B. and Bahar, M. (2003) Analysis of genetic diversity in pomegranate cultivars of Iran, using random amplified polymorphic DNA (RAPD) markers. In: *Proceedings of the Third National Congress of Biotechnology*. 2. Iran, pp. 343–345.
Talebi Bedaf, M., Bahar, M., Sharifnabi, B. and Yamchi, A. (2011) Evaluation of genetic diversity among Iranian pomegranate (*Punica granatum* L.) cultivars, using ISSR and RAPD markers. *Taxonomy and Biosystematics* 8, 35–44.
Tavousi, M., Kaveh, F., Alizadeh, A., Babazadeh, H. and Tehranifar, A. (2016) Effect of salinity and deficit irrigation on quantity and quality of pomegranate (*Punica granatum* L). *Iranian Journal of Irrigation and Drainage* 4(10), 499–507.
Teixeira da Silva, J.A., Rana, T.S., Narzary, D., Verma, N., Meshram, D.T. *et al.* (2013) Pomegranate biology and biotechnology: a review. *Scientia Horticulturae* 160, 85–107. DOI: 10.1016/j.scienta.2013.05.017.
Terakami, S., Matsuta, N., Yamamoto, T., Sugaya, S., Gemma, H. *et al.* (2007) Agrobacterium-mediated transformation of the dwarf pomegranate (*Punica granatum* L. var. Nana). *Plant Cell Reports* 26(8), 1243–1251. DOI: 10.1007/s00299-007-0347-2.
Thongtham, C. (1986) Germplasm collection and conservation of pomegranate in Thailand. *IBPGR Newsletter* 10(8), 8–10.
Ue, B.S., Weng, R.F. and Zhang, M.Z. (1992) Chromosome numbers of Shanghai plants I. *Investigation et Studium Naturae* 12, 48–64.
Ulukapi, K. and Gul Nasircilar, A. (2015) Developments of gamma ray application on mutation breeding studies in recent years. *International Conference on Advances in Agricultural, Biological & Environmental Sciences (AABES-2015)*, London, 22-23 July 2015. DOI: 10.15242/IICBE.C0715044.
USDA (2007) Repository inventory of available accessions for *Punica granatum*. *ARS, The National Clonal Germplasm Repository (NCGR) at Davis*. Available at: www.ars.usda.gov/Main/docs.htm?docid¼12856
Valizadeh Kaji, B. and Abbasifar, A. (2017) Transformation of pomegranate (*Punica granatum* L.) a difficult-to-transform tree. *Biocatalysis and Agricultural Biotechnology* 10, 46–52. DOI: 10.1016/j.bcab.2017.02.007.
Valizadeh Kaji, B., Ershadi, A. and Tohidfar, M. (2014) Agrobacterium-mediated transformation of pomegranate (*Punica granatum* L.) 'Yousef Khani' using the gus reporter gene. *International Journal of Horticultural Science and Technology* 1, 31–41.
Valizadehkaji, B., Ershadi, A. and Tohidfar, M. (2013) In vitro propagation of two Iranian commercial pomegranates (*Punica granatum* L.) cvs. 'Malas Saveh' and 'Yusef Khani'. *Physiology and Molecular Biology of Plants* 19(4), 597–603. DOI: 10.1007/s12298-013-0193-3.

Van Harten, A.M. (1998) *Mutation Breeding: Theory and Practical Applications*. Cambridge University Press, Cambridge, UK, pp. 137–158.

Vazifeshenas, M., Khayyat, M., Jamalian, S. and Samadzadeh, A. (2009) Effects of different scion-rootstock combinations on vigor, tree size, yield and fruit quality of three Iranian cultivars of pomegranate. *Fruits* 64(6), 343–349. DOI: 10.1051/fruits/2009030.

Verma, V., Kanwar, K., Tufchi, M. and Kashyap, M. (2014) Agrobacterium-mediated Cry1A(b) gene transfer in *Punica granatum* L. cv. Kandhari Kabuli using different in vitro regeneration pathways. *Journal of Crop Science and Biotechnology* 17(1), 1–10. DOI: 10.1007/s12892-013-0033-6.

Villeux, R. (1985) Diploid and polyploid gametes in crop plants: mechanisms of formation and utilization in plant breeding. In: Janick, J. (ed.) *Plant Breeding Review*. 3. AVI Publishing Company, Westport, Connecticut, pp. 253–258.

Vorsa, N. and Bingham, E.T. (1979) Cytology of 2N pollen formation in diploid alfalfa, *Medicago sativa*. *Canadian Journal of Genetics and Cytology* 21(4), 525–530. DOI: 10.1139/g79-057.

Wang, J.X., Sun, Y., Cui, G.M. and Hu, J.J. (2001) Transgenic maize plants obtained by pollen mediated transformation. *Actor Botanica Sinica* 43(3), 275–279.

Wang, Y., Yin, X.L. and Yang, L.F. (2006) Breeding of Zaoxuan 018 and 027 pomegranate selections. *China Fruits* 4, 6–8.

Wang, W.L., Tsay, J.S., Yeuh, C.S., Chen, R.S. and Tsay, J.G. (2007) Phylogenetic relationships of pomegranate (*Punica granatum* L.) accessions by using rDNA, ITS, RAPD and ISSR molecular markers.. *Journal of Taiwan Society for Horticultural Science* 53, 157–172.

Wang, L., Zhang, Y., Qi, X., Gao, Y. and Zhang, X. (2012) Development and characterization of 59 polymorphic cDNA-SSR markers for the edible oil crop *Sesamum indicum* (Pedaliaceae). *American Journal of Botany* 99(10), e394–e398. DOI: 10.3732/ajb.1200081.

Wang, Z., Kang, M., Liu, H., Gao, J., Zhang, Z. *et al*. (2014) High-level genetic diversity and complex population structure of Siberian apricot (*Prunus sibirica* L.) in China as revealed by nuclear SSR markers. *PLoS ONE* 9(2), e87381. DOI: 10.1371/journal.pone.0087381.

Xue, H., Cao, S., Li, H., Zhang, J., Niu, J. *et al*. (2017) De novo transcriptome assembly and quantification reveal differentially expressed genes between soft-seed and hard-seed pomegranate (*Punica granatum* L.). *Plos One* 12(6), e0178809. DOI: 10.1371/journal.pone.0178809.

Yan, M., Zhao, X., Zhou, J., Huo, Y., Ding, Y. *et al*. (2019) The complete chloroplast genomes of *Punica granatum* and a comparison with other species in Lythraceae. *International Journal of Molecular Sciences* 20(12), 2886. DOI: 10.3390/ijms20122886.

Yang, R.P., Long, W.H., Zhang, H., Xu, B. and Li, W.X. (2007) RAPD analysis of 25 *Punica granatum* germplasm resources collected in Yunnan province. *Journal of Fruit Science* 24, 226–229.

Yang, H., Zhang, T., Masuda, T., Lv, C., Sun, L. *et al*. (2011) Chitinase III in pomegranate seeds (*Punica granatum* Linn.): a high-capacity calcium-binding protein in amyloplasts. *Plant Journal* 68, 765–776.

Yang, H., Li, M., Qi, X., Lv, C., Deng, J. *et al*. (2012) Identification of seven water-soluble non-storage proteins from pomegranate (*Punica granatum* Linn.) seeds. *Food Science and Technology International* 18(4), 329–338. DOI: 10.1177/1082013211428008.

Yang, Y.L., Guo, X.Y., Di, Q. and Sun, Y. (2016) Observation of GFP expression in pomegranate (*Punica granatum* L.) via pollen-mediated transformation. *Plant* 4(4), 23–28. DOI: 10.11648/j.plant.20160404.11.

Yazici, K. and Ercişli, S. (2017) Characterization of hybrid pomegranate genotypes based on sunburn and cracking traits related to maturation time. *Journal of Applied Botany and Food Quality* 90, 132–139.

Yezhov, V.N., Smykov, A.V., Smykov, V.K., Khokhlov, S.Y., Zaurov, D.E. *et al*. (2005) Genetic resources of temperate and subtropical fruit and nut species at the Nikita botanical gardens. *HortScience* 40(1), 5–9. DOI: 10.21273/HORTSCI.40.1.5.

Yuan, Z., Yin, Y., Qu, J., Zhu, L. and Li, Y. (2007) Population genetic diversity in Chinese pomegranate (*Punica granatum* L.) cultivars revealed by fluorescent-AFLP markers. *Journal of Genetics and Genomics* 34(12), 1061–1071. DOI: 10.1016/S1673-8527(07)60121-0.

Yuan, Z., Fang, Y., Zhang, T., Fei, Z., Han, F. *et al*. (2018) The pomegranate (*Punica granatum* L.) genome provides insights into fruit quality and ovule developmental biology. *Plant Biotechnology Journal* 16(7), 1363–1374. DOI: 10.1111/pbi.12875.

Zamani, Z. (1990) Evaluation of pomegranate genotypes in Saveh-Iran. MSc Thesis in Horticultural Science. Faculty of Agriculture, University of Tehran, Karaj, Iran.

Zamani, Z., Sarkhosh, A., Fatahi, R. and Ebadi, A. (2007) Genetic relationships among pomegranate genotypes studied by fruit characteristics and RAPD markers. *The Journal of Horticultural Science and Biotechnology* 82(1), 11–18. DOI: 10.1080/14620316.2007.11512192.

Zamani, Z., Zarei, A. and Fatahi, R. (2010) Characterization of progenies derived from pollination of pomegranate cv. Malase-Tourshe-Saveh using fruit traits and RAPD molecular marker. *Scientia Horticulturae* 124(1), 67–73. DOI: 10.1016/j.scienta.2009.12.021.

Zamani, Z., Adabi, M. and Khadivi-Khub, A. (2013) Comparative analysis of genetic structure and variability in wild and cultivated pomegranates as revealed by morphological variables and molecular markers. *Plant Systematics and Evolution* 299(10), 1967–1980. DOI: 10.1007/s00606-013-0851-5.

Zarei, A., Zamani, Z. and Fatahi-Moghadam, M.R. (2009) Evaluation of genetic relationships among some Persian cultivated and wild pomegranate accessions using RAPDs and SSRs molecular markers. *Horticulture, Environment and Biotechnology* 50, 224–232.

Zarei, A., Zamani, Z., Fatahi, R., Mousavi, A. and Salami, S.A. (2013) A mechanical method of determining seed-hardness in pomegranate. *Journal of Crop Improvement* 27(4), 444–459. DOI: 10.1080/15427528.2013.790867.

Zarei, A., Zamani, Z., Fatahi, R., Mousavi, A., Salami, S.A. *et al.* (2016a) Differential expression of cell wall related genes in the seeds of soft- and hard-seeded pomegranate genotypes. *Scientia Horticulturae* 205, 7–16. DOI: 10.1016/j.scienta.2016.03.043.

Zarei, A., Zamani, Z., Fattahi, R., Salami, A. and Mousavi, A. (2016b) Analysis of the phenylpropanoid enzyme activities and products in the Soft- and Hard-Seeded pomegranate genotypes during fruit development. *International Journal of Fruit Science* 16(3), 242–258. DOI: 10.1080/15538362.2015.1089814.

Zarei, A. (2017) Biochemical and pomological characterization of pomegranate accessions in Fars Province of Iran. *SABRAO Journal of Breeding and Genetics* 49(2), 155–167.

Zarei, A. and Ebrahimi, A. (2017) EST-SSR identification in pomegranate expressed sequence tags. *2nd International and 10th National Biotechnology Congress of Islamic Republic of Iran*, Tehran, Iran, 29–31 August.

Zarei, A. and Sahraroo, A. (2018) Molecular characterization of *pomegranate (Punica granatum* L.) accessions from Fars Province of Iran using microsatellite markers. *Horticulture, Environment, and Biotechnology* 59(2), 239–249. DOI: 10.1007/s13580-018-0019-x.

Zhang, Y.P., Tan, H.H., Cao, S.Y., Wang, X.C., Yang, G. *et al.* (2012) A novel strategy for identification of 47 pomegranate (*Punica granatum*) cultivars using RAPD markers. *Genetics and Molecular Research* 11(3), 3032–3041. DOI: 10.4238/2012.May.30.1.

Zhang, B.L. and Stolz, L.P. (1991) *In vitro* shoot formation and elongation of dwarf pomegranate. *Horticultural Science* 26(8), 1084.

Zhao, C.L. (2007) Breeding of new early pomegranate cultivars 'Taihanghong'. *China Fruits* 3, 5–6.

Zhao, Y.L., Feng, Y.Z., Li, Z.H. and Cao, Q. (2006) Breeding of the new pomegranate cultivar 'Yushiliu 4'. *China Fruits* 2, 8–10.

Zhao, L., Li, M., Cai, G., Pan, T. and Shan, C. (2013) Assessment of the genetic diversity and genetic relationships of pomegranate (*Punica granatum* L.) in China using RAMP markers. *Scientia Horticulturae* 151, 63–67. DOI: 10.1016/j.scienta.2012.12.015.

Zhao, X., Yuan, Z., Feng, L. and Fang, Y. (2015) Cloning and expression of anthocyanin biosynthetic genes in red and white pomegranate. *Journal of Plant Research* 128(4), 687–696. DOI: 10.1007/s10265-015-0717-8.

Zoccatelli, G., Dalla Pellegrina, C., Consolini, M., Fusi, M., Sforza, S. *et al.* (2007) Isolation and identification of two lipid transfer proteins in pomegranate (*Punica granatum*). *Journal of Agricultural and Food Chemistry* 55(26), 11057–11062. DOI: 10.1021/jf072644x.

Zolfagharieh, H., Vafaeishoushtari, R., Farazmand, H., Ardakani, M.R., Babaii, M. *et al.* (2008) Application of nuclear technique for determination controlling dose of pomegranate fruit moth, *Ectomyelois ceratoniae* Zeller (Lep Pyralidae). *Journal of Entomological Research* 1(1), 35–42.

5 World Pomegranate Cultivars

J.M. Chater[1*], A. Yavari[2*], A. Sarkhosh[3], Z. Jia[1], D.J. Merhaut[1], J.E. Preece[4], F. Cossio[5], G. Qin[6], C. Liu[6], J. Li[6], P. Shilpa[7], K.D. Babu[7], J. Sharma[7], C. Yilmaz[8], J. Bartual[9], Z. Mustafayeva[10], M.A. Saeedi[11], N.A. Awd[12], J. Moersfelder[3] and L. Hou[13]

[1]Department of Botany and Plant Sciences, University of California, Riverside, California, USA; [2]Pomologist, private consultant, Tehran, Iran; [3]Horticultural Sciences Department, University of Florida, Gainesville, Florida, USA; [4]National Clonal Germplasm Repository, USDA-ARS, University of California, Davis, California, USA; [5]Fruit crops breeder, private consultant, Verona, Italy; [6]Institute of Horticultural Research, Anhui Academy of Agricultural Sciences, Hefei, China; [7]ICAR-National Research Centre on Pomegranate, Solapur, India; [8]Department of Horticulture, Osmangazi University, Eskisehir, Turkey; [9]Conselleria d'Agricultura, Medi Ambient, Canvi Climàtic i Desenvolupament Rural. Estació Experimental Agrària d'Elx., Elx, Spain; [10]Institute of Genetics Resources, Baku, Azerbaijan; [11]Afghanistan National Horticulture Development Organization, Kabul, Afghanistan; [12]Horticultural Research Institute, Agricultural Research Center, Giza, Egypt; [13]Research Center of Pomegranate in Zaozhuang, Zaozhuang, China

5.1 Introduction

Interesting and important pomegranate cultivars have been reported from several locations all over the world, including Europe, Asia and North Africa. There are perhaps thousands of domesticated pomegranate cultivars on Earth, with traits ranging from very sour to very sweet in taste and from white, yellow, pink, red, purple, multicoloured to jet-black in colour. Some are used for fresh fruit or juice; others for medicine, syrups, sauces, ornamental purposes and pastes. Some pomegranate cultivars are earlier to mature than others and some have short harvest windows, while others have longer ones. Many believe peel thickness impacts storability and postharvest shelf-life. Others believe those with soft seeds are best for fresh market, whereas those with hard seeds are best suited for juice and other purposes. Cultures from around the world celebrate the pomegranate tree for its prized fruits, which have been providing nourishment for our species for thousands of years. Many of these cultivars were selected, propagated and distributed during domestication and the movement of people and their plants. Several others were selected in breeding programmes, some ongoing, others from another era. Regardless of the origins of the world's most beloved pomegranate cultivars, it is important to acknowledge the germplasm repositories of all nations,

*Corresponding author: john.chater@ucr.edu; alimohammadyavari@gmail.com

Fig. 5.1. 'Pand post'. (Photo: Khoshal Basharat.)

Fig. 5.2. 'Nazak post'. (Photo: Khoshal Basharat.)

which allow for the brilliant diversity of pomegranates, and so many other important crops, to exist in perpetuity. The important and more famous cultivars of countries that are major pomegranate origins, producers and exporters are introduced in alphabetical order in the text that follows.

5.2 Afghanistan

5.2.1 'Pand post' ('Kandahari')

This is a local cultivar of Kandahar province, Afghanistan (Fig. 5.1). There are two Kandahari cultivars grown commercially: 'Pand post' (thick peel) and 'Nazak post' (thin peel). To clarify, the famous Kandahari cultivar is the 'Pand post' accession. It is common among growers and traders to call it 'Pand post' rather than 'Kandahari', but outside Kandahar this cultivar is called 'Kandahari'. 'Pand post' is a vigorous tree with low thorniness. Average yield is 40 kg per tree. The height of the tree is up to 5 m. It has a large fruit size, the average size is 400–500 g and the largest reaches 900–1000 g. The colour of the skin is red. Arils are dark red in colour. The taste is sweet and the seeds are very hard. Total soluble solids (TSS) content is 19–20% at maturity. This is a mid-season cultivar, which typically matures on 5 October. This cultivar is the one commercially grown for exports because of its relatively thick rind, which allows it to be exported over long distances and provides for a longer shelf-life. The shelf-life is about 3 months.

5.2.2 'Nazak post'

This cultivar is a local cultivar of Kandahar province, Afghanistan (Fig. 5.2). It has a sweet taste and hard seeds. The plant is vigorous with no thorniness. Average yield is 45 kg per tree. The height of the tree is up to 5 m. The fruit peel colour is red and thin. The aril colour is pink to red. The average fruit size is 400 g and the largest reaches 700–800 g. The TSS is 20.3% and titratable acidity (TA) 1.3% at maturity stage. It is an early cultivar, which matures around 25 September in Afghanistan. The tree's tendency to produce suckers is low. It is very susceptible to cracking and for this reason it is grown for local consumption in Kandahar.

5.2.3 'Bedana'

The 'Bedana' (seedless) cultivar is the most famous pomegranate cultivar grown in Kapisa province, Tagab district, Afghanistan (Fig. 5.3). As its name implies, this cultivar is very soft seeded, so much so that people call it 'Bedana', meaning seedless. This cultivar is the most expensive pomegranate in Afghanistan. Most of the product is exported to Pakistan and it fetches a very high price. The plant's average yield is low, about 15 kg per tree. The height of the tree is up to 5–6 m. The fruit peel colour is cream yellow and it is very thin. The aril colour is light pink and it has very soft seeds. The average fruit weight is 235 g. The TSS is 20% and TA is 1% at maturity. It matures at the end of October.

Fig. 5.3. 'Bedana'. (Photo: Khoshal Basharat.)

5.2.4 'Surkhak'

'Surkhak' is a local cultivar of Kandahar province, Afghanistan (Fig. 5.4). The taste is sweet, average yield is 35kg per tree. The height of the tree is up to 3m. The fruit peel colour is red and the peel is thick. The aril colour is dark red with semi-hard seeds. The average fruit size is 300–400 g and the largest reaches 700g. The TSS content is 18.9%, and TA is 1.1% at maturity. It is a mid-season cultivar, which matures on 25 September in Afghanistan. The shelf-life is about 2 months.

5.2.5 'Tor Anar'

This is a local cultivar of Farah and Kandahar provinces in Afghanistan (Fig. 5.5). It is an ornamental cultivar, so the flavour is poor, and it is relatively less juicy. The plant is of medium vigour with medium thorniness. The average yield is 35–40kg per tree. The height of the tree is up to 5m. The fruit peel colour is black and dark brown. The aril colour is red and purple with hard seeds. The average fruit weight is 100–200 g and the largest weigh 300g. The TSS is 17.7%, and TA is 1.2% at maturity. It is a late-maturing cultivar, which matures on 10 October in Afghanistan.

Fig. 5.4. 'Surkhak'. (Photo: Khoshal Basharat.)

Fig. 5.5. 'Tor Anar'. (Photo: Khoshal Basharat.)

5.2.6 'Sherinak'

'Sherinak' is a local cultivar of Balkh province in Afghanistan (Fig. 5.6). It has a sweet taste. The average yield is 30kg per tree. The tree height is up to 6m. The fruit peel colour is light red. The aril colour is red with hard seeds. The average fruit weight is 300–400 g and the largest are 500g. The TSS content is 20%, TA is 1.2% at maturity. It is a late-maturing cultivar that is ready for harvest around 10 October in Afghanistan. The shelf-life is about 2 months.

5.3 Azerbaijan

5.3.1 'Bala Mursal'

'Bala Mursal' is a popular cultivar of Azerbaijan (Fig. 5.7). The tree is compact and large in size. The vegetative development is moderate. It is widely grown in the Aran region. Its fruits are large and dark red with brown spots on the peel. The average weight of the fruit is 350g; large fruits are 600–700g. The standard fruit production of a tree is 80–90% marketable fruit. The juice ratio is 38–45%, aril length is 11.4mm and width is 10.5mm. The weight of 100 arils is 58g. The degree of aril septation is very weak.

Fig. 5.6. 'Sherinak'. (Photo: Khoshal Basharat.)

Fig. 5.7. 'Bala Mursal'. (Photo: Ziyafet Mustafayeva.)

The peel ratio of fruits is 50%. The thickness of the peel is 6 mm. The TSS and TA are 17.4% and 1.3–1.5%, respectively. The juice is cherry red-coloured, the taste is sour-sweet. The juice ratio is 38–50%. The average yield of the tree is 30–35 kg. The storage life is about 3–4 months. It is one of the most important fruit cultivars in Azerbaijan. Its appearance and taste show major differences from other cultivars. One of the advantages is that the peel is thick. It is mainly used as fresh fruit.

5.3.2 'Girmiz Gabig'

This is one of the cultivars of the national selection of Azerbaijan (Fig. 5.8). The height of the tree is 3–7 m at the age of 10 years and the width is 2.3 m. It is most common in the Absheron peninsula, Ganja and Gazakh. The average weight of the fruit is 180–200 g, with a maximum weight of 400 g. The colour of the peel is red. It is thin-skinned and peel thickness is 1.25 mm, aril length is 10.25 mm, width is 5.75 mm. Seed length is 6.9 mm and width is 2 mm. The degree of aril septation is very weak. The ratio of peel to fruit mass is 25.6%. Suberization is common on the crown section of the fruit. The colours of peel and juice are dark red. The weight of 100 arils is 23–35 g. The ratio of the peel of the fruit mass is 25.6%. The TSS and TA are 16.0–17.4% and 1.45–1.94%, respectively. The taste is sour-sweet. Acidity is higher than other cultivars. It matures in mid-October. The average yield per tree is 35–40 kg. The storage life is about 4–5 months. The main uses of fruits of this cultivar are juice and narsharab (pomegranate sauce).

Fig. 5.8. 'Girmiz Gabig'. (Photo: Ziyafet Mustafayeva.)

Fig. 5.9. 'Shirin Girmizi'. (Photo: Ziyafet Mustafayeva.)

5.3.3 'Shirin Girmizi'

The tree is tall and large in size. The plant can be grown about 4–7 m in height and will occupy 4.7 m in diameter. The branches are raised and straight upright. 'Shirin Girmizi' is one of the cultivars of the national selection of Azerbaijan. The fruit is dark red (Fig. 5.9). It is a middle-season cultivar. It is found in many regions of Azerbaijan. The average fruit weight is 250 g, maximum weight is 500–600 g. The fruit peel is very thin and it has about 1 mm thickness. The length of the arils is 9.75 mm and they are 7 mm wide. Seed length is 7 mm, width is 3 mm. The weight of 100 arils is 26.8–35 g. The weight of 100 seeds is 6 g. The juice ratio is 40–49% and juice colour is red. The ratio of the peel to the fruit mass is 27–28%. The TSS and TA are 12.75% and 0.65%, respectively, at maturity. The taste is sweet. The storage life is about 3 months. It is possible to get 20–30 kg of fruit from one bush. It is mainly used as fresh fruit.

Fig. 5.10. 'Valas'. (Photo: Ziyafet Mustafayeva.)

5.3.4 'Valas'

'Valas' is a popular cultivar of Azerbaijan (Fig. 5.10). Its cultivation has spread mainly in the Absheron peninsula and Goychay region. The colour of the fruit peel is red. The average fruit weight is 110–150 g; maximum weight is 500–580 g. Thickness of the peel is 2 mm. The colour of the arils is red, the juice is a red (raspberry) colour. Aril length is 10 mm, width 7 mm, seed length is 7 mm, seed width is 3.8 mm. Juice ratio is 37–49%. The TSS is 15.5%, TA is 1.65–1.77% at maturity. It is sour-sweet cultivar. The storage life is about 4–5 months. The avarage yield of a tree is 30–35 kg. The fruits of this cultivar are used for table fruits, juice or cider.

5.3.5 'Azernaijan Guleyshasi'

This is a popular cultivar of Azerbaijan (Fig. 5.11). It is most common in many regions of Azerbaijan, mainly in the Absheron peninsula, Ganja and Goychay. The colour of the peel of the fruit is red. Suberization is common on the crown section of

Fig. 5.11. 'Azernaijan Guleyshasi'. (Photo: Ziyafet Mustafayeva.)

Fig. 5.12. 'Baiyushizi'. (Photo: Lefeng Hou.)

fruit. This cultivar is also found in pink ('Çəhrayı Güleyşə'). The maturing period is the end of October. The average fruit weight is 250 g; maximum weight is 400–460 g. The fruit peel is about 2–2.8 mm in thickness. The ratio of peel to fruit mass is 37.52%. The average weight of 100 arils is 30–40 g. The weight of 100 seeds is 3.4–4.5 g. The juice ratio is 45–55%. The colour of the aril and juice is from red to dark red. The TSS and TA at maturity are 16.6% and 1.5–2.9%, respectively, and the taste is pleasant. Its postharvest shelf-life is about 2–3 months in cold storage. The yield is 25–30 kg per tree. Fruits of this cultivar are mainly used for canned fruit and juice.

5.4 China

5.4.1 'Baiyushizi'

This is a well-known superior cultivar in China, which originates in Anhui province and is planted in Anhui, Henan, Shandong, Sichuan and Yunnan provinces (Fig. 5.12). This is a bud mutation of local cultivar 'Sanbai'. The colour of flower, peel and arils are yellow white to milky white, but it has bigger fruit and seed than the parent. The average fruit size is about 450 g and the maximum size reaches 1200 g. Hundred-aril weight is 84 g. The flavour is sweet and seeds are semi-soft. The TSS is 16.4%, sugar content is 12.6%, the TA is 0.32% and the vitamin C content is 1.49 mg/100 g fresh weight (Zhu *et al.*, 2009). Generally, it matures in late September. It has a desirable flavour but poor storability.

Fig. 5.13. 'Dabenzi'. (Photo: Lefeng Hou.)

5.4.2 'Dabenzi'

This is a local cultivar of Anhui province in China and has been cultivated for more than 300 years (Fig. 5.13). It is vigorous and productive. It has a large fruit size, sweet-sour flavour and hard seeds. Average fruit size is above 400 g (Cao and Hou, 2013). Generally, it ripens in October. It has good tolerance to drought and diseases, and excellent storability. This cultivar is of major importance because it was the first pomegranate to have its genome sequenced and published as a reference (Qin *et al.*, 2017), which is a most important contribution to punicology.

5.4.3 'Dongyan'

The cultivar originates in Henan province of China (Fig. 5.14). The fruit is near round and has a thick rind. The colour of skin is bright red and arils are crimson red in colour. Average fruit size is above 360 g. Hundred-aril weight is 52.4 g. The flavour is sour-sweet and the seeds are semi-soft. Mature fruit TSS is 16% (Chen *et al.*, 2012). Generally, it matures in late October. Fruits have excellent transportability.

Fig. 5.14. 'Dongyan'. (Photo: Lefeng Hou.)

Fig. 5.15. 'Huaibeiqingpiruanzi' (Qingpiruanzi). (Photo: Lefeng Hou.)

5.4.4 'Huaibeiqingpiruanzi' ('Qingpiruanzi')

This is a local cultivar of Anhui province in China (Fig. 5.15). It is vigorous and productive (Cao and Hou, 2013). The average height of the plant is more than 4 m. Its fruits are very large, reaching 1500 g in weight, and average fruit size is above 600 g (Cao and Hou, 2013). The skin colour is yellow-green and slightly reddish on the sunny/blush side. Arils are pinkish to bright red in colour. The flavour is sweet and seeds are semi-soft. It is late maturing and matures generally in October. This cultivar has good drought- and disease tolerance and it is planted as the leading cultivar in Huaibei region, Anhui province.

5.4.5 'Lintongjingpitian' ('Jingpitian')

The cultivar originates in Shanxi province (陕西省), China (Fig. 5.16). It is also known as 'Fenhongshiliu' and 'Fenpitian' due to its pink, bright and clean skin. It is a vigorous small tree. Fruits are near round and average fruit size is 300 g. Arils are crimson red in colour and have a sweet flavour. The seeds are hard. Mature fruit TSS

Fig. 5.16. 'Lintongjingpitian' (Jingpitian). (Photo: Lefeng Hou.)

is 16% (Cao and Hou, 2013). This cultivar has excellent tolerance to drought and cold, and grows well in infertile soils. Generally, it matures in late September. Fruit split occurs frequently when it rains during the period of maturity.

5.4.6 'Mengzitianlvzi' ('Tianlvzi')

This is a local and leading commercial pomegranate cultivar of Mengzi, Yunnan province, China (Fig. 5.17). It is a shrub growing in warm regions and has poor cold tolerance. The shape of its fruit is round, being 7 cm in length and 7.8 cm in diameter. The fruit skin colour is yellow-green, and the arils are light red to bright red in colour. Fruit size is above 250 g on average. The fruit has a sweet flavour and semi-soft seeds. For mature fruits, TSS is 13.8%, sugar content is 14.4% and TA is 0.45% (Cao *et al.*, 2017). It matures in late September.

Fig. 5.17. 'Mengzitianlvzi' (Tianlvzi). (Photo: Lefeng Hou.)

Fig. 5.18. 'Tunisiruanzi' (Tunisia). (Photo: Lefeng Hou.)

5.4.7 'Tunisiruanzi' ('Tunisia')

This is an introduced cultivar in China, and it originates in Tunisia (Fig. 5.18). It is one of the main soft-seeded pomegranate cultivars widely planted in China. The cultivar is vigorous and productive. Fruits are large and seeds are soft. Average fruit size is above 400 g. The skin colour is bright red and arils are ruby-red in colour. It is very juicy and has a sweet taste. Mature fruit TSS is 15%, sugar content is 15.5%, the TA is 0.29% and the vitamin C content is 1.87 mg/100 g fresh weight (Cao and Hou, 2013). This cultivar matures early and is generally mature in late August. This cultivar has moderate tolerance to drought and diseases but has poor cold tolerance.

5.4.8 'Taishanhong'

This is a well-known cultivar in Shandong province, China (Fig. 5.19). It has big fruits with bright red peel and crimson red arils. Average fruit size is above 500 g. Seeds are semi-soft. It is juicy and has a sweet taste. The TSS is 17%, sugar content is 14.98%, the TA is 0.28% and the vitamin C content is 5.26 mg/100 g fresh weight (Cao and Hou, 2013). It is a mid- to late-season cultivar that matures in late September to early October. This cultivar is drought tolerant, disease resistant and has poor cold tolerance.

Fig. 5.19. 'Taishanhong'. (Photo: Lefeng Hou.)

5.4.9 'Huaibeiruanzi 3'

This is a well-known commercial cultivar of Anhui province, China, with a sweet flavour and semi-soft seeds (Fig. 5.20). The fruit is near round. The skin colour is yellow, and the arils are white. The average fruit size is 267.2 g, and the maximum weight is 557 g. It is juicy and the average 100-aril weight is 63.5–70.0 g. For mature fruit, TSS is 15.0%, sugar content is 15.5% and the TA is 0.62% (Cao *et al.*, 2017). Generally, it matures at the end of September.

5.4.10 'Jianshuihongzhenzhu'

This is a commercial cultivar from Yunnan province, China (Fig. 5.21). It has poor cold tolerance and is always cultivated in warm regions. The shape of the fruit is near round, 7.4 cm long and 8.4 cm in diameter. Fruit size is 326–574 g. The skin colour of the cultivar is crimson red and the arils are red or pinkish-red in colour. The taste is sweet-sour and seeds are semi-soft. Mature fruit TSS is 14–16% (Cao and Hou, 2013). It is mature by the end of August to September in Yunnan province. It is better for juice and fresh eating.

Fig. 5.20. 'Huaibeiruanzi 3'. (Photo: Lefeng Hou.)

Fig. 5.21. 'Jianshuihongzhenzhu'. (Photo: Lefeng Hou.)

5.4.11 'Lintongsanbaitian'

The cultivar originates in Shanxi province (陕西省), China (Fig. 5.22). The name is derived from the colour of the flower, skin and arils: they are all yellow white to milky white. The fruit flavour is sweet and the seeds are semi-soft. Average fruit size is 300 g. Mature fruit TSS is 16% (Cao and Hou, 2013). Generally, it is mature in late September. Fruit splitting occurs when it rains during late development.

Fig. 5.22. 'Lintongsanbaitian'. (Photo: Lefeng Hou).

Fig. 5.23. 'Qiuyan'. (Photo: Lefeng Hou).

5.4.12 'Qiuyan'

This is a commercial cultivar that originates in Shandong province, China (Fig. 5.23). It is a small tree having a moderate growth rate. The fruit shape is near round. Its average size is 425 g, and the maximum size is 725 g. Skin colour is red and arils are pinkish red. It is juicy and has a sweet-sour taste. Mature fruit TSS is 16.4%. Seeds are semi-soft. It matures in mid- to late-October. The fruit has excellent resistance to splitting and to cold and it has good storability (Cao *et al.*, 2017). This cultivar has wide adaptability and can grow well in other pomegranate-growing areas.

5.5 Egypt

5.5.1 'Assuity'

This cultivar, native to the area of the Assuit governorate, is located in upper Egypt. The plant is moderately vigorous. The fruits are roundish to flattened; medium to small in size (diameter 7.5 cm and length 6.6 cm), weight 185 g. Fruit firmness is about 77 Newtons (N). Peel thickness 0.3 mm and its weight is 97 g. The arils are medium to large, weighing 87.5 g. It has an acid taste, arils produce juice with 12.6% TSS and a very high acidity (2.6% TA), and total vitamin C content of about 14 mg/100 ml juice. Anthocyanin is about 3.9 mg/100 fresh arils. The arils are highly juicy, with a medium to large volume (67 ml). This cultivar has the trait of earliness. The period of maturation is towards the end of July. There is little sunburn and cracking of the fruit (Abou El-Khashab *et al.*, 2005). It is sold at a high price because it is one of the first cultivars to mature in the Assiut governorate. There are unknown numbers of strains for this important cultivar, estimated to be at least three.

5.5.2 'Manfalouty'

The most common cultivar in the Assiut governorate, the fruit is very large, the weight more than 600 g, with round-shaped ribs and a small, short tubular calyx. The colour of the fruit is pinkish red with a beautiful dark crimson colour, the soft skin is sometimes very thick, the seeds are large in size, the juice is red and ruby, the juice is sweet with some sweet acidity. It matures in early September. The plant is almost vigorous. The fruits are roundish and medium to large in size (diameter 8.6 cm and length 8.1 cm), fruit weight is 353 g. Fruit volume is about 303 cm^3. Fruit firmness is about 71 N and the fruit has five carpels. It is the most common local cultivar used in breeding programmes due to elite traits (Nahla *et al.*, 2014). Peel thickness is 0.5 mm. The arils are medium–large, weighing 210 g. It has a sweet taste, juice of 15.5% TSS and the acidity is low; total vitamin C content is about 13 mg/100 ml juice, anthocyanin content is about 9.5 mg/100 fresh arils. 'Manfalouty' has large, juicy, dark red arils with volume (67 ml) and matures from the end of August to the beginning of September. 'Manfalouty' is very sensitive to salt stress (Abo-Taleb *et al.*, 1998; Saeed, 2005). Authors' note: 'Manfalouty' has various spellings in the scientific literature.

5.5.3 'Nab El-Gamal'

'Nab El-Gamal' is the best Egyptian cultivar with respect to loss of chlorophyll in response to elevated salt concentration in irrigation water (Saeed, 2005). Fruit colour tends to be slightly yellow and it has a rough shape.

The aril is very large, and seeds are hard. It matures in the middle of the season, late August to early September. The advantage of this cultivar is that the fruits are of large size and have a sweet taste, with low acid content, increasing sweetness (Mohamed *et al.*, 2015). It is very similar to 'Manfalouty'. The fruits are roundish and large in size (diameter 9.5 cm and length 8.9 cm); fruit weight is 479 g. Fruit volume is about 416.5 cm^3. Fruit firmness is about 79 N and the fruit has five carpels. Peel thickness is 0.5 mm, and the weight is 186 g in fruit. The arils are medium–large, weighing 215 g. It has a sweet taste, 14.4% TSS with low acidity, about 3.2 mg/100 ml juice vitamin C content and anthocyanin content is about 6.1 mg/100 fresh arils. It has large, juicy, pink arils with a volume of 71.8 ml/100 arils.

5.5.4 'Hegazy'

This cultivar is native to the upper Egypt area. It is a vigorous cultivar with fruit weight ranging from 334–500 g and a diameter of 8.5–11.5 cm. The fruit volume is about 281 cm^3. Fruit firmness is about 76 N and the fruit has five carpels. Peel thickness is 0.5 mm and its weight is 140 g/fruit. The arils are medium–large, ranging from 181–215 g/ fruit. It has a sweet taste, with juice having 16% TSS and low acidity. Total vitamin C content is about 11.4 mg/100 ml juice and anthocyanin content is about 10.1 mg/100 fresh arils. It has large, juicy arils with a volume of 71.8 ml/100 arils. The fruit is round in shape, ribbed yellow or green. The shape of the fruits is reported to be unattractive (Sawarsan *et al.*, 2011).

5.5.5 'Wardy'

This cultivar is also called the 'Head of the Mule' due to its large fruit size. The fruit is round and not ribbed and its weight ranges from 226.5–300 g. The diameter is 8.5–10 cm. Fruit volume is about 214 cm^3, firmness is about 74.4 N and each fruit has five carpels. The calyx is long and thick. Peel thickness is 0.63 mm and its weight is 84.5 g/fruit. The rind is soft, the colour is pale yellow and the shape of the fruit is a round non-polygon. The seed is big, with the arils coloured pink and juicy. The juice taste is acid-free. The arils are medium–large, and range from 132–165 g/ fruit. The soft seeds are easy to chew. It has a sweet taste, with juice of 14.4% TSS and a total vitamin C content of about 4 mg/100 ml juice. Anthocyanin is about 10.1 mg/100 fresh arils. It has large, juicy arils with a volume of 62.5 ml. This cultivar matures in late July, which makes it an early cultivar.

5.5.6 'Araby'

Its trees are medium in size, with light flowers and medium-sized fruits showing the phenomenon of ribbing. The neck of the fruit is very short. The colour of the fruit is light yellow, slightly green with light pink spots; it has large seeds that are easy to chew and that contain a lot of juice in the arils. The taste is sweet and free of acids, matures early in late July to early August and it is suitable for cultivation in warm climates (Hassan *et al.*, 2012). The fruit weight ranges from 233–290 g. Fruit volume is about 230 cm^3, with a fruit firmness of about 75 N. Each fruit has five carpels. The calyx is long and thick. Peel thickness is 0.61 mm and its weight is 96.5 g/fruit. The fruit shape is rough, the colour is yellow to green and shiny with patches of light pink. The seed is large with pink to light red arils. The arils are medium–large and the weight ranges from 132–165 g/fruit. It has a sweet taste, with 12.8% TSS and a total vitamin C content of about 3.8 mg/100 ml juice. Anthocyanin is about 1.5 mg/100 fresh arils. This cultivar has large, juicy arils with a volume of 66.6 ml/100 arils.

5.6 Greece

5.6.1 'Ermioni'

The cultivar 'Ermioni' originated from a local population in the area of Ermioni in the north-east region of the Peloponnese peninsula of Greece (Fig. 5.24). It is an important

Fig. 5.24. 'Ermioni'. (Photo: Ferdinando Cossio.)

cultivar with good productivity, although it is not considered as a true single cultivar but as a population, since there are some differences in its appearance and fruit characteristics from orchard to orchard. Today it is also cultivated in some other regions of Greece, even in the northern region of the country. The tree is not vigorous. It ripens from 20 September to 10 October in Peloponnese. The fruit is medium to large in size. The external fruit colour is red, but the surface is not always covered completely by red colour. The arils are big (around 0.4 g), have a pink to red colour and the seeds are relatively soft. It is a very sweet cultivar, aromatic, suitable mainly for table consumption but in Greece is also used for juice production. The TSS is usually 14% (but also 16–17% some seasons); the average TA is 0.25–0.30% at maturity.

5.7 India

5.7.1 'Ganesh'

'Ganesh' (also known as 'GBG-1' earlier) is one of the oldest popular pomegranate cultivars in India, cultivated extensively in the Pune, Solapur and Satara districts of Maharashtra (Fig. 5.25). The cultivar

Fig. 5.25. 'Ganesh'. (Photo: Shilpa Parashuram.)

was developed by Dr Cheema through seedling selection from open pollinated fruits of 'Alandi' at Ganeshkhind Fruit Experiment Station, Pune in 1936. 'Ganesh' has revolutionized the cultivation of pomegranate in Maharashtra, but after release of 'Bhagawa', its acreage has decreased in India and now it is holding second position after 'Bhagawa'. The height of the tree is up to 2.37 m. The fruits are large in size with a yellowish smooth surface with a red tinge. The average fruit size is 315 g. The seeds are soft with light pinkish arils. It has moderate lateness in terms of the maturing period. The fruits are mature about 150 days after blooming. The plant average yield is about 30 kg per tree. The TSS is 16.1%, TA is 0.35–0.40% at maturity. The fruit start maturing in the month of November for mrig bahar-treated trees (India). Light pink arils have comparatively less preference in the export market. This cultivar is highly prone to the physiological disorder known as fruit cracking. The cultivar is susceptible to bacterial blight, fruitborer and fungal diseases.

5.7.2 'Bhagawa'

This is the most popular and ruling commercial cultivar of pomegranate in India (Fig. 5.26). It occupies the largest area under cultivation in India and has high demand for export. It was selected from the F_2 population of the cross made between 'Ganesh' × 'Gul-e-Shah Red' at Mahatma Phule Krishi Vidyapeeth (MPKV), Rahuri, Maharashtra. In 2003, 'Bhagawa' was released for commercial cultivation by MPKV, Rahuri. The tree is evergreen with medium height (2.5 m)

Fig. 5.26. 'Bhagawa'. (Photo: Shilpa Parashuram.)

and spreading habit. This cultivar blooms throughout the year under Indian conditions with the maturity period of 170–180 days after blooming. Its average fruit weight is 280 g. It has an attractive red-coloured rind and sweet, juicy red arils with soft seeds. The fruit has 35–40% juiciness, 15.8% TSS and 0.4–0.5% TA at maturity. It is suitable for long-distance markets as it has a medium–thick rind and better keeping quality, relatively. It is suitable for both table and processing purposes. The cultivar has disadvantages of long duration for maturity (180 days) and it is susceptible to major insect pests and diseases.

5.7.3 'Phule Arakta'

'Phule Arakta' was developed and released by MPKV, Rahuri for commercial cultivation in 2003 (Fig. 5.27). It is a selection from F_2 progeny of a cross between 'Ganesh' × 'Gul-e-Shah Red'. The trees are of medium height with spreading habit. It is under commercial cultivation in very selected pockets of production. The fruit is medium size, weighs about

Fig. 5.27. 'Phule Arakta'. (Photo: Shilpa Parashuram.)

Fig. 5.28. 'Mridula'. (Photo: Shilpa Parashuram.)

266 g with a deep red peel, thin rind and dark red arils. It is known to be earlier maturing (140 days after bloom) than 'Bhagawa' (180 days after bloom) with very soft seeds. The tree yields an average of 21–26 kg fruits. Arils are juicy (38–43% fruit juiciness), sweet (15.45% TSS) and low acidity (0.4–0.45% TA). Average length and width of aril is 10.26 mm and 6.63 mm, respectively. It is suitable for both fresh consumption as well as postharvest processing. The thin rind affects its storability and transportability to longer distances. It is susceptible to major insect pests, diseases and fruit cracking.

5.7.4 'Mridula'

In 1994, an F_2 promising plant (No. 61) with most of the desirable commercial attributes had been released by MPKV, Rahuri as 'Mridula' after their continuous breeding efforts, which started 18 years before the date of release (Fig. 5.28). It is also a seedling selection made with an open pollinated F_2 population derived through hybridization between 'Ganesh' × 'Gul-e-Shah Red' cultivars. It is cultivated in limited areas. Fruits weigh an average of 257 g. The fruits are of medium size with rough surface and deep red-coloured, thin rind, having medium-sized, dark red arils. Fruit maturity period ranges between 135 and 140 days after bloom. Seeds are much softer than 'Bhagawa'. Average fruit juiciness is about 40–45%. Arils are sweeter (15.5% TSS) and less acidic (0.4–0.45% TA). Average fruit yield per tree is 21–26 kg. It is an early maturing cultivar (140 days after bloom). It is not suitable for transportation over longer distances. It is

Fig. 5.29. 'Ruby'. (Photo: Shilpa Parashuram.)

Fig. 5.30. 'Solapur Lal'. (Photo: Shilpa Parashuram.)

susceptible to fruit cracking, bacterial blight and fungal diseases.

5.7.5 'Ruby'

'Ruby' is the most popular cultivar in Karnataka, India, and was released by IIHR, Bengaluru, in 1997 (Fig. 5.29). Hybrid no. 15-9-94 was derived from crossing between {[('Ganesh × Kabul') × 'Yercaud'] F_1–F_2} × {[('Ganesh × Gul-e-Shah Rose Pink')] F_1–F_2}. With appealing bright red fruit colour, sweet arils and soft seeds, it was named 'Ruby'. The tree height measures about 2.3 m with spreading habit. The fruit weighs around 270 g with thin rind. Average fruit juiciness (%) is 40–45%. Juicy arils contain 15.53% TSS and 0.4–0.45% TA at maturity. Fruits are ready to harvest after 165–175 days of bloom. Fruit yield per tree is about 23–28 kg. It is most suitable for the domestic market. Fruits are smaller in size than 'Bhagawa' and prone to cracking. The cultivar is also susceptible to major insect pests and diseases.

5.7.6 'Solapur Lal'

This is the first 'Biofortified cultivar' in pomegranate, newly released by ICAR-National Research Centre on Pomegranate, Solapur in 2017 (Fig. 5.30). It is a hybrid obtained from a cross between 'Bhagawa' × [('Ganesh' × 'Nana') × 'Daru']. Tree growth is vigorous with medium height (2 m) and a spreading nature. Fruits are attractive with red colour having medium rind thickness. Average fruit weight is around 265 g, containing red, sweet arils with 17.6% TSS and 0.4% TA at maturity. Mean aril length and width is about 10.21 mm and 6.55 mm, respectively. Fruit yield per plant varies between 30 and 36 kg. It has 25–35% more yield than 'Bhagawa'. Fruit juiciness is about 45–48%. Fruit maturity takes place about 160–165 days after bloom. It has 60% more iron and 25% more zinc compared with 'Bhagawa'. Also, it is rich in anthocyanin and ascorbic acid content, suitable for fresh market and juice purposes. It is a medium–hard-seeded cultivar, and is not resistant to bacterial blight.

5.7.7 'Solapur Anardana'

'Solapur Anardana' was developed specifically for the purpose of making anardana and was newly released by ICAR-National Research Centre on Pomegranate, Solapur in 2017 (Fig. 5.31). It is also a hybrid selection derived from crossing 'Bhagawa' × [('Ganesh' × 'Nana') × 'Daru']. The fruit weighs 270 g on average. Fruits are of medium size with a red, medium-thick rind, medium-size red arils and soft seeds. TSS and TA content in arils at maturity are 16.6% and 4.85%, respectively. This is a highly sour cultivar with 40–45% fruit juiciness. The fruit matures 145–155 days after bloom. Fruit yield per plant ranges between 28 and 34 kg. Red arils with high TA content, high anthocyanin (460 mg/100 g aril) and the fact that it is high yielding (20–22 metric tonnes/ha) make it suitable for anardana. However, it is susceptible to bacterial blight.

Fig. 5.31. 'Solapur Anardana'. (Photo: Shilpa Parashuram.)

5.7.8 'Phule Bhagawa Super'

'Phule Bhagawa Super' is a promising selection (Selection-4) from 62 collections of 'Bhagawa' evaluated at MPKV, Rahuri (Fig. 5.32). The cultivar was recommended for commercial cultivation in 2013. Fruits are medium size (278 g) with smooth, attractive and red-coloured rind of medium thickness. It has red, sweet, juicy arils with soft seeds. Fruit juiciness is 35–40%. The TSS and TA content of the juice is 15.71% and 0.4–0.45%, respectively. Fruits are ready to harvest 165–170 days after bloom. It is suitable for domestic and export markets. Fruit yield per plant ranges between 28 and 33 kg. It matures about 2 weeks earlier than 'Bhagawa', with high yield and better quality compared with 'Bhagawa'. It is susceptible to bacterial blight.

5.7.9 'Phule Anardana'

'Phule Anardana' was released by MPKV Rahuri in 2015 for anardana-making purposes (Fig. 5.33). Average fruit weight is 278.5 g. Fruits are of medium size with red rind and arils. At maturity, the juice contains 14.55% TSS and 4.18% TA. This highly sour cultivar gives a fruit yield of 18–22 kg per plant. High TA makes it suitable for anardana. It is rich in vitamin C and mineral nutrients.

Fig. 5.32. 'Phule Bhagawa Super'. (Photo: Shilpa Parashuram.)

Fig. 5.33. 'Phule Anardana'. (Photo: MPKV Rahuri.)

5.7.10 'Kandhari Seedless'

'Kandhari Seedless' (Hybrid-301) is a hybrid between 'Kandhari Kabuli' × 'Bhagawa' and was identified and developed by Dr Y.S. Parmar at the University of Horticulture and Forestry, Kullu, Himachal Pradesh (Fig. 5.34). It is known to be suitable for cultivation in the hills of Himachal Pradesh. It matures in mid-October at 'Bajaura'. The fruit weighs between 350–400 g with a glossy, red rind and arils. Arils are sweet with a slight acidic blend in taste containing 15% TSS and 0.52% TA at maturity. It is a soft-seeded

Fig. 5.34. 'Kandhari Seedless'. (Photo: Jayant Kumar.)

Fig. 5.35. 'Yellow Nana'. (Photo: Shilpa Parashuram.)

cultivar that yields about 38–42 kg fruits per plant. It has relatively better keeping quality, less cracking and it is suitable for long-distance transport, fresh market and processing purposes. It is less susceptible to *Alternaria* fruit spot, *Phomopsis* rot and Anthracnose. It has common insect pest problems (Kumar *et al.*, 2017).

5.7.11 'Yellow Nana'

'Yellow Nana' is a new ornamental pomegranate cultivar that was developed at ICAR-Indian Institute Horticultural Research, Bengaluru, Karnataka by crossing 'Kabul Yellow' × 'Nana' (Fig. 5.35). It is a dwarf cultivar that produces profuse, miniature, yellow-coloured flowers and fruits with highly acidic light yellow arils with hard seeds. The fruit weighs about 20 g having a thin rind. TSS and TA content in arils is about 15.05 and 5.5% at maturity, respectively. Aril length and width is about 6.31 mm and 4.08 mm. It is suitable for decoration in indoor and outdoor applications, particularly in potting containers.

5.8 Iran

5.8.1 'Rabab-e-Neyriz'

'Rabab-e-Neyriz' is the most famous commercial Iranian pomegranate cultivar, grown in Fars province, Neyriz region (Fig. 5.36) (Mohseni, 2009). The tree has a vigorous growing habit producing a high yield where the growing conditions are optimal. The height of the tree is up to 7 m. The chilling and heat (growing degree hours, GDH) requirement of Rabab is 600 h and 4920 ± 264 GDH (°C), respectively (Soloklui *et al.*, 2017). Fruit size varies from medium to large, ranging from 350–1000 g. The peel is thick with light red colour but it can also become dark red in areas with cool nights during fruit development. Arils are deep red in colour, medium, with a delicious sour-sweet taste and juicy. Seeds are moderately hard and small to medium. The harvesting period of 'Rabab' starts from mid November and continues until the end of November. TSS is 18.26% and TA is 1.08% at maturity (Varasteh *et al.*, 2009). It has good postharvest storage life (more than 4 months) and, owing to its thick peel, is suitable for long shipments. It is used for fresh consumption, fresh juice, juice concentrate and paste purposes.

Fig. 5.36. 'Rabab-e-Neyriz'. (Photo: Alimohammad Yavari.)

Fig. 5.37. 'Malas-e-Yazdi'. (Photo: Alimohammad Yavari.)

Fig. 5.38. 'Malas-e- Saveh'. (Photo: Alimohammad Yavari.)

5.8.2 'Malas-e-Yazdi'

This is one of the most common Iranian pomegranate cultivars, originally from Yazd province, and it is mainly cultivated in this area (Fig. 5.37) (Mohseni, 2009). The term 'Malas' means sour-sweet in the Persian language (Farsi) and refers to fruits that have a sour-sweet taste. The plant is moderately vigorous and productive. The fruits are medium–large in size, with light red peel and a thin rind. The arils are large, juicy, light red (or pink) and the percentage of the arils is higher than its peel percentage with a high juice content. The seed is of medium hardness. The juicy arils contain 16.6% TSS and 1.35% TA (Varasteh *et al.*, 2009). It matures in the middle of October. It is suitable for fresh consumption, juice, paste, anardana and short-term cold storage. The chilling requirement of 'Malas-e-Yazdi' is about 634 h (Soloklui *et al.*, 2017). This cultivar is moderately tolerant to salinity (Okhovatian-Ardakani *et al.*, 2010) but it is sensitive to aril browning.

5.8.3 'Malas-e-Saveh'

'Malas-e-Saveh' is one of the most famous commercial Iranian pomegranate cultivars grown in the Saveh region (Markazi province) (Fig. 5.38) (Mohseni, 2009). It is a medium-vigorous cultivar with a medium-sized canopy. The fruits are roundish, medium–large in size with an average fruit weight of 400 g. Arils are red and seed hardiness is medium. The taste is sour-sweet and TSS is 18% and TA is 1.19% at maturity (Varasteh *et al.*, 2009). The fruit peel is thick and red in colour, and it is a suitable cultivar for long-term cold storage and export. It is a late-maturing cultivar, with maturation from mid- to the end of October. It is suitable for the fresh market, juice, paste, concentrate and the sauce industries.

5.8.4 'Yousef Khani'

'Yousef Khani' is one of the most important cultivars grown successfully in Saveh (Fig. 5.39). The tree is moderately vigorous and productive. The fruits are roundish, medium–large in size and average fruit weight is 350 g. The fruit peel has medium thickness and red colour. Aril colour is red to deep red. Seeds are of medium hardness. It is a late-maturing cultivar with harvest starting in the end of October. Its taste is sour-sweet containing 17% TSS and 0.7% TA. This cultivar is suitable for fresh consumption and juice purposes. 'Yousef Khani' is moderately tolerant to salinity and drought.

Fig. 5.39. 'Yousef Khani'. (Photo: Alimohammad Yavari.)

Fig. 5.40. 'Shishe-Kab'. (Photo: Zahra Zare.)

5.8.5 'Shishe-Kab'

'Shishe-Kab' is the most common Iranian pomegranate cultivar in the east of Iran, originally from Ferdows, a city located in south Khorasan province (Fig. 5.40). It is mainly cultivated in the south and north of Khorasan province (Mohseni, 2009). It is a cultivar of moderate vigour. One of the distinctive morphological features of this cultivar is its tall fruit crown. The aril percentage is low and the peel percentage is high. Fruits have a thick peel with red colour. Arils are light red in colour with moderately hard seeds. The average fruit weight is 350 g and the largest fruit of this cultivar can weigh more than 500 g. Its taste is sour-sweet, containing 17.96% TSS and 0.79% TA (Varasteh *et al.*, 2009). It is a late-maturing cultivar that can be harvested at the end of October. This cultivar is suitable for long-term storage. It is moderately tolerant to salinity (Karimi and Hasanpour, 2014) but is sensitive to aril browning.

5.8.6 'Naderi-e-Badrud'

'Naderi' is one of the most important commercial Iranian cultivars and is grown mainly in the Badrud region (Isfahan province) (Fig. 5.41) (Mohseni, 2009). It is a medium-vigorous cultivar. Fruit shape is round, peel colour is red, aril and juice colour are red, and its taste is sweet-sour with hard seeds. The average fruit weight is 300 g. The harvesting period of 'Naderi' starts from the end of October and the harvest window continues until the middle of November. The TSS is 16.98% and TA is 0.79% at maturity (Varasteh *et al.*, 2009). The fruit peel is thick, which is believed to make it suitable for long-term storage. This cultivar is used for fresh consumption, juice, concentrate and paste purposes.

Fig. 5.41. 'Naderi-e-Badrud'. (Photo: Alimohammad Yavari.)

5.8.7 'Shirin-e-Shahvar'

The term Shirin means 'sweet', and Shahvar means 'king's favour' in Persian and implies the quality of this cultivar. This cultivar grows in most pomegranate production areas in Iran, especially in central and southern cities of Iran (Fig. 5.42) (Behzadi Shahrbabaki, 1998). The peel is thin and white-yellow in colour. Arils are large, juicy, sweet and white to pink in colour. The seeds are soft. 'Shirin-e-Shahvar' is a mid-season cultivar and its harvesting period depends on the region in which it is grown, typically starting from the end of August and continues until the end of September. The TSS is 16.67% and TA is 0.58% at maturity. Due to its short postharvest storage life, it is suitable for fresh consumption.

5.8.8 'Poost Siah'

'Poost Siah', meaning black skin, is one of the medicinal and ornamental pomegranate cultivars grown in different regions in Iran (Fig. 5.43). In traditional medicine, this cultivar has been used to treat pertussis, diarrhoea and jaundice. However, it is not a commercially cultivated cultivar at present. The plant has medium vigour. The fruit peel colour is dark violet

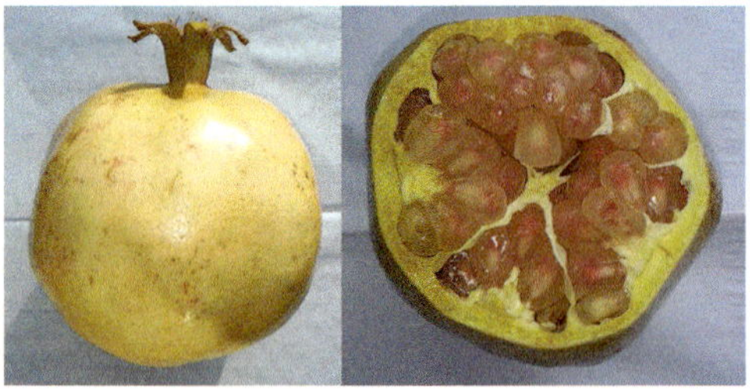

Fig. 5.42. 'Shirin-e-Shahvar'. (Photo: Alimohammad Yavari.)

to black. The aril colour is light red and purple with hard seeds. The average fruit size is 150–200 g. It is a late-maturing cultivar and is ready to harvest at the end of October. The fruit peel is thick. The taste is sweet and less juicy. The TSS is 13.97% and TA 0.5% at maturity stage. 'Poost Siah' is resistant to some pests such as carob moth and this cultivar is resistant to fruit cracking (Ranjbar *et al.*, 2004).

5.8.9 'Golnar Farsi Sarvestan'

Owing to its beautiful flowers 'Golnar' is used as a drought-resistant ornamental plant, and its flowers are used in both traditional and modern medicine for curing some diseases and in cosmetics (Fig. 5.44). In Iran, ornamental pomegranates that do not produce fruit and only produce flowers with many petals are called 'Golnar' (Behzadi Shahrbabaki, 1998). There are many ornamental cultivars with different flower colours. 'Golnar Farsi Sarvestan' is one of the Iranian ornamental cultivars grown in Fars province. It has double

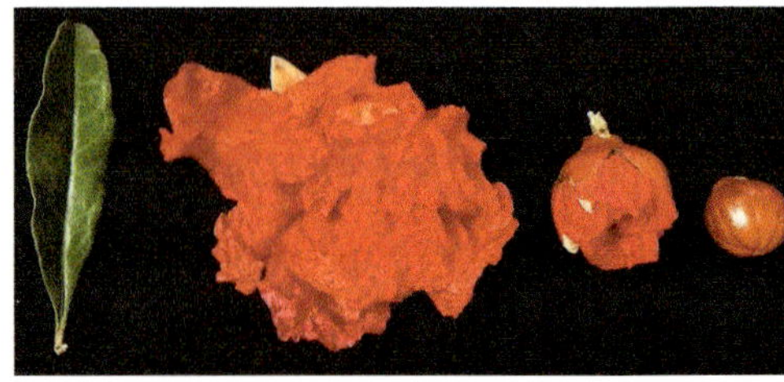

Fig. 5.44. 'Golnar Farsi Sarvestan'. (Photo: Alimohammad Yavari.)

red flowers, which open early in May. This cultivar is a 'double flower' cultivar because all the flower stamens are modified into petals. This cultivar has intermediate cold hardiness (Ghasemi Soloklui *et al.*, 2018).

5.8.10 'Bihaste Ravar'

This is one of the soft-seeded cultivars of Iran, originally from the Ravar region (Kerman province) (Fig. 5.45). As its name indicates, this cultivar is so very soft-seeded that people call it 'Bihaste' – 'seedless' in the Persian language. The fruit peel colour is yellow and thin. Arils are coloured white with very soft seeds. The taste is sweet and the fruit matures early in September. The fruit yield is low, shelf-life is poor and it is suitable for fresh consumption. The chilling and heat (GDH) requirement of 'Bihaste Ravar' is 234 h and 5,328 ± 346 GDH (°C), respectively. (Soloklui *et al.*, 2017).

Fig. 5.43. 'Poost Siah'. (Photo: Alimohammad Yavari.)

Fig. 5.45. 'Bihaste Ravar'. (Photo: Abdolkarim Zarei.)

Fig. 5.46. 'Acco'. (Photo: Doron Holland.)

Fig. 5.47. 'Shani Yonay'. (Photo: Doron Holland.)

5.9 Israel

5.9.1 'Acco' ('Aco', 'Ako', 'Akko')

This cultivar was identified in Israel, and takes its name from the homonymous city. It was selected for the earliness, the bright red colour of the peel, the sweetness of the arils and the softness of the seeds. The fruits have a bright light red skin (the red colour appears early on the peel) homogeneous over the whole surface (Fig. 5.46). The plant is of medium vigour with good productivity. The fruits are medium-sized (250–450 g). Arils weigh about 0.23 g, are deep red in colour, and have a well-balanced sweet and sour taste. The TSS ranges are 15–16% and TA is about 0.49% at maturity. The seeds are small and moderately soft. The juice is bright red. Maturation is very early. In Italy, it is harvested in mid-September; in Israel or southern Spain, 15 days earlier (end of August). Fruits have a limited shelf-life of about 1 month. It is relatively resistant to fruit cracking and sunburn.

5.9.2 'Shani Yonay'

This is a recently registered Israeli cultivar, identified in 1995 and patented in Israel in 2003 (Fig. 5.47). Very early, in Israel it matures towards the end of August and the fruits remain on the plant until the end of September, so it matures more than 1 month before 'Wonderful'. The plant is of medium vigour with reduced spinescence (thorniness). It is self-fertile with good productivity, but is less productive than 'Wonderful'. The shape of the roundish fruit is more polygonal and flatter than 'Acco'. It produces medium-sized fruits, with a homogeneous bright red skin (the red colour appears early on the peel). The arils are medium-sized (0.23 g compared with 0.37 g for 'Wonderful'), ruby red and intensely coloured. The seeds are small and soft. The juice is sweet with low acidity. The TSS reaches 14–15%, and TA is 0.51% at maturity (Holland *et al.*, 2007). The peel is thin, and the fruit has a limited shelf-life of about 1–2 months. It is not susceptible to cracking and sunburn damage.

5.9.3 'Emek'

This is a recent Israeli cultivar, identified in 2003, protected by a patent (Fig. 5.48).The plant is of medium vigour, assurgent but with open insertion angles. Maturation is very early; in Israel it matures in mid-August until mid-September (about 6 weeks before 'Wonderful'). It is self-fertile, of good productivity, with rounded fruits and a prominent calyx. It has homogeneous bright red skin (the red colour appears early on the peel), and is of large size. The peel has a medium thickness. Arils (around 0.36 g) are intensely red coloured. The seeds are mildly soft. The juice is sweet with low acidity and is intensely red. The TSS averages 14%, and TA is 0.42% at maturity (Holland *et al.*, 2014). In general, the fruits have a limited shelf-life, about 1 month. It is not very susceptible to cracking and sunburn damage.

Fig. 5.48. 'Emek' (Photo: Doron Holland.)

5.10 Italy

5.10.1 'Dente di Cavallo' ('Horse's tooth')

This is one of the main and historically important Italian cultivars, spread throughout the territory. In Sicily four types have been described: typical, late, hard-seeded and soft-seeded. Generally, it is a very vigorous and hardy plant, slow to enter into production, and of average productivity. The spinescence is average. The fruits are large, with a yellow-red peel (Fig. 5.49). The arils are large, pink, of sweet and sour taste (but also sour or sweet), with seeds that can be both hard and soft. The maturation period is medium late (mid-October). The peel is consistent, 4–5 mm thick. The fruits can be well preserved. It is a polyclonal genotype (Cossio and Vitelli, 2018).

Fig. 5.49. 'Dente di Cavallo'. (Photo: Ferdinando Cossio.)

Fig. 5.50. 'Grossa of Faenza'. (Photo: Ferdinando Cossio.)

5.10.2 'Grossa of Faenza'

This is a great cultivar of the Italian countryside (from Faenza, Emilia Romagna) and is interesting for the large size of its fruit, which can easily exceed the weight of 1 kg, but which can even reach 2 kg (Fig. 5.50).The plant is vigorous and assurgent, with little spiny branches and good productivity, hardy and resistant to diseases. The period of maturation is late, towards the end of October and the beginning of November. The fruits are moderately sensitive to cracking. The peel is pinkish yellow, very thick and resistant; the calyx closure is semi-closed or closed.

Fig. 5.51. 'Primosole'. (Photo: Ferdinando Cossio.)

The arils are large (0.44 g) and juicy, easily distinguishable, rosy or reddish. The TSS averages 17%, and TA is 1.4% at maturity. The seed is hard/medium hard. The taste of the juice is very pleasant sweet and sour. It has a long shelf-life (Cossio and Vitelli, 2018).

5.10.3 'Primosole'

This has been identified among the ancient Sicilian cultivars by researchers at the University of Catania, in Italy. The plant is vigorous, with the characteristic of a reduced spinescence. It has a medium productivity, with large fruits of marketable size (Fig. 5.51).The peel is pinkish yellow. The arils are large, of an intense pink colour, with a very aromatic flavour and a high sugar to acid ratio, it tastes sweet. The TSS is 16.2% and TA is 0.19% at maturity. The seeds are moderately soft. The period of maturation is very late, at the beginning of November after 'Mollar de Elche' and after 'Wonderful' are mature. The fruits are moderately sensitive to cracking and have a good shelf-life.

5.11 Morocco

5.11.1 'Sefri'

A typical Moroccan pomegranate, it is a cultivar with moderate vigour with a semi-erect and compact growth form. The plant is assurgent and of good productivity. The maturation of fruit takes place at the beginning of October. The fruits are large, with yellow skin, with a thick peel. The arils are light pink, semi-soft, juicy and very sweet. Average TSS is 15–17% and TA is 0.24%. The fruits are of good storage quality. It is a polyclonal genotype (Cossio and Vitelli, 2018).

5.12 Spain

5.12.1 'Valenciana'

This cultivar is native to the area of the Valencian Community (Spain), which is located along the Mediterranean coast on the east side of the Iberian Peninsula. It is considered the typical Spanish cultivar with soft seeds, starting the pomegranate season in Spain. The plant is moderately vigorous with few thorns. The fruits are roundish to flattened; medium–large in size (diameter 86–91 mm), weight is 290–310g (Fig. 5.52).The skin is light pink, and of average consistency. If the fruits are left on the tree they become pink-reddish. Peel thickness is 2.9 mm. The arils are medium–large, and pink. The taste is very sweet, juice has 14.5–16% TSS and very low acids (0.20–0.26% TA). Maturity index (MI) is 62–72 (Bartual and Valdés, 2011). The arils, which are medium size (0.31 g), have high juiciness. The tegument is extremely soft and narrow. This cultivar has the benefit of earliness but has poor shelf-life. The period of maturation is the end of August to mid-September. The yield in juice is 41% v/w (volume/weight of fruit) with not less than 74% v/w of aril.

5.12.2 'Mollar de Elche'

This is the most widely known Spanish pomegranate, with a very sweet taste and

Fig. 5.52. 'Valenciana'. (Photo: Julian Bartual.)

Fig. 5.53. 'Mollar de Elche'. (Photo: Julian Bartual.)

remarkably soft seeds (Fig. 5.53). Genetically, it is a cultivar-population native to the Elche area (Alicante province). Selected clones are available as IVIAGRANA49 and IVIAGRANA55. The European Union recognized the 'Mollar de Elche' pomegranate from Alicante as Protected Designation of Origin. The plant is vigorous, medium to low thorniness and produces roundish to flattened fruits of excellent size and weight, 93 mm diameter, 350 g fruit. The peel of the fruit is pinkish yellow, but it can also become deep pinkish with a yellow background and short crown. Furthermore, the relatively aromatic arils are pinkish-red, medium–large (0.37–0.44 g), very sweet in taste, 15.5–17.6 TSS and very low acid levels, 0.18–0.27% TA (Bartual and Valdés, 2011). The tegument is extraordinarily soft, almost not felt in the mouth. It is a late cultivar with maturation in the first week of October, it has very good storage life, and can run commercially for 4 months. Rustic and tolerant to diseases in semi-arid conditions, it is not normally subject to 'black heart' caused by *Alternaria alternata*. In the semi-arid area of Spain, sunburn and cracking are the most significant physiological disorders affecting 'Mollar de Elche'.

5.12.3 'Tendral'

This popular pomegranate cultivar is mainly cultivated in the Valencia province. It is an upright tree with vigorous growth, lightly weeping. The branches have few to no thorns. The bud break is 1 week earlier and its leaves are lighter green than 'Mollar de Elche'. This cultivar has moderate yield (40 kg per tree). The fruits are ready for commercial harvesting from 15 to 25 September. It has a yellowish to pink fruit peel and is medium to large in size (average weight about 320 g, maximum diameter 85–90 mm) (Fig. 5.54). Mature fruits are very sweet and have a good eating quality and flavour and medium thickness of skin (3.78 mm). Pink arils are medium- to large-sized (11 mm length and 7.3 mm diameter) and heavy in weight (0.41 g). Juice yield is 65–70% v/w (juice/aril). Some traits at maturity are low acids (0.40% TA) and sweet taste (15.9% TSS, with an MI of 36). It has extremely soft tegument, which is useful for local markets, germplasm conservation and

Fig. 5.54. 'Tendral'. (Photo: Julian Bartual.)

breeding. This cultivar seems to have low susceptibility to pests and diseases.

5.12.4 'Zafari'

This cultivar is also named 'Safari'. This cultivar is cultivated in the Andalusia region (south of Spain). It was probably introduced from Syria during the 8th century (Garcia Sanchez, 2011). Harvesting begins in the second week of September. With yellowish to orange skin, this cultivar usually does not develop red colour in the peel (Fig. 5.55). The fruit weight is 360 g, spherical and slightly rounded, with a thin rind. The colour of the aril is between pearl white to pink they are extremely large in size (12.26 × 7.7 mm) and weight (0.50 g). The yield of juice is moderate. The internal characteristics of arils and juice are 15.3% TSS, 3.89 pH and the acids are very low (0.30–0.35% TA), even at early stages of fruit development. The seeds have low to medium hardness. The fruits have a tendency to be affected by fruit cracking, more frequently during periods of high humidity following rains. This cultivar is also susceptible to sunburn.

Fig. 5.55. 'Zafari'. (Photo: Julian Bartual.)

Fig. 5.56. 'MR-100'. (Photo: Julian Bartual.)

5.12.5 'MR-100'

A patented cultivar, with traits very similar to 'Mollar de Elche', but larger average fruit weight and it matures a week earlier (Fig. 5.56). Develops vigorous shoots with a moderately weeping growth habit. Trees are almost thornless. Peel is pink to reddish, fruit colour progresses gradually and this characteristic serves for harvesting purposes. In warm areas, external colouration will not be complete. Average yield is 40–60 kg per tree, depending on the agronomic conditions. The fruit size is 250–400g (Melgarejo, 2014). The juice content reaches 33% of the fruit weight. It is a cultivar with low acids (0.3% TA), and it tastes very sweet (15–17% TSS). It has pink arils, medium to large in size. Harvest begins in early October in Spain. As with the majority of the pinkish late pomegranate cultivars, this cultivar is affected by sunburn and cracking. It has a light tendency to weep in the juvenile phase, training the young plant against a support is recommended in the early stages of growth.

5.12.6 'Purple Queen'

This cultivar shares common characteristics with some new early maturing hybrids registered in Spain; these include sweet taste (16.0% TSS) and low acids (0.4% TA). The stage at which the crop should be harvested is when the juice has a maturity index of 23. It has small to medium fruit size (300 g) with a juice/fruit yield of about 33% (v/w). The red fruits are round with medium thickness of rind

Fig. 5.57. 'Purple Queen'. (Photo: Julian Bartual.)

(Fig. 5.57). Aril size is medium, with a dark red juice colour. The seeds are semi-soft. For horticulture, fruit thinning should be performed, since fruit set is high. Fruit matures early, from the second week of August to September (Melgarejo, 2014). It is not recommended for long periods of storage. Average yield is similar to 'Acco'.

5.12.7 'Kingdom'

This cultivar is registered by Spanish breeders. It is a late cultivar with sour-sweet taste and large sized fruit (550–600 g) and appealing appearance (Fig. 5.58). This cultivar has dark red skin and aril colour. Peel thickness is medium–high. It is a cultivar with medium-sized arils (0.3 g) and semi-hard tegument. Compositional analyses of juice include this cultivar within the 'sweet-sour' group (no less than 1.2–1.64% TA) and TSS values are in the range from 18–20%. For juiciness, juice/fruit is 30%. The highest yield of marketable fruits is from 10 October, but cropping could be achieved in November (Melgarejo, 2014). Storability: 90–120 days. Average yield is 40–60 kg/tree depending on ecological conditions. Physiological disorders such as sunburn and cracking can affect this cultivar.

Fig. 5.58. 'Kingdom'. (Photo: Julian Bartual.)

5.12.8 'Iliana'

'Iliana' is a hybrid from a controlled cross made in the breeding programme at the Agricultural Experiment Station of Elche, Valencian Agricultural Research Institute (IVIA) and Cambayas Coop V. This cultivar resulted from crossbreeding between 'Mollar de Elche' and an unknown red cultivar. 'Iliana' is a vigorous tree with spreading to upright growth habit; it has moderately thorny branches. The fruit is round and medium to large in size, weighing more than 300 g (82 mm diameter, 3.7 mm rind thickness) with good sensory fruit quality. It has bright, dark red peel, which makes this cultivar very attractive in appearance (Fig. 5.59). Arils are medium in size (0.29 g) with relatively soft to medium seeds. This cultivar has good eating quality with a juice yield of 69% v/w juice per

Fig. 5.59. 'Iliana'. (Photo: Julian Bartual.)

Fig. 5.60. 'Rugalate'. (Photo: Julian Bartual.)

aril, and a sweet tasting low-acid juice (0.25–0.33% TA, pH 3.3–3.8, 14–15%TSS). The maturity index is particularly high for an early red cultivar, ranging from 40 to 51. It matures in the last week of August; if not picked, fruit could last on the tree until the end of September. Thinning needs to be done once fruit development begins. This cultivar has poor postharvest shelf-life. *Botrytis* sp. can affect fruit crowns.

5.12.9 'Rugalate'

This cultivar is a new hybrid obtained by the Agricultural Experiment Station of Elche (Alicante) and Valencian Agricultural Research Institute (IVIA, Moncada) by crossing 'Mollar de Elche' and 'Wonderful' (Bartual *et al.*, 2012). It has sweet and soft seeds with large fruits (350–400 g), 85–90 mm in diameter, 3.8 mm rind, morphologically spherical and as flattened as 'Mollar de Elche' (Fig. 5.60). Consistently closed and short to medium crown (17.6 mm). The juice has about 16.1% TSS, a pH of 3.5–3.8 and low acids, with 0.41% TA and b23–38 MI. Seeds are softer than other red cultivars, such as 'Wonderful' (Mena *et al.*, 2011). It has excellent sensory quality with intense red skin colouration, and maintains good quality during cold storage for 3–4 months, which is comparable to 'Wonderful' (Villamón *et al.*, 2019). It is a late-maturing cultivar, with harvests from 7–14 October. The fruits contain large red seeds 0.29–0.38 g; yielding 68.6% of juice per fruit (v/w). The fruits can be affected by physiological disorders, such as sunburn and cracking.

5.12.10 'Tastem'

This accession is the result of a controlled breeding programme from the Agricultural Experiment Station of Elche (Alicante), Valencian Agricultural Research Institute (IVIA, Moncada) and Cambayas Coop V. It was bred through crossbreeding 'Mollar de Elche' with an unknown red cultivar (Bartual *et al.*, 2015). It is self-fertile, with medium–high vigour, an early cultivar with red skin colour that appears very early during fruit development (Fig. 5.61). The fruits mature from the end of August to the first week of September and have a storage life of 2–3 months in refrigeration. It has good eating quality, arils are light red, soft seeds with a weight of 0.33 g. Yield of juice is about 32%, taste is sweet-sour with 16.7% TSS, pH 3.55, 0.30–0.38% TA and 27–35 MI. This cultivar is currently widespread in the Elche area. Plants start to produce fruit in their second or third year.

5.12.11 Author's note

It is reported that there are other cultivars on the market, for instance, pink cultivars such as 'Gorda de Xativa' or red cultivars such as 'Bigful' 'Mely' or 'Crucial', with limited agronomic data available.

5.13 Tunisia

5.13.1 'Gabsi'

This is the typical Tunisian cultivar originating from the Gabes region in the south-east of Tunisia. It is cultivated mainly in the Gabes region, but also in other parts of the country. It is a productive cultivar with rounded shape and large-sized fruits (Fig. 5.62). At maturity the skin is yellow pink. The arils are red, or reddish pink, very sweet, with low acid and soft seeds.

Fig. 5.61. 'Tastem'. (Photo: Julian Bartual.)

Average TSS is about 17% (14–18%). TA at maturity ranges from 0.15–0.25%. It reaches maturity in mid-September (at Gabes). It is a polyclonal genotype.

5.13.2 'Tounsi'

This is an important Tunisian cultivar, originating from the Tetour region in the north-west of Tunisia. It is cultivated mainly in north Tunisia. Maturation takes place around mid-October. It produces medium to large fruits, with red or yellowish red peel and ruby-red arils (Fig. 5.63). The arils are juicy, sweet and have low acids. The seeds are soft. Average TSS is from 14–16% and TA is approximately 0.2–0.3% at maturity.

Fig. 5.62. 'Gabsi'. (Photo:Faten Boussaa.)

Fig. 5.63. 'Tounsi'. (Photo: Fateh Aljane.)

5.14 Turkey

5.14.1 'Hicaznar'

'Hicaznar' (pronunciation in English, Hijaznar) is the main commercial cultivar in production in Turkey. The cultivar was selected in 1983 from Kemer, Antalya by Dr Caner Onur. The tree is moderately vigorous and is high yielding. It is self-fertile with medium thorniness. Fruit shape is round, peel colour is dark red, aril and juice colour are dark red (Fig. 5.64). The average fruit weight is 417.1–598.0 g, the aril ratio is 53.4%, 100 aril weight is 40.9 g, juice ratio is 43.9% and it has a sweet-sour taste and moderate seed hardness. It grows well in the Mediterranean region. The harvesting period of the Hicaznar cultivar starts from the beginning of October and continues until the end of October. The TSS is 16.2% and TA is 1.26% at maturity. It can be stored for up to 6 months. 'Hicaznar' is the main exported cultivar in Turkey. 'Hicaznar' is a pomegranate cultivar similar to 'Wonderful'. 'Hicaznar' pomegranate is a regular, highly yielding cultivar in different climates; at the same time, pollinators increase the yield and fruit quality. 'Hicaznar' is also the main cultivar for the pomegranate

Fig. 5.64. 'Hicaznar'. (Photo: Cenap Yilmaz.)

Fig. 5.65. 'Lefan'. (Photo: Cenap Yilmaz.)

food industry in Turkey. It is suitable for juice, sauce and dried fruit purposes because of its sour sweet taste, dark-coloured juice, high TSS content and aroma. 'Hicaznar' fruits have moderate tolerance to fruit cracking and sunburn. Insufficient colouration on rind of 'Hicaznar' can be a problem in warmer climates (Onur, 1982; Özgüven and Yılmaz, 2000; Yılmaz, 2007; Yılmaz *et al.*, 2009; Özgüven *et al.*, 2011; Anonymous, 2014; Holland *et al.*, 2014).

5.14.2 'Lefan'

This cultivar was selected in Iskenderun, Hatay. 'Lefan' is generally grown in the eastern Mediterranean and south-eastern Anatolian regions. The tree has a moderate growth rate and is high yielding. It is self-fertile with moderate thorniness. Fruit shape is round, peel colour is yellow-red, aril and juice colour are red (Fig. 5.65). The average fruit weight is 582.5–635.0 g, the aril ratio is 54.69%, 100 aril weight is 58.38 g, juice ratio is 45.91% and it has a sweet-sour taste with hard seeds. The fruit peel is very thick. The fruits of 'Lefan' can be harvested from the first week of October. The TSS is 16% and TA is 0.98% at maturity. This cultivar tends to form bigger fruits. It is suitable for storage and it can be stored for up to 5–6 months. The cultivar is moderately resistant to fruit cracking and sunburn. It is suitable for the juice and sauce industries (Onur, 1982; Özgüven and Yılmaz, 2000; Yılmaz, 2007; Ak *et al.*, 2009; Yılmaz *et al.*, 2009; Özgüven *et al.*, 2011).

5.14.3 'Katirbasi'

This cultivar (pronunciation in English, Katirbashi) was selected in Dörtyol, Hatay, and is widely grown in eastern Mediterranean and south-eastern Anatolian regions. The tree has a vigorous growth rate and is high yielding. It is self-fertile with moderate thorniness. Fruit shape is round, peel colour is yellow-red, aril and juice colour is red (Fig. 5.66). The average fruit weight is 330.4–525.0 g, the aril ratio is 64.6%, 100-aril weight is 50.58 g, the juice ratio is 52.16%, and its taste is sweet-sour and seed hardness is moderate. The harvest time is from the beginning of September. The TSS is 15.2% and TA is 1.37% at maturity. 'Katirbasi' is suitable for storage for up to 4–5 months. It is suitable for the juice and sauce industries. This cultivar is moderately resistant to fruit cracking

Fig. 5.66. 'Katirbasi'. (Photo: Cenap Yilmaz.)

Fig. 5.67. 'Silifke Asisi'. (Photo: Cenap Yilmaz.)

and sunburn (Onur, 1982; Özgüven and Yılmaz, 2000; Yılmaz, 2007; Ak *et al.*, 2009; Yılmaz *et al.*, 2009; Özgüven *et al.*, 2011).

5.14.4 'Silifke Asisi'

The cultivar (pronunciation in English, Silifke Ashisi) was selected in Silifke, Mersin. The tree has a high growth rate and is high yielding. It is self-fertile with medium thorniness. Fruit shape is round, peel colour is attractive yellow-pink, aril and juice colour is pink-red (Fig. 5.67). The average fruit weight is 413.0–636.5 g, the aril ratio is 57.2%, 100-aril weight is 58.37 g, juice ratio is 46.1%. The taste is sweet-sour and it is hard-seeded. The peel of the fruit is thick. The cultivar is suitable for long-term storage (more than 5–6 months). The maturation time of 'Silifke Asisi' is around the middle of October. The TSS is 15.2% and TA is 1.10% at maturity. This cultivar tends to form bigger fruits and is relatively resistant to fruit cracking and sunburn. It is suitable for the juice and sauce industries. The cultivar may have a low yield in certain conditions, especially in seaside regions (Onur, 1982; Özgüven and Yılmaz, 2000; Yılmaz, 2007; Ak *et al.*, 2009; Yılmaz *et al.*, 2009; Özgüven *et al.*, 2011).

5.14.5 'Izmir-23'

This cultivar was selected in Çesme (pronounced Cheshme), Izmir, and is grown mainly in the Aegean region of Turkey. The tree has a low to moderate growth rate and yield. The tree height of this cultivar is generally less than 2.5–3 m. It is self-fertile. The cultivar has less thorns than typical for a pomegranate. Fruit shape is round, peel colour is pink, aril and juice colour is red (Fig. 5.68). The average fruit weight is 292.0–385.0 g, the aril ratio is 57.23%, 100-aril weight is 49.35 g, seed to aril ratio is 14.28%, juice ratio is 47.85%, and it is a sweet and soft-seeded cultivar. The fruit are not suitable for long-term storage. The harvest begins during the third week of September. The TSS is 15.4% and TA is 0.22% at maturity. The cultivar is moderately sensitive to fruit cracking and sunburn (Ercan *et al.*, 1991; Özgüven and Yılmaz, 2000; Yılmaz, 2007; Yılmaz *et al.*, 2009).

5.14.6 'Izmir-26'

This cultivar was selected in Çeşme, Izmir, and is grown mainly in the Aegean region of Turkey. The tree has a low to moderate growth

Fig. 5.68. 'Izmir-23'. (Photo: Cenap Yilmaz.)

rate and yield. The tree height of this cultivar is generally less than 2.5–3 m. It is self-fertile. The cultivar is less thorny than average. Fruit shape is round, peel and aril colour is pink, juice colour is red (Fig. 5.69). The average fruit weight is 285.6–307.0 g, the aril ratio is 60.62%, 100-aril weight is 46.21 g, juice ratio is 53.08%, and it is a sweet and soft-seeded cultivar. The fruits of this cultivar are not suitable for long-term storage. The harvest time starts from the third week of September. The TSS is 16.0% and TA is 0.21% at maturity. The cultivar is moderately sensitive to fruit cracking and sunburn (Ercan *et al.*, 1991; Özgüven and Yılmaz, 2000; Yılmaz, 2007; Yılmaz *et al.*, 2009).

Fig. 5.69. 'Izmir-26'. (Photo: Cenap Yilmaz.)

Fig. 5.70. 'Izmir-1264'. (Photo: Cenap Yilmaz.)

5.14.7 'Izmir-1264'

This cultivar was selected in Marmaris, Muğla, and is grown mainly in the Aegean region of Turkey. The tree has a moderate growth rate and yield, is self-fertile and has average thorniness. 'Izmir-1264' is similar to 'Hicaznar'. Fruit shape is flat and round, peel colour is dark red, aril and juice colour is dark red (Fig. 5.70). The average fruit weight is 451.0–511.0 g, the aril ratio is 43.44%, 100-aril weight is 34.04 g, juice ratio is 36.47%, and it is a sweet-sour and hard-seeded cultivar. The cultivar is suitable for long-term storage. 'Izmir 1264' can be harvested from the second week of October. The TSS is 16.5% and TA is 1.35% at maturity. The cultivar is moderately sensitive to fruit cracking and sunburn (Ercan *et al.*, 1991; Özgüven and Yılmaz, 2000; Yılmaz, 2007; Yılmaz *et al.*, 2009).

Fig. 5.71. 'Izmir-1513'. (Photo: Cenap Yilmaz.)

5.14.8 'Izmir-1513'

The cultivar was selected in Nazilli, Aydin, and is grown mainly in the Aegean region of Turkey. The tree has a moderate growth rate and has high yield, is self-fertile and has medium thorniness. 'Izmir-1513' is similar to 'Hicaznar'. Fruit shape is round, peel colour is dark red, aril and juice colour is dark red (Fig. 5.71). The average fruit weight is 299.8–458.0 g, the aril ratio is 56.99%, 100-aril weight is 41.78 g, juice ratio is 43.56%, and it is a sweet-sour and hard-seeded cultivar. The cultivar is suitable for long-term storage. 'Izmir 1513' can be harvested from the second week of October. The TSS is 16.3% and TA is 1.40% at maturity. The cultivar is moderately sensitive to fruit cracking and sunburn (Ercan *et al.*, 1991; Özgüven and Yılmaz, 2000; Yılmaz, 2007; Yılmaz *et al.*, 2009).

5.14.9 'BATEM Esinnar'

The origin of 'BATEM Esinnar' is a result of the pomegranate breeding programme in the West Mediterranean Research Institute, Antalya. It is a new cultivar for Turkey and was registered in 2009. The tree has a high growth rate and is a cultivar with moderate yield. It is self-fertile. Fruit shape is round, peel colour is dark red, aril and juice colour is dark red (Fig. 5.72). The average fruit weight is 518.5 g, the aril ratio is 53.78%, 100-aril weight is 38 g, and it is a sweet and soft-seeded cultivar. The cultivar can be harvested starting from the last week of September. The TSS is 15.6% and TA is 0.45% at maturity (Yıldız Turgut, 2012; Anonymous, 2014).

5.14.10 'Efenar 35'

The 'Efenar 35' is a result of the pomegranate breeding programme in the Aegean Agricultural Research Institute, Izmir. It is a new cultivar for Turkey and was registered in 2015. The cultivar is a hybrid between 'Izmir 1445' × 'Izmir 23'. The tree has a moderate growth rate and yield (35–40 kg/tree). It is self-fertile. It is harvested is the middle of the Turkish pomegranate season.

Fig. 5.72. 'BATEM Esinnar'. (Photo: Alpaslan Sahin.)

Fig. 5.73. 'Efenar 35'. (Photo: Deniz Aksoy.)

Fruit shape is round, peel and aril colour is dark red (Fig. 5.73). The average fruit weight is 350 g, the aril ratio is 60.0%, 100-aril weight is 35 g, and it is a sweet and soft-seeded cultivar. The TSS is 15.3% and TA is 0.40% at maturity (Anonymous, 2015; Aksoy, 2017).

5.14.11 'Tezeren 35'

'Tezeren 35' is a result of the pomegranate breeding programme at the Aegean Agricultural Research Institute, Izmir. It is a new cultivar for Turkey, registered in 2015. 'Tezeren 35' is a hybrid between 'Izmir 1513' × 'Izmir 23'. The tree has a moderate growth rate and moderate yield (30–35 kg/tree). It is self-fertile. It is harvested in early season (third week of August). The TSS is 16.1% and TA is 0.50% at maturity. Fruit shape is round, peel colour and aril colour is dark red (Fig. 5.74). The average fruit weight is 300.0 g, the aril ratio is 56.0%, 100-aril weight is 35 g, and it is a sweet and soft-seeded cultivar (Anonymous, 2015; Aksoy, 2017).

Fig. 5.74. 'Tezeren 35'. (Photo: Deniz Aksoy.)

Fig. 5.75. 'Canernar'. (Photo: Cenap Yilmaz.)

5.14.12 'Canernar'

The 'Canernar' is a result of the pomegranate breeding programme at the West Mediterranean Research Institute, Antalya. The tree has a moderately strong growth rate and is high yielding. It is self-fertile. The cultivar has few thorns. Fruit shape is round, peel colour is dark red, aril and juice colour is purple (Fig. 5.75). The average fruit weight is 321.2 g, the aril ratio is 46.86%, 100-aril weight is 30.97 g, juice ratio is 34.9%, and it is a sweet-sour and soft-seeded cultivar. 'Canernar' is a very early cultivar. The harvesting period of 'Canernar' starts the from middle of August and the period continues until the middle of September. The TSS is 17.6% and TA is 1.50% at maturity. 'Canernar' is not suitable for storage. 'Canernar' is very sensitive to diseases and insect damage especially in humid conditions. The cultivar is resistant to fruit cracking and sunburn (Yılmaz *et al.*, 2009).

5.15 United States

5.15.1 'Wonderful'

This is currently the most important cultivar in the USA. It is a medium to large, sweet-tart, red pomegranate (Chater *et al.*, 2018a) (Fig. 5.76),

Fig. 5.76. 'Wonderful'. (Photo: Ferdinando Cossio.)

but can be rather sweet in arid regions with high summer temperatures and quite sour in coastal regions with mild summers. It originated in Florida, USA and its parentage is unknown (Stover and Mercure, 2007). Some believe the cultivar is originally from Spain. The fruit is shiny and deep red when light conditions are perfect, yellow and light red with red speckles are common in fruit deep in the canopy with less light exposure or on the shaded side of the fruit. Peel is leathery and tough with medium peel thickness. Juice is sweet-tart to tart, or even sour, depending on climate (Chater *et al.*, 2018b), and can be quite sweet (TA < 1.0%) in inland arid regions of California, such as Kern County. Seeds are moderately hard and small. A proportion of American consumers report disliking the texture of the seeds, with many refusing to eat them. Fruit are mature in mid to late October, but growers are known to harvest earlier in late September to avoid autumn rains; this reduces fruit quality in the USA markets. Average TA reported has ranged from 1.1–1.6% and average TSS has ranged from 16.8–17.1%, although growers will often pick when TSS is 15%. 'Wonderful' pomegranate arils have measured as low as 14% in the Californian market (unpublished data). TA can be less than 1%. Fruit are considered mature when TA is lower than 1.85% (Kader, 2006) and TSS is at or above 15%. Average weight of 100 arils is 32.5 g and the average weight of a fruit is 443.1 g (Chater *et al.*, 2018a). The tree is medium to large and its branches hang or droop when fruit develop and fill with water (Brooks and Olmo, 1997). The tree suckers profusely and requires frequent pruning. It is a hardy tree that can produce an average of over 300 fruits per tree when cultivated by a master grower (e.g. The Wonderful Company). Other growers may average yields closer to 50–75 fruits per tree (Chater and Garner, 2018). 'Wonderful' is the industry-standard cultivar in America and many other countries and it has been the premier cultivar for the past several decades, although plantings of 'Wonderful' have recently declined significantly over the past 5 years. 'Wonderful' was originally brought to California in 1896 (Stover and Mercure, 2007) and its name is believed by some to be derived from Greek mythology.

5.15.2 'Early Wonderful'

A medium to large early season red pomegranate, this cultivar is a bud mutation of 'Wonderful' and was selected by L.K. Wileman *et al.* near Visalia, California and patented in 1974, patent number 3520. The fruit is uniformly bright red, shiny and matures in mid-September or 2 weeks before 'Wonderful' (Fig. 5.77). Flavour is similar to, yet thinner than 'Wonderful' with seeds being medium hard. This has historically been a minor commercial cultivar in California (Stover and Mercure, 2007) and can be found in nurseries in California. The tree is vigorous and has a similar growth habit to 'Wonderful'. It is reported to have good pest resistance and can be strip-harvested in September (Brooks and Olmo, 1997).

Fig. 5.77. 'Early Wonderful'. (Photo: John M. Chater.)

Fig. 5.78. 'Foothill'. (Photo: John M. Chater.)

5.15.3 'Early Foothill'

A medium, red, early-eason commercial pomegranate (Stover and Mercure, 2007), which originated in Porterville, California and is a bud mutation of 'Wonderful'. Selected by R.J. Owen and patented in 1976, the plant patent number 3884. It colours up earlier than 'Wonderful' and has redder flowers. The fruits are smaller than 'Wonderful' and have less acids. Arils and juice are red, and seeds are moderately hard. Flavour is mild and delicate and less tart than 'Wonderful' and 'Foothill' (AKA 'Foothill Late'). 'Early Foothill' matures late August or early September and is earlier than 'Wonderful' and 'Foothill'. The tree form is similar to 'Wonderful' (Brooks and Olmo, 1997).

5.15.4 'Foothill' AKA 'Foothill Late'

Medium to large red, commercial pomegranate, like 'Early Foothill', originated in Porterville, California and is also a bud mutation of 'Wonderful' (Fig. 5.78). It was selected by R.J. Owen and it was patented in 1976 with plant patent number 3883. The fruit is similar to 'Wonderful', but it has redder flowers and it is more acidic than 'Wonderful'. This cultivar is later than 'Wonderful' and its tree form is similar to 'Wonderful'. This has been a commercial cultivar in California and to complicate matters, some growers market 'Early Foothill' as 'Foothill', which has a sweeter and milder flavour (Brooks and Olmo, 1997).

Fig. 5.79. 'Granada'. (Photo: Ferdinando Cossio.)

5.15.5 'Granada'

Red, early-season, commercial pomegranate (Stover and Mercure, 2007), this cultivar is a bud mutation of 'Wonderful' and originated in Lindsay, California and was patented by M.J. Slayman in 1966 with plant patent number 2618. The fruit is fully red to dark red and has small moderately hard seeds (Fig. 5.79). This is a commercial cultivar in California and is much earlier than 'Wonderful', with some growers starting shipping in early August and ending in early September. The flavour of 'Granada' is significantly less acidic than 'Wonderful' and it can be considered a sweet cultivar. The fruit have a tendency to russet lightly during development. The tree form is similar to 'Wonderful' (Brooks and Olmo, 1997).

5.15.6 'Eversweet'

A small to medium-sized, pink pomegranate with sweet soft seeds, originated in Camarillo, California and was selected by S.J. Chater in the late 20th century (Fig. 5.80). Its parentage is believed to be a chance seedling from a Lebanese cultivar. This cultivar was patented in the 1980s with plant patent

Fig. 5.80. 'Eversweet'. (Photo: Ferdinando Cossio.)

number 5418 (Brooks and Olmo, 1997). The rind is pink with red blush and soft and it does not have a long postharvest shelf-life. Fruit can be yellow if set under the canopy and away from direct sunlight. Average fruit weight is 250 g and average weight of 100 arils is 36 g. Average TA is 0.36% and average TSS is 16.0%. It has medium-sized soft seeds with pink to reddish pink colour depending on climate (Chater *et al.*, 2018a). The tree is smaller and blooms two or three times per season, making it an indeterminant cultivar with multiple cohorts of fruit (Brooks and Olmo, 1997). It has an excellent fruity, sweet flavour in coastal regions and has less citric acid than malic acid (Chater *et al.*, 2019). Fruit mature over a long harvest window due to multiple blooms and fruit sets. Maturity can be as early as late August and continue to October. Fruit are usually desiccated and/or split by late October and November.

5.15.7 'Al-Sirin-Nar'

Medium to large, sweet-tart, reddish pomegranate, originated in either Azerbaijan or Turkmenistan and came to the USDA/ARS Southeastern Fruit and Tree Nut Research

Fig. 5.81. 'Al-Sirin-Nar'. (Photo: Jeff Moersfelder.)

Laboratory in Byron, Georgia from Michael Hothchkiss (Fig. 5.81). Peel is soft and thick. The seeds are hard and arils and juice are red. Juice of 'Al-Sirin-Nar' pomegranates had an average of 1.10% TA in citric acid equivalent. TSS averaged 17.3% (Chater *et al.*, 2018a). The tree is vigorous and has relatively high yields. Due to hard seeds, this cultivar would be a good candidate for juice. The fruit is mature by late October.

5.15.8 'Blaze'

A medium-sized, sweet-tart, red pomegranate, with a shape similar to 'Wonderful'. It originated in Camarillo, California and was bred and selected by S.J. Chater in the late 20th century (Fig. 5.82). Peel is leathery and thick (Chater, 2014). Average TA is 1.1% and average TSS is 17.4%. Arils and juice are red to dark red. Seeds are medium-sized and moderately hard (Chater *et al.*, 2018a). The tree grows in a similar way to 'Wonderful'; tree size is medium to large and its branches hang or droop when fruit develop. This cultivar has high yield in optimal conditions and is a candidate for fresh market or juice (Chater, 2014). The fruit mature by mid- to late October (Chater *et al.*, 2018a).

Fig. 5.82. 'Blaze'. (Photo: John M. Chater.)

Fig. 5.83. 'Desertnyi'. (Photo: John M. Chater.)

5.15.9 'Desertnyi'

A medium-sized, sweet-tart pomegranate with soft seeds, which originated in Turkmenistan and was developed by N.J. Zaktreger, O.F. Mizgireva and A.D. Strebkova (Preece *et al.*, 2016). The parentage is reported to be the following cross: [('Kazake' × 'Wonderful') × 'Wonderful'] (Levin, 2006). It came to USDA germplasm in 1999. It is a pinkish red to red fruit with red arils (Fig. 5.83). The rind is soft and the fruit has poor shelf-life and does not transport well (Preece *et al.*, 2016). The seeds are small and soft and it has a flavour that some have described as citrusy. Fruit quickly perish on the tree if left on beyond harvest date, making the harvest window for this cultivar rather narrow. Average fruit weight is 383 g and average 100-aril weight is 37.3 g. Average TA is 1.5% and average TSS is 16.7% (Chater *et al.*, 2018a). The tree is small and grows slowly and may be a semi-dwarf type. Branches are thorny and the tree grows as a spreading bush if not properly pruned (Preece *et al.*, 2016). The branches may be weaker than other cultivars. This cultivar has poor pest resistance.

5.15.10 'Haku Botan'

A yellow, ornamental, sour pomegranate with white double-flower blooms (Preece *et al.*, 2016) (Fig. 5.84). This cultivar originated in Japan Shibamichi Nursery in Saitama, Japan. It was introduced to the USDA germplasm in 1990. The fruit are small to large, white to yellow, with calyx lobes that often come together at the tips (Preece *et al.*, 2016). The arils and seeds are small. The seeds are medium hard to hard.

Fig. 5.84. 'Haku Botan'. (Photo: Jeff Moersfelder.)

The juice has a lemony flavour and is sour to sweet-sour. Average TA is 3.9% and average TSS is 15.8% (Chater *et al.*, 2018a). Arils are white to yellow, but sometimes some pink can be observed very late in the season (November). The fruit are susceptible to sunburn (Preece *et al.*, 2016), wind damage and split, and will almost turn inside-out late season, exposing the arils. The tree form is upright and up to 4 m in height. The tree produces bright white double flowers that have commercial potential and the branches generally lack thorns.

5.15.11 'Parfianka'

Originally from Turkmenistan, the plant is moderately vigorous, spinescent, very productive, with medium to large-sized fruits (Kennedy, 2010; Chater *et al.*, 2018a). The fruits have a homogeneous, intense, bright red colour (Fig. 5.85). The arils are ruby-red, medium–large, and the juice is dark red. Some have argued that it is superior to 'Wonderful' in taste, being sweeter and more aromatic. Soluble solids 18%, total acids 0.83% in Italian systems. Average TA is 1.22% and average TSS is 15.9% in California, USA (Chater *et al.*, 2018a). The seed is small and soft. The plant has less vigour in comparison with 'Wonderful', but is readily propagated via hardwood cuttings (Kennedy, 2010; Chater *et al.*, 2017). The fruit matures in mid-October, or about 2 weeks before 'Wonderful'. The fruits are sensitive to botrytis and insects. The shelf-life is poor.

5.15.12 'Purple Heart'

Medium-sized, tart-sweet, red pomegranate, marketed as 'Sharp Velvet' in California, USA. Fruit shape is similar to 'Wonderful', with medium to large fruit (Fig. 5.86). It originated in Camarillo, California and was bred and

Fig. 5.85. 'Parfianka'. (Photo: Jeff Moersfelder.)

Fig. 5.86. 'Purple Heart'. (Photo: Jeff Moersfelder.)

selected by S.J. Chater. It was introduced to the USDA germplasm in 1990. The peel is leathery and thick (Preece *et al.*, 2016). Average TA is 1.4% and average TSS is 16.8% (Chater *et al.*, 2019). Arils and juice are red to dark red. Seeds are medium-sized and moderately hard. Seeds are crunchy with a nutty flavour (Preece *et al.*, 2016). The tree grows in a similar way to 'Wonderful'; tree form is medium to large and its branches droop. The cultivar has high yield in optimal conditions and is a candidate for fresh market or juice. The fruit mature by mid- to late October in Davis, California.

5.15.13 'Ink'

A medium-sized, sweet-tart, dark red pomegranate, which originated in Chico, California at the California Plant Introduction Station. The fruit

Fig. 5.87. 'Ink'. (Photo: Jeff Moersfelder.)

Fig. 5.88. 'Ariana'. (Photo: Jeff Moersfelder.)

shape is similar to 'Wonderful' (Fig. 5.87). The arils are large and are solid to dark red. The seeds are medium hard. This cultivar scored well in USDA sensory panels (Moersfelder, 2008).

5.15.14 'Ariana'

A medium- to large-sized, tart-sweet, red pomegranate with soft seeds, which came to USDA germplasm in 1999 and originated from Turkmenistan (Moersfelder, 2008). The fruits are round-flat. The arils are red and large, the juice is dark red (Fig. 5.88). The seeds are small and soft, TSS 14.1% and TA 1.25% at maturity. The plant has been described as a low-growing shrub with multiple trunks that grows more horizontally. The fruit mature in late October (Preece *et al.*, 2016).

References

Abo-Taleb, S.A., Moaman, V.F. and El-Deen, S.S. (1998) Growth of pomegranate transplants as affected by different water regimes. *Annals of Agricultural Science Moshtohor Journal* 36, 1073–1091.

Abou El-Khashab, A.M., El-Iraqy, M.A. and Essa, K.B. (2005) Evaluation of some pomegranate (*Punica granatum* L.) cultivars under new reclaimed region. *Egyptian Journal of Agricultural Research* 2(1), 59–69.

Ak, B.E., Ikinci, A., Parlakci, H., Özgüven, A.I. and Yılmaz, C. (2009) Some pomological traits of different pomegranate varieties grown in Sanliurfa-Turkey. *Acta Horticulturae* 818,115–120. DOI: 10.17660/ActaHortic.2009.818.15.

Aksoy, D. (2017) Determination of blooming, pollen and fruit set characteristics on some pomegranate cultivars. MSc Thesis. Adnan Menderes University, Graduate School of Applied and Natural Sciences, Department of Horticulture, Aydın, Turkey.

Anonymous (2014) *The Catalog of Variety*. Batı Akdeniz Agricultural Research Institute, Antalya, Turkey.

Anonymous (2015) *The Catalog of Registered Varieties 2015*. Republic of Turkey, Ministry of Food, Agriculture and Livestock, Variety Registration and Seed Certification Center, Ankara, Turkey.

Bartual, J. and Valdés, G. (2011) The pomegranate: state of the art, problems and future of this species in the Valencian region. *Agricultura y Cooperación* 316, 24–27.

Bartual, J., Valdés, G., Escartin, N., Lozoya, A., Andreu, J. *et al*. (2012) Pomegranate improvement through clonal selection and hybridization in Elche. *Options Mediterraneennes* 103, 71–74.

Bartual, J., Palou, L. and Pérez-Gago, M.B. (2015) Characterization of fruit traits from 'Mollar de Elche' pomegranate progenies. *Acta Horticulturae* 1106, 25–30. DOI: 10.17660/ActaHortic.2015.1106.5.

Behzadi Shahrbabaki, H. (1998) *Genetic Diversity of Pomegranate Genotypes in Iran*. Nashr Amoozesh Keshavarzi Publications, Tehran, Iran.

Brooks, R.M. and Olmo, H.P. (1997) *Brooks and Olmo Register of Fruit and Nut Varieties*, 3rd edn. ASHS Press, Alexandria, Virginia.

Cao, S.Y. and Hou, L.F. (2013) *China Fruit Monograph – Punica granatum*, 1st edn. China Forestry Publishing House, Beijing, China.

Cao, S.Y., Li, X.H. and Hao, Z.X. (2017) *Monograph of Local Fruit Varieties in China – Punica granatum*, 1st edn. China Forestry Publishing House, Beijing, China.

Chater, J.M. (2014) Register of new fruit and nut cultivars list 47: pomegranate. *HortScience* 49(4), 396–421.

Chater, J.M. and Garner, L.C. (2018) Foliar nutrient applications to 'Wonderful' pomegranate (*Punica granatum* L.). II. Effects on leaf nutrient status and fruit split, yield and size. *Scientia Horticulturae* 242, 207–213. DOI: 10.1016/j.scienta.2018.07.015.

Chater, J.M., Merhaut, D.J., Preece, J.E. and Blythe, E.K. (2017) Rooting and vegetative growth of hardwood cuttings of 12 pomegranate (*Punica granatum* L.) cultivars. *Scientia Horticulturae* 221, 68–72. DOI: 10.1016/j.scienta.2017.04.025.

Chater, J.M., Merhaut, D.J., Jia, Z., Mauk, P.A. and Preece, J.E. (2018a) Fruit quality traits of ten California-grown pomegranate cultivars harvested over three months. *Scientia Horticulturae* 237, 11–19. DOI: 10.1016/j.scienta.2018.03.048.

Chater, J.M., Merhaut, D.J., Jia, Z., Arpaia, M.L., Mauk, P.A. *et al*. (2018b) Effects of site and cultivar on consumer acceptance of pomegranate. *Journal of Food Science* 83(5), 1389–1395. DOI: 10.1111/1750-3841.14101.

Chater, J.M., Mathon, C., Larive, C.K., Merhaut, D.J., Tinoco, L.W. *et al*. (2019) Juice quality traits, potassium content, and 1H NMR derived metabolites of 14 pomegranate cultivars. *Journal of Berry Research*, 1–17.

Chen, Y.H., Hu, Q.X., Li, H.T., Zheng, X.B., Tan, B. *et al*. (2012) A new pomegranate cultivar 'Dongyan'. *Acta Horticulturae Sinica* 39(7), 1411–1412.

Cossio, F. and Vitelli, V. (2018) *Il Melograno: Botanica, Varietà, Impianto, Cure colturali, Difesa e Utilizzi (Guida Illustrata)*. Supplemento a Vita in Campagna, Ed. L'Informatore Agrario, Verona, Italy.

Ercan, N., Özvardar, S. and Baldıran, E. (1991) *Nar Çeşit Araştırma Projesi Ara Sonuç Raporu*. T.C. Tarım ve Köyişleri Bakanlığı Ege Tarımsal Araştırma Enstitüsü Müdürlüğü, Menemen, İzmir, Turkey.

Garcia Sanchez, E. (2011) Fruit production in al-Andalus: an example of biodiversity. *Estudio Avanzados* 16, 51–70.

Ghasemi Soloklui, A.A., Gharaghani, A., Yavari, A.M., Eshghi, S., Nasrabadi, M.E. *et al*. (2018) Genetic variations of cold hardiness within Iranian ornamental pomegranates. *Acta Horticulturae* 1190, 53–58. DOI: 10.17660/ActaHortic.2018.1190.9.

Hassan, N.A., Halwagi, A.A. and Sayed, H.A. (2012) Phytochemicals, antioxidant and chemical properties of 32 pomegranate accessions growing in Egypt. *World Applied Sciences Journal* 16(8), 1065–1073.

Holland, D., Hatib, K., Bar-Ya'akov, I., Yonay, E. and Abd El Hadi, F. (2007) 'Shani-Yonay' pomegranate. *HortScience* 42(3), 710–711. DOI: 10.21273/HORTSCI.42.3.710.

Holland, D., Bar-Ya'akov, I. and Hatib, K. (2014) 'Emek', a red and very early-ripening new pomegranate cultivar. *HortScience* 49(7), 968–970. DOI: 10.21273/HORTSCI.49.7.968.

Kader, A.A. (2006) Postharvest biology and technology of pomegranates. In: Seeram, N.P., Schulman, R.N. and Heber, D. (eds) *Pomegranates: Ancient Roots to Modern Medicine*. CRC Press, Boca Raton, Florida, pp. 211–220.

Karimi, H.R. and Hasanpour, Z. (2014) Effects of salinity and water stress on growth and macro nutrients concentration of pomegranate (*Punica granatum* L.). *Journal of Plant Nutrition* 37(12), 1937–1951. DOI: 10.1080/01904167.2014.920363.

Kennedy, C.T. (2010) Register of new fruit and nut cultivars list 45: pomegranate. *HortScience* 45(5), 746–747.

Kumar, J., Sharma, S.D., Thakur, D. and Khosla, K. (2017) New variety of pomegranate: 'Kandhari Seedless' for mid-hills of Himachal Pradesh. *Souvenir 2nd National Seminar-cum-Farmers' Fair on 'Pomegranate for Health, Growth & Prosperity' organized by ICAR-National Research Centre on Pomegranate, Solapur and Society for Advancement of Research on Pomegranate*, Solapur, India, 28–30 April, pp. 21–23.

Levin, G.M. (2006) *Pomegranate*, 1st edn. Texas A&M Press, College Station, Texas.

Melgarejo, P. (2014) El granado. La fruticultura del siglo XXI en España. ED. Cajamar. *Serie Agricultura* 10, 225–240 (in Spanish).

Mena, P., García-Viguera, C., Navarro-Rico, J., Moreno, D.A., Bartual, J. *et al.* (2011) Phytochemical characterisation for industrial use of pomegranate (*Punica granatum* L.) cultivars grown in Spain. *Journal of the Science of Food and Agriculture* 91(10), 1893–1906. Available at: https://doi.org/ DOI: 10.1002/jsfa.4411.

Moersfelder, J. (2008) Pomegranates: interesting selections in the USDA-ARS Davis, National Clonal Germplasm Repository collection. United States Department of Agriculture. Available at: https://crec.ifas.ufl.edu/extension/pomegranates/pdfs/Moersfelder_Jeff_NCGR_091412.pdf (accessed 29 July 2019).

Mohamed, A.K.A., Ibrahim, R.A., Abdel-Salam, M.M. and Abd-El- Ghany, A.M.M. (2015) Physico-chemical and antioxidant contents during developmental stages in three pomegranates cultivars under Assiut condition. *Assiut Journal of Agricultural Sciences* 46(4), 77–96.

Mohseni, A. (2009) The situation of pomegranate orchards in Iran. *Acta Horticulturae* 818, 35–42. DOI: 10.17660/ActaHortic.2009.818.3.

Nahla, A., El-Taweel, A.A. and Aly, A.A. (2014) Studies on cross pollination between Manfaloty pomegranate and some evaluated import cultivars. *British Journal of Applied Science & Technology* 4(25), 3701–3715. DOI: 10.9734/BJAST/2014/10518.

Okhovatian-Ardakani, A.R., Mehrabanian, M., Dehghani, F. and Akbarzadeh, A. (2010) Salt tolerance evaluation and relative comparison in cuttings of different pomegranate cultivars. *Plant, Soil and Environment* 56(4), 176–185. DOI: 10.17221/158/2009-PSE.

Onur, C. (1982) Akdeniz Bölgesi Narlarının Seleksiyonu. Doctoral Thesis. Çukurova Üniversitesi, Fen Bilimleri Enstitüsü, Sarıçam/Adana, Turkey.

Özgüven, A.I., Yılmaz, M. and Yılmaz, C. (2011) The fruit traits of some pomegranate cultivars in Adana ecological conditions. *Acta Horticulturae* 890, 197–200.

Özgüven, A.I. and Yılmaz, C. (2000) Pomegranate growing in Turkey. In: Melgarejo-Moreno, P., Martínez-Nicolás, J.J. and Martínez-Tomé, J. (eds) *Production, Processing and Marketing of Pomegranate in the Mediterranean Region: Advances in Research and Technology*. CIHEAM-IAMZ, Zaragoza, pp. 41–48.

Preece, J.E., Chater, J.M. and Moersfelder, J. (2016) Register of new fruit and nut cultivars list 48: pomegranate. *HortScience* 51(6), 620–620.

Qin, G., Xu, C., Ming, R., Tang, H., Guyot, R. *et al.* (2017) The pomegranate (*Punica granatum* L.) genome and the genomics of punicalagin biosynthesis. *The Plant Journal* 91(6), 1108–1128. DOI: 10.1111/tpj.13625.

Ranjbar, V., Asadi, Y., Hoseininia, M. and Behzadi Shahrbabaki, H. (2004) *Pomegranate Guide (Plantation, Cultivation and Harvesting)*. Nashr Amoozesh Keshavarzi Publication, Tehran, Iran.

Saeed, W.T. (2005) Pomegranate cultivars as affected by paclobutrazol, salt stress and change in fingerprints. *Bulletin of Faculty of Agriculture, Cairo University* 56, 581–615.

Sawarsan, M.R., El-Bolok, K.T. and Abou-Taleb, S.A. (2011) Comparative study on some pomegranate cultivars grown under the ecological conditions of Souhag governorate. *Agriculture Research Journal* 11(2), 101–116.

Soloklui, A.A.G., Gharaghani, A., Oraguzie, N., Eshghi, S. and Vazifeshenas, M. (2017) Chilling and heat requirements of 20 Iranian pomegranate cultivars and their correlations with geographical and climatic parameters, as well as tree and fruit characteristics. *HortScience* 52(4), 560–565. DOI: 10.21273/HORTSCI11614-16.

Stover, E. and Mercure, E.W. (2007) The pomegranate: a new look at the fruit of paradise. *HortScience* 42(5), 1088–1092. DOI: 10.21273/HORTSCI.42.5.1088.

Varasteh, F., Arzani, K., Zamani, Z. and Mohseni, A. (2009) Evaluation of the most important fruit characteristics of some commercial pomegranate (*Punica granatum* L.) cultivars grown in Iran. *Acta Horticulturae* 818, 103–108. DOI: 10.17660/ActaHortic.2009.818.13.

Villamón, D., Palou, L., Bartual, J., Taberner, V., de la Fuente, B. *et al.* (2019) Fruit quality attributes of a new Spanish pomegranate cultivar at harvest and during cold storage. *Acta horticulturae* 1254(41), 275–282. DOI: 10.17660/ActaHortic.2019.1254.41.

Yıldız Turgut, D. (2012) Determination of phenolic composition and antioxidant activities of some pomegranate (*Punica granatum* L.) cultivars and genotypes grown in the Mediterranean region of Turkey. MSc Thesis. Süleyman Demirel University, Graduate School of Applied and Natural Sciences, Department of Food Engineering, Isparta, Turkey.

Yılmaz, C. (2007) *Pomegranate*. Hasad Publisher, Istanbul.

Yılmaz, C., Özgüven, A.I., Gülşen, O., Canan, İ. and Yılmaz, M. (2009) The conservation and molecular characterization of the pomegranate genotypes in Turkey. Project final report. Project no: 1110045. The Scientific and Technological Research Council of Turkey (TUBITAK), Ankara, Turkey.

Zhu, L.W., Zhang, S.M., Jia, B. and Ye, Z.F. (2009) A new pomegranate cultivar 'Baiyushizi'. *Acta Horticulturae Sinica* 36(3), 460.

6 Propagation Techniques and Nursery Management

N.V. Singh[1]*, H.R. Karimi[2], J. Sharma[1] and K.D. Babu[1]
[1]ICAR-National Research Centre on Pomegranate, Solapur, India; [2]Department of Horticultural Science, Vali-e-Asr University of Rafsanjan, Rafsanjan, Iran

6.1 Introduction

Pomegranate is a powerhouse of therapeutic and nutritional properties and belongs to the coveted category of the 'super fruits'. The versatile adaptability of this wonder fruit crop to a wide range of climatic and edaphic conditions makes it one of the choicest fruit crops in climatically challenged areas (Singh *et al.*, 2012a; Maity *et al.*, 2014). It is a relatively hardy plant and responsive to improved cultural practices; plants produce attractive ruby-red fruits and refreshing red arils packed tightly inside a leathery rind ensuring keeping qualities (Levin, 2006). These features have made pomegranate cultivation a highly lucrative agriculture business in water-scarce semi-arid tropics and subtropics. The high returns on investment per unit area from this crop have resulted in a rapid increase in area, production and export of pomegranate during the recent past (Sharma *et al.*, 2014).

Although there are no systematic data available to strongly support the pomegranate revolution globally, rough estimates suggest that the global production and acreage of pomegranate have registered an unprecedented growth of about twofold since 2007–2008. The global pomegranate acreage increased from roughly 0.22 million ha to 0.5 million ha and production rose from 2.3 million metric tonnes (t) to 6.0 million t over the same period (Holland and Bar-Ya'akov, 2008; Chandra *et al.*, 2010). The global production is led by India with 2.8 million t, followed by China, Iran, Turkey, the USA, Tunisia, Morocco, Spain and Israel (National Horticultural Board, 2018 http://nhb.gov.in; Singh *et al.*, 2017a). This unprecedented expansion demands the availability of quality planting material to ensure sustainability and profitability from the established orchards.

Healthy planting material is the first step for successful crop production, and this becomes all the more important in horticultural crops that are perennial. The rapid pace with which pomegranate expansion is under progress necessitates huge availability of elite planting material, and this requires confluence of traditional propagation methods with modern propagation techniques and logistics to make available the required quantities. Taking into consideration the area expansion during the past 10 years, on average more than 20 million elite pomegranate saplings are globally required per year for plantation to meet the pace of expansion. Traditionally, pomegranate is propagated through hardwood cuttings and air layering but in the recent past *in vitro*-raised disease-free and bio-hardened pomegranate saplings are becoming popular among

*Corresponding author: nripendras72@gmail.com

pomegranate farmers to avoid the spread of diseases like bacterial blight through planting material carrying latent infection. There are also a few reports in the recent past on exploring possibilities of grafting, budding and other innovative propagation methods in pomegranate (Karimi *et al.*, 2011; Karimi, 2011; Chandra and Jadhav, 2012; Pal and Singh, 2017; Singh *et al.*, 2017b, c; Ahire *et al.*, 2017). This chapter examines various propagation methods, techniques and nursery management practices in pomegranate.

6.2 Nursery Management in Pomegranate with Special Reference to India (Sharma and Srivastav, 2004; Ahire *et al.*, 2017; Sharma *et al.*, 2014)

Until recently, before commercialization of micropropagation protocols, the greatest bottleneck in the fast expansion of pomegranate plantation was limited availability of healthy and elite saplings. Traditionally, pomegranate planting material is propagated through air layering and stem cuttings, which may carry latent infections of threatening diseases like bacterial blight and wilt and act as inoculum sources in new areas of pomegranate cultivation. Hence, procuring healthy planting material from a reliable certified nursery is of paramount importance. Therefore it is high time to establish stringent nursery guidelines and ensure their strict execution to avoid the spread of diseases in new areas through infected planting material.

Essential steps for establishing an ideal commercial nursery for production of elite planting material:

- The mother plants should be maintained by the recognized organizations or the protector of the variety. Relevant morphological/genetic/molecular markers or Distinctness, Uniformity and Stability (DUS) characteristics should be established for maintaining varietal identity and purity.
- Healthy mother plants should be used to establish progeny orchards/increase block and should be monitored by a team of experts regularly for any kind of serious infection.
- Approved/certified nurseries / tissue culture laboratories must use elite planting material and follow only standard propagation methods and protocols like air layering / hardwood cuttings / micropropagation, budding, grafting, etc.
- Suspected propagating material has to be tested regularly through diagnostic symptoms, ooze tests, microscopy, molecular diagnostics, *in vitro* culturing, etc.
- Each selected source tree should be registered, and nurserymen should use permanent and separate tags to label saplings of each propagation group.
- Non-certified nursery stock cannot be grown or introduced into the same greenhouse structure.

Nursery sanitation and treatment of mother plant and saplings:

- The nursery should be kept clean.
- Drenching of nursery with bleaching powder solution (a.i. 33% Cl) every 3 months at the rate of 25 kg/1000 l water/ha on ground.
- Pruning tools – secateurs, etc., if used should be sterilized after handling each plant with sodium hypochlorite (2.5%).
- Keep nursery free from weeds, which may be latent carriers or multiplication ground for several diseases, nematodes and insect pests.
- Apply Bordeaux paste (10%) to the cut ends of the mother plant and air-layered cuttings.
- Treat the roots of air layered cuttings with copper oxychloride (COC) at 2.5 g/l to protect against soil-borne diseases at the early stage and plant them in the standard size polyethylene bags filled with potting mixture containing plant-beneficial microbes.

Nursery certification:

- The issuing of a phytosanitary certificate should be mandatory.
- Periodical inspection of the specified/nurseries should be mandatory for the presence of any disease/insect pests before the planting material is lifted.
- Mother plants and saplings should be regularly monitored for:

– bacterial blight leaf spots and nodal blight;

– *Cercospora* leaf spots (common);

– *Alternaria* leaf spots and heart rot;

– wilt – *Ceratocystis fimbriata*, *Macrophomina*, *Rhizoctonia*, *Fusarium*;

– Sucking pests – thrips and aphids (most common nursery pest in pomegranate), mealybugs, scale insects and mites;

– nematodes (root knot nematode);
– other insect pests and diseases.

Nursery site and facilities:

- The nursery site must be well connected to roads.
- All nursery stock must be propagated only at the approved site.
- There must be facilities such as shadenet houses, mist houses, low tunnel poly houses, a place for the sterilization of potting mixture and other climate-controlled structures.
- Pest monitoring tools such as yellow sticky traps or other detection devices for insects should be used by the nursery.
- Commercial nursery sites must be about 1 km away from commercial pomegranate plantations unless and until propagation is carried out in highly sanitized protected structures.
- The site should have an adequate quality water supply and the area should be well drained.
- The site should have adequate parking outside the facility, and a dedicated area for delivery of saplings.
- The nursery site should incorporate natural or artificial windbreaks and must be fenced with secured entrances.
- Site access should have permitted entry only through an area that incorporates decontamination areas for personnel and equipment.
- Either a soilless planting mixture such as different proportions of cocopeat, perlite, compost and vermiculite should be used, or planting mixtures should be sterilized to render them free from soil pathogens, insect pests and nematodes. Sterilization may be done through (i) solarization, (ii) aerated steam, (iii) using chemicals or (iv) using electric soil sterilizers.
- The potting mixture should be biofortified with plant-beneficial microbes.
- Well-rooted plants either in nursery bags with sterile potting mixture or well-established and trained plants in plastic pots should be encouraged for orchard planting.
- Proper record keeping should exist for sale of saplings and mother plant maintenance.

6.3 Sexual Propagation

Pomegranate is propagated commercially using cuttings, air layering and *in vitro* propagation methods, but seed-raised plants are required for crop improvement programmes. Though pomegranate is not commercially propagated through seeds owing to morphological and yield variability among plants raised through seeds, information on seed germination of different genotypes is useful to plant breeders and pathologists to screen large populations of seedlings and develop new hybrids (Singh *et al.*, 2016a). In addition, Indian local varieties used for anardana preparation are mostly grown through seeds, so, germination and growth studies using wild Daru type pomegranate will be useful for increasing wild pomegranate plantations in the areas of their natural habitat. It is also possible to raise seedlings and graft or bud with desired cultivars, although this is a more expensive and time-consuming process and is not a usual task in commercial pomegranate propagation. There are few reports of sexual propagation of pomegranate. Previous reports have stated that the seed germination percentage varies from 7–98% and the time taken for germination ranges from 10 to more than 100 days depending on variety, seed hardiness and storage period (Levin, 2006; Chandra *et al.*, 2014). In dwarf pomegranate, seed germination is very low due to the presence of water-soluble inhibitors in the fleshy seedcoat (Cervelli and Belletti, 1994; Jalikop, 2007). Generally, varieties with the hardest seeds have a lower germination percentage than soft-seeded ones, which indicates a mechanical and physical dormancy in pomegranate seeds (Rawat *et al.*, 2010; Karimi *et al.*, 2011; Silva *et al.*, 2016). Although Noda *et al.* (2002) reported the cause of delay in the germination of pomegranate seeds is the presence of sarcotesta that surrounds the seeds and contains phenolic compound, anthocyanin and tannins. Silva *et al.* (2016) identified the main reasons for the lack of uniformity in germination of pomegranate seeds as physical and physiological dormancy. There are several reports related to the removal of dormancy in the seeds of some pomegranate cultivars by warm and cold moist chilling treatments (Rawat *et al.*, 2010; Karimi *et al.*, 2011; Shalimu *et al.*, 2016; Silva *et al.*, 2016). Karimi *et al.* (2011) reported that in 'Malas-e-Yazdi' cultivar, the highest percentage of germination was obtained by 40 days' treatment of stratification, while in another study, Olmez *et al.* (2007) identified that 60 days of stratification is suitable for seed germination of wild genotypes of pomegranate. In summary, it can be postulated that the amount of stratification for breaking seed dormancy in pomegranate, depending on the variety, is between 25 and 60 days at 5°C (Olmez *et al.*, 2007; Rawat *et al.*, 2010; Karimi *et al.*, 2011). It has also been reported that in seeds of pomegranate, higher germination percentages are achieved by the combination of warm and cold stratification rather than by cold stratification alone (Shalimu *et al.*, 2016). A combination of warm and cold stratification-induced H_2O_2 change led to greater changes (elevation followed by attenuation) in activities of the scavenging enzymes than that induced by cold stratification alone.

6.4 Vegetative Propagation

Clonal or vegetative propagation ensures production of genetically identical saplings, thus it ensures uniformity of trees and keeps all horticultural traits intact and similar to the mother plant or original cultivar. Hence, pomegranate is commercially propagated through different methods of vegetative propagation.

6.4.1 Stem cutting

This is a commercial method of pomegranate propagation and pomegranate orchards worldwide use saplings propagated through stem cuttings (Levin, 2006; Day and Wilkins, 2009; Holland *et al.*, 2009). Propagation by stem cutting requires formation of an adventitious root system, as the potential shoot system is already there in the form of shoot buds. After observing cutting success from different methods of stem cutting, it can be concluded that success varies with intrinsic factors like genetic material, nutrient reserve, hormonal levels of stems, age dimension and maturity of stem, position of the branch on the mother plant and extrinsic factors like pruning, fertilization, irrigation, stem cutting time, rooting environment and hormone applications (Chadha, 2001; Polat and Caliskan, 2009; Chandra *et al.*, 2014; Singh, 2017b). Hardwood, semi-hardwood and softwood cuttings have been reported for propagation in pomegranate (Table 6.1). Though Saroj *et al.* (2008) and Hussain *et al.* (2012) reported high rooting success under controlled conditions in semi-hardwood owing to higher endogenous protein levels but in general the use of hardwood cuttings is the preferred method of pomegranate propagation due to high cutting success (Reddy and Reddy, 1989, 1990; Sandhu *et al.*, 1991; Panwar *et al.*, 2001; Chadha, 2001; Rajan and Markose, 2007; Chandra *et al.*, 2014). It is easy to observe initial sprouting in pomegranate cuttings to the tune of more than 85%, but due to the inability of some of the cuttings to form an adventitious root system, the ultimate cutting success reduces considerably (Singh *et al.*, 2015). The maturity of wood plays a vital role in rooting of different types of stem cuttings used for propagation in pomegranate. Six to 18-month-old basal portions of stem, during or immediately after rest or dormancy should be used for propagation through hardwood cutting, and 20–25-cm-long, 6–12-mm-thick cuttings with four nodes lead to high cutting success (Purohit and Shekharappa, 1985; Reddy and Reddy, 1990; Dhillon and Sharma, 1992; Rajan and Markose, 2007; Chandra *et al.*, 2014). Stem cuttings in pomegranate are generally low in root-promoting cofactors, and preconditioning by girdling, ringing, etiolation and basal wounding of cuttings have been reported to promote rooting cofactors and rooting of cuttings (Reddy and Reddy, 1989; Yesiloglu *et al.*, 1997; Pandey and Bisen, 2010). Use of plant growth regulators, mainly auxins, induces rooting of cuttings

Table 6.1. Different types of cuttings and growth regulators used in pomegranate propagation.

Cultivar	Type of cutting	Duration of growth regulator treatment	Concentration of growth regulator (ppm)	Reference
'Bassein Seedless'	Hardwood cutting	Quick dip	2500 indole-3-butyric acid (IBA) + 2500 paclobutrazol	Reddy and Reddy, 1989
'Bassein Seedless'	Hardwood cutting	Quick dip	2500 IBA + 2500 1-naphthaleneacetic acid (NAA)	Reddy and Reddy, 1990
'Ganesh'	Hardwood, semi-hardwood and softwood cuttings	Quick dip	5000 IBA	Ghosh *et al.*, 1988
'Bedana'	Hardwood cutting	Quick dip	1000 polyhydroxybutyrate (PHB) + 2500 NAA	Hore and Sen, 1993
'Ganesh', 'Dholka'	Hardwood cutting	Quick dip	1000 PHB + 5000 IBA	Tripathi and Shukla, 2004
'Jalore Seedless'	Hardwood and semi-hardwood cutting	Quick dip	2500 IBA	Saroj *et al.*, 2008
'Bhagawa'	Hardwood cutting	Quick dip	2500–5000 IBA	Singh *et al.*, 2014; Sharma *et al.*, 2014
'Bhagawa'	*In situ* hardwood cutting	Quick dip	2500–5000 IBA	Singh, 2017
Pomegranate ('ME12', 'CR02' and 'PT08')	Hardwood cutting	Quick dip and basal wounding	12,000 IBA	Melgarejo *et al.*, 2000
Pomegranate (multiple varieties)	Hardwood cutting	Quick dip	1000 IBA	Polat and Caliskan, 2009
'Wonderful'	Hardwood cutting	Quick dip	1000 IBA + 500 GA_3	Sarrou *et al.*, 2014

and improves cutting success. The increased hydrolytic activity, enhanced callus and tissue formation, and vascular differentiation in the presence of auxin might be the reason for improved rooting (Singh, 2014). Both quick dip and prolonged dip methods have been used by different researchers but quick dip of cuttings for 30 s to 5 min in a solution with 1000–5000 ppm indole-3- butyric acid (IBA) is found to be the best to induce rooting in pomegranate cuttings (Ghosh *et al.*, 1988; Dhillon and Sharma, 1992; Hore and Sen, 1993; Panwar *et al.*, 2001; Tripathi and Shukla, 2004; Saroj *et al.*, 2008; Chater *et al.*, 2017). In addition to IBA, 1-naphthalene acetic acid, p-hydroxybenzoic acid, paclobutrazol, ferulic acid, indole-3-acetic acid, Ethrel®, salicylic acid, GA_3, H_2O_2 and melatonin at different concentrations either alone or in combination have been reported to promote rooting in pomegranate (Tripathi and Shukla, 2004; Saroj *et al.*, 2008; Karimi *et al.*, 2012; Sarrou *et al.*, 2014). Many researchers have reported the use of plant-beneficial microorganisms like *Trichoderma harzianum*, *Pseudomonas fluorescence*, *Azospirilum* sp., *Azotobacter* sp., etc. to improve rooting of cuttings in pomegranate due to their growth-promoting properties (Patil *et al.*, 2001; Jaganath *et al.*, 2009).

The medium used for planting of cuttings has a great influence on root proliferation and cutting success (Chandra *et al.*, 2014). Different types of rooting media are used for propagation

Fig. 6.1. Stages of hardwood cuttings (HWCs) in pomegranate including sanitation protocol. Upper left to right: bundles of HWCs of desired length and thickness; treatment of HWC in lukewarm solution of fungicide and bactericide; indole-3-butyric acid (IBA) treatment of basal portion of cuttings. Bottom left to right: sprouted cuttings on sterile cocopeat medium; rooted cuttings in nursery bags with potting mixture; root biomass of HWC-raised plants. (Photos: Nripendra Singh.)

of pomegranate through cuttings, which include river silt, cocopeat, biopeat, saw dust, vermiculite, perlite, sand and other potting mixtures (Deol and Uppal, 1990; Batista *et al.*, 2011). Hu and Wang (1993), reported 98% rooting success in hardwood cutting on fly ash medium. Ansari (2013) observed maximum rooting in hardwood cuttings on sand and vermiculite rooting medium. Sand and vermiculite medium along with bottom heat treatment gave 85% rooting success in pomegranate cuttings when planted in a mist chamber (Khalil, 2013). An ideal rooting medium for rooting of pomegranate cuttings should have unrestricted gaseous exchange and sufficient aeration to allow air to reach to the newly forming roots, and can hold enough and uniform moisture to avoid drying or moisture stress (Rajkumar *et al.*, 2017).

Singh *et al.* have standardized hardwood cutting and sanitation protocols for high cutting success and production of healthy saplings in pomegranate cv. 'Bhagawa' (Fig. 6.1) (Singh *et al.*, 2014) as follows:

- Pruned wood/stem immediately after rest phase/dormant phase preferred for a high success rate.
- Take 6–18-month-old shoots for the hardwood cuttings for high success; lateral shoots, which usually flower and fruit heavily, should not be used for making cuttings.
- Stem cuttings ranging in length from 20–25 cm and thickness 0.6–1.2 cm perform best.
- Before planting it is desirable to sanitize the cuttings by giving them a 15-min dip in 2-bromo-2-nitro-1,3-propanediol @ 500 mg/l + carbendazim (50% WP) @ 1.0 g/l dissolved in lukewarm water at 45°C to get rid of non-systemic surface pests and disease pathogens.
- Surface sterilize the cuttings using NaOCl at 2% for 15 min.
- Dip the lower part (basal end, about 2–3 cm) of stem cuttings for 1 min in a solution of IBA at 2.5 g/l for inducing roots in stem cuttings.
- Plant the cuttings in sterilized/biofortified cocopeat for faster rooting in a polyhouse/shadenet house.
- Well-rooted cuttings should be transferred after 45–60 days to nursery bags filled with presterilized cocopeat in the upper three-quarters (place rooted cuttings in the top

cocopeat portion) and a sterilized mixture of sand, soil and farmyard manure (1:1:1) in the bottom quarter.

- At the time of transfer, place premultiplied plant-beneficial microflora formulation containing *P. fluorescens*, arbuscular mycorrhizal fungi (AMF) (only multiplies in/on root system) and *Aspergillus niger* AN27 in the root zone.
- Plant in the field after 45 days of growth in bags. Before shifting to the field keep plants for 1 week in the shade for acclimatization.

6.4.2 *In situ* hardwood cutting

Pomegranate orchards are sometimes established by direct planting of unrooted freshly harvested cuttings (Blumenfeld *et al.*, 2000; Chandra and Babu, 2010; Holland *et al.*, 2009). *In situ* hardwood cutting offers an easy, excellent and cost-effective way to expand the already existing orchard. Preparation of hardwood cuttings is similar to stem cutting, but after making cuttings ready, they are directly planted in the field. While refilling the pit, a central portion of 20 cm (depth) × 15 cm (diameter) should be left unfilled, half of this portion should be filled with cocopeat followed by pit treatment with fungicides and insecticides. Each pit is planted with four cuttings and temporarily covered with old shade net, to maintain optimum moisture the pit should be supplied with two drippers. (Fig. 6.2) (ICAR-NRC on Pomegranate, 2016; Singh, 2017).

Fig. 6.2. *In situ* hardwood cutting. Left to right: planted cutting covered with discarded shade net and other material; sprouted *in situ*-planted cuttings. (Photos: Nripendra Singh.)

6.4.2.1 Advantages

- No transplantation shock to the plants.
- Very cost effective and saves manpower and nursery costs.
- Easy and farmer friendly.

6.4.3 Air layering

Air layering is the most prevalent method of pomegranate propagation in the Deccan Plateau region of India (Chandra *et al.*, 2014). For air layering, upright branches of 0.8–1.5 cm in diameter from a healthy tree are selected in the rainy season and girdled 2–3 cm in length. Rooting hormone (2000–3000 ppm IBA) is applied on the upper cut of the girdle and wrapped with moist rooting medium (sphagnum moss) with the help of a small, black/white polythene strip (200–300 gauze), and both the sides are tied with coir/jute thread or string (Fig. 6.3) (Chandra and Babu, 2010; Tomar, 2011; Sharma *et al.*, 2014). In general, sphagnum moss is used as a substrate for covering the girdled portion of air layers and IBA either alone or in combination with p-hydroxybenzoic acid at a concentration ranging from 1000–5000 ppm is used to induce early rooting in air layers for high success (Hore and Sen, 1995; Bhosale *et al.*, 2009; Tomar, 2011; Chandra *et al.*, 2014). However, sawdust, fly ash, coconut coir and pond soil have also been reported as substrates for covering the girdled stem for pomegranate air layers (Allioli *et al.*, 2001). Rooting takes place between 30–40 days, and well-rooted layers are detached from the mother plants after about 60–75 days. These air layers, after 70–80% defoliation, can be planted in the nursery or in polythene bags containing soil, sand and well-rotted farmyard manure in a 1:1:1 ratio. The optimum time for air layering is June–August under the Deccan Plateau conditions of India (Chandra *et al.*, 2014; Tayade *et al.*, 2017). Well-developed layered plants can be used for establishment of orchards.

6.4.4 Mound layering

Multiplication of pomegranate through mound layering is a novel, convenient and cheap technique for obtaining planting material. Considering the tendency of pomegranate to produce profuse

Fig. 6.3. Air layering in pomegranate. Clockwise: ringing; air layers on mother plant; detached air layers. (Photos: Nripendra Singh.)

suckers as needed for stool layering, this propagation method has been tested at ICAR-National Research Centre in India. This method of propagation has been successfully used for propagation of apple rootstocks, quince, guava and some ornamentals (Rymbai and Reddy, 2010; Chandra *et al.*, 2014). Exploitation of ground layering as an alternative method of pomegranate propagation has earlier been suggested by Chadha (2001). In mound layering of pomegranate under the climate conditions of Solapur, India, mother plants are headed back at 5 cm height above ground level in March. In late June, lanolin paste with 2500 ppm IBA is applied after wounding/girdling at the base of stool shoots, when stool shoots attain approximately 45 cm in height and are covered up manually with field soil up to 20 cm above ground level. The shoot tip is removed at the height of 1 m of stool shoot from ground level (Chandra and Babu, 2010; Chandra *et al.*, 2014; Singh *et al.*, 2017a).

The per unit (m^2) production of air layered planting material from a well-established pomegranate orchard of cv. 'Bhagawa' planted at standard spacing of 4.5 × 3.0 m is about 11.1 rooted plants as compared with rooted stool shoots, in which it is as high as 19.5 per m^2, if progeny plant spacing is 0.5 × 0.5 m (Singh *et al.*, 2017b). Thus, it is possible to produce more saplings per unit using stool or mound layering and this can be an option for plant propagation for small and marginal farmers in semi-arid areas for expansion of their own pomegranate orchard (Chandra *et al.*, 2014; Singh *et al.*, 2017b).

6.5 Grafting and Budding

Grafting and budding are horticultural techniques, practised for many years and in many parts of the world. It is mainly practised to obtain the advantages of rootstocks for soil and water salinity, soil-borne pests and diseases, and for benefits like precocity and dwarfing of fruit trees and other biotic and abiotic challenges in a climate change scenario. Pomegranate is a fruit tree that has been propagated traditionally by stem cuttings and air layering, and no serious attempt has been made to exploit grafting and budding. Grafting in pomegranate has been reported earlier by Asadov (1987), but Kar *et al.* (1989) explored the possibility of top working through budding and grafting in pomegranate. Recently, in some areas of Iran, farmers used this technique to change the pomegranate cultivars in their orchards. Environmental and non-environmental stresses such as drought, salinity of water and soil, soil-borne diseases, frost and nutrient imbalances are among the factors limiting pomegranate cultivation across the globe (Karimi and EiniTari, 2016). Screening of pomegranate germplasm, particularly wild genotypes, against various biotic and abiotic stresses should be carried out to identify potential rootstocks. Thus, non-commercial genotypes that are resistant to environmental and non-environmental stresses can be used as rootstock (Karimi and Farahmand, 2011; Karimi and Mirdehghan, 2013; Nowrozy *et al.*, 2016; Karimi and Hasanpour, 2017). One of the characteristics of pomegranate is the ease of propagation of clonal rootstocks in comparison with other fruit trees. Thus, uniform pomegranate orchards can be raised with grafted plants. There are reports on pomegranate propagation through cleft and bench grafting, wedge grafting, omega grafting, patch budding, shield budding, ring budding and stenting methods (Karimi, 2011; Karimi and Farahmand, 2011; Chandra and Jadhav, 2012; Hasanpour *et al.*, 2015; Nowrozy *et al.*, 2016; Karimi and Nowrozy, 2017Karimi and Hasanpour, 2017). Table 6.2 shows the methods of grafting and budding reported on the pomegranate and the ideal time to do it.

6.5.1 Shield and ring budding

Shield and ring budding methods are commonly used to change cultivars in the pomegranate orchards. Both methods are performed when the rootstock bark is slipping during the early summer and the scion buds are matured and hardened. These budding methods are performed on suckers of the trees. Ring budding during late June gave 75% graft success on suckers of mature trees of 'Ali-Agaei' and 'Shirin-e-Shahvar' cultivars as rootstocks (Nowrozy *et al.*, 2014).

6.5.2 Patch budding

Chandra *et al.* (2013) have standardized patch budding in pomegranate, which can be utilized effectively for *in situ* budding. More than 90% success has been achieved when 'Bhagawa' scion bud was budded on wild pomegranate rootstocks during November to February under Solapur conditions (Chandra *et al.*, 2014). One-year-old rootstock and a patch bud of 20 mm × 10 mm had been found ideal for patch budding. The patch of bark containing the bud is cut from the bud stick similar to the bark patch removed from the rootstock. After

Table 6.2. Methods of budding and grafting in pomegranate.

Budding and grafting method	Ideal time[a]	References
Cleft grafting	Early March	Nowrozy *et al.*, 2016 Nowrozy *et al.*, 2014
Bench grafting	Early March	Karimi and Farahmand, 2011
Wedge grafting	January	Chandra *et al.*, 2009; Chandra and Jadhav, 2012
Omega grafting	Early March	Karimi and Nowrozy, 2017
Stenting	Early March	Karimi, 2011 Karimi and Nowrozy, 2017
Chip budding	Late March	Nowrozy *et al.*, 2016 Nowrozy *et al.*, 2014
Shield budding	Late June	Nowrozy *et al.*, 2014
Ring budding	Late June	Nowrozy *et al.*, 2014
Patch budding	November to February	Chandra *et al.*, 2013, Chandra *et al.*, 2014

[a]May vary according to geographical location

Fig. 6.4. Patch budding in pomegranate: scion bud and rootstock (left); sprouted bud after union (right). (Photos: Nripendra Singh.)

the bud patch is removed from the bud stick, it must be placed immediately on the rootstock to avoid loss of moisture from the bud. The patch from the bud stick should fit snugly in the rootstock from where the bark of similar size has been removed. Polythene tape, polythene strips or grafting tape are used for covering the budded portion to enhance budding success by maintaining high moisture at the budded portion (Fig. 6.4) (Singh, 2017).

6.5.3 Bench grafting

This method can only be used in the nursery to produce grafted pomegranate plants. One-year-old rooted cuttings or 2-year-old seedlings are used as rootstock. In early March, the rootstocks are removed from the nursery and grafted. In this method, first, the rootstocks are headed back at a height of 10–15 cm in length, and a vertical cut is made on them. Then the bottom part of the scion is cut as a wedge and placed in the vertical slit of the rootstock and tightly fitted with yarn (Fig. 6.5). The grafted plants are then planted in a substrate of moist sawdust, and the temperature is kept at 18 ± 2°C until callus formation and it is finally planted in the nursery bed. The most important point in this method is the preparation of scions from young trees, since older scions produce callus later (Karimi and Farahmand, 2011; Nowrozy *et al.*, 2016).

6.5.4 Cutting and grafting (stenting method)

In pomegranate, propagation using bench and cleft grafting requires skill and time to produce rooted rootstocks (Karimi and Farahmand, 2011). An alternative method

Fig. 6.5. Bench grafting in pomegranate. Left to right: grafted rootstocks for cultivation in nursery bed; growing grafted plants. (Photos: Hamidreza Karimi.)

is simultaneous cutting and grafting, which is sometimes called the stenting method. Stenting is a method for quick propagation of plants. Cutting and grafting are performed simultaneously. The scion is grafted on to a non-rooted rootstock. The formation of the graft union and adventitious roots on the rootstock occur simultaneously. Stenting is now being used worldwide by rose growers and is also a valuable technique in propagating species of conifers and also rhododendron, apple, plum and pear (Hartmann *et al.*, 2002; Izadi *et al.*, 2013). In this method, stem cuttings of 1–1.2 cm diameter should be harvested from the mother plant and cut back to the desired length of 15–20 cm for using as rootstocks. Also, scions 1–1.2 cm in diameter and 5–7 cm in length are selected. Rootstocks and scions are prepared n a similar way to bench or cleft grafting, and then the scions are placed into the rootstock gap in such a way that the cambium layers of scion and rootstock have the maximum area of contact. The grafted area is wrapped using a cotton band, and the bottom of rootstocks are treated with IBA hormone in 500–1000 mg/l concentration and are planted in moist perlite medium. Then the whole system (scion and rootstock) is covered in a layer of perlite or cocopeat. Grafted cuttings are kept under these conditions for 4 weeks, and then the perlite or cocopeat layer is removed to avoid adventitious root formation on the scion (Karimi, 2011). The best time to perform stenting is early March.

6.5.5 Omega grafting

There are hand machines for grafting fruit trees, and one of them is the omega grafting machine. The machine can be used for the production of grafted plants of pomegranate in a nursery or to change a cultivar in the orchard. Ease of use of grafting tools and increased efficiency are the benefits of omega grafting. Additionally, the percentage of successful grafts increases due to better fitting of the rootstock and the scion. In this method, as for the stenting method, stem cuttings of 1–1.2 cm in diameter and 15–20 cm in length and scions of 1–1.2 cm in diameter and 5–7 cm in length are removed from mother plants. In the first stage, rootstocks and scions are cut with the omega grafting machine, and then the scions are placed on rootstocks and they are tied with special plastic tape (Fig. 6.6).This machine can be used for increasing the efficiency of the bench and stenting methods (Nowrozy *et al.*, 2014; Karimi and Nowrozy, 2017).

Fig. 6.6. Stages of omega grafting in pomegranate. Upper left to right, respectively: rootstock and scion cuts; placement of scion on the rootstock; wrapped grafted area using cotton. Bottom left to right, respectively: cultivation of grafted cuttings in the bed; the growth of scions; plants grafted after 3 months. (Photos: Hamidreza Karimi.)

6.5.6 Effective factors for graft success

Compatibility of rootstock and scion, grafting methods, scion and rootstock type, the age of scion and rootstock, environmental conditions (temperature, humidity and oxygen), worker skill and active compounds in rootstock and scion are the most important factors for success, survivability and graft growth in pomegranate (Karimi, 2011). The favourable temperature for cellular activity, in general, varies from 12°C to 35°C. Therefore grafting should be carried out when the temperatures are favourable for cambial activity and there is high humidity in the vicinity of the cambial region of graft union. Pomegranate is one of the plants in which the amount of cell division in the cambium layer is dependent upon the amount of oxygen, so the site of grafting should be treated with a lower amount of grafting wax. Type of rootstock also affects the graft success and the growth of the scion (Karimi, 2011; Karimi and Nowrozy, 2017). In a study, the highest graft success percentage using the stenting method was obtained with 'Gorch-e-Dadashi' rootstock (Karimi and Nowrozy, 2017). Reports on the effect of available constitutive biochemicals of rootstock and scion on the graft success of pomegranates are limited. In a study, Karimi and Nowrozy (2017) investigated the effect of carbohydrates, phenolic compounds and biochemical constituents in the rootstock and scion on the graft success and mortality percentage of pomegranate, and reported that in some rootstocks a negative correlation was observed between phenolic compounds of rootstock wood and graft success percentage, whereas in all rootstocks, mortality percentage of grafted plants was correlated positively with phenolic compounds of rootstock wood. It has also been reported that, owing to the role of carbohydrates in cell division, scions with higher carbohydrates have higher graft success.

6.5.7 Interaction of rootstock and scion

Selection of an appropriate graft combination is crucial for the production of quality fruit because the rootstock and scion combination influences photosynthetic parameters, mineral uptake, plant size, water relations, fruit quality and yield efficiency (Goncalves *et al.*, 2006). There are very few reports available on rootstock effects on growth, yield and physiological parameters of pomegranate (Vazifeshenas *et al.*, 2009; Chandra and Jadhav, 2012; Karimi and EiniTari, 2016; Ahire *et al.*, 2016, 2017; Karimi and Nowrozy, 2017). Table 6.3 shows the traits that have been influenced by the rootstock type in two cultivars of pomegranate namely 'Rabab-e-Neyriz' and 'Khafr-e-Jahroom' when they were grafted on 'Gorch-e-Dadashi', 'Gorch-e-Shahvar' and 'Post Ghermaz-e-Aliaghai' rootstocks.

6.5.8 Effect of rootstock on shoot growth and vigour

One of the most important factors affecting scion growth and vigour is rootstock. In order to confirm the dwarfing effects of rootstocks, it is necessary to examine the effects of rootstock on scion growth. In a study, Karimi and Nowrozy (2017) studied the effect of three rootstocks, 'Gorch-e-Dadashi', 'Gorch-e-Shahvar' and 'Post Ghermaaz-e-Aliaghai', on growth parameters of 'Rabab-e-Neyriz' and 'Khafr-e-Jahroom' in field conditions and reported that the highest shoot length and leaf number were obtained when 'Rabab-e-Neyriz' was grafted on to 'Gorch-e-Shahvar', whereas the lowest shoot length was found when 'Rabab-e-Neyriz' was grafted on to 'Post Ghermaz-e-Aliaghai'.

6.5.9 Effect of rootstock on echo physiological parameters

Previous studies concluded that the genotype of rootstocks affects physiological parameters of scion in pomegranate. Cultivars grafted on vigorous rootstocks exhibited higher stem water potential, stomatal conductance, intercellular CO_2 assimilation and*Fv*/*Fm* than those grafted on dwarfing rootstocks (Goncalves *et al.*, 2006; Fotouhi-Ghazvini *et al.*, 2007). In pomegranate the type of rootstock also influences soil–plant analysis development (SPAD) index, leaf chlorophyll content and relative water content of leaves of grafted plants (Karimi and EiniTari, 2016; Biniyaz *et al.*, 2017). In general, in pomegranate, it can be postulated that rootstocks with a robust and efficient root system for absorption of minerals have a higher photosynthetic rate (Biniyaz *et al.*, 2017).

6.5.10 Effect of rootstock on nutrient composition of leaf

Rootstock has a significant effect on the nutrient concentrations of scion leaves. Rootstocks have selective absorption properties of mineral nutrients, which means that some rootstocks have more absorption of mineral nutrients, which makes it possible to provide the elements needed for the scion. The rootstock effect on the nutrition concentration of a leaf is related to the characteristics of the root system, so that rootstocks with a stronger root system are more capable of absorbing mineral nutrients from

Table 6.3. Interaction of scion and rootstock on some growth and physiological traits of pomegranate.

		Traits													
Rootstock	Scion	Shoot length	Shoot diameter	Sucker number	Leaf area	Leaf number	SPAD	Ch. a	Ch. b	Total ch	*Fv/Fm*	Ca	Mg	Cu	Zn
GSH	KH		+	–						–		+	+	+	+
	RN	–				+						+		+	+
GD	KH	+	+	+			+	+	+				+	+	+
	RN						+						+		+
PZ	KH		+					+	+	+			+		+
	RN	+		+			+							+	+

[a] + and – means a significant and non-significant effect, respectively.
GD, GSH and PZ are 'Gorch-e-Dadashi', 'Gorch-e-Shahvar' and 'Post Ghermaz-e-Aliaghai' rootstocks, respectively; RN and KH are 'Rabab-e-Neyriz' and 'Khafr-e-Jahroom', respectively.

the soil. There are several reports about the effects of rootstock on the nutrient concentrations of leaves in fruit trees, but very few reports are available in pomegranate due to its vegetative propagation by cuttings. Biniyaz *et al.* (2017) reported that vigorous rootstocks such as 'Gorch-e-Dadashi' and 'Gorch-e-Shahvar' increased the zinc, manganese and iron concentrations in the shoots of 'Rabab-e-Neyriz' cultivar under field conditions.

6.5.11 Effects of rootstock on tolerance of scion to salinity stress

Most pomegranate cultivation is in arid and semi-arid regions of the world, where soil salinity and water stress are the main limitations for optimum yield. Pomegranate is considered to be moderately tolerant to salinity (Naeini *et al.*, 2006; Hasanpour *et al.*, 2014: Karimi and Hasanpour, 2014). Pomegranate cutting cv. 'Malas-e-Shirin' was tolerant of up to 40 mM of NaCl in potted cultures of 1:1 sand-perlite medium irrigated with complete Hoagland's solution (Naeini *et al.*, 2006). There is little evidence for rootstock effect on increasing tolerance to salinity in grafted pomegranate plants. Hasanpour *et al.* (2015) and Karimi and Hasanpour (2017) evaluated two grafted combinations of pomegranate 'Gabri'/ 'Tab-o-Larz', and 'Gabri'/'Malas-e-Yazdi' to salinity stress created through irrigation water. Based on echo-physiological and nutrient concentrations of leaves they reported that 'Tab-o-Larz' rootstock either restricted the uptake/transport of Na from root to shoot or maintained sufficient levels of K to enhance salinity tolerance in 'Gabri' cultivar.

6.5.12 Effects of rootstock on sucker production

Suckers continuously emerge from pomegranate trunk and crown during the growing season, which need to be removed regularly. Suckers can be removed by hand or by chemical treatments during the growing season. These practices offer a temporary solution to the problem. To overcome this problem, rootstocks with low sucker production characteristic can be exploited. It has been found that the type of rootstock can affect sucker length, sucker diameter and sucker number in pomegranate (Vazifeshenas *et al.*, 2009; Biniyaz *et al.*, 2017). Vazifeshenas *et al.* (2009) reported that when 'Shahvar' was grafted on to 'Golnar-e-Fars' it produced fewer suckers compared with 'Poost Siyah', 'Malas-e-Esfahan' and 'PostSefid-e-Jahrom' rootstocks. In a similar study, Biniyaz *et al.* (2017) showed that when 'Ghermez-e-Aliaghai' was used as rootstock for 'Rabab-e-Neyriz', it produced fewer suckers as compared with non-grafted plants.

6.6 Micropropagation and Bio-hardening

Availability of a large quantity of disease-free, genuine planting material is a major constraint in expanding the area under pomegranate cultivation. Micropropagation can be explored as an alternative tool to fulfil the demand for good-quality, disease-free planting material in bulk. Both direct and indirect organogenesis, including somatic embryogenesis, have been successfully attempted in pomegranate for plant propagation and other purposes. Micropropagation in pomegranate can be initiated either through proliferation of existing meristems or regeneration from adventitious meristems or by somatic embryogenesis (Kajla *et al.*, 2013). Since, the primary objective of the chapter is to describe pomegranate propagation methods, we concentrate mainly on *in vitro* propagation through direct organogenesis. Micropropagation has five distinct stages. Stage 0: pretreatment of explants to disinfect and surface sterilize them. Stage I: establishment of infection-free cultures. Stage II: *in vitro* proliferation/direct or indirect organogenesis (Kumar *et al.*, 2017). Stage III: rooting of microshoots. Stage IV: hardening/acclimatization of rooted plantlets (Fig. 6.7).

6.6.1 Selection of parent material

The selection of the mother plant for excising explant is a critical step for the success of micropropagation in pomegranate. Only healthy mother plants with proven horticultural traits should be selected for excising the explants (Debergh and Maene, 1981; Kumar *et al.*, 2017). However,

Fig. 6.7. Stages of *in vitro* pomegranate propagation. Upper left to right: explant pretreatment; *in vitro* culture establishment; *in vitro* proliferation; *in vitro* rooting. Bottom left to right: primary hardened *in vitro*-raised plants; secondary bio-hardened *in vitro*-raised saplings. (Photos: Nripendra Singh.)

most of the time an apparently healthy looking mother plant may harbour latent or symptomless infections of fungal, bacterial and viral pathogens. Thus, attention should be given to making sure that the stock plant is healthy, vigorously growing and preferably maintained under protected structures with controlled conditions (Torres, 1988; Guranna *et al.*, 2018).

6.6.2 Selection of explants

The source and type of explant have been considered as important factors for *in vitro* propagation of pomegranate. Many types of explants have been reported in pomegranate tissue culture by various researchers, but to minimize the chances of somaclonal variations and other physiological abnormalities, direct organogenesis is preferred; thus, explants having preformed meristematic buds like shoot tips and nodal segments or axillary buds should be used for *in vitro* propagation (Bhojwani and Razdan, 1983; Singh *et al.*, 2007, 2011; Kumar *et al.*, 2017). Among various explants for *in vitro* establishment in pomegranate, shoot tips and nodal segments have been extensively used by many researchers in cultivars like 'Bhagawa', 'Mridula' and 'Ganesh' for direct regeneration (Naik *et al.*, 1999; Murkute *et al.*, 2004; Chaugule *et al.*, 2007; Singh *et al.*, 2007, 2011; Patil *et al.*, 2011; Singh *et al.*, 2016b). However, protocols for *in vitro* regeneration of *Punica granatum* L. plantlets using explants like leaf segments and cotyledons (Murkute *et al.*, 2004; Raj and Kanwar, 2008; Kanwar *et al.*, 2010), anthers (Naik *et al.*, 1999) or through embryogenesis from various seedling explants, petals and immature zygotic embryos (Kanwar *et al.*, 2010) have also been well documented. Seed-based explants like cotyledonary nodes were used by Naik *et al.* (1999) and Singh *et al.* (2013) in pomegranate cv. 'Bhagawa' for *in vitro* propagation. Shoot tips (2–3 cm long), nodal segments (3–4 cm), leaf segments (1–2 cm^2) and petal segments (2–3 cm^2) were taken as explants by Guranna *et al.* (2018) .

6.6.3 Pretreatment of explants

Establishment of infection-free cultures has often proved to be the deciding factor for cost-effective micropropagation and treatment of explants prior to ingression using standard decontamination protocols is mandatory to achieve this goal. Contamination through explant in tissue culture might originate either from the surface contamination of explants or endophytically from the tissue of explants (Singh *et al.*, 2010). Washing the plant material with clean water before initiating the sterilization process drastically reduces the initial pathogen load. The excised explants are first washed with running tapwater for 5 min followed by treatment with an aqueous solution of fungicides and bactericides/antibiotics for 20–30 min, and repeatedly rinsed four or five times with autoclaved water followed by surface sterilization with NaOCl or mercuric chloride. Though $HgCl_2$ and NaOCl are the two most widely employed surface sterilants, other chemicals have also been tried with varying success. Damiano *et al.* (2008) successfully sterilized axillary bud segments using a combination of NaOCl and Na methiolate for 20 min, which gave good explant survival (65%). Patil *et al.* (2011) soaked nodal segments in fungicide (M-45) solution (1 g/l) for 45 min, followed by streptomycin solution (100 mg/l) treatment for 20 min. Finally 1 g/l mercuric chloride solution for 10 min was used to treat these explants followed by rinsing three times with sterile distilled water for complete sterilization of nodal explants. Singh and Patel (2016) treated nodal explants by keeping them in a solution of 0.2% carbendazim (50% WP) + 0.05% streptomycin for 1 h followed by 10 min treatment with 10% Teepol solution and repeated rinsing to remove traces of Teepol. This was followed by surface sterilization in a laminar hood with 0.1% $HgCl_2$ solution.

6.6.4 Phenol exudation and its management

Phenol exudation and resultant browning of media is a serious impediment in the successful establishment of pomegranate cultures. The oxidized products of phenols, that is quinones, are known to be highly reactive and inhibit enzyme activity leading to the necrosis/death of explants (Fig. 6.8) (Hu and Wang, 1993). Strategies have been adopted to tackle phenolic browning aiming at avoiding exudation or neutralizing or avoiding its build-up in the media. The different approaches are culture of juvenile explants, or new growth flushes during the active growth period, culture in darkness, transfer of explant to fresh medium at short intervals, culture in liquid medium, inclusion of antioxidants in the culture media, soaking explants in water or solutions containing antioxidants prior to ingression, or use of adsorbing agents such as activated charcoal (AC), polyvinyl pyrrolidone (PVP), etc. (Weatherhead *et al.*, 1978; Wang *et al.*, 1994). In addition use of low salt medium and optimum concentration of growth regulators, sealing the cut ends with paraffin wax (Bhat and Chandel, 1991; Singh *et al.*, 2007, 2011), keeping the culture medium in the dark and drying the explant under laminar airflow (Krishna *et al.*, 2006) have all been tried. Wang *et al.* (1994) suggested

Fig. 6.8. Phenol exudation and media browning. Left to right: browning of media due to phenol exudation; necrosis of nodal explant, necrosis of shoot tip. (Photos: Nripendra Singh.)

the use of nodal segments collected during active growth and keeping the culture in the dark at 5°C for 6–8 days to reduce phenol exudation. Singh and Khawale (2003) reported that explants excised from glasshouse-grown plants exude fewer phenols, and liquid medium with filter paper wicks was also helpful in establishing cultures. In pomegranate, fast subculture of explants on to fresh medium, use of antioxidants, culture of juvenile plant parts etc. are employed to manage the phenolic problem in tissue culture. Murkute *et al.* (2004) and Chaugule *et al.* (2007) suggested fast subculturing to control media browning. Singh *et al.* (2011) showed that sealing of the cut ends of explants with presterilized paraffin minimized the media browning by preventing phenol exudation and leading to high culture establishment (83.95%) compared with the control (16.55%). Jones and Saxena (2013) incorporated 2-aminoindane-2- phosphonic acid (AIP), a competitive inhibitor of phenylalanine amonialyase (PAL) enzyme responsible for oxidative browning of culture, into the medium to reduce media and tissue browning and enhance shoot growth. Singh and Patel (2016) reported that the intensity of media browning was influenced by the position and length of nodal explants.

6.6.5 Media composition for culture establishment and shoot proliferation

The media for culture establishment and shoot proliferation should be composed of basal nutrients, growth regulators and other media supplements. Most of the reports on pomegranate micropropagation have recommended the use of Murashige and Skoog (MS), modified MS, Woody Plant Medium (WPM), Quoirin and Lepoivre (QL) or Nitsch and Nitsch (NN) basal media for culture establishment and shoot proliferation (Kumar *et al.*, 2017). The ratio of cytokinin to auxin determines rooting and shooting in pomegranate micropropagation. Among cytokinins, 6-benzylaminopurine (6-BAP), kinetin, zeatin riboside (ZR), thidiazuron (TDZ) and cytokinin supplements like adenine sulfate, 2iP, are mostly used at different concentrations for culture establishment and shoot proliferation, and among auxins most of the researchers used 1-naphthaleneacetic acid (Singh *et al.*, 2010; Naik and Chand, 2011; Teixeira da Silva *et al.*, 2013; Kumar *et al.*, 2017). However, there are some reports of the use of GA_3 for pomegranate culture establishment. Use of media supplements like coconut milk, silver nitrate, cycloheximide, etc. is also documented for culture establishment and *in vitro* proliferation of pomegranate. Yang *et al.* (1991) suggested supplementing the medium with 1.0 µM NAA and 2.0 µM BA for maximum shoot production on terminal shoot explants in a dwarf pomegranate genotype. Drazeta (1997) compared the micropropagation in different pomegranate cultivars 'Slatki Barski', 'Serbetas', 'Konjski Zubi' and 'Dividis' and obtained the best production of shoots on medium containing 1 mg/l BAP and 0.1 mg/l NAA, but shoots exhibited vitrification, hence subsequent cultures were suggested to be transferred on to a medium containing BAP at 0.5 mg/l and 0.1 mg/l NAA. Fougat *et al.* (1997) also obtained success on MS medium supplemented with 0.5 mg/l kinetin, 1.0 mg/l BA and 500 mg/l CH (cycloheximide). A comparative study on shoot proliferation was attempted by Naik *et al.* (1999) on an elite pomegranate cultivar 'Ganesh' using nodal stem segments excised from a mature tree. They tried three cytokinins, namely BA, zeatin riboside (ZR) and thidiazuron (TDZ), and found that the highest number of shoots developed on medium containing 2.0 mg/l ZR, while TDZ was the least effective. Later, Naik *et al.* (2000) obtained high-frequency axillary shoot proliferation and plant regeneration from established cotyledonary nodes. Shoot development was induced on nodes upon culture on MS medium supplemented with 2.3 to 23.0 µM BA or kinetin. Both the type and concentration of cytokinin significantly influenced the shoot proliferation. The maximum number of shoots (9.8 shoots/explant) was obtained on medium containing 9.0 µM BA.

Shoot culture was established by repeatedly subculturing the original cotyledonary nodes on a fresh batch of the same medium after each harvest of the newly formed shoots. Murkute *et al.* (2004) obtained the highest shoot proliferation of pomegranate cv. 'Ganesh' on MS basal medium supplemented with 1.0 mg/l BAP + 0.5 mg/l NAA. Singh and Khawale (2003) found half-strength MS medium supplemented with 1.0 mg/l IBA + 1.0 mg/l kinetin along with 200 mg/l AC to be the best medium for *in vitro* establishment of nodal segments of pomegranate cv. 'Jyoti'. Axillary buds were aseptically removed from the established cultures and transferred on to proliferation medium consisting of MS basal medium supplemented with various concentrations of BA, kinetin and 40 mg/l adenine sulfate. For shoot elongation and rooting, MS medium containing 2.0 mg/l IBA, 200 mg/l AC and 40 mg/l sucrose was best. Singh *et al.* (2007) carried out *in vitro* clonal propagation of pomegranate cv. 'G137' using nodal segments and shoot tips of mature trees. They reported maximum culture establishment on MS medium supplemented with 2.0 mg/l BAP + 0.1 mg/l NAA and 0.5 mg/l GA_3. The same basal medium supplemented with 1.0 mg/l BAP, 1.0 mg/l kinetin and 0.1 mg/l NAA produced the highest number of shoots per explant alongwith the longest shoots. Chaugule *et al.* (2007) used shoot tips and nodal segment for *in vitro* propagation of pomegranate cv. 'Mridula' and reported the highest percentage of shoot proliferation on MS medium supplemented with 0.4 mg/l NAA + 1.0 mg/l BAP. Damiano *et al.* (2008) showed good shoot multiplication on basal QL medium supplemented with BA (0.4 mg/l) and IBA (0.05 mg/l). Kanwar *et al.* (2010) reported when cotyledonary explants, excised from *in vitro*-germinated seedlings of pomegranate were inoculated on solid MS medium supplemented with 21 μM NAA and 9 μM 6-benzyladenine (BA), 80% of explants developed callus. A high frequency of shoot organogenesis was obtained when explants were incubated on MS medium supplemented with 8 μM BA, 6 μM NAA and 6 μM GA_3. However, adding 24 μM silver nitrate ($AgNO_3$) to this medium markedly enhanced the shoot regeneration frequency (63%) and mean number of shoots per explant (11.26) and length of shoots (2.22 cm). The nodal explants grown on MS medium containing 1.8 mg/l BAP, 0.9 mg/l NAA, 1 mg/l silver nitrate and 30 mg/l adenine sulfate had the highest proliferation rate (10–15 shoots/explants) in the establishment stage (Patil *et al.*, 2011). Singh *et al.* (2013) found that the cotyledonary explant on MS basal medium fortified with 1 mg/l BAP, 1 mg/l kinetin and 200 mg/l activated charcoal gave the maximum multiplication rate (86.33 %). Valizadehkaji *et al.* (2013) recommended the use of 4.7–9.2 μM kinetin and 0.54 μM NAA on WPM medium for proliferation of shoot tips and nodal segments of Iranian pomegranate cultivars.

6.6.6 *In vitro* rooting

Various types of auxins, basal media and media supplements like activated charcoal have been reported by many researchers to promote *in vitro* rooting of pomegranate microshoots. In most cases, NAA and IBA at concentrations ranging from 0.5 mg/l to 2.0 mg/l were used to induce rooting in *P. granatum* L. (Omura *et al.*, 1987; Mahishni *et al.*, 1991; Yang and Ludders, 1993; Fougat *et al.*, 1997; Drazeta, 1997; Naik *et al.*, 2000; Amin *et al.*, 2003). Among basal media, half-strength MS medium, MS medium with half-strength macro salts and WPM are generally used for rooting of microshoots in pomegranate (Singh *et al.*, 2010; Naik and Chand, 2011; Kumar *et al.*, 2017). Naik *et al.* (1999) reported good induction of roots in *in vitro*-derived shoots (80%) with half-strength MS medium containing 1.0 mg/l IBA. From each shoot, three or four roots developed to form a complete plantlet. Again, Naik *et al.* (2000) observed root initiation within 10–15 days in half-strength MS medium supplemented with 0.54 to 5.4 μM NAA and the highest number of roots (10.33 roots/shoot) was formed in medium containing 0.54 μM NAA. Kantharajah *et al.* (1998) showed that lower salt level in culture medium had a beneficial effect on *in vitro* rooting. They obtained both highest rooting and a higher number of roots per microshoot on WPM supplemented with 2 mg/l NAA. Singh and Khawale (2003) suggested the role of AC during rooting with a higher level of sucrose, that is MS medium containing 2.0 mg/l IBA, 200 mg/l AC and 40 g/l sucrose, which also resulted in improved root quality. Murkute *et al.* (2004) reported efficient rooting on half-strength MS basal medium supplemented with either NAA or IAA at 0.5 mg/l. In contrast, Chaugule *et al.* (2007) suggested supplementation of auxin at 0.5 mg/l to be optimum irrespective of their type. Singh *et al.* (2007) could induce good rooting (70.37%) with dual auxins, that is IBA and NAA both supplemented to the rooting medium. However, Damiano *et al.* (2008) suggested that the effective auxin level ranged from 0.75–2.0 mg/l either synthetic or natural (IBA or IAA) for 'Mridula'. Kanwar *et al.*

(2010) reported that the highest frequency of *in vitro* rooting, mean number of roots/shoot (4.32) and mean root length (2.71 cm) were obtained when proliferated shoots were transferred to half-strength MS medium supplemented with 0.02% AC. The plantlets grown on MS medium were found to have better survival compared with WPM medium. Concentrations of 0.5 mg/l NAA and 0.5 mg/l IBA showed an equal rooting response in both media, whereas thick root formation was observed in the medium containing IBA (Patil *et al.*, 2011). Naik and Chand (2011) reported varying success in half-strength WPM and MS medium for *in vitro* rooting of pomegranate. Singh *et al.* (2013) reported the maximum number of roots per shoot (4.17) and root length (3.87 cm) when shoots were rooted on MS basal medium supplemented with 0.5 mg/l NAA, and 200 mg/l AC. Valizadehkaji *et al.* (2013) used 0.54 µM NAA and 4.9 µM IBA on half-strength WPM for *in vitro* rooting of pomegranate microshoots.

6.6.7 Acclimatization of plantlets

Commercialization of *in vitro* propagation has been limited due to high field mortality and slow growth rate of tender plantlets during the acclimatization phase. Increasing sucrose level, use of growth retardants, use of antitranspirants, photoautotrophic micropropagation, use of different colours of shading, hardening under protected structures with very high humidity, and simultaneous rooting and acclimatization are some of the strategies adopted by various workers to enhance *ex vitro* survival and improve acclimatization of *in vitro*-raised plants (Maene and Debergh, 1985; Driver and Suttle, 1987; Graebe, 1987; Wainwright and Scrace, 1989). Different types of potting media like vermicompost, sand, cocopeat, perlite and vermiculite have been utilized for the hardening period of *in vitro*-raised pomegranate saplings. However, a mix of cocopeat, perlite and vermiculite in different combinations is mostly used as the primary hardening medium to achieve good *ex vitro* survival (Mahishni *et al.*, 1991; Yang *et al.*, 1991; Naik *et al.*, 1999, 2000; Murkute *et al.*, 2004; Singh *et al.*, 2007, 2013). Various types of hardening containers/vessels with different types of capping strategies have been used in pomegranate hardening. Singh and Khawale (2003) used a glass jar with a polypropylene (PP) cap filled with moistened peat Soilrite® (1:1) medium for hardening and found 86.5% plantlet survival. *In vitro*-rooted plantlets are transferred to a sterilized potting mixture comprising cocopeat +perlite + vermiculite (1:1:1) in glass jars either with a PP cap or plastic pots with a polythene cover and survival as high as 89% was achieved in pomegranate cv. G-137 (Singh *et al.*, 2007). Before transfer to the medium, the basal portion of the plants are washed thoroughly with 0.1% Bavistin® (Carbendazim 50% WP) to remove adhered medium (Singh *et al.*, 2007). Valizadehkaji *et al.* (2013) used pots containing a sterilized cocopeat–perlite mixture covered with polythene bags for hardening of *in vitro*-rooted pomegranate plantlets. Plantlets were kept at 25 ± 1°C in artificial light (50 µmol m^2/s) for 3–4 weeks for proper hardening. Commercially, primary hardening of *in vitro*-raised rooted pomegranate plantlets in India is carried out in nursery trays or net pots and hardened under low tunnel polyhouses with very high relative humidity.

6.6.8 Plantlet bio-hardening

The micropropagated plants are exposed to various stresses upon transfer from *in vitro* to *ex vitro* greenhouse conditions because of the sudden environmental changes and certain incomplete physiological developments resulting in sudden wilting and high field mortality of the saplings (Singh *et al.*, 2012a, Singh *et al.*, 2016b). These *in vitro*-raised plantlets have poorly formed and weak root systems (Hazarika, 2003), poorly developed cuticle and/or non-functional stomata, being unfavourable for nutritional and environmental conditions (Louro *et al.*, 1999; Hazarika, 2003). Bio-hardening or biopriming is one of the ways to improve field performance of these saplings and can be defined as utilization of plant-beneficial microbes and their formulations and inoculating them into the rhizosphere and/or phyllosphere of *in vitro*-raised plants or saplings during *in vitro* or *ex vitro* or nursery stage to improve field survival and performance of saplings (Kajla *et al.*, 2013; Singh *et al.*, 2016b). The root endophyte *Piriformospora indica* promotes explant hardening

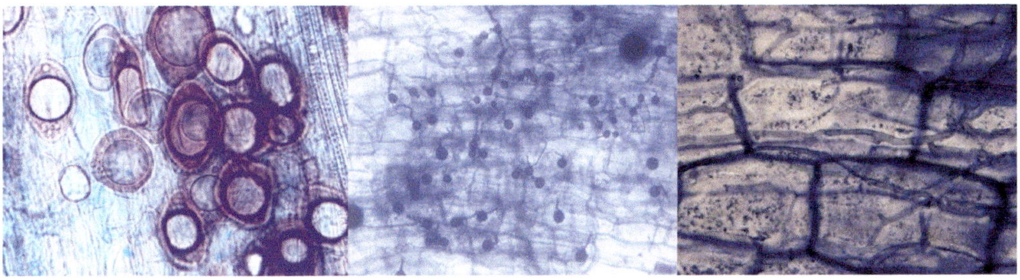

Fig. 6.9. Arbuscular mycorrhizal fungi in root cortical cells of pomegranate. Left to right: vesicles; vesicles and hyphae; arbuscules and hyphae. (Photos: Nripendra Singh.)

(Sahay and Varma, 1999); *Pseudomonas* spp. can reduce hyperhydricity (Bela *et al.*, 1998) and *Bacillus pumilus*, *Alcaligenes faecalis* and *Pseudomonas* spp. improve shoot multiplication (Monier *et al.*, 1998). Mycorrhization in micropropagation has been reported to have significant impact on post-transplant performance of *in vitro*-grown plants (Lovato *et al.*, 1996) (Fig. 6.9). The *ex vitro* performance of tissue culture-raised plantlets can effectively be improved by utilizing plant-beneficial microbes or bioagents like AMF and other beneficial fungi in the hardening process (Mondal *et al.*, 2000; Rupnawar and Navale, 2000; Aseri *et al.*, 2008; Singh *et al.*, 2012a, Singh *et al.*, 2012b, Singh *et al.*, 2016b). Though both, *in vitro* and *ex vitro* bioprimings have been reported, generally primary hardened *in vitro*-raised plants are used for biopriming and plant-beneficial microbes are placed near the root zone of plants planted in sterile potting mixture. Plant-beneficial microbes like *Glomus aggregatum*, *Glomus intraradices*, *Glomus mossae*, *Glomus manihotis*, *Trichoderma harazianum*, *Aspergillus niger* strain AN-27, *Pseudomonas fluorescence*, actinomycetes, etc. can be used in pomegranate to improve *ex vitro* establishment, before transferring them to sterilized soil mixture (soil:sand:organic manure, 1:1:1) in the glasshouse (Fig. 6.8) (Mathur *et al.*, 2008; Singh *et al.*, 2012a; Kajla *et al.*, 2013; Singh *et al.*, 2016b). These biohardening agents establish themselves in the roots and rhizosphere of *in vitro*-raised plants and are extremely efficient in mobilizing nutrients in deficient soils and increasing soil exploration capacity through their mycelia for better uptake of various nutrients. Inoculated and bio-primed/bio-hardened plantlets exhibit higher survival, more root and shoot biomass production, enhanced photosynthesis and better nutrient uptake (Puthur *et al.*, 1998; Rupnawar and Navale, 2000; Singh *et al.*, 2012a, b). The higher survival rates and biomass production of *in vitro*-raised mycorrhized plantlets might be due to an increased rhizosphere-exploring area through the differentiation of extra radical mycorrhiza mycelia, which initially improves water uptake and consequently uptake of some nutrients and optimizes photosynthesis (Mathur and Vyas, 1999; Krishna *et al.*, 2006). Rhizosphere interactions between soil microorganisms such as nitrogen-fixing bacteria, plant growth-promoting rhizobacteria and AMF help in maintaining plant nutrient balances. AMF colonization increases phenol content and also plays a role in enhancing induced systemic resistance (ISR) and systemic acquired resistance (SAR), thus protecting plants against pathogens (Paula *et al.*, 1993; Singh, 2017). It interacts with heavy metals/micronutrients and can restore the equilibrium of nutrient uptake that is unbalanced by heavy metals. It can alleviate Al toxicity, and contribute to soil aggregation and structure stability.

6.7 Somatic Embryogenesis, Embryo Culture and Shoot Bud Organogenesis

Induction of somatic embryos *in vitro* depends upon the morphogenic potential of different vegetative tissues. Jaidka and Mehra (1986) reported indirect shoot organogenesis and somatic embryogenesis in pomegranate. They found that for callus induction from seedling explants of pomegranate cv. 'Kandhari', MS medium supplemented with 4.0 mg/l NAA + 2.0 mg/l kinetin and 15% coconut water was most effective. Embryo-like structures were observed with 2.0 mg/l NAA + 2.0 mg/l BAP. One year later, Omura *et al.* (1987) reported adventitious shoot bud formation on leaf segments of dwarf pomegranate var. 'Nana'. They were also successful in obtaining plantlet regeneration from suspension culture derived from leaf callus on MS medium supplemented with 2.0 µM BAP and 1.0 µM NAA (Omura *et al.*, 1990). Bhansali (1990) found vigorous proliferation of cotyledonary tissue-based embryogenic cell clusters with regular subculturing on RBM-II (half-strength MS medium containing 1 µM kinetin, 2 µM BAP and 5 µM 2,4-dichlorophenoxyacetic acid). Developmental stages of somatic embryos were expressed on subculturing with a low level of 2,4-D (2.5 µM). Embryo maturation was obtained on RBM-III (half-strength MS medium containing 2 µM kinetin, 2 µM BAP and 2.5 µM 2,4-dichlorophenoxyacetic acid) and IV media (half-strength MS medium containing 2 µM kinetin, 2 µM BAP and 0.5 µM 2,4-dichlorophenoxyacetic acid). Calli from leaf segments and stem explants of *P. granatum* L. var. 'Nana' were initially cultured on modified MS basal medium supplemented with BA, zeatin, kinetin or 2-iso pentanyladenine (2iP) at 0.1–1.5 mg/l and IAA, IBA or NAA at 0.1–1.0 mg/l for organogenesis (Yang and Ludders, 1993). Adventitious shoot elongation was stimulated on MS basal medium supplemented with 0.5 mg/l BA and 0.1 mg/l IBA. Elongated shoots rooted easily on half-strength MS medium. Nataraja and Neelambika (1996) were successful in getting somatic embryos from cultured pomegranate petals on MS medium supplemented with 5.0 mg/l BAP and 5.0 mg/l IAA. Fougat *et al.* (1997) compared different types of explants for *in vitro* regeneration in cv. 'Ganesh' on MS medium supplemented with various growth regulator combinations and found callus growth and induction from cotyledon and leaf explants were best on MS medium supplemented with 4.0 mg/l NAA, 2.0 mg/l kinetin and 15% coconut water. Enhanced axillary branching from nodal segments and proliferation of shoot tip meristems was achieved on MS medium supplemented with 0.5 mg/l kinetin, 1.0 mg/l BA and 500 mg/l casein hydrolysate. Rooting was best in shoots derived from all explant sources, on MS medium supplemented with 4.0 mg/l NAA, 2.0 mg/l kinetin and 15% coconut water. Kantharajah *et al.* (1998) studied the combination effect of media, plant growth regulators and explant source on *in vitro* culture of pomegranate cv. 'Wonderful'. Callus cultures were initiated from leaf and nodal explants obtained from aseptically cultured shoots. Callus initiation and growth were best on MS basal medium containing either 1 mg/l BAP (leaf explant) or 1 mg/l BAP + 0.4 mg/l NAA (nodal explant). The minimum duration for callus induction from leaf segments was 8.8 days, whereas cotyledons recorded 10.0 days. Callus weight and proliferation was highest on MS medium containing 0.4 mg/l NAA + 1.0 mg/l BAP. The cotyledon was found to be most responsive (78.94%) for callus induction on MS medium fortified with 0.4 mg/l NAA + 1.0 mg/l BAP. However, the calli derived from the leaf segments showed greater dry weight than those derived from the cotyledon. On nodal cuttings, the best adventitious bud formation was observed on half-strength MS medium supplemented with 0.5 mg/l BAP + 0.1 mg/l NAA. This medium also promoted good proliferation along with the formation of the longest shoots. The highest rooting and average number of roots/explant were induced on WP medium supplemented with 2 mg/l NAA.

There was poor somatic embryo induction confirming the recalcitrant nature of pomegranate. Kanwar *et al.* (2010) excised zygotic embryos from seeds collected at 16 weeks after full bloom, and when these embryos inoculated on MS medium containing 21 µM NAA, 9 µM BA, 30 g/l sucrose and 15% coconut water produced the highest frequency of embryogenic callus, clumps with globular embryos, and mean number of both globular and heart-shaped embryos per callus clump. Subjecting zygotic embryo explants to a 6-week dark incubation period was essential for embryogenic callus induction, and these were subsequently transferred to a 16 h photoperiod for further growth and development of somatic embryos. Germination of somatic embryos was best on MS medium supplemented with 60 g/l sucrose. Guranna *et al.* (2018) obtained rapid regeneration of plants through indirect organogenesis in pomegranate cv. 'Bhagawa'. They reported nodal segments as the best explant for induction of callus on MS medium supplemented with 5 mg/l BAP and 0.4 mg/l NAA and shoot induction from calli was best on MS basal medium with 2 mg/l BAP + 0.1 mg/l NAA + 0.5 mg/l GA_3.

6.8 Synthetic Seeds

Synthetic seeds are similar to seeds in function. The synthetic seed technology encapsulates somatic embryos and/or axillary shoot buds, apical shoot tips, embryogenic calli, protocorm or protocorm-like bodies or other micropropagules in a nutrient-rich hydrogel (Singh *et al.*, 2010; Sharma *et al.*, 2013; Teixeira da Silva *et al.*, 2013). These synthetic seeds are used as seed analogues for mechanical planting at a commercial level or air sowing. In pomegranate, Naik and Chand (2006) have successfully encapsulated nodal segments from *in vitro*-proliferated shoot cultures or axenic cotyledonary nodes in calcium alginate hydrogel containing MS medium supplemented with 4.44 μM BAP and 0.54 μM NAA. For this, sodium alginate (1–6%) and the consolidation solution of calcium chloride (50–125 mM) is usually used, and a combination of 3% sodium alginate and 100 mM calcium chloride was most suitable for formation of ideal synthetic seed. The morphogenic response of encapsulated nodal segments was the highest in MS medium augmented with 4.44 μM BAP and 0.54 μM NAA. Encapsulated nodal segments stored up to 30 days at 4°C were capable of sprouting.

References

Ahire, D.B., Ranpise, S.A. and Shirsath, H.A. (2016) Study of pomegranate (*Punica granatum* L.) propagation using wedge grafting. *Advances in Life Sciences* 5, 8823–8826.

Ahire, D.B., Ranpise, S.A. and Shete, M.B. (2017) Assessment of graft compatibility of different rootstocks of pomegranate. *International Journal of Minor Fruits, Medicinal and Aromatic Plants* 3, 21–26.

Allioli, T.B., Reddy, P.N., Hussain, S.A. and Patil, C.V. (2001) Fly ash: new medium for induction of rooting in air layers of dry land fruits. *Karnataka Journal of Agricultural Sciences* 14, 536–536.

Amin, M.N., Rahman, M.M. and Manik, M.S. (2003) In vitro clonal propagation of *Paederia foetida* L. – a medicinal plant of Bangladesh. *Plant Tissue Culture and Biotechnology* 13, 117–123.

Ansari, S. (2013) Effect of different collecting time and different medium on rooting of pomegranate Malas-Torsh cv. cuttings. *Bulletin of Environment Pharmacology and Life Sciences* 2(164), 168.

Asadov, K.S. (1987) Forest orchard in Azerbaidzhan. *Lesnoe Khozyaistov* 9, 59–60.

Aseri, G.K., Jain, N., Panwar, J., Rao, A.V. and Meghwal, P.R. (2008) Biofertilizers improve plant growth, fruit yield, nutrition, metabolism and rhizosphere enzyme activities of pomegranate (*Punica granatum* L.) in Indian Thar desert. *Scientia Horticulturae* 117(2), 130–135. DOI: 10.1016/j.scienta.2008.03.014.

Batista, P.F., Maia, S.S.S., Coelho, M.F.B., Benedito, C.P. and Guimaraes, I.P. (2011) Vegetative propagation of pomegranate in different substrates. *Revista Verde De Agroecologiae Desenvolvimento Sustentavel* 6, 96–100.

Bela, J., Ueno, K.I. and Shetty, K. (1998) Control of hyper-hydricity in anise (*Pimpinella anisum*) tissue culture by *Pseudomonas* spp. *Journal of Herbs, Spices & Medicinal Plants* 6(1), 57–67. DOI: 10.1300/J044v06n01_08.

Bhansali, R.R.A.J. (1990) Somatic embryogenesis and regeneration of plantlets in pomegranate. *Annals of Botany* 66(3), 249–253. DOI: 10.1093/oxfordjournals.aob.a088022.

Bhat, S.R. and Chandel, K.P.S. (1991) A novel technique to overcome browning in tissue culture. *Plant Cell Reports* 10(6–7), 358–361. DOI: 10.1007/BF00193159.

Bhojwani, S.S. and Razdan, M.K. (1983) *Plant Tissue Culture: Theory and Practice*. Elsevier Science Publishers, the Netherlands.

Bhosale, V.P., Jadhav, R.G., Masu, M.M., Sitapara, H.H. and Patel, H.C. (2009) Response of different media and plant growth regulators on rooting and survival of air layers in pomegranate (*Punica granatum* L.) cv. Sinduri. *Proceedings of 2nd International Symposium on Pomegranate and Minor Including Mediterranean Fruits*. University of Agricultural Sciences, Dharwad, India, pp. 72–73.

Biniyaz, N., Karimi, H.R., MohamadiMirik, A.A. and Esmaelzadeh, M. (2017) Study of rootstock effects on growth and eco-physiological parameters of two pomegranate cultivars in Rafsanjan environment conditions. Msc Thesis. Vale-e-Asr University of Rafsanjan, Iran.

Blumenfeld, A., Shaya, F. and Hillel, R. (2000) Cultivation of pomegranate. *Options Mediteraneennes* 42, 143–146.
Cervelli, C. and Belletti, P. (1994) Effect of thermic treatments and seed manipulations on emergence of dwarf pomegranate (*Punica granatum* L. 'Nana'). *Acta Horticulturae* 362,189–196. DOI: 10.17660/ActaHortic.1994.362.23.
Chadha, K.L. (2001) *Handbook of Horticulture*, 1st edn. Indian Council of Agricultural Research, New Delhi, India.
Chandra, R. and Babu, K.D. (2010) Propagation of pomegranate – a review. *Vegetable, Cereal Science and Biotechnology* 4, 51–55.
Chandra, R. and Jadhav, V.T. (2012) Grafting method and time in pomegranate (*Punica granatum* L.) under semi-arid agro-climatic conditions of Maharashtra. *Indian Journal of Agricultural Sciences* 82, 990–992.
Chandra, R., Jadhav, V.T., Sharma, J. and Marathe, R.A. (2009) Effect of grafting method and time on scion sprouting, graft success and subsequent growth of grafter plants of pomegranate (*Punica garanatum* L.) Bhagawa. *Acta Horticulturae* 82, 990–992.
Chandra, R., Jadhav, V.T. and Sharma, J. (2010) Global scenario of pomegranate culture with special reference to India. *Fruit, Vegetable and Cereal Science and Biotechnology* 4, 7–16.
Chandra, R., Singh, N.V. and Pal, R.K. (2013) Patch budding in pomegranate – an easy *in situ* budding technique. *ICAR News* 19, 1–2.
Chandra, R., Pal, R.K., Rigveda, D., Singh, N.V. and Maity, A. (2014) Propagation practices in pomegranate: a review. *Indian Journal of Arid Horticulture* 9, 1–6.
Chater, J.M., Merhaut, D.J., Preece, J.E. and Blythe, E.K. (2017) Rooting and vegetative growth of hardwood cuttings of 12 pomegranate (*Punica granatum* L.) cultivars. *Scientia Horticulturae* 221, 68–72. DOI: 10.1016/j.scienta.2017.04.025.
Chaugule, R.R., More, T.A., Kamble, A.B. and Karale, A.R. (2007) Studies of micropropogation in pomegranate (*Punica granatum* L.). *Recent Trends in Horticultural Biotechnology (Vol I) and II ICAE National Symposium on Biotechnological Interventions for Improvement of Horticultural Crops: Issues and Strategies*, Vellanikkara, Kerala, pp. 195–199.
Damiano, C., Padro, M.D.A. and Frattarelli, A. (2008) Propagation and establishment in vitro of mytle (*Myrtus communis* L.), pomegranate (*Punica granatum* L.) and mulberry (*Morus alba* L.). *Propagation of Ornamental Plants* 8, 3–8.
Day, K.R. and Wilkins, E.D. (2009) Commercial pomegranate production in California. *Proceedings of the 2nd International Symposium on Pomegranate and Minor Including Mediterranean Fruits*, Dharwad, India, pp. 33–41.
Debergh, P.C. and Maene, L.J. (1981) A scheme for commercial propagation of ornamental plants by tissue culture. *Scientia Horticulturae* 14(4), 335–345. DOI: 10.1016/0304-4238(81)90047-9.
Deol, I.S. and Uppal, D.K. (1990) Effect of different rooting media on rooting and growth of hard wood and semi hardwood cuttings of pomegranate (*Punicag ranatum* L.). *Punjab Horticultural Journal* 30, 140–144.
Dhillon, W.S. and Sharma, K.K. (1992) Effect of indole butyric acid on rooting of cuttings in pomegranate (*Punica granatum* L). *Punjab Agricultural Journal of Research* 29, 350–353.
Drazeta, L. (1997) Pomegranate (*Punica granatum* L.) propagation by *in vitro* method of tissue culture. *Review of Research Work at the Faculty of Agriculture, Belgrade* 42, 49–59.
Driver, J.A. and Suttle, G.R.L. (1987) Nursery handling of propagules. In: Bonga, J.M. and Durzan, D.J. (eds) *Cell and Tissue Culture in Forestry*. Martinus Nijhoff, Dordrecht, Germany, pp. 320–335.
Fotouhi-Ghazvini, R., Sajadian, H., Hokmabadi, H. and Ahmad, S. (2007) Effects of pistachio rootstocks on echo-physiological characteristics of commercial pistachio cultivars. *International Journal of Agriculture and Biology* 2, 352–354.
Fougat, R.S., Pandya, S.B., Ahmed, T. and Godhani, P.R. (1997) *In vitro* studies in pomegranate (*Punica granatum* L.). *Journal of Applied Horticulture Navsari* 3, 23–29.
Ghosh, D., Bandyopadhyay, A. and Sen, S.K. (1988) Effect of NAA and IBA on adventitious root formation in stem cuttings of pomegranate (*Punica granatum* L.). *Indian Agriculturist* 32, 239–243.
Goncalves, B., Moutinho-Pereira, J., Santos, A., Silva, A.P., Bacelar, E. *et al.* (2006) Scion–rootstock interaction affects the physiology and fruit quality of sweet cherry. *Tree Physiology* 26(1), 93–104. DOI: 10.1093/treephys/26.1.93.
Graebe, J.E. (1987) Gibberellin biosynthesis and control. *Annual Review of Plant Physiology* 38(1), 419–465. DOI: 10.1146/annurev.pp.38.060187.002223.

Guranna, P., Hosmani, I., Sathyanarayana, R., Hegde, R. and Hipparagi, K. (2018) Micropropagation in pomegranate (*Punica granatum* L.) cv. Bhagwa through indirect organogenesis and assessment of genetic fidelity by RAPD marker. *Biotechnology Journal International* 20, 1–8.

Hartmann, H.T., Kester, D.E., Davies, J.F.T. and Geneve, R.L. (2002) *Plant Propagation: Principles and Practices*, 7th edn. Prentice-Hall Inc., Englewood Cliffs, New Jersey.

Hasanpour, Z., Karimi, H.R. and Mirdehghan, S.H. (2014) Effects of salinity and water stress on echo-physiological parameters and micronutrients concentration of pomegranate (*Punica granatum* L.). *Journal of Plant Nutrition* 38, 1–13.

Hasanpour, N., Karimi, H.R. and Mirdehghan, S.H. (2015) Effects of rootstock and salinity stress on the vegetative, biochemical and eco-physiological characteristics of pomegranate cv. 'Gabri'. *Journal of Horticultural Science and Technology* 16, 47–62.

Hazarika, B.N. (2003) Acclimatization of tissue-cultured plants. *Current Science* 85, 12–25.

Holland, D. and Bar-Ya'akov, I. (2008) The pomegranate: new interest in ancient fruit. *Chronica Horticulturae* 48, 12–15.

Holland, D., Hatib, K. and Bar-Ya'akov, I. (2009) Pomegranate: botany, horticulture, breeding. In: Janick, J. (ed.) *Horticultural Reviews*. 35. Wiley and Sons, Hoboken, New Jersey, pp. 127–191.

Hore, J.K. and Sen, S.K. (1993) Root formation in pomegranate (*Punica granatum* L.) stem cuttings with NAA and auxin synergists under intermittent mist. *Crop Research* 6, 252–256.

Hore, J.K. and Sen, S.K. (1995) Role of non auxinic compounds and IBA on root regeneration in air layers of root formation in pomegranate (*Punica granatum* L.). *Current Research, University of Agricultural Science* 24, 83–85.

Hussain, I., Khattak, A.M., Amin, M.U., Aman, F. and Sajid, M. (2012) Response of different pomegranate cuttings types to different environmental conditions. *Sarhad Journal of Agriculture* 28, 15–18.

Hu, C.Y. and Wang, P.J. (1993) Meristem, shoot tip and bud culture. In: Evans, D.A., Sharp, D.R., Ammirato, P.V. and Yamada, Y. (eds) *Handbook of Plant Cell Culture*. 1. MacMillan, New York, pp. 177–216.

ICAR-NRC on Pomegranate (2016) Annual report 2015–16. ICAR-National Research Centre on Pomegranate, Solapur, India.

Izadi, Z., Zarei, H. and Alizadeh, M. (2013) Role of grafting technique on the success of stenting propagation of two rose (*Rosa* sp.) varieties. *American Journal of Plant Sciences* 4, 41–44.

Jaganath, S., Meenakshi, H.C., Harinikumar, K.M. and Nachegowda, V. (2009) Effect of microbial inoculants on rooting of pomegranate (*Punica granatum* L.) cvs. Bhagwa and Ganesh stem cutting. *Proceedings of 2nd International Symposium on Pomegranate and Minor Including Mediterranean Fruits*. University of Agricultural Sciences, Dharwad, India, p. 72.

Jaidka, K. and Mehra, P.N. (1986) Morphogenesis in *Punica granatum* (pomegranate). *Canadian Journal of Botany* 64(8), 1644–1653. DOI: 10.1139/b86-220.

Jalikop, S.H. (2007) Linked dominant alleles or inter-locus interaction results in a major shift in pomegranate fruit acidity of 'Ganesh' × 'Kabul Yellow'. *Euphytica* 158(1–2), 201–207. DOI: 10.1007/s10681-007-9443-1.

Jones, A.M.P. and Saxena, P.K. (2013) Inhibition of phenylpropanoid biosynthesis in *Artemisia annua* L.: a novel approach to reduce oxidative browning in plant tissue culture. *PLoS ONE* 8(10), e76802. DOI: 10.1371/journal.pone.0076802.

Kajla, S., Poonia, A., Kharb, P. and Duhan, J.S. (2013) Role of biotechnology for commercial production of fruit crops. In: Salar, R.K., Gahlawat, S.K., Siwach, P. and Duhan, J.S. (eds) *Biotechnology: Prospects and Applications*. Springer, New Delhi, India, pp. 27–36.

Kantharajah, A.S., Dewitz, I. and Jabbari, S. (1998) The effect of media, plant growth regulators and source of explants on *in vitro* culture of pomegranate (*Punica granatum* L.). *Erwerbsobstbau* 40, 54–58.

Kanwar, K., Joseph, J. and Deepika, R. (2010) Comparison of *in vitro* regeneration pathways in *Punica granatum* L. *Plant Cell, Tissue and Organ Culture* 100(2), 199–207. DOI: 10.1007/s11240-009-9637-4.

Kar, P.L., Singh, R.P. and Chadha, T.R. (1989) Studies on the standardization of top working technique in wild pomegranate trees. *Horticulture Journal* 2, 68–70.

Karimi, H.R., Esmaelizadeh, M. and Abolipour, M. (2011) Preliminary study of stratification treatment on germination of pomegranate (*Punica granatum* L.). *Iranian Horticultural Science Congress*, Isfahan University of Technology, 5–8 September.

Karimi, H.R., Afzalifar, M. and Mansouri, M.Z. (2012) Effect of IBA and salicylic acid on rooting and vegetative parameters of pomegranate cuttings. *International Journal of Agriculture: Research and Reviews* 2, 1085–1091.

Karimi, R. (2011) Stenting (cutting & grafting): a new technique for propagation pomegranate (*Punica granatum* L.). *Journal of Fruit and Ornamental Plant Research* 19, 73–79.

Karimi, H.R. and EiniTari, F. (2016) Effect of $NaHCO_3$ on photosynthetic characteristics, iron and sodium transfer in pomegranate. *Journal of Plant Nutrition* 40, 1–12.

Karimi, R. and Farahmand, H. (2011) Study of pomegranate (*Punica granatum* L.) propagation using bench grafting. *Journal of Fruit and Ornamental Plant Research* 19, 67–72.

Karimi, H.R. and Hasanpour, Z. (2014) Effects of salinity and water stress on growth and macro nutrients concentration of pomegranate (*Punica granatum* L.). *Journal of Plant Nutrition* 37(12), 1937–1951. DOI: 10.1080/01904167.2014.920363.

Karimi, H.R. and Hasanpour, N. (2017) Effects of salinity, rootstock, and position of sampling on macro nutrient concentration of pomegranate cv. Gabri. *Journal of Plant Nutrition* 40(16), 2269–2278. DOI: 10.1080/01904167.2016.1263324.

Karimi, H.R. and Mirdehghan, S.H. (2013) Correlation between the morphological characters of pomegranate (*Punica granatum* L.) traits and their implications for breeding. *Turkish Journal of Botany* 37, 355–362.

Karimi, H.R. and Nowrozy, M. (2017) Effects of rootstock and scion on graft success and vegetative parameters of pomegranate. *Scientia Horticulturae* 214, 280–287. DOI: 10.1016/j.scienta.2016.11.047.

Khalil, A. (2013) Effect of different collecting time and different medium on rooting of pomegranate Malas Torsh cv. cuttings. *Bulletin of Environment, Pharmacology and Life Sciences* 2, 164–168.

Krishna, H., Singh, S.K. and Patel, V.B. (2006) Screening of arbuscular mycorrhizal fungi for enhanced growth and survival of micropropagated grape (*Vitis vinifera*) plantlets. *Indian Journal of Agricultural Sciences* 76, 297–301.

Kumar, R., Kumar, R., Jakhar, M.L. and Verma, R. (2017) Pomegranate micropropagation - A review. *International Journal of Pure & Applied Bioscience* 5(5), 1138–1149. DOI: 10.18782/2320-7051.5124.

Levin, G.M. (2006) *Pomegranate,* 1st edn. Third Millennium Publishing, Tempe, Arizona.

Louro, R.P., Dos Santos, A.V. and Machado, R.D. (1999) Ultrastructure of *Eucalyptus grandis* × *Eucalyptus urophylla* . I. shoots cultivated *in vitro* in multiplication and elongation-rooting media. *International Journal of Plant Sciences* 160(2), 217–227. DOI: 10.1086/314118.

Lovato, P.E., Gianinazzi-Pearson, V., Trouvelot, A. and Gianinazzi, S. (1996) The state of mycorrhizas and micropropagation. *Advances in Horticulture Science* 10, 46–52.

Maene, L. and Debergh, P. (1985) Liquid medium additions to established tissue cultures to improve elongation and rooting *in vivo*. *Plant Cell, Tissue and Organ Culture* 5(1), 23–33. DOI: 10.1007/BF00033566.

Mahishni, D.M., Muralikrishna, A., Shivshankar, G. and Kulkarni, R.S. (1991) Shoot tip culture method for rapid clonal propagation of pomegranate (*Punica granatum* L.) *Horticulture – New Technologies and Applications*. *Proceedings of the International Seminar on New Frontiers in Horticulture*, Indo-American Hybrid Seeds, Bangalore, India, pp. 215–216.

Maity, A., Pal, R.K., Chandra, R. and Singh, N.V. (2014) *Penicillium pinophilum* – a novel microorganism for nutrient management in pomegranate (*Punica granatum* L.). *Scientia Horticulturae* 169, 111–117. DOI: 10.1016/j.scienta.2014.02.001.

Mathur, A., Mathur, A.K., Verma, P., Yadav, S., Gupta, M.L. *et al.* (2008) Biological hardening and genetic fidelity testing of micro-cloned progeny of *Chlorophytum borivilianum* Sant. et Fernand. *African Journal of Biotechnology* 7, 1046–1053.

Mathur, N. and Vyas, A. (1999) Improved biomass production, nutrient uptake and establishment of *in vitro* raised *Ziziphus mauritiana* by VA mycorrhiza. *Journal of Plant Physiology* 155(1), 129–132. DOI: 10.1016/S0176-1617(99)80153-9.

Melgarejo, P., Martinez, J., Martinez, J.J., Martinez, V.R. and Amoros, A. (2000) Study of the rooting capacity of eleven pomegranate (*Punica granatum* L.) clones, using plastic to cover the soil. *Options Mediterraneennes* 42, 169–173.

Mondal, G., Dureja, P. and Sen, B. (2000) Fungal metabolites from *Aspergillus niger* AN27 related to plant growth promotion. *Indian Journal of Experimental Biology* 38, 84–86.

Monier, C., Bossis, E., Chabanet, C. and Samson, R. (1998) Different bacteria can enhance the micropropagation response of *Cotoneaster lacteus* (Rosaceae). *Journal of Applied Microbiology* 85(6), 1047–1055. DOI: 10.1111/j.1365-2672.1998.tb05270.x.

Murkute, A.A., Patil, S. and Singh, S.K. (2004) *In vitro* regeneration in pomegranate cv. Ganesh from mature trees. *Indian Journal of Horticulture* 61, 201–208.

Naeini, M.R., Khoshgoftarmanesh, A.H. and Fallahi, E. (2006) Partitioning of chlorine, sodium, and potassium and shoot growth of three pomegranate cultivars under different levels of salinity. *Journal of Plant Nutrition* 29(10), 1835–1843. DOI: 10.1080/01904160600899352.

Naik, S.K. and Chand, P.K. (2006) Nutrient-alginate encapsulation of *in vitro* nodal segments of pomegranate (*Punica granatum* L.) for germplasm distribution and exchange. *Scientia Horticulturae* 108(3), 247–252. DOI: 10.1016/j.scienta.2006.01.030.

Naik, S.K. and Chand, P.K. (2011) Tissue culture-mediated biotechnological intervention in pomegranate: a review. *Plant Cell Reports* 30(5), 707–721. DOI: 10.1007/s00299-010-0969-7.

Naik, S.K., Pattnaik, S. and Chand, P.K. (1999) *In vitro* propagation of pomegranate (*Punica granatum* L. cv. Ganesh) through axillary shoot proliferation from nodal segments of mature tree. *Scientia Horticulturae* 79(3-4), 175–183. DOI: 10.1016/S0304-4238(98)00218-0.

Naik, S.K., Pattnaik, S. and Chand, P.K. (2000) High frequency axillary shoot proliferation and plant regeneration from cotyledonary nodes of pomegranate (*Punica granatum* L.). *Scientia Horticulturae* 85(4), 261–270. DOI: 10.1016/S0304-4238(99)00149-1.

Nataraja, K. and Neelambika, G.K. (1996) Somatic embryogenesis and plantlet regeneration from petal culture of pomegranate, *Punica granatum* L. *Indian Journal of Experimental Biology* 34, 719–721.

National Horticultural Board (2018) Horticultural statistics at a glance – 2017. Available at: http://nhb.gov.in (accessed 25 May 2020).

Noda, Y., Kaneyuki, T., Mori, A. and Packer, L. (2002) Antioxidant activities of pomegranate fruit extract and its anthocyanidins: delphinidin, cyanidin, and pelargonidin. *Journal of Agricultural and Food Chemistry* 50(1), 166–171.

Nowrozy, M., Karimi, H.K. and Mirdehghan, S.H. (2014) Effect of rootstock, scion and some methods of grafting on graft success and vegetative parameters of pomegranate. MSc Thesis. Vale-e-Asr University of Rafsanjan, Iran, Rafsanjan, Iran.

Nowrozy, M., Karimi, H.K. and Mirdehghan, S.H. (2016) Effect of rootstock, scion and grafting method on vegetative propagation of pomegranate. *Iranian Journal of Horticultural Science* 47, 337–350.

Olmez, Z., Temel, F., Gokturk, A. and Yahyaoglu, Z. (2007) Effects of sulphuric acid and cold stratification pretreatments on germination of pomegranate (*Punica granatum* L.) seeds. *Asian Journal of Plant Science* 6, 427–430.

Omura, M., Matsuta, N., Moriguchi, T. and Kozaki, I. (1987) Adventitious shoot and plantlet formation from cultured pomegranate leaf explants. *HortScience* 22, 133–134.

Omura, M., Matsuta, N., Moraguchi, T. and Kozaki, I. (1990) Suspension culture and plantlet regeneration in dwarf pomegranate (*Punica granatum* var. *nana* Pers.). *Bulletin of the Fruit Tree Research Station* 17, 19–33.

Pal, R.K. and Singh, N.V. (2017) *Pomegranate for Nutrition, Livelihood Security and Entrepreneurship Development*. Daya Publishing House, New Delhi, India.

Pandey, S.K. and Bisen, A. (2010) Effect of mechanical treatments on rooting in cuttings of guava, lemon and pomegranate. *Journal of Horticulture and Forestry* 2, 95–98.

Panwar, R.D., Kaushik, R.A., Singh, S., Gupta, R.B. and Singh, S. (2001) Effect of indole butyric acid (IBA) on rooting of hard wood cuttings in pomegranate (*Punica granatum* L.). *Haryana Journal of Horticulture Science* 30, 72.

Patil, P.B., Patil, C.P. and Kumar, S. (2001) Impact of inoculation microorganism on rootability of pomegranate cuttings. *Karnataka Journal of Agricultural Sciences* 14, 1020–1024.

Patil, V.M., Dhande, G.A.D., Thigale, D.M. and Rajput, J.C. (2011) Micropropagation of pomegranate (*Punica granatum* L.) 'Bhagava' cultivar from nodal explant. *African Journal of Biotechnology* 10, 18130–18136.

Paula, M.A., Siqueira, J.O. and Dobereiner, J. (1993) Occurrence of vesicular arbuscular mycorrhizal fungi and diazptrophic bacteria associated with sweet potato. *Revista Brasileira de Ciência do Solo* 17, 349–356.

Polat, A.A. and Caliskan, O. (2009) Effect of IBA on rooting cutting in various pomegranate genotypes. *Acta Horticulturae* 818, 187–192.

Purohit, A.G. and Shekharappa, K.E. (1985) Effect of type of cutting and indole butyric acid on rooting of hard wood cuttings of pomegranate (*Punica granatum* L.). *Indian Journal of Horticulture* 42, 30–36.

Puthur, J.T., Prasad, K.V.S.K., Sharmila, P. and Pardha Saradhi, P. (1998) Vesicular arbuscular mycorrhizal fungi improves establishment of micropropagated *Leucaena leucocephala* plantlets. *Plant Cell, Tissue and Organ Culture* 53(1), 41–47. DOI: 10.1023/A:1006068026377.

Rajan, S. and Markose, B.L. (2007) Propagation of horticultural crops. In: Peter, K.V. (ed.) *Horticultural Science Series (6)*. New India Publishing Agency, New Delhi, India, pp. 81–84.

Rajkumar, R., Gora, J.S., Kumar, R., Singh, A., Kumar, A. *et al.* (2017) Effect of different growing media on the rooting of pomegranate (*Punica granatum* L.) cv. 'Phulearakta' cuttings. *Journal of Applied and Natural Science* 9(2), 715–719. DOI: 10.31018/jans.v9i2.1263.

Raj, D. and Kanwar, K. (2008) *In vitro* regeneration of *Punica granatum* L. plants from different juvenile explants. *Journal of Fruit and Ornamental Plant Research* 18, 5–22.

Rawat, J.M.S., Tomar, Y.K. and Rawat, V. (2010) Effect of stratification on seed germination and seedling performance of wild pomegranate. *Journal of American Science* 6, 97–99.

Reddy, Y.T.N. and Reddy, Y.N. (1989) Rooting of pomegranate cuttings as influenced by paclobutrazol, polythene covering and basal wounding. *Progressive Horticulture* 21, 106–110.

Reddy, Y.T.N. and Reddy, Y.N. (1990) Effect of basal wounding, growth regulator and polythene covering on rooting of pomegranate cuttings. *Journal of Maharashtra Agricultural University* 15, 153–155.

Rupnawar, B.S. and Navale, A.M. (2000) Effect of VA-mycorrhizal inoculation on growth of pomegranate layers. *Journal of Maharashtra Agricultural University* 25, 44–46.

Rymbai, H. and Reddy, G.S. (2010) Effect of IBA concentrations on guava stooling and plantlets survival under open and polyhouse conditions. *Indian Journal of Horticulture* 67, 443–446.

Sahay, N.S. and Varma, A. (1999) *Piriformospora indica* : a new biological hardening tool for micropropagated plants. *FEMS Microbiology Letters* 181(2), 297–302. DOI: 10.1111/j.1574-6968.1999.tb08858.x.

Sandhu, A.S., Minhas, P.P.S., Singh, S.N. and Kambhoj, J.S. (1991) Studies on rhizogenesis in hard wood cuttings of pomegranate. *Indian Journal of Horticulture* 48, 302–304.

Saroj, P.L., Awasthi, O.P., Bhargawa, R. and Singh, U.V. (2008) Standardization of propagation by cutting under mist system in hot arid region. *Indian Journal of Horticulture* 65, 25–30.

Sarrou, E., Therios, I. and Dimassi-Theriou, K. (2014) Melatonin and other factors that promote rooting and sprouting of shoot cuttings in *Punica granatum* cv. Wonderful. *Turkish Journal of Botany* 38, 293–301. DOI: 10.3906/bot-1302-55.

Shalimu, D., Sun, J., Baskin, C.C., Baskin, J.M., Sun, L. *et al.* (2016) Changes in oxidative patterns during dormancy break by warm and cold stratification in seeds of an edible fruit tree. *AoB Plants* 8, 1–13. DOI: 10.1093/aobpla/plw024.

Sharma, S., Shahzad, A. and Teixeira da Silva, J.A. (2013) Synseed technology – a complete synthesis. *Biotechnology Advances* 31(2), 186–207. DOI: 10.1016/j.biotechadv.2012.09.007.

Sharma, J., Chandra, R., Babu, D., Meshram, D.T. and Maity, A. (2014) Pomegranate: cultivation, marketing and utilization. Technical Bulletin No. NRCP/2014/1. ICAR-NRCP, Solapur, India.

Sharma, R.R. and Srivastav, M. (2004) *Plant Propagation and Nursery Management*, 1st edn. International Book Distribution Co., Lucknow, India.

Silva, J.G., Lopes, K.P., Cavalcante, J.R., Pereira, N.A.E. and Barbosa, R.C.A. (2016) Pre-germinate treatments in pomegranate seeds (*Punica granatum* L.): effect on physiological quality. *Revista Brasileira Fruticultura* 39, 1–5.

Singh, N.V., Singh, S.K. and Patel, V.B. (2007) *In vitro* axillary shoot proliferation and clonal propagation of 'G-137' pomegranate (*Punica granatum* L.). *Indian Journal of Agricultural Science* 77, 509–511.

Singh, S.K., Singh, A., Singh, N.V. and Ramajayam, D. (2010) Pomegranate tissue culture and biotechnology. *Fruit, Vegetable, Cereal Science and Biotechnology* 4, 35–44.

Singh, N.V., Singh, S.K. and Patel, V.B. (2011) *In vitro* culture establishment studies in pomegranate. *Indian Journal of Horticulture* 68, 307–311.

Singh, N.V., Singh, S.K., Singh, A.K., Meshram, D.T., Suroshe, S.S. *et al.* (2012a) Arbuscular mycorrhizal fungi (AMF) induced hardening of micropropagated pomegranate (*Punica granatum* L.) plantlets. *Scientia Horticulturae* 136, 122–127. DOI: 10.1016/j.scienta.2012.01.005.

Singh, N., Singh, S. and Meshram, D. (2012b) *Pomegranate: In Vitro Propagation and Biohardening*. Lambert Academic Publishing, Saarbrucken, Germany.

Singh, P., Patel, R.M. and Kadam, S. (2013) *In vitro* mass multiplication of pomegranate from cotyledonary nodal explants. *African Journal of Biotechnology* 12, 2863–2868.

Singh, K.K. (2014) Effect of IBA concentration on the rooting of pomegranate (*Punica granatum* L.) cv. Ganesh hard wood cuttings under mist house condition. *Plant Archives* 14, 1111–1114.

Singh, K.K. (2017) Vegetative propagation of pomegranate (*Punica granatum* L.) through cutting – a review. *International Journal of Current Microbiology and Applied Sciences* 6(10), 4887–4893. DOI: 10.20546/ijcmas.2017.610.458.

Singh, N.V., Chandra, R. and Pal, R.K. (2014) Two stage hardwood cutting protocol for pomegranate. *ICAR News* 20, 20–22.

Singh, N.V., Abburi, V.L., Ramajayam, D., Kumar, R., Chandra, R, Chandra, Sharma, K.K. *et al.* (2015) Genetic diversity and association mapping of bacterial blight and other horticulturally important traits with microsatellite markers in pomegranate from India. *Molecular Genetics and Genomics* 290(4), 1393–1402. DOI: 10.1007/s00438-015-1003-0.

Singh, N.V., Singh, S., Chandra, R., Babu, K.D. and Pal, R.K. (2016a) Comparative evaluation of seed germination and parameters of seedling growth in pomegranate genotypes (*Punica granatum* L.). *Research on Environment and Life Science* 9, 282–284.

Singh, N.V., Sharma, J., Chandra, R., Babu, K.D., Shinde, Y.R. *et al.* (2016b) Bio-hardening of *in-vitro* raised plants of Bhagwa pomegranate (*Punica granatum*). *Indian Journal of Agricultural Science* 86, 132–136.

Singh, N.V., Gaikwad, N.N. and Pal, R.K. (2017a) Pomegranate – a potential fruit crop for doubling income and ensuring livelihood security of farmers in natural resource deficit regions. In: Saroj, P.L., Sharma, B.D. and Jatav, M.K. (eds) *Training Manual*. ICAR sponsored Winter School on Doubling Income through Advance Approaches for Fruit and Vegetables in the Arid Region at ICAR-CAIH, Bikaner, India, pp. 418–426.

Singh, N.V., Chandra, R., Awachare, C.M., Babu, K.D. and Pal, R.K. (2017b) A novel method of propagation in pomegranate: mound layering. *Progressive Horticulture* 49(1), 92–94. DOI: 10.5958/2249-5258.2017.00020.3.

Singh, N.V., Babu, K.D., Chandra, R., Sharma, J., Sahu, P. *et al.* (2017c) Quality planting material production in pomegranate. In: Pal, R.K. and Singh, N.V. (eds) *Pomegranate for Nutrition, Livelihood Security and Entrepreneurship Development*. Daya Publishing House, New Delhi, India, pp. 69–79.

Singh, N.V. (2017b) Horticultural practices and propagation methods and techniques for climate resilient pomegranate cultivation. In: Wakchaure, G.C., Gaikwad, B.B., Meena, K.K., Singh, Y., Nangare, D.D., *et al.* (eds) *Technical Compendium of Model Training Course on Climate Smart Agriculture for Enhancing Crop and Water Productivity under Abiotic Stress Conditions*. ICAR – National Institute of Abiotic Stress Management, Baramati, India, pp. 46–53.

Singh, S.K. and Khawale, R.N. (2003) Plantlet regeneration from the nodal segments of pomegranate (*Punica granatum* L.) cv. Jyoti. In: Kumar, A., Roy, S. and Sopory, S.K. (eds) *Plant Biotechnology and its Applications in Tissue Culture*. I.K. International Pvt. Ltd, New Delhi, pp. 105–113.

Singh, P. and Patel, R.M. (2016) Factors affecting *in vitro* degree of browning and culture establishment of pomegranate. *African Journal of Plant Science* 10, 43–49.

Tayade, S.A., Joshi, P.S., Raut, H.S. and Shete, M.M. (2017) Effect of time and air layer shoot on rooting and survival of air layers in pomegranate cv. *Bhagwa. International Journal of Minor Fruits, Medicinal and Aromatic Plants* 3, 20–24.

Teixeira da Silva, J.A., Rana, T.S., Narzary, D., Verma, N., Meshram, D.T. *et al.* (2013) Pomegranate biology and biotechnology: a review. *Scientia Horticulturae* 160, 85–107. DOI: 10.1016/j.scienta.2013.05.017.

Tomar, K.S. (2011) Effect of different concentrations of growth regulators on rooting and survival percentage of pomegranate air layers. *Progressive Agriculture* 11, 431–433.

Torres, K. (1988) *Tissue Culture Techniques for Horticultural Crops*. Nostrand Reinhold Co., New York.

Tripathi, S.N. and Shukla, H.S. (2004) Propagation of pomegranate (*Punica granatum* L.) cultivars by stem cuttings with indole butyric acid and p-hydroxybenzoic acid. *Indian Journal of Horticulture* 61, 362–365.

Valizadehkaji, B., Ershadi, A. and Tohidfar, M. (2013) *In vitro* propagation of two Iranian commercial pomegranates (*Punica granatum* L.) cvs. 'Malas Saveh' and 'Yusef Khani'. *Physiology and Molecular Biology of Plants* 19(4), 597–603. DOI: 10.1007/s12298-013-0193-3.

Vazifeshenas, M., Khayyat, M., Jamalian, S. and Samadzadeh, A. (2009) Effects of different scion-rootstock combinations on vigor, tree size, yield and fruit quality of three Iranian cultivars of pomegranate. *Fruits* 64(6), 343–349. DOI: 10.1051/fruits/2009030.

Wainwright, H. and Scrace, J. (1989) Influence of *in vitro* preconditioning with carbohydrates during the rooting of microcuttings on *in vivo* establishment. *Scientia Horticulturae* 38(3–4), 261–267. DOI: 10.1016/0304-4238(89)90073-3.

Wang, Q., Tang, H., Quan, Y., Tang, Y.Q. and Zhou, G. (1994) Phenol induced browning and establishment of shoot-tip explants of 'Fuji' apple and 'Jinhua' pear cultured *in vitro*. *Journal of Horticultural Science* 69(5), 833–839. DOI: 10.1080/14620316.1994.11516519.

Weatherhead, M.A., Burdon, J. and Henshaw, G.G. (1978) Some effects of activated charcoal as an additive to plant tissue culture media. *Zeitschrift für Pflanzenphysiologie* 89(2), 141–147. DOI: 10.1016/S0044-328X(78)80054-3.

Yang, Q.G., Chen, X.J., Zhang, L.F., Gou, Z.L. and Zhang, Q.X. (1991) Micropropogation and transplantation of the valuable and rare pomegranate cultivar Ruanzi. *Plant Physiology Communications* 1, 14–16.

Yesiloglu, T., Gübbük, H., Polat, E. and Erkan, M. (1997) The effects of girdling and scoring of cuttings on the rooting rate and quality of nursery plants of pomegranate. *Acta Horticulturae* 441,407–410. DOI: 10.17660/ActaHortic.1997.441.62.

Yang, Z.H. and Ludders, P. (1993) Organogenesis of *Punica granatum* L. var. *Nana. Angewandte Botanik* 67, 151–156.

7 Environmental Requirements and Site Selection

C. Yilmaz[1]*, M. Rezaei[2] and A. Sarkhosh[3]
[1]Department of Horticulture, Eskisehir Osmangazi University, Eskisehir, Turkey; [2]Horticultural Sciences Department, Shahrood University of Technology, Shahrood, Iran; [3]Horticultural Sciences Department, University of Florida, Gainesville, Florida, USA

7.1 Introduction

The pomegranate (*Punica granatum* L.) is grown over a very wide geographical area, from desert areas to very humid regions, from sea level to high plateaus in the subtropics and tropical climate zone between 41 northern and 41 southern latitudes (Hodgson, 1917; Levin, 2006b). It is native to Iran and grown extensively in arid and semi-arid regions worldwide (Sarkhosh *et al.*, 2006). Wild forms of pomegranate are found in the Near-East Transcaucasia, Dagestan, central Asia (Kopet-Dag, Pamiro-Alaj), and also in Asia Minor, Iran, Afghanistan and India (Ozguven *et al.*, 2012). *Punica protopunica* Balf. is the only other species of *Punica*, an endemic wild species in Socotra (Yemen), which has an arid tropical climate (Brown and Mies, 2012). Pomegranate is generally known as a subtropical climate fruit. It is a partially or completely deciduous species in subtropical areas and in tropical regions, it is evergreen. It can be also locally grown in special microclimate regions in temperate climate zones and it grows well in arid and semi-arid climates. Favourable growth takes place where winters are cool and summers are hot. It can withstand frosty conditions, but below −12°C the hardiness is poor (Levin, 2006a). A day temperature of up to 38°C and dry climate during fruit development produce the best quality fruits (Ikinci *et al.*, 2014). Areas with high relative humidity or rain are unsuitable for its cultivation as fruits produced under such conditions tend to taste less sweet (Kumar, 1990; Ozguven *et al.*, 2012). Pomegranate, whose production is growing rapidly throughout the world in recent years, is now mainly produced in India, Iran, China, Turkey, the USA, Azerbaijan, Israel, Afghanistan, Pakistan, Tunisia, Egypt, Spain and Syria, and it is also grown in some other countries (South Africa, Cyprus, Italy, France, Lebanon, Saudi Arabia, Yemen, Oman, Armenia, Georgia, Kazakhstan, Turkmenistan, Tajikistan, Kirghistan, Bangladesh, Myanmar, Vietnam, Thailand, Chile, Mexico, Argentina, Brazil and Australia) at minor production levels (Holland *et al.*, 2009).

7.2 Temperature Requirements

Pomegranate is a subtropical climate fruit species that can be grown in the tropics, some temperate zones and microclimate areas. The

*Corresponding author: cenapyilmaz@yahoo.com

tree is mainly deciduous in the subtropical climate zone but some cultivars grow evergreen in the tropics, especially in the Deccan Plateau in India. The pomegranate has the ability to adapt to a wide range of temperature regimes and the most suitable climate for pomegranate is the Mediterranean regions of Asia, Africa, America and Europe (Kumar, 1990; Holland *et al.*, 2009). In Table 7.1 the temperature and precipitation levels of some of the main pomegranate production areas in the world are presented. The highest yield and fruit quality are obtained in the climate type typified by hot, dry summers and mild, rainy winters with minimum temperatures not lower than −12°C (Levin, 2006b). Pomegranate is not a frost-resistant species even it is grown in some microclimate regions within the temperate zone. Pomegranate cultivation is restricted mainly by winter freezing but also by spring late frosts and autumn early freeze. Melgarejo *et al.* (1997) reported that pomegranate trees can be grown in areas where the winter temperatures drop to −15°C and certain acidic and central Asian cultivars even survive temperatures of −25°C or −30°C (Aleksandrov, 1950). However, central Asian pomegranate cultivations overcome the lower temperatures (−25°C, −30°C) by covering (with soil or other material) the trees in winter (Levin, 2006b; Yuan and Zhao, 2019).

In cold areas, pomegranate may be damaged by late spring frosts. The most critical of spring frosts for pomegranates is when buds are awakened and new leaflets are formed, the plant can be damaged by −1.5 to −2°C. The damage of late spring frosts is dependent on how late it occurs and the severity of frost level. In subtropical conditions and temperate microclimate areas, usually fresh shoots are damaged primarily. Pomegranate flowers are not damaged because of late flowering (Levin, 2006a).

In autumn, pomegranate trees can be damaged as a result of autumn early frosts before the fall of leaves. In this condition, temperatures of −2 to −3°C are harmful (Levin, 2006a). This can result in completely dried out leaves and death of branches. If fruits are not harvested, fruits are also damaged by cracking. The damage of early frost in autumn depends on how early it occurs and the severity of the frost. Autumn cold, especially if it is in the early morning and with rainfall, is considered to be the most hazardous cold for pomegranate fruit. In this case, most pomegranate fruits are heavily damaged by cracking, and if the intensity of the cold leads to freezing of the fruit, the internal texture of the flesh becomes frozen, and the fruit becomes unusable.

7.2.1 Winter cold hardiness

Ghasemi Soloklui *et al.* (2012), in their research, showed that acclimation and deacclimation did not occur simultaneously in different pomegranate cultivars and this phenomenon had a crucial role in cold tolerance of cultivars, especially in autumn and late winter. 'Post Sefid Bafgh' cultivar showed high cold tolerance early in autumn, but it was susceptible to cold during winter. 'Naderi', 'Yusef Khani', 'Malas Saveh' and 'Robab Neyriz' had the highest mid-winter cold hardiness; 'Mahabadi' showed intermediate hardiness, whereas 'Post Sefid Bafgh' and 'Shishe Kap' were found to be cold-susceptible in this period. The minimum temperature that pomegranate can tolerate depends mainly on the degree of dormancy winter hardiness. Levin (1995) differentiated between winters with respect to the pomegranate cold-hardiness after a long time working in the biggest pomegranate gene bank at Garrygala, Turkmenistan as follows:

1. Mild winters: air temperature lowers gradually, frosts are no lower than −10 to −11°C, and the pomegranate does not suffer from the frosts.
2. Mildly cold winters: frosts of −12 to −15°C, insignificant (considerable in some cultivars) damage.
3. Cold winters: air temperatures reach −15 to −17°C, some cultivars are destroyed on the ground surface, and others are considerably damaged.
4. Severe winters: frosts reach down to −17°C and lower, and practically all cultivars have the aboveground part destroyed.

Sepahvand *et al.* (2011) reported that the Karaj region of Iran is located in a temperate area, where, usually, winter temperatures fall to sub-zero temperatures. Winter freezing killed all of the pomegranate trees (10 cultivars) three

Table 7.1. The mean values for minimum and maximum temperatures and precipitation of the main pomegranate-growing regions in the world.

		Jan	Feb	Mar	Apr	May	Jun	Jul	Aug	Sept	Oct	Nov	Dec	Mean (temp.)/total(Rain)
Solapur, India	Min. temp.	5.7	17.4	21	24.4	25.5	23.6	22.4	22	21.6	20.7	17.2	14.9	19.7
	Max. temp.	30.8	33.7	37.2	39.3	40.4	35.1	31.5	31.3	31.5	32.3	31	30	33.7
	Rainfall (mm)	1	3	5	13	30	110	125	120	194	82	24	6	71.3
Antalya, Turkey	Min. temp.	5.8	6.4	7.6	10.7	14.8	18.9	22.3	21.9	18.8	14.5	10.5	7.4	13.3
	Max. temp.	14.8	15.4	17.8	21.2	25.9	30.2	34.1	33.5	30.8	26.6	21.2	16.5	24
	Rainfall (mm)	212	159	103	48	29	8	3	3	11	66	126	241	100.9
Shiraz, Iran	Min. temp.	–0.3	1	3.7	7	12.7	15.9	19.3	17.9	13.9	8.8	3.1	1.1	8.7
	Max. temp.	12.4	14.5	18.9	21.9	30.2	34.9	37	35.9	33.1	27.3	19.9	14.6	25.1
	Rainfall (mm)	85	55	51	27	5	1	1	0	0	8	29	54	31.6
Jinan, China	Min. temp.	–5.3	–3	2.6	9.8	16.1	20.9	23.1	22	16.8	10.5	3.4	–3	9.5
	Max. temp.	3.4	6.2	13.3	21.5	27.8	32.2	32.4	31	27.1	21.4	13	5.6	19.6
	Rainfall (mm)	6	8	12	25	35	73	188	159	54	28	17	8	61.3
Kandahar, Afghanistan	Min. temp.	–0.7	1.5	6.4	12.2	16.1	20.1	22.8	20	13.6	8.2	2.5	0	10.2
	Max. temp.	12.2	14.4	21.7	28.1	34.5	39.9	40.8	39	34.4	28.3	21	15.3	27.5
	Rainfall (mm)	62	43	29	11	4	0	2	0	0	1	3	21	17.6
Tulare, California, USA	Min. temp.	2.4	4.4	6.2	8.3	11.6	15	17.5	16.8	14.2	9.9	5.2	2.1	9.5
	Max. temp.	12.6	16.7	20	24.1	28.9	33.3	36.3	35.4	32.2	26.9	18.8	12.5	24.8
	Rainfall (mm)	45	41	38	24	7	2	0	0	6	11	29	34	23.7

times during a 15-year period up to the soil level, but in some years with moderate winters, some of these cultivars have shown better tolerance. They reported that after a −9.8°C minimum temperature in winter, moderate freezing injuries were found in most of the cultivars and the lowest level of injury was in 'Shirine Saveh' while the highest was in 'Poost Siahe Shirin'. But in the winter of the next year, all of the 10 cultivars were seriously damaged and dried up to soil level after a −16.4°C minimum temperature. Singh *et al.* (2011b) found that in an arid tropical environment the upper growing portions of pomegranate trees were damaged by low temperature/freezing during the winter season when the temperature only fell below 0°C. Fruits on the trees became hard, and arils went black. However, the plants recovered after employing standard cultural practices and irrigation.

7.2.2 Chilling requirement

Pomegranate orchard establishment is expensive, time-consuming and requires much care. This requires special attention of scientists and growers directed towards understanding climatic elements in a given location in order to establish a successful orchard. It appears that cultivar placement in a locality must be based on the chilling and heat requirement of the cultivars. Accordingly, cultivars should be selected carefully for the particular climatic condition in order to to meet their chilling demands. The chilling requirement of pomegranate is very low, and most pomegranate trees do not need winter chilling, since pomegranates are also grown in semitropical areas that do not have cold winter temperatures. It has been suggested that some pomegranate cultivars require 100 to 150 h to fulfil their chilling requirement (Ashton, 2006). Westwood (1999) suggested 100–300 h of chilling were required for pomegranate. It has been reported that the chilling requirement of some pomegranate cultivars in Iran ranges from 233–633 h until buds break. The higher chilling requirement (450–633 h) of some cultivars in this experiment highlighted the risk of growing some of the Iranian pomegranate cultivars ('Poost Nazok Torosh Abarkuh','Malas Yazdi', 'Rabab Poost Ghermez Neyriz' and 'Makhmal Malas Shahreza') in warmer climates (Day and Wilkins, 2011; Soloklui *et al.*, 2017). Chilling requirement showed a moderate correlation with stomatal density, seed hardness and regional wind speed (Soloklui *et al.*, 2017). Biao (2007) assessed the chilling requirement and growth and development of a pomegranate cultivar, 'Mudanhua', in a solar greenhouse. 'Mudanhua' only needed to pass through 17 days in a 0–7.2°C low-temperature treatment to break dormancy, and it grows and develops normally in greenhouse conditions after 408 h of low temperature. Cui *et al.* (2009) reported a 431-h chilling requirement at 0–7.2°C for two varieties of pomegranate in China. Deng *et al.* (2019) reported that variation of the chilling period in Florida results in the yield alteration of some pomegranate varieties. Inadequately low temperatures to meet the chilling requirements for pomegranate in tropical climates forced the growers to forcibly defoliate trees after harvest or apply synthetic abiotic stresses to artificially induce dormancy (Edwards, 1987; Griesbach, 2007). If this is practised, trees appear to be able to resume their annual cycle without requiring chill. Luedeling *et al.* (2009) reported that pomegranate cultivation was carried out at high oases in spite of the risk of insufficient chilling in Oman.

7.2.3 Flowering in tropical climates

Several pomegranate cultivars have adapted to both tropical and subtropical climates. They are evergreen, semi-deciduous or deciduous. The flowering habit, fruiting and flower physiology alter depending on the habitat. In the tropical climate of south India with its mild winters, growth and flowering are continuous processes, while in the subtropical climate of north India, the trees remain dormant during the winter and flowering occurs in the following spring, while in temperate climates, flowering occurs later and even in summer. In India, 'Ganesh', 'Jaloreseedless', 'Dholka', 'P-13', 'P-16', 'P-23', 'P-26', 'G-137', 'Musket', 'Kandhari', 'Kabul' and 'Mridula' pomegranates have evergreen natures. In the middle and south of India where the climate is tropical, pomegranates flower throughout the year producing small fruits,

Table 7.2. Flower regulation (Bahar treatment) of pomegranate in tropical conditions in India.

Flowering season	Flowering time	Harvest time	Fruit colour quality	Yield
Ambe Bahar	January–February	June–September	Excellent	Good
Mrig Bahar	June–July	December–February	Poor	Good
Hasta Bahar	September–October	March–April	Good	Low

which may not be desirable commercially; hence, flowering needs to be concentrated in a shorter period so as to enable a prolific harvest at a given time. To avoid this problem, trees are given Bahar treatment. Looking at patterns of precipitation, flowering can be induced during June–July (Mrig Bahar), September–October (Hasta Bahar) and January–February (Ambe Bahar) (Table 7.2). In this treatment, the plants are given a rest period by withholding irrigation. Generally, a 45–60-day stress period is found to be sufficient to induce flowering. The rest period facilitates the shedding of leaves. Trees can also be forced to defoliate with chemical applications (Ethrel®, Curacron or thiourea). At the end of the rest period, trees are pruned. The recommended doses of fertilizers are applied immediately after pruning and irrigation is resumed. This leads to profuse flowering and fruiting. The fruits are ready for harvest 4–5 months after flowering. The time of flowering treatment (Bahar treatment) is regulated taking into consideration the availability of irrigation water, market demand and pest/disease incidence in a given locality (Chandra and Meshram, 2010; Hiwale, 2015; Lal *et al.*, 2017).

The fruits of Ambe Bahar are ready for harvest from June to September. As the fruit development takes place during dry months, they develop an attractive colour and quality, and are thus suitable for export. Similarly, due to dry weather, the incidences of pest and disease attacks are limited. However, Ambe Bahar can only be carried out in areas with assured irrigation facilities. The Mrig Bahar crop is harvested from December to February. Usually, this condition is favoured as the flowering and fruiting period coincides with rainy season or immediately after rains, and the crop is taken without much irrigation. As the fruits develop during the rainy season and mature during winter, the colour and sweetness of the fruit are affected. The fruits from Hasta Bahar are harvested from March to April. They have attractive rinds with dark-coloured arils. Since the availability of the fruits during this season is limited, they fetch high value. Optimum water stress cannot be developed during this period as withholding of irrigation coincides with the rainy season. This leads to poor flowering and thus affects the yield (Chandra and Meshram, 2010; Hiwale, 2015; Lal *et al.*, 2017).

7.2.4 Heat requirement

Pomegranate is a plant that loves high temperatures and the highest quality fruits are obtained in arid regions with long, warm, dry summers without rainy days. Ikinci *et al.* (2014) showed that the best-quality pomegranate fruits were obtained in temperatures of 38°C and with a dry climate during fruit development; however, the accumulated total heat requirement together with the high temperature is also important for fruit quality.

The heat requirements were calculated as the growing degree hours (GDH) and growing degree days (GDD). GDH is mostly used for bud flowering heat requirements from breaking of dormancy to 50% of flower opening, but it is also used to calculate accumulated heat from breaking of dormancy to the leaf fall date in some experimental orchards. GDD is the sum of temperatures that fruit needs to grow in a region to reach the expected quality level. GDD is determined by subtracted threshold temperature from the mean daily temperature (Jackson, 1999), daily and based on the number of days between each phenological phase, the sum of effective temperature (°C-day) 'day-degree' is calculated. Temperatures below the threshold temperature have not been taken into account. The growth temperature threshold for pomegranate (mean daily temperature) is considered to be 10°C (Melgarejo *et al.*, 1997; Samani, 2014). GDH can also be measured following the models proposed by Richardson *et al.* (1975) and Anderson *et al.* (1986). Richardson *et al.* (1975) defined one GDH as 1 h at a temperature 1°C above the base temperature of 4.5°C. GDH is calculated by subtracting 4.5°C from each hourly temperature between 4.5 and 25°C. All temperatures above 25°C are assumed equal to 25°C. Anderson *et al.* (1986) proposed a growing degree model using an asymmetric curvilinear model with GDH accumulation between 4 and 25°C (base and optimum).

Ikinci *et al.* (2014) found that the effective heat summation requirement from bud swelling to 50% flowering stage for 'Suruc', 'Katirbasi' and 'Hicaznar' pomegranates was 643, 655 and 718 GDD or 25,000, 25,270 and 28,000°C GDH, respectively, and from bud swelling to harvest the values are 2734, 2802 and 3289 GDD or 73,670, 74,105 and 88,052°C GDH, respectively, in Sanliurfa province, south-eastern Turkey. They concluded that the effective heat summation of Sanliurfa province is enough for commercial growing of all pomegranate cultivars. Soloklui *et al.* (2017) reported that the heat requirement of flowering buds of 20 pomegranate cultivars in Yazd, Iran on the basis of GDH ranged from 4096 to 7928°C with a base temperature of 10°C. Meshram *et al.* (2016) found that total GDD accumulations of 10 pomegranate cultivars grown in Maharashtra, India ranged from 2948 to 4105°C from defoliation to harvest time.

The vegetative budbreak time of pomegranates in Shirvan region (Azerbaijan) is in the first half of April. The growing duration is nearly the same in different cultivars and is up to 230–235 days. According to total required temperatures for completion of all phenomenal phases, the studied cultivars were close to each other, needing 5040 to 5140°C, and formation up to the maturation of the fruits needed 1475–1560°C (Mammadov, 2015). Cao *et al.* (2015) reported that the accumulated temperature in pomegranate-growing regions in China is between 4133 and 6532°C. A study at the National Research Centre on Pomegranate (Solapur, India) determined the total GDD, photo-thermal index and heat use efficiency for 12 pomegranate varieties during 2015–2016 in Mrig Bahar. Total GDD accumulations of all the varieties ranged from 2390.30 to 3575.10°C from defoliation to the harvesting period. The GDD ranged from 932.80–1753.20°C at flowering stage and 284.90–903.50°C at the reproductive stage. The lowest and highest GDD from defoliation to harvesting period build-up was 2390.20°C for 'IC-318707' and 3575.10°C for 'Bhagwa' . Photo-thermal index (PTE) and heat use efficiency (HUE) of 10 varieties ranged from 17.20–19.9°C /day and 0.70–7.9 metric tonnes (t)/ha/degree at flowering and reproductive stages (Anonymous, 2016). A study under the Jodhpur climate revealed that 'Jalor Seedless' required an accumulation of 41 degree days, whereas 'Jodhpur Red' required only 33 degree days per mm increase in fruit size after fruit setting indicating that growth rate of fruit size is faster in 'Jodhpur Red' in comparison with 'Jalor Seedless' cultivar. It has also been found that an accumulation of 13 degree days and 164 degree days, respectively, is required to increase the fruit weight by 1 g and total soluble solids (TSS) by 1 degree Brix, after fruit setting in 'Jalor seedless' cultivar (Rao and Singh, 1998). It appears that the heat requirement of pomegranate cultivars is affected by climate, altitude, cultivar, age of the tree, models used for calculation and the year in which the experiment was conducted.

Pomegranate is tolerant to high temperatures and can withstand dry hot winds at 46–48°C; however, in some cases, high temperature may cause damage. Lye (2010) reported that Alice Springs and northwards in Australia is just too hot for pomegranate, as the fruit boils on the tree, and colour is also poor in these regions. Onur (1988) reported that the growing season for the vegetative period of pomegranate in Mediterranean climate conditions is 180–215 days and for the fruit development period is 120–160 days. Fruit ripening time varies between 80–90 days and 180–200 days according to cultivar and climate condition. Fruits of some sweet cultivars mature 30–50 days before the fruit of sour cultivars (Yılmaz, 2007). Fawole and Opara (2013) indicated that the 'Ruby' pomegranate reached maturity between 132 and 139 days after full bloom (DAFB) and 'Bhagwa' between 140 and 165 DAFB in Western Cape, South Africa. Babu *et al.* (2017) reported the period of fruit set to maturity is 180 DAFB for 'Bhagwa', 175 DAFB for 'Ruby', 150 DAFB for 'Ganesh' and 145 DAFB for 'Jalore Seedless' in Solapur ecological conditions with Mrig Bahar treatment. The period for fruit set to maturity takes 120–155 days according to cultivar during Ambe Bahar in New Delhi (Meena *et al.*, 2011). Čizmović *et al.* (2014) reported that the duration of the growing season in studied cultivars ranged from 241 days in 2003 on the site of Dobra Voda to 275 days in 2004 on the site of Tomba in Montenegro.

The difference in temperature between night and day enhances the colouring of the rind and arils. Higher temperatures during maturation retard colouration of fruit peel and arils. Manera *et al.* (2012) found that the colourimetric values of pomegranate rind were highly correlated with the air temperature during fruit development and ripening. The higher correlation coefficients (0.9) indicated a significant effect of air temperature on rind colour development in pomegranates. Borochov-Neori *et al.* (2011) have also shown that there is an inverse relationship between anthocyanin accumulation and the high season temperatures. Cyanidins were generally more abundant, but delphinidin accumulation was enhanced in cooler seasons. Shulman *et al.* (1984) and Schwartz *et al.* (2009) have shown that pomegranate fruits grown in Mediterranean coastal areas have higher anthocyanin content than those grown in desert climate conditions. Borochov-Neori *et al.* (2009) found that the red colour intensity of the arils was inversely related ($R^2=0.89–0.94$) to the sum of heat units accumulated during fruit ripening, and Fawole and Opara (2013) found that the low heat condition is known to be an optimum factor for the biosynthesis of red anthocyanin compounds in pomegranate fruit. Chater *et al.* (2018) indicate that there were significant site and cultivar effects on many traits as well as site–cultivar interactions. In California, the coastal trial grew significantly faster than the semi-arid inland site. However, the inland site was more productive than the coastal site for the first 3 years, although production during year four of establishment was similar at both sites. Franck (2012) reported that many pomegranate plantations in Chile have been placed in zones with suboptimal climatic conditions, such as high autumn rain probability (regions south of Valparaíso) that may cause a high incidence of fruit cracking (especially 'Wonderful' cultivar). Also the maturation period of plantations near the coast, with milder summer and winter temperatures and frequent foggy days in spring, tends to spread in time, resulting in a a significant proportion of fruit that does not fulfil its ripening process or develops very poor skin colour.

7.3 Light

Pomegranate is a light-loving plant. To obtain high yields and high-quality pomegranate fruit, a long, hot, dry and extremely sunny summer period after fruit set is required. Excessive shading adversely affects the development of the tree. However, direct sunlight on the fruit in hot places is harmful as it causes overheating and sunburn (Yılmaz, 2007). Obviously, light intensity affects photosynthesis and therefore the tree's vegetative growth, flower production, and the size, colour and composition of the fruit. These factors in turn determine the quantity and quality of production. Mditshwa *et al.* (2013) observed an association between light intensity and vitamin C content. Fruit grown in locations with high light intensity had higher vitamin C content and lower phenolic content than those of fruit from low-altitude locations and low light intensity. Pomegranates require at least 6 h of direct sunlight a day in order to ensure good fruit colour and productivity. When selecting a pomegranate orchard location, choosing the south direction for the orchard will increase the probability of receiving a better amount of radiation. The tree spacing of pomegranate is very important. Narrow tree spacing causes crowding of the orchard. After several years trees shade each other and photosynthesis declines, leading to a reduction in tree growth and development. Insufficient light also affects flower induction and differentiation, flowering, fruit formation and fruit quality. Therefore, tree rows should be preferably located in the south—north direction for more light exposure when the orchard is being installed.

In training and pruning of plants, the aim is to create more leaf area, which is better able to absorb more solar radiation leading to good productivity. The most effective training system appropriate to the ecological conditions must be chosen and summer pruning should be applied. Pomegranates are especially sensitive to sunburn because they are terminal-bearing plants on generally thin branches that bend with the increase in fruit weight as the season progresses. This exposes fruit parts that had developed previously in the shade, and are extremely sensitive to sunburn (Melgarejo *et al.*, 2004). Yazici and Kaynak (2006) observed that sunburn damage occurred on 'Hicaznar' pomegranate when temperatures were higher than 30°C and solar radiation was higher than 610 Wm^{-2}. In this study sunburn damage on pomegranate fruit called 'blacking' occurred on the sun-exposed side of the fruit when the surface temperature reached up to 45–50°C. Therefore, July, August and September (in the northern hemisphere) were determined to be the months with a high risk of sunburn damage. Meanwhile, an effect of air humidity on solar radiation leading to a decrease in fruit surface temperature was determined (Yazici and Kaynak, 2006). Schrader *et al.* (2003) observed that factors such as clouds, wind and precipitation caused rapid fluctuations in fruit surface temperature (FST). For example, the appearance of a few clouds markedly decreased solar radiation, and quickly decreased FST below the threshold temperature required to induce sunburn browning in apple. Yazici and Kaynak (2006) reported that sunburn browning in pomegranate was induced when the FST reached 35 ± 1°C, sunburn browning with black necrosis occurred when the FST reached 40 ± 1°C and blacking was induced when the FST reached 45 ± 1°C. Photoselective netting, kaolin application, evaporative cooling and bagging of the fruits should be used to prevent sunburn damage in pomegranate (Shahak *et al.*, 2004; Yazici and Kaynak, 2006; Shlomo, 2015; Meena *et al.*, 2016). Yazici and Kaynak (2006) found that shading treatments decreased sunburn damage on pomegranate fruits compared with controls. However, fruit cracking increased when the humidity was high. In addition, shading plus kaolin application was more effective for decreasing sunburn compared with shading treatment alone. Sunburn damage of fruits was reduced from 21.9% in the untreated control to 9.4% in the Surround® WP-treated fruits (Melgarejo *et al.*, 2004).

The oasis environment confers an advantage for fruit growing in the deserts in Tunisia. Typically, the oasis is based on the date palm grove, an essential element for creating its microclimate that allows the cultivation of other fruit trees (pomegranate, fig, citrus, etc.). Pomegranate trees were intercropped with date palm trees and other fruit species. In this case, pomegranate trees were shaded by date palm and other trees under the severe light conditions of the desert. It was found that the oasis pomegranates exhibited better-quality fruit, high total anthocyanin content, high hydrophilic antioxidant activity and high levels of other mineral and biochemical content (Boussaa *et al.*, 2018, 2019).

7.4 Soil

Pomegranate grows on heavy clay soils, clay loess-like loamy soils, chestnut soils, loamy soils, loamy-pebble soils, sandy loam soils, rich with humus, black earth (chernozem) soils, light humus soils with pebble inclusions, yellow soils (zheltozem), podzolclay lime, alluvial soils, red Mediterranean laterite, red laterite, seaside sands, gravel talus, dry rocky hills, alkali soils and lime-rich soils, as well as on limestone formations on arid hills. In Ethiopia, pomegranate grows on red soils (krasnozem) (Levin, 2006b; Yılmaz, 2007). Thus, it has a wide adaptability for soil type. The most suitable soil for pomegranate is fertile, humus-rich, deep, medium structure, with good drainage and especially alluvial soils. The best yield and quality is achieved in such soils. It is known that pomegranates require well-drained soils with good water-holding capacity and a reliable water source for irrigation. In choosing the land, in addition to factors such as soil type, proper soil texture and depth, slope, cultivation method, tree distance, and row direction should be considered.

The soil depth of the land to be cultivated should be at least 50 cm deep (Toledo and Albujer, 2000). The pomegranate is a species sensitive to waterlogging or root hypoxia (Olmo-Vega *et al.*, 2017). Pomegranate trees in heavy soils that hold water should be planted on raised beds (Fig. 7.1) (Dagar *et al.*, 2001a). Pomegranate cultivation will not be a problem unless there is a drainage problem in heavy soils. Fruit colouration will be less in heavy soils. Dagar *et al.* (2001b) reported that pomegranate tree was damaged by 45 days of waterlogging caused by flood water entry into the experimental area during the 1995 monsoon season. If adequate

Fig. 7.1. Raised-bed planting of pomegranate tree in regions with heavy soil or waterlogging problem. (Photo: Cenap Yilmaz.)

drainage provision exists, pomegranate can be cultivated in alkaline soils successfully. Under waterlogging stress, pomegranate cannot produce fruit, and in the next season the tree may die (Singh *et al.*, 1997). Olmo-Vega *et al.* (2017) tested three pomegranate cultivars in a flooding pot experiment in Spain and showed that 'Mollar de Elche' was the most tolerant and 'Valenciana' was the most sensitive cultivar to flooding.

In natural conditions, pomegranate does not grow in salty soils but pomegranate is considered a salt-tolerant species. Pomegranate has a xylopodium – an underground woody stem that is a storage organ – to supply nutrients and act as a buffer for absorption of toxic sodium ions, in particular to restore the aboveground bush if it dies back after a cold winter, and as the plant ages. The xylopodium is formed during the first years of the pomegranate's life (Levin, 2006b; Chandra *et al.*, 2010). The amount of salt in the root zone that pomegranate tree can tolerate is 0.2% and it can be grown with 1.5–2.0 meq/l of salt in irrigation water if the soil drainage is good. The electrical conductivity (EC) value of the soil should not be higher than 3.5 mmho/cm. This value should be at lower levels (1.7 mmho/cm) in high-clay soils. Although the pH limits for pomegranates are between 4.5 and 7.2, the range for best tree growth is pH 5.5–7.2 (Jain and Neetin, 2018). The recommended active lime ratio in the soil is between 12 and 15% (Jain and Neetin, 2018). The organic matter content of the soil should not be lower than 1%. The organic matter content should be checked to make provisions for its improvement. Well-composted farm manure should be used in the orchard. In the first years, 10–15 m^3/ha and the following years 25–30 m^3/ha per year farm manure is generally recommended. Farm manure should be given in the autumn–winter months to cover the whole root area and mixed with the soil. The use of green manure also gives beneficial results by considering the organic matter demand of pomegranate. For this purpose, plants such as sun hemp (*Crotalaria juncea* L.), glyricidia (*Gliricidia sepium*), karanj (*Pongamia pinnata*) and neem (*Azadirachta indica*) are planted in the entire pomegranate orchard and are buried in the soil during flowering periods (Marathe *et al.*, 2017). These plants have benefits in terms of enriching the soil with nitrogen. Organic fertilization should be continued every year or every couple of years in pomegranate orchards, as organic matter is very difficult to accumulate in soils of hot and dry regions. Applications of organics have been shown to be effective in decreasing pH and EC values of the soil (Marathe *et al.*, 2017). Green manuring through sun hemp decreased pH (6.27) and EC (0.38 dS/m) of soil (Marathe *et al.*, 2017). Green manure crops, pulses and vegetables are recommended as intercrops till the main crop attains 4–5 years of age (Jain and Neetin, 2018). Intercropping leads to economic benefits from either both of the component crops or at least one in comparison with a monocrop system. Intercropping improves mobilization of nutrients in the rhizosphere, thus resulting in better growth and yield from component crops (Wasaki *et al.*, 2003).

7.5 Topography

Natural pomegranate forests are mostly found around rivers and lakesides, valley slopes and dried river beds. These locations are very suitable for pomegranate cultivation, but pomegranate groves can be built up to the altitudes of 800–1000 m in plains and plateaus. Especially in regions with cold winters, the build-up of cold air patches in the hollows should be considered. If the terrain is very inclined, a terraced contour system suitable for inclined areas should be established. Franck (2012) reported that the best conditions for growing pomegranates in

Chile are in the interior of the transversal valleys (Andes foothills) of the northern cultivation zones: especially the Atacama and Coquimbo Regions, which therefore constitute 83% of the country's pomegranate production area.

7.6 Water (Quality and Quantity)

Pomegranate is a drought-tolerant plant and tree life can continue for several years, although it is not very efficient under dry conditions. Under irrigation conditions, the tree develops strongly and gives plenty of fruit. In the summer period, if the precipitation is abundant and regular, the trees develop very well and give plenty of fruit, but the fruits tend to soften and have low transport and storage quality. In the Mediterranean climate zone, pomegranate can survive by natural precipitation, but in these conditions fruit yield and quality is extremely low. A soil water shortage leads to less photosynthesis and this may cause the tree's growth to decrease. Pomegranate is a relatively winter-hardy and drought-tolerant plant and can thrive under desert conditions but bears well only under irrigation (Phule, 2002).

Although pomegranate has a long-lasting tolerance to drought, the yield and fruit quality are negatively affected under water stress. In this condition, fruits cannot develop and remain small. Burmistrov (1993) reported that pomegranates reproduce by seed in the wild. Seedlings appear in April. Growth is slow at first, especially in dry conditions, then speeds up somewhat. On dry slopes, a seedling may take 10–12 years to reach a height of 1 m; if irrigated it may reach the same height in 2 years. The water requirement of pomegranate depends on the climatic condition, especially the rainfall regime and temperature of the region where it is grown. Pomegranate needs approximately 1200 mm/year of water. A part of this water requirement is compensated for by winter and spring rainfalls. However, it is necessary to irrigate the trees during the summer period when there is little or no rain. For the best fruit quality, it is necessary to provide the soil with adequate moisture. In fact, pomegranate water requirements are high as a study in Israel showed that water application for pomegranate tree cultivation is around 5000–6000 m^3/ha (Holland *et al.*, 2009). More specifically, water use of control trees under no soil water limitations changed during the growing season from 0.23 to 5 mm/day under class 'A' pan evaporation values of 3.06–9.19 mm/day (Bhantana and Lazarovitch, 2010), although Sulochanamma *et al.* (2005) did not find any significant increase in fruit yield when water was applied at 0.6, 0.8 or 1.0 of pan evaporation. Pomegranate irrigation is mostly done by surface irrigation (pan and furrow) and drip irrigation systems. Some farmers also use the mini sprinkler irrigation system. In order to obtain high quantities of good-quality fruit, the tree should be irrigated regularly with the same amount each time. Adequate soil moisture must be provided, especially during the fruit development period. The number of cracked fruits will be reduced if sufficient soil moisture is provided, especially at the end of summer and autumn when the fruit is maturing. It is possible to irrigate on one side of the tree, but it is preferable to irrigate two sides of the tree because of the root system, which is shallow and weak, in order to enlarge the root system and to ensure the anchoring of the tree.

The level of water requirement depends on environmental factors that drive evaporative demand and transpiration, and also salinity and electrolyte composition in the soil solution, soil type and structure, soil aeration, root penetration, tree hydraulic architecture, tree canopy size and crop load (Meshram *et al.*, 2011). Bhantana and Lazarovitch (2010) estimated k_c values by using a lysimeter for two pomegranates 'Wonderful' and 'SP-2' from 30 days after bud burst (DAB) to 240 DAB with various levels of salinity of irrigation water: 0.8, 1.4, 3.3, 4.8 and 8 dS/m (Table 7.3). There were slight differences in k_c between the two cultivars and when the salinity level increased, the k_c values mostly decreased. Bhantana (2009) observed that fruit yield was reduced from 42,000 to 6000 kg/ha when EC of irrigation water was increased from 0.8 to 7.5 dS/m.

Use of salty water in agriculture is not recommended, but saline and recycled water could be used for irrigating pomegranate orchards. In Israel, several desert orchards in the Negev Highlands and in the southern Arava area are irrigated with saline water (between 2.5 and 4.0 dS/m). Some pomegranate cultivars ('Israeli'

Table 7.3. Crop coefficient (k_c) values of two pomegranate cultivars under various irrigation water salinity conditions at days after bud burst (DAB) (Bhantana and Lazarovitch, 2010)

DAB	EC = 0.8 dS/m		EC = 1.4 dS/m		EC = 3.3 dS/m		EC = 4.8 dS/m		EC = 8 dS/m	
	'Wonderful'	'SP-2'	'Wonderful'	'SP-2'	'Wonderful'	'SP-2'	'Wonderful'	'SP-2'	'Wonderful'	'SP-2'
30	0.16	0.15	0.15	0.15	0.14	0.14	0.12	0.09	0.09	0.09
60	0.19	0.19	0.19	0.19	0.16	0.15	0.13	0.13	0.09	0.09
90	0.49	0.44	0.45	0.41	0.33	0.31	0.23	0.21	0.13	0.13
120	0.64	0.58	0.52	0.53	0.38	0.35	0.25	0.24	0.12	0.11
150	0.53	0.6	0.42	0.54	0.43	0.37	0.29	0.29	0.17	0.17
180	0.39	0.45	0.32	0.44	0.41	0.32	0.3	0.3	0.21	0.17
210	0.22	0.28	0.19	0.27	0.32	0.23	0.25	0.25	0.16	0.12
240	0.2	0.28	0.17	0.23	0.18	0.33	0.28	0.22	0.15	0.15

and 'Turkmen') under these conditions grew to produce normal yields and fruit qualities without apparent damage to the trees in Israel (Holland *et al.*, 2009). Pomegranate can tolerate saline water up to concentration levels of 40 mM NaCl in the water (Naeini *et al.*, 2004; Naeini and Fallahi, 2006). Şafak and Pırlak (2016) reported that better development of pomegranate plants was observed up to 3000 ppm of NaCl solution. The amount of soil moisture interacts indirectly with the susceptibility of plant to diseases and pests and winter hardiness (Meshram *et al.*, 2011; Nasrabadi *et al.*, 2019).

The water requirement of a 5-year-old pomegranate tree varied from 5.62–47.81 l/day in the western part of Maharashtra, India and the seasonal water requirement of pomegranate was 14057.33 l/year/tree (Chandra and Meshram, 2010). Regardless of the amount of water required in pomegranate orchards, one of the important issues is providing soil moisture with irrigation at regular periods. Regular irrigation is essential during the reproductive growth stage as irregular irrigation causes dropping of flowers and small fruits (Patil *et al.*, 2002). A sudden change in soil moisture causes stress, which affects adversely the fruit development and leads to fruit cracking. For further information on pomegranate water requirements, see Chapter 11.

7.7 Environmental Limiting Factors for Pomegranate Production

7.7.1 Rain

The amount and distribution of rainfall is an important factor in growth and development of pomegranate crop. Water is required at different stages of plant growth. Water shortage at the time of early growth, bud differentiation, blossoming, and fruit set and development results in an undesirable effect. However, rain at the time of flowering may wash out pollen grains and greatly reduce the fruit set. It is generally observed that fruits are more juicy when they mature during the rainy season due to high atmospheric humidity. Fruits that mature during the rainy season contain less sugar and more acid than fruits maturing during the dry season. Maintaining the quality of fruits after harvest that have developed under high atmospheric humidity may not be easy. Mditshwa *et al.* (2013) reported that fruit from locations with low rainfall had high TSS and low titratable acidity (TA). Rainfall in autumn can affect the yield as the pomegranate fruit will crack, and this can be a more severe problem than when the crop is affected by a shortage of irrigation water (Fig. 7.2). Areas that receive regular summer rain are probably not suitable as potential production zones, as fruit tends to be soft, with poor transport and storage characteristics. Heavy rains may cause flooding, resulting in substantial fruit damage. However, in monsoon precipitation with a dry season, pomegranate can be grown using some flowering treatments (Singh *et al.*, 2011a). Planting on borders (ridges) can improve drainage and root aeration. Plastic mulching can also be used to improve drainage and soil water potential.

Fig. 7.2. Soil moisture fluctuation or sudden rain before or during harvesting results in pomegranate fruit cracking injury. (Photo: Cenap Yilmaz.)

7.7.2 Humidity

The quality of pomegranate fruit is adversely affected by a humid climate. It has been shown that relative humidity is negatively correlated with fruit surface temperature (Yazici and Kaynak, 2006). Air circulation is essential, especially in the spring during bloom. Flowers may not set or will abort if conditions are too humid. Thus, open areas free from shade and with a

Fig. 7.3. Russeting damage on pomegranate peel due to high humidity and precipitation during fruit development stages. (Photo: Cenap Yilmaz.)

gentle slope to promote natural air drainage are encouraged for pomegranate cultivation. High atmosphere humidity and the dew formed on fruit peel cause russeting on fruit towards maturity (Fig. 7.3). The russeting occurs because of damage to the tissue of the epidermis of the peel by dew in humid conditions. This damage in pomegranate fruit is generally seen in advanced stages of fruit development and towards maturity. On the whole or part of the peel surface of the damaged fruit, a suberin layer is formed. Thus, the fruit cannot become the normal colour and loses its attractiveness completely. This damage also triggers fruit cracking (Yılmaz, 2007).

Special care should be taken in areas with high relative humidity, as fruit quality may be affected and there may be high incidence of fungal diseases. Polat (2013) reported that *Alternaria* infections are more common in pomegranate leaves in areas with high humidity, and Day and Wilkins (2011) found that low relative humidity in the San Joaquin Valley precludes common development of foliar diseases in pomegranates. Bacterial blight (*Xanthomonas axonopodis* pv. punicae) build-up is rapid during the rainy and humid season. High disease severity is observed from July to October in the monsoon region of India. Temperatures between 25 and 35°C coupled with humidity above 50%, rains and wind favour rapid disease development. The disease affects all plant parts, but is most destructive on the fruits (Anonymous, 2013).

7.7.3 Elevation

The best elevation for pomegranate production is up to 1000 m above sea level in subtropical climate zones and 1800 m in tropical climate zones. Most of the pomegranate areas in Afghanistan are located between 700 and 1500 m above sea level and it can be grown in such elevation ranges commercially (Samadi, 2011). In China, the pomegranate-growing area has the lowest altitude of 50 m (Anhui Huaiyuan region), and the highest is up to 1800 m (Cao *et al.*, 2015). In Nepal, both cultivated and wild forms grow on open and dry slopes of warm valleys and outer hills ranging between 700 and 2700 m (Chaudhary, 2000; Lama *et al.*, 2001; Joshi and Joshi, 2001). Pomegranate trees grow native in well-defined forest patches in the northern areas of Pakistan. The plant is commonly found at altitudes of 1000–2000 m in the north-west of Pakistan and wild pomegranates frequently appear, especially in Khyber Pakhtunkhwa (Chitral, Dir and Kurrum regions), Baluchistan, south Waziristan, and some areas of Kashmir and Hazara (Jurenka, 2008; Ali and Begum, 2015).

High altitudes affect pomegranate yield and fruit quality negatively. High altitude increases the frost risk for pomegranate production. The fruit size, aril yield and yield per tree decrease in 'Hicaznar' in cultivation areas above 350 m altitudes in Mediterranean climate. It has been determined that the increase in both peel and aril colour density of 'Hicaznar' happened as the elevation increased (Yaman *et al.*, 2015). '33N26 Çekirdeksiz VI' cultivar has soft seeds, yellow skin and light pink arils when growing in a coastal area in Mersin, Turkey; however, in higher elevation mountainous areas it is red skinned and arilled (Onur, 1982). Mphahlele *et al.* (2016) reported that fruit from a lower altitude (222 m), characterized by the Mediterranean climate, had higher TSS, glucose and fructose contents than fruit from semi-arid (662 m) and subtropical (898 m) climates at commercial harvest. Mphahlele *et al.* (2014) also found that pomegranate fruits grown in an area with lower altitude were associated with higher bioactive compounds at full ripe stage according to principal component analysis.

7.7.4 Salinity

In natural conditions, pomegranate does not grow in salty soils. However, pomegranate is considered a salt-tolerant species. When the salt accumulation in the soil exceeds 0.5%, the pomegranate roots begin to die. This situation affects tree development and efficiency. A negative relationship has been found between the amount of water soluble minerals in the soil (including sulfates and chlorides) and root development. Importable sodium content in soil affects root formation (Yılmaz, 2007). The maximum amount of salts permitted in the soil in the root zone is 0.2%, and the 1.5–2.0 meq/l limit is acceptable if the drainage is good. Pomegranate yields can be reduced at least 10% or more when the amount of salts in the soil exceeds 5 dS/m. It has been found that the pomegranate plant is very tolerant to saline water irrigation up to an EC of 15 dS/m with little foliar salt damage and a slight growth reduction without considering fruit production (Sun *et al.*, 2018).

7.7.5 Wind

Although severe winds hamper the development of pomegranate plants, there is no significant damage at normal winds. However, especially in thorny varieties, the peel of the fruits is damaged by spines because of the wind and the market value is negatively affected (Onur, 1988). Pomegranate trees are especially sensitive to strong winds and extremely windy areas should be avoided when planning a cultivation site. The mechanical effect of wind is especially noted in young plantations in zones with dominant winds from one direction. The trunk is at an angle and the crown is deformed, leaving the tree unbalanced. In places where there is a risk of wind and frost damage, a windbreak should be installed in the direction of the prevailing winds to protect from wind and frost damage. The effect of the wind varies according to its intensity, temperature and humidity. Hodgson (1917) indicated that pomegranates seldom ripen well in California near the Pacific coast or cool regions, usually remaining very sour and tart, as well as poorly coloured. Moreover, the ratio of acid to total solids runs high, and when exposed to cool sea breezes, the plant does not bloom well or set much fruit.

7.7.6 Frost and freezing

Among all the limiting factors of pomegranate cultivation, frosts, especially winter freezing, cause the most losses to pomegranate growers in colder regions. The main producers of pomegranates in the marginal desert regions such as Iran, over a period of about 10 years, encountered extremely low winter temperatures at least once, resulting in general damage to the orchards. In this case, severe winter freezing in 2007 (–21°C for 3 consecutive days in winter) and 2016 (–10°C for 48 h in late autumn) destroyed 36,931 ha and 35,000 ha of pomegranate orchards in Iran, respectively (Ministry of Agriculture of Iran, 2017). The rate of winter freezes has increased in recent decades with global climate change, and it is very difficult to predict areas without freezing. However, considering long-term meteorological data and using cold-tolerance cultivars will greatly help to minimize the losses.

Types of winter injury include blackheart, cambium injury, bark splitting, trunk splitting, shoot death (dieback), leaf bud death and root death. Frost burns of trunks are often seen in open areas of pomegranate-growing zones. The dead bark is separated from the wood and callus is produced at the border of this injured area. Bark splitting of the trunk is more common in young trees (Ghasemi Soloklui *et al.*, 2012; Levin, 2006b). In most cases, farmers are forced to cut the main scaffold trunks from the surface of the soil (Fig. 7.4). In these conditions, cultivars that can grow faster and reach the fruit-bearing stage sooner are useful. Cold hardiness of the same cultivars varies in different winters, which may be related to age, yield status, the physiological state of the plants, the agro-technical level and the time of winter minimum temperatures. However, some genotypes show the same level of frost tolerance in different winters; for example, 'Kazake', a soft-seeded cultivar, and its progenies are not usually frost tolerant (Levin, 1995). Hodgson (1917) reported that the freeze of 1886 in the

Fig. 7.4. The freezing injury of pomegranate trees on 25 November 2016 in Semnan, Iran at −10°C: (a) browning and dying of the main tree trunk from the soil surface and the emergence of the new suckers in the following growing season; (b) injury of main branches of the tree. (Photos: Marzieh Vafadar.)

south of the USA killed all the sweet pomegranate trees to the ground, but did little damage to sour varieties and this greater frost resistance of the sour varieties has been noted several times. Levin (1995) also reported that there are differences between pomegranate varieties in terms of resistance to frosts in winter. He reported that soft-seeded cultivars are less frost hardy and are damaged at −11 to −12°C, while hard-seeded cultivars are damaged at −15 to −16°C and lower. Levin (2006b) reported that pomegranate is not a resistant plant to extreme cold. Dwarf pomegranate varieties can withstand up to −6 to −7°C and soft-seeded varieties up to −12 to −13°C. At temperatures of −14 to −15°C, annual branches of pomegranate trees and some buds are killed, and the main branches are damaged at −16 to −17°C. The sub-surface parts of the plant are damaged at temperatures of −18°C and lower, and the tree is completely killed. Tolerance to frost and winter cold varies between cultivars. At the same time, this tolerance affects the rest period and the swelling of the buds.

The amount of tree damage by freezing depends on several factors such as cultivar, nutritional status, topographic features of the area, tree acclimation stage and irrigation levels (Ghasemi Soloklui *et al.*, 2012; Nasrabadi *et al.*, 2019). Freezing injuries to trees can be increased with occurrence of low temperatures prior to dormancy in the autumn, and in mid-winter during full dormancy trees have more tolerance to freezing injury (Ghasemi Soloklui *et al.*, 2012). Starch content at the end of the growing season seems to be positively related to pomegranate cold hardiness (Nasrabadi *et al.*, 2019). The faster the tree enters the dormancy stage, the more tolerance it has to low temperatures. In the autumn, the usual irrigation of trees is cut off as soon as possible so the trees enter the dormancy stage sooner. In 2016, the pomegranate orchards in Semnan region, Iran, which traditionally stopped irrigation earlier, had earlier leaf abscission, and the loss of trees owing to late autumn freezing was much lower or did not occur (Mehdi Rezaei, personal observation).

Spring frost is rarely a major problem for pomegranate orchards. Flowering of pomegranate trees is later than other fruit trees and their flowering period is longer. However, late spring frost may injure the terminal shoots and flower buds. Low temperatures in the early autumn and during fruit harvesting time can also increase the percentage of fruit cracking, and if the temperature drops, it can completely destroy the fruits on trees. Day and Wilkins (2011) reported that both winter freezes and spring frosts caused damage to pomegranate trees in the San Joaquin Valley in early March 2008, and again in 2009 spring frost events occurred where temperatures dropped to as low as −3°C. Mature trees were unaffected, but some young orchards suffered from shoot dieback and growers were forced to regrow trees from the resulting ground suckers (Fig. 7.5). An unusually hard freeze occurred in December 1990 in California, USA, where the temperature dropped to −6 to −8°C. A number of mature orchards were damaged severely with death of scaffolds, and some young orchards were killed outright (Day and Wilkins, 2011).

Levin (2006a) found that the nature of pomegranate blossoming at the northern border of its cultural habitat varies a lot from year to year and greatly depends on the level of harm caused by winter frosts, spring weather conditions, etc. Slight plant freezing affects the nature of blossoming mainly indirectly, which results in mainly forming functionally male flowers during the second year and demonstrates a late start of the blossoming period.

Fig. 7.5. Frost damage is one of the major problems in pomegranate orchards during late winter or early spring. Types of frost injury are (a) cambium injury, (b) bark splitting and trunk splitting, (c) shoot death (dieback) and leaf bud death. (Photos: Cenap Yilmaz.)

Fig. 7.6. Drought stress after water shortage before and during harvest time can significantly reduce pomegranate fruit quality. (Photo: Cenap Yilmaz.)

7.7.7 Water

The major pomegranate orchards are established in arid and semi-arid regions of the world, which often experience drought. Pomegranate trees are tolerant to soil water deficit and in fact the best quality of crop is obtained from moderately drought-prone areas (Galindo *et al.*, 2017); however, long periods of drought during the last stage of fruit development can cause significant economic damage to the growers (Fig. 7.6). This crop tolerates extreme drought for long periods during which the trees put on very little vegetative growth with no fruit production. However, the trees tend to recover and thrive after being irrigated and given subsequent care, and for commercial pomegranate production, orchards need regular irrigation especially in the dry season and during fruit development (Holland *et al.*, 2009; Intrigliolo *et al.*, 2011; Khattab *et al.*, 2012). In addition to water quantity, the quality of irrigation water can also be a limiting factor in the cultivation of pomegranate. Pomegranate trees are moderately tolerant to salinity and salinity stress, and pomegranate injuries from salt stress are enhanced by increasing irrigation intervals (Karimi and Hasanpour, 2014). It is noteworthy that salinity of irrigation water for pomegranate trees should not exceed 2 mS/cm, due to the fact that pomegranate is moderately sensitive to salinity (salt effects at EC = 3 mS/cm) (Amri *et al.*, 2011).

Pomegranate commercial orchards in arid and semi-arid regions are rarely exposed to prolonged waterlogging, and high soil moisture and waterlogging in pomegranate orchards are rarely considered a serious problem, so little research has been carried out on this issue. However in tropical regions and semi-arid climates with high precipitation in the summer season, such as the south-east of Spain, waterlogging is a limiting factor, especially in heavy soils (Olmo-Vega *et al.*, 2017). Olmo-Vega *et al.* (2017) showed that after only 6 days of waterlogging, the pomegranate tree had more than 30% growth reduction. Long-term flooding in the growing season can destroy pomegranate trees, especially in heavy soils (Fig. 7.7). However, pomegranate forests grow in northern Iran in heavy soils with high soil moisture content. Some pomegranate cultivars have more tolerance than others (Olmo-Vega *et al.*, 2017).

Fig. 7.7. Waterlogging damage to pomegranate tree; a few days can cause foliage wilt and sudden heavy leaf drop follows. (Photo: Cenap Yilmaz.)

7.7.8 Sunburn

Although pomegranates prefer hot and dry weather with sunny days, in severe light conditions and high temperatures, a large percentage of the tree's fruit is damaged by sunburn (Fig. 7.8). Summer temperatures may rise above 45°C in pomegranate-growing regions, resulting in sunburn damage to the fruits, which may inflict losses of up to 40% of the total yield (Melgarejo *et al.*, 2004). Nutritional status, proper irrigation, shading and kaolin treatment can reduce sunburn damage (Melgarejo *et al.*, 2004; Yazici and Kaynak, 2006). For more information on sunburn, see Chapter 12.

Fig. 7.8. Sunburn damage to pomegranate fruit can significantly reduce fruit quality and causes aril discolouration. (Photo: Cenap Yilmaz.)

References

Anonymous (2013) Ahuja, D.B. (ed.) *Pests of Fruits (Banana, Mango, and Pomegranate) 'E' Pest Surveillance and Pest Management Advisory*. National Centre for Integrated Pest Management, New Delhi and State Department of Horticulture, Commissionerate of Agriculture, Pune, India.

Aleksandrov, A.D. (1950) The pomegranate. *Sad I Ogorod* 2, 9–40.

Ali, K. and Begum, H.A. (2015) A comparative assessment of climate change effect on some of the important tree species of Hindu-Kush Himalayas, using predictive modelling techniques. *International Journal of Advanced Research* 3, 1230–1240.

Amri, E., Mirzaei, M., Moradi, M. and Zare, K. (2011) The effects of spermidine and putrescine polyamines on growth of pomegranate (*Punica granatum* L. cv Rabbab) in salinity circumstance. *International Journal of Plant Physiology and Biochemistry* 3, 43–49.

Anderson, J.L., Richardson, E.A. and Kesner, C.D. (1986) Validation of chill unit and flower bud phenology models for 'Montmorency' sour cherry. *Acta Horticulturae* 184,71–78. DOI: 10.17660/ActaHortic.1986.184.7.

Anonymous (2016) *ICAR-NRCP Annual Report 2015–16*. ICAR-National Research Centre on Pomegranate, Solapur, India.

Ashton, R. (2006) *The Incredible Pomegranate: Plant and Fruit*. Third Millennium Publishing, Tempe, Arizona.

Babu, K.D., Singh, N.V., Gaikwad, N., Maity, A., Suryavanshi, S.K. *et al.* (2017) Determination of maturity indices for harvesting of pomegranate (*Punica granatum*). *Indian Journal of Agricultural Sciences* 87, 1225–1230.

Bhantana, P. (2009) Evapotranspiration, growth, yield and quality of pomegranate (*Punica granatum* L.) under combined drought and salinity stress. MSc Thesis. Ben-Gurion University, Negev, Israel.

Bhantana, P. and Lazarovitch, N. (2010) Evapotranspiration, crop coefficient and growth of two young pomegranate (*Punica granatum* L.) varieties under salt stress. *Agricultural Water Management* 97(5), 715–722. DOI: 10.1016/j.agwat.2009.12.016.

Biao, W.A.N.G. (2007) Brief report on the chilling requirement and characteristic of *Punica granatum* L. growth and development in solar greenhouse. *Journal of Anhui Agricultural Sciences* 35(12), 3517–3520.

Borochov-Neori, H., Judeinstein, S., Tripler, E., Harari, M., Greenberg, A. *et al.* (2009) Seasonal and cultivar variations in antioxidant and sensory quality of pomegranate (*Punica granatum* L.) fruit. *Journal of Food Composition and Analysis* 22(3), 189–195. DOI: 10.1016/j.jfca.2008.10.011.

Borochov-Neori, H., Judeinstein, S., Harari, M., Bar-Ya'akov, I., Patil, B.S. *et al.* (2011) Climate effects on anthocyanin accumulation and composition in the pomegranate (*Punica granatum* L.) fruit arils. *Journal of Agricultural and Food Chemistry* 59(10), 5325–5334. DOI: 10.1021/jf2003688.

Boussaa, F., Zaouay, F., Hernandez, F., Noguera-Artiaga, L., Carbonell-Barrachina, Á. *et al.* (2018) Cropping system contributes largely to fruit composition and sensory properties of pomegranate (*Punica granatum* L. var. Gabsi). *South African Journal of Botany* 115, 170–178. DOI: 10.1016/j.sajb.2018.01.016.

Boussaa, F., Zaouay, F., Burlo-Carbonell, F., Nuncio-Jáuregui, N., Gmati, M. *et al.* (2019) Combined effects of cropping system and harvest date determine quality and nutritional value of pomegranate fruits (*Punica granatum* L. cv. Gabsi). *Scientia Horticulturae* 249, 419–431. DOI: 10.1016/j.scienta.2019.02.007.

Brown, G. and Mies, B. (2012) *Vegetation Ecology of Socotra*. Springer Science & Business Media, New York.

Burmistrov, L.A. (1993) *Pomegranate Culture in Central Asia*. 17. WANATCA Yearbook, pp. 3–13.

Cao, S.Y., Li, H., Yuan, P., Tan, H.-H., Zhao, D. *et al.* (2015) An overview of cultivation, scientific research, and industrialization of Chinese pomegranate. *Acta Horticulturae* 1089, 369–373.

Chandra, R. and Meshram, D.T. (2010) Pomegranate culture in Deccan plateau of India. *Fruit, Vegetable and Cereal Science and Biotechnology* 4, 113–119.

Chandra, R., Babu, K.D., Jadhav, V.T. and Silva, J.A.T. (2010) Origin, history, and domestication of pomegranate. *Fruit, Vegetable and Cereal Science and Biotechnology* 4, 1–6.

Chater, J.M., Santiago, L.S., Merhaut, D.J., Jia, Z., Mauk, P.A. *et al.* (2018) Orchard establishment, precocity, and eco-physiological traits of several pomegranate cultivars. *Scientia Horticulturae* 235, 221–227. DOI: 10.1016/j.scienta.2018.02.032.

Chaudhary, R.P. (2000) Forest conservation and environmental management in Nepal: a review. *Biodiversity and Conservation* 9(9), 1235–1260. DOI: 10.1023/A:1008900216876.

Čizmović, M., Popović, R. and Džubur, A. (2014) Phenological characteristics of the major pomegranate (*Punica granatum* L.) cultivars grown in different agro-ecological conditions of Montenegro. *Agriculture and Forestry* 60, 61–66.

Cui, P., Cao, S.Y. and Jiang, J.F. (2009) Preliminary report about chilling requirement determination of jujube, pomegranate and fig. *Nonwood Forest Research* 27(1), 91–93.

Dagar, J.C., Sharma, H.B. and Shukla, Y.K. (2001a) Raised and sunken bed technique for agroforestry on alkali soils of northwest India. *Land Degradation & Development* 12(2), 107–118. DOI: 10.1002/ldr.442.

Dagar, J.C., Singh, G. and Singh, N.T. (2001b) Evaluation of forest and fruit trees used for rehabilitation of semiarid alkali-sodic soils in India. *Arid Land Research and Management* 15(2), 115–133. DOI: 10.1080/15324980151062742.

Day, K. and Wilkins, E. (2011) Commercial pomegranate (*Punica granatum* L.) production in California. *Acta Horticulturae* 890, 275–286.

Deng, Z., Castle, W., Vallad, G.E., Agehara, S., Thetford, M. *et al.* (2019) Pomegranate: an emerging fruit crop in southeast United States? *Acta Horticulturae* 1254,149–156. DOI: 10.17660/ActaHortic.2019.1254.23.

Edwards, G.R. (1987) Producing temperate-zone fruit at low latitudes-avoiding rest and the chilling requirement. *HortScience* 22, 1236–1240.

Fawole, O.A. and Opara, U.L. (2013) Fruit growth dynamics, respiration rate and physico-textural properties during pomegranate development and ripening. *Scientia Horticulturae* 157, 90–98. DOI: 10.1016/j.scienta.2013.04.004.

Franck, N. (2012) The cultivation of pomegranate cv. Wonderful in Chile. In: Melgarejo, P. and Valero, D. (eds) *II International Symposium on the Pomegranate*. Série A. Séminaires Méditerranéens; n. 103. CIHEAM/Universidad Miguel Hernández, Zaragoza, Spain, pp. 97–99.

Galindo, A., Calín-Sánchez, Á.., Griñán, I., Rodríguez, P., Cruz, Z.N. *et al.* (2017) Water stress at the end of the pomegranate fruit ripening stage produces earlier harvest and improves fruit quality. *Scientia Horticulturae* 226, 68–74. DOI: 10.1016/j.scienta.2017.08.029.

Ghasemi Soloklui, A.A., Ershadi, A. and Fallahi, E. (2012) Evaluation of cold hardiness in seven Iranian commercial pomegranate (*Punica granatum* L.) cultivars. *HortScience* 47(12), 1821–1825. DOI: 10.21273/HORTSCI.47.12.1821.

Griesbach, J. (2007) *Growing Temperate Fruit Trees in Kenya*. World Agroforestry Center (ICRAF), Nairobi, Kenya.

Hiwale, S. (2015) *Sustainable Horticulture in Semiarid Dry Lands*. Springer, New Delhi.
Hodgson, R.W. (1917) *The Pomegranate. University of California Publications Bulletin No. 276*. University of California Publications Bulletin No. 276. University of California, Berkeley, California.
Holland, D., Hatib, K. and Bar-Ya'akov, I. (2009) Pomegranate: botany, horticulture, breeding. In: Janick, Jules. (ed.) *Horticultural Reviews*. 35. John Wiley & Sons. Inc, New Jersey, pp. 127–192.
Ikinci, A., Mamay, M., Unlu, L., Bolat, I. and Ercisli, S. (2014) Determination of heat requirements and effective heat summations of some pomegranate cultivars grown in southern Anatolia. *Erwerbs-Obstbau* 56(4), 131–138. DOI: 10.1007/s10341-014-0220-8.
Intrigliolo, D.S., Puerto, H., Bonet, L., Alarcón, J.J., Nicolas, E. *et al.* (2011) Usefulness of trunk diameter variations as continuous water stress indicators of pomegranate (*Punica granatum*) trees. *Agricultural Water Management* 98(9), 1462–1468. DOI: 10.1016/j.agwat.2011.05.001.
Jackson, D. (1999) *Climate and Fruit Plants*. In: Jackson, D. and Earl Looney, N. (eds) *Temperate and Subtropical Fruit Production*, 2nd edn. CAB International, Wallingford, UK.
Jain, K. and Neetin, D. (2018) Pomegranate the cash crop of India: a comprehensive review on agricultural practices and diseases. *International Journal of Health Sciences and Research* 8, 315–336.
Joshi, K.K. and Joshi, S.D. (2001) *Genetic Heritage of Medicinal and Aromatic Plants of Nepal Himalayas*. Buddha Academic Publishers, Kathmandu.
Jurenka, J. (2008) Therapeutic applications of pomegranate (*Punica granatum* L.). A review. *Alternative Medicine Review* 13(2), 128–144.
Karimi, H.R. and Hasanpour, Z. (2014) Effects of salinity and water stress on growth and macro nutrients concentration of pomegranate (*Punica granatum* L.). *Journal of Plant Nutrition* 37(12), 1937–1951. DOI: 10.1080/01904167.2014.920363.
Khattab, M.M., Shaban, A.E., El-Shrief, A.H. and Mohamed, A.E.-D. (2012) Effect of humic acid and amino acid on pomegranate trees under deficit irrigation. I: growth, flowering, and fruiting. *Journal of Horticultural Science & Ornamental Plants* 4, 253–259.
Kumar, G.N.M. (1990) Pomegranate. In: Nagy, S., Shaw, P.E. and Wardowski, W.F. (eds) *Fruits of Tropical and Subtropical Origin*. Florida Science Source, Ocala, pp. 328–347.
Lal, N., Sahu, N., Marboh, E.S., Gupta, A.K. and Patel, R.K. (2017) A review on crop regulation in fruit crops. *International Journal of Current Microbiology and Applied Sciences* 6(7), 4032–4043. DOI: 10.20546/ijcmas.2017.607.418.
Lama, Y.C., Ghimire, S.K. and Aumeeruddy-Thomas, Y. (2001) *Medicinal Plants of Dolpo. Amchis' Knowledge and Conservation*. WWF Nepal Program, Kathmandu.
Levin, G.M. (1995) *Aspects of pomegranate culture in Turkmenistan. Bulletin des Ressources Phytogenetiques (IPGRI/FAO)*. Noticiario de Recursos Fitogeneticos (IPGRI/FAO).
Levin, G.M. (2006a) *Pomegranate*. Third Millennium Publications, Tempe, Arizona.
Levin, G.M. (2006b) *Pomegranate Roads: A Soviet Botanist's Exile from Eden*. Floreant Press, Forestville, California.
Luedeling, E., Gebauer, J. and Buerkert, A. (2009) Climate change effects on winter chill for tree crops with chilling requirements on the Arabian Peninsula. *Climatic Change* 96(1-2), 219–237. DOI: 10.1007/s10584-009-9581-7.
Lye, C. (2010) Pomegranate – a new option for irrigated areas of the Murray-Darling Basin. *IREC Farmers' Newsletter* 183, 27–30.
Mammadov, J.S. (2015) The influence of natural environmental conditions on fruiting of subtropical crops. *Global Journal of Biology, Agriculture & Health Sciences* 4, 69–71.
Manera, F.J., Legua, P., Melgarejo, P., Martínez, R., Martínez, J.J. *et al.* (2012) Effect of air temperature on rind colour development in pomegranates. *Scientia Horticulturae* 134, 245–247. DOI: 10.1016/j.scienta.2011.11.016.
Marathe, R.A., Sharma, J., Murkute, A.A. and Babu, K.D. (2017) Response of nutrient supplementation through organics on growth, yield and quality of pomegranate. *Scientia Horticulturae* 214, 114–121. DOI: 10.1016/j.scienta.2016.11.024.
Mditshwa, A., Fawole, O.A., Al-Said, F., Al-Yahyai, R. and Opara, U.L. (2013) Phytochemical content, antioxidant capacity and physicochemical properties of pomegranate grown in different microclimates in South Africa. *South African Journal of Plant and Soil* 30(2), 81–90. DOI: 10.1080/02571862.2013.802033.
Meena, K., Singh, R., Pareek, S., Kashyap, P., Sheikh, M. *et al.* (2011) Evaluation of pomegranate (*Punica granatum* L.) genotypes for morphological and flowering characteristics under semi-arid climate. *Acta Horticulturae* 890, 233–237.

Meena, V.S., Kashyap, P., Nangare, D.D. and Singh, J. (2016) Effect of coloured shade nets on yield and quality of pomegranate (*Punica granatum*) cv. Mridula in semi-arid region of Punjab. *Indian Journal of Agricultural Sciences* 86(4), 500–505.

Melgarejo, P., Martínez Valero, R., Guillamón, J.M. and Amorós, M.M. (1997) Phenological stages of the pomegranate tree (*Punica granatum* L.). *Annals of Applied Biology* 130(1), 135–140. DOI: 10.1111/j.1744-7348.1997.tb05789.x.

Melgarejo, P., Martínez, J.J., Hernández, F., Martínez-Font, R., Barrows, P. *et al.* (2004) Kaolin treatment to reduce pomegranate sunburn. *Scientia Horticulturae* 100(1-4), 349–353. DOI: 10.1016/j.scienta.2003.09.006.

Meshram, D., Gorantiwar, S., Mittal, H.K. and Purohit, R.C. (2011) Water requirement of pomegranate (*Punica granatum* L.) for Solapur district of Maharashtra State. *Acta Horticulturae* 890, 311–322. DOI: 10.17660/ActaHortic.2011.890.43.

Meshram, D.T., Chandra, R., Singh, N.V. and Pal, R.K. (2016) Thermal requirement of pomegranate varieties growing in Maharashtra. *Indian Journal of Horticulture* 73(3), 327–333. DOI: 10.5958/0974-0112.2016.00072.4.

Ministry of Agriculture of Iran (2017) Agricultural statistics, Executive Committee on Management of Environmental Stresses for Horticultural Products. Available at: https://www.maj.ir (accessed 26 May 2020).

Mphahlele, R.R., Stander, M.A., Fawole, O.A. and Opara, U.L. (2014) Effect of fruit maturity and growing location on the postharvest contents of flavonoids, phenolic acids, vitamin C and antioxidant activity of pomegranate juice (cv. Wonderful). *Scientia Horticulturae* 179, 36–45. DOI: 10.1016/j.scienta.2014.09.007.

Mphahlele, R.R., Caleb, O.J., Fawole, O.A. and Opara, U.L. (2016) Effects of different maturity stages and growing locations on changes in chemical, biochemical and aroma volatile composition of 'Wonderful' pomegranate juice. *Journal of the Science of Food and Agriculture* 96(3), 1002–1009. DOI: 10.1002/jsfa.7186.

Naeini, M.R., Khoshgoftarmanesh, A.H. and Fallahi, E. (2006) Partitioning of chlorine, sodium, and potassium and shoot growth of three pomegranate cultivars under different levels of salinity. *Journal of plant nutrition* 29(10), 1835–1843.

Naeini, M.R., Khoshgoftarmanesh, A.H., Lessani, H. and Fallahi, E. (2004) Effects of sodium chloride-induced salinity on mineral nutrients and soluble sugars in three commercial cultivars of pomegranate. *Journal of Plant Nutrition* 27(8), 1319–1326. DOI: 10.1081/PLN-200025832.

Nasrabadi, M., Ramezanian, A., Eshghi, S., Kamgar-Haghighi, A.A., Vazifeshenas, M.R. *et al.* (2019) Biochemical changes and winter hardiness in pomegranate (*Punica granatum* L.) trees grown under deficit irrigation. *Scientia Horticulturae* 251, 39–47. DOI: 10.1016/j.scienta.2019.03.005.

Olmo-Vega, A., García-Sánchez, F., Simón-Grao, S., Simón, I., Lidón, V. *et al.* (2017) Physiological responses of three pomegranate cultivars under flooded conditions. *Scientia Horticulturae* 224, 171–179. DOI: 10.1016/j.scienta.2017.06.013.

Onur, C. (1982) Pomegranate selection in Mediterranean region. PhD Thesis. Çukurova University, Institute of Applied and Natural Sciences, Adana, Turkey.

Onur, C. (1988) Nar (pomegranate). *Derim* 5(4), 147–190.

Ozguven, A.I., Yilmaz, C. and Keles, D. (2012) Pomegranate biodiversity and horticultural management. *Acta Horticulturae* 940, 21–27.

Patil, A. V., Karale, A.R. and Bose, T.K. (2002) Pomegranate. In: Bose, T.K., Mitra, S.K. and Sanyal, D. (eds) *Fruits: Tropical and Subtropical*. 2. Naya Udyog, Calcutta, India, pp. 125–162.

Phule, B. (2002) Pomegranate cultivation in Solapur district: a Geo economical analysis. Ph.D. Thesis. Department of Geography, Shivaji University, India, Kolhapur.

Polat, P. (2013) Isolation, morphological and molecular characterization of *Alternaria* species affecting some crops grown in Çukurova region. MSc Thesis. Çukurova University, Institute of Applied and Natural Sciences, Department of Plant Protection, Adana, Turkey.

Rao, A.S. and Singh, R.S. (1998) Climatic features and crop production. In: Faroda, A.S. and Singh, M. (eds) *Fifty Years of Arid Zone Research in India*. Central Arid Zone Research Institute, Jodhpur, India, pp. 17–37.

Richardson, E.A., Seeley, S.D., Walker, R.D., Anderson, J. and Aschcroft, G. (1975) Pheno-climatography of spring peach bud development. *HortScience* 10, 236–237.

Şafak, C. and Pırlak, L. (2016) A research on salt tolerance of some pomegranate (*Punica granatum* L.) cultivars. *Selçuk Tarım Bilimleri Dergisi* 3(1), 30–36.

Samadi, G.R. (2011) Status of pomegranate (*Punica granatum* L.) cultivation in Afghanistan. *Acta Horticulturae* 890, 55–60.

Samani, Z. (2014) *Evaluating and Managing Crops Water Requirement. Handbook of Plant and Crop Physiology*. CRC Press, Boca Raton, Florida.

Sarkhosh, A., Zamani, Z., Fatahi, R. and Ebadi, A. (2006) RAPD markers reveal polymorphism among some Iranian pomegranate (*Punica granatum* L.) genotypes. *Scientia Horticulturae* 111(1), 24–29. DOI: 10.1016/j.scienta.2006.07.033.

Schrader, L., Zhang, J. and Sun, J. (2003) Environmental stresses that cause sunburn of apple. *Acta Horticulturae* 618,397–405. DOI: 10.17660/ActaHortic.2003.618.47.

Schwartz, E., Tzulker, R., Glazer, I., Bar-Ya'akov, I., Wiesman, Z. *et al.* (2009) Environmental conditions affect the color, taste, and antioxidant capacity of 11 pomegranate accessions' fruits. *Journal of Agricultural and Food Chemistry* 57(19), 9197–9209. DOI: 10.1021/jf901466c.

Sepahvand, E., Zamani, Z., Askari, M.A. and Khademi, O. (2011) Phenotypic characterization of ten pomegranate cultivars at karaj condition of Iran. *Acta Horticulturae* 890,243–249. DOI: 10.17660/ActaHortic.2011.890.34.

Shahak, Y., Gussakovsky, E.E., Gal, E. and Ganelevin, R. (2004) ColorNets: crop protection and light-quality manipulation in one technology. *Acta Horticulturae* 659,143–151. DOI: 10.17660/ActaHortic.2004.659.17.

Shlomo, M. (2015) Efficiency of bagging pomegranate fruits. *Acta Horticulturae* 1089,485–488. DOI: 10.17660/ActaHortic.2015.1089.66.

Shulman, Y., Fainberstein, L. and Lavee, S. (1984) Pomegranate fruit development and maturation. *Journal of Horticultural Science* 59(2), 265–274. DOI: 10.1080/00221589.1984.11515196.

Singh, G., Dagar, J.C. and Singh, N.T. (1997) Growing fruit trees in highly alkali soils—a case study. *Land Degradation & Development* 8(3), 257–268. DOI: 10.1002/(SICI)1099-145X(199709)8:3<257::AID-LDR259>3.0.CO;2-Q.

Singh, N.P., Dhillon, W.S. and Gill, P.P.S. (2011a) Quality improvement studies in pomegranate under subtropics of India. *Acta Horticulturae* 890, 363–369.

Singh, R., Sharma, B., Bhargava, R. and More, T. (2011b) Introduction and evaluation of anardana type pomegranate under hot arid conditions. *Acta Horticulturae* 890, 239–242.

Soloklui, A.A.G., Gharaghani, A., Oraguzie, N., Eshghi, S. and Vazifeshenas, M. (2017) Chilling and heat requirements of 20 Iranian pomegranate cultivars and their correlations with geographical and climatic parameters, as well as tree and fruit characteristics. *HortScience* 52(4), 560–565. DOI: 10.21273/HORTSCI11614-16.

Sulochanamma, B.N., Yellamanda Reddy, T. and Subbi Reddy, G. (2005) Effect of basin and drip irrigation on growth, yield and water use efficiency in pomegranate cv. Ganesh. *Acta Horticulturae* 696, 277–279.

Sun, Y., Niu, G., Masabni, J.G. and Ganjegunte, G. (2018) Relative salt tolerance of 22 pomegranate (*Punica granatum*) cultivars. *HortScience* 53(10), 1513–1519. DOI: 10.21273/HORTSCI13362-18.

Toledo, J. and Albujer, E. (2000) Project of technical standards for pomegranate integrated production in Valencia. *Options Méditerranéennes. Série A: Séminaires Méditerranéens*, 149–155.

Wasaki, J., Yamamura, T., Shinano, T. and Osaki, M. (2003) Secreted acid phosphatase is expressed in cluster roots of lupin in response to phosphorus deficiency. *Plant and Soil* 248(1/2), 129–136. DOI: 10.1023/A:1022332320384.

Westwood, M.N. (1999) *Temperate Zone Pomology: Physiology and Culture*. Timber Press, Portland, Oregon.

Yaman, S., Toprak, Z. and Bayazit, S. (2015) Determination of fruit quality characteristics of Hicaznar cultivar grown in different elevations. *Fruit Science* 2(2), 9–15.

Yazici, K. and Kaynak, L. (2006) Effects of kaolin and shading treatments on sunburn on fruit of Hicaznar cultivar of pomegranate (*Punica granatum* L. cv. Hicaznar). *Acta Horticulturae* 818, 167–174.

Yuan, Z.H. and Zhao, X. (2019) Pomegranate genetic resources and their utilization in China. *Acta Horticulturae* 1254,49–56. DOI: 10.17660/ActaHortic.2019.1254.8.

Yılmaz, C. (2007) *Pomegranate*. Hasad Publishing, Istanbul.

8 Orchard Establishment and Tree Management

G. Ferrara[1]*, M. Palasciano[1], A. Sarkhosh[2], F. Cossio[3], K.D. Babu[4] and A. Mazzeo[1]

[1]Department of Soil, Plant and Food Sciences, University of Bari 'Aldo Moro', Bari, Italy; [2]Horticultural Sciences Department, University of Florida, Gainesville, Florida, USA; [3]Fruit crops breeder, Private consultant, Verona, Italy; [4]ICAR-National Research Centre on Pomegranate, Kegaon, Solapur, India

8.1 Introduction

The pomegranate is a rustic tree that can survive in hard conditions with almost no care, but for commercial cultivation an adequate supply of fertilizers, water, application of specific phytosanitary treatments, training, winter and summer pruning operations, and some support or protection structures are essential. Before planting an orchard, it is necessary to consider various aspects:

- Characteristics of the chosen site in terms of both climate (winter cold, wind, exposure, length of the vegetative season, temperature, etc.) and soil (texture, fertility, depth, etc.).
- Type of machinery and equipment available, as well as the experience of the workforce (in land management, cultivation operations, plant protection, harvesting methods, etc.).
- Preferred type of cultivation (conventional, integrated, or organic).
- Greater or lesser complexity of the chosen plantation system (as free shape or with more or less sophisticated and expensive support structures).
- Availability of water.
- Size of the orchard.

In the following sections, several aspects of the orchard establishment and the tree management operations will be described.

8.2 Selecting the Orchard Site

Before planting a pomegranate orchard, it is advisable to carefully consider and then act in such a way to prepare all the operations in time: gathering information, choosing the nursery material, preparing the soil and installing any necessary support structures. The cost of both establishment and management is high and significantly influenced by several factors such as the choice of the site, the variety and the training system (simple or 'supported').

In choosing the site, the climate and sun exposure must be taken into consideration, as well as the type of soil, water availability and quality, the drainage, pre-planting soil operations (deep

*Corresponding author: giuseppe.ferrara@uniba.it

tillage, rock fragmentation, ripping, organic matter, air circulation, etc.) The wide adaptation of pomegranate to different climates has favoured the differentiation of several genotypes/ landraces in the world areas of its cultivation (Ferrara *et al.*, 2011, 2014a). Pomegranates require a prolonged and warm summer season, with many hours of light to ensure a good fruit skin colour, high productivity and regular fruit ripening. Other factors for site selection include the cost of land (in some countries), proximity to facilities (fruit processing) and markets. The optimal site selection will require very few corrections or integrations in the successive operations, from soil preparation (fertilization, pH correction, organic matter, etc.) onward (training system, mulching, etc., for more information, refer to Chapter 7).

8.3 Site Preparation and Orchard Planning and Establishment

8.3.1 Basic fertilization

Prior to planting trees, it is essential to carry out a soil analysis of a representative sample of the future orchard and make any needed integration/correction. Soil analysis is useful for knowing the actual levels of the macroelements (nitrogen, phosphorus, potassium, etc.) and microelements (iron, boron, manganese, etc.), but also to determine the ratios (i.e. Ca/K and Mg/K) and highlight any imbalances. An excess of potassium, for example, can inhibit the absorption of calcium, which is very useful for the pomegranate in particular before flowering (Maity *et al.*, 2019). Proper necessary fertilization involves the supply of organic substances via mature manure or compost or organic or mixed organic fertilizers, preferably slow-release (Cossio and Vitelli, 2018).

Organic matter in the soil generally acts as a natural source of nutrients and can improve water- and nutrient-holding capacity in light soils and increase drainage in heavier soils. Usually, soil organic matter of 2–5% is considered suitable for fruit crop production (Neilsen *et al.*, 2014). Before planting, some soil chemical and physical operations are recommended; for example, for possible pH corrections, adding lime to acid soils (with pH values lower than 6.5), and calcium sulfates (agricultural gypsum) to alkaline soils (pH higher than 7.5), as well as adding organic substances (manure, compost, etc.) to soils with a low level of carbon.

Traditionally, during soil preparation up to 50 metric tonnes (t) per hectare (ha) of manure is applied as well as 100–150 kg of phosphorus and 200–250 kg of potassium. Apart from particular soil conditions (poor or problematic soils), organic and mineral elements are better supplied during the most sensitive phenological stages of the season after the orchard establishment, as is done for species such as grape (Ferrara *et al.*, 2018) to better meet the plants' needs (Bindraban *et al.*, 2015). Before transplanting, to make amendments and corrections to the soil, around 4 t/ha of agricultural gypsum can be applied to improve the porosity of clay soils and reduce the pH, and the green manure with legumes and/or crucifers can be added to improve the organic carbon content. If small plants are planted (just rooted in peat pots or from trays) a limited quantity of controlled-release fertilizer is recommended to obtain a correct rooting, applied around the plant at about a 20-cm distance from the stem; for example, a fertilizer such as 'Osmocote Smart-Release' (19-6-12 formula) at a dose of 80–100 g/plant (for more information, refer to Chapter 9).

8.3.2 Soil preparation

The need for deep tillage is based on the fact that for this fruit tree, which has a production cycle of at least 25 years (some trees live up to 500–600 years!), the only possibility for tilling the soil in depth is before planting. This will favour the development of a sound root system capable of supporting the tree in terms of water and nutrient absorption. The soil is subjected to subsoiling at a depth of about 70–90 cm carried out at least 4–5 months before the time of planting. Subsoiling is recommended to be performed in two steps in two perpendicular directions. Then the cultivation bed is prepared by tilling the soil with a tiller (grubber), a disc harrow and a cutter. Successive steps include the squaring and staking of the surface destined to become the orchard, defining the position and orientation of

Fig. 8.1. Pomegranate orchard with very wide ridges. (Photo: Vito Vitelli.)

the rows with a picket placed at the beginning and at the end of each row.

Raised beds in correspondence with the rows are essential to avoid stagnation of water near the plant collar, particularly in areas where heavy rains often occur or where soils are heavy. The ridges must have a triangular section with a vast base, over 4 m (Fig. 8.1); they are achieved with a three- or four-blade plough, with passages 'to fill it'. The height varies depending on the type of soil: from 30 cm in loose soils up to 60 cm in clay soils (Fig. 8.2).

8.3.3 Orchard floor management

Weed control in a pomegranate orchard is essential because weeds exert intense competition for water and nutrients towards plants; in the case of newly planted trees, competition can occur even for sunlight. Weeds can be managed by mulching, mowing, tilling or applying herbicides. Where possible, the adoption of permanent sod in the inter-row (Fig. 8.3) and successive mowing is the most advisable practice, both for eco-compatibility and for the positive effects on the characteristics of the soil, as well as against erosion (Fig. 8.4). Perennial meadow grasses such as fescue and ryegrass can be used. In small family orchards weeds can be easily controlled by using a brush cutter (see successive paragraph for more detailed information on weed control).

8.3.4 Planting

The choice of the most suitable time for planting is in relation to the state of preparation of the soil and the seasonal conditions, which should be planned and organized carefully. If the nursery supplies bare-rooted plants, the ideal time

Fig. 8.2. Ridges in a clay soil covered with black polyethylene mulch. (Photo: Giuseppe Ferrara.)

Fig. 8.3. Inter-rows with natural permanent sod. (Photo: Alimohammad Yavari.)

for planting is late winter or early spring, as soon as the risk of frost is overcome, generally from the beginning of February to mid-March in the northern hemisphere, and from August to September in the southern hemisphere. If the nursery supplies the plants in pots, planting can instead – theoretically – be done at any time of year, but it is advisable to avoid the full summer in case of water stress problems. Before planting, it is necessary to prepare planting holes of at least 20 × 20 × 20 cm size. At the time of making the holes the soil must not be too wet, because the soil on the hollow wall could compact and prevent the development of the roots of

Fig. 8.4. Ridges and inter-rows with sod before mowing. (Photo: Giuseppe Ferrara.)

Fig. 8.5. Pomegranate saplings with stakes and protectors. (Photo: Ali Sarkhosh.)

Fig. 8.6. Young Ypsilon orchard. (Photo: Vito Vitelli.)

the young plants outside the hole. Plants require irrigation immediately after transplanting in order to reduce the planting stress and the drying of the roots. It is advisable to use stakes and plant protectors at the moment of planting (Fig. 8.5).

8.4 Planting Density

One of the positive aspects of pomegranate is the earliness with which the plants bear fruit. In the second year, the first fruits can be harvested. As a consequence, one of the problems with the establishment of the orchard is the tendency of the young branches to bend down, preventing the regular and optimal formation of the plant structure. For this reason, it is common practice to tie the branches to supports or shorten them with adequate pruning in order to strengthen them. Because of the heavy fruit load once the trees become established, it is often necessary to support the branches by employing special wood/metal structures, as in the transversal Ypsilon (Fig. 8.6), and with horizontal wires supported by poles (Fig. 8.7). In commercial orchards, the optimal distances between plants need to be defined and optimized and may show some variability. The recommended plant density is between 500 and 750 plants/ha (Cossio and Vitelli, 2018).

In the transversal Ypsilon trellis, traditionally adopted in Israel, the planting layout is 6 m space

Fig. 8.7. The metal structure of the Y-trellis. (Photo: Alimohammad Yavari.)

Fig. 8.8. Orchard spacing in vase training. (Photo: Ferdinando Cossio.)

between the rows × 3.5 m space between trees on the row, but can be reduced to 6 × 3 m or 5 × 3 m or 4.5 × 2 m depending on the variety, soil, etc. Some fruit growers adopt the so-called 'variable layout' planting to obtain higher production in the early years; successively, alternate plants are removed, in order to reach the final distances (5 × 4 m). In the 'vase' training system, tree layout is 5.5 × 3 m or 5.5 × 3.5 m; however, spacing of 5 × 3 and 6 × 4 m are also adopted (Fig. 8.8). In all training systems, the tree height should not exceed 3 m to facilitate harvesting and pruning operations.

Planting density should be chosen in the appropriate way for several reasons: sufficient sunlight to penetrate into the canopy since light is essential for fruit to ripen properly; adequate aeration between the trees to reduce phytosanitary problems; and efficient movement of tools, machinery and people during cultivation practices (pruning, harvesting). In some areas, planting is at higher densities (3 × 2 m), with some success in moving towards a superintensive system such as is used for other fruit crops.

North–south orientation is recommended for the rows, because the sun moves across the sky from east to west, and a north–south row orientation captures maximum sunlight and allows for even light distribution across the orchard. However, the penetration of light in the inter-row depends on the distance between the rows and the distance between plants on the row. A square or rectangular system of planting is well suited for pomegranate.

8.5 High-Density Plantations and Management

High-Density Planting (HDP) is a method of fruit tree cultivation involving dense planting of fruit trees, allowing small or dwarf trees with a modified canopy for better light interception and distribution and ease of mechanized field operation. HDP gives higher yield and returns/unit area due to increased number of trees/unit area. It is possible, by regular pruning and use of bioregulators, to maintain the size and shape of trees.

In recent years, in California and other countries, growers have been willing to grow pomegranate trees as fruiting walls in the same way as pome and stone fruits. A vertical canopy development that rejects traditional multidimensional tree development (the natural habit of pomegranate) in favour of dense, flat canopies (Day and Wilkins, 2011) could be an alternative for growing pomegranate trees (Fig. 8.9). These fruiting walls are developed by narrow pruning and hedging, thus increasing tree density. The higher plant density can decrease the time required to reach profitable yield when compared with lower densities (open vase or Ypsilon). Concentrating the pomegranate trunk, scaffold or branch growth into a zone that can be more easily pruned, sprayed, thinned and harvested has potential for improved efficiency and mimics the approaches already deployed for apple and pear production in many countries of the world (Day and Wilkins, 2011) and recently also developed for olive, sweet cherry and almond. Vertical wire trellises fit well within the context of the fruiting wall strategy. It is possible to either use vase multiple trunks or single trunks in a fruiting wall approach, but these forms would seem less apt to garner many of the benefits as quickly or easily. However, in a very high-density orchard the production could be unsatisfactory due to lack of light, with fruits that tend to form only in the upper and outer part of the canopy, the colour and size of the fruit are generally not optimal and there are difficulties in the application of pesticides.

Plants of the cultivar Mridula trained to ultra-high density planting (3 × 2 m) treated with some plant growth regulators (PGRs) showed impressive results (Shanmugasundaram

Fig. 8.9. High-density orchard. (Photo: Ali Sarkhosh.)

and Balakrishnamurthy, 2017). The application of 1-naphthaleneacetic acid (NAA) at 10 ppm + gibberellic acid (GA_3) at 50 ppm in this intensive training system increased the fruits per plant (65.11 fruit), average fruit weight (272.5 g), fruit volume (290 cc), fruit length (7.7 cm), fruit diameter (7.9 cm), number of arils per fruit (670), total aril weight per fruit (186.2 g), 100-aril weight (25.85 g), aril recovery (25.85%), total seed weight (20.23 g) and fruit yield/plant (17.74 kg) with respect to control treatment (water application) (Shanmugasundaram and Balakrishnamurthy, 2017). With regard to quality parameters, application of GA_3 50 ppm + KNO_3 1% increased total sugar (8.01%), reducing sugar (7.35%), total soluble solids (16.2%) and anthocyanin content (18.85%). Three sprays of NAA at 10 ppm + GA_3 at 50 ppm starting from 150 days after pruning at 30-day intervals resulted in the highest yield attributes, whereas the quality attributes were ameliorated by application of GA_3 at 50 ppm + KNO_3 1% (Shanmugasundaram and Balakrishnamurthy, 2017).

In India pomegranate cultivar 'Bhagwa' is planted at a spacing of 4.5 × 3.0 m. However, there are different HDP systems that adopt various spacings, as mentioned in Table 8.1.

8.6 Time of Planting

Pomegranates are usually planted in late winter or at the beginning of spring. The best months in the northern hemisphere are from December to February, whereas in the southern hemisphere it should be done from August to September. Pomegranates can also be planted later in the spring season when taken rooted in pots from the nursery (Fig. 8.10), but an early planting is preferred to optimize the length of the growing season in particular in cool areas. Late planting times reduce the length of the growing season, in particular for young and small pomegranate plants, which need more time to develop shoots, new roots and accumulate reserves.

Table 8.1. Planting 'Bhagwa' at different spacings in India.

Spacing	Row–row distance (m)	Plant–plant distance (m)	Area occupied by a plant (m^2)	Density/ha
5 × 5 m	5	5	25	400
5 × 4 m	5	4	20	500
5 × 3 m	5	3	15	666
5 × 2.5 m	5	2.5	12.5	800
5 × 2 m	5	2	10	1000
4 × 4 m	4	4	16	625
4 × 3 m	4	3	12	833
4 × 2.5	4	2.5	10	1000
4 × 2 m	4	2	8	1250
3 × 3 m	3	3	9	1111
3 × 2.5 m	3	2.5	7.5	1333
2.5 × 2.5 m	2.5	2.5	6.25	1600

Fig. 8.10. Potted pomegranate trees in the nursery. (Photo: Alimohammad Yavari.)

8.7 Variety and Rootstock Selection

Growing the right varieties in the region to meet the market demand is a critical decision. Many pomegranate cultivars and landraces are cultivated in the world-growing regions (for more information, refer to Chapter 5). DNA marker analysis indicates that landraces/ varieties cultivated in some countries (Italy, Israel, Turkmenistan, the USA, Japan) appeared less similar concerning the accessions within the same country than between countries (Giancaspro *et al.*, 2017). Using rootstock in pomegranate is not common, although in some parts of India different genotypes of rootstocks for different purposes (Table 8.2) may be used for grafting/budding.

Table 8.2. Rootstocks used for pomegranate in India.

S. no	Biotic/abiotic stress	Rootstocks
1	Bacterial blight	IC-1181, IC-1253, IC-1256, IC-1259, IC-1272
2	Salinity	IC-318706, IC-318707
3	Fruit cracking	IC-318712

8.8 Post-Planting Care

The period post-planting is a very difficult and delicate one. Poor attention devoted to either water and nutrient requirements or pest care of the young and growing pomegranate plants can have negative consequences for pomegranate orchard establishment (long juvenile phase, increased costs, death of young plants, etc.). Staking of the young plants is done during or soon after planting. Usually, a bamboo (wood) pole of 1.5–2 m in length is used for staking the plants. The main growing shoot is tied with the bamboo pole with the help of a coir rope/jute string. Staking helps the plants to grow straight and protects the plants from bending of the canopy due to the velocity of the wind. A tree tube or guard around the young plants is recommended to protect them from animal browsing, extreme temperatures and herbicide drifts (Fig. 8.5).

8.8.1 Irrigation

Irrigation is done immediately at the time of planting to keep the rhizosphere moist. Then, regular irrigation is provided using a drip/ microsprinkler irrigation system. In the early stages of root development, the pomegranate root system occupies only a small portion of the soil, in particular in the case of small rooted plants in pots. In these early stages, the amount of water required is limited, but the frequency of irrigation should be high, since these young plants are less able to face water scarcity during the vegetative season. Irrigation is necessary to stimulate the development of a good and extensive root system. Initially, irrigation may be provided using a single drip-tape or drip-pipe with two drippers placed on either side of the pomegranate trees and microtubes or a microsprinkler about 30 cm from the trunk attached to the pipe (Fig. 8.11). In some climatic conditions (hot climates) two irrigation pipelines can be used. One or two irrigation pipelines can be positioned beneath the mulching material or even above it.

Fig. 8.11. Single irrigation pipeline. (Photos: Vito Vitelli and Alimohammad Yavari.)

As the canopy size increases, the drippers may be placed away from the plants, as direct trickling of water on the root surface may increase the chance of wilt disease such as phytophthora crown rot. Use of drip irrigation should be preferred to other systems and emitters should not be too close to the collar of the tree to avoid excessive moisture (to avoid fungal diseases) and stimulate root growth to reach the water (for more information, refer to Chapter 10).

8.8.2 Nutrition

Mineral nutrition is necessary for optimal plant growth. In the establishment years, it is fundamental to supply nitrogen for the development of the shoots, and nitrogen is generally applied in spring in order to be readily available to the root system for plant growth. Mineral nutrients require a localized application because of the small development of the root system and their application through fertigation is preferred when possible (for more information, refer to Chapter 9).

8.8.3 Weeds and wind

In the first years, competition of young pomegranate trees with weeds can be extreme (Fig. 8.12). Young pomegranates have a small root system and weeds growing close can compete for nutrients and water. Growth of weeds on the row should be controlled by using mulching materials or mowing or herbicides. In order to avoid breakage of the young trees caused by the wind, support poles can be used to sustain the first steps of the establishment, in particular in windy areas.

Fig. 8.12. Weeds mowed in the inter-rows and competing with pomegranate trees on the row (top), mulching on the row and tillage in the inter-rows (below). (Photos: Giuseppe Ferrara and Vito Vitelli.)

8.8.4 Pests and diseases

In the early years, shoots and leaves need to develop well in order to establish a good orchard. Registered pesticides must be used to protect the young trees from pests and diseases. Although pomegranate is considered a resistant species, the following pests are reported to cause financial loss to pomegranate trees: aphids (*Aphis punicae*, *Aphis gossypii*), leopard moth (*Zeuzera pyrina*), moths (*Cryptoblabes gnidiella*, *Cydia pomonella*), Mediterranean fruit fly (*Ceratitis capitata*), mealybugs (*Planococcus citri*), fungi (*Phytophthora* spp., *Pilidiella*/*Coniella granati*), thrips (*Scirtothrips dorsalis*), fruit and leaf spots (*Colletrotrichum*/*Gloeosporium gloesporioides*,

Gloromella cingulata, *Sphaceloma/Gloeosporium punicae*, *Alternaria* spp., *Cercospora punicae*, *Xanthomonas axonopodis* pv. *punicae*), rot (*Botrytis cinerea*, *Penicillium* spp., *Pilidiella granati*), black heart (*Alternaria* spp., *Aspergillus niger*), anthracnose (*Colletotrichum gloeosporioides*), etc., (for more information, refer to Chapters 12 and 13).

8.9 Training Systems

8.9.1 Multiple trunks

Pomegranate is basically a deciduous tree; however, it can behave as deciduous, semi-deciduous and evergreen in temperate, subtropical and tropical climates, respectively. Looking into the fruiting behaviour and growth of trees, an open centre or vase-shaped training system can be followed to develop a suitable framework for optimum growth, flowering and fruiting. Well-grown saplings of 5–6 months old may be planted and three or four healthy suckers arising from the ground level can be kept, and the extra ones should be removed. These suckers can develop a proper canopy within a 2-year period. Thus, this system allows three or four strong stems with six to eight fruiting branches for producing quality fruits from the third year of planting.

In some countries, pomegranate trees are trained as bushes with many stems arising from the ground, a sort of multi-trunk system following the natural tendency of this species. The multi-trunk system (Fig. 8.13) favours the maximum development of tree canopy and results in a bush-like appearance; broken trunks can be regrown from the crown or adjoining trunks to replenish the productive canopy, and provide some indemnification against frost damage (LaRue, 1977). This training system should be avoided in commercial orchards because of many problems for the different horticultural practices (pesticide application, harvesting, pruning, machinery movement, etc.). The feature of this training system is mainly an increased number of trunks in the orchard and some assume a consequent higher number of fruits. A four stem multi-trunk training system resulted in the highest yield under Solapur conditions compared with single-, double-, triple-, five- and more than five-stem (control) systems. However, drawbacks of the multi-trunk training system are numerous and include factors such as more suckers for trimming and time needed for pruning; difficulty in supporting fruiting branches to avoid excessive bending or breaks; difficult harvesting and longer harvesting time; low level of mechanization; and lower marketable fruit rate. Although the bush or multi-trunk training systems are the natural growth habit of the pomegranate, these systems can be adopted in small-scale orchards, backyards, parks and gardens. However, the multi-trunk system is widely adopted for cultivation of pomegranate in countries like India, Iran, Afghanistan, etc.

Fig. 8.13. Pomegranate tree trained as multi-trunk. (Photo: Ali Sarkhosh.)

8.9.2 Single trunk

This training system is advantageous over the multi-stem training system for production of larger numbers of better-grade fruits for fresh market. This system is practised by relatively few farmers around the world and can be seen in countries like Israel, Spain, Italy and the USA. It facilitates easy intercultural operations. However, pomegranate is susceptible to shot hole borer and stem borer under tropical conditions, and trees with a single-stem training system are at the risk of wilt disease.

Fig. 8.14. Removal of sucker and water sprouts in a single trunk trellis. (Photo: Ali Sarkhosh.)

The single-trunk strategy with a tree like pomegranate, which has a bush-type habit, requires significant management for branch development. Delays in production capacity often occur while the trunk is grown to a sufficient diameter to bear total tree crop load on scaffolds and branches (Day and Wilkins, 2011). The benefits of this system include the ease of sucker and water-sprout removal (Fig. 8.14) and the efficiency with which farming practices can be made close to the treeline (lower labour hours). In this training system a higher level of mechanization can be applied (pruning, harvesting, pesticide and foliar applications, etc.) and quality of fruit is generally higher than in the multi-trunk system. Suckering in a single-trunk system can either be done manually through removing by pulling (when shoots are not lignified yet), or cutting with pruning shears, or chemically, since lower branches and canopy are removed in a process referred to as skirting (Day and Wilkins, 2011). However, care must be used when using chemicals in order to avoid toxicity to the young pomegranate trees. The single-trunk system also offers better penetration of sunlight into the inner layers of the canopy and water sprouts can be easily trimmed manually (green and soft tissues) or mechanically. Some drawbacks include the need for use of support systems (wires, poles, bars, etc.) necessary to sustain the fruit load (higher establishment costs) and avoid breaks and the higher number of trees to be planted in the orchard (trees/ha). The most common and traditional tree training for the single-trunk system is the open centre with some variants.

8.9.3 Open centre training systems (Spanish system)

As mentioned above, the presence of a single trunk compared with the multi-trunk training system reduces the summer pruning operations because of the lower number of suckers. The vase, with a single trunk and a rounded canopy, is adopted in Spain for cultivars such as 'Mollar' and 'Valenciana' (Fig. 8.15) and

Fig. 8.15. Trees of 'Mollar' in Spain. (Photos: Alimohammad Yavari.)

Fig. 8.16 Open centre system in Tunisia. (Photo: Vito Vitelli.)

also in other countries of the Mediterranean basin (Italy, Greece) with some variants depending on the location (Fig. 8.16). The free vase is the training system resembling that of other fruit species, with a single trunk and three or four branches. Growing pomegranate as a single trunk with a height of 50–80 cm and split into four to six primary scaffolds facilitates the management of the soil underneath the tree and other horticultural practices. In the first years after planting it is necessary to support the young trees with horizontal wires. For training the tree as open centre, the shoots from the trunk need to be selected and either headed back or thinned during the winter pruning in the first years of shaping. It is a training system with low cost because it does not require supporting structures, but sunburn or mechanical damage to fruits (scratches) or shoots (breaks) can be significant.

8.10 Modern Orchard Systems (Y, T and Central Leader)

The most innovative training system for pomegranate is the system that has a transversal Y-shaped structure, with a 3.5 m space between trees on the row and 6 m space between rows (Fig. 8.6). It is an expensive system to install but very valid in modern orchards, which allows forming an expanded canopy with a large production area (Fig. 8.17) and, above all, significantly reduces sunburn and mechanical damage (Cossio and Vitelli, 2018). Pomegranate is a species suitable for different shape and training systems since it can continuously produce new shoots because of its natural bushy growing habit. Because of this habit, other training systems have been adopted in the past decade for the cultivation of pomegranate in various areas of the world, and some of them are described below.

Fig. 8.17. Positioned and tied branches in the Y-trellis. (Photos: Vito Vitelli.)

8.10.1 Y-trellis

This training system has a single trunk with 6–12 branches positioned as an upside-down umbrella or in a double-inclined wing (Fig. 8.17). This system requires poles and horizontal wires in order to sustain the canopy (Fig. 8.17) and has gained increasing popularity in modern pomegranate orchards (Atzmon, 2015). In Israel, the system is made of Y-shaped, 40-mm-thick metal bars with a distance between the bars of 10 m in the row (Fig. 8.18). On each arm of the Y there are three metal wires, 3 mm thick and 0.5 m spaced. The splitting of the Y is at the height of 1.2 m,the total height is 1.8 m and the width of the bar is 3 m (Atzmon, 2015). This system has many advantages compared with the traditional training system (multi-trunks): almost no broken branches (20% in traditional); lower fruit temperature and less direct sun exposure (Fig. 8.19); better fruit quality (i.e. sunburn, scratches, colour, etc.); fewer damaged fruits (20 vs. 50%); higher yield (40–60 vs. 20–40 t/ha); easy and effective in the application of pesticides; and better accessibility for desuckering, pruning and harvesting (Atzmon, 2015).

8.10.2 Central leader training systems

There are several trellis systems belonging to this type, such as the fuse, the pyramid and the spindle. These systems can be adopted in the case of high-density plantations for better management of the fruiting wall. There is a continuous main stem with branches distributed at different heights, which form the canopy (fuse, pyramid, etc.).

8.10.3 Espalier training systems

The tree structure is created in two dimensions and the main trunk is tied to the wires of the supporting structure. The shoots are fastened to training wires and can be horizontal or bent, either regular or irregular. The canopy is continuous and vertical, similar to the Palmette training system in stone fruit species.

8.10.4 Pergola training system

In this trellis system, the trees form a continuous horizontal canopy with a single trunk and the branches and shoots in the higher part; this trellis also requires poles and wires for supporting the horizontally developed canopy (Fig. 8.20).

8.11 Effects of Training System and Planting Density on Yield

In many fruit crops, such as apple, pear and grape, much is known about the effects of the training system on yield and related parameters, but such information is lacking in the case of pomegranate (Gill *et al.*, 2011). The yield of pomegranate is influenced by the training system adopted in the orchard. Aliev (1979) reported the maximum number of fruits per tree was achieved when pomegranate trees were trained to a multi-trunk system. Another work also reported a positive effect on yield with the

Fig. 8.18. Y-trellis in winter. (Photo: Ferdinando Cossio.)

multi-trunk training system (Balasubramanyan *et al.*, 1997). Training systems favouring light penetration in the canopy can stimulate photosynthetic activity during the fruit growth period and might increase yield (Durand, 1997). In a more recent trial conducted in India for 3 years (2006–2008), data showed that the training system influenced the number of harvested

Fig. 8.19. Fruits in the Y-trellis. (Photo: Giuseppe Ferrara.)

fruits with a significantly higher number of fruits harvested from multi-trunks and single-trunk 30 cm height-trained trees compared with single-trunk 45 cm height-trained ones (Gill *et al.*, 2011). Moreover, the highest yield and the biggest fruits were obtained with the multi-trunk trellis, followed by single-trunk 15 cm height. However, training systems also had inconsistent effects on fruit quality (Gill *et al.*, 2011). Although some positive aspects have been reported for multi-trunk training in India, there are some negative aspects that have been reported that limit use of this training system in modern orchards (pruning, harvesting, mechanization, etc.) in other countries of the world (Israel, Spain, Italy, the USA). However, the quality of fruits was found to be higher in the single-stem training system than for the multi-trunk system. Hence, for export purposes, a single-stem training system may be advantageous, whereas for the domestic market and juice production four-stem training (multi-trunk) system might be advisable.

Fig. 8.20. Pergola-type system (left) in summer, and (right) winter. (Photos: Vito Vitelli.)

8.12 Pollination and Pollinizer

Pomegranate flowers develop into either hermaphrodite or staminate flowers. Staminate (male) flowers are bell-shaped, with a poorly developed atrophied ovary containing few ovules, and are infertile. These flowers senesce without fruit set. Hermaphrodite (bisexual) flowers are vase-shaped with a normal ovary capable of developing fruit (Holland *et al.*, 2009). In addition, a minor form of intermediate flower has also been reported (for more information, refer to Chapter 2). They are tubular in shape and on rare occasions they set malformed fruits (Babu, 2010). Pollen germination in fertile and infertile flowers of pomegranate under *in vitro* culture in different media ranged from 34.88 to 60.07% and from 26.27 to 45.73%, respectively (Eshghi *et al.*, 2010). A medium containing 5 mg/l boric acid without molybdic acid had the highest pollen germination percentage of fertile (60.07%) and infertile flowers (45.73%). Flower retention, pollen formation, pollen tube growth or germination are affected by the nutrient boron (Camacho-Cristóbal *et al.*, 2008; Hänsch and Mendel, 2009).

8.13 Canopy Management

8.13.1 General considerations

Canopy management for pomegranate is a cultivation practice adopted in both traditional and modern orchards. It is important for growing trees because of the canopy microclimate, which affects several aspects such as flower differentiation, fruit yield and quality, and disease incidence. Canopy management is necessary to improve light and air in the canopy in order to optimize the growth of the shoots and the ripening of the fruits but also to reduce diseases. The canopy of a pomegranate tree includes the shoots and the leaves distributed in the surrounding space according to the different training systems adopted (from multi-trunk to single-trunk systems). In high-density orchards, canopies are almost continuous, whereas in traditional orchards the canopies are separated from each other and thus defined as discontinuous. However, canopy management includes many techniques applied in the pomegranate orchard, the most important including summer and winter pruning. Canopy management may also include the shoot positioning technique (with wires and poles), which is applied for many new training systems (Y-trellis) in order to facilitate production, phytosanitary treatments and harvesting. In general, any cultivation technique aiming at modifying the position, size and the number of shoots within the canopy can be considered as part of canopy management. These changes tend to modify the canopy microclimate, which is the climate within and in close proximity to the canopy. The canopy microclimate is influenced by several factors apart from cultivation techniques, such as sunlight, temperature, humidity, wind, rainfall, evaporation, etc.

Fig. 8.21. Orchard netting to reduce fruit sunburn. (Photo: Ali Sarkhosh.)

8.13.2 Sunlight and temperature

The sun's radiation, essential for plant photosynthesis (photosynthetically active radiation, PAR) is within the range of 400–700 nm and is abundant at the external part of the canopy. Values decrease in the inner layers of the canopy with possible consequences for flower bud differentiation (number of flowers and sex ratio). This should be kept in mind for training systems such as the multi-trunk system.

The sun's radiation affects both leaf and fruit temperature, especially in the case of sun-exposed organs. Excessive sun exposure can cause sunburn on the fruits leading to exterior (brown skin) and interior alterations (white arils), which cause a loss of the commercial product. The excessive temperature on the fruit can cause poor colouration of the skin. Photoselective netting, kaolin application, evaporative cooling and bagging of the fruits can be used to prevent sunburn damage in pomegranate fruit (Fig. 8.21).

8.13.3 Humidity

An increase of humidity within the canopy is caused either by transpiration of leaves or fruitlets, in particular in the case of dense canopies of high-density pomegranate orchards (superintensive systems). High humidity may favour the spread of some pests and diseases in pomegranate trees.

8.13.4 Wind

Wind increases the evapotranspiration of plants, and also high-velocity winds may cause the breakage of shoots and damage to the fruits. Wind velocity can be reduced by the use of windbreaks where prone to it. In general wind velocity is reduced in the middle of dense pomegranate canopies because of the presence of different plant organs, thus reducing also the transpiration in the inner layers.

8.13.5 Transpiration

As mentioned above, the solar radiation intensity in the inner canopy is lower compared with the outside, and this lower value reduces transpiration from leaves and fruits located in the middle of the canopy. In general, external leaves are more stressed than the internal leaves because of the difference in the transpiration rate.

8.13.6 Rainfall

Plants can catch part of the falling rain within the canopy, thus either reducing the amount and speed of rainwater reaching the soil beneath the canopy, or increasing the humidity in the inner canopy.

8.13.7 Effects of canopy management on tree productivity

Canopy management has significant effects on yield and quality of pomegranate fruits. When considering the leaf area exposed to sun, it is clear that a greater area means a higher yield because of the greater amount of photosynthesis accomplished by the leaves. However, there is also a potential effect of light on bud differentiation, with more flower buds in the external canopy and a lower differentiation in the internal layers. In this latter case, reflective mulches are often used in order to increase sunlight in the middle/inside of the canopy for better effects (Fig. 8.19) on buds (differentiation) and fruits (ripening). A balance of light and nutrient interception can also be pursued with two important practices for pomegranate: desuckering and removal of water sprouts. Almost all the varieties of pomegranate can produce many suckers (Fig. 8.22) during the season, which need to be removed to reduce competition for the nutrients, in particular in young pomegranate orchards during establishment. Removal of water sprouts or excessive vigorous shoots can help to maintain a balanced canopy space between fruits and shoots by reducing unfruitful and sucking branches. Shoot removal and desuckering should be performed early in the season in order to either facilitate the operation (green and soft tissues) or to reduce the competition for water and nutrients. Shoot positioning is currently adopted in modern pomegranate orchards in order to better sustain the shoots carrying lots of heavy fruits and facilitate the penetration of sunlight into the internal canopy. In vase-trained or multi-trunk training systems shoot positioning is generally not adopted, and the consequence is the bending or breaking of many branches.

All the operations of canopy management tend to reduce the density of the vegetation in order to favour more sunlight in the inner layers; increase air movement and balance the humidity; increase flower bud differentiation; improve fruit ripening; reduce incidence of diseases; and promote easier harvesting and other cultivation practices.

Fig. 8.22. Suckers in summer. (Photo: Vito Vitelli.)

8.14 Types of Pruning

8.14.1 General considerations

Pruning is a complex task of cutting operations to regulate the growth, development and production of the tree. Pruning is a cultivation practice dating back to thousands of years ago, such that it is reported in the Greek and Latin texts as an essential horticultural practice in the orchard for obtaining an almost constant yield. The main objective of pruning is to achieve better results in terms of yield and quality or, in the case of ornamental pomegranates, better aesthetic results. Pruning includes a very large and diversified number of operations involving the aerial organs and the root system, performed both during rest (dormancy) and the vegetative season (Igles *et al.*, 2002). Pomegranate has been always considered as a minor fruit such as loquat, prickly pear, etc. and little attention has been devoted to its horticultural practices such as pruning; as a consequence, pruning techniques commonly adopted for other important fruit species (stone and pome fruits) have also been used for pomegranate with no particular considerations of the vegetative and productive behaviour of this species. However, in later years the growing interest in pomegranate cultivation increased activities to ameliorate some horticultural practices in order to develop the best training systems for the species. Pruning is necessary for shaping the tree in the orchard to reach its final tree shape, and to regulate both vegetative and productive activity with respect to soil and climatic conditions, cultivation practices and production goals. The recent development of more superintensive planting for this almost forgotten species has not yet allowed the definition of the best pruning system. It should not be forgotten that the pomegranate is naturally a bushy species and its canopy form must also be adapted to the specific area or cultivation region, considering the optimization of the canopy and its vegetative–productive balance, and taking into account the different cultivars and the pruning experience of the workers of each region.

8.14.2 Winter pruning

Pruning can be split into winter pruning or summer pruning. The first one is generally carried out during the dormancy period, whereas the latter is done during the growing season and is also defined as green pruning. The winter pruning of the tree is necessary to: create a strong and robust structure; have the right shape for the different training systems adopted; intercept the light and ensure there is air circulating within the canopy; stimulate vegetative activity; facilitate the application of pesticide; produce fruits of good quality and size; reduce alternate bearing; and balance the aerial and the root systems. The right light intensity is fundamental for the optimal vegetative–productive activity of the tree. Shaded leaves are not able to produce the optimal amount of photosynthates necessary to sustain shoot and fruit growth. Shaded leaves have a negative balance since they catabolize more photosynthates than they are able to synthesize. The optimal availability of light also acts positively on the formation of flower buds and fruit ripening (i.e. colour) (Fig. 8.19).

There is not much information available on pomegranate winter and summer pruning compared with other fruit crops such as grape, peach or apple. However, a little information is available on the new training systems and the behaviour of recently introduced varieties, in terms of data on qualitative, technological and health aspects (Pontonio *et al.*, 2019). In the case of pomegranate, particularly scarce information is available on the vegetative–productive balance and appropriate canopy management. Heavy or light/absent pruning can have different effects: in the absence of pruning, there is an increase of small shoots, whereas intense pruning stimulates the strong growth of the shoots, in particular, water sprouts (Chakma, 2014).

Winter pruning can be accomplished for trellising in the first years (pruning of young trees) and then for production (pruning of mature trees). Trellis pruning is performed in 2–3 years in order to achieve as soon as possible the final trellis to be kept in the orchard. Once the tree has achieved the final shape, pruning should be done each year to regulate vegetative and productive activity. In the past, when pruning was very light or almost inconsistent for pomegranate, the trees were easily trained to multi-trunks or were better known as bush-shaped trees. Now this system is not generally adopted in modern mechanized orchards where single-trunk systems are commonly adopted.

Time of pruning can also affect some parameters in pomegranate production. A trial in India on 'Ruby' cultivar indicated that yield and time of harvesting were optimal when pruning was done at the earliest date of 15 November. Later pruning dates decreased production and delayed harvest time. Fruit quality was also highest on trees pruned at the earliest date (fruit weight, TSS, reducing sugars and sugar/acid ratio); however, late pruning times improved the skin colour of the fruits (Ghosh *et al.*, 2012). In the northern hemisphere winter pruning can be done sometime during January, February or March, and in the southern hemisphere it can be done during June, July or August, depending on the climate conditions – the colder the winter, the later the pruning.

8.14.3 Summer pruning

Summer pruning aims to: keep the trellis and the size of the tree; control the vigour of the tree (trimming/heading of shoots and water sprouts); regulate the canopy microclimate (leaf removal); increase carbohydrates and nitrogen content in the fruiting shoots (desuckering); and improve the fruit quality (fruit thinning, girdling). In California, pre-harvest summer pruning is typically accomplished by removing non-fruiting shoots and suckers from the middle of the canopy (Day and Wilkins, 2011). The increase of light in the canopy stimulates the development of the exterior skin red colour, and sometimes fruit size can be improved slightly since photosynthetic activity is enhanced on newly illuminated leaves adjacent to fruit, so they are better able to transport photosynthates to the nearby growing fruits (Day and Wilkins, 2011). Girdling consists of removing a ring of bark (phloem tissue) from shoots (interruption of the phloem vessels) to restrict the movement of assimilates from the aerial portion of the tree to the basal portion (trunk and roots), so the sap flow is directed to buds, flowers and fruits present above the girdled zone. The phloem vessel interruption is temporary (few days), and subsequently the tree restores the vessels through the production of a callus (Ferrara *et al.*, 2014b). Girdling may be performed after fruit set to increase size, and also at the beginning of fruit ripening to improve fruit skin colour and advance fruit maturation. Girdling reduced fruit splitting and improved quality in a trial conducted on the cultivar 'Manfalouty' (Ahmed-Amin *et al.*, 2000). El-Kholy (2005) reported that the use of girdling increased both carbohydrates and C/N ratio in shoots, similar to that found by Abdel-Galil (2008). In particular, desuckering together with girdling of 3 mm increased fruit set, yield, reducing sugars and TSS of 'Manfalouty' pomegranate, while non-desuckering plus girdling of 9 mm decreased yield and fruit quality over the season (Abdel-Galil, 2008) probably because of the competition of suckers and the wider girdled zone to be restored. The practice of desuckering, which is fundamental for pomegranate, is helpful to increase carbohydrates and nitrogen content and C/N ratio in the fruiting shoots of pomegranate cultivar 'Manfalouty' (Abdel-Galil, 2008).

8.14.4 Flower and fruit thinning and vegetative growth (fruiting balance)

The non-bearing period of pomegranate trees is about 1 year. Pomegranate plants put forward blooms from 12–14 months after planting. Though the farmers' practice involves obtaining a crop from 1 year after planting, scientifically/physiologically it is not recommended as explained in the following paragraphs. In order to maintain the vigour of bearing trees and enhance orchard longevity, mechanical/manual deblossoming is advised at 1-year planting. In this method, nipping of flowers is done by removing the flowers emerging during the undesirable/unwanted period. This helps to conserve the food reserves for the next season and for shoot development. The plants are allowed to flower and set fruits only at the end of the second year or beginning of the third year. As the tissue culture plants are vigorous and precocious in nature, crop regulation for onset of flowering and fruiting is done at the end of the second year, in months 18–24. Flower and fruit thinning is usually done by hand.

Fruit thinning is an important horticultural practice for producing good sized pomegranates for fresh consumption. A study conducted in Iran on the cultivar 'Malase Torshe Saveh' with different intensities of hand thinning (0, 10, 20, 30 and 40%), applied when fruitlet diameter was about 30 mm, showed interesting results (Jafari *et al.*, 2014). Mean fruit weight, length, diameter and volume increased with increasing thinning severity and fruit thinning generally increased fruit quality (TSS and TSS/TA). Intense hand thinning ameliorated skin fruit colour compared with control non-thinned trees, probably because of the increased leaf-to-fruit ratio (Jafari *et al.*, 2014). In general, fruit thinning increased commercial value and marketability by increasing fruit quality; meanwhile, yield was almost unaffected in all treatments (Jafari *et al.*, 2014). Similar non-significant effects of fruit thinning on yield have been reported for 'Jyoti' pomegranate, when only 25 fruits/tree were retained after hand thinning (Padmavathamma and Hulamani, 1996) and for 'Manfalouty' pomegranate (Hussein *et al.*, 1994). Fruit thinning intensity can be applied according to market demands, since consumers are attracted by big fruits but excessively large size should be avoided because consumers can find these fruits difficult to manage. In Spain, fruit thinning is usually done in the first week of June and is repeated after 20–30 days (end of June or early July). Depending on the phenological stage of the fruits at thinning and the crop load, from 7–8 to 12–15 kg fruit/tree could be removed (Melgarejo Moreno *et al.*, 2010). Too many fruits on trees may have a negative effect on the next season's yield (bud differentiation) and will also impact the current season's fruit size. To avoid this, thinning of fruits can be performed 4–5 weeks after flowering. It is also important to thin fruits in clusters as fruit clusters usually create an ideal environment for pests and diseases. A general rule will also be to thin out the fruits borne on weak spurs, as well as deformed and damaged fruits. Generally, the early flowers will result in the larger fruits, thinning should be applied on the late fruits and a warmer spring will enhance the development of large fruits.

Deleafing is also a cultivation practice that can be done a few weeks before harvest time, if the lack of sunlight is an issue in the orchard at that time of the season. Deleafing about 2–3 weeks before harvest will improve the evenness of peel colour, increasing fruit marketability. It should be applied by just removing a few leaves on the branch (lateral, spur, etc.) where the fruit is connected to the tree.

8.14.5 Intensity of pruning

The intensity of pruning depends upon several factors such as: training system, cultivar, soil and climatic conditions, irrigation, fertilization, vigour and age of the tree, etc. The pruning can be defined as light, moderate or severe depending on the amount of wood removed. Shukla *et al.* (2007) defined the intensity of the pruning for pomegranate using the reduction in the length of the shoot, that is, light (25%), moderate (50%) and intense (75%). Masalkar *et al.* (2009) reported the effects of different intensities of pruning on canopy development, with the greatest tree growth either for unpruned or moderate pruning (both thinning and heading back 30 cm shoots), while trees with intense pruning and thinning of shoots showed a slightly smaller canopy. The highest number of fruits (81) was obtained with the treatment of 20 cm heading back without thinning, whereas the highest fruit yield (18.76 kg/tree) was recorded with the treatment of 20 cm heading back but with thinning (Masalkar *et al.*, 2009). Chakma (2014) reported, in a trial performed in India on the cultivar 'Kandhari Kabuli', the noteworthy effects of different intensities of pruning on several vegetative and productive parameters. Heading back the shoots to 15 cm in length induced the longest shoot growth (56.34 cm) with respect to heading back to 60 cm (41.23 cm); however, unpruned trees (control) reached the highest height and the biggest canopy volume (Chakma, 2014). The highest growth of shoots was registered for heavily pruned trees, which can be attributed to the availability of nutrient reserves for fewer vegetative points, because of the removal of other competing shoots. Intense pruning, that is, heading back to 15 cm, reduced fruit set, fruit drop and number of fruits/tree compared with light (heading back to 60 cm) or no pruning (control). However, intense pruning slightly reduced the yield/tree (7.88 kg) compared with light or no pruning (≈10 kg), but significantly increased fruit weight (≈280 vs. 210 g) with a more intense and uniform red colour (Chakma, 2014). Heavy pruning significantly increased size and weight of pomegranate fruits (Masalkar *et al.*, 2009). Intensity of pruning also affected the organoleptic quality of pomegranate fruits, with the highest TSS and ascorbic acid content and lowest TA for the trees with the intense pruning (heading back to 15 cm) compared with trees with light or no pruning at all (Chakma, 2014). This was probably because of a lesser number of competitive sinks (fruits), the lower fruit load and higher availability of photosynthates for the remaining fruits. Light or no pruned pomegranate trees (loading more fruits) showed lower fruit cracking (1.7%) compared with heavy pruned tress (2.5%), probably because big fruits absorbed much more water and the skin was less elastic than in small fruits (Chakma, 2014).

For weak cultivars, the intensity of pruning should be high, whereas in the case of vigorous cultivars the intensity should be kept low in order not to stimulate excessive growth of water sprouts and suckers. Suckers and water sprouts are the response of the tree to intense trimming in order to retrieve, as soon as possible, the missing parts. A recent study conducted in India revealed that heading back the fruiting shoots to 15 cm showed the best results in terms of shoot extension (56.34 cm), fruit size (diameter 8.66 cm and length 8.65 cm), fruit weight (278.50 g), marketable yield (10.25 kg), TSS (°Brix 13.60) and TSS/TA (42.10); good results were also achieved by heading back to 30 cm, a moderate intensity of pruning (Sharma and Singh, 2018).

Severe winter pruning on pomegranate trees showed a decrease in the population of both *A. punicae* and its natural enemies (Mdellel *et al.*, 2015). The intensity of pruning can also affect rainfall reaching the soil in rainfed pomegranate orchards because of the trapping of the water on leaves, branches and shoots (Hakimi *et al.*, 2018). A huge canopy is able to save and use a significant amount of water in this way. Decreases in tree height, canopy cover, crown length and leaf area index (LAI) were correlated with a significant increase in rainfall reaching the orchard floor for intensity of 40% pruning compared with the control (no pruning) (Hakimi *et al.*, 2018).

8.14.6 Pruning young trees

The pomegranate tree has elastic branches, which allows growers to adopt different training systems from vase to central leader or pergola. The pruning for training of young trees aims to give a better shape to the tree for each particular orchard. The main aspects to be considered, when pruning young trees, are the following: accordance with the tree habitus; stimulating early production; stimulating light interception; keeping the tree healthy; providing a strong scaffold; and facilitating orchard practices (pruning, pesticide treatments, harvesting, etc.). Following these criteria will lead to the final trellis of the tree, which will be kept during the whole life of the orchard.

8.14.7 Pruning mature trees

Once the training system has been given to the orchard, trees need to be pruned annually during the dormant period. This type of pruning stimulates the growth of fruiting shoots and avoiding an excessive development of the canopy with water sprouts and suckers. The dead, overlapping, thin and broken shoots need to be removed with this pruning in order to obtain a canopy with optimal health and light interception. The main aspects to be considered when pruning mature and productive trees are the following: balancing the vegetative and productive activities; delaying the senescence of the tree with rejuvenation cuts; elimination of damaged and dead parts; keeping optimal light and air in the canopy; and making horticultural practices easier. Here are the pruning steps to be adopted from young to mature trees in the case of two training systems, Ypsilon and open vase, in a temperate-warm climate of the northern hemisphere.

8.14.8 Y-trellis training system

In the first year, at the planting time of the trees (winter), there is the heading cut at 50–60 cm (February/March) from the collar of the tree (Cossio and Vitelli, 2018) (Fig. 8.23a). Since the tree is very delicate at this time, it is better to use poles (12–14 mm) and horizontal wires (80 cm above the ground) to support the young trees. The Y-structure should be also placed at this time together with the horizontal metal wires (2–3 mm). In the northern hemisphere, after bud sprouting (April/May), the basal shoots are trimmed in order to have a free trunk of 50–60 cm high (Fig. 8.23b). The shoots are subsequently (May/June–June/July) headed (Fig. 8.23c) to stimulate the development of several branches (6–12). The fruits in the first year, if present, should be dropped, in order to stimulate the vegetative activity and speed up the training of the tree. In summer (June/July–July/August), the upper shoots are tied up around the tutor/pole to reduce the bending of the branches (Fig. 8.23d). In the second year, before bud opening, the shoots are thinned to leave only the main branches (Fig. 8.23e), which are subsequently tied to the horizontal wires of the Y-structure (Fig. 8.23f). Summer pruning is essential at this time of season either for trimming or heading back some water sprouts (10–15 cm) in the basal part of the canopy for the renewal of the vegetation; poorly positioned water sprouts should be trimmed leaving the ones to possibly be used as new branches (Fig. 8.23g and Fig. 8.23h). Another important operation to be done in summer is desuckering because of the natural tendency of pomegranate to develop many suckers. Second-year fruits need to be thinned leaving only a few to complete the training system of the tree or can be completely dropped for better development of the trellis (Fig. 8.23i). During the third year the training system is generally completed. There is the thinning of the shoots, trimming either the weak and broken ones or the vigorous ones; the remaining shoots to be used as fruiting branches are tied up to

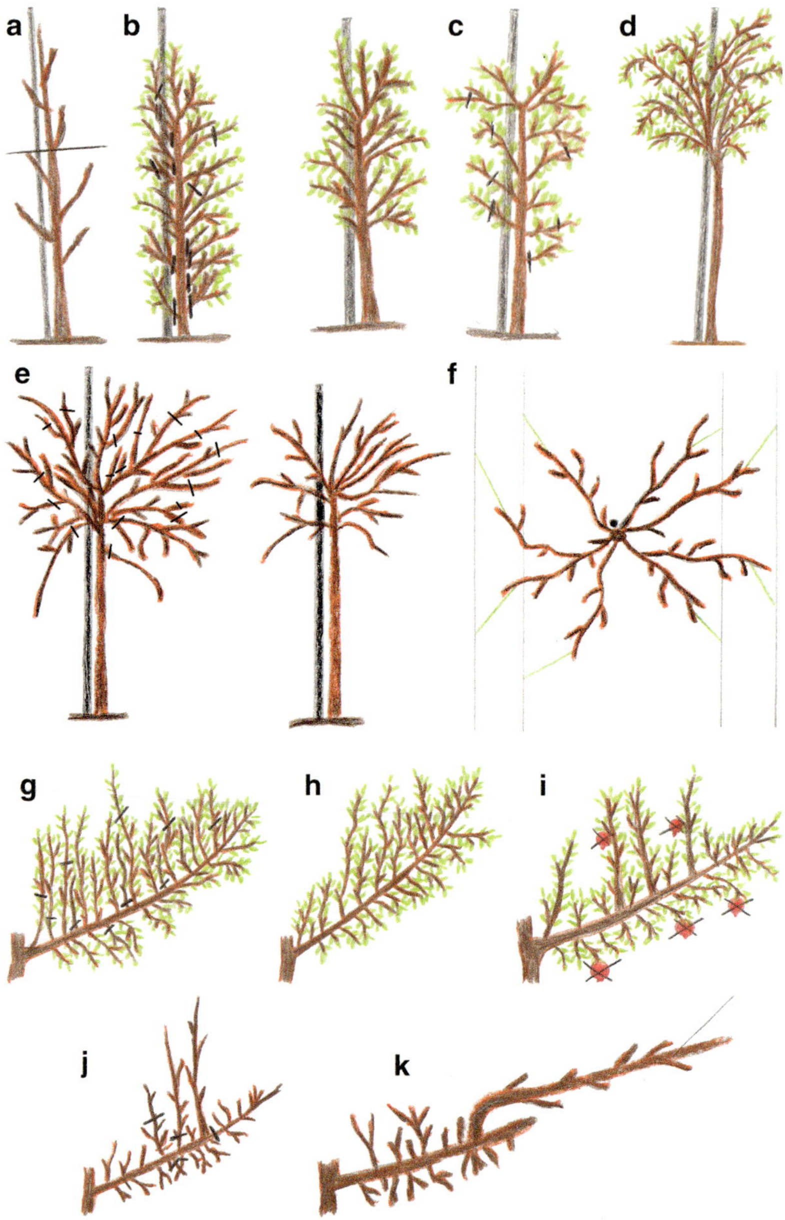

Fig. 8.23. Pruning tree in Y training system. (a) Young pomegranate tree at planting tied to a pole and cut at 50–60 cm. (b) Young pomegranate tree before (left) and after (right) the trimming of basal shoots in order to have a free trunk. (c) Heading of the shoots in spring–summer. (d) The upper shoots are tied up around the pole in summer to reduce the bending of the branches. (e) Before (left) and after (right) the thinning/heading back of the shoots. (f) The upper shoots are tied to the horizontal wires of the Y-structure. (g) The water sprouts and shoots before trimming/heading back. (h) The branch after the trimming/heading back of the shoots/water sprouts. (i) The fruits can be completely dropped or heavily thinned. (j) The trimming/heading back of the shoots, either the weak and broken ones or the vigorous ones. (k) Bending and tying the shoots to the structure. (Drawing: Mariarosa Mincuzzi and Giuseppe Ferrara, adapted from Cossio and Vitelli (2018).)

Fig. 8.24. (a) Mature Ypsilon trained pomegranate tree in winter. (b) Branches tied to wires. (Photos: Vito Vitelli.)

the wires of the structure often after bending (Fig. 8.23j and Fig. 8.23k). Some water sprouts can be used as new branches, and the old ones are trimmed. In summer, desuckering and trimming of some water sprouts should be done (to reduce competition) together with fruit thinning to obtain a good quality of fruits. From the fourth year the tree has the final shape and a generally good yield (20–30 kg/tree); from now onward it is important to carry out a summer pruning practice (i.e. fruit thinning, desuckering) for better quality fruits (size, colour, shape, organoleptic properties), and winter pruning to replace old branches with new fruiting ones (Fig. 8.24a). It is important to control the water sprouts (thinning) and renew the shoots both for vegetative and productive aspects (Cossio and Vitelli, 2018).

The Y-trellis training system has the advantage that it sustains the canopy with its metal structure, in particular in the presence of a heavy fruit load, because in traditional pomegranate orchards up to 20% of branches can break, whereas with this system branches are tied to wires (Fig. 8.24b) and can sustain the fruits with almost no broken branches (Atzmon, 2015). This system better protected the fruits of 'Wonderful' from sunburn (shading) compared with traditional trellising, when the internal temperature of sun-exposed fruits can reach up 47°C, much higher than that of shadowed fruits of 35°C (Atzmon, 2015). Moreover, this trellising system significantly reduces the scratches on the fruits caused by rubbing against shoots, thorns and branches. All horticultural practices are better facilitated with this system, such as the application of pesticides, which is more efficient, fruit thinning, pruning or harvesting. Fruit quality and yield of 'Wonderful' (Atzmon, 2015) are improved when trellising the trees with wires and poles compared with traditional systems, in particular with reduction of damaged fruits (20% vs. 50%), better red colour (80% vs. 50% coloured fruits) and yield (40–60 t/ha vs. 20–40 t/ha).

8.14.9 Open vase training system (Spanish system)

In the first year, at the planting of the trees (winter), there is a heading cut at 50–60 cm from the collar of the tree (Fig. 8.25a). The basal shoots are all trimmed up in the first 30–40 cm, and the tree is tied up to tutors or poles (3–4 cm) in order to avoid breakages (Fig. 8.25a). In summer (May/June), three or four vigorous shoots are selected and headed back to 25–30 cm for reinvigoration and stimulating new branching; basal shoots and suckers are trimmed (Fig. 8.25b). These shoots will be the primary branches of the trees, whereas weak and broken shoots are trimmed. The heading back can be repeated some weeks later to stimulate further branching (Fig. 8.25c and Fig. 8.25d). The fruits of the first year, if present, should be dropped in order to favour vegetative activity and speed up the training of the tree (Fig. 8.25d). In the second year, before buds opening, there is the thinning of the shoots to

Fig. 8.25. Pruning tree in open vase training system. (a) Trimming of basal shoots are all trimmed up in the first 30–40 cm. The young tree is tied up to a pole in order to avoid breakages. (b) Before (left) and after (right) both the thinning/heading back of the shoots and trimming of basal shoots/suckers. (c) Before (left) and after (right) heading back of shoots in summer time. (d) Pomegranate tree after heading back (seen from above). Fruits are completely dropped. (e) Before (left) and after (right) thinning/heading back of shoots during winter pruning. (f) Pomegranate tree after winter pruning seen from above. (g) The branch before summer pruning operations. (h) The branch with the cuts of summer pruning. (i) The fruits are heavily thinned or completely discarded. (j) The shoots before trimming/heading back in summer time. Fruits are thinned. (k) The shoots after trimming/heading back in summer time. (l) Heading back of the branch during winter pruning. (m) Bending and tying of the branch during winter pruning when a supporting structure is utilized. (Drawing: Mariarosa Mincuzzi and Giuseppe Ferrara, adapted from Cossio and Vitelli (2018).)

leave only six to eight secondary branches to stimulate shoot development (Fig. 8.25e and Fig. 8.25f). In summer, the heading back of water sprouts (10–15 cm) in the basal part of the branches has to be performed together with trimming of unnecessary shoots and desuckering (Fig. 8.25g and Fig. 8.25h). Second-year fruits need to be heavily thinned or completely discarded (Fig. 8.25i) to complete the training system of the tree (no support structure for this training system). In the third year there is the final development of the four to six branches arranged in a radial pattern; the apical shoots are used to keep the growth whereas the basal shoots are generally thinned or headed back. In summer (July), water sprouts in the internal canopy are trimmed or headed back (Fig. 8.25j and Fig. 8.25k), whereas the suckers developed are generally all trimmed. Fruits are thinned to three or four fruits per branch or even less depending on both the development and fruit load of the tree (Fig. 8.25j). From the fourth year onward, the new shoots are left in the outer part of the canopy (for the growth) with trimming of internal shoots (weak, old, exhausted). The summer pruning aims at leaving the shoots in the apical position of the branches (outer canopy) but also some water sprouts in the internal canopy (renewal of the fruiting branches). In the successive years, summer pruning operations will be important for both keeping the growth in the external part of the canopy and trimming the internal shoots; fruit thinning will be also done. The heading back and tying of some branches will also be necessary with winter pruning for the renewal of the vegetation (Fig. 8.25l and Fig. 8.25m). This is a straightforward, low-cost maintenance training system adopted in many growing areas of the world (Fig. 8.15 and Fig. 8.16). However, trees yield later than with the Y-system and there are more problems with sunburn and mechanical damage and with some horticultural practices (pesticide application, pruning, harvesting).

In general, some secondary branches should be allowed to develop from each main branch, but excessive growths that would lead to overcrowding should be removed, as should any suckers that develop from the base of the tree. Short fruiting spurs appear primarily on 2- or 3-year-old wood and are found growing mostly on the outer layer of the canopy. Light annual pruning encourages the growth of new fruiting spurs, while more aggressive pruning will significantly impact yield. Pruning of the fruiting tree will consist mainly of removal of excessive overcrowded growths, deadwood and suckers (Fig. 8.16). Adequate fruit-bearing wood should be retained. The height of pomegranate trees can be maintained to values easily managed by mechanical topping machinery.

8.14.10 Tree rejuvenation/reshaping

Apart from the types of pruning described above, there are also other types of pruning such as rejuvenation and reshaping. The rejuvenation pruning consists of intense pruning (cutting of branches) of old and/or damaged trees in order to stimulate new shoot formation, thus reducing the old unproductive wood. It is a type of pruning sometimes adopted to recover old trees. The reshaping pruning is adopted to change the trellis of the tree when it is no longer considered appropriate for optimal productivity as well as for its management. There are generally intense cuts (branches and even the trunk) and often the suckers are used to grow a new tree according to the new trellis.

These types of pruning are normally applied to old trees, which are still healthy with the aim of reinvigorating the tree to improve cropping potential. Even severely bacterial blight-infected or freeze-damaged orchards can be rejuvenated by proper pruning and orchard management practices to replace the lost trees. Major limbs (thick fruiting branches) are normally removed to encourage replacement with new young branches. This type of pruning should be carried out over a number of years depending on orchard condition and age of trees. Rejuvenation can be achieved via cutting back the tree to the crown during the first month of spring and by training and pruning of emerging shoots later on, by keeping three to five shoots for multi-trunk system or one shoot for single-trunk systems (Fig. 8.26). Pruning plants to 30 cm from ground level has been found to be the most effective method to rejuvenate the old pomegranate orchards and can be recommended as standard practice in rejuvenating old orchards (Hiwale *et al.*, 2006).

8.15 High-Density Orchard Management

In recent years, in California (USA) and other regions, growers have been willing to grow pomegranate trees as fruiting walls as is done

Fig. 8.26. Tree rejuvenation. (Photo: Alimohammad Yavari.)

for pome and stone fruits, olive, grape, almond and sweet cherry. This involves a vertical canopy development that rejects traditional bushy tree development in favour of dense, flat canopies (Day and Wilkins, 2011). These fruiting walls are developed by narrow pruning and hedging, with an increased tree density. The higher plant density can reduce the time required to reach a profitable yield when compared with lower densities (open vase/Ypsilon). Condensing pomegranate trunk, scaffold or branch growth into a narrower zone that can be more easily pruned, sprayed, thinned and harvested has the potential for improved efficiency and mimics the approaches already deployed for apple and pear production in many countries of the world (Day and Wilkins, 2011). Vertical wire trellises fit well within the context of the fruiting wall strategy as they are also used for the Y-system. It is possible to use either free-form multiple trunks or single trunks in a fruiting wall approach, but these forms would seem less apt to garner many of the benefits as quickly or as easily as is gained with the central leader system.

8.16 Intercropping

In modern orchard systems, pomegranate is grown as a standalone crop, as shown in the previous paragraphs on training systems and pruning. However, in some countries of the world pomegranate trees are intercropped with either other fruit tree species or vegetables in order to harvest more crops. In particular, in Mediterranean countries, before becoming an important crop, pomegranate was grown with other fruit crops such as olive, grape or almond.

Intercropping is one of the important cropping systems in both temperate and tropical Indian conditions and is adopted in small farming systems as it can yield more crops (fruits, vegetables, etc.) compared with a single-crop system (Bhatti *et al*., 2006). A recent work in India (Sharma *et al*., 2015) showed that intercropping systems with pomegranate and garlic, pea, urad bean, broccoli and turmeric significantly improved growth attributes of the pomegranate trees and physical, chemical and biological attributes of the rhizosphere compared with pomegranate trees grown as a single crop. In particular, among the intercropping systems tested, intercropping with urad bean and pea improved water and carbon content of the soil together with an increased concentration of nutrients and microbial biomass (Sharma *et al*., 2015).

Moreover, in some African countries of the Mediterranean basin pomegranate is often intercropped with different species. In Tunisia, the oasis is a particular horticultural environment and can be considered an intercropping system that is one of the most effective human adaptation strategies in an area with significant temperature variations (in particular high temperatures) and low precipitation (Cheneval, 2016). The oasis is a horticultural system based on date palm, which creates a microclimate that allows the cultivation of other fruit tree species (pomegranate, fig, citrus, etc.), forage crops and vegetables (Sellami and Sifaoui, 1998). In arid regions, the oasis has a vegetation cooling effect because of extensive evapotranspiration compared with the dry outer environment (Hao *et al*., 2016). The different canopy layers in the oasis horticultural system can reduce the incoming solar radiation, with cooler temperatures at the lower layer of the vegetation, such as vegetables and small fruit trees (Middel *et al*., 2014). Data have demonstrated that the summer air temperature of an oasis could be 2–7°C cooler than the surrounding desert environment, and as a result the relative humidity is generally 5–15% higher than the surroundings (Saaroni *et al*., 2004; Potchter *et al.*, 2008). Intercropping pomegranate in the oasis influences the quality of pomegranate fruits, and the arils of the fruits grown in the oasis present either a more intense red colour or higher acid content than the arils outside the oasis from trees grown as a single crop (Boussaa *et al.*, 2018). Moreover, fruits grown in the oasis in full shade have higher anthocyanin content and lower sugars than the fruits outside the oasis in full sun conditions (Boussaa *et al.*, 2018). The cooler and shaded microclimate of the oasis favours a higher content of volatile compounds, in particular, hexanal and limonene (Boussaa *et al.*, 2018).

A recent work (Boussaa *et al.*, 2019) showed that the oasis environment was more favourable to produce large-sized fruits with both higher aril yield and higher juice content compared with full sun conditions (outside the oasis) of the single-crop systems. Moreover, oasis pomegranates exhibited higher total anthocyanin content, hydrophilic antioxidant activity, and higher levels of magnesium, potassium and manganese content than fruits grown outside the oasis.

8.17 Endogenous Effects of Plant Growth Regulators (PGRs) on Flowering

Application of PGRs can influence the sex expression and distribution of flower types in pomegranate. Gibberellic acid induced more male flowers and reduced hermaphrodite flowers, whereas Ethrel® and maleic hydrazide induced more hermaphrodite and fewer male flowers (Abubakar *et al.*, 2012). Application of plant biostimulants on 'Kandhari Kabuli' cultivar significantly improved flowering, yield, return bloom and reduced the fruit drop. The highest flowering, yield/plant and minimum fruit drop were recorded on trees treated with Spic cytozyme (4 ml/l), whereas the highest return bloom was observed with the application of Vipul (15 ml/l) (Abubakar *et al.*, 2012). Application of GA_3 at 100 ppm (45, 90, 103 and 135 days after fruit set) plus $CaCl_2$ 2% + borax 0.2% + $MgSO_4$ 0.5% increased the total sugar content (15.73%) and the juice percentage (44.66%) of the fruits and favoured a longer shelf-life and the lowest loss in weight (Deepa *et al.*, 2018). Application of NAA at 50 ppm and Ethrel® at 200 ppm were found effective for fruit set, yield and other quality components. In particular, NAA at 50 ppm was found to be effective in increasing the number of fruits per tree, fruit weight, yield, number of hermaphrodite flowers and reduced fruit drop; Ethrel® at 200 ppm was able to advance the first harvest of fruits, and changed the flower sex/ratio by reducing the number of male flowers compared with hermaphroditic ones (Goswami *et al.*, 2013). The cultivar 'Kandhari' responded to the application of different PGRs (NAA, GA_3, 2,4-D) applied at various concentrations (Phawa *et al.*, 2017). The application of GA_3 at 75 ppm showed maximum plant height (194.90 cm), canopy volume (3.81 m^3), internode length (8.07 cm), shoot length (15.57 cm) and leaf area (7.37 cm^2) (Phawa *et al.*, 2017). The maximum number of flowers (28.01) was recorded with the application of 75 ppm GA_3 followed by 40 ppm NAA (28.58). An increase in number of flowers following GA_3 application might be due to the fact that the trees sprayed with GA_3 and NAA remained physiologically more active to build up sufficient food reserves for developing new flowers. Auxins are also known to stimulate flower bud initiation. Hence, the increase in flowering may be due to enhanced photosynthesis, which increased the potential of trees to develop more flower buds (Phawa *et al.*, 2017). Application of NAA at 40 ppm advanced the time of flowering as days (23.67) compared with the control (26.21), but earlier flowering by 2–3 days was also obtained with the other PGRs (Phawa *et al.*, 2017).

In order to reduce fruit drop in pomegranate, some ethylene-inhibiting chemicals have been used, such as cobalt chloride and KNO_3 (Reddy *et al.*, 2011). Application of 2.5 ppm cobalt chloride recorded a significant increase in number of fruits and yield per tree followed by 5 ppm cobalt chloride and KNO_3 treatments, and the effects were more evident with foliar applications compared with soil application (Reddy *et al.*, 2011).

Quality and yield parameters were improved in 'Kandhari' cultivar after the application of NAA, GA_3, 6-BA and their combination at different concentrations compared with a control treatment (Thakur and Sharma, 2018). Among the PGRs applied, NAA at 30 ppm after flowering (May) was the most effective in improving fruit weight and size, aril weight, juice and ascorbic acid content (Thakur and Sharma, 2018).

Fig. 8.27. Spring frost damage. (Photo: Giuseppe Ferrara.)

8.18 Freezing and Frost Protection

Pomegranate can withstand frosty conditions during winter time but will not survive long below −15°C (Soloklui *et al.*, 2012). Cold temperatures can severely damage pomegranate trees; either winter freezes or spring frosts (Fig. 8.27) can cause damage to trees in Mediterranean climates. In California (San Joaquin Valley) frost events occurred where temperatures dropped to as low as −3°C in March 2008 and 2009 (Day and Wilkins, 2011) causing damage to pomegranate trees. During spring time, even a few hours (4–6 h) below 0°C can irreversibly damage the trees. Mature trees were unaffected but a number of young orchards suffered shoot dieback and growers were forced to regrow trees from resulting ground suckers. However, low temperatures in winter time can cause damages even to mature trees. In December 1990 in California temperatures dropped to −6 to −8°C and a number of mature orchards were damaged severely, with death of scaffolds occurring for several years afterwards, and some young orchards were killed outright (Day and Wilkins, 2011). In order to face these events, frost protection similar to strategies used in citrus during critical periods can be used also for pomegranate (Day and Wilkins, 2011). Frost could be a real threat to young pomegranate plantings and special care should be taken to reduce the danger of frost in certain areas. Trees planted in open lower areas and trees exposed to prevailing cold winds are most likely to suffer frost damage. Freezing tolerance can vary among the different pomegranate cultivars as reported in a trial in Iran (Soloklui *et al.*, 2012). Iranian pomegranate cultivars 'Naderi', 'Yusef Khani', 'Malas Saveh', and 'Robab Neyriz' showed the highest midwinter cold hardiness; 'Mahabadi' showed intermediate hardiness, whereas 'Poost Sefid Bafgh' and 'Shishe Kap' were found to be cold susceptible. Cold tolerance of the cultivars, from autumn to midwinter, was significantly correlated with soluble carbohydrate content, which plays an important role in osmotic adjustment of cells to withstand frost (Soloklui *et al.*, 2012), but the natural habitat of the variety (genetic characteristics) is also a very important factor that determines its degree of cold hardiness.

Paclobutrazol (PBZ) is a member of the triazole plant growth inhibitors group able to induce tolerance to a number of biotic and abiotic stresses (Moradi *et al.*, 2017). Application of PBZ on seedlings of the pomegranate cultivar 'Robab' subjected to freezing stress (−3°C for 7 h) improved the growth rate and increased relative leaf chlorophyll content, chlorophyll fluorescence ratio, relative water content, soluble carbohydrate content, and enzyme activity of ascorbate peroxidase and guaiacol peroxidase compared with the control (Moradi *et al.*, 2017). PBZ at a concentration of 75 mg/l ameliorated the injury caused by freezing stress by reducing proline content and leaf electrolyte leakage (Moradi *et al.*, 2017).

8.19 Salinity Management

Salinity is one of the most important environmental stresses that severely limits plant growth and productivity in different ways, such as leading to water deficiency, specific ion toxicity and/or ionic imbalance, or a combination of these factors (Karimi and Hassanpour, 2017).

Pomegranate is considered moderately tolerant to salinity (Naeini *et al.*, 2005), but, as for other physiological aspects, the tree response to salinity is variety dependent (Naeini *et al.*, 2006; Okhovatian-Ardakani *et al.*, 2010). Holland *et al.* (2009) noted that pomegranate was amenable to irrigation with saline water (between 2.5 and 4.0 dS/m) and produced normal yield. Based on studies on evapotranspiration, crop coefficient and growth of two young pomegranate cultivars under salt stress, the term moderately sensitive to salinity has been suggested to more adequately describe the response of pomegranate to irrigation with saline water (0.8–8.0 dS/m) (Bhantana and Lazarovitch, 2010). Increasing irrigation with saline water showed positive effects on phenolic accumulation in the arils of two pomegranate accessions but with negative effects on anthocyanins to an extent depending on the accession (Borochov-Neori *et al.*, 2013). The growth of pomegranate cultivars 'Alak Torsh' and 'Malas Torsh', irrigated with saline waters, showed a decline from 0 down to 40 mM NaCl, as indicated by the number of internodes, length of the main stem, length of internodes and leaf surface area (Naeini *et al.*, 2006). In an 80-day experimental period on three pomegranate cultivars ('Malas Shirin', 'Alak Torsh' and 'Malas Torsh'), irrigation with waters containing 0, 40, 80 and 120 mM NaCl increased Na, Cl and K concentrations and decreased Ca, Mg and N concentrations in the plant tissues; and soluble sugar content was also reduced with increasing NaCl concentrations (Naeini *et al.*, 2005). Doring and Ludders (1986) reported that the chlorophyll concentration and photosynthetic efficiency of pomegranate leaves decreased with increasing NaCl concentrations in culture solution and also a positive correlation existed between concentrations of Na and Cl in plant tissue with those in the culture solutions.

The irrigation of 'Wonderful' and 'SP-2' accession with waters at increasing EC (1.2–8.0 dS/m) showed various effects on fruit quality (Neori *et al.*, 2014). The content of phenolics in the fruit peel of both pomegranate accessions increased considerably when the irrigation water salinity was raised above EC 3 dS/m and at high salinity levels (EC 6 and 9 dS/m). The 'SP-2' accession accumulated significantly higher level of phenolics compared with 'Wonderful' (Neori *et al.*, 2014). A function in plant defence against salinity-induced oxidative stress was suggested for phenolics, significantly increasing at high EC values (Gould and Lister, 2006; Di Ferdinando *et al.*, 2012). A higher salt tolerance of the 'SP-2' accession compared with 'Wonderful' may be related to the more abundant accumulation of phenolics in this accession (Neori *et al.*, 2014). Increasing irrigation water salinity favoured anthocyanin accumulation in both pomegranate accessions, but 'Wonderful' fruit peel was significantly richer in anthocyanins compared with the 'SP-2' accession up to EC 6 dS/m, whereas at EC 9 dS/m pigment concentrations in both accessions were similar (Neori *et al.*, 2014). The anthocyanins detected in fruit peel of 'Wonderful' and 'SP-2' were mono- and diglucosides of cyanidins, pelargonidins and delphinidins; with high salinity water, 'Wonderful' fruit peel accumulated more of the purple delphinidins, whereas 'SP-2' had more of the orange-coloured pelargonidins (Neori *et al.*, 2014). The accumulation of anthocyanins in the peel and a reduction in the arils (Borochov-Neori *et al.*, 2013) could be explained by the existence of different pathways of anthocyanin synthesis in the pomegranate fruit peel and aril (Neori *et al.*, 2014). High salinities, EC 6 and 9 dS/m, were associated with significantly elevated concentrations of gallotannins, ellagic acid derivatives and flavonols in 'Wonderful', and punicalagins, ellagic acid derivatives and flavonols in 'SP-2' (Neori *et al.*, 2014).

Salinity of water can affect the absorption and translocation of nutrients within the pomegranate tree (Hasanpour *et al.*, 2015). In a potted experiment with the Iranian pomegranate cultivars 'Rabab' and 'Shishegap', water salinity (sodium and calcium chloride salts) affected the concentration of iron (Fe^{2+}), zinc (Zn^{2+}), copper (Cu^{2+}) and manganese (Mn^{2+}) in leaves and roots (Hasanpour *et al.*, 2015). The concentration of zinc (Zn^{2+}), copper (Cu^{2+}) and manganese (Mn^{2+}) in roots and manganese (Mn^{2+}) in shoots increased depending on the salinity of the water (0, 30 and 60 mM) with the highest concentrations detected at 60 mM. The high salinity negatively affected the soil plant analysis development (SPAD) values and the Fm/Fv ratio of the two pomegranate cultivars (Hasanpour *et al.*, 2015). As expected, water salinity increased the concentration of sodium (Na^{+}), chloride (Cl^{-}), calcium (Ca^{2+}) and potassium (K^{+}) both in roots and shoots (Karimi and Hasanpour, 2014). The two pomegranates showed a different behaviour towards the relative water content (RWC) of leaves; the RWC of 'Rabab' decreased in an irrigation interval of 6 days with 60 mM of salinity, whereas 'Shishegap' was not affected, which could be related to the higher tolerance to salinity of this cultivar (Hasanpour *et al.*, 2015). Differences between the two pomegranate cultivars for the salinity effects were also observed for the leaf fresh and dry weight, since in 'Shishegap' the 30 mM salinity treatment increased leaf fresh and dry weight compared with the control, whereas in 'Rabab' there was a significant decrease in the leaf weight (Karimi and Hasanpour, 2014). The 'Shishegap' cultivar was able to restrict either the uptake/transport of Cl or maintain sufficient levels of K, better than 'Rabab' (Karimi and Hasanpour, 2014). Okhovatian-Ardakani *et al.* (2010) investigated the salt tolerance of different pomegranate cultivars and reported that 'Malas-e-Yazdi' and 'Tab-o-Larz' are salinity-tolerant cultivars, while 'Gabri' and 'Khafr-e-Jahrom' are the most sensitive ones. Salinity tolerance of 10 commercial Iranian cultivars in pots was reported by Tabatabaei and Sarkhosh (2006) with significant differences among them.

In general, the effect of salinity stress on pomegranate cultivars is enhanced by increasing irrigation intervals, from 2–6 days (Hasanpour *et al.*, 2015). The presence of a rootstock can also affect the concentration of nutrients in pomegranate; the low sodium concentration in shoots of 'Gabri' cultivar grafted on 'Tab-o-Larz' as rootstock was possibly related to the higher ability of 'Tab-o-Larz' compared with 'Malas-e-Yazdi' to accumulate/exclude Na in roots and prevent the transportation to shoots (Karimi and Hassanpour, 2017). Pomegranate cultivars/rootstocks lacked the mechanisms to control Cl levels in leaves, and the negative salinity effect was associated with the accumulation of Cl more than Na in this organ (Karimi and Hassanpour, 2017). Na and Cl accumulated more in leaves located in the lower part of the canopy compared with the upper part, which is probably a mechanism of pomegranate to prevent the harmful effects of these ions on the young outer fruiting shoots and leaves together with the abscission of the leaves of the lower part in order to remove the accumulated toxic ions (Karimi and Hassanpour, 2017). 'Gabri' cultivar grafted onto 'Tab-o-Larz' and 'Malas-e-Yazdi' as rootstocks showed a lower concentration of Cl and Na in shoots compared with non-grafted 'Gabri' trees. The lower Cl and Na levels in leaves could be due to the capacity of roots of 'Tab-o-Larz' and 'Malas-e-Yazdi' rootstocks to either store these ions or reduce their transportation to the scion, and the 'Tab-o-Larz' as rootstock was more efficient in this activity (Karimi and Hassanpour, 2017). Grafted pomegranates had more K than non-grafted ones probably because of a more selective uptake of K compared with Na by the rootstock, which could partially explain the higher tolerance to salinity (Karimi and Hassanpour, 2017). Salinity decreased the concentration of Mg in shoots and this effect was less evident in plants grafted on 'Tab-o-Larz' (Karimi and Hassanpour, 2017). A recent study conducted on pomegranate cultivars 'Malas-e-Saveh' and 'Shishe Kab' in greenhouse and field using irrigation water with EC values ranging from 1.5–12 dS/m and 1.05–7.46 dS/m for greenhouse and field, respectively, showed differences between cultivars, and 'Malas-e-Saveh' had more tolerance compared with 'Shishe Kab' (Khayyat *et al.*, 2016). Salinity of water reduced the chlorophyll content in pomegranate leaves but significantly increased chloride and sodium content as a consequence of the higher EC of water (Khayyat *et al.*, 2016). Moreover, use of high EC water was negatively correlated with the content of several nutrients such as potassium, nitrogen, calcium and magnesium. Non-photochemical quenching, effective quantum yield of photochemical energy conversion in PSII reduced under the highest salinity level in the field; however, basal quantum yield of non-photochemical processes in PSII increased in the highest salinity irrigation (Khayyat *et al.*, 2016).

8.20 Weed Management (Mechanical, Chemical, Mulching)

Weed control in pomegranate orchards is important as weeds create competition for water and nutrients. Weeds can be managed by mowing (Fig. 8.3), tillage (Fig. 8.28), cover crop in the alleys or application of herbicides. Weeds are controlled mainly with pre-emergence herbicides,

Fig. 8.28. Tilled pomegranate orchard. (Photo: Giuseppe Ferrara.)

while, after germination, in the plant rows, glyphosate can be applied as a post-emergence herbicide. Chemical control can include application of post-emergence herbicides such as glyphosate (Roundup®) and oxyfluorfen (Goal®) on all pomegranates or flumioxazin (Chateau®) on non-bearing pomegranates. Pre-emergence herbicides including oryzalin (Surflan®) and napropamide (Devrinol®) are also used. In organic farming hand labour plus mechanical tillage/mowing is typically employed (Day and Wilkins, 2011).

Weeds could also host insect pests as well as diseases that could create problems in the orchard, but cover crops (living mulch) could be seeded in the alleys to reduce weeds and act as hosts for natural enemies of potential pests and diseases. Mulching, in addition to preventing weed development, preserves the soil moisture around the young plants. Mulching with a black plastic sheet is not recommended due to the considerable heat it causes in the soil but can help in controlling the growth of the weeds on the row (Fig. 8.5). Alternatively, a two-coloured plastic sheet can be used for mulching, with one side black (inside) and the other side white (outside) (Fig. 8.4); the latter must be set upwards to decrease the temperature accumulation of the soil and reflect the light to the lower part of the canopy for a better bud differentiation and fruit development and ripening (Fig. 8.19).

Equally valid is mulching with breathable mesh sheets, better if white or green (anti-algae), available in nurseries, which allow the passage of water (Fig. 8.29). This type of material for mulching limits the problems related to the reduced ventilation that leads to trunk rot and galls. Mulching plays a very important role with respect to moisture conservation, and reduction of weeds and nematode control. Deficiency of

Fig. 8.29. Mulched row with breathable mesh sheets. (Photo: Giuseppe Ferrara.)

moisture in the soil during the fruit development stage is responsible for fruit cracking (following rainy events) and reduces fruit quality in pomegranate. Besides controlling the weed population, mulching with black polythene reduces the population of nematodes, which is one of the main causes of wilting in pomegranate (Warade *et al.*, 2008). Warade *et al.* (2008) reported that mulching pomegranate trees with black polythene showed better performance than other treatments (sugarcane trash, soybean straw and wheat straw) with respect to yield and number of fruits. Highest marketable fruits and total yield (15.55 and 17.09 kg/tree, respectively) were recorded by providing mulching with a black polythene sheet. The average number of fruits per tree (58.75) and average weight of fruits (268.75 g) was also found to be maximal with black polythene mulch treatment.

Chattopadhyay and Patra (1992) investigated the effect of different mulches (black polythene, banana trash and sawdust) on pomegranate yield in west Bengal. They found the maximum number of fruits (60 fruits/tree) with black polythene followed by sawdust (51 fruits/tree), banana trash (44 fruits/tree) and the control (no mulch) (39 fruits/tree), respectively. The maximum increase in plant height (12.90 cm) was obtained with black polythene and minimum (8.52 cm) in the 'no mulch' treatment (Chattopadhyay and Patra, 1992). Singh *et al.* (1990) studied the influence of different cultural practices on premature fruit cracking and found that mulching with dry grass significantly reduced fruit cracking of pomegranate over all the other practices. This reduced fruit cracking may be the consequence of higher soil moisture content in mulched orchards with respect to clean ones, thus reducing the adverse effects of rains leading to rapidly increasing soil moisture with a consequent higher accumulation of water in fruits thus causing the cracking.

Studies in west Bengal (Chattopadhyay and Patra, 1997) revealed that the influence of soil cover in conserving soil moisture did not follow any pattern. However, the variation in soil moisture at 25 cm and 50 cm depths below the mulch was significant in November, December, January and February. In general, the black polythene resulted in better moisture conservation in the months of no or low rainfall. Chakma (2014) studied the effects of different mulching treatments on the row and reported positive effects of black polythene and grass mulch (10 cm) on shoot growth (≈50 vs. 40 cm) and tree volume (8–9 vs. 6 m^3; Aliev, 1979) compared with herbicides or no mulching at all. Black polythene and grass mulch exerted positive effects also on fruit set, fruit weight and size, TSS and yield (Chakma, 2014), probably because of higher soil moisture facilitating absorption of nutrients by the root system.

Apart from these effects on the tree, the use of mulching can also have positive effects on soil management in order to conserve soil moisture in regions facing water scarcity. The use of black polythene and grass mulch was able to keep higher moisture in the soil compared with herbicides or no mulching during almost the whole growing season (Chakma, 2014). Harvest equipment moving through commercial orchards requires dry soil and the use of inter-row living mulching (cover crops) can facilitate early footing in the orchards. Autumn heavy rain can result in severe damage to the fruits, such as cracking, in particular for late-ripening cultivars. In this case, the presence of a cover crop for mulching in the inter-row can reduce such a problem either by facilitating harvest operations or reducing fruit cracking from almost 100 to 40% (Fathi Abd Elhadi, unpublished data).

Mulching exerted positive effects on 7-year-old trees of cultivar 'Kandhari Kabuli' in India. In particular, black polythene proved to be most effective in increasing plant growth (51.80 cm), yield (13.05 kg/tree) and in conservation of soil moisture. Grass mulch was better when compared with other orchard floor management practices in terms of growth (47.00 cm) and yield/tree (12.35 kg). Black polythene and grass mulching also improved quality parameters such as TSS, TA, TSS/TA and ascorbic acid content (Sharma *et al.*, 2017). These data indicated that mulching can be a sustainable tool for modern management of pomegranate orchards for different goals: control of weeds, sustainable environment, conservation of soil moisture, better yield and quality of fruits, no use of herbicides and ease of passage in the orchards after rainy events. The use of olive pomace mulch was able to control the growth of weeds and reduce evapotranspiration compared with control in a pomegranate orchard of 'Manfalouty' in desert conditions in Egypt (Seidhom and

Abd-El-Rahman, 2011). The application of olive pomace had positive effects also on yield and water use efficiency with values significantly higher than control and bitumen mulch, in particular when a 6-day irrigation interval was adopted (Seidhom and Abd-El-Rahman, 2011). A recent experiment showed the interesting results of application of pecan (*Carya illinoensis* (Wangenh.) K.Kock) hulls in controlling weed growth beneath the pomegranate trees close to the collar; moreover, application of almond hulls on the row applied as a layer of 4–5 cm or even thicker lasted for at least 2 years and completely inhibited the growth of weeds thus avoiding the use of chemical herbicides in the row (Ferrara, unpublished data). Organic mulching improves water retention capacity and prevents carbon losses from the soil; moreover, there is the chance to reuse the organic materials produced on the farm (pecan and almond hulls, pomegranate rinds, grape pruning materials, olive pomace, etc.).

8.21 Tillage and cover crops

The soil of a pomegranate orchard can be covered with permanent vegetation or managed with tillage, mulching or use of herbicides as described in the previous paragraph. Tillage in pomegranate orchards is accomplished with year-round use of disc harrows. Clean cultivation (Fig. 8.28) improves air circulation and reduces the competition with weeds, but reduces organic matter with potential adverse effects on soil structure and erosion. In the past, when pomegranate was considered a low-importance tree fruit species, there was no use of irrigation and clean cultivation was generally adopted to control weeds and maximize the use of water. Recent interest in growing pomegranate and the environmental aspects have induced some changes in soil management, with reduced or no-tillage at all. The orchard floor of pomegranate trained to the Ypsilon or vase system is kept clean of weeds by using mulching material such as coloured polythene, TNT fabrics, organic materials, etc. Ridges (raised beds) are commonly adopted to improve drainage and root air circulation and are covered with the mulching material; beneath the mulches the irrigation pipes are generally placed. The alleys can be either tilled or covered with living mulch such as *Trifolium* species or a mix of leguminous and monocot species in order to improve organic matter and N content in the soil. The soil cover of the alleys exerts a very limited or null competition with the pomegranate trees because of the distance of the cover crops from the root system of the pomegranate, in particular when ridges are used in the orchard. The competition can be better managed and almost discounted depending on the species, the distance from the root system, irrigation and fertilization management. The use of fertigation and application of water close to the pomegranate root system and the presence of leguminous species in the alleys can reduce the competition and improve soil characteristics. Moreover, ground cover of the alleys affects the sustainability of the orchard by storing the nutrients in the soil and facilitating machine access. Ground cover can be used in autumn–winter–spring (seeded or natural) with two or three mowings to control the height; in late spring cover crops are mowed to the ground or incorporated into the soil.

Permanent vegetation can also be used beneath the pomegranate trees with dichondra or clover, with positive effects on N in the soil but can be competitive for nutrients and water. Weeds on the tree row may be controlled with herbicides, or more sustainably with organic (straw, mowed material, pruning residues, hulls, etc.) or living mulches (leguminous species) successively mowed. The use of organic mulch in the tree row can improve the soil moisture retention and release nutrients, in particular during the dry season; the layer should be at least 8–10 cm deep to last for a couple of seasons. Cost and transportation of the organic material to be used as mulch can limit its application, but the use of available material from the farm can overcome this problem, such as mowed grass, hulls, pruning materials, residues of fruits (pomegranate, orange, olive, grape, etc.). Inorganic mulches such as black/white plastic and geotextile sheets can also be applied, but their cost is generally higher than the organic ones, and they also present a problem of disposal although they generally last longer. Some reflective inorganic mulches can also be used to improve bud differentiation, flowering and fruit quality such as the red colour of the skin.

Management of the pomegranate orchard could include a clean tree row (or mulched) and the alleys with cover crops or natural weeds or tilled. Subterranean clover (*Trifolium subterraneum* L.) can be used as a cover crop to improve soil physical and chemical properties, improving water retention, nitrogen and organic matter content, and has a good reseeding ability. Subterranean clover reseeds in early summer and dies in full summer when temperatures are high, thus not being competitive with the pomegranate tree but making a layer of organic mulch beneath the tree. The use of spontaneous weeds as the ground cover can favour the development of a wide fauna community useful for the control of some pests and thus stimulating the biological activity in the soil.

However, young pomegranate trees are very susceptible to competition with weeds compared with mature pomegranate trees. Weeds on the tree row can be mechanically controlled but this should be done with care in order to avoid damage to the trunk; alternatives could be fire and high-pressure water, but the use of these devices can have some drawbacks, such as fuel hazards (flame) or limited efficacy (water). However, in arid areas, the land can be kept tilled, as done in some areas of Spain; surface tillage in the alleys with a harrow has the advantage of interrupting the capillarity and preserving the soil water more efficiently.

References

Abdel-Galil, H.A. (2008) Effect of suckering and girdling on yield and fruit quality of 'Manfalouty' pomegranate under Assiut environments. *Assiut Journal of Agricultural Sciences* 39, 83–100.

Abubakar, A.R., Ashraf, N. and Ashraf, M. (2012) Effect of plant biostimulants on flowering, fruit drop, yield and return bloom of pomegranate cv Kandhari Kabuli. *The Asian Journal of Horticulture* 7, 473–477.

Ahmed-Amin, K.I., El-Salhy, A.M., Abdel Galil, H.A. and El-Kholy, M.S. (2000) Effect of girdling, fruit thinning and uniconazole spraying on fruiting of 'Manfalouty' pomegranate. *Assiut Journal of Agricultural Sciences* 31, 177–189.

Aliev, M.A. (1979) Flowering and fruiting of pomegranate in relation to number of stems per bush. *Subtropicheskie Kultury* 6, 63–65.

Atzmon, I. (2015) Growing pomegranates using a trellising system. *Acta Horticulturae* 1089, 427–430. DOI: 10.17660/ActaHortic.2015.1089.59.

Babu, D.K. (2010) Floral biology of pomegranate (Punica granatum L.). In: Chandra, R. (ed.) *Fruit, Vegetable and Cereal Science and Biotechnology: Pomegranate*, Special Issue 2. Volume 4. Global Science Books, pp. 45–50.

Balasubramanyan, S., Anbu, S., Bangarusamy, U. and Chokalingam, P. (1997) Effect of pruning and training on growth, yield and quality of pomegranate in black soil under rainfed conditions. *South Indian Horticulture* 45, 271–273.

Bhantana, P. and Lazarovitch, N. (2010) Evapotranspiration, crop coefficient and growth of two young pomegranate (*Punica granatum* L.) varieties under salt stress. *Agricultural Water Management* 97(5), 715–722. DOI: 10.1016/j.agwat.2009.12.016.

Bhatti, I.H., Ahma, R., Jabbar, A., Nazir, M.S. and Mahmood, T. (2006) Competitive behavior of component crops in different sesame–legume intercropping systems. *International Journal of Agriculture and Biology* 8, 165–167.

Bindraban, P.S., Dimkpa, C., Nagarajan, L., Roy, A. and Rabbinge, R. (2015) Revisiting fertilisers and fertilisation strategies for improved nutrient uptake by plants. *Biology and Fertility of Soils* 51(8), 897–911. DOI: 10.1007/s00374-015-1039-7.

Borochov-Neori, H., Lazarovitch, N., Judeinstein, S., Patil, B.S. and Holland, D. (2013) Climate and salinity effects on color and health promoting properties in the pomegranate (*Punica granatum* L.) fruit arils. *ACS Symposium Series, Tropical and Subtropical Fruits: Flavors, Color, and Health Benefits* 1129, 43–61.

Boussaa, F., Zaouay, F., Hernandez, F., Noguera-Artiaga, L., Carbonell-Barrachina, A.. *et al*. (2018) Cropping system contributes largely to fruit composition and sensory properties of pomegranate (*Punica granatum* L. var. Gabsi). *South African Journal of Botany* 115, 170–178. DOI: 10.1016/j.sajb.2018.01.016.

Boussaa, F., Zaouay, F., Burlo-Carbonell, F., Nuncio-Jáuregui, N., Gmati, M. *et al.* (2019) Combined effects of cropping system and harvest date determine quality and nutritional value of pomegranate fruits (*Punica granatum* L. cv. Gabsi). *Scientia Horticulturae* 249, 419–431. DOI: 10.1016/j.scienta.2019.02.007.

Camacho-Cristóbal, J.J., Rexach, J. and González-Fontes, A. (2008) Boron in plants: deficiency and toxicity. *Journal of Integrative Plant Biology* 50(10), 1247–1255.

Chakma, J. (2014) Effect of different orchard management practices on the growth, productivity and rejuvenation of declining trees of pomegranate (*Punica granatum* L cv. Kandhari Kabuli). MSc Thesis. Yashwant Singh Parmar University of Horticulture & Forestry, Solan, India.

Chattopadhyay, P.K. and Patra, S.C. (1992) Effect of soil covers on growth, flowering and yield of pomegranate (*Punica granatum*). *South Indian Horticulture* 40, 309–312.

Chattopadhyay, P.K. and Patra, S.C. (1997) Effect of mulches on soil temperature in pomegranate. *Indian Journal of Horticulture* 54, 280–282.

Cheneval, J.B. (2016) How to enhance resilience for oasis ecosystems in Maghreb? *CIHEAM Watch Letter* 36, 90–92.

Cossio, F. and Vitelli, V. (2018) Il Melograno: Botanica, Varietà, Impianto, Cure Colturali, Difesa e Utilizzi (Guida Illustrata). In: *Supplemento a Vita in Campagna*. 42. L'Informatore Agrario, Verona, Italy.

Day, K.R. and Wilkins, E.D. (2011) Commercial pomegranate (*Punica granatum* L.) production in California. *Acta Horticulturae* 890, 275–286.

Deepa, M.G., Patil, S.N., Gollagi, S.G., Patil, D.R., Suma, R. *et al.* (2018) Effect of foliar application of gibberellic acid and nutrients on physiology and quality of pomegranate (*Punica granatum* L.) cv. Bhagwa under northern dry zone of Karnataka. *International Journal of Chemical Studies* 6, 3403–3407.

Di Ferdinando, M., Brunetti, C., Fini, A. and Tattini, M. (2012) Flavonoids as antioxidants in plants under abiotic stresses. In: Ahmad, P. and Prasad, M.N.V. (eds) *Abiotic Stress Responses in Plants: Metabolism, Productivity and Sustainability*. Springer Science and Business Media, LLC, New York, pp. 159–180.

Doring, J. and Ludders, P. (1986) Effect of different salt treatment on *Punica granatum* L. at different root treatments. *Die Gartenbauwissenschaft* 52, 92–96.

Durand, G. (1997) Effects of light availability on the architecture of canopy in mango (*Mangifera indica* L.) cv. Manzana trees. *Acta Horticulturae* 455, 217–227.

El-Kholy, M.S. (2005) Histological and physiological studies on flowering and fruiting of pomegranate (*Punica granatum*) under Assiut environments. PhD Thesis. Assiut University, Assiut, Egypt.

Eshghi, S., Teixeira da Silva, J.A. and Ranjbar, R. (2010) Molybdenum and boron affect pollen germination of strawberry and fertile and infertile flowers of pomegranate. *Fruit, Vegetable and Cereal Science and Biotechnology* 4(2), 148–150.

Ferrara, G., Cavoski, I., Pacifico, A., Tedone, L. and Mondelli, D. (2011) Morpho-pomological and chemical characterization of pomegranate (*Punica granatum* L.) genotypes in Apulia region, southeastern Italy. *Scientia Horticulturae* 130(3), 599–606. DOI: 10.1016/j.scienta.2011.08.016.

Ferrara, G., Giancaspro, A., Mazzeo, A., Giove, S.L., Matarrese, A.M.S. *et al.* (2014a) Characterization of pomegranate (*Punica granatum* L.) genotypes collected in Puglia region, southeastern Italy. *Scientia Horticulturae* 178, 70–78. DOI: 10.1016/j.scienta.2014.08.007.

Ferrara, G., Mazzeo, A., Netti, G., Pacucci, C., Matarrese, A.M.S. *et al.* (2014b) Girdling, gibberellic acid, and forchlorfenuron: effects on yield, quality, and metabolic profile of table grape cv. Italia. *American Journal of Enology and Viticulture* 65(3), 381–387. DOI: 10.5344/ajev.2014.13139.

Ferrara, G., Malerba, A.D., Matarrese, A.M.S., Mondelli, D. and Mazzeo, A. (2018) Nitrogen distribution in annual growth of 'Italia' table grape vines. *Frontiers in Plant Science* 9, 1374. DOI: 10.3389/fpls.2018.01374.

Ghosh, S.N., Bera, B., Roy, S., Kundu, A. and Bhattacharyya, A. (2012) Effects of crop management factors on pomegranate cultivation in West Bengal. *Acta Horticulturae* 940,163–170. DOI: 10.17660/ActaHortic.2012.940.20.

Giancaspro, A., Mazzeo, A., Giove, L.S., Zito, D., Marcotuli, I. *et al.* (2017) Exploiting DNA-based molecular tools to assess genetic diversity in pomegranate (*Punica granatum* L.) selections and cultivars. *Fruits* 72(5), 292–305. DOI: 10.17660/th2017/72.5.5.

Gill, P.P.S., Dhillon, W.S. and Singh, N.P. (2011) Influence of training systems on growth, yield and fruit qualiy of pomegranate 'Kandhari'. *Acta Horticulturae* 890, 305–310.

Goswami, J.D., Patel, N.M., Bhadauria, H.S. and Wankhade, V.R. (2013) Effect of plant growth substances on growth, fruit setting and yield of pomegranate cv Sinduri. *International Journal of Agricultural Sciences* 9, 332–334.

Gould, K.S. and Lister, C. (2006) Flavonoid functions in plants. In: Andersen, Ø.M. and Markham, K.R. (eds) *Flavonoids: Chemistry, Biochemistry and Applications*. CRC Press, Boca Raton, FL, pp. 397–441.

Hänsch, R. and Mendel, R.R. (2009) Physiological functions of mineral micronutrients (Cu, Zn, Mn, Fe, Ni, Mo, B, Cl). *Current Opinion in Plant Biology* 12(3), 259–266.

Hakimi, L., Sadeghi, S.M.M., Van Stan, J.T., Pypker, T.G. and Khosropour, E. (2018) Management of pomegranate (*Punica granatum*) orchards alters the supply and pathway of rain water reaching soils in an arid agricultural landscape. *Agriculture, Ecosystems & Environment* 259, 77–85. DOI: 10.1016/j.agee.2018.03.001.

Hao, X., Li, W. and Deng, H. (2016) The oasis effect and summer temperature rise in arid regions – case study in Tarim Basin. *Scientific Reports* 6(1), 35418. DOI: 10.1038/srep35418.

Hasanpour, Z., Karimi, H.R. and Mirdehghan, S.H. (2015) Effects of salinity and water stress on echo-physiological parameters and micronutrients concentration of pomegranate (*Punica granatum* L.). *Journal of Plant Nutrition* 38(5), 795–807. DOI: 10.1080/01904167.2014.944711.

Hiwale, S.S., Dhandhar, D.G. and Bagle, B.G. (2006) Rejuvenation of pomegranate through non-selective pruning.*In: Ghosh, S.N., Mitra, S.K., Banik, B.C., Hasan, M.A., Sarkar, S.K., Dhua, R.S., Kabir, J. and Hore, J.K. (eds) Proceedings of the national symposium on production, utilization and export of underutilized fruits with commercial potentialities, Kalyani, Nadia, West Bengal, India, 22-24 November, 2006*, pp. 174–176.

Holland, D., Hatib, K. and Bar-Yaakov, I. (2009) Pomegranate: botany, horticulture, breeding. In: Janick, J. (ed.) *Horticultural Reviews*. Wiley-Blackwell, New York, pp. 127–191.

Hussein, M.A., El-Sese, A.M., El-Mahdy, T.K. and Abd-El-Sabour, B. (1994) Physiological studies on thinning effects on the yield and fruit quality of 'Manfalouty' pomegranate. B – Sevin, NAA and hand thinning influences on some fruit physical and chemical properties. *Assiut Journal of Agricultural Sciences* 25, 41–50.

Ingles, C., Geisel, P.M. and Unruh, C.L. (2002) Fruit trees: training and pruning deciduous trees. ANR, University of California, Davis, California.

Jafari, A., Arzani, K., Fallahi, E. and Barzegar, M. (2014) Optimizing fruit yield, size, and quality attributes in 'Malase Torshe Saveh' pomegranate through hand thinning. *Journal of the American Pomological Society* 68, 89–96.

Karimi, H.R. and Hasanpour, Z. (2014) Effects of salinity and water stress on growth and macro nutrients concentration of pomegranate (*Punica granatum* L.). *Journal of Plant Nutrition* 37(12), 1937–1951. DOI: 10.1080/01904167.2014.920363.

Karimi, H.R. and Hassanpour, N. (2017) Effects of salinity, rootstock, and position of sampling on macro nutrient concentration of pomegranate cv. Gabri. *Journal of Plant Nutrition* 40(16), 2269–2278. DOI: 10.1080/01904167.2016.1263324.

Khayyat, M., Tehranifar, A., Davarynejad, G.H. and Sayyari-Zahan, M.H. (2016) Effects of NaCl salinity on some leaf nutrient concentrations, non-photochemical quenching and the efficiency of the PSII photochemistry of two Iranian pomegranate varieties under greenhouse and field conditions: preliminary results. *Journal of Plant Nutrition* 39(12), 1752–1765. DOI: 10.1080/01904167.2016.1201686.

LaRue, J.H. (1977) Growing pomegranates in California. In: *Pamphlet 2458*. University of California, Division of Agricultural Sciences.

Maity, A., Babu, K.D. and Sarkar, A. (2019) Guidelines for fertilizer use in pomegranate orchards based on seasonal uptake and partitioning of nutrients. *Scientia Horticulturae* 252, 138–148. DOI: 10.1016/j.scienta.2019.03.047.

Masalkar, S.D., Joshi, V.R., Masalkar, S.D., Kulkarni, S.R. and Chavan, S.D. (2009) Effect of different pruning levels on yield and quality of pomegranate. In: *2nd International Symposium on 'Pomegranate and Minor including Mediterranean Fruits'*. 101. University of Agricultural Sciences, Dharwad, India.

Mdellel, L., Halima Kamel, M.B. and Assadi, B. (2015) Impact of winter pruning of pomegranate trees on *Aphis punicae* (Hemiptera, Aphididae) and its natural enemies in Tunisia. *Annales de la Société entomologique de France* 51(3), 266–271. DOI: 10.1080/00379271.2015.1114425.

Melgarejo Moreno, P., Hernández García, F. and Legua Murcia, P. (2010) El Granado. In: Melgarejo Moreno, P., Hernández García, F. and Legua Murcia, P. (eds) *Proceedings of I Jornadas Nacionales sobre el Granado: Producción, Economía, Industrialización, Alimentación y Salud*. SPE3, Valencia, Spain, pp. 36–37.

Middel, A., Häb, K., Brazel, A.J., Martin, C.A. and Guhathakurta, S. (2014) Impact of urban form and design on mid-afternoon microclimate in Phoenix local climate zones. *Landscape and Urban Planning* 122, 16–28. DOI: 10.1016/j.landurbplan.2013.11.004.

Moradi, S., Baninasab, B., Gholami, M. and Ghobadi, C. (2017) Paclobutrazol application enhances antioxidant enzyme activities in pomegranate plants affected by cold stress. *The Journal of Horticultural Science and Biotechnology* 92(1), 65–71. DOI: 10.1080/14620316.2016.1224605.

Naeini, M.R., Khoshgoftarmanesh, A.H., Lessani, H. and Fallahi, E. (2005) Effects of sodium chloride-induced salinity on mineral nutrients and soluble sugars in three commercial cultivars of pomegranate. *Journal of Plant Nutrition* 27(8), 1319–1326. DOI: 10.1081/PLN-200025832.

Naeini, M.R., Khoshgoftarmanesh, A.H. and Fallahi, E. (2006) Partitioning of chlorine, sodium, and potassium and shoot growth of three pomegranate cultivars under different levels of salinity. *Journal of Plant Nutrition* 29(10), 1835–1843. DOI: 10.1080/01904160600899352.

Neilsen, G., Forge, T., Angers, D., Neilsen, D. and Hogue, E. (2014) Suitable orchard floor management strategies in organic apple orchards that augment soil organic matter and maintain tree performance. *Plant and Soil* 378(1–2), 325–335. DOI: 10.1007/s11104-014-2034-8.

Neori, H.B., Judeinstein, S., Tripler, E., Holland, D. and Lazarovitch, N. (2014) Salinity effects on colour and health traits in the pomegranate (*Punica granatum* L.) fruit peel. *International Journal of Postharvest Technology and Innovation* 4(1), 54–68. DOI: 10.1504/IJPTI.2014.064145.

Okhovatian-Ardakani, A.R., Mehrabanian, M., Dehghani, F. and Akbarzadeh, A. (2010) Salt tolerance evaluation and relative comparison in cuttings of different pomegranate cultivar. *Plant, Soil and Environment* 56(4), 176–185. DOI: 10.17221/158/2009-PSE.

Padmavathamma, A.S. and Hulamani, N.C. (1996) Effect of fruit thinning on yield, weight, physical and chemical parameters of pomegranate (*Punica granatum* L. cv. Jyoti). *Journal of Research ANGRAU* 24, 73–76.

Phawa, T., Prasad, V.M. and Rajwade, V.B. (2017) Effect of plant growth regulators on growth and flowering of pomegranate (*Punica granatum* L.) cv. Kandhari in Allahabad agro-climatic conditions. *International Journal of Current Microbiology and Applied Sciences* 6(8), 116–121. DOI: 10.20546/ijcmas.2017.608.015.

Pontonio, E., Montemurro, M., Pinto, D., Marzani, B., Trani, A. *et al.* (2019) Lactic acid fermentation of pomegranate juice as a tool to improve antioxidant activity. *Frontiers in Microbiology* 10(1550). DOI: 10.3389/fmicb.2019.01550.

Potchter, O., Goldman, D., Kadish, D. and Iluz, D. (2008) The oasis effect in an extremely hot and arid climate: the case of southern Israel. *Journal of Arid Environments* 72(9), 1721–1733. DOI: 10.1016/j.jaridenv.2008.03.004.

Reddy, P., Ch., R., Narayanareddy, P, Y.N. and Chandrasekhar, R. (2011) Effect of ethylene inhibiting chemicals on yield and quality of pomegranate. *Acta Horticulturae* 890, 353–358.

Saaroni, H., Bitan, A., Dor, E.B. and Feller, N. (2004) The mixed results concerning the 'oasis effect' in a rural settlement in the Negev Desert, Israel. *Journal of Arid Environments* 58(2), 235–248. DOI: 10.1016/j.jaridenv.2003.08.010.

Seidhom, S.H. and Abd-El-Rahman, G. (2011) Prediction of traditional climatic changes effect on pomegranate trees under desert condition in El-Maghara, Egypt. *Journal of American Science* 7, 268–280.

Sellami, M.H. and Sifaoui, M.S. (1998) Measurements of microclimatic factors inside the oasis: interception and sharing of solar radiation. *Renewable Energy* 13(1), 67–76. DOI: 10.1016/S0960-1481(97)00064-5.

Shanmugasundaram, T. and Balakrishnamurthy, G. (2017) Exploitation of plant growth substances for improving the yield and quality of pomegranate under ultra high density planting. *International Journal of Current Microbiology and Applied Sciences* 6(3), 102–109. DOI: 10.20546/ijcmas.2017.603.011.

Sharma, S.D., Kumar, P., Bhardwaj, S.K. and Chandel, A. (2015) Agronomic performance, nutrient cycling and microbial biomass in soil as affected by pomegranate based multiple crop sequencing. *Scientia Horticulturae* 197, 504–515. DOI: 10.1016/j.scienta.2015.10.013.

Sharma, D.P., Chakma, J., Sharma, N. and Singh, N. (2017) Effect of different orchard management practices on the growth and production of rejuvenated of pomegranates (*Punica granatum* L.) cv. Kandhari Kabuli. *Journal of Applied and Natural Science* 9(1), 577–581. DOI: 10.31018/jans.v9i1.1233.

Sharma, D.P. and Singh, N. (2018) Effect of rejuvenation pruning on the growth, productivity and disease incidence in declining trees of pomegranate (*Punica granatum* L.) cv. Kandhari Kabuli. *Journal of Applied and Natural Science* 10(1), 358–362. DOI: 10.31018/jans.v10i1.1630.

Shukla, A.K., Singh, D., Shukla, A.K. and Meena, S.R. (2007) Pruning and training of fruit crops. In: Yadav, P.K. (ed.) *Fruit Production Technology*. International Book Distributing Company, Lucknow, India, pp. 135–148.

Singh, R.P., Sharma, Y.P. and Awasthi, R.P. (1990) Influence of different cultural practices on premature fruit cracking of pomegranate. *Progressive Horticulture* 22, 96–98.

Soloklui, A.A.G., Ershadi, A. and Fallahi, E. (2012) Evaluation of cold hardiness in seven Iranian commercial pomegranate (*Punica granatum* L.) cultivars. *HortScience* 47(12), 1821–1825. DOI: 10.21273/HORTSCI.47.12.1821.

Tabatabaei, S.Z. and Sarkhosh, A. (2006) Analysis and comparison of salinity tolerance among 10 Iranian commercial pomegranate cultivars. In: *1st International Symposium on 'Pomegranate and Minor Mediterranean Fruits'*. 54. Adana, Turkey.

Thakur, C. and Sharma, C.L. (2018) Effect of plant growth regulators on physico-chemical parameters of pomegranate (*Punica granatum* L cv. Kandhari). *International Journal of Chemical Studies* 6, 1849–1855.

Warade, S.D., Joshi, V.R., Masalkar, S.D. and Kulkarni, S.R. (2008) Effect of different mulches on yield and quality of pomegranate. In: *2nd International Symposium on 'Pomegranate and Minor including Mediterranean Fruits'*. 104. University of Agricultural Sciences, Dharwad, India.

9 Soil and Nutrition

A. Maity[1], M. Khayyat[2]*, F. Azarmi-Atajan[3], S. Agehara[4] and A. Sarkhosh[5]
[1]ICAR-National Research Centre on Pomegranate, Kegaon, Solapur, India; [2]Department of Horticultural Science, College of Agriculture, University of Birjand, Birjand, Iran; [3]Department of Soil Science and Engineering, College of Agriculture, University of Birjand, Birjand, Iran; [4]Gulf Coast Research and Education Center, University of Florida, Wimauma, Florida, USA; [5]Horticultural Sciences Department, University of Florida, Gainesville, Florida, USA

9.1 Introduction

Soil nutrient availability greatly affects plant growth and development. An article written by Jan Baptista van Helmont, which was published by his son in Amsterdam in 1652, reported that a potted willow cutting irrigated with rainwater gained 76.7 kg in 5 years, whereas the soil weight decreased by 56 g. He attributed the weight gain of willow to water and ignored the small weight reduction in the soil. About 200 years later, Justus von Liebig, who is considered as the father of modern plant nutrition, analysed plant tissues and soil. He discussed the consequence of cropping and addressed the need to replace essential nutrients with fertilizers. For example, 10 metric tonnes (t) of pomegranate fruit may remove about 14 kg of nitrogen (N), 15.5 kg of potassium (K) and 1.43 kg of phosphorus (P). The discovery of the importance of nutrients in plant growth, flowering and fruiting is a noteworthy achievement of science. Most of the mineral nutrients are from the soil solution. The essential roles of minerals in plant growth has been repeatedly demonstrated.

Pomegranate (*Punica granatum* L.) requires 17 elements for plant growth and development (Al-Maiman and Ahmad, 2002; Mirdehghan and Rahemi, 2007; Korkmaz and Aşkın, 2013; Marathe *et al.*, 2016a). These elements are called essential elements and include carbon (C), hydrogen (H), oxygen (O), which mainly come from air and water, as well as N, P, K, sulfur (S), calcium (Ca), magnesium (Mg), boron (B), zinc (Zn), copper (Cu), iron (Fe), manganese (Mn), chloride (Cl), molybdenum (Mo) and nickel (Ni), which are present either in the soil or supplied as fertilizers. N, P, K, S, Ca and Mg are called macronutrients. Other elements including B, Zn, Cu, Fe, Mn, Cl, Mo and Ni are required in much smaller amounts and are called micronutrients (Table 9.1). Irrespective of the amount required, a deficiency of any of these elements will impair vegetative growth, flowering, fruiting and the tolerance to both abiotic and biotic stresses including salinity, drought, heat, frost, pests and diseases.

It is well-known that selection of the proper soil is of paramount importance in the success or failure of a fruit crop. Similarly, the effect of plant nutrition on quality fruit production has been well recognized. In the past, not much attention was paid to fruit quality parameters, as the priority was just to produce a higher yield.

*Corresponding author: mhdkhayat@birjand.ac.ir

Table 9.1. Forms of nutrient elements in the soil solution.

Element	Cationic	Anionic
Nitrogen (N)	Ammonium (NH_4^+)	Nitrate (NO_3^-), Nitrite (NO_2^-)
Phosphorus (P)	-	Hydrogen phosphate (HPO_4^{2-}), Dihydrogen phosphate ($H_2PO_4^-$)
Potassium (K)	Potassium ion (K^+)	-
Calcium (Ca)	Calcium ion (Ca^{2+})	-
Magnesium (Mg)	Magnesium ion (Mg^{2+})	-
Sulfur (S)	-	Sulfate $(SO_4)^{2-}$
Iron (Fe)	Ferrous ion (Fe^{2+}), Ferric ion (Fe^{3+})	-
Manganese (Mn)	Manganese ion (Mn^{2+})	-
Zinc (Zn)	Zinc ion (Zn^{2+})	-
Copper (Cu)	Cupric ion (Cu^{2+})	-
Boron (B)	-	Borate ($H_2BO_3^-$), Boric acid (H_3BO_3)
Molybdenum (Mo)	-	Molybdate (MoO_4^{2-})
Chlorine (Cl)	-	Chloride ion (Cl^-)

Source: Marschner (1995); Rashna (1997).

However, recently, there has been a paradigm shift from increasing production to improving fruit quality. It is especially true in the case of pomegranate, which is commonly cultivated as a cash crop in many countries for export purposes. Hence, the selection of suitable soil, as well as the application of balanced nutrients, are prerequisites to harness higher fruit yield with required quality standards.

9.2 Suitable Soil

Soil is a major source of nutrients required for crop production. Plant nutrient uptake capacity is influenced by many factors including soil physical and chemical properties, soil moisture, root growth and development, rootstocks, cultivars, climatic conditions, and biotic and abiotic stresses. Soil nutrient availability is critical for efficient plant nutrient uptake. Generally, the availability of nutrients in agricultural soils is less than the optimum level for plant growth and yield production. Although the total amount of nutrients in the soil may be high, their chemical properties and chemical reaction with other soil components can make them unavailable for plant use (Osman, 2013; Miransari, 2013).

Soils are classified based on their origin, mineral or organic content, the proportion of clay, silt and sand particles, depth, colour, the soil reaction (pH) (Truog, 1947), water-holding capacity and cation exchange capacity (CEC). Surface soils are generally more fertile than subsoils because of the accumulation of organic residues. The decomposition rates of plant and animal-derived organic matter in cultivated lands depend on climatic conditions. However, especially in arid climate zones, its amount rarely reaches 5%. Such soils consisting predominately of inorganic constituents are termed mineral or inorganic soils (Mengel *et al.*, 2001; Osman, 2013).

Soil particles are classified as clay (less than 0.002 mm), silt (0.002–0.02 mm) or sand (0.02–2.0 mm) based on the diameter. According to these soil particle proportions, the soils are classified into different texture types by using the soil texture triangle. Soil texture affects many properties of the soils. Although particles larger than 2 mm in diameter, such as gravel and stones, may constitute part of the soil mass and help trees with anchorage and rapid drainage of water, they do not play a role in plant nutrition (Dudley *et al.*, 2008b). Soils containing a high proportion of these large particles are known as gravelly soils. Pomegranate has the potential to grow on a variety of soils. In India, it is mostly grown on very shallow (10–25 cm), light-textured and rocky soils with low moisture and poor nutrient-holding capacity. The best soil for trees is sandy loam soil because its overall characteristics are optimum. Although this soil has a smaller water-holding capacity, it has more rapid water infiltration than loamy clay soils. More clay particles in the soil can lead to soil compaction, which inhibits root penetration and exploration. In heavily compacted soils, roots are unable to extract water and nutrients efficiently, even if they are present in adequate quantities. Sandy soils have better drainage than clay soils, but they are usually less fertile because of low cation and anion exchange capacity (Mengel *et al.*, 2001; Osman, 2013).

Sandy soils require more frequent irrigation and fertilization than loamy soils. Smoke (2018) stated that irrigation should be provided daily for light-textured soils because of the low water-retention capacity. El-Desouky and El-Hamied (2014) suggested that amendment with organic fertilizers improves the growth and development of pomegranate trees under dune soil conditions. Furthermore, they found improved plant growth by subsurface irrigation compared with surface irrigation. Performance of pomegranate is equally good on heavy black soils when raised beds are used to promote proper drainage. It is reported that 'Jyoti', 'Ganesh' and 'Raichur-1' clones of pomegranate performed well under the Vertisols of the Tungabhadra project area in Karnataka, India.

9.3 Soil pH and Different Processes of Buffer Action

Soils are categorized as acidic, neutral or alkaline according to their hydrogen ion concentration or their negative logarithm-pH (Mengel *et al.*, 2001; Taiz and Zeiger, 2002; Osman, 2013). The formation of limited-soluble compounds plays an important role in controlling the availability of some nutrients. Under alkaline soil conditions, zinc and copper are not readily available to plants. Under neutral to alkaline soils, calcium phosphate precipitates, and its solubility decreases with increasing alkalinity. Excessively alkaline soils are usually the results of accumulation of sodium carbonate in the soil or Na replacement on a large percentage of the ion exchange sites, or both. Under acidic to neutral conditions, P is specifically adsorbed on to the surface of iron oxide, aluminium oxide and clay mineral particles, and is not displaced from these surfaces to any appreciable extent by other components of the soil solution. In addition to specific adsorption of phosphate on the oxide surface, iron phosphates and aluminium phosphates may precipitate. In highly acidic soils, aluminium oxide is dissolved to create toxic levels of Al. Conversion of insoluble manganese dioxide to the soluble Mn^{2+} also leads to toxic concentrations of Mn. Organic matter in the soil is mineralized (decomposed) by microorganisms. On the other hand, using organic materials with low levels of these elements (straw and so on) leads to incorporation of these elements from the soil into organic matter. Under the condition of lacking O_2 in the soil (waterlogging), soil microorganisms may make chemical transformations that produce toxic levels of some substances. Hydrogen sulfide, methane gas, Mn^{2+} and other organic materials are examples of toxins that occur in these soils. Toxin production strongly depends on temperature and they form much more rapidly when the soil is warmer. Because of adsorption, precipitation and cycling of organic matter, the soil can buffer the soil solution. Soil pH also affects the extent to which most nutrients are bound by soil particles. However, the pH of the soil near the root is not the same as the pH of the bulk of the soil. Many substances such as carbon dioxide, which forms carbonic acid in soil solution and organic matter, are excreted from the roots. Thus, the pH of soil adjacent to a root alters and reduces and leads to an increase in the availability of nutrients to the plants, except in acid soil. Soil pH more than 7.5 greatly limits the solubility of many elements including Zn, Cu, Mn and Fe. Low pH (less than 5.5) may lead to deficiencies in Ca, Mg, P and Mo, and results in excessive amounts of Mn, Fe or Al (Fig. 9.1).

The solution to these problems lies in correction of the soil condition including the addition of lime (for raising pH), the addition of sulfur or gypsum (for lowering pH) and the addition of organic matter (to improve soil structure and to maintain optimal water) or adding fertilizers in some instances (Mengel *et al.*, 2001; Osman, 2013).

Cation exchange occurs primarily on the surface of clay minerals and active sites of

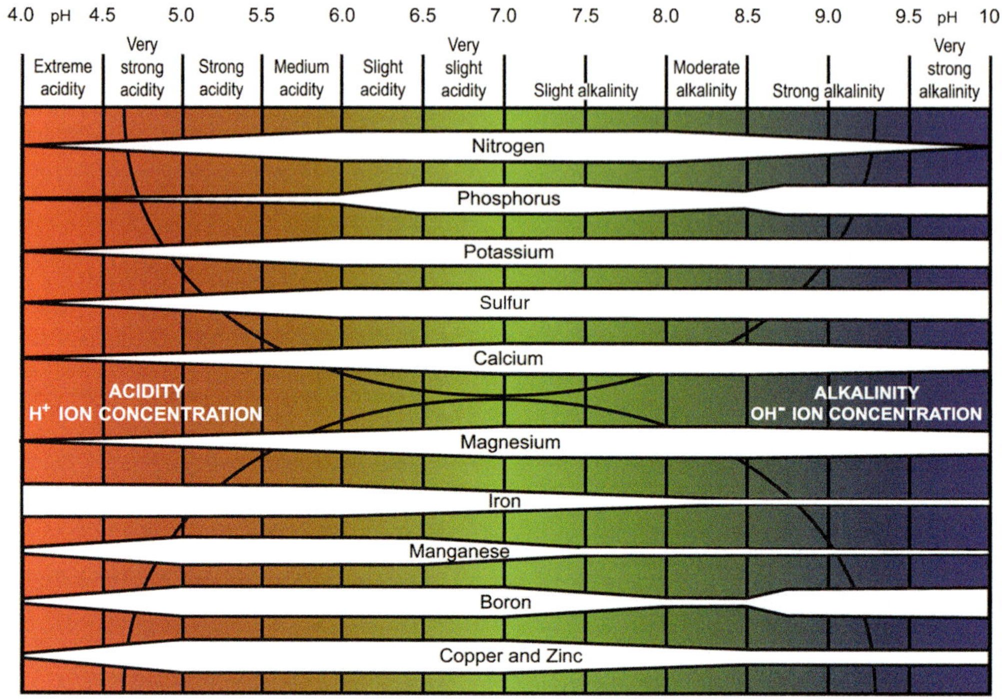

Fig. 9.1. Availability of nutrients as influenced by soil pH. (Source: https://www.pda.org.uk/pda_leaflets)

organic matter, because there are negative electrical charges that must be equilibrated by positively charged cations. The amounts of Ca, Mg, K and Na are affected by the amount of clay and organic matter in the soil. Negatively charged soil particles hold some cations more firmly than others. Specific adsorption is a highly selective phenomenon (Mengel *et al.*, 2001; Osman, 2013).

9.4 Mineral Nutrients

All nutritional elements are necessary for healthy growth and development, flowering and fruiting of plants. Macronutrients are required for synthesis and activity of organic compounds or used for osmoregulation and cellular pH maintenance. Micronutrients are the constituents of enzyme molecules and are involved in photosynthesis, cell membrane structure, and disease resistance or susceptibility. For example, sufficient amounts of Ca, B or Zn or an excess of nitrogen are very important in these cases.

9.4.1 Nutrient source and their availability in soil

The distribution of plant mineral nutrients in a natural soil is extremely heterogeneous. Soil nutrient concentrations and availability in a pomegranate orchard may thus vary both spatially and temporally, which greatly affect the acquisition of nutrients by plant roots. There are several sources of nutrients for crop production as follows (Mengel *et al.*, 2001; Osman, 2013):

- soil organic matter (plant residues, microorganisms, etc.);
- mineral fertilizers;
- organic fertilizers;
- release from parent rocks by weathering and release from fixed sites in clay minerals (e.g. K from illite);
- atmospheric deposition;
- N-fixation by specialized microorganisms.

The use of chemical fertilizers is the most common and quick method to supply both macronutrients and micronutrients. Although chemical fertilizers have largely contributed to improved crop productivity, excessive use has led to environmental and groundwater pollution. Therefore, it is essential to optimize the application of chemical fertilizers to minimize their adverse environmental effects. Various methods have been proposed and tested to increase nutrient availability in soils. The use of beneficial soil microbes is one of the novel ways to increase nutrient availability and reduce environmental stresses. Soil contains various microorganisms such as bacteria, fungi, algae and protozoa. In general, bacteria and fungi play more important roles in soil processes than other microorganisms. It is estimated that there are about 9×10^7 bacteria and about 2×10^5 fungi in 1 g of a typical soil. The population of these microorganisms is greater in the plant root zone (rhizosphere) than in bulk soil (Glick, 2018). The useful interactions between microorganisms and plants, as well as the role of microorganisms in improving plant growth have been reported (Cosme and Wurst, 2013; Bompadre *et al.*, 2014). Moreover, microorganisms play an essential role in increasing plant resistance to environmental stresses. The plant growth-promoting rhizobacteria (PGPR) and mycorrhizal fungi are the main soil microorganisms with plant growth-promoting potential under stressed conditions. The PGPR are able to promote plant growth via various mechanisms including production of beneficial chemical compounds, such as phytohormones (indole-3-acetic acid, gibberellins and cytokinins), siderophores (Fe-chelating agent), and hydrogen cyanide (HCN), atmospheric nitrogen fixation and increasing nutrient availability (such as K, P, Fe, Zn) by production of organic acids, proton and phosphatase enzymes. Arbuscular mycorrhizal fungi (AMF) increase the uptake of nutrients required for photosynthesis (Marschner and Dell, 1994). Also, in the mycorrhizal symbiosis, AMF increase water transport more efficiently than the non-mycorrhizal plants. This process improves plant growth under environmental stress conditions such as water deficit and salinity. Improvement of photosynthesis in pomegranate inoculated with AMF under different water availability conditions has been reported by Bompadre *et al.* (2015). Also, Maity *et al.* (2014) reported that *Penicillium pinophilum* isolated from the pomegranate rhizosphere significantly increased the growth, photosynthetic rate and nutrient (N, P and K) uptake in pomegranate. However, the combined application of *P. pinophilum* with insoluble K was more effective. Inoculation of pomegranate plants with AMF (*Rhizophagus intraradices* strains) increased antioxidant defences, including reactive oxygen species (ROS)-scavenging enzymes superoxide dismutase (SOD), catalase (CAT) and ascorbate peroxidase (APX) in shoots and roots under water stress conditions. Aseri *et al.* (2008) indicated that the use of *Azotobacter chroococcum* and *Glomus mosseae* alone or in combination improved the growth and biomass production of pomegranate in both nursery and field experiments. Mir and Sharma (2012) reported that inoculation of pomegranate cuttings with biofertilizers (N_2-fixing bacteria, PSB and AM) increased the growth and biomass production of cuttings, the rhizosphere microbial activity, and concentration of metabolites and nutrients.

9.4.2 Nitrogen

The nitrogen sources may occur as soluble in water, and insoluble N mainly as proteinaceous materials, or as conjugates with other compounds that are not readily utilized by plants. Insoluble nitrogenous sources are broken down into simpler molecules by various fungi and bacteria. These simpler molecules are nothing but ammonic compounds. *Nitrosomonas* bacteria oxidize ammonium salts to nitrite and *Nitrobacter* oxidize the nitrite to nitrate. These processes are collectively known as nitrification. As a result of the activities of these soil microorganisms, the complex N-containing substances become available for absorption by roots. Primary forms of nitrogen in soil solution are nitrate (NO_3^-) and ammonium (NH_4^+) ions. Because nitrate, as a negatively charged ion is not attracted by soil particles, it is freely mobile in the soil and is susceptible to leaching. Nitrate can readily move to plant roots by mass flow.

On the other hand, ammonium is held firmly as an exchangeable cation on the surface of negatively charged clay particles. Thus, its mobility in a soil is relatively restricted. Gaseous nitrogen (N_2) makes up to 78% of the atmosphere by volume. The plant can use this atmospheric

N_2 only after its reduction to ammonium. The reduction of N_2 into ammonium is carried out by certain microorganisms in the process of N fixation in the soil. Pomegranate trees are unable to fix gaseous nitrogen because they do not have symbiotic nitrogen-fixing bacteria on their roots, but they have associative symbiosis with *Azospirillum* sp., which fixes atmospheric N_2 and supplies N to pomegranate plants (Ramos, 1997; Mengel *et al.*, 2001; Taiz and Zeiger, 2002; Osman, 2013).

9.4.3 Phosphorus

Phosphorus (P) occurs in soils in various forms, which have considerably different availability to plants. The concentration of P ion in most soils varies from 0.1 to 10 μM (Raghothama and Karthikeyan, 2005). In soils, P is present as inorganic (mineral) and organic forms, and it is absorbed by roots as phosphate. Generally, there is no reason to use P fertilizer in deciduous orchard trees, except that planting of cover crops are supplied. Phosphorus availability to plant roots depends on soil pH and the amount of P present. Approximately, 80% of inorganic P added to soil in the form of chemical fertilizers is rapidly precipitated, becoming unavailable for plant uptake. The optimal pH range for maximum phosphorus availability is 6.0–7.0. When soils are too acidic, P is tied up on the surface of iron oxide and aluminium oxide resulting in precipitation by extremely low soluble iron phosphate and aluminium phosphate.

On the other hand, the increase in pH in alkaline soils leads to the formation of low-soluble calcium phosphate. Moreover, there is a slow conversion of iron phosphate, aluminium phosphate and calcium phosphate to a slightly soluble form of phosphate, which may take place over months or years. The relatively low immobility of phosphate in soils and its presence in very low concentrations in the soil solution means that plants can very easily suffer from a deficiency of this nutrient. The insoluble forms of P in soil such as $Ca_3(PO_4)_2$, Al_3PO_4 and Fe_3PO_4 can be solubilized by a group of microorganisms. Thus, the use of phosphate solubilizing bacteria (PSB) can enhance P availability for plant nutrient uptake. Several studies have indicated that inoculation with PSB improved the growth and yield of various crops (Ramos, 1997; Mengel *et al.*, 2001; Babana and Antoun, 2007; Osman, 2013; Valetti *et al.*, 2018).

9.4.4 Potassium

Potassium (K) can be present in abundance in the soil, occurring in various forms: as a structural element in soil minerals, associated with organic matter in exchangeable form, in a readily exchangeable or slowly exchangeable form in clay minerals and as a cation in the soil solution. By far the largest fraction of K that presents as a structural component of the soil minerals is virtually inaccessible to plants. On the other hand, the clay minerals, which constitute only a few percent of the total K in soils, usually provide the major source of K by buffering the soil solution, which provides the plant roots with K. During drought periods, K may become deficient due to (i) the enhanced fixation within the layers of clay minerals (e.g. illite), which is favoured by dry conditions, and (ii) by interruption of the diffusion pathways by which K is usually transported to tree roots (Ramos, 1997; Mengel *et al.*, 2001; Osman, 2013).

9.4.5 Calcium and magnesium

Calcium and magnesium are closely related and have similar chemistry in the soil. Their occurrence in the soil solution is in the greatest abundance compared with all other essential elements. The cation exchange process significantly affects their availability, which amuses 80–90% of the negatively charged exchange sites on soil particles of rich soils. Under the acidic condition of soils, they are usually washed out. Thus the application of lime (calcium carbonate) or dolomite lime (a mixture of calcium carbonate and magnesium carbonate) is recommended to neutralize the acidity and refill Ca and Mg. In sandy and neutral soils, Mg deficiency occurs, which is corrected via Epsom salts (magnesium sulfate) application (Ramos, 1997; Mengel *et al.*, 2001; Osman, 2013).

9.4.6 Sulfur

The most abundant sulfur (S) component in most agricultural soils is organic sulfur. In the solution of soil, S occurs as sulfate ion $(SO_4)^{2-}$. Generally, it is quite mobile in soil, and there is a small tendency for S to be adsorbed by soil particles in certain acid soils. Different S sources are atmospheric S that is delivered to the soil through rainfall, and many common fertilizers including ammonium sulfate, super phosphate and mixed fertilizers are used as a source of N. In an experiment in Washington State, it was stated that water containing more than 1.2 ppm S from sulfate (equivalent to 3.6 ppm sulfate) provides sufficient S for crops. Even with insufficient annual S addition, deficiency may not develop until the store of S in soil organic matter is exhausted (Ramos, 1997; Mengel *et al.*, 2001; Osman, 2013).

9.4.7 Micronutrients

Micronutrient availability is affected greatly by soil pH (Ramos, 1997; Taiz and Zeiger, 2002). Most micronutrients are only poorly accessible to plant roots at high pH, but availability increases as soil pH decreases.

The solubility of iron oxide and manganese oxide significantly control their availability. Both of them are affected by oxidation and reduction processes. Via an oxidation reaction ferrous iron (Fe^{2+}) transforms to ferric iron (Fe^{3+}); the reverse is the reduction. In well-aerated soils, the oxidized ferric form is dominant, which in oxide form has low solubility. In the Mn ion (Mn^{2+}), oxidation occurs after it has precipitated, and the final product is called manganese dioxide, in which Mn has a 4^+ charge. The soil pH condition affects the solubility of both ferric oxide and manganese dioxide: their solubility decreases rapidly as soil pH increases, and despite enough Fe and Mn content in most soils for plant growth and development, deficiency occurs because of their unavailability to the plants. Lowering soil pH improves the availability of these elements, which can be achieved by the soil incorporation of elemental sulfur. Microorganisms convert sulfur to sulfuric acid. It is recommended to apply the S in a band rather than broadcasting because banding concentrates the acidification and only part of the soil needs to be acidified to improve Fe and Mn availability. Depletion of soil O_2 increases reduction reactions leading to the reduction of ferric oxide and manganese dioxide to the high concentrations of ferrous and manganous ions that become toxic to plant roots (Ramos, 1997; Mengel *et al.*, 2001; Osman, 2013).

The concentration of zinc (Zn) and copper (Cu) elements in soils is much less than Fe and Mn. The pH-dependent adsorption process probably dominates in controlling Zn and Cu availability. A soil pH of 6.5 is something of a dividing point: as pH increases above this level, their availability is severely limited. Moreover, the amounts of Zn and Cu present in the soil and also the extent of surface adsorbtion are additional factors, contributing to their availability in the soil. These are immobile, so fertilizing with Cu or Zn under deficiency conditions is not useful. As with Fe and Mn, it is recommended to amend the soil to make it less alkaline (Ramos, 1997; Mengel *et al.*, 2001; Taiz and Zeiger, 2002; Osman, 2013).

Boron is held on soil particles to some extent and occurs in the soil solution as borate anion $(H_2BO_3)^-$ in alkaline soils or as neutral boric acid (H_3BO_3). A narrow concentration range of B in soil solution produces healthy growth and development, and below or above this range, deficiency and toxicity may arise. For correction of the deficiency, application of borax or boric acid is useful. Toxicity is often due to using irrigation water high in B, usually from well sources (Ramos, 1997; Mengel *et al.*, 2001; Taiz and Zeiger, 2002; Osman, 2013).

Molybdenum (Mo) occurs as the molybdate ion $(MoO_4)^{2-}$, which is required by plants only in small amounts. In acid soils, and via specific adsorption, Mo is removed from the soil solution similar to the way phosphate is removed (Ramos, 1997; Mengel *et al.*, 2001; Taiz and Zeiger, 2002; Osman, 2013).

Chlorine (Cl) occurs in soils as the highly mobile chloride ion (Cl^-), which is only needed by plants in small amounts. The wave action of the ocean adds Cl to the atmosphere, supplying the small plant requirement via annual rainfall, and by importation with surface irrigation (Ramos, 1997; Mengel *et al.*, 2001; Osman, 2013).

9.4.8 Orchard practices influencing nutrient availability

Pomegranate trees are long-lived plants that can remain productive for several years. With the exception of the establishment year, orchards soils should not be disturbed by ploughing or tillage as happens in arable land. In orchards, all soil-related activities are focused on the control of herbaceous flora growing between the tree rows. In general, there are three options for managing the orchard floor (Ramos, 1997; Kumar, 1997; Bose *et al.*, 2001):

- Keeping the soil bare without any grass/herbaceous vegetation or other coverage.
- Covering the soil with either living vegetation or with natural or artificial materials.
- The mixed system of (i) spatial variation of bare and covered areas or (ii) temporarily covered areas.

In practice, the choice of the appropriate system is more or less a question of water availability in a particular region. In dry climatic areas, the soil has to be kept free from any vegetation other than pomegranate trees in order to eliminate the competition for water as well as nutrients. In more humid areas, only young pomegranate trees have to be protected from herbal vegetation. It is well accepted that under undisturbed grass/herb vegetation, soils develop good structure, optimum aeration and organic matter content. On the other hand, the competition of herbs with pomegranate trees for water and nutrient resources is not negligible. Tree roots are very sensitive to the biological status of the soil surface so fine root distribution between different soil layers can vary. In contrast, under grass vegetation, the fine roots are found at greater depth where there is less competition with the grass roots. Annual plants are much more effective than pomegranate trees in absorbing nutrients including those that have been applied as fertilizers. Many studies have shown that the mineralized N content (N_{min}) in the 0–30 cm soil layer under grass or organic mulches is higher than at lower levels (Sánchez *et al.*, 2007).

For pomegranate orchard management, supplying of fertilizers, the release of biochemicals (e.g. from organic substances or humic matters) produced from their incorporation into the soil and the fixation of mineral nutrients are the most important factors influencing the nutrient balance in the soil. The availability of nutrients to plant roots is a function of conditions in the soil and of its biological activity. In general, plant roots only absorb nutrients that are dissolved in the soil solution and present as ions. In their acquisition by tree roots, nutrients must be transported in the soil solution to the root surface. Two main processes are involved in this movement: mass flow and diffusion. The mass flow of nutrients to roots is facilitated by transpiration. For most of the nutrients that are present in relatively high concentrations (NO_3^-, Ca^{2+}, Mg^{2+}), mass flow is how they are transported to the root surface. However, for K and particularly for P the ionic concentrations in soil solution are comparatively low, so that the amounts transported by mass flow are not high enough to meet the plant nutrient requirements. These two nutrients move to the roots by the physical process of diffusion, that is, down the concentration gradient induced by the removal of these nutrients at the plant root surface during uptake. Factors that disturb this process, therefore, induce deficiency.

9.4.9 Function of nutrients

There are numerous publications on the function of mineral nutrients in fruit trees, which deal mainly with growth and physiological aspects (Ferree and Warrington, 2003; Gradziel, 2017). Today, mineral nutrition has to be considered more in relation to aspects of fruit quality rather than to yield. Fruits are regarded as healthy food, and thus fertilization of pomegranate trees is not only a means of increasing productivity of the plant, but also of promoting the formation of valuable components within the fruit.

Nitrogen (N) is the driving force for vegetative and generative development of the trees. Besides its promoting effects on shoot growth, N is necessary for flower bud formation, fruit set and fruit development. The influence of N on fruit properties that are related to human health is ambiguous. On the one hand N promotes the development of a growing fruit, but, on the other hand, the formation of some vitamins, colour pigments (carotenoids and anthocyanins) and aroma compounds can be suppressed by an excess of N and the optimum sugar–acid ratio can be impaired. Nitrate as an N source is more effective than NH_4 or urea in maintaining a physiologically adequate level of Ca in fruits. Nitrogen is also effective in the distribution and absorption of other nutrients in the plant (Zekri and Obreza, 2003).

Phosphorus (P) is mostly related to flowering and fruiting as well as to the energy metabolism of the plant. Phosphorus promotes yield by increasing the number of flowers, fruit set and fruit size. Phosphorus level seems to be the regulator of meristematic activity. It has a beneficial effect on fruit quality parameters such as fruit and skin firmness. Also, P is a vital nutrient for several plant functions such as carbohydrate metabolism and transport (Vance *et al.*, 2003).

Potassium (K) is the second most important nutrient for pomegranate trees in terms of the requirement. Being very mobile within the plant, K is not directly involved in the structural components of the tree but it plays a major role in a number of physiological processes. These involve water relations of the tree, raising frost tolerance and lowering the susceptibility of plants to attack by pests and diseases, as well as a wide range of biochemical processes in the developing fruit. Potassium participates in numerous enzymatic reactions and is an important factor in the development of fruit colour, total soluble solids (TSS) and vitamin C content (Khayyat *et al.*, 2012). K is the key nutrient in osmoregulation and the maintenance of cell turgor and therefore closely related to the firmness of the fruit. Fruits rich in K are more resistant to sunscald (Ebert, 2009). However, too much K in relation to Ca can induce fruit disorders.

Calcium (Ca) unlike K is the typical structural nutrient element in the tree and the fruit. Calcium forms bonds in the middle lamellae and the microfibrils of the fruit tissue, and is therefore crucial for fruit firmness. Ca also has an essential function in the maintenance of cell membrane integrity: Ca deficiency causes membrane leakage and favours the onset of diseases (Orlov *et al.*, 2005; Marathe *et al.*, 2016a).

Magnesium (Mg) functions as the central atom of the chlorophyll molecule in the plants; however, its role in the growth and development of pomegranate trees is often underestimated. More than 60% of Mg is located elsewhere in the cell and is involved in numerous metabolic reactions. In terms of fruit quality, Mg improves fruit size and colour, increases sugar content, and promotes the formation of aroma compounds and acidity (Marschner, 1995; Marathe *et al.*, 2016a).

Sulfur (S) is essential for protein synthesis and the formation of aromatic compounds in the fruit. Sulfur containing substances can either enhance the plant's tolerance to diseases or act as repellents to pests. The tripeptide glutathione plays a key role in the formation of substances that are involved in stress alleviation of both biotic and abiotic stresses (i.e. heat and cold stress) (Marschner, 1995; El-Rauof and Dawoud, 2015; Marathe *et al.*, 2016a; Cannon and Ho, 2018).

The significance of micronutrients in fruit quality has not been evaluated in detail. However, most of the physiological processes depend on the action of iron (Fe), manganese (Mn), copper (Cu), zinc (Zn) and boron (B), which are involved in numerous enzymatic reactions. Micronutrient deficiencies in trees appear mainly on soils with a high pH, where their availability is very low. Deficient trees develop characteristic leaf symptoms (Bose *et al.*, 2001; Ramos, 1997; Ferree and Warrington, 2003; Marathe *et al.*, 2016a).

Iron (Fe) is an abundant element in soils, mostly present in its oxidized form Fe^{3+} but to be available to the roots of most plant species including pomegranate trees it has to be chemically reduced to Fe^{2+}. The role of Fe in the formation of fruit quality compounds is not clear. However, the element is closely linked to CO_2 assimilation in the leaf with 80% of total Fe in plants being localized in the chloroplasts. In addition to that, Fe is an activator of many biochemical processes such as regulation of oxidation/reduction pathways. The mobility of Fe in the tree is very low, as is also true of Zn, therefore young leaves and shoots show typical deficiency symptoms (Marathe *et al.*, 2016a).

Manganese (Mn) is also involved in CO_2 assimilation and respiratory pathways. The element plays a crucial role in N assimilation and Mg uptake. It activates enzymes and regulates membrane permeability. Over a very narrow concentration range, Mn favours the formation of green colour pigments in fruits. Manganese improves fruit size and also fruit yield attributes (Marschner, 1995).

Zinc (*Zn*) is one of the most deficient micronutrients in soils worldwide, and Zn deficiency is present in many calcareous or alkaline soils. The element is essential for fruit set of trees. It has a strong influence on elongation growth. Zn-deficient trees have very short internodes resulting in rosette-like, stunted shoots known as 'little leaf'. This appears to relate to the requirement of Zn for the synthesis of the growth hormone indole acetic acid (IAA) (Marschner, 1995; Arteca, 1996; Khorsandi *et al.*, 2009; Marathe *et al.*, 2016a).

Copper (*Cu*) is of crucial importance for growth-related processes such as in meristematic tissues as well as in xylem development (Marschner, 1995; Adrees *et al.*, 2015).

Boron (*B*) imparts a beneficial effect on fruit set and yield, which is demonstrated by the physiological evidence of a high requirement of B during the reproductive phase of growth. B can suppress postharvest disorders and/or influence the uptake and deficiency of Ca. Recent studies have pointed out the role of B in alleviating water deficiency stress in plants and in enhancing frost resistance of trees (Cooling, 1967; Cooling and Jones, 1970; Hanson and Breen, 1985). B deficiency impairs Ca transport in trees and may lead to Ca deficiency in fruits. This microelement is most strongly connected to fruit quality (Marschner, 1995; Bose *et al.*, 2001; Korkmaz and Aşkın, 2013).

9.4.10 Nutrient uptake pattern and their removal with harvest

Plant nutrient uptake is quite variable because it is dependent on specific site conditions during the growing season and plant developmental stages (Fig. 9.2). As a result, uptake of nutrients from the soil may vary considerably from year to year and from orchard to orchard. The yield and quality of fruit trees are influenced by the nutrient dynamics of the plant. Although, in the early stages of fruit development carbon and nutrient demand of fruit and leaves may be met, in part by redistribution from storage pools in perennial tree organs, in the later stages, developing fruit may draw on labile nutrient pools in pre-senescent leaves (Ramos, 1997; Ferree and Warrington, 2003). Thus, it is necessary to understand the redistribution pattern of different nutrients from foliar parts to fruit as the major sink during its developmental stages, so that the nutrient may be supplemented at the right time and with the proper perspective. The concentration of N, P, K, S, Fe, Zn and B in leaves were shown to decrease while Ca, Mg, Mn, and Cu concentrations increased during fruit growth and development (Raghupathi and Bhargava,

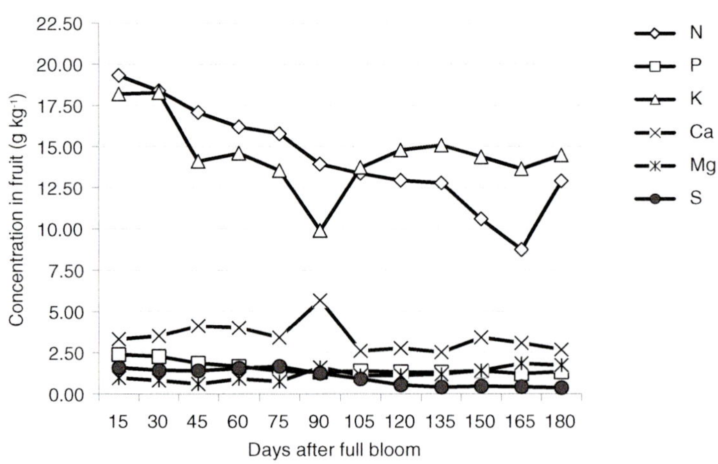

Fig. 9.2. Seasonal changes of macronutrients in pomegranate fruit. (Source: Maity *et al.*, 2017.)

1998). Mirdehghan and Rahemi (2007) stated that the concentration of most elements in arils and peel decreased during fruit growth and development. Major nutrient (N, P and K) concentrations in fruit declined at two rates, very sharply during 15–75 days after full bloom (DAFB) and then gradually in the following days until harvest (Fig. 9.2).

On the contrary, the concentration of Mg in fruit continued to increase during the fruit growth period, while Ca concentration increased up to 90 DAFB and then sharply declined at 105 DAFB and remained almost stable during the rest of the period. However, S concentration decreased very fast during 75–135 DAFB and stabilized before harvest. The relative order of macronutrient concentrations in fruit at fruit set (i.e. 15 DAFB) and the fruit enlargement stage (15–60 DAFB) was N > K > Ca > P > S > Mg, while at the fruit development stage (60–120 DAFB) the order of macronutrients was K > N > Ca > P > Mg > S. Likewise, at harvest the relative order of concentrations of macronutrients in fruit was K > N > Ca > Mg > P > S. Considerable accumulation of nutrients occurred during the early fruit development stage and continued until harvest for most of the primary nutrients. For some elements, including Ca, N, P and K, there was a good linear relationship ($R^2 = 0.90–0.94$) between the growth period and nutrient accumulation (Al-Maiman and Ahmad, 2002; Mirdehghan and Rahemi, 2007).

Concentrations of micronutrient cations (Fe, Mn, Zn and Cu) in fruits also decreased from a maximum just after fruit set to the minimum at fruit harvest (Fig. 9.3). Manganese concentration dropped very fast within 30 DAFB, while the same happened with Fe and Zn throughout 75 DAFB (Al-Maiman and Ahmad, 2002; Mirdehghan and Rahemi, 2007).

In comparison, Cu concentration declined gradually throughout 105 DAFB, and B concentration fell at a slower pace throughout the fruit growth and development period to a minimum at harvest. By harvest, the micronutrient present at the highest concentration in fruit was Fe (130.93 mg/kg), followed by Mn (32.87 mg/kg), B (24.07 mg/kg), Zn (17.47 mg/kg) and Cu (3.17 mg/kg), respectively. During fruit enlargement, demand for P, K, Fe, Mn and Zn were high, while during fruit development the requirement for N, Ca, Mg, S and Cu were high. Therefore, it is necessary to supplement with a good balance of macro and micronutrients according to their

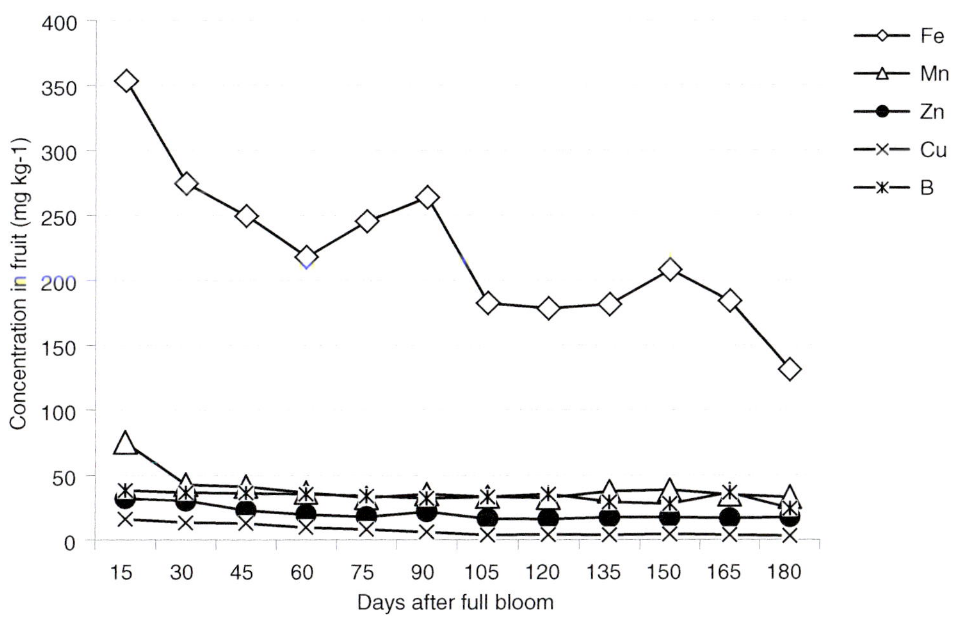

Fig. 9.3. Seasonal changes of micronutrients in pomegranate fruit. (Source: Maity *et al.*, 2017).

Table 9.2. Nutrient removal with harvest for producing 10-t pomegranate fruit yield per hectare.

Cultivar	N	P	K	Ca	Mg	S	Fe	Mn	Zn	Cu
	kg						g			
'Ganesh'	11.20	2.00	17.40	4.54	0.66	1.46	184.00	9.50	26.00	13.00
'Bhagwa'	13.87	1.43	15.54	2.87	1.85	0.40	140.60	35.30	18.76	3.40

Source: Raghupathi and Bhargava (1998).

demand before fruit set and growth in order to achieve desirable fruit quality.

The removal of nutrients by fruits is usually much lower than field crops (e.g. cereals, tuber crops) because the nutrient element concentrations in fruits are relatively low (Al-Maiman and Ahmad, 2002; Mirdehghan and Rahemi, 2007). Nitrogen recommendations for pomegranate trees are higher than the amount present in the fruit because N and other nutrients are also incorporated into leaves, stems and roots. It has also to be taken into account that minerals stored in leaves, stems and roots are not exported from the orchard but are stored in the plant to be redistributed to the major sink of the tree during the following years. Potassium is by far the most abundant nutrient element in fruits and is exported from the orchards with the harvested fruits in substantial amounts, followed by nitrogen and calcium. The cultivar 'Ganesh' removes more calcium from orchard soil than 'Bhagwa' (Table 9.2). Among the micronutrients, iron is the most abundant in the fruit and is removed from orchard soil in considerable amounts with the harvested fruit. It should also be pointed out that the K/Ca ratio is higher in 'Bhagwa' in comparison with 'Ganesh' as this cultivar removes high amounts of K and relatively low amounts of Ca, which may induce a K-Ca imbalance. It also suggests that 'Bhagwa' is more susceptible to Ca deficiency in the fruit and requires more careful monitoring of Ca nutrition (Raghupathi and Bhargava, 1998).

9.5 Nutrient Deficiency Symptoms

Both nutrient deficiency and toxicity are reflected by characteristic symptoms on leaves. Although a bit late, the occurrence of visible deficiency symptoms can be used as a simple indication of nutrient imbalances in the plant. Different behaviours are seen among nutrient elements. Most of the macronutrients, including N, P, K and Mg, are supplied from soil or remobilization from stores in the older leaves. Thus, symptoms of deficiency appear first in these older leaves. On the other hand, the main source of Ca, B, Fe, Zn, Cu and Mo for the growing plants is the soil, and they cannot be remobilized from older leaves. Therefore, deficiency symptoms for these elements are seen first in youngest leaves. Roots are very a important organ in acquiring nutrients for their growth and development and also for nutrient transport to aerial parts. Many factors including soil nutrient availability, the size and the health of the root system, the soil area that is occupied by roots, chemical and physical characteristics of the soil, root diseases, soil moisture and temperature, competition between roots and stress conditions around the root zone affect nutrient acquisition by roots. Certain nutrient deficiencies strongly affect root growth and branching. Deficiencies in K, P and N reduce branching of the root system and also root length (except for N). The leaf deficiency symptoms as described in Table 9.3 are indicative of progressive and severe deficiency, which should be rectified with proper fertilization practices (Marschner, 1995).

9.5.1 Management of nutrient supply

Like many other fruit trees, high yields and fruit quality of pomegranate are achieved with soils and irrigation water that are good quality, with desirable levels of salts and low salinity. For example, salinity often creates some field problems including decreased soil-water availability, reduced rates of water infiltration into the soil and plant tissue toxicity (Ashraf, 1994; Marschner, 1995; Ashraf and Harris, 2004; Silva-Ortega *et al.*, 2008). To diagnose

Table 9.3. Mineral nutrient deficiency symptoms in pomegranate.

Nitrogen (N)	Uniform yellowing of whole leaves appears initially on lower and mature leaves. These leaves become stiffer and break into pieces on folding. At the advanced stage, leaves are light green to yellow (chlorosis). Early leaf drop, retarded growth of the plant organs including shoots and roots. Massive number of flower buds and hermaphrodite flowers per plants, but small fruit. Reduction in the plant biomass (Fig. 9.4a)
Phosphorus (P)	Deficiency symptoms prominently appear on young leaves. Small, but dark green leaves, sometimes with red margins. Leaf margins turned upward and with tunnel-like shape. The growth of the plant is retarded. Delayed bud burst. Poor flowering and fruiting. Reduction in the plant biomass (Fig. 9.4b)
Potassium (K)	Deficiency symptoms appear initially on older leaves. Older leaves with chlorotic margins, later turning necrotic, beginning from the leaf tip. Many brown spots appear on the dorsal side of leaves along the leaf margin starting from the tip. Leaves sometimes curled. Weak branches, poor quality of fruits, shorter postharvest life (Fig. 9.4c)
Calcium (Ca)	Chlorosis of young leaves initiates with the purplish colouration of the interveinal area, midribs lighter in colour. Veins remain green during initial stages and become yellow at later stages. The yellow portion of the leaf tip acquires inverted 'V' shape. At an advanced stage, yellow portion of leaves turns dark brown in colour, and half of the leaves from the tip dry up. Leaf drop, dieback of branches. Twisted and deformed tissues at the growing tips. Reduction of root dry weight (Fig. 9.4d)
Magnesium (Mg)	Mature leaves become chlorotic, starting from areas in the region of the midrib (interveinal chlorosis), then progressing later from the centre to the leaf margins. Leaves are sometimes curled. Initially, grey patches appeared on side margin of the leaves and subsequently spread on the whole leaf (Fig. 9.4e)
Sulfur (S)	Similar to N deficiency, chlorotic, pale green leaves. In contrast to N deficiency, young leaves are more affected. Yellowing starts in the middle of the leaf around the midrib and interveinal areas turn yellow in colour, and the whole leaf becomes pale yellow. The intensity of yellowing is very low as compared with nitrogen deficiency (Fig. 9.4f)
Iron (Fe)	Young leaves become yellow, but their veins remain green. Leaves may lose all pigments and turn white, later necrotic (Fig. 9.4g)
Manganese (Mn)	Similar to Fe deficiency, but veins have a green seam, chlorotic interveinal areas turn pale green to yellow, sometimes necrotic spots (Fig. 9.4h)
Zinc (Zn)	Similar to Fe deficiency, leaves are small, twigs are stunted and rosette-like. The deficiency is first pronounced as interveinal chlorosis in young and mid-shoot leaves. Poor flowering and fruiting (Fig. 9.4i)
Copper (Cu)	Young leaves turn yellow or pale, necrotic leaf tips and margins. Dieback of young shoots and necrosis of the apical meristems, small sized fruits. Leaves on top of the plant show unusual puckering with veinal chlorosis
Boron (B)	Leaf chlorosis or yellow spots, smaller leaves with a hard leaf texture. Stems and leaves distorted. Malformed and cracked fruits. Reduction of root dry weight (Fig. 9.4j)
Molybdenum (Mo)	It could reveal itself as interveinal yellow spotting and mottling of older leaves. Reduction of root dry weight

Source: Marathe *et al.* (2016a).

Fig. 9.4. Nutrient deficiency symptoms in pomegranate trees (a) nitrogen deficiency, (b) phosphorus deficiency, (c) potassium deficiency, (d) calcium deficiency, (e) magnesium deficiency, (f) sulfur deficiency, (g) iron deficiency, (h) manganese deficiency, (i) zinc deficiency and (j) boron deficiency. (Photos: Ashis Maity.)

deficiencies or toxicities, soil and plant analysis and also water analysis in some cases are essential for the determination of the cause of the deficiency or toxicity. For precise analysis, good sampling should be conducted. Two types of sampling can be done: annual or biannual sampling for routine monitoring of soil and/or water. Marginal soil/water quality is often found in the area under development, and thus sampling prior to planting an orchard is essential and should be continued annually or biannually. Sampling is necessary when problems appear in conditions where soil and water are mostly uniform and of good quality, and previous crops have shown no toxicity symptoms. However, annual sampling may be more useful than biannual sampling, because it allows salinity conditions to be checked and rectified before and during the early stages of development using the easiest and cheapest methods with also fewer adverse effects on the orchard.

9.5.2 Soil test

Like other tree fruit crops, fertilization of pomegranate trees should always be based on the soil nutrient test for the orchard soil (Marathe *et al.*, 2016b; Gawada *et al.*, 2018; Thanari and Suma, 2018). Information about the nutrient reserves can prevent the producers from under- or overestimation of the trees' demand; meanwhile, other influencing factors such as climatic condition, physiological stage and the expected yield level should always be considered. To assess the nutrient levels in the soil, representative samples need to be analysed using laboratory equipment. Thus, the grower himself often collects the samples, but chemical analysis is carried out in a specialized laboratory. Soil analysis alone has limitations, because regarding root growth and development, the sample may not accurately represent the soil that roots are feeding from. It is possible to take a soil sample at any time of the year, because the level of elements does not change significantly throughout the year. However, some elements, such as nitrogen, may be lost because of the denitrification process, and under prolonged wet soil, leaching of NO_3, Cl and Na, and to some extent B, may happen. Thus, when looking at these elements, time of soil sampling and also time and volume of prior irrigation or rain should be taken into account. The following factors must be considered while sampling soil for chemical analysis:

- Sampling from different spots around the tree may lead to better representation, because soil may not be homogeneous throughout the orchard. In many cases, nutrient deficiencies are associated with localized soil differences such as those associated with old riverbeds, differences in topography, sand deposits and so on. In some situations, deep sampling should be performed to determine the depth of a high concentration of a toxic element. Generally, soil samples should be collected from the areas in which the roots are active, although this is often difficult to determine. Salinity may vary considerably throughout the orchard; thus, achieving representative soil samples is challenging (Pennock *et al.*, 2007).

- Considering the rainfall patterns and irrigation scheduling and water distribution (drip-wetted area in drip irrigation), soil samples should be prepared at the same time each year or every couple of years. Thus, the effect of water volume and evapotranspiration on salinity distribution and accumulation is removed. The method of irrigation and uniform watering must be considered when sampling is done. A small quantity of water (5–10 cm depth per application) frequently used over several days is more useful for moving the salinity below the root zone than an equal depth of water applied in one large flood irrigation. Drip irrigation that leaves a minimum of wetted soil on the orchard floor is more useful for leaching during summer than flood irrigation (Mahmoud and Sheren, 2014). However, it is important to avoid long periods of saturation. In flood-irrigated orchards, sampling 1.5–3 m to the side of the tree row will be indicative. In sprinkler-irrigated trees, samples should be collected across the sprinkler including soils from the centre of the wetting pattern (these receive the most applied water) and soils from the edges of the wetting pattern where salinity is accumulated.

It is recommended to sample after harvest time, because irrigation is commonly delayed or restricted during harvest and this leads to salinity accumulation in the root zone. Moreover, autumn sampling gives better results than winter sampling if additional irrigation water is needed for reducing salt contents during winter, because the trees are dormant in winter and least sensitive to overirrigation.

The area sampled should be restricted to a uniform soil type or condition within the orchard. Sample the same depth of soil in at least nine locations within a section of land that is considered to have a similar soil type. One large sample is needed; thus mix the samples together. For each soil depth and type, separate samples are combined and drying is recommended. If there are several distinctly different soil types, soil textures, drainage conditions or depths to impervious layer, or if different fertilizer or crop histories exist in the orchard, they should be sampled separately. In micro-irrigation systems, after winter rainfall in spring check the soil salinity. Uniform tree growth is a useful index of uniform soil type. For large orchards, the area included in any one soil sample collection generally should not exceed 10 acres (4 ha). In a 10-acre orchard, a minimum of 10–20 subsamples is suggested.

Sample topsoil and subsoil separately. Most surface-applied fertilizers move slowly into the soil or may be bound to soil particles near the surface. As a result, surface soil samples often more closely reflect the accumulation of nutrients from recent fertilizer applications, while subsoil samples may indicate either inherent soil fertility or the effect of a long-term fertilization programme. Separate soil samples of surface and subsurface provide a means of evaluating both of these factors. For a topsoil sample, scrape away the surface 2–3 cm of soil, then collect samples from the 0–30 cm depth and a separate sample from the 30–60 cm depth. A 0–8 cm depth crusting problem may be observed, because high salinity and excess sodium/chloride levels in irrigation water lead to restricted water penetration. Subsoils at 25 cm to 150 cm depths will show the typical depth of water penetration and salt accumulation. If infiltration is a problem or the irrigation period is too short, salinity accumulation in the top 30–60 cm will often be higher than the 90–120 cm depths. Although samples may be collected with a spade, a soil auger is usually a more convenient tool. Sampling should be done thoroughly throughout the area being tested (Ramos, 1997; Ferree and Warrington, 2003). Thoroughly mix the subsamples together to provide a sample for the soil test. Take about 800 g of the sample to the laboratory the same day; however, if you have to wait a few days to submit the sample, then air dry it by exposing the sample on a paper under wind flow (Ramos, 1997; Gregorich and Carter, 2007).

Standardized analytical procedures have been developed for the assessment of macro- and micronutrients. The results quantify the pool of nutrients in the soil that is available for plant roots. For both K and P, the available fraction is much lower than the respective total contents of these nutrients in the soil. In addition to the nutrient concentration in soil or water, the salinity condition also may be seen. Three types of salinity conditions are present in the field: excess root zone salinity, poor water penetration rates and accumulation of specific elements to toxic levels, which are diagnosed by salinity analysis. Excess root zone salinity leads to a reduction of soil-water availability and decreases its absorption by roots. These soils have a greater potential to retain water and the plant must exert more energy to absorb its required water.

Moreover, if these soils contain sufficient amounts of water, plants may show water stress symptoms. Leaching of salts is very important to cope with this problem. Poor water penetration can reduce water uptake and root zone aeration, which results from a high proportion of Na. Under this status, and when irrigated, soil aggregates and structures will be dispersed into particles. Both salinity (via EC_e or EC_w) and sodicity (via SAR) should be measured. High amounts of Mg and/or K cause the soil to become less stable and more impermeable. Moreover, when the Mg to Ca ratio exceeds 1:1, infiltration may occur (Quirk and Schofield, 1955).

In terms of toxic accumulation, Na, Cl and B are the primary ions of concern. Under their toxicity, trees accumulate these elements in the woody tissues and eventually in leaves. Leaf burn on margins often results from excess Cl or Na in leaf tissue. In B toxicity, the margins of foliage develop leaf burn that develops into interveinal necrosis with twisting and curling. Toxicity of nitrate-nitrogen (NO_3-N) becomes a concern when too much N fertilizer is applied, leading to large, curled or cup-shaped leaves and temporary defoliation before a tremendous regrowth. The concentrations of NO_3-N in irrigation water and soil are very important Ramos, 1997; Erez, 2000; Wallender and Tanji, 2011).

9.5.3 N-min method

This method allows the mineralized fraction of the soil to be assessed, thereby indicating the current amount of nitrogen that is available for the tree. Usually, shortly before flowering, soil samples are taken from a depth of 0–30 cm and 30–60 cm within the tree rows. N-min values, which include extracted nitrate and some ammonium N, can vary markedly due to specific site conditions and microclimatic factors. However, this method provides valuable data in estimating the appropriate N application. The requirement of pomegranate trees for vegetative growth and fruit yield has to be estimated taking into account both the current status of available N and the potentially available N from the soil reserves in order to obtain maximum quality fruit and to avoid N losses by nitrate leaching into the groundwater or by denitrification.

Available P, K, Ca, Mg and S are analysed from mixed soil samples using different extraction methods, which simulate the activity of tree roots in nutrient removal from soil. For meaningful interpretation of soil test data, soil fertility norms developed for pomegranate (Table 9.4) are used. If the soil test value for any nutrient is below the optimum range, the application of that nutrient will result in yield response. If it is within the optimum range, the amount of nutrients removed by the harvested fruit of the previous crop need to be supplemented, although many studies show a poor relationship between soil test results and leaf analysis in the orchard. The major limitation in using soil testing to evaluate the nutrient status of orchards is associated with problems in obtaining samples representative of conditions throughout the root zone and throughout the orchard. In existing orchards, soil testing provides the additional information necessary for interpreting the results of leaf analysis and in formulating fertilization programmes.

9.5.4 Plant analysis

Soil nutrient analysis does not always provide sufficient information about the bioavailability of nutrients to trees. Therefore, additional plant analysis is recommended in order to assess the uptake of nutrients from the soil. Leaf analysis provides a complete picture of the current nutritional status of the trees. Many factors that influence leaf composition must be considered when interpreting leaf analysis results. Leaf sampling technique forms the basis for any leaf analysis programme for judicious

Table 9.4. Soil fertility norms for pomegranate.

Nutrients	Diagnostic recommendation integrated system (DRIS) norms				
	Very low	Low	Optimum	High	Excess
Available N (kg/ha)	<114.30	114.30–161.40	161.50–255.50	255.60–302.50	>302.50
Available P (kg/ha)	<6.14	6.14–11.77	11.78–23.01	23.02–28.63	>28.63
Available K (kg/ha)	<127.90	127.90–232.20	232.30–440.90	441.00–545.20	>545.20
Exchangeable Ca (cmol (p^+)/kg)[a]	<18	18.00–27.64	27.65–46.92	46.93–56.55	>56.55
Exchangeable Mg (cmol(p^+)/kg)	<5.26	5.26–9.80	9.81–21.89	21.90–27.43	>27.43
Available S (mg/kg)	<1.92	1.92–6.93	6.94–16.95	16.96–21.96	>21.96
Diethylenetriaminepentaacetic acid (DTPA) Fe (mg/kg)	<1.12	1.12–3.70	3.71–8.85	8.86–11.43	>11.43
DTPA Mn (mg/kg)	<3.59	3.59–7.71	7.72–15.96	15.97–20.08	>20.08
DTPA Zn (mg/kg)	<0.31	0.31–0.51	0.52–0.92	0.93–1.12	>1.12
DTPA Cu (mg/kg)	<0.14	0.14–2.11	2.12–6.06	6.07–8.03	>8.03
Available B (mg/kg)	<0.04	0.04–0.16	0.17–0.40	0.41–0.52	>0.52

Source: Gosavi *et al.* (2017).

[a] A few laboratories report exchangeable cations as mg/kg (ppm). It is more useful to express them as centimoles of positive charge per kilogram of soil (cmol (+)/kg), numerically equal to milliequivalents per 100 g of soil (me/100 g).

fertilization. The levels of most elements vary with leaf age, either during the season or along the shoot. Different sampling methods may be chosen. If the aim is to determine the problem in an isolated tree or area, sampling a few poor and some good trees should suffice. If a determination of nutrient status in a large orchard is required, a survey containing trees from many sites within the orchard is needed. Looking for abnormal symptoms in foliage or growth is useful to diagnose nutrition problems. Often the location and shape of the symptom can be used to identify a nutrient deficiency. To compare leaf samples to standard, the leaves sampled must be comparable in physiological age to those used in developing the standards (Jones Jr, 2001).

Thus, sampling should be done in the month of April or August for February or June fruiting, respectively. In the northern hemisphere, August sampling, and for the southern hemisphere, April sampling should be considered. Leaf analysis standards are based on leaf pair samples collected from the eighth node from the growing tips (Maity *et al.*, 2017) as it was found that most of the nutrients are stabilized in the eighth leaf pair from the growing tips (Fig. 9.5). Samples should consist of 100 leaves collected from several trees (5 trees, 20 leaves from each) in the area being sampled. Trees may be selected at random or by following a predetermined pattern. No more than two leaves should be taken from an individual terminal shoot.

Fig. 9.5. Pomegranate leaves sampled for tissue nutrient analysis. (Photo: Shinsuke Agehara.)

9.5.5 Interpretation of leaf analysis results

Foliar diagnosis is very difficult and requires years of experience, because symptoms vary with time and plant species, and they can be affected by environmental conditions and the occurrence of multiple deficiencies in the same tree. Identifying the marginal deficiencies is also very difficult, and conditions such as low temperatures, waterlogging, the presence of toxic elements, mechanical and spray damage may be misinterpreted as nutrient deficiency (Ramos, 1997). The first step in evaluating the nutritional status of orchards is to compare results of leaf analysis (Gosavi *et al.*, 2017) with a set of standard values (Table 9.5).

Nitrogen (N): The most desirable N management programme provides a relatively high nitrogen status early in the season to encourage rapid leaf development, flower bud formation and fruit set and then allows nitrogen to decline gradually as the season progresses to favour fruit colour development. Optimum growth of young trees is associated with leaf nitrogen values of approximately 1.31 to 2.15% (Gosavi *et al.*, 2017). As the tree matures, less vegetative growth is required and the satisfactory level of nitrogen is generally reduced to improve colour development and fruit firmness. It is clear that optimum levels of copper improve the uptake of N by the plants (Malvi, 2011).

Phosphorus (P): leaf P level above 0.13% usually indicates an adequate supply of P within the trees (Gosavi *et al.*, 2017). Since the availability of phosphorus is strongly influenced by soil pH, low leaf phosphorus values frequently indicate a low soil pH condition that is limiting the availability of soil P and thus plant P uptake. On the other hand, high values frequently result from the accumulation of phosphorus when growth and leaf expansion is limited by deficiencies of other nutrients such as zinc (Marathe *et al.*, 2016a).

Potassium (K): Values in the range of 1.30–2% are generally considered to be adequate for pomegranate crop (Gosavi *et al.*, 2017). Visual symptoms of potassium deficiency are usually evident with leaf potassium values of 0.20% or less. Leaf potassium shows an inverse relationship with crop load. Thus a value of 0.61%

Table 9.5. Leaf nutrient norms for pomegranate.

	Diagnostic recommendation integrated system (DRIS) norms				
Nutrient	Very low	Low	Optimum	High	Very high
N (%)	<0.89	0.89–1.31	1.32–2.15	2.16–2.57	>2.57
P (%)	<0.13	0.13–0.17	0.18–0.24	0.25–0.28	>0.28
K (%)	<0.92	0.92–1.28	1.29–1.99	2.00–2.35	>2.35
Ca (%)	<0.34	0.34–0.63	0.64–1.20	1.21–1.48	>1.48
Mg (%)	<0.10	0.10–0.22	0.23–0.45	0.46–0.57	>0.57
S (%)	<0.09	0.09–0.15	0.16–0.26	0.27–0.31	>0.31
Fe (mg/kg)	<79.99	79.99–103.03	103.04–149.12	149.13–172.16	>172.16
Mn(mg/kg)	<22.96	22.96–39.59	39.60–72.85	72.86–89.47	>89.47
Zn (mg/kg)	<9.92	9.92–15.98	15.99–26.10	26.11–31.16	>31.16
Cu (mg/kg)	<4.57	4.57–6.15	6.16–9.32	9.33–9.90	>9.90
B (mg/kg)	<15.11	15.11–23.37	23.38–39.88	39.89–48.14	>48.14

Source: Gosavi *et al.* (2017).

potassium may be adequate in a sample from a heavily cropping orchard but might indicate marginal supply in a lightly cropping or non-bearing orchard. Leaf potassium levels of 2.0% or greater are not uncommon with young non-bearing trees. Leaf K levels decline as trees mature and the level of cropping increases (Marathe *et al.*, 2016a).

The nitrogen–potassium ratio often provides additional information in judging potassium status, in terms of fruit quality. High N–K ratios (e.g. more than 1.5% in apple) usually indicate that potassium supply is inadequate, while low ratios (less than 1.5%) might indicate either that the nitrogen supply is too low or that potassium supply is too high (Spectrum Analytic Inc., 2006). In other fruit, Hammami *et al.* (209) stated that a N–K ratio more than 0.9% is required for production of qualified mandarin fruits.

In addition to tree age and level of cropping, soil moisture and soil management practices also affect leaf potassium status. Even if the soil potassium supply is adequate, water-deficit conditions may limit the availability of potassium in the soil and thus results in a low leaf potassium level. Soil management practices such as the use of clean cultivated or herbicide strips along the tree rows or mulching, which reduces moisture stress, generally result in a higher leaf potassium level. It is reported that orchards with low potassium levels are more prone to diseases especially bacterial blight disease, which is the major factor for the pomegranate decline in Maharashtra state of India (Patil, 2014).

Calcium (Ca): Calcium content in leaf samples is considered to be adequate in the range from 0.64 to 1.20% (Gosavi *et al.*, 2017). Low leaf calcium is often, but not always, associated with low soil Ca supply and low pH, particularly in the subsoil. When adequate soil Ca is available, low leaf Ca may be the result of boron or zinc deficiency. Normal application of potassium or magnesium has little effect on calcium unless soil Ca supply is low. Leaf Ca and N levels generally have a positive correlation under normal growing conditions. This relationship exists because both Ca and N uptake by roots are largely regulated by transpiration-driven mass flow (Marathe *et al.*, 2016a).

Higher N, by increasing growth and leaf surface, enhances total transpiration. Excessively high nitrogen supply frequently promotes the development of high leaf to fruit ratios, which accentuates the problems associated with low Ca in the fruit. This is particularly important when soil moisture is inadequate because Ca is removed from the fruit as water moves from fruit to leaves under moisture stress condition (Ramos, 1997; Marathe *et al.*, 2016a).

Magnesium (Mg): Magnesium concentration within the range of 0.23–0.45% is usually satisfactory (Gosavi *et al.*, 2017), but should be considered in relation to potassium. The requirement for magnesium increases as the potassium status of the tree increases. For practical purposes, a ratio of the percentage of K to Mg in the leaf sample of 4:1 or greater usually indicates that magnesium supply is inadequate.

Boron (B): Boron shortages frequently occur in orchards, particularly on coarse textured soils and during the dry season. Leaf concentrations of 20–40 ppm are required for normal tree performance (Gosavi *et al.*, 2017). Low boron levels are often associated with a calcium deficiency problem. Interpretation of leaf boron values must recognize past boron application practices. If no foliar sprays of boron were used before leaf sample collection, a level of 20–40 ppm B usually indicates an adequate boron supply. However, if post-petal fall boron sprays were used, leaf levels in the 20–40 ppm range indicate a need to continue boron applications as a combination of soil and foliar applications.

Zinc (Zn): Interpretation of leaf Zn level is complicated by Zn-containing materials in foliar applications and by interaction with phosphorus. If no foliar sprays containing Zn have been applied, 15.99–26.10 ppm indicates adequate Zn; 9.92–15.98 ppm indicates low zinc status; and less than 9.92 ppm indicates Zn deficiency (Gosavi *et al.*, 2017). Relying strictly on these levels to assess zinc status may be misleading for two reasons: (i) Growth is reduced as Zn becomes limiting. This limited growth results in accumulation of Zn to a higher concentration than would occur with healthy growth. (ii) A high level of phosphorus tends to reduce the availability of Zn within the tree, as a result of the formation of insoluble zinc phosphate precipitates. When zinc is limited, the reduced growth also tends to result in a higher concentration of phosphorus within the leaf tissue, further aggravating the problem. An evaluation of the ratio of phosphorus to zinc in the leaf tissue provides a second means of assessing relative Zn status. Examining the foliage for visual symptoms of zinc deficiency provides a third means of verifying the adequacy of the zinc supply (Marathe *et al.*, 2016a).

Manganese (Mn): Manganese deficiency is found more frequently on high pH soils and coarse-textured soils. A concentration of 39.60–72.85 ppm of Mn indicates an adequate amount of this element in the leaf samples from the trees that have not been sprayed with manganese-containing materials. Concentrations below 22 ppm are usually accompanied by Mn deficiency symptoms, which include interveinal yellowing (chlorosis). Leaf samples from trees that have been sprayed with Mn-containing fungicides may show a high level of Mn. In such cases, the leaf samples need to be thoroughly washed with ethylenediaminetetraacetic acid (EDTA) solution before analysis. This eliminates the contamination by physiologically inactive Mn (Gosavi *et al.*, 2017).

Iron (Fe): Iron content of leaf samples fluctuates over a considerable range, often in response to variations in soil and weather conditions and with contamination of samples by dust (Gosavi *et al.*, 2017).

Copper (Cu): Copper shortage can be a problem on coarse-textured soils and soils with a pH of 6.3 or higher. Levels of 6.16–9.32 ppm in leaf samples generally indicate a satisfactory copper level (Gosavi *et al.*, 2017). Symptoms of Cu deficiency are associated with leaf content of 4.57 ppm or less and may appear as roughening and enlargement of lenticels on shoots followed by necrosis, shoot dieback during the season of growth and limited fruit set despite heavy bloom (Hippler *et al.*, 2017).

9.5.6 Nutrient recommendations, fertilizer application and manuring

Roots respond to localized nutrient enrichment and rapidly proliferate to uptake nutrients from that area. This response can be seen in band fertilization, especially N banding, in which root concentrations are seen around the band, compared with the bulk of the soil. Commercial cultivation of pomegranate was started with the realization of its nutraceutical value and market potentials, while it was considered as a minor fruit until 1986. Thus, nutrient management aspects have attained prime importance in enhancing productivity and quality of fruits. Research done in the late 1950s in Tulare County, California showed that mature 'Wonderful' pomegranate required only about 40–60 kg of N annually while P and K were of no benefit for improving yield, size and quality of fruit. During those days 1.2–4.0 kg ammonium sulfate along with 15–40 kg farmyard manure (FYM), 2.5–3.5 kg cake, sometimes 5.0 kg wood ash and 1.0 kg lime were recommended to meet the nutrient requirement of pomegranate per acre in India. But with the commercialization of pomegranate cultivation and more research results, it has been found that availability of all three major nutrients (N, P and K) is essential for satisfactory root growth, higher photosynthesis rates and higher yield with best fruit quality (Singh *et al.*, 1988; Padmavathamma and Hulamani, 1998). Based on fertilizer response, several recommendations have been worked out for various soil and crop situations in India and abroad. The following suggestions have been made: 250–625 g *N* + 125–250 g P_2O_5 + 125–250 g K_2O per plant for 'Ganesh' grown on black soils; 120 kg N + 90 kg P + 60 kg K/ha for 'Kzyr Anar'; 240 kg N + 160 kg P + 60 kg K/ha for 'Jodhpur Red'; and 375g *N* + 375 g P + 375 g K per plant for 'Dholka'. Dhillon *et al.* (2011) also recommended 180 g *N* + 60 g P + 120 g K per plant for 'Kandhari'. Prasad and Mali, 2000, 2003 also stated that increase in N levels from 350 to 500 g per plant improves fruit growth and size; however, increments of N levels up to 750 g per plant decreased the fruit weight. Hasani *et al.* (2016) stated that application of urea as a deep soil placement was better than foliar applied urea for pomegranate fruit yield and quality. Pomegranate responds particularly well to sulfate of potash as compared with muriate of potash. Foliar application of potassium chloride or potassium sulfate, both at 0.5, 1.0, 1.5 and 2.0%, from bud to harvesting stage at 15-day intervals increased leaf K level and fruit quality. In another study on pomegranate, Hasani *et al.* (2012) suggested the foliar spray of 0.6% $MnSO_4$ and 0.3% $ZnSO_4$ to achieve better quality in fruits.

Manures are organic materials originating as animal waste or vegetation that are incorporated into soil to enrich it and improve the structure and water-holding capacity. Under a warm, arid climate the beneficial effects of manures are not long lasting. Fully rotted manure should be used because microbial activity in fresh manure generates enough heat to kill roots and young seedlings. It is recommended to apply manure lightly or compost it prior to use. In some conditions, continuous application of manure has disadvantages. Under poor drainage of water or where less irrigation water is applied, large applications of chicken manure lead to sodium accumulation to toxic levels (Ryugo, 1988). Some organic constituents of manures tend to bind or chelate certain essential elements, such as Zn, making them unavailable to plants. There are many sources of manure including products from dairy and beef cattle and poultry industries, vermicompost, blood from slaughterhouses, sediment from sewage treatment facilities, the residues of seed oils after oil has been extracted and green manure.

Chemical fertilizers have some advantages over organic fertilizers, in that organic fertilizers come from unknown sources with varying compositions, thus the exact amount of any essential element supplied for maintaining trees in a healthy state may not be clear (Table 9.6; Van Slyke, 1950). In a study done by Kurer *et al.* (2017), it was found that vermicompost and poultry manure are the best sources of nitrogen for pomegranate trees. Mir *et al.* (2013) found that the microbial biomass pool in terms of *Pseudomonas* sp., soil fungi, *Azotobacter chroococcum*, *Actinomycetes* and arbuscular mycorrhizal fungi increased by 385.57, 60.26, 134.19, 168.02 and 39.87%, respectively, by manuring compared with the control. Olyaie Torshiz *et al.* (2017) also stated that manuring led to yield and fruit quality improvement of pomegranate.

Table 9.6. Percentage of composition of fresh solid animal excrement.

Species	Water	Nitrogen	Phosphorus	Potassium
Horse	75	0.55	0.13	0.33
Cow	85	0.40	0.08	0.08
Sheep	60	0.75	0.22	0.37
Swine	80	0.55	0.22	0.33
Hen	55	1.00	0.35	0.33

Source: Van Slyke (1950).

It was found that application of compost (25 kg/tree) or humic acid (25 g/tree) accompanied by recommended NPK doses improved fruit yield and quality of pomegranate trees (Abd-Ella *et al.*, 2009). Sandor (2011) also reported that application of humic acid for pomegranate nursery improves sapling quality and balances nutrient uptake. Humic acid improves root growth, soil physical condition and microbial activity, and thus could improve fertilizer use. Moreover, it was observed that organic matter decomposition and the mineralization processes in the soil increase. Results of a large number of experiments on manures and fertilizers conducted countrywide confirmed that neither chemical fertilizers alone nor organic sources exclusively can achieve production sustainability (Ryugo, 1988; Baviskar *et al.*, 2011; Khachi *et al.*, 2015; Wani *et al.*, 2017). However, integrated application of inorganic fertilizers along with organics (FYM at 25 kg/tree) increased tree spread and yield of pomegranate. It is reported that application of 10 kg FYM per plant alone or in combination with recommended NPK, or poultry manure 5 kg and bone meal 1 kg along with recommended NPK were found to be effective to boost the all-round growth of pomegranate plants. It was also observed that 50% supplementation of inorganic fertilizer with cattle dung manure increased fruit yield and improved fruit quality parameters such as TSS, ascorbic acid and sugar contents compared with application of organics alone (Mir *et al.*, 2013). Similarly, supplementation of inorganic fertilizers with vermicomposting in a 50:50 ratio increased plant height, canopy volume and fruit yield in sandy soils of a hot arid region (Meena, 2010). An integrated approach consisting of 10 kg vermicomposting, 25% recommended dose of NPK, 5 kg neem cake and 20g phosphate solubilizing bacteria per plant was found to produce large number of flowers (15.35/shoot), higher fruit setting (45.64%) and fruit retention (45.7%), and consequently higher yield as compared with yield obtained from recommended dose of NPK through inorganic fertilizers (Meena, 2010). The integrated supply approach not only increases fruit yield but also improves fruit qualities such as TSS, TSS:acid ratio, ascorbic acid content, reducing and non-reducing and total sugars.

Fertilization programmes must be developed based on the requirements of the crop and the characteristics of the soil on which it is being grown. Effects of soil management practices on nutrient availability and of cultivation practices such as pruning on tree vigour and nutrient requirements are significant factors that must be accounted for in adopting the programme to the individual orchard. Physical soil conditions throughout the root zone affect not only the depth of rooting, but also the distribution and type of root system that develops. Slowly drained or imperfectly drained soils are subjected to low oxygen conditions that may result in damage to tree roots, impairment of root function and alteration in the availability of various nutrient elements. Likewise, coarse-textured soils are more subject to moisture stress and may require special attention in dealing with elements such as nitrogen, boron and magnesium. Many nutritional problems of orchards are often more directly attributable to poor soil physical conditions than to fertilization programmes (Ryugo, 1988).

9.5.7 Type of fertilizers and the method of application

Various fertilizers are in use for the nutrition of pomegranate trees (Table 9.7). According to the crop management system and site conditions, solid (granular or crystalline) or liquid fertilizers are applied as straight (a single type of nutrient) or as complex NPK products. Today the use of micronutrient fertilizers is very common in fruit production. Nanofertilizers aim to make nutrients more available to leaves, consequently increasing nutrient use efficiency (Suppan, 2013). Some characteristics of nanoparticles, including the large specific surface area, unique magnetic/optical properties, electronic states and catalytic behaviour, confer nanoparticles a better reactivity than the equivalent bulk materials (Agrawal and Rathore, 2014). According to origin, nutrient sources can be classified as:

- Mineral fertilizers and organic manures (composts, farmyard manure, residues from the processing of organic materials).
- Soil amendments and products with low nutrient content (e.g. gypsum, lime, rock flour).

Table 9.7. Properties of mineral fertilizers used in fruit tree production.

Main nutrients	Fertilizer	Chemical composition	Nutrient content $N-P_2O_5-K_2O$ (%)	Properties
N	Urea	$(NH_2)_2CO$	46-0-0	Acidic
	Calcium nitrate	$Ca(NO_3)_2$	16-0-0 19% Ca	Basic
	Calcium ammonium nitrate (CAN)	$NH_4NO_3 + CaCO_3$	(21–27)-0-0 10% Ca	Basic
	Ammonium nitrate (AN)	NH_4NO_3	34-0-0	Acidic Quick acting
	Ammonium sulfate (AS)	$(NH_4)_2SO_4$	21-0-0 24% s	Acidifying
	Urea ammonium nitrate solution	$(NH_2)_2CO + NH_4NO_3$	(28–32)-0-0	Slightly acidifying effect in soil
P	Monoammonium phosphate (MAP)	$NH_4H_2PO_4$	11-50-0	Acidic
	Diammonium phosphate (DAP)	$(NH_4)_2HPO_4$	18-46-0	Acidic
	Phosphoric acid	H_3PO_4	0-61-0	Acidic Quick acting
	Single super phosphate (SSP)	$Ca(H_2PO_4)_2$	0-(16–18)-0 20% Ca	Neutral Quick acting
	Triple super phosphate (TSP)	$Ca(H_2PO_4)_2.H_2O$	0-(44–52)-0 14% Ca	Neutral
K	Muriate of potash (MOP)	KCl	0-0-60	Neutral
	Sulfate of potash (SOP)	K_2SO_4	0–0-(50–52) 18% s	Neutral
	Nitrate of potash (NOP)	KNO_3	13-0-44	Basic Quick reacting
	Monopotassium phosphate (MKP)	KH_2PO_4	0-52-35	Acidic
Ca	Gypsum	$CaSO_4.2H_2O$	18% s, 23% Ca	Not soluble
	Lime	$CaCO_3$ $Ca(OH)_2$	50–65% Ca	Hardly soluble Slow acting
Mg	Epsom salt	$MgSO_4.7H_2O$	16% MgO 13% s	Water soluble
	Kieserite	$MgSO_4.H_2O$	25% MgO 20% s	Water soluble
	Dolomitic limestone	$MgCO_3$ $CaCO_3$	5–20% MgO 14–32% Ca	Slow acting
S	Elemental sulfur	S	100% s	Strongly acidifying
	Sulfate types of other nutritional elements	$(NH_4)_2SO_4$ K_2SO_4 $MgSO_4$	24% s 18% s 13–20% s	Acidifying

Source: Ramos, 1997.

Nitrogen can be applied as NO_3 or NH_4 or other forms like amides in urea. Nitrate (NO_3-N) is immediately available to plant roots; however, due to its high mobility in the soil it is also subject to leaching after rainfall or irrigation. Under anaerobic soil conditions, it can also be lost to the atmosphere by denitrification in the form of oxide of N. Ammonium-N (NH_4-N) can be taken up directly by tree roots but it is also converted in the soil into nitrate by nitrification. The mobility of NH_4-N in the soil is low, so it is not readily leached. However, N losses via gaseous NH_3 are possible particularly in high pH soils. Urea-N applied to the soil has to be converted into NO_3 or NH_4 before it becomes available to the tree. Thus, NO_3 containing fertilizers are fast acting, while NH_4-containing or amide fertilizers are slower (Singh and Singh, 2015). The choice of the form or source of N for soil application is most frequently made on the basis of cost per unit of nitrogen, as nitrogen application does increase soil acidity and hence lime requirement. Nitrogen-containing fertilizers are susceptible to leaching and other losses. In order to reduce leaching:

- Do not use nitrate-N forms in sandy soils (coarse-textured soils).
- Do manage irrigation level and remove excessive rainfall from the orchard.
- Use ammonium-N fertilizer as broadcasting.
- Incorporate fertilizers containing ammonia (NH_3) or ammonium some depth below the soil surface shortly after application.
- Do adjust the soil pH to below 7.0 (to reduce volatilization).
- Increase the cation exchange capacity of the soil.
- Apply ammonium-containing fertilizers in coarse-textured soils in the winter.

Different methods for application of nitrogen-containing fertilizers are broadcasting, drilling, banding, application through the irrigation system (fertigation) and foliar spray. Banding in a 40-cm-wide strip allows less N use by weeds on the orchard floor, compared with broadcasting. Marathe *et al.* (2017) found that root growth and density or root activity of pomegranate plants is confined to a 0–60 cm radial distance and 0–45 cm vertical distance. Therefore, fertilizers and water may be applied in this zone for better utilization of inputs by the tree roots. Many growers do not want to disturb the soil surface and the orchard is not tilled, so nitrogen is seldom drilled into the soil. If the irrigation water and the applied nitrogen are uniform, N fertigation through the irrigation system saves money. Sprinkler and low-volume (drip or mini-sprinkler) irrigation systems can be used to supply nitrogen in the fertigation method. Compared with conventional systems, injection of N fertilizers through low-volume irrigation systems results in higher N recovery, because of the stimulation of a higher density of fibrous roots in a restricted volume of soil, which effectively intercepts the N fertilizer to the area. Moreover, the use of a high-frequency drip irrigation/fertigation method minimizes soil water saturation, which causes soil anaerobic conditions and leaching losses of NO_3-N. There are many factors affecting relative need for N-containing fertilizers. It is recommended that under sandy and coarse-textured soils the nitrogen application times are changed from two large doses once or twice per year to small doses six to eight times over the growing season. Leaves also respond to fertilizers via resumption of leaf expansion and meristematic growth, which provides new sites for photosynthesis. After expansion, more nutrients are needed to supply proteins and membranes and also younger leaves. Under deficient conditions, these nutrients are remobilized from the old leaves and transported through the phloem to the young tissues such as fruits and seeds. Davarpanah *et al.* (2017) evaluated two forms of nitrogen fertilizers as spraying: nano- N and urea in two different concentrations and at different times of growth season. They found that nano-N fertilizers are more efficient compared with urea because of the low rate needed for application and also better effects on pomegranate yield and quality.

Ordinary superphosphate (0-16-0) and triple superphosphate (0-48-0) are the sources most frequently used in orchards. Monoammonium phosphate (11-48-0), diammonium phosphate (18-46-0), various other phosphate compounds and animal manures are additional sources of phosphorus used in pre-plant soil preparation or in established orchards. The greatest attention should be given to pre-plant incorporation of phosphate throughout the rooting zone, at least in the upper 40 cm depth of the soil. Application of phosphates to the soil surface in established orchards is inefficient in meeting crop needs. A high rate of phosphate application can enhance Zn and Cu deficiencies. Incorporation of the appropriate amount of phosphate during pre-plant soil preparation should provide ample phosphorus for the life of the orchard, provided that soil pH is maintained in the range of 6.0–6.5 throughout the root zone.

The form of potassium to be applied should be determined by both the amount of potassium required and the available soil magnesium. Muriate of potash is suitable material if the magnesium supply is high. When both potassium and magnesium supplies are low, both elements should be applied using a material containing both of them such as sulfate of potash-magnesia (0-0-22-11). When the magnesium supply is adequate, and only a small amount of potassium and nitrogen is needed, nitrate of potash (13-0-44) may be a suitable material. As with nitrogen, placement influences the efficiency of potassium fertilizer use. Applying potassium fertilizers in a narrow 15–20 cm band on both sides of the row approximately one-half the distance from the trunk to the outer spread of the branches is effective. The foliar spray of potassium chloride at 1.0% on 'Yercaud-1', 'Ganesh' and 'Jyoti' pomegranates improved the fruit size significantly over the control. This also led to increased fruit yield in all the cultivars (Muthumanickam and Balakrishnamoorthy, 1999). Khayyat *et al.* (2012) also found that application of the sulfate form of potassium when fruits are 30 mm in diameter is very useful for quality and growth improvement of pomegranate fruit.

Limestone is the primary and most economical source of calcium in acidic soil. Consistent soil testing and a liming programme are basic requirements in managing soil Ca supply in acidic soil. Gypsum (24% Ca) has been used as a source of Ca in alkaline soils after working out the gypsum requirement of the soil. Other materials commonly used as soil-applied fertilizers containing calcium include ordinary superphosphate (20% Ca) and triple superphosphate (14% Ca). Calcium nitrate (24% Ca) has also been considered as an additional source for orchards (Ryugo, 1988; Osman, 2013). Kamal *et al.* (2017) stated that spraying some nutrients including calcium chloride and potassium oxide at the early stages of fruit growth and development (approximately 8 weeks after full bloom) has beneficial effects on improving fruit characteristics.

Dolomitic limestone is the most used source, but magnesium content of dolomitic limestone from different sources varies considerably. Other sources of magnesium for soil application include kieserite (a kind of magnesium sulfate with 17.3% Mg), magnesium oxide (49–56% Mg), sulfate of potash-magnesia (11% Mg) and Epsom salt (a kind of magnesium sulfate, called hepta-hydrate sulfate with 10% Mg). Soluble forms such as sulfate of potash-magnesia or kieserite are preferred to magnesium oxide for surface application, but if thoroughly incorporated into the soil, as in pre-plant preparation, magnesium oxide can be used effectively.

Soil applications of boron are essential in managing the supply of this element. Boron is readily mobile within the soil and can be effectively supplied through soil surface applications in established orchards. When a new site is being prepared for planting, it is recommended that an appropriate amount of boron be thoroughly mixed into the top soil. Rates of boron to be applied are determined on the basis of soil texture, boron already present as indicated by the soil test and crop needs. Granular fertilizer grade borate (14.3% B) can be used for soil application blended with other fertilizer materials. Solubor (20% B) applied as a foliar spray has also been effective. A complete boron programme frequently includes both a soil application to meet the basic need of the crop, and one or more foliar applications to supply additional boron at critical stages of crop development. Nano-B chelate fertilizer and nano-Zn chelate fertilizer also tested and were effective on pomegranate trees (Davarpanah *et al.*, 2016). There are significant differences between cultivars when their roots are exposed to high levels of boron. Sarafi *et al.* (2017) found that 'Ermioni' is more tolerant to high boron level (10 mg/l) in nutrient solution compared with 'Wonderful'. They stated that the highest B concentrations were observed in roots followed by stems and apical and basal leaves. Brown *et al.* (1998) reported that *Punica granatum* 'Nana' plants were very tolerant of B toxicity, and no clear B toxicity symptoms were observed even when treated with a nutrient solution containing 25 mg/l B for 5 months. Nable *et al.* (1997) and Reid *et al.* (2004) stated that high B concentrations may inhibit cell wall expansion and root growth and reduce leaf area, photosynthetic rate and overall plant growth. Mirzapour and Khoshgoftarmanesh (2013) in evaluation between foliar and soil applications of zinc and iron found that soil application of Fe-EDDHA +$ZnSO_4.7H_2O$, particularly as localized placement, was an effective approach to increase fruit yield and quality of pomegranate in calcareous soils.

9.5.8 The timing of fertilizer application

The timing of application of nutrients in modern orchard system is an important prerequisite for high yield and quality of fruit. Pomegranate trees undergo a rest period after harvesting of the previous crop followed by rapid development of new leaves, flowers and shoots. The tree's demand for water during the growing period is a function of leaf area, fruit load and environmental conditions, with vapour saturation deficit as the dominant factor. At the start of new season, flowers and leaves develop with virtually no nutrient supply from the roots, this new flush being more or less completely dependent on nutrients stored in buds and woods. Dependence on this storage for early growth and development in trees explains the importance of an adequate supply of nutrients before the initiation of the rest period. A high level of nitrogen reserves in the tree favours the development of flowers and shoots in the following year's bud break. Bearing in mind that N is a strong promoter of vegetative growth, this nutrient should be available for the tree throughout the entire growing season. Thus small split applications of N are more favourable to the trees than only a single application per year. Davarpanah *et al.* (2017) stated that split application of N fertilizers as spraying at full bloom and 1 month later is very useful for increasing pomegranate yield. Slow-release N-sources such as organic manures or coated mineral fertilizers are suitable alternatives to fast-acting N fertilizers based on NO_3.

Abdel-Sattar and Mohamed (2017) found that a combination of fast- and slow-release N fertilizers and different levels of soil moisture led to the improvement of vegetative growth traits, that is, tree height, shoot length, dry weight of leaves and chlorophyll content. In comparison with N and K, P is required by the tree in relatively small amounts. Like Ca, but to a lesser extent, P accumulates in the frame wood of the tree mainly as phytate, showing its importance for structural processes. Under normal soil conditions, a single application per year (or over even longer periods) is sufficient for pomegranate trees.

The function of K is strongly related to the water status of the tree. Besides that, K is the dominant nutrient in the fruit. The relatively high mobility of K in light-textured soils requires a continuous supply to the tree throughout the season on these soils. In clay soils, a single application is sufficient. However, split doses during the growing period have been shown to be more effective. Singh *et al.* (1988) and Bhujbal (1990) reported that split application of NPK increased individual fruit weight, soft seeds and good taste in pomegranate, and four split applications were more useful. They found better results when they fertilized the plants four times including March, April, May and June, at an interval of 1 month. The same is true for Mg, being very mobile in light-textured soils and easily leached from the root zone. Potassium is antagonistic to Ca, competing for binding sites of the clay minerals in the soil as well as in the apoplast of the tree root and in cellular membranes. Towards the end of the season, the K/Ca ratio has to be maintained at an optimum range concerning fruit quality. Increase in leaf NPK and Mg contents and decrease in Mn and Zn contents with the increasing doses of NPK fertilizers was reported in the case of 'Arabi' pomegranate (Haggag and El-Shamy, 1987).

Micronutrients (e.g. B) have to be applied early enough to the soil to reach the target organs (flower buds) or applied with foliar sprays. Micronutrient supply should be based on the appearance of deficiency symptoms and on foliar analysis. In conclusion, the timing of nutrient supply is dependent on the developmental stage of the tree as well as on orchard factors such as soil type, water availability and microclimate. In this context, fertigation has an advantage over spreading of solid fertilizers, because nutrients can be applied in tiny amounts and precisely at the time they are needed. Also, foliar sprays are by far the fastest method of nutrient application in order to correct deficiencies.

9.5.9 Salinity management and its effects on pomegranate trees

Accumulation of sodium chloride, sodium carbonate or both, and B compounds lead to development of salty soils. Excessive B, Na and Cl are toxic to plants. Moreover, accumulated Na tends to displace exchangeably Ca and Mg from soil particles, and by destruction of the soil structure leads to make the soil impermeable to water (Shabala, 2012).

Similar to other fruit trees, pomegranate also suffers from salinity levels higher than it can tolerate. There are many reports around the world about pomegranate responses to this stress (Wang *et al.*, 1995Asrey and Shukla, 2003; Naeini *et al.*, 2005; Kulkarni *et al.*, 2007; Bhantana and Lazarovitch, 2010; Okhovatian-Ardakani *et al.*, 2010; Bonyanpour and Khosh-Khui, 2013; Khayyat *et al.*, 2014; Sarafi *et al.*, 2014; Karimi and Hasanpour, 2014; Neori *et al.*, 2014; Sun *et al.*, 2015; Hasanpour *et al.*, 2015; Tavousi *et al.*, 2015; Khayyat *et al.*, 2016; Ibrahim, 2016; Bidabadi *et al.*, 2017; Karimi and Hassanpour, 2017).

Nowadays, we know that salinity stress strongly affects the plant growth and development via water stress (i.e. by lowering osmotic potential of the soil solution and thus reducing water uptake) or ionic stress (i.e. by nutritional imbalance and/or toxicity) or by a combination of them (Ashraf, 1994; Marschner, 1995; Ashraf and Harris, 2004; Silva-Ortega *et al.*, 2008). It has been accepted that pomegranate plants can tolerate salinity under arid and semi-arid zones. However, there are many differences among varieties and cultivars of pomegranate, because different aspects of growth, development and physiology are affected by salt stress. Bhantana and Lazarovitch (2010) suggested the pomegranate should be categorized as a moderately sensitive crop. On the other hand, some researchers clustered this plant as moderately tolerant to salinity; thus, it was recommended that pomegranate should not grow in soils with electrical conductivity (EC) of saturation paste more than 10 dS/m (Maas *et al.*, 1993; Allen *et al.*, 1998; Jain and Dass, 1988; Patil and Waghmare, 1982; Holland *et al.*, 2009; Bhantana and Lazarovitch, 2010; Khayyat *et al.*, 2014). Based on previous studies, it was also recommended that pomegranate plants should not be grown with salinized waters with approximate salinity level more than 4 dS/m (Holland *et al.*, 2009; Khayyat *et al.*, 2014).

The tolerance of pomegranate to Cl ranges from 1 to 2 milliequivalents per litre (meq/l) in saturation extract, and for Na it ranges from 15–70 meq/l in saturation extract of 0–40 cm depth of the soil around the roots (Khayyat *et al.*, 2014). Regarding irrigation water, the tolerable concentration of bicarbonate, sodium and chlorine for pomegranate trees is between 2.3–3.9, 13–14 and 8 –23.8 meq/l, respectively (Khayyat *et al.*, 2014). One of the main causes for increasing salinity of root zone is irrigation water. The best way is to use more tolerant cultivars/rootstocks under salinity conditions. Khayyat *et al.* (2014) stated that significant differences exist among the cultivars under salt stress. They found that pomegranate root had a strong role in preventing Na and Cl transportation to aerial parts, although there was a positive correlation between Cl and Na concentration in soil-saturated paste and their concentrations within the leaf. Moreover, they observed differential behaviour between Cl and Na ions. Chloride movement to aerial parts increases as salinity level increases; however, Na entrance is inhibited. There are differences between cultivars for chlorine accumulation within the leaf. Khayyat *et al.* (2014) suggested that Cl^- may be useful as osmoticum in leaf tissues of some pomegranate cultivars. Doering and Luedders (1987) and Ludders (1987) also found that sodium strongly accumulated in roots and was prevented from moving aboveground. Doering and Luedders (1986) also noted that root pressure significantly decreased under salinity, which led to lower transpiration flow and entrance of calcium into the plant organs. All those changes might be related to any change that occurred in cuticle thickening of leaf, production of parenchymal cells for water accumulation and salt accumulation in those cells (Zarinkamar and Asfa, 2005). Accumulation of potassium in leaf tissue is significantly affected by salinity. The K/Na ratio is a strong characteristic for screening under salt stress. Khayyat *et al.* (2014) found a significant difference between pomegranate cultivars with regard to this trait under salinity. Munns (1985) stated that this ratio should not be reduced to lower than 1.1 and, if reduced, it may lead to disturbance in metabolic activities of leaf. A reducing trend was also observed in the accumulation of nitrogen, calcium and iron in leaves. Karimi and Hassanpour (2017) used a grafting technique to study the influence of rootstock under salinity conditions. They found that the mentioned rootstock ('Tab-O-Larz' cultivar, from Iran) significantly decreased transportation of sodium and chlorine from root to shoot. Another way of reducing this problem is transporting salts out of the root zone with deep permeation, leading to a leaching process. In other words, it means to supply more water than crops need. Incorporation of saline water in irrigation causes a reduction in transpiration (Dudley *et al.*, 2008a), which subsequently results in reduced evapotranspiration (Bhantana and Lazarovitch, 2010). An increase in crop evapotranspiration and increment in concentration of salts in the water lead to more salinity transported into the orchard. There is a good correlation between relative yield and relative evapotranspiration. Thus, more leaching is required to transport salts beyond the root zone. Effective salinity management is easier to achieve in soils with deep, well-drained profiles. This condition provides a place to accumulate unwanted salts, far from the root zone. On the other hand, salinity management is difficult in poorly drained soils, because the only place where salts can accumulate is near the root systems. The depth of the water table is also very important. It is closest to the soil surface in the spring and farthest in the autumn. As a result, salinity that is leached in the autumn can be transported back into the root zone in the spring, when the water table rises. It is recommended to increase the soil-water content more than field capacity throughout the root zone before leaching can occur. Moreover, it is known that small amounts of water applied frequently via sprinkler or winter rainfall more effectively transport salinity below the root zone compared with an equal volume of water applied in one large flood application. Applications of large quantities of water lead to more water movement through the large pores and are not able to wash out the salinity from the small pores. Finally, soil sampling provides a view about the level of salinity or the specific ion concentration that exceeds the critical level for the plant. Based on these data, the required leaching frequency is decided.

Soil reclamation is needed when production of orchards is limited by salinity (Ramos, 1997). For this goal, up to three times leaching is needed. The depth of leaching required for reclamation is an estimate of the depth in which water reduces salinity to a level that does not reduce the plant yield. This parameter is shown as the volume of water used in the root zone. Table 9.8 shows the depth of leaching water for reclamation of salinized soils.

Regarding the table, it is assumed that leaching should be done in several small amounts of irrigation or rain with 2 or more days in between for drainage. Care must be taken to avoid complete saturation of the root zone, because after saturation it is easier for salts from the lower depths to rise back up into the root zone via capillary action, thus winter reclamation is preferred. Evaporation also should be considered, and any water lost this way must be added to the amounts presented in this table. Approximately as much as five

Table 9.8. The depth of leaching water required per 30 cm depth of root zone to be reclaimed given the initial average salinity and final desired salinity.

Desired root zone salinity (dS/m)	Initial salinity (dS/m)			
	6	8	10	12
	Water required for leaching (cm of water per 30 cm of root zone)			
3	3.5	5.0	7.0	9.0
5	0.5	1.8	8.9	4.3
7	0.0	1.3	1.3	2.3

Source: Hoffman (1986).

times as much water is needed to leach the other salts, depending on soil type and the residual of precipitated B in the profile.

Soils with poor permeability rates are candidates for treatment with amendments. This process supplies exchangeable Ca that displaces Na, Mg and K in some instances. Sodium (and Mg and K to a lesser degree) causes swelling and dispersion of soil particles when it is irrigated, and Ca improves soil aggregation and porosity. Soils with high SAR and low EC value need amendment. After the amendment, sufficient leaching is needed to remove salinity from the root zone. Soil amendment is done in two ways. In one way, Ca is added to the soil directly using Ca salts including gypsum, lime, dolomite, calcium chloride and calcium nitrate. The highest solubility rate is for $Ca(NO_3)_2$ and $CaCl_2$. Gypsum is moderately soluble and dolomite and lime are very slowly soluble, when pH is greater than 7.2 (Ramos, 1997; Osman, 2013).

Data presented here are applicable for all irrigation waters with less than 1.0 dS/m. It is adapted from a report by Hoffman (1986).

Addition of gypsum to irrigation water is simple, convenient and may be less expensive than $CaCl_2$ and $Ca(NO_3)_2$. Application of lime and dolomite to acidic waters with pH less than 7.0 is useful. Addition of $CaCl_2$ or $Ca(NO_3)_2$ does not affect much soil pH, but lime and dolomite increase this characteristic, especially when applied to acidic soil. In another method of soil amendment, known as acid-forming, amendments, namely sulfur (S), sulfuric acid (H_2SO_4), urea-sulfuric acid, ammonium polysulfide and lime sulfur are used. These cause dissolution of lime that is native to the soil, so they supply exchangeable Ca indirectly. The S compounds contribute to microbiological reactions, but the acid dissolves soil-lime to form gypsum, which then dissolves in the irrigation water to provide exchangeable Ca. Moreover, the acid-forming amendment can also increase the availability of Ca in irrigation water by neutralizing HCO_3^- and CO_3^{2-} that react with Ca to form lime precipitates.

Selection of soil amendment method is largely dependent on the presence or absence of lime and the type and relative cost of materials. Lime is abundant in the soil surface, so consider either the Ca salt or an acid-forming method. The amendment materials added to the root zone are very important. For example, some materials add SO_4, and others may add chloride or nitrate. The two latter materials should be used with care, because Cl^- and NO_3^- may be harmful for plant growth, when in the active growing stage. When very low amounts of lime exist in the soil, application of acidic materials is not recommended. These soils are neutral or acidic in pH. Thus, application of Ca salts is appropriate. The acid-forming amendments may be more useful on very alkaline soils with a pH above 8.4 for reducing the pH. There are various ways of amendment including applying the materials in the water, applying to the soil surface and irrigating the soil, broadcasting the materials and tilling it into the soil and applying the materials in a band in the soil. In soils with penetration problems, caused by surface crusting, amendment materials can be directly added to the water. This crust is on the soil surface and is often thinner than 2 cm. It is created because of the water quality used for irrigation. The small amount of amendment is needed as a frequent application. If trying acid-forming amendment, be sure that soil-lime is present in the soil surface or that water contains high Ca and HCO_3^- levels. It is possible to broadcast amendments such as gypsum on to the soil surface and introduce it into the soil by irrigation, which is an alternative to water treatments. With this method, the amount of gypsum used is lesser than that used for water amendment. However, the time of application is very important. It is suggested to use amendment before the soil infiltration problem arises. Applying the treatments too early will also lead to poor results. Calcareous soils often have more than 2% lime content and acid-forming amendments in irrigation water can be effective as long as water is applied with drip or microjet irrigation. This type of watering limits the volume of soil that is irrigated and concentrates the amendment (Ramos, 1997; Osman, 2013).

References

Abd-Ella, E.E., Mervate, S.S. and Wafaa, A.Z. (2009) Effect of some organic and mineral fertilizer applications on growth and productivity of pomegranate trees. *Alexandria Science Exchange Journal* 31(3), 296–304.

Abdel-Sattar, M. and Mohamed, Y. (2017) Pomegranate trees productivity in response to three levels of irrigation and slow or fast nitrogen release fertilizer as well as their combinations. *Journal of Plant Production* 8(8), 813–820. DOI: 10.21608/jpp.2017.40873.

Adrees, M., Ali, S., Rizwan, M., Ibrahim, M., Abbas, F. *et al.* (2015) The effect of excess copper on growth and physiology of important food crops: a review. *Environmental Science and Pollution Research* 22(11), 8148–8162. DOI: 10.1007/s11356-015-4496-5.

Agrawal, S. and Rathore, P. (2014) Nanotechnology pros and cons to agriculture: a review. *International Journal of Current Microbiology and Applied Sciences* 3(3), 43–55.

Al-Maiman, S.A. and Ahmad, D. (2002) Changes in physical and chemical properties during pomegranate (*Punica granatum* L.) fruit maturation. *Food Chemistry* 76(4), 437–441. DOI: 10.1016/S0308-8146(01)00301-6.

Allen, R.G., Pereira, L.S., Raes, D. and Smith, M. (1998) Crop apotranspiration: guidelines for computing crop waterrequirements. In: *FAO Irrigation and Drainage*, Paper 56. Food and Agriculture Organization of the United Nations, Rome.

Arteca, R.N. (1996) *Plant Growth Substances: Principles and Applications*. Springer Science & Business Media.

Aseri, G.K., Jain, N., Panwar, J., Rao, A.V. and Meghwal, P.R. (2008) Biofertilizers improve plant growth, fruit yield, nutrition, metabolism and rhizosphere enzyme activities of pomegranate (*Punica granatum* L.) in Indian Thar desert. *Scientia Horticulturae* 117(2), 130–135. DOI: 10.1016/j.scienta.2008.03.014.

Ashraf, M. (1994) Organic substances responsible for salt tolerance in *Eruca sativa*. *Biologia Plantarum* 36(2), 255–259. DOI: 10.1007/BF02921095.

Ashraf, M. and Harris, P.J.C. (2004) Potential biochemical indicators of salinity tolerance in plants. *Plant Science* 166(1), 3–16. DOI: 10.1016/j.plantsci.2003.10.024.

Asrey, R. and Shukla, H.S. (2003) Salt stress and correlation studies in pomegranate (*Punica granatum* L.). *Indian Journal of Horticulture* 60(4), 330–334.

Babana, A.H. and Antoun, H. (2007) Biological system for improving the availability of Tilemsi phosphate rock for wheat (*Triticum aestivum* L.) cultivated in Mali. *Nutrient Cycling in Agroecosystems* 76(2–3), 285–295. DOI: 10.1007/s10705-006-9067-1.

Baviskar, M.N., Bharad, S.G., Dod, V.N. and Barne, V.G. (2011) Effect of integrated nutrient management on yield and quality of sapota. *Plant Archives* 11(2), 661–663.

Bhantana, P. and Lazarovitch, N. (2010) Evapotranspiration, crop coefficient and growth of two young pomegranate (*Punica granatum* L.) varieties under salt stress. *Agricultural Water Management* 97(5), 715–722. DOI: 10.1016/j.agwat.2009.12.016.

Bhujbal, U.G. (1990) *Ganesh Dalimb*. Continental Prakashan, Vijayanagar, India.

Bidabadi, S.S., Dehghanipoodeh, S. and Wright, G.C. (2017) Vermicompost leachate reduces some negative effects of salt stress in pomegranate. *International Journal of Recycling of Organic Waste in Agriculture* 6(3), 255–263.

Bompadre, M.J., Silvani, V.A., Bidondo, L.F., Ríos de Molina, M.D.C., Colombo, R.P. *et al.* (2014) Arbuscular mycorrhizal fungi alleviate oxidative stress in pomegranate plants growing under different irrigation conditions. *Botany* 92(3), 187–193. DOI: 10.1139/cjb-2013-0169.

Bompadre, M.J., Fernández Bidondo, L., Silvani, V.A., Colombo, R.P., Pérgola, M. *et al.* (2015) Combined effects of arbuscular mycorrhizal fungi and exogenous cytokinins on pomegranate (*Punica granatum*) under two contrasting water availability conditions. *Symbiosis* 65(2), 55–63. DOI: 10.1007/s13199-015-0318-2.

Bonyanpour, A. and Khosh-Khui, M. (2013) Effects of salt and drought stress conditions on callus growth, proline content and antioxidant enzyme activity of *Punica granatum* 'Nana'. *Biotechnology, an Indian Journal* 7(7), 257–262.

Bose, T.K., Mitra, S.K. and Sanyal, D. (2001) *Fruits: Tropical and Subtropical*, 3rd edn. Naya Udyog, Kolkata, India.

Brown, P.H., Hu, H. and Roberts, W.G. (1998) Redefining boron toxicity symptoms in some ornamentals. *Slosson Report* 95, 1–7.

Cannon, R.J. and Ho, C.-T. (2018) Volatile sulfur compounds in tropical fruits. *Journal of Food and Drug Analysis* 26(2), 445–468. DOI: 10.1016/j.jfda.2018.01.014.

Cooling, E.N. (1967) Frost resistance in *Eucalyptus grandis* following the application of fertilizer borate. *Rhodesia, Zambia and Malawi Journal of Agricultural Research* 5, 97–100.

Cooling, E.N. and Jones, B.E. (1970) The importance of boron and NPK fertilizers to eucalyptus in the southern province, Zambia. *East African Agricultural and Forestry Journal* 36(2), 185–194. DOI: 10.1080/00128325.1970.11662459.

Cosme, M. and Wurst, S. (2013) Interactions between arbuscular mycorrhizal fungi, rhizobacteria, soil phosphorus and plant cytokinin deficiency change the root morphology, yield and quality of tobacco. *Soil Biology and Biochemistry* 57, 436–443. DOI: 10.1016/j.soilbio.2012.09.024.

Davarpanah, S., Tehranifar, A., Davarynejad, G., Abadía, J. and Khorasani, R. (2016) Effects of foliar applications of zinc and boron nano-fertilizers on pomegranate (*Punica granatum* cv. Ardestani) fruit yield and quality. *Scientia Horticulturae* 210, 57–64. DOI: 10.1016/j.scienta.2016.07.003.

Davarpanah, S., Tehranifar, A., Davarynejad, G., Aran, M., Abadía, J. *et al.* (2017) Effects of foliar nano-nitrogen and urea fertilizers on the physical and chemical properties of pomegranate (*Punica granatum* cv. Ardestani) fruits. *HortScience* 52(2), 288–294. DOI: 10.21273/HORTSCI11248-16.

Dhillon, W.S., Gill, P.P.S. and Singh, N.P. (2011) Effect of nitrogen, phosphorus and potassium fertilization on growth, yield and quality of pomegranate 'Kandhari'. *Acta Horticulturae* 890, 327–332.

Doering, J. and Luedders, P. (1986) Effect of different salt treatments on *Punica granatum* L. at different root temperatures. *Gartenbauwissenschaft* 52(2), 92–96.

Doering, J. and Luedders, P. (1987) Influence of sodium salts on Na, Cl and SO4 content in leaves, shoots and roots of *Punica granatum* L. *Gartenbauwissenschaft* 52, 26–31.

Dudley, L.M., Ben-Gal, A. and Shani, U. (2008a) Influence of plant, soil, and water on the leaching fraction. *Vadose Zone Journal* 7(2), 420–425. DOI: 10.2136/vzj2007.0103.

Dudley, L.M., Ben-Gal, A. and Lazarovitch, N. (2008b) Drainage water reuse: biological, physical, and technological considerations for system management. *Journal of Environmental Quality* 37(S5), S25–S35. DOI: 10.2134/jeq2007.0314.

Ebert, G. (2009) Fertilizing for high yield and quality: pome and stone fruits of the temperate zone. *Ipl Bulletin* 19, 74.

El-Desouky, M.I. and El-Hamied, S.A.A. (2014) Improving growth and productivity of pomegranate fruit trees planted on sandy dunes slopes at Baloza district (N. Sinai) using different methods of drip irrigation, organic fertilization and soil mulching. *IOSR Journal of Agriculture and Veterinary Science* 7(12), 86–97.

El-Rauof, F.A. and Dawoud, H.D. (2015) Improving nutritional status, yield and fruit quality of Barhi date palm cultivar by using different levels of elemental sulphur fertilization under soba conditions. *Journal of Network Communications and Emerging Technologies* 2(2), 16–20.

Erez, A. (ed.) (2000) *Temperate Fruit Crops in Warm Climates*. Springer Science & Business Media, Dordrecht, The Netherlands463 pp.

Ferree, D.C. and Warrington, I.J. (eds) (2003) *Apples: Botany, Production, and Uses*. CAB International, Wallingford, UK.

Gawada, S.N., Kale, A.P., Shaikh, J.A. and Sharma, R.C. (2018) Study on nutrient package for pomegranate. *Indian Journal of Agriculture and Research* 52(2), 199–202.

Glick, B.R. (2018) Soil microbes and sustainable agriculture. *Pedosphere* 28(2), 167–169. DOI: 10.1016/S1002-0160(18)60020-7.

Gosavi, A.B., Deshpande, A.N. and Maity, A. (2017) Identifying nutrient imbalances in pomegranate (cv. Bhagwa) at different phonological stages by the diagnosis and recommendation integrated system. *Journal of Plant Nutrition* 40(13), 1868–1876. DOI: 10.1080/01904167.2016.1267209.

Gradziel, T.M. . (ed.) (2017) *Almonds: Botany, Production and Uses*. CAB International, Wallingford, UK.

Gregorich, E.G. and Carter, M.R. (2007) *Soil Sampling and Methods of Analysis*. CRC Press, Boca Raton, Florida.

Haggag, M.N. and El-Shamy, H.A. (1987) Response of fig and pomegranate fruit trees to NPK fertilization [Egypt]. *Alexandria Journal of Agricultural Research* 32, 199–208.

Hammami, A., Rezgui, S. and Hellali, R. (209) Leaf nitrogen and potassium concentrations for optimum fruit production, quality and biomass tree growth in clementine mandarin under Mediterranean climate. *Journal of Horticulture and Forestry* 2(7), 161–170.

Hanson, E.J. and Breen, P.J. (1985) Effects of all boron sprays and environmental factors on fruit set and boron accumulation in ‘Italian’ prune flowers. *Journal of American Society and Horticulture Science* 110, 389–392.

Hasani, M., Zamani, Z., Savaghebi, G. and Fatahi, R. (2012) Effects of zinc and manganese as foliar spray on pomegranate yield, fruit quality and leaf minerals. *Journal of Soil Science and Plant Nutrition* 12471–480. DOI: 10.4067/S0718-95162012005000009.

Hasani, M., Zamani, Z., Savaghebi, G. Sofla, H.S. and Sheikh Sofla, H. (2016) Effect of foliar and soil application of urea on leaf nutrients concentrations, yield and fruit quality of pomegranate. *Journal of Plant Nutrition* 39(6), 749–755. DOI: 10.1080/01904167.2015.1047525.

Hasanpour, Z., Karimi, H.R. and Mirdehghan, S.H. (2015) Effects of salinity and water stress on echo-physiological parameters and micronutrients concentration of pomegranate (*Punica granatum* L.). *Journal of Plant Nutrition* 38(5), 795–807. DOI: 10.1080/01904167.2014.944711.

Hippler, F.W.R., Boaretto, R.M., Quaggio, J.A. Mattos, D. and Mattos Jr, D. (2017) Copper in citrus production: required but avoided. *Citrus Research & Technology* 38(1), 99–106. DOI: 10.4322/crt.ICC067.

Hoffman, G.J. (1986) Guidelines for reclamation of salt-affected soils. *Applied Agricultural Research* 1(2), 65–72.

Holland, D., Hatib, K. and Bar-Ya'akov, I. (2009) Pomegranate: botany, horticulture, breeding. *Horticultural Reviews* 35(2), 127–191.

Ibrahim, H.I.M. (2016) Tolerance of two pomegranates cultivars (*Punica granatum* L.) to salinity stress under hydroponic culture conditions. *Journal of Basic and Applied Scientific Research* 6(4), 38–46.

Jain, B.L. and Dass, H.C. (1988) Effect of saline water on performance of saplings of jujube (*Ziziphus mauritiana*), Indian cherry (*Cordia dichotoma* var Wallichii) and pomegranate (*Punica granatum*) at nursery stage. *Indian Journal of Agricultural Sciences* 58(5), 420–421.

Jones Jr, J.B. (2001) *Laboratory Guide for Conducting Soil Tests and Plant Analysis*. CRC Press, Boca Raton, Florida.

Kamal, H.M., Elisa, M.A. and Mohammed, A.A. (2017) Effect of some mineral compounds on yield and fruit quality of pomegranate. *Bioscience Research* 14(4), 1197–1203.

Karimi, H.R. and Hasanpour, Z. (2014) Effects of salinity and water stress on growth and macro nutrients concentration of pomegranate (*Punica granatum* L.). *Journal of Plant Nutrition* 37(12), 1937–1951. DOI: 10.1080/01904167.2014.920363

Karimi, H.R. and Hassanpour, N. (2017) Effects of salinity, rootstock, and position of sampling on macro nutrient concentration of pomegranate cv. Gabri. *Journal of Plant Nutrition* 40(16), 2269–2278. DOI: 10.1080/01904167.2016.1263324.

Khachi, B., Sharma, S.D., Vikas, G., Kumar, P. and Mir, M. (2015) Study on comparative efficacy of bio-organic nutrients on plant growth, leaf nutrient contents and fruit quality attributes of kiwi fruit. *Journal of Applied and Natural Science* 7(1), 175–181. DOI: 10.31018/jans.v7i1.584.

Khayyat, M., Tehranifar, A., Zaree, M., Karimian, Z., Aminifard, M.H. *et al.* (2012) Effects of potassium nitrate spraying on fruit characteristics of ‘Malas Yazdi’ pomegranate. *Journal of Plant Nutrition* 35(9), 1387–1393. DOI: 10.1080/01904167.2012.684130.

Khayyat, M., Tehranifar, A., Davarynejad, G.H. and Sayyari-Zahan, M.H. (2014) Vegetative growth, compatible solute accumulation, ion partitioning and chlorophyll fluorescence of ‘Malas-e-Saveh’ and ‘Shishe-Kab’ pomegranates in response to salinity stress. *Photosynthetica* 52(2), 301–312. DOI: 10.1007/s11099-014-0034-9.

Khayyat, M., Tehranifar, A., Davarynejad, G.H. and Sayyari-Zahan, M.H. (2016) Effects of NaCl salinity on some leaf nutrient concentrations, non-photochemical quenching and the efficiency of the PSII photochemistry of two Iranian pomegranate varieties under greenhouse and field conditions: preliminary results. *Journal of Plant Nutrition* 39(12), 1752–1765. DOI: 10.1080/01904167.2016.1201686.

Khorsandi, F., Yazdi, F.A. and Vazifehshenas, M.R. (2009) Foliar zinc fertilization improves marketable fruit yield and quality attributes of pomegranate. *International Journal of Agriculture and Biology* 11(6), 766–770.

Korkmaz, N. and Aşkın, M.A. (2013) Effects of calcium and boron foliar application on pomegranate (*Punica granatum* L.) fruit quality, yield, and seasonal changes of leaf mineral nutrition. *Acta Horticulturae* 1089, 413–422.

Kulkarni, T.S., Desai, U.T., Kshirsagar, D.B. and Kamble, A.B. (2007) Effects of salt regimes on growth and mineral uptake of pomegranate. *Annals of Arid Zone* 46(1), 77–82.

Kumar, D.N. (1997) *Introduction to Horticulture*. Rajalakshmi Publications, Nagercoil, India.

Kurer, B.S., Patil, D.R., Gandolkar, K., Mesta, R.K., Nagaraj, M.S. *et al.* (2017) Response of pomegranate to different organic manures under northern dry zone of Karnataka, India. *International Journal of Current Microbiology and Applied Sciences* 6(11), 86–90. DOI: 10.20546/ijcmas.2017.611.011.

Maas, E.V., Baligar, B.V., Duncan, R.R. and Yohe, J.M. (1993) Testing crops for salinity tolerance. *Proceedings of a Workshop on Adaptation of Plants to Soil Stresses* 234, 247

Mahmoud, I.E.D. and Sheren, E.H. (2014) Improving growth and productivity of pomegranate fruit trees planted on sandy dunes slopes at Baloza district (N. Sinai) using different methods of drip irrigation, organic fertilization and soil mulching. *IOSR Journal of Agriculture and Veterinary Science* 7(12), 86–97.

Maity, A., Pal, R.K., Chandra, R. and Singh, N.V. (2014) *Penicillium pinophilum*—A novel microorganism for nutrient management in pomegranate (*Punica granatum* L.). *Scientia Horticulturae* 169, 111–117. DOI: 10.1016/j.scienta.2014.02.001.

Maity, A., Babu, K.D., Sarkar, A. Pal, R.K. and Dinesh Babu, K. (2017) Seasonality of nutrients *vis-à-vis* fruit quality of pomegranate cv. Bhagwa on vertisol. *Journal of Plant Nutrition* 40(9), 1351–1363. DOI: 10.1080/01904167.2016.1267750.

Malvi, U.R. (2011) Interaction of micronutrients with major nutrients with special reference to potassium. *Karnataka Journal of Agricultural Science* 24, 106–109.

Marathe, R.A., Murkute, A.A. Dhinesh Babu, K. and Babu, K.D. (2016a) Mineral nutrient deficiencies and nutrient interactions in pomegranate. *National Academy Science Letters* 39(6), 407–410. DOI: 10.1007/s40009-016-0487-4.

Marathe, R.A., Babu, K.D. and Shinde, Y.R. (2016b) Soil and leaf nutritional constraints in major pomegranate growing states of India. *Agricultural Science Digest* 36(1), 52–55.

Marathe, R.A., Chaudhary, D.T. and Shinde, Y.R. (2017) Roots density and activity of pomegranate grown in light textured soil of semi-arid region. *Vegetos – An International Journal of Plant Research* 30(3), 48–50. DOI: 10.5958/2229-4473.2017.00154.9.

Mirdehghan, S.H. and Rahemi, M. (2007) Seasonal changes of mineral nutrients and phenolics in pomegranate (*Punica granatum* L.) fruit. *Scientia Horticulturae* 111(2), 120–127. DOI: 10.1016/j.scienta.2006.10.001.

Mir, M. and Sharma, S.D. (2012) Influence of biofertilizers on plant growth, fruit yield, nutrition and rhizosphere microbial activity of pomegranate (*Punica granatum* L.) cv. Kandhari Kabuli. *Journal of Applied Horticulture* 14(02), 124–128. DOI: 10.37855/jah.2012.v14i02.24.

Mirzapour, M.H. and Khoshgoftarmanesh, A.H. (2013) Effect of soil and foliar application of iron and zinc on quantitative and qualitative yield of pomegranate. *Journal of Plant Nutrition* 36(1), 55–66. DOI: 10.1080/01904167.2012.733049.

Marschner, H. (1995) *Mineral Nutrition of Higher Plants*, 2nd edition. Academic Press Ltd, London.

Marschner, H. and Dell, B. (1994) Nutrient uptake in mycorrhizal symbiosis. *Plant and Soil* 159(1), 89–102. DOI: 10.1007/BF00000098.

Meena, C.L. (2010) Integrated nutrient management in pomegranate (*Punica granatum* L.) cv. 'Ganesh'. Doctoral Dissertation. Maharana Pratap University of Agriculture and Technology, Udaipur, India.

Mengel, K., Kirkby, E.A., Kosegarten, H. and Appel, T. (2001) *Principles of Plant Nutrition*. Springer, Dordrecht, the Netherlands.

Mir, M., Hassan, G.I., Mir, A., Hassan, A. and Sulaimani, M. (2013) Effects of bio-organics and chemical fertilizers on nutrient availability and biological properties of pomegranate orchard soil. *African Journal of Agricultural Research* 8(37), 4623–4627.

Miransari, M. (2013) Soil microbes and the availability of soil nutrients. *Acta Physiologiae Plantarum* 35(11), 3075–3084. DOI: 10.1007/s11738-013-1338-2.

Munns, R. (1985) Na^+, K^+ and Cl^- in xylem sap flowing to shoots of NaCl-treated barley. *Journal of Experimental Botany* 36(7), 1032–1042.

Muthumanickam, D. and Balakrishnamoorthy, G. (1999) Spraying of potassium solution on the yield and quality of pomegranate (*Punica granatum* L). *South Indian Horticulture* 47(1/6), 152–154.

Nable, R.O., Bañuelos, G.S. and Paull, J.G. (1997) Boron toxicity. *Plant and Soil* 193(2), 181–198. DOI: 10.1023/A:1004272227886.

Naeini, M.R., Khoshgoftarmanesh, A.H., Lessani, H. and Fallahi, E. (2005) Effects of sodium chloride-induced salinity on mineral nutrients and soluble sugars in three commercial cultivars of pomegranate. *Journal of Plant Nutrition* 27(8), 1319–1326. DOI: 10.1081/PLN-200025832.

Neori, H.B., Judeinstein, S., Tripler, E., Holland, D. and Lazarovitch, N. (2014) Salinity effects on colour and health traits in the pomegranate (*Punica granatum* L.) fruit peel. *International Journal of Postharvest Technology and Innovation* 4(1), 54–68. DOI: 10.1504/IJPTI.2014.064145.

Okhovatian-Ardakani, A.R., Mehrabanian, M., Dehghani, F. and Akbarzadeh, A. (2010) Salt tolerance evaluation and relative comparison in cuttings of different pomegranate cultivar. *Plant, Soil and Environment* 56(No. 4), 176–185. DOI: 10.17221/158/2009-PSE.

Olyaie Torshiz, A., Goldansaz, S.H., Motesharezadeh, B., Asgari Sarcheshmeh, M.A. and Zarei, A. (2017) Effect of organic and biological fertilizers on pomegranate trees: yield, cracking, sunburning and infestation to pomegranate fruit moth *Ectomyelois ceratoniae* (Lepidoptera: Pyralidae). *Journal of Crop Protection* 6(3), 327–340.

Orlov, S.N., Aksentsev, S.L. and Kotelevtsev, S.V. (2005) Extracellular calcium is required for the maintenance of plasma membrane integrity in nucleated cells. *Cell Calcium* 38(1), 53–57. DOI: 10.1016/j.ceca.2005.03.006.

Osman, K.T. (2013) *Soils: Principles, Properties and Management*. Springer Science & Business Media, Dordrecht, The Netherlands.

Padmavathamma, A.S. and Hulamani, N.C. (1998) Effect of N and K nutrition on growth and yield of pomegranate cv. Jyoti and RCR-1. *Karnataka Journal of Agricultural Science* 11(4), 1126–1128.

Patil, R.J. (2014) Soil potassium level is the key factor for control of *Xanthomonas axanopodis* pv. Punicae infection of pomegranate in field. *National Symposium cum exhibition on 'Pomegranate for Nutrition, Livelihood Security and Entrepreneurship Development'*, Solapur, India, 5–7 December 2014, pp. 147–148.

Patil, V.K. and Waghmare, P.R. (1982) Salinity tolerance of pomegranate. *Journal of Maharashtra Agricultural University* 7, 268–269.

Pennock, D., Yates, T. and Braidek, J. (2007) Soil sampling design. In: Carter, M.R. and Gregorich, E.G. (eds) *Soil Sampling and Methods of Analysis*. CRC Press, Boca Raton, Florida, pp. 1–14.

Prasad, R.N. and Mali, P.C. (2000) Effect of different levels of nitrogen on quality characters of pomegranate fruit cv Jalore Seedless. *Haryana Journal of Horticultural Science* 29, 186–187.

Prasad, R.N. and Mali, P.C. (2003) Effect of different levels of nitrogen on quality characters of pomegranate fruit cv Jalore Seedless. *Scientific Horticulturae* 8, 35–39.

Quirk, J.P. and Schofield, R.K. (1955) The effect of electrolyte concentration on soil permeability. *Journal of Soil Science* 6(2), 163–178. DOI: 10.1111/j.1365-2389.1955.tb00841.x.

Raghothama, K.G. and Karthikeyan, A.S. (2005) Phosphate acquisition. *Plant and Soil* 274(1-2), 37–49.

Raghupathi, H.B. and Bhargava, B.S. (1998) Leaf and soil nutrient diagnostic norms for pomegranate (*Punica granatum* L). *Journal of Indian Society for Soil Science* 46, 412–416.

Ramos, D.E. (ed.) (1997) *Walnut Production Manual*. Vol. 3373. UC ANR Publications, Davis, California.

Reid, R.J., Hayes, J.E., Post, A., Stangoulis, J.C.R. and Graham, R.D. (2004) A critical analysis of the causes of boron toxicity in plants. *Plant, Cell & Environment* 27(11), 1405–1414. DOI: 10.1111/j.1365-3040.2004.01243.x.

Ryugo, K. (1988) *Fruit Culture: Its Science and Art*. John Wiley and Sons Inc, Hoboken, New Jersey.

Sánchez, E.E., Giayetto, A., Cichón, L., Fernández, D., Aruani, M.C. *et al.* (2007) Cover crops influence soil properties and tree performance in an organic apple (*Malus domestica* Borkh) orchard in northern Patagonia. *Plant and Soil* 292(1-2), 193–203. DOI: 10.1007/s11104-007-9215-7.

Sandor, F. (2011) The effect of humic substances on pomegranate nursery production in Nangarhar, Afghanistan. *Journal of Environmental Science and Engineering* 5, 214–226.

Sarafi, E., Chatzissavvidis, C.H. and Therios, I. (2014) Effect of calcium and boron on the ion status, carbohydrate and proline content, gas exchange parameters and growth performance of pomegranate cv. 'Wonderful' plants grown under NaCl stress. *Turkish Journal of Agricultural and Natural Sciences* (Special Issue 2), 1606–1617.

Sarafi, E., Chatzissavvidis, C. and Therios, I. (2017) Response of two pomegranate (*Punica granatum* L.) cultivars to six boron concentrations: growth performance, nutrient status, gas exchange parameters, chlorophyll fluorescence, and proline and carbohydrate content. *Journal of Plant Nutrition* 40(7), 983–994. DOI: 10.1080/01904167.2016.1262403.

Shabala, S. (ed.) (2012) *Plant Stress Physiology*. CAB International, Wallingford, UK.

Silva-Ortega, C.O., Ochoa-Alfaro, A.E., Reyes-Agüero, J.A., Aguado-Santacruz, G.A. and Jiménez-Bremont, J.F. (2008) Salt stress increases the expression of p5cs gene and induces proline accumulation in cactus pear. *Plant Physiology and Biochemistry* 46(1), 82–92. DOI: 10.1016/j.plaphy.2007.10.011.

Singh, Y.S. and Singh, B. (2015) Soil and fertilizer nitrogen. In: Rattan, R.K., Katyal, J.C., Dwivedi, B.S., Sarkar, A.K. and Bhattacharayya, T., *et al.* (eds) *Soil Science: An Introduction. Indian Society of Soil Science*. New Delhi, pp. 541–569.

Singh, R.R., Singh, H.G. and Chauhan, K.S. (1988) Effect of N, P and K on physico chemical composition of pomegranate fruit local selection. *Progressive Horticulture* 20(1–2), 77–79.

Smoke, W. (2018) Nutrient uptake, growth and yield of pomegranate as influenced by irrigation frequencies under light textured soils. *Journal of Environmental Biology* 39, 143–148.

Spectrum Analytic Inc. (2006) Fertilizing appleswww.spectrumanalytic.com/support/library/pdf/fertilizing g_apple_trees.pdf (accessed 20 April 2015).

Sun, Y., Niu, G., Iglesias, J., Altland, J. and Cabrera, R.I. (2015) Salt tolerance of 22 pomegranate cultivars. *ASHS 2015 annual conference*, New Orleans, Louisiana, 5 August.

Suppan, S. (2013) *Nanomaterials in Soil: Our Future Food Chain?* The Institute of Agriculture and Trade Policy, Minneapolis, Minnesota.

Taiz, L. and Zeiger, E. (2002) *Plant Physiology*, 3rd edn. Sinauer Associates, Sunderland, Massachusetts.

Tavousi, M., Kaveh, F., Alizadeh, A., Babazadeh, H. and Tehranifar, A. (2015) Effects of drought and salinity on yield and water use efficiency in pomegranate tree. *Journal of Materials and Environmental Science* 6(7), 1975–1980.

Thanari, N. and Suma, R. (2018) Effect of fertigation and soil application of major nutrients on growth and yield of pomegranate cv. *Bhagwa. International Journal of Chemical Studies* 6(5), 3062–3065.

Truog, E. (1947) Soil reaction influence on availability of plant nutrients. *Soil Science Society of America Journal* 11(C), 305–308. DOI: 10.2136/sssaj1947.036159950011000C0057x.

Valetti, L., Iriarte, L. and Fabra, A. (2018) Growth promotion of rapeseed (*Brassica napus*) associated with the inoculation of phosphate solubilizing bacteria. *Applied Soil Ecology* 132, 1–10. DOI: 10.1016/j.apsoil.2018.08.017.

Van Slyke, L.L. (1950) *Fertilizers and Crop Production*. Orange Judd Publishing Company, New York.

Vance, C.P., Uhde-Stone, C. and Allan, D.L. (2003) Phosphorus acquisition and use: critical adaptations by plants for securing a nonrenewable resource. *New Phytologist* 157(3), 423–447. DOI: 10.1046/j.1469-8137.2003.00695.x.

Wallender, W.W. and Tanji, K.K. (eds) (2011) *Agricultural Salinity Assessment and Management*. American Society of Civil Engineers, New York.

Wang, L., Na, K., Jiang, W., Ling, Z. and Wang, Y. (1995) Study on contents of sodium and potassium ions of pomegranate and peach plants under sodium chloride stress and their salt tolerrance. *Acta Horticulturae Sinica* 22(4), 336–340.

Wani, I., Mehraj, S., Ali, M., Hassan, A., Wani, S. *et al.* (2017) Effect of inorganic and organic fertilisers on yield and soil nutrient status of walnut orchard. *International Journal of Plant & Soil Science* 16(2), 1–13. DOI: 10.9734/IJPSS/2017/32310.

Zarinkamar, F. and Asfa, A. (2005) The effect of salinity on anatomical structure and alkaloid production in pomegranate. *Rostaniha* 6(2), 97–106.

Zekri, M. and Obreza, T.A. (2003) *Plant Nutrients for Citrus Trees*. University of Florida Cooperative Extension Service, Institute of Food and Agricultural Sciences, EDIS, Florida.

10 Water Requirements and Responses to Irrigation Restrictions

D.S. Intrigliolo[1]*, D. Wang[2], M.B. Pérez-Gago[3], L. Palou[3], J. Ayars[2], H. Puerto[4] and J. Bartual[5]

[1]*Spanish National Research Council (CSIC), Centre for Applied Biology and Soil Sciences (CEBAS), Campus Universitario Espinardo, Murcia, Spain;* [2]*United States Department of Agriculture (USDA), Agricultural Research Service (ARS), San Joaquin Valley Agricultural Sciences Center, Water Management Research Unit, Parlier, California, USA;* [3]*Institut Valencià d'Investigacions Agràries (IVIA), Centre de Tecnologia Postcollita (CTP), Valencia, Spain;* [4]*Department of Engineering, Miguel Hernández University of Elche, Desamparados, Spain;* [5]*Conselleria de Agricultura, Desarrollo Rural, Emergencia Climática y Transición Ecológica, Estació Experimental Agrària d'Elx, Elx, Spain*

10.1 Pomegranate Water Relations

Pomegranate is a versatile perennial woody plant that grows and produces fruit in a wide range of soil and climatic conditions. The optimum growth conditions include relatively high temperatures and low humidity during the growing season and cool but not cold winters during dormancy. The plant grows well in arid and semi-arid environments where it can withstand periodic drought without causing permanent damage. Because of the origin of the plant, it is most adapted to the Mediterranean-like climate under mild winters and hot summers with the majority of the rain occurring during the winter. Therefore, major producing countries in the world include India, Iran, Turkey, Spain, Israel, China and the USA, among many other countries across all continents (see Chapter 3 for more details).

Depending on the local weather conditions and cultural practices, irrigation is often required to provide supplemental water to meet plant needs, especially in arid and semi-arid climates. Despite the assumed ability of pomegranate to tolerate drought and water stress, sufficient hydration via irrigation or precipitation is needed to maintain essential physiological functions such as photosynthesis, carbohydrate metabolism and protein synthesis.

Surface gravity flow in the form of flood, basin or furrow irrigation has been the conventional method of irrigation for most agricultural and horticultural crops until recent conversion to pressurized systems, such as sprinkler and micro-irrigation systems. In most production areas, pomegranate is commonly irrigated by either surface or drip irrigation. Meshram *et al.* (2010) compared pomegranate yield and water use efficiency (WUE) between surface drip, subsurface drip and surface irrigation, and found

*Corresponding author: dintri@cebas.csic.es

Fig. 10.1. Installation of subsurface drip irrigation tubing prior to planting a pomegranate orchard in California (USA). (Photo: Claude J. Phene *et al.*, 2015, University of California Agriculture and Natural Resources.)

that subsurface drip (Fig. 10.1) resulted in the highest yield and WUE and the lowest under surface irrigation.

Drip irrigation has shown positive effects on pomegranate growth parameters, such as tree height and canopy size (Sulochanamma *et al.*, 2005). Significant water savings were achieved with drip irrigation compared with surface irrigation in Iran (Behnia, 1999) and India (Chopade *et al.*, 2001). In a recent study by Ayars *et al.* (2017) pomegranate yield and WUE showed no significant difference between surface drip and subsurface drip, indicating the possibility of using subsurface drip on this crop.

The efficiency for most well managed surface irrigation systems ranges from 65–75%, whereas drip and micro-sprinkler irrigation systems can achieve irrigation efficiency of 85–90%. The higher country average pomegranate yield for Israel and the USA (>20 t/ha) can probably be attributed in large part to the wide adoption of advanced irrigation technology, such as drip. The choice of irrigation system can also depend on financing availability to convert gravity-based irrigation systems to more efficient pressurized systems and site-specific conditions at the location such as soil type and field topography.

Pomegranate is generally believed to have evolved to a drought-tolerant plant (Rodríguez *et al.*, 2012). Many woody plants have adapted to drought by two broad mechanisms: drought avoidance and drought tolerance. For drought avoidance, the plants tend to possess an annual growth cycle that flourishes during the rainy part of the year such as winter to late spring and remain semi-dormant during the hot and dry summer and autumn seasons. For drought tolerance, the plants often follow the seasonal patterns of growth but have developed specific physiological mechanisms to cope with water stress. The drought-tolerance mechanisms can be expressed in the whole plant including proliferation in root systems, differentiation in xylem structure and function that resists liquid water flow, dehydration tolerance through osmotic adjustment, drought hardening and control of transpiration.

Deep and wide-spreading root systems can increase the plant's ability to take up water from larger soil volumes compared with smaller root systems. Despite its relatively small canopy size compared with most deciduous fruit trees, Hiwale *et al.* (2009) reported that the root distribution of pomegranate was concentrated in the top 60 cm soil depth and within 60 cm radial distance from the tree trunk (85% of root biomass). Plants minimize the water flow resistance between roots and leaves for a given degree of transpiration by adjusting the xylem structure so that water is easily transported to the active sites for photosynthesis and other important physiological functions. In a drought environment, where there is a lack of water, plants still need to make available the limited water taken up by the roots to active locations of the plant. Drought tolerance through osmotic adjustment is achieved, when experiencing drought stress, by actively increasing solute content in plant cells, which lowers the osmotic potential, thus maintaining turgor pressure in the plant tissue. Drought hardening is a feedback mechanism for plants to develop lower rates of stomatal and cuticular transpiration after exposure to water stress during the early stages of growth. Control of transpiration for drought tolerance in plants may be expressed in reduced overall growth rate, reduced leaf size and shape, and stomatal control. For woody plants, it is believed that up to 90% of the variation in tree trunk diameter growth can be attributed to variations in plant water stress in arid regions (Ortuño *et al.*, 2010). Intrigliolo *et al.* (2011a) found the maximum diurnal trunk shrinkage of mature pomegranate trees was strongly related to plant water status measured as midday stem water potential. This is a good indicator of pomegranate tree water status, and it can be further used for managing irrigation.

Compared with most deciduous fruit trees, pomegranate leaves are small and narrow, which enables more efficient heat dissipation and minimizes water loss under hot and dry weather conditions. Rodríguez *et al.* (2012) determined diurnal and seasonal pomegranate leaf water relations under deficit irrigation regimes to simulate drought conditions to characterize leaf stomatal control. They found that leaf stomatal conductance was reduced proportionally from 400 to 100 mmol/m^2/s both diurnally and seasonally, which clearly indicates a stomatal control mechanism responding to the imposed water stress.

Given the growth and plant-water characteristics of pomegranate, certain irrigation strategies such as deficit irrigation, and technologies such as remote sensing can be applied to enhance water management effectiveness and efficiency. Deficit irrigation has been applied in fruiting trees because fruit yield and quality at harvest may not be sensitive to a certain degree of water stress (Fereres and Soriano, 2007). For example, similar pomegranate fruit yield was obtained by Intrigliolo *et al.* (2013) under different deficit irrigation managements with drip irrigation, by comparing trees with 50% of crop water use and trees with 100% of crop water use. Additional details regarding the deficit irrigation of pomegranates will be discussed later in the chapter.

Thanks to recent advances in electronics, remote sensing technology has evolved rapidly in both sensors and platforms that can be used affordably for agricultural purposes. In an attempt to estimate pomegranate water requirement using remote sensing data, for example, multispectral images of the tree canopy (Fig. 10.2), Zhang *et al.* (2017) found strong relationships between pomegranate tree canopy cover and crop coefficient.

The preliminary findings are promising because spatial distribution of crop evapotranspiration (ET_c) can be estimated from the crop coefficient (K_c) and reference evapotranspiration (ET_o) for any given time of the growing season, thus providing a near real-time guide for irrigation scheduling. In addition, canopy images can be obtained from different mobile platforms from the ground, to air, to space.

Fig. 10.2. Determination of canopy vigour in a young pomegranate orchard using a multispectral camera at the University of California Kearney Agricultural Center (KAC). (Photo: Claude J. Phene *et al.*, 2015, University of California Agriculture and Natural Resources.)

10.2 Water Requirements

The determination of water requirements for any crop is difficult, particularly for a perennial crop that relies primarily on irrigation to meet its water demand. Plant water use is estimated using many different techniques, for example, lysimeters, eddy covariance, water or energy balance and sap-flow sensors, each with its advantages and drawbacks (Wullschleger *et al.*, 1998).

There are several studies of the water requirement of pomegranate in the literature that describe the measured water requirement for developing trees (Bhantana and Lazarovitch, 2010; Mittal *et al.*, 2011, 2012). These methods include estimating water balance based on applied water and changes in the soil water content. Often ET_o is not available to characterize K_c as the ratio of ET_c and ET_o (Allen *et al.*, 1998). Here we have summarized the main research carried out specifically to determine tree water needs of either 'Wonderful' in California or 'Mollar de Elche' in Spain.

10.2.1 Water requirements for 'Wonderful' trees

'Wonderful' is the primary cultivar grown in California with approximately 13,000 ha under cultivation (Day and Wilkins, 2011), as the primary production area in the USA. This area is considerably smaller than the production areas found in Iran, Turkey and China (see Chapter 3 for details).

Factors that complicate the determination of water requirements include the irrigation methods, for example, micro-irrigation, sprinkler and surface irrigation, the orchard configuration and the pruning methods. Irrigation scheduling and the availability of water during critical growth periods will affect the measured crop water use (Allen *et al.*, 1998). Pomegranate has been characterized as a drought-tolerant plant, which further complicates a definitive determination of the water requirement and a crop coefficient.

Experiments were conducted at the University of California Kearney Agricultural Center (KAC), and the USDA-ARS San Joaquin Valley Agricultural Sciences Center (SJVASC) in California (USA) to determine the water requirements of pomegranate (*Punica granatum* L. cv. 'Wonderful') grown with both surface and subsurface drip irrigation (Ayars *et al.*, 2017). The main treatments were surface drip irrigation and subsurface drip irrigation with the laterals installed at a depth of 50–55 cm. There were two laterals per tree row with a lateral located at 1.1 m on each side of the tree row. Both locations were equipped with automated weighing lysimeters (Phene *et al.*, 1991; Ayars *et al.*, 1996; Schneider *et al.*, 1996). A California Irrigation Management Information System (CIMIS) located approximately 1 km from the site provided meteorological data for the determination of the ET_o. The trees were planted in the spring of 2010 at both locations. Both sites were planted using a randomized complete block design for statistical purposes.

On the KAC site, the trees were planted with a between-row spacing of 4.9 m and a within-row spacing of 3.6 m. There was one tree planted on the lysimeter (Fig. 10.3) that was subsurface drip irrigated with the equivalent number of emitters per tree as found in the surrounding field.

Fig. 10.3. A young pomegranate tree growing in the large weighing lysimeter facility of the University of California Kearney Agricultural Center (KAC). (Photo: Claude J. Phene *et al.*, 2015, University of California Agriculture and Natural Resources.)

The irrigations were controlled using the lysimeter and irrigation was initiated when 1 mm of water loss was measured. The trees were pruned to multi-stem bush configuration with the height limited to approximately 3 m. After 2012 an additional 10% of water was applied to the surface-irrigated plots to compensate for weed growth and evaporation.

The trees on the SJVASC site were planted with a 5 m between-row spacing and a 2.75 m within-row spacing. Surface drip with a single lateral on each row was used to irrigate the trees with 4 l/h emitters positioned 0.5 m on each side of the tree with an additional 8 l/h emitter placed halfway between the trees. There were two trees planted on the lysimeter and irrigation was initiated when a total of 4 mm of water loss was measured by the lysimeter. This site was used for a deficit irrigation trial with 35, 50, 75 and 100% replacement of water as measured by the lysimeter. These trees were pruned to a vase-type configuration with four main leaders for fruiting starting at approximately 0.3 m above ground. The tree height was limited to approximately 3 m, as well, for harvest convenience.

Applied irrigation water was measured with flow meters, and soil water content in the KAC lysimeter was measured using heat dissipation matric potential sensors. The yield was determined with a single harvest, and all the fruit from each tree was weighed and evaluated for quality (Ayars *et al.*, 2017).

Recent studies have developed methods to use remotely sensed data to characterize fractional ground cover (f_c) to estimate crop water use (Trout and Gartung, 2006; Bartual *et al.*, 2019). Zhang *et al.* (2015) demonstrated a remote sensing technology to determine normalized difference vegetation index (NDVI) to estimate (f_c) for use in determining the crop coefficient K_c.

Zhang *et al.* (2017) collected remotely sensed and ground cover data in both fields during the 2012–2014 growing seasons. From the NDVI data, they were able to determine fractional ground cover. The crop coefficients determined from the lysimeter data were then correlated to the fractional canopy cover.

The resulting equations were:

$K_c = 0.86(+/-0.11)f_c + 0.22\ (+/-0.06)\ R^2 = 0.90$ (10.1) for the KAC site

$K_c = 0.51\ (+/-0.18)f_c + 0.27(+/-0.05)\ R^2 = 0.57$ (10.2) for the SJVASC site

For more detail, the reader is referred to Zhang *et al.* (2017). Of note is the difference between the slope of the equations resulting from the differences in planting density and canopy structure. These equations should represent a tool for estimating 'Wonderful' pomegranate water use on a regional scale.

On the KAC site, the applied irrigation increased from 25.4 mm in 2010 to 932.2 mm for the surface drip irrigation system and from 25.4 mm to 843.3 mm for the subsurface drip irrigation system in response to the crop water requirement increasing from 53.3 mm to 952.5 mm (Ayars *et al.*, 2017). The ET_c and ET_0 data from 2015 were used to calculate the crop coefficient for a mature multi-trunk pomegranate tree. The data were fitted using a fifth-order polynomial equation. The resulting equation is:

$$K_c = -0.0125 - 0.000585x + 0.0001624x^2 - 1.452\,E^{-6}x^3 + 5.314\,E^{-9}x^4 - 7.13\,E^{-12}x^5 \quad (10.3)$$

Where x is the day of the year (DOY). The crop coefficient was also calculated using the data from 2013 after 3 years of growth. The resulting equation is:

$$K_c = -3.904 + 0.087\,x - 0.00088\,x^2 + 3.821\,E^{-6}x^3 - 8.59\,E^{-9}x^4 + 7.203\,E^{-12}x^5 \quad (10.4)$$

Where x is the DOY. These equations represent a starting point for calculating K_c for multi-trunk pomegranate trees.

10.2.2 Water requirements for 'Mollar de Elche' trees

In Europe, the major pomegranate production area is located in the eastern part of Spain, mainly in the province of Alicante (Valencian Community) where the commercial cultivar 'Mollar de Elche' is a local variety with a Protected Denomination of Origin (Granada de Elche Food Wine Spain, 2018). It belongs to the sweet cultivars group (Melgarejo *et al.*, 2000) and is most suitable for the fresh market. Its fruit has a cream-yellow to red-coloured skin, a deep pink to red arils, a sweet taste and soft seeds. As a deciduous tree, in local conditions bud break usually takes place at the end of February with a staggered bloom that begins about the second fortnight of April (10–20 April) and continues throughout May. Fruit set of the first flowering begins about 10 May. Fruit growth progresses through June, July and August, fruit ripening begins in September and harvest follows in the last week of September and October. Leaf drop takes place during November.

An experiment was conducted in a commercial mature pomegranate orchard (*P. granatum* L. cv. 'Mollar de Elche') at Elche, Alicante, Spain (38°N, 0°W, elevation 97 m). Trees were planted at a spacing of 5 × 4 m and average tree-shaded area was 56%. The drip irrigation system had a single drip line per row with eight 4.0 l/h nominal discharge emitters per tree. The soil was sandy loam with an effective depth greater than 120 cm. Climatic conditions during the experimental periods are reported in Fig. 10.4.

The seasonal variation of plant water status, leaf gas exchange, soil water content, sap flow rates and the ratio between irrigation + effective rainfall, and ET_0 were used to derive irrigation recommendations. Sap flow of trees was measured using the heat pulse technique, since this method provides direct estimates of plant transpiration, can be easily automated and allows continuous long-term records. None the less, this method also presents some uncertainties because it is based on some theoretical assumptions used to convert measured parameters to mass flow. Therefore, sap flow determinations were calibrated by soil water balance performed with capacitance probes with sensors installed at different depths (10, 30, 50 and 70

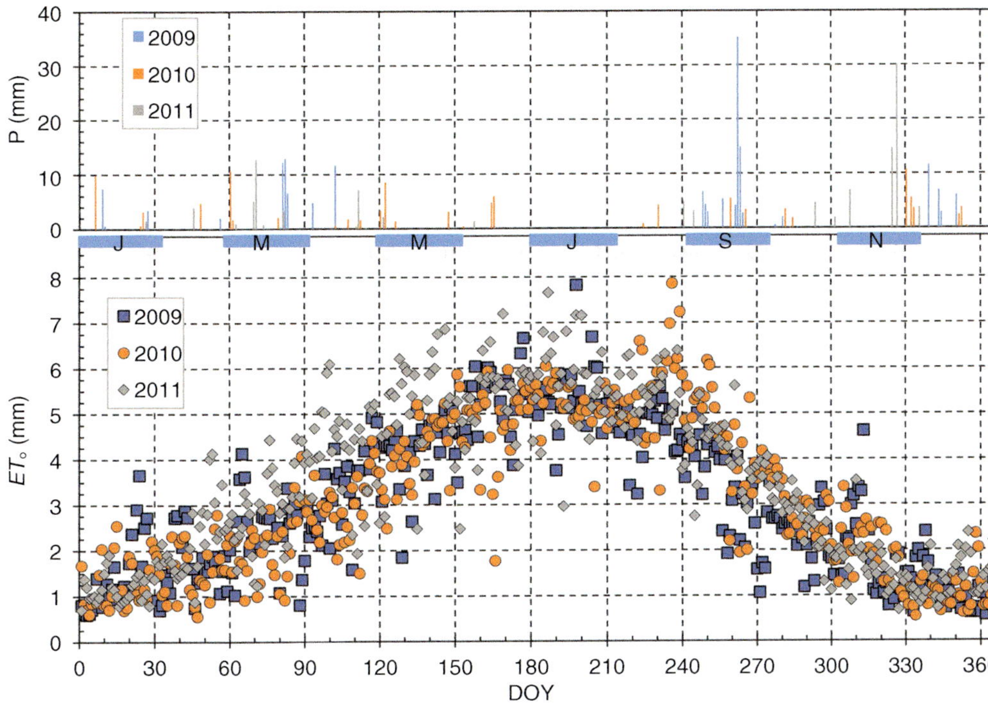

Fig. 10.4. Seasonal variation of daily rainfall (P) and reference evapotranspiration (ET_o) for years 2009, 2010 and 2011. DOY, the day of the year. Measured 4 km from the experimental plot. (From: Elx Agro-Weather Station, available at http://riegos.ivia.es/listado-de-estaciones/elx)

cm) and about 20 cm distant from the emitters. Probe readings at 70 cm depth showed no deep percolation was occurring which confirmed the calibration of the sap flow sensors (Buesa *et al.*, 2011).

The results obtained from several studies (Intrigliolo *et al.*, 2011a, b, 2013) shown in Fig. 10.5 demonstrate that the maximum K_c values can reach 0.75–0.95 implying that despite pomegranate's resistance to drought stress, potential water use can be quite large.

Though pomegranate is a crop that can thrive in arid and salty soil conditions, the results provided here validate the idea provided by Holland *et al.* (2009) and Bhantana and Lazarovitch (2010) that potential pomegranate water needs can be high, particularly during summer. Thus, the results of regulated deficit irrigation (RDI) of pomegranate discussed later in the chapter show great potential to reduce irrigation water use and increase the water use efficiency. As expected, the K_c results obtained with mature Elche trees are higher than the results obtained by Bhantana and Lazarovitch (2010) with young trees in Israel. Nevertheless, considering the shaded area and the time to bud break in relation to predominant climate characteristics (e.g. degree-day calculations) could help to obtain a general K_c for all growing areas.

Pomegranate trees are considered moderately tolerant to salinity (Holland *et al.*, 2009), though the results obtained by Bhantana and Lazarovitch (2010) suggest it should be listed as a moderately sensitive crop rather than a moderately tolerant one. Comparing both experiments again, irrigation water salinity in Elche was considerably higher during the 2009 and 2010 seasons than the 1 dS/m threshold provided for 'Wonderful' cultivar. There is still scope for investigating the ET_c reduction due to irrigation water salinity and whether there is a differential response to salinity between the 'Mollar de Elche' and 'Wonderful' cultivars.

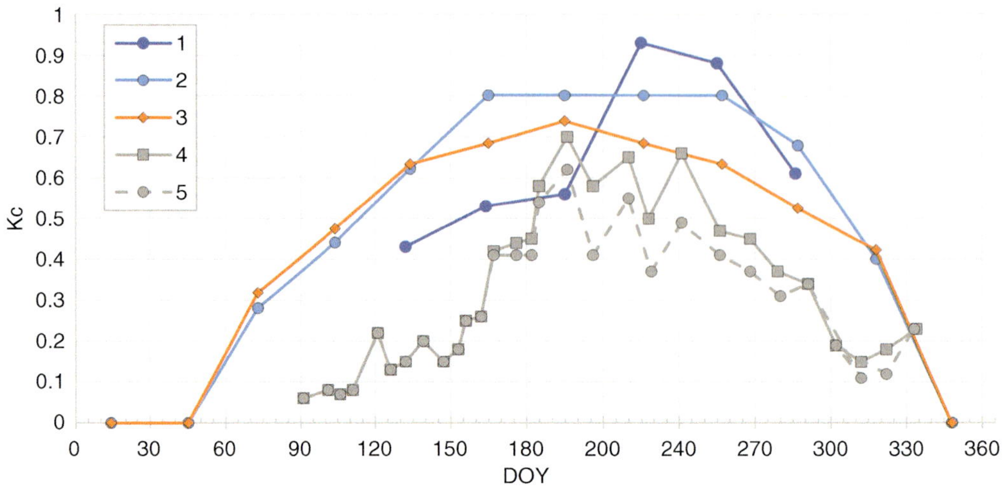

Fig. 10.5. Values of K_c reported for Mollar de Elche (ME). (1) ME 2009 shaded area 56% (Intrigliolo *et al.*, 2011a); (2) ME 2010 and (3) ME 2011 (Buesa *et al.*, 2011); (4) 'Wonderful' CE 0.8 dS/m and (5) 'Wonderful' CE = 1.4 dS/m (Bhantana and Lazarovitch, 2010).

10.3 Responses to Water Stress

10.3.1 Yield and tree performance

The chronic scarcity of water reserves for irrigation in the Mediterranean basin might result in limits on irrigation – the main limiting factor for pomegranate production in the area. In conditions of water shortage, RDI has been shown as a suitable technique that can be used by growers to reduce the amount of water applied to some crops with no or minimal reductions in yield (Ruiz-Sánchez *et al.*, 2010). RDI was developed in the 1980s as a strategy to reduce the growth of vigorous trees and to save water (Chalmers *et al.*, 1981; Bausch, 1995). Water restrictions must be applied in phenological periods when fruit growth is less sensitive to soil water deficit (i.e. non-critical) ensuring water availability during the rest of the season to meet full tree water requirements. Therefore, the use of RDI techniques requires knowledge of the periods of crop sensitivity to deficit irrigation, which differ from crop to crop depending on their agronomic and physiological characteristics.

RDI has been studied in many citrus species (García-Tejero *et al.*, 2010; Ballester *et al.*, 2011, 2013). The sensitive periods of citrus to deficit irrigation have been clearly identified and the effects of water stress on vegetative and reproductive growth are well documented (Ruiz-Sánchez *et al.*, 2010).

In pomegranate, on the other hand, the literature about RDI effects on crop performance is poor, and just a few recent studies report the effects of RDI on pomegranate yield and its components (Intrigliolo *et al.*, 2013; Ayars *et al.*, 2017; Selahvarzi *et al.*, 2017).

In a study performed on mature pomegranate trees, cv. 'Mollar de Elche', grown in a drip-irrigated commercial orchard located in Elche (Alicante, Spain), Intrigliolo *et al.* (2013) showed that pomegranate appears to be a crop that is tolerant to deficit irrigation. Fruit weight was not drastically reduced by RDI applied at any time during the season. RDI enabled water savings of up to 20% without any reduction in yield. The increase in the number of fruits per tree compensated for the smaller average fruit weight. Since the number of fruits harvested is determined by the number of fruits remaining on the tree after the fruit drop period, this period seems to be of particular importance in pomegranate trees for the success of an RDI strategy. Results obtained from this study showed that in pomegranate trees, RDI is a strategy that can be potentially applied in seasons when the crop level is low.

Table 10.1. Suggested crop coefficients (K_c) for scheduling regulated deficit irrigation (RDI) in mature pomegranate trees under Mediterranean conditions. Threshold values of midday stem water potential (Ψ_{stem}) are also reported for ensuring either optimum plant water status during the entire growing season or for RDI applications optimizing water use efficiency under conditions of water scarcity.

Month	K_c for deficit irrigation	Ψ_{stem} for optimum water status (MPa)	Ψ_{stem} for RDI applications (MPa)
March	0.28	−0.5	−0.5
April	0.31	−0.6	−0.7
May	0.30	−0.7	−0.9
June	0.35	−0.8	−1.2
July	0.65	−0.9	−0.9
August	0.66	−1.0	−1.0
September	0.70	−1.0	−1.0
October	0.70	−0.9	−1.0
November	0.60	−0.85	−1.0

This same 4-year study conducted by Intrigliolo *et al.* (2013) showed that moderate plant water stress (midday stem water potential (Ψ_{stem}) not lower than −1.2 MPa) during early spring (mid-May to end of June, coinciding with fruit drop episodes) decreased fruit drop. Water restrictions applied later in the season, at the end of the fruit drop period, had less of an effect on reducing fruit drop. Similarly, Selahvarzi *et al.* (2017) also found that RDI applied during flowering and fruit set has positive effects on pomegranate flowering and fruits. This knowledge of the water stress effects on pomegranate trees, as well as the appropriate timing to apply RDI strategies in the cv. 'Mollar de Elche', may be of great utility in semi-arid areas to increase the fruit crop level.

Recommended K_c values for the application of RDI strategies have been obtained (Table 10.1) based on the results from the aforementioned studies. When applying deficit irrigation, it is also of crucial importance to evaluate the plant water status to ensure that a moderate potentially beneficial water stress does not result in a severe water deficit detrimental to fruit production. Based on the results reported by Intrigliolo *et al.* (2013) the suggested threshold Ψ_{stem} values under either full or deficit irrigation conditions are shown in Table 10.1.

However, because Ψ_{stem} measurement cannot be easily automated, there is a need to look for other tools for continuously monitoring plant water status. In this sense, trunk dendrometers have been widely assessed in fruit trees to monitor plant water status (Ortuño *et al.*, 2010). From trunk diameter variations (TDV) two indexes are typically obtained: the maximum diurnal trunk shrinkage (MDS) and the trunk growth rate (TGR). Particularly, MDS has been shown to have the potential to serve as a plant water stress indicator (Fernández and Cuevas, 2010). This is because MDS is usually higher in plants with soil water deficit than in well-irrigated trees. In pomegranates Intrigliolo *et al.* (2013) showed that MDS is a sensitive indicator of pomegranate tree water status and it can be further used for managing irrigation. However, the seasonal changes in the MDS-Ψ_{stem} relationship should be taken into account when attempting to use threshold MDS values for scheduling irrigation.

Deficit irrigation can also be applied using partial root zone drying (PRD), which alternates water application between the two sides of the tree in order to maintain part of the root system in contact with dry soil, while the rest of the root zone is in a wet condition. Recent studies conducted with the 'Rabab' cultivar by Parvizi *et al.* (2016) showed that PRD irrigation applied at a rate equivalent to 75% of the estimated water needs resulted in water savings with respect to the fully irrigated trees. This is most likely because of some indirect beneficial effect due to the smaller soil volume wetted by PRD, and hence a higher efficiency in water application was obtained.

10.3.2 Fruit quality at harvest

Although, in general, irrigation has a positive effect on pomegranate vegetative growth, yield and fruit weight, the effect of deficit irrigation strategies on fruit quality at harvest differs across studies. Early reports in India indicated that the application of different irrigation regimes allowed controlling the desired ripening time. Thus, full irrigation applied with drip systems in pomegranate trees increased fruit weight and juice content and reduced fruit cracking by around 60% in comparison with deficit irrigated fruit, but chemical parameters, such as soluble solids content (SSC), total sugars, reducing sugars and pH of juice, decreased with increasing water under drip irrigation, while titratable acidity (TA) and the amount of non-reducing sugars increased (Prasad and Mali, 2002).

More recent research on this subject has been conducted in Spain, particularly with the autochthonous cultivar 'Mollar de Elche'. Comparison among sustained deficit irrigation (SDI) (water constantly applied at 50% of control regime) and three RDI regimes with severe water restrictions applied during different phenological periods of 'Mollar de Elche' pomegranates (25% of the control irrigation amount during flowering, fruit set and early fruit growth [$RDI_{Fl.-Fr.set}$], linear fruit growth [$RDI_{Fr.-gr}$] and last part of fruit growth and ripening period [RDI_{Ripe}]) showed that SDI, and to a lesser extent RDI_{Ripe}, can be used to accelerate fruit ripening and advance the harvest date due to earlier rind colouration and sugar accumulation (Laribi *et al.*, 2013). This feature has important implications for the commercialization of 'Mollar de Elche' pomegranate fruit as this cultivar is most often picked based on fruit external colouration. Therefore, an increase in rind red colour and the subsequent harvest advance will allow an increase in grower benefits, since the first fruit reaching the market have in general higher commercial prices. Similarly, Galindo *et al.* (2014) reported that 'Mollar de Elche' pomegranate fruit subjected to SDI (33% ET_o) during the second half of the rapid fruit growth period to the last harvest advanced the optimal harvest time about 7–8 days compared with irrigated fruit (105% ET_o). These fruit exhibited a higher maturity index (MI, as the ratio of SSC/TA) and a darker and more intense garnet colour than control fruit. Moreover, Mellisho *et al.* (2012) showed considerable differences in the response of 'Mollar de Elche' pomegranates to deficit irrigation treatments that depended on fruit harvest date. Arils from fruit subjected to moderate water stress (32% ET_o from the beginning of the season to the end of the first half of the linear fruit growth phase, 74% ET_o during the second half of the linear fruit growth phase and 36% ET_o during the end of the fruit growth and ripening phase) showed higher a* and lower h° (hue angle) values and a significant increase in SSC and MI compared with those from control fruit, which reflected earlier ripening. This was later confirmed by Peña *et al.* (2013). On the contrary, a more pronounced water stress level during the second half of the fruit growth phase (32% ET_o up to the first half of the linear fruit growth phase, irrigation was withheld during the second half of the linear fruit growth phase and reirrigated at the levels of control plants during the end of the fruit growth and ripening phase) was more critical for fruit size than for juice chemical characteristics, probably because under this situation carbon assimilation should be allocated to the synthesis of primary metabolites, which did not exceed the amount used for fruit growth to the detriment of the synthesis of carbon-based secondary metabolites (Mellisho *et al.*, 2012). Furthermore, the differences between both irrigation regimes and full irrigation were only observed in fruit harvested at the beginning of the season, but not at the end of the season. In contrast, Mena *et al.* (2013) indicated that SDI strategies that induced moderate (43% ET_o throughout the experimental period) and severe water stress (12% ET_o) led to pomegranate juice of lower visual attractiveness (yellowish colour), with this being more pronounced under severe water stress, and there was no effect on SSC, TA, and MI among treatments.

A more recent work suggested that the sensitivity of 'Mollar de Elche' pomegranate fruit to water stress during the critical phenological period of fruit ripening is not constant, and that for productivity and fruit quality to be adversely affected it is necessary to exceed a threshold level of water stress (Galindo *et al.*, 2017). In this work, increased water stress induced by withholding irrigation during different-length periods before harvest increased redness and darkness of the fruit peel, whereas more than 25 days of water restrictions were required to observe significant changes in the colour of pomegranate juice. On the other hand, the modification of SSC, TA, MI and pH values in response to irrigation withholding was not very clear and showed some differences compared with the results presented by other authors in similar experimental conditions in which water restrictions in the final phase of fruit growth and ripening showed a significant increase in SSC and TA of 'Mollar de Elche' juice (Mellisho *et al.*, 2012; Laribi *et al.*, 2013).

Several reasons could explain the colour increase in fruit from trees subjected to some water stress during the ripening period, although in some cases the results are ambiguous. Rind colouration in pomegranates is predominantly due to anthocyanins, and sugars are known to have an important role in anthocyanin biosynthesis (Laribi *et al.*, 2013). In fact, different studies have showed that moderate and severe deficit irrigation improves total anthocyanin content of 'Mollar de Elche' fruit (Mellisho *et al.*, 2012; Laribi *et al.*, 2013; Mena *et al.*, 2013). Furthermore, Laribi *et al.* (2013) concluded that the timing of water stress is an important factor to take into account for anthocyanin accumulation in the rind and aril tissues, and the application of RDI in the linear fruit growth phase was the condition that induced the highest amount of anthocyanin in the juice. In this sense, Galindo *et al.* (2017) also found that a short period (6 days) of irrigation restriction at the end of the ripening period enhanced the content of bioactive compounds in fruit of this cultivar, mainly anthocyanins, punicalagin and ellagic acid, whereas longer periods had no effect on the production of these compounds. On the other hand, Mena *et al.* (2013) reported that severe SDI at 12% ET_0 throughout the experimental period dramatically reduced the content of phenolic compounds, especially anthocyanins and punicalagin, whereas moderate (43% ET_0) water stress maintained the total content of anthocyanins. In these works, juice antioxidant capacity was not affected by water stress and did not correlate with the red colour intensity and the total anthocyanin content, suggesting that phenolics other than anthocyanins may be the major contributors to pomegranate juice antioxidant activity. Other workers reported no effect of water deficit irrigation on either total phenolics, total antioxidant capacity or total anthocyanin and punicalagin content of 'Mollar de Elche' pomegranate, probably because the actual level of water stress was different (Galindo *et al.*, 2014; Cano-Lamadrid *et al.*, 2018). Contrary to these results, Peña *et al.* (2013) reported that total anthocyanins, gallic acid and total vitamin C were significantly correlated with the antioxidant capacity of 'Mollar de Elche' pomegranates. In this work, anthocyanin, catechin and caffeic acid contents of RDI fruits were lower, under similar water restrictions to those applied by Mellisho *et al.* (2012), and vitamin C content was higher than those of control samples.

Within a particular growing region, the cultivar is also an important factor affecting the influence of water restrictions on fruit quality. In recent work conducted in Spain, Cano-Lamadrid *et al.* (2018) reported that the juice colour density of 'Wonderful' pomegranates, which has a high correlation with catechin-phloroglucinol and monomeric anthocyanins such as cyanidin-3-glucoside, was reduced when water deficit (60% ET_c from fruit set to harvest) was applied, whereas the opposite trend was observed in 'Mollar de Elche' fruit grown under similar Mediterranean conditions and irrigation regimes. This was explained by an increase in the degree of anthocyanin polymerization and a consequent deterioration of the red colour as water stress increased. This was clearly influenced by the differences between cultivars in the composition of phenolic compounds and their stability. On the other hand, 'Wonderful' fruit were more sensitive to changes in the sugar profile than 'Mollar de Elche' fruit, and their values of glucose and fructose increased in fruit grown under deficit irrigation strategies. Furthermore, water stress caused a reduction of total aldehydes (mainly hexanal) and terpenoids in both cultivars, losing vegetable aroma notes. Contrary to these results, Centofanti *et al.* (2017) reported no changes in fruit colour, pH, SSC, mineral content and bioactive compounds of 'Wonderful' pomegranates grown in California subjected to different SDI conditions (35, 50, 75 and 100% ET_c). Therefore, the discrepancies among the different studies regarding fruit quality at harvest could be attributed not only to differences in irrigation regimes, but also to cultivar, agroclimatic conditions, harvest time, the age of trees and other environmental factors.

In research conducted in Turkey, Dinc *et al.* (2018) reported that irrigation strategies that combined irrigation levels (0.50, 0.75, 1.00 and 1.25 times the pan evaporation data) and intervals (3- and 6-day interval) did not affect the SSC, TA and pH of young 'Hicaznar' pomegranates. However, an increase in SSC and TA and a decrease in pH were observed as the amount of irrigation water applied decreased. In Iran, Parvizi and Sepaskhah (2015) investigated the effect on quality attributes of 'Rabab' pomegranates due to deficit irrigation including partial root drying (irrigation was applied on one side of the tree during flowering and fruit set and withheld during the growing season) with the application of 50 and 75% of water requirement. Partial root drying strategies increased MI and juice content percentage, and decreased TA in comparison with full irrigation. Furthermore, the irrigation strategies with a higher level of water stress increased SSC and decreased vitamin C in comparison with other irrigation strategies. 'Rabab' cultivar had a higher tolerance to drought stresses than 'Shishehgap', mainly as a consequence of defence mechanisms such as a higher accumulation of soluble sugars and a greater activity of antioxidant enzymes in the leaves, showing the importance of plant species and cultivars in the tolerance to abiotic stress (Ebtedaie and Shekafandeh, 2016). In further studies conducted in Iran, Selahvarzi *et al.* (2017) did not recommend SDI (50% ET_c throughout the growing season) for 'Shahvar' pomegranate trees in arid and semi-arid areas, especially for long time periods, as the severe water stress affected the flowering period and significantly reduced yield, despite the fact that this treatment increased the total phenolic content and the antioxidant activity of the fruit. For this cultivar, a mild water stress early in the season and later recovery (no watering until fruit set followed by watering similar to the control) seemed more appropriate in terms of yield maintenance, although no benefit was observed in terms of increasing the amounts of bioactive compounds.

10.3.3 Fruit storage potential

In addition to inducing changes in fruit quality attributes at harvest, deficit irrigation of pomegranate trees can also modify the postharvest performance of the fruits and influence their storage potential. General physicochemical changes during cold storage and shelf-life of pomegranate fruit include an increase in SSC, MI and weight loss, and a decrease in TA and firmness, as well as changes in rind colour that tend towards a decrease in red tonalities or an increase of yellowish surfaces. In general, weight loss, chilling injury and decay are the most important problems limiting the storability of pomegranate.

Research in Spain by several authors working with 'Mollar de Elche' pomegranates showed that the fruits resulting from SDI and RDI treatments had better postharvest behaviour than those from full irrigation because of retarded chilling injury incidence (Laribi *et al.*, 2013; Peña *et al.*, 2013), higher sensory and nutritional quality, and longer shelf-life (Laribi *et al.*, 2013; Peña *et al.*, 2013; Peña-Estévez *et al.*, 2016a). However, the effect depended on irrigation and storage conditions. Peña *et al.* (2013) reported that SDI fruit stored at 5°C had earlier SSC increases than control fruit, while the SSC increases were retarded during the shelf-life period at 15°C. This was explained by a possible acceleration of fruit metabolism caused by SDI-induced stress. In addition, SDI reduced chilling injury symptoms, water and firmness loss, and increased phenolic and anthocyanin contents during shelf-life, suggesting a relationship between high levels of antioxidants and the delayed development of chilling injury. Laribi *et al.* (2013) also concluded that deficit irrigation, depending on the phenological period when water shortage is applied, could be used as a field practice to improve the postharvest performance of 'Mollar de Elche' pomegranate fruits. Overall, fruit from trees subjected to SDI or water withholding only during ripening (RDI_{Ripe}) showed lower weight loss and chilling injury symptoms, and maintained higher SSC and superior reddish colouration compared with the control treatment during up to 19 weeks of cold storage at 5°C with or without a shelf-life period of 7 days at 20°C. However, water withholding during

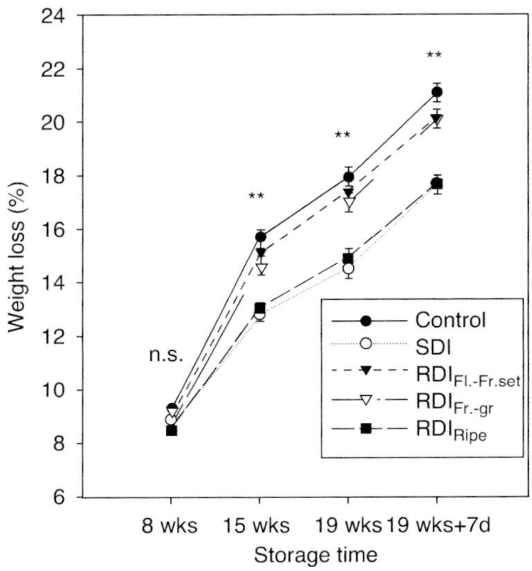

Fig. 10.6. Weight loss of 'Mollar de Elche' pomegranates after 8, 15 and 19 weeks of cold storage at 5°C and 19 weeks at 5°C plus a shelf-life period of 7 days at 20°C in response to different irrigation treatments. Control irrigated at 100% of crop evapotranspiration (ET_c), sustained deficit irrigation (SDI) irrigated at 50% of the ET_c during the entire season. In the regulated deficit irrigation (RDI), severe water restrictions (25% ET_c) were applied during one of three phases: flowering and fruit set ($RDI_{Fl.-Fr.set}$), fruit growth ($RDI_{Fr.-gr.}$) or the final phase of fruit growth and ripening (RDI_{Ripe}). n.s. and ** indicates no significant and significant differences at $P < 0.01$, respectively, using the least significant difference (LSD) test. Error bars indicate the magnitude of LSD. (From: Laribi *et al.*, 2013, Elsevier.)

other phenological periods (fruit set, early fruit growth and linear fruit growth) did not affect these quality attributes during storage. Due to its direct economic implications, the reduction of weight loss may have an special significance for 'Mollar de Elche' pomegranate growers and marketers (Fig. 10.6). In this work, postharvest decay incidence of pomegranate increased with storage time and no effect of irrigation treatments was observed. On the other hand, water withholding during the linear phase of fruit growth and during ripening helped maintain higher juice anthocyanin content after 19 weeks of cold storage plus 7 days of the shelf-life period at 20°C.

In other research in Spain, Peña-Estévez *et al.* (2016a) reported that 'Mollar de Elche' pomegranates subjected to SDI maintained higher SSC than control samples after 14 days of storage at 5°C, but no significant differences were observed for longer storage periods, most likely due to the reduction of fruit metabolism throughout cold storage. On the other hand, although the juice of SDI fruit had a lower total antioxidant capacity and total phenol content at harvest, the reduction of these parameters during cold storage was lower for SDI than for control fruit. When evaluating the shelf-life of fresh arils from control and SDI-treated fruit in modified atmosphere packaging, these authors concluded that arils could be processed from whole fruit stored up to 90 days at 5°C with good sensory and health-promoting properties. However, storage of whole fruit longer than 60 days reduced the shelf-life of the obtained arils from 14 to 10 days at 5°C. In a study of the effect of different irrigation regimes and postharvest treatments (vapour heat application at 95°C for 7–10 s vs. NaClO as sanitizing agents) on the quality and shelf-life of fresh 'Mollar de Elche' pomegranate arils, Peña-Estévez *et al.* (2015) observed a synergistic effect between water deficit and postharvest heat treatments. The best results were found when vapour heat treatment was applied to pomegranates cultivated under SDI conditions, with irrigation withheld for 16 or 26 days prior to harvest. This represented water savings of 6–11% and a shelf-life of arils of 18 days at 5°C. These results were confirmed in a subsequent study in which the effect of deficit irrigation and postharvest heat treatments on phenolic content, antioxidant capacity and activity of the main oxidative enzymes during storage of fresh pomegranate arils was assessed (Peña-Estévez *et al.*, 2016b). Arils from fruit grown under SDI had higher initial content of phenolic compounds and total antioxidant capacity than control arils, probably due to the stressing conditions generated by SDI. Phenolic compound content, however, decreased during storage to a greater extent for SDI samples. Furthermore, 46–58% lower phenylalanine ammonia lyase (PAL) and 39% higher polyphenol oxidase (PPO) values were registered in arils from SDI fruit compared with control arils, while peroxidase (POD) activity was similar in arils from both irrigation treatments. During cold storage, PAL and PPO activities generally increased while POD activity decreased, and the maximum variation in enzymatic activity was retarded in arils from SDI fruit compared with control arils.

The most important findings for fruit quality and storage potential described in the above two sections are summarized in Table 10.2.

10.4 Fertigation

Recent trends in pomegranate tree planting have been towards high-density orchards and localized irrigation systems (Haneef *et al.*, 2014). This fact has stimulated an interest in improving fertigation techniques, which allow frequent additions of small amounts of fertilizer that are more carefully timed to meet tree demands. Nowadays, new drip irrigation plantations are also equipped with a fertilization system (Fig. 10.7). As a consequence, irrigation scheduling should be jointly considered with the appropriate application of fertilizers using the drip irrigation systems.

There have been few attempts to date to assess the application of fertigation strategies on pomegranate. Plants normally take up the nutrients dissolved in water from the soil through their roots. For this reason, even if there is a sufficient quantity of nutrients in the soil, it is not possible for plants to take up the required nutrients when water is limited. Taking into account that water levels and optimal irrigation regime depend on soil type, tree size, the physiological phase of the tree and, particularly, the potential evaporation, optimized irrigation is necessary for healthy plant growth in dry and particularly hot seasons in most of the pomegranate cultivation areas. Tree water and nutrient relations are crucial factors affecting not only tree performance and bioactive compounds, but also fruit physiological disorders (such as fruit cracking and sunburn) and fruit postharvest performance in pomegranate (Intrigliolo *et al.*, 2013; Laribi *et al.*, 2013; Mphahlele *et al.*, 2014).

Synchronization between the fertilizer application and crop demand during growth increases nutrient use efficiency, is cost-effective and protects the environment. However, little is known about the pattern of nutrient uptake in pomegranate fruit. This understanding is required to schedule fertilizer applications to coincide with periods of high nutrient demand, thereby maximizing nutrient use efficiency. Fertigation has to accommodate the growth of shoots, flowers and fruits. Flowering occurs about 1 month after bud break on newly developed branches of the same year, mostly on spurs or short branches. The synchronous growth of shoots during the blooming and fruit setting stage in pomegranate has to take into account the competition, and implies interactions among nutrients and its effects on fruit yield. The growth curve for pomegranate fruit (Shulman *et al.*, 1984; Varasteh *et al.*, 2008) follows a single sigmoid pattern with three development stages. Stage I is an initial phase of rapid fruit growth by cellular division and stage II is characterized by slower growth. Fruit expansion also occurs during stage III when vegetative growth is almost complete.

Fertigation management refers to providing a specific quantity of water and fertilization at an appropriate time to the effective root zone of the crop, thus deriving maximum water and nutrient efficiency. Good fertigation management is needed to minimize fertilizer leaching. In general, N, P and K crop uptakes are correlated with crop biomass accumulation. In order to manage irrigation and mineral content, it is important to know the dynamics of nutrient accumulation in developing fruits (Tagliavini *et al.*, 2000). In mature orchards (30 t/ha), dry matter allocation to fruit reaches up to 55% of the total tree biomass produced per year (García-Gómez, 2011). At harvest the relative order of concentration of macronutrients both in arils and peel is K > N > Ca > P > Mg >Na (Mirdehghan and Rahemi, 2007). Some differences have been reported among cultivars. Al-Maiman and Ahmad (2002) reported high contents of Cu, Fe, Zn, Mg, P, Na, Ca and K in the seeds and juice in the arils of Iranian cultivar 'Malas Yazdi' and in the 'Taifi' cultivar cultivated in Saudi Arabia. Among seven cultivars the highest amounts of N, P, S and Cl were found in 'Bhagwa', while 'Arakta', 'Ruby' and 'Wonderful' had the highest amount of Mg, Ca and Na, respectively (Amos Fawole *et al.*, 2012). Considering the nutrient content (nitrogen, phosphorus, potassium, calcium and magnesium) in 'Mollar de Elche' pomegranate

Table 10.2. Effect of deficit irrigation strategies on pomegranate fruit quality at harvest and during cold storage.

Cultivar	Deficit irrigation strategy	Effect on fruit quality or storage potential	References
'Mollar de Elche'	SDI (50% of control irrigation): 32% ET_o beginning of the season to end of the first half of linear fruit growth; 74% ET_o second half of linear fruit growth; and 36% ET_o end of fruit growth and ripening phase SDI (32% of control irrigation): water was withheld during the second half of linear fruit growth	Moderate water stress produced earlier fruit ripening, whereas a more pronounced stress had little effect on juice characteristics Deficit irrigation increased total anthocyanin content, but did not affect the total antioxidant capacity and the main sugar and acid contents Moderate water stress reduced chilling injury, weight and firmness loss, and increased phenolics and anthocyanins during cold storage and shelf life	Mellisho *et al.* (2012) Peña *et al.* (2013)
	SDI: 50% of control regime throughout the season $RDI_{Fl.-Fr.set}$: 25% of control during flowering, fruit set and early fruit growth $RDI_{Fr.-gr}$: 25% of control in the linear fruit growth RDI_{Ripe}: 25% of control in the last part of fruit growth and ripening	SDI, and to a lesser extent RDI_{Ripe}, increased SSC, rind red colouration and total anthocyanins at harvest and during cold storage These treatments also reduced weight loss and chilling injury during cold storage and shelf-life	Laribi *et al.* (2013)
	SDI: 43% ET_o SDI: 12% ET_o	SDI provided a dramatic decrease in bioactive phenolic compounds, especially anthocyanins and punicalagin, and a reduced rind red colour density, being more pronounced under severe water stress. No effect on SSC, TA and MI	Mena *et al.* (2013)
	SDI: 33% ET_o throughout the season	SDI advanced the optimal harvest time about 7–8 days with respect to fully irrigated trees	Galindo *et al.* (2014)
	SDI: water was withheld for 6, 15, 25 and 36 days before harvest	A very short period of irrigation restriction (6 days) at the end of the ripening period induced earlier harvest and enhanced the bioactive compound content (anthocyanins, phenolic compounds, punicalagin and ellagic acid)	Galindo *et al.* (2017)

Continued

Table 10.2. Continued

Cultivar	Deficit irrigation strategy	Effect on fruit quality or storage potential	References
	SDI: 78% less water than the ET_0 SDI: Water was withheld for 16 and 26 days before harvest	SDI-treated fruits had lower total antioxidant capacity and total phenols at harvest and showed higher SSC during cold storage Regardless of SDI treatment, whole fruit stored up to 90 days at 5°C rendered fresh arils with good sensory and health-promoting properties. However, storage of whole fruit longer than 60 days reduced the shelf life of the arils from 14 days to 10 days	Peña-Estévez *et al.* (2016a, b)
'Wonderful' and 'Mollar de Elche'	SDI: 60% ET_c during fruit growth and ripening	The effect of deficit irrigation was cultivar dependent Water deficit reduced rind colour density in 'Wonderful' pomegranates but increased it in 'Mollar de Elche' fruit. Water stress caused a reduction of total aldehydes and terpenoids in both cultivars, losing vegetable aroma notes	Cano-Lamadrid *et al.* (2018)
'Wonderful'	SDI: 35, 50, 75 and 100% ET_o	SDI conditions had no effect on physicochemical quality, mineral content and bioactive compounds	Centofanti *et al.* (2017)
'Rabab'	PRD and deficit irrigation with 50 and 75% of the water requirement	PRD increased fruit MI and juice content and decreased TA compared with control fruit. Higher level of water stress increased SSC and decreased vitamin C	Parvizi and Sepaskhah (2015)
'Shahvar'	SDI: 50% ET_c throughout the growing season RDI: no watering until fruit set stage	SDI-treated fruit had higher values of total phenolic compounds and greater antioxidant activity. SDI negatively affected fruit yield	Selahvarzi *et al.* (2017)
'Hicaznar'	Combined irrigation levels (0.50, 0.75, 1.00 and 1.25 times the evaporation data) and intervals (3- and 6-day interval)	An increase in SSC and TA was observed as water deficit increased	Dinc *et al.* (2018)

SDI, sustained deficit irrigation; ET_o, reference evapotranspiration; RDI, regulated deficit irrigation; PRD, partial root drying; SSC, soluble solid content; TA, titrable acidity; MI, maturation index.

fruit, the yield implies removing from the soil approximately 1.33 N, 0.46 P_2O_5, 2.51 K_2O, 0.28 CaO and 0.16 MgO fertilizer units (kg) per metric tonne (t). Nutrient content of fertilizers is expressed as the element or combinations with other elements. Nitrogen is generally expressed in the elemental form, phosphates and potash fertilizers may be expressed either as the oxide form (P_2O_5, K_2O) or as the element (P, K), and calcium and magnesium are normally expressed in their oxide form (CaO, MgO) (Table 10.3).

The main categories of available recommendation schemes and tools for N management for use with crops are: soil testing approaches; soil solution analysis; N balance calculations; decision support systems; and crop-plant testing

Fig. 10.7. Irrigation and fertilization equipment for applying fertigation in a pomegranate orchard. (Photo: Claude J. Phene *et al.*, 2015, University of California Agriculture and Natural Resources.)

approaches (Thompson *et al.*, 2017). Knowledge of the soil nutrient availability is critical for optimized fertigation management. The soil sampling procedure varies depending on the type of orchard: (i) collect soil cores just inside the drip line of the canopy; (ii) collect soil cores from at least 15–20 locations to form a representative composite sample; and (iii) mix the soil taken into one composite sample. There are a number of different sampling patterns that can be used to give reliable results for soil testing. If the orchard is made up of large areas with different characteristics, then these areas should be sampled separately. Sampling should occur at the same time each year. Sample to the same depth every time the orchard is sampled. To determine N lixiviates and salinity in a pomegranate orchard, two sampling depths of 0–30 cm and 30–60 cm are recommended. For an orchard of about 1 ha, at least 10 individual cores should be collected.

Several novel approaches to optimize N fertilizer application using soil and plant nutrient testing, as well as simplified decision support systems (DSSs) have been shown to be effective tools to improve nitrogen use efficiency (harvested yield per unit of nutrient used) in crop production (Thompson *et al.*, 2015). In a study conducted by Ayars *et al.* (2017) on 'Wonderful' investigating the effect of different irrigation strategies and levels of nitrogen fertilization on yield and quality in a commercial pomegranate orchard, it was shown that the N requirement is in the range of 62–112 kg/ha (109–198 g/tree) for a mature pomegranate orchard, similar to other N recommendations. However, Dhillon *et al.* (2011) found that lower N applications at a rate of 24 kg/ha would be sufficient if the tree crop level does not reach more than 20 t/ha.

The effect of different combinations of irrigation, and NPK fertilization on yield and fruit quality at harvest of 'Mollar de Elche' pomegranates was studied (Intrigliolo *et al.*, 2019) and a recommendation of P_2O_5 application of between 45 and 50 kg/ha should be enough to maintain the yield.

The pattern of increasing most of the nutrient accumulations by fruit with age is similar to those described for other fruits. Potassium requirements of fruits progressively increase as fruit maturity approaches. Approximately 70% of the calcium (Ca) requirements of fruit may be taken up during the first stages of growth and development, when supply by xylem is likely to predominate. Significant accumulation during early fruit growth identifies this as a period when

Table 10.3. Macronutrients contained in pomegranate fruit (% dry mass), amount of the element (kg per metric tonne of fresh fruit) and fertilizer units removed per tonne of fruit. Dry matter percentage is 20% of fresh fruit.

Macronutrient	Element content % (dry mass)	Element (in kg per tne of fruit)	Conversion of element to fertilizer units	Fertilizer units (kg) removed from soil per tne of fruit
N (nitrogen)	0.65–0.68	1.33	1.0	1.33 (N)
P (phosphorus)	0.08–0.12	0.20	2.29	0.46 (P_2O_5)
K (potassium)	0.94–1.15	2.09	1.20	2.51 (K_2O)
Ca (calcium)	0.05–0.15	0.20	1.40	0.28 (CaO)
Mg (magnesium)	0.03–0.07	0.10	1.67	0.16 (MgO)

adequate supply of Ca to the plant is crucial, and when the Ca nutrition of fruit might most easily influence those properties where fruit quality is known to suffer because of a Ca imbalance later in the season (Mirdehghan and Rahemi, 2007).

The best indication of successful fertilizer management practices for fruit trees is having leaf nutritional concentrations within the optimum ranges. Proper tissue sampling is needed for commercial nutrient monitoring and valid conclusions for fertilizing the orchards. Mature leaves (3–4-month-old spring cycle) from the middle third of the current season branch must be collected for analysis from at least 15–25 trees, in all four quadrants of the tree, away from shoots containing fruit, fruitlets, flowers and points of expansion. Because of the seasonal changes in plant leaf nutrients, the concentration content of leaf nutrients in pomegranate varies depending on the phenological stage (Hepaksoy *et al.*, 2016). Leaf N, P and K contents are expected to decrease with vegetation because plants take up a large portion of phosphorus and potassium, which is needed during the first periods of development. Therefore, it is strongly recommended to collect leaf samples for analysis at the same time every year, preferably in the first part of the summer (July in the northern hemisphere), during the first stage of fruit growth, when the highest leaf K content will be obtained (Bartual *et al.*, 2015b).

In semi-arid conditions the average values of leaf analyses in 'Mollar de Elche' showed a wide range of variation; N 1.22–1.68%, P 0.12–0.23%, K 0.53–0.78%, Ca 1.67–2.33%, Mg 0.44–0.65%, Na 0.03–0.08%, S 0.12–0.14%, B 5.7–20.9 mg/kg, Fe 51–102 mg/kg, Cu 4–10.7 mg/kg, Mn 6.5–29 mg/kg and Zn 9–17 mg/kg (Bartual *et al.*, 2015b). These values are similar to those found in the leaves of 'Hicaznar' by Ozkan (2005), where nitrogen content ranged from 1.38–1.82%, phosphorus 0.15–0.25%, potassium 0.87–1.43%, calcium 0.84–2.58% and magnesium 0.21–0.44% during the vegetative period. Under rainfed conditions, Mir *et al.* (2015) found the nutrient content of pomegranate leaves was approximately: N (2.63%), P (0.25%), K (1.57%), Fe (197.8 mg/kg), Cu (14.6 mg/kg), Zn (59.3 mg/kg) and Mn (200.4 mg/kg).

The interaction between soil, plant nutrients and irrigation has been studied, and it has been shown that the nature of the crop response to fertilization is markedly modified by the soil moisture regime and vice versa. Hepaksoy *et al.* (2016) reported that irrigation frequency and quantity affected nitrogen, calcium and magnesium nutrient uptake of the trees of 'İzmir 1513' pomegranate. In addition, the diffusion of K^+ ions in the soil is much slowed down when it dries out. Ca and Mg leaf concentrations are negatively correlated with leaf K, and when soil becomes drier, there is a competitive interaction for plant uptake between K (which moves in the soil mainly by diffusion) and Ca and Mg (which are transported by mass flow to the root surface) (Nielsen *et al.*, 1986).

The required amount of annual fertilizer should be applied during the plant's vegetative growth season. The optimal timing for nitrogen application is when the tree is starting a new vegetative growth. Seventy per cent of the tree's annual fertilizer rate should be applied from late winter to late spring, during the flowering and fruit set period. Parvizi and Sepaskhah (2015) reported variable results of the effect of irrigation strategies and fertilization on particular fruit parameters. Adding an excessive amount of nitrogen has disadvantages such as delayed fruit maturation and, at harvest, decreased red aril colour (Bartual *et al.*, 2015a, b, c).

The presence of sufficient total quantities of essential nutrients in soil does not guarantee the availability of these nutrients to plants, because of other factors, such as the presence of salts and soil moisture content (Fageria *et al.*, 2011). Fertilizers applied by injection into the irrigation water are strongly connected to salinity in the wetted area where most active roots are located. Nutrient additions, on the other hand, have been more successful in improving crop quality, such as the correction of Na-induced Ca^{2+} deficiencies by the addition of supplemental calcium. Although pomegranate is considered to be moderately tolerant to salinity (Maas and Hoffmann, 1976; Maas, 1993); Kulkarni *et al.* (2007) reported that pomegranate tolerates sodium chloride up to EC_e 6 dS/m without mortality and with satisfactory growth, but this significantly decreased the concentration of N, P, K, Ca and Mg in leaves. Karimi and Hasanpour (2014) reported different tolerance to salinity among cultivars. Fertilizers added for fertigation should not increase salinity of the irrigation water higher than EC_e 4.0 dS/m since a decline in growth rate occurred at salinity levels of NaCl concentration higher than 40 mM. Naeini *et al.* (2005) demonstrated pomegranate can tolerate salinity up to 40 mM without any symptoms of ion toxicity (Naeini *et al.*, 2005, 2006). In Spain, pomegranate grew to produce standard yield (20–30 t/ha) and fruit qualities without apparent damage to the tree with similar salinity levels (Naeini *et al.*, 2005,2006).

There are two main stresses imposed by salinity on plant growth. One is water stress imposed by the increase in osmotic potential of the rhizosphere as a result of high salt concentration. If the salt concentration is not this high, the stress is ion stress and may be caused by one particular type of ion.

Water salinity and drought affect the concentration of Fe^{2+}, Zn^{2+}, Cu^{2+} and Mn^{2+} in pomegranate leaves and roots (Hasanpour *et al.*, 2015). Salinity could affect micronutrient concentrations in the plant differently, depending on plant species and the salinity level.

10.5 Agronomic Practices to Cope with Water Quantity Restrictions

Water is consumed by horticultural crop systems owing to soil evaporation and plant transpiration. While this last component is needed to ensure optimum plant productivity, soil water evaporation should be reduced as much as possible in order to achieve high irrigation efficiency. Woody perennial crops often have incomplete ground cover, leaving part of the soil directly receiving a high radiation regime that increases the evaporative component of the orchard evapotranspiration (Fereres *et al.*, 2003). Subsurface drip irrigation systems eliminate most of the soil evaporation component from the water balance calculation, and they could be used in cases of water scarcity as was previously highlighted. Additionally, when the volume of applied water needs to be restricted in relation to potential ET_c, decreasing the volume of soil wetted by the drip system, by reducing the number of emitters per tree, can be a useful way to increase the irrigation efficiency.

Weeds and particularly cover crops compete with the main crop of an agricultural ecosystem for water and nitrogen. Besides subsurface drip irrigation (Fig. 10.8), plastic covers and mulches (either organic or plastic) can be used to reduce competition from weeds and evaporation.

Fig. 10.8. Difference in area of weeds between surface drip (left) and subsurface drip (right). (Photo: Claude J. Phene *et al.*, 2015, University of California Agriculture and Natural Resources.)

Where cover crops are used for reducing soil erosion, less competitive cover crops species can be adopted, such as legume species, with growth rates and/or cycles that minimize competition with the main crop.

In fruit trees, there is evidence that high fruit load may enhance the sensitivity of fruit growth to water stress (Berman and DeJong, 1996). Hence, reducing fruit load has been used to mitigate the adverse effects of plant water stress, though with important yield penalties (López *et al.*, 2006; Marsal *et al.*, 2008). Under low crop demand conditions, a reduced plant photosynthesis rate due to water stress is less detrimental since fruit are the major sink for carbohydrates, particularly during stage III of fruit growth. Also, lowering fruit load has been shown to reduce plant water use because of a reduction in stomatal conductance via feedback mechanisms (Hansen, 1971). However, specific research in pomegranates should be conducted because the general recommendations given here are mainly based on studies on stone fruit trees. Eliminating part of the actively transpiring canopy surface area (such as whole branches) can also be used to help tree survival under extreme drought conditions (Marsal *et al.*, 2006). It is obvious that this practice has major consequences for the current year tree performance, but at least it can guarantee plant survival. Also, innovative canopy forms can be designed in order to optimize light interception, reducing tree transpiration under soil water-limiting conditions. In this sense, Intrigliolo and Lakso (2010) have shown that in north–south oriented vineyard rows, leaning vines towards the west can be used to slightly (+8%) increase the overall water use efficiency because vine light interception decreases during the warmer hours of the day when evaporative demand is higher. Under a future scenario of reduced water availability, new orchard designs might alleviate the impact of drought and heat spells. In a pomegranate study Gill *et al.* (2011) showed how a trellis system can be employed to train pomegranate trees. However, the study focused on fruit growth and quality and more emphasis should be given in the future to determining the effects of the training systems on tree water use and the orchard WUE.

In citrus trees, for instance, the use of shading nets has been proven to be useful for increasing water use efficiency (Alarcón *et al.*, 2006) and even crop performance (Cohen *et al.*, 1997). The installation of a shading net is expensive, but for high-value crops its use may be profitable. Specific research using shading nets in pomegranates has also been conducted as a tool for improving fruit composition and reducing incidence of fruit disorders (Kale *et al.*, 2018).

10.6 Recommendations for Further Research

The studies conducted on the assessment of RDI strategies in pomegranate trees highlight that further research is still needed in order to elucidate the physiological mechanisms involved in fruit drop and flowering, as well as to explore how different water stress levels might affect fruit drop in 'Mollar de Elche' and other cultivars. More research is also encouraged to provide growers with tools for predicting the tree bearing capacity before the second wave of fruit abscission. If this could be quantified, growers would have more information to decide if and when to reduce fruit load by inducing a certain degree of plant water stress. Other topics that, based on the literature reviewed, require more attention for an efficient nutrient management of pomegranate trees are: (i) study of the effects of specific mineral nutrition programmes on fruit quality and disorders; and (ii) study of the physiological responses of pomegranate trees to climate change (heat spells, low winter chilling).

References

Al-Maiman, S.A. and Ahmad, D. (2002) Changes in physical and chemical properties during pomegranate (*Punica granatum* L.) fruit maturation. *Food Chemistry* 76(4), 437–441. DOI: 10.1016/S0308-8146(01)00301-6.

Alarcón, J.J., Ortuño, M.F., Nicolás, E., Navarro, A. and Torrecillas, A. (2006) Improving water-use efficiency of young lemon trees by shading with aluminised-plastic nets. *Agricultural Water Management* 82(3), 387–398. DOI: 10.1016/j.agwat.2005.08.003.

Allen, R.G., Pereira, L.S., Raes, D. and Smith, M. (1998) Crop evapotranspiration – guidelines for computing crop water requirements. FAO Irrigation and Drainage paper 56. FAO, Rome, Italy.

Amos Fawole, O. Linus Opara, U, Fawole, O.A. and Opara, U.L. (2012) Composition of trace and major minerals in different parts of pomegranate (*Punica granatum*) fruit cultivars. *British Food Journal* 114(11), 1518–1532. DOI: 10.1108/00070701211273009.

Ayars, J.E., Mead, R.M., Soppe, R.W., Clark, D.A. and Schoneman, R.A. (1996) Weighing lysimeters for shallow ground water management studies. *Proceedings of the International Conference on Evapotranspiration and Irrigation Scheduling*, St. Joseph, Michigan, pp. 825–837.

Ayars, J.E., Phene, C.J., Phene, R.C., Gao, S., Wang, D. *et al.* (2017) Determining pomegranate water and nitrogen requirements with drip irrigation. *Agricultural Water Management* 187, 11–23. DOI: 10.1016/j.agwat.2017.03.007.

Ballester, C., Castel, J., Intrigliolo, D.S. and Castel, J.R. (2011) Response of clementina de Nules citrus trees to summer deficit irrigation: yield components and fruit composition. *Agricultural Water Management* 98(6), 1027–1032. DOI: 10.1016/j.agwat.2011.01.011.

Ballester, C., Castel, J., Intrigliolo, D.S. and Castel, J.R. (2013) Response of navel Lane late citrus trees to regulated deficit irrigation: yield components and fruit composition. *Irrigation Science* 31(3), 333–341. DOI: 10.1007/s00271-011-0311-3.

Bartual, J., Fernandez-Zamudio, M.A. and De-Miguel, M.D. (2015a) Situation of the production, research and economics of the pomegranate industry in Spain. *Acta Horticulturae* 1089, 345–349. DOI: 10.17660/ActaHortic.2015.1089.45.

Bartual, J., Laribi, A.I., Palou, L., Pérez-Gago, M.B., Nortes, P.A. *et al.* (2015b) Improving pomegranate fruit quality by means of watering management in semi-arid eastern Spain. *Acta Horticulturae* 1089, 431–436. DOI: 10.17660/ActaHortic.2015.1089.60.

Bartual, J., Pérez-Gago, M.B., Pomares, F., Palou, L. and Intrigliolo, D.S. (2015c) Nutrient status and irrigation management affect anthocyanins in 'Mollar de Elche' pomegranate. *Acta Horticulturae* 1106, 85–92.

Bartual, J., García-González, J.F., Guerra, D., Parra, J., Bonet, L. *et al.* (2019) Combined effects of regulated deficit irrigation and fertilization regime on Mollar de Elche pomegranate tree performance in eastern Spain. *Acta Horticulturae* 1254, (in press)

Bausch, W.C. (1995) Remote sensing of crop coefficients for improving the irrigation scheduling of corn. *Agricultural Water Management* 27(1), 55–68. DOI: 10.1016/0378-3774(95)01125-3.

Behnia, A. (1999) Comparison of different irrigation methods for pomegranate orchards in Iran: irrigation under conditions of water scarcity. *17th International Congress Irrigation and Drainage*, Granada, Spain, pp. 207–217.

Berman, M.E. and DeJong, T.M. (1996) Water stress and crop load effects on fruit fresh and dry weights in peach (*Prunus persica*). *Tree Physiology* 16(10), 859–864. DOI: 10.1093/treephys/16.10.859.

Bhantana, P. and Lazarovitch, N. (2010) Evapotranspiration, crop coefficient and growth of two young pomegranate (*Punica granatum* L.) varieties under salt stress. *Agricultural Water Management* 97(5), 715–722. DOI: 10.1016/j.agwat.2009.12.016.

Buesa, I., Badal, E., Guerra, D., García, J., Lozoya, A. *et al.* (2011) Development of an irrigation scheduling recommendation for pomegranate trees (*Punica granatum*). II Symposium Internacional Sobre el Granado. *Proceedings Options Méditerranéennes*, 141–145.

Cano-Lamadrid, M., Galindo, A., Collado-González, J., Rodríguez, P., Cruz, Z.N. *et al.* (2018) Influence of deficit irrigation and crop load on the yield and fruit quality in Wonderful and Mollar de Elche pomegranates. *Journal of the Science of Food and Agriculture* 98(8), 3098–3108. DOI: 10.1002/jsfa.8810.

Centofanti, T., Banuelos, G.S., Wallis, C.M., Ayars, J.E. and USDA (2017) Deficit irrigation strategies and their impact on yield and nutritional quality of pomegranate fruit. *Fruits* 72(1), 47–54. DOI: 10.17660/th2017/72.1.5.

Chalmers, D.J., Mitchell, P.D. and Vanheek, L. (1981) Control of peach tree growth and productivity by regulated water supply, tree density and summer pruning. *Journal of the American Society for Horticultural Sciences* 106, 307–312.

Chopade, S.Q., Gorantiwar, S.D., Pampattiwar, P.S. and Supe, V.S. (2001) Response of pomegranate to drip, bubbler and surface irrigation methods. *Advances in Horticulture and Forestry* 8, 53–59.

Cohen, S., Moreshet, S., Guillou, L.L., Simon, J.-C. and Cohen, M. (1997) Response of citrus trees to modified radiation regime in semi-arid conditions. *Journal of Experimental Botany* 48(1), 35–44. DOI: 10.1093/jxb/48.1.35.

Day, K.R. and Wilkins, E.D. (2011) Commercial pomegranate (*Punica granatum* L.) production in California. *Acta Horticulturae* 890, 275–285.

Dhillon, W.S., Gill, P.S. and Singha, N.P. (2011) Effect of nitrogen, phosphorus and potassium fertilization on growth, yield and quality of pomegranate 'Kandhari'. *Acta Horticulturae* 890, 327–332.

Dinc, N., Aydinsakir, K., Isik, M., Bastug, R., Ari, N. *et al.* (2018) Assessment of different irrigation strategies on yield and quality characteristics of drip irrigated pomegranate under Mediterranean conditions. *Irrigation Science* 36(2), 87–96. DOI: 10.1007/s00271-017-0565-5.

Ebtedaie, M. and Shekafandeh, A. (2016) Antioxidant and carbohydrate changes of two pomegranate cultivars under deficit irrigation stress. *Spanish Journal of Agricultural Research* 14(4), e0809. DOI: 10.5424/sjar/2016144-9317.

Fageria, N.K., Gheyi, H.R. and Moreira, A. (2011) Nutrient bioavailability in salt affected soils. *Journal of Plant Nutrition* 34(7), 945–962. DOI: 10.1080/01904167.2011.555578.

Fereres, E., Goldhamer, D.A. and Parsons, L.R. (2003) Irrigation water management of horticultural crops. *HortScience* 38(5), 1036–1042. DOI: 10.21273/HORTSCI.38.5.1036.

Fereres, E. and Soriano, M.A. (2007) Deficit irrigation for reducing agricultural water use. *Journal of Experimental Botany* 58, 147–159.

Fernández, J.E. and Cuevas, M.V. (2010) Irrigation scheduling from stem diameter variations: a review. *Agricultural and Forest Meteorology* 150, 135–151.

Galindo, A., Calín-Sánchez, A., Collado-González, J., Ondoño, S., Hernández, F. *et al.* (2014) Phytochemical and quality attributes of pomegranate fruits for juice consumption as affected by ripening stage and deficit irrigation. *Journal of the Science of Food and Agriculture* 94(11), 2259–2265. DOI: 10.1002/jsfa.6551.

Galindo, A., Calín-Sánchez, Á., Griñán, I., Rodríguez, P., Cruz, Z.N. *et al.* (2017) Water stress at the end of the pomegranate fruit ripening stage produces earlier harvest and improves fruit quality. *Scientia Horticulturae* 226, 68–74. DOI: 10.1016/j.scienta.2017.08.029.

García-Gómez, K.I. (2011) Estimación de la acumulación de biomasa y extracción estacional de nitrógeno, fósforo, potasio, calcio y magnesio en plantas de granado (*Punica granatum* L). Master's Thesis. Universidad de Chile.

García-Tejero, I., Romero-Vicente, R., Jiménez-Bocanegra, J.A., Martínez-García, G., Durán-Zuazo, V.H. *et al.* (2010) Response of citrus trees to deficit irrigation during different phenological periods in relation to yield, fruit quality, and water productivity. *Agricultural Water Management* 97(5), 689–699. DOI: 10.1016/j.agwat.2009.12.012.

Gill, P.P.S., Dhillon, W. and Singh, N. (2011) Influence of training systems on growth, yield and fruit quality of pomegranate 'Kandhari'. *Acta Horticulturae* 890, 305–310.

Granada de Elche Food Wine Spain (2018) PDO Granada Mollar de Elche. Available at: https://www.foodswinesfromspain.com/spanishfoodwine/global/food/news/new-detail/NEW2016610309.html (accessed 14 October 2018).

Haneef, M., Kaushik, R.A., Sarolia, D.K., Mordia, A. and Dhakar, M. (2014) Irrigation scheduling and fertigation in pomegranate cv. Bhagwa under high density planting system. *The Indian Journal of Horticulture* 71, 45–48.

Hansen, P. (1971) The effect of fruiting upon transpiration rate and stomatal opening in apple leaves. *Physiologia Plantarum* 25(2), 181–183. DOI: 10.1111/j.1399-3054.1971.tb01424.x.

Hasanpour, Z., Karimi, H.R. and Mirdehghan, S.H. (2015) Effects of salinity and water stress on echo-physiological parameters and micronutrients concentration of pomegranate (*Punica granatum* L.). *Journal of Plant Nutrition* 38(5), 795–807. DOI: 10.1080/01904167.2014.944711.

Hepaksoy, S., Bahaulddin, A. and Kukul Kurttas, Y.S. (2016) The effects of irrigation on leaf nutrient content in pomegranate 'İzmir 1513'. *Acta Horticulturae* 1139, 581–586.

Hiwale, S.S., More, T.A. and Bagle, B.G. (2009) Root distribution pattern in pomegranate 'Ganesh' (*Punica granatum* L.). *Acta Horticulturae* 890, 44–49.

Holland, D., Hatib, K. and Ya'akov, B.I. (2009) Pomegranate: botany, horticulture, breeding. In: *Horticultural Reviews*. Wiley, In, pp. 127–191.

Intrigliolo, D.S., Nicolás, E., Bonet, L., Ferrer, P., Alarcón, J.J. *et al.* (2011a) Water relations of field grown pomegranate trees (*Punica granatum*) under different drip irrigation regimes. *Agricultural Water Management* 98(4), 691–696. DOI: 10.1016/j.agwat.2010.11.006.

Intrigliolo, D.S., Puerto, H., Bonet, L., Alarcón, J.J., Nicolás, E. *et al.* (2011b) Usefulness of trunk diameter variations as continuous water stress indicators of pomegranate (*Punica granatum*) trees. *Agricultural Water Management* 98(9), 1462–1468. DOI: 10.1016/j.agwat.2011.05.001.

Intrigliolo, D.S., Bonet, L., Nortes, P.A., Puerto, H., Nicolás, E. *et al.* (2013) Pomegranate trees performance under sustained and regulated deficit irrigation. *Irrigation Science* 31(5), 959–970. DOI: 10.1007/s00271-012-0372-y.

Intrigliolo, D.S. and Lakso, A.N. (2010) Effects of amount of light interception and canopy orientation to the sun on grapevine water status and canopy gas exchange. *Acta Horticulturae* 889, 99–104.

Intrigliolo, D.S., Bartual, J., García-González, J.F., Guerra, D., Parra, J. *et al.* (2019) Quantifying pomegranate tree responses to water and nutrients for a sustainable fertirrigation. *Acta Horticulturae* 1254, 193–198.

Kale, S.J., Nath, P., Meena, V.S. and Singh, R.K. (2018) Semi-permanent shadenet house for reducing the sunburn in pomegranates (*Punica granatum*). *International Journal of Chemical Studies* 6, 2053–2057.

Karimi, H.R. and Hasanpour, Z. (2014) Effects of salinity and water stress on growth and macro nutrients concentration of pomegranate (*Punica granatum* L.). *Journal of Plant Nutrition* 37(12), 1937–1951. DOI: 10.1080/01904167.2014.920363.

Kulkarni, T.S., Desai, U.T., Kshirsagar, D.B. and Kamble, A.B. (2007) Effects of salt regimes on growth and mineral uptake of pomegranate (*Punica granatum* L) cv. Mrudula. *Annual Arid Zone* 46, 77–82.

Laribi, A.I., Palou, L., Intrigliolo, D.S., Nortes, P.A., Rojas-Argudo, C. *et al.* (2013) Effect of sustained and regulated deficit irrigation on fruit quality of pomegranate cv. 'Mollar de Elche' at harvest and during cold storage. *Agricultural Water Management* 125, 61–70. DOI: 10.1016/j.agwat.2013.04.009.

López, G., Mata, M., Arbonés, A., Solans, J.R., Girona, J. *et al.* (2006) Mitigation of effects of extreme drought during stage III of peach fruit development by summer pruning and fruit thinning. *Tree Physiology* 26(4), 469–477. DOI: 10.1093/treephys/26.4.469.

Maas, E.V. (1993) Testing crops for salinity tolerance. In: Maranville, J.W., Baligar,B.V., Duncan, R.R. and Yohe, J.M. (eds) *Proceedings of a Workshop on Adaptation of Plants to Soil Stresses, l–4 August 1993*. INTSORMIL Pub. No. 94-2. University of Nebraska, Lincoln, NE, pp. 234–247.

Maas, E.V. and Hoffmann, G.J. (1976) Crop salt tolerance: evaluation of existing data. *Proceedings International Conference Texas Technical University*, Lubbock, Texas, pp. 187–197.

Marsal, J., López, G., Mata, M. and Girona, J. (2006) Branch removal and defruiting for the amelioration of water stress effects on fruit growth during stage III of peach fruit development. *Scientia Horticulturae* 108(1), 55–60. DOI: 10.1016/j.scienta.2006.01.008.

Marsal, J., Mata, M., Arbones, A., Del Campo, J., Girona, J. *et al.* (2008) Factors involved in alleviating water stress by partial crop removal in pear trees. *Tree Physiology* 28(9), 1375–1382. DOI: 10.1093/treephys/28.9.1375.

Melgarejo, P., Salazar, D.M. and Artés, F. (2000) Organic acids and sugars composition of harvested pomegranate fruits. *European Food Research and Technology* 211(3), 185–190. DOI: 10.1007/s002170050021.

Mellisho, C.D., Egea, I., Galindo, A., Rodríguez, P., Rodríguez, J. *et al.* (2012) Pomegranate (*Punica granatum* L.) fruit response to different deficit irrigation conditions. *Agricultural Water Management* 114, 30–36. DOI: 10.1016/j.agwat.2012.06.010.

Mena, P., Galindo, A., Collado-González, J., Ondoño, S., García-Viguera, C. *et al.* (2013) Sustained deficit irrigation affects the colour and phytochemical characteristics of pomegranate juice. *Journal of the Science of Food and Agriculture* 93(8), 1922–1927. DOI: 10.1002/jsfa.5991.

Meshram, D.T., Gorantiwar, S.D., da Silva, J.A.T., Jadhav, V.T. and Chandra, R. (2010) Water management in pomegranate. *Fruit, Vegetable and Cereal Science and Biotechnology* 4, 106–112.

Meshram, D.T., Gorantiwar, S.D., Mittal, H.K., Singh, A.K. and Lohkare, A.S. (2012) Water requirement of pomegranate (*Punica granatum* L.) plants up to five-year age. *Journal of Applied Horticulture* 14, 47–50.

Mir, M., Sharma, S.D. and Kumar, P. (2015) Nutrient dynamics: effect on cropping behavior, nutrient profile and quality attributes of pomegranate (*Punica granatum* L.) under rainfed agroclimatic conditions. *Journal of Plant Nutrition* 38, 83–95.

Mirdehghan, S.H. and Rahemi, M. (2007) Seasonal changes of mineral nutrients and phenolics in pomegranate (*Punica granatum* L.) fruit. *Scientia Horticulturae* 111, 120–127.

Mittal, H.K., Meshram, D.T., Purohit, R.C. and Gorantiwar, S.D. *et al.* (2011) Water requirement of pomegranate (*Punica granatum* L.) for Solapur district of Maharashtra State. *Acta Horticulturae* 890, 311–322. DOI: 10.17660/ActaHortic.2011.890.43.

Mphahlele, R.R., Fawole, O.A., Stander, M.A. and Opara, U.L. (2014) Preharvest and postharvest factors influencing bioactive compounds in pomegranate (*Punica granatum* L.): a review. *Scientia Horticulturae* 178, 114–123. DOI: 10.1016/j.scienta.2014.08.010.

Naeini, M.R., Khoshgoftarmanesh, A.H., Lessani, H. and Fallahi, E. (2005) Effects of sodium chloride-induced salinity on mineral nutrients and soluble sugars in three commercial cultivars of pomegranate. *Journal of Plant Nutrition* 27(8), 1319–1326. DOI: 10.1081/PLN-200025832.

Naeini, M.R., Khoshgoftarmanesh, A.H. and Fallahi, E. (2006) Partitioning of chlorine, sodium, and potassium and shoot growth of three pomegranate cultivars under different levels of salinity. *Journal of Plant Nutrition* 29(10), 1835–1843. DOI: 10.1080/01904160600899352.

Nielsen, D.R., van Genuchten, M.T. and Biggar, J.W. (1986) Water flow and solute transport processes in the unsaturated zone. *Water Resources Research* 22(9), 89–108. DOI: 10.1029/WR022i09Sp0089S.

Ortuño, M.F., Conejero, W., Moreno, F., Moriana, A., Intrigliolo, D.S. *et al.* (2010) Could trunk diameter sensors be used in woody crops for irrigation scheduling? A review of current knowledge and future perspectives. *Agricultural Water Management* 97(1), 1–11. DOI: 10.1016/j.agwat.2009.09.008.

Ozkan, Y. (2005) Investigations on physical and chemical characteristics of some pomegranate genotypes (*Punica granatum* L.) of Tokat province in Turkey. *Asian Journal of Chemistry*17, 939–942.

Parvizi, H., Sepaskhah, A.R. and Ahmadi, S.H. (2016) Physiological and growth responses of pomegranate tree (*Punica granatum* (L.) cv. Rabab) under partial root zone drying and deficit irrigation regimes. *Agricultural Water Management* 163, 146–158. DOI: 10.1016/j.agwat.2015.09.019.

Parvizi, H. and Sepaskhah, A.R. (2015) Effect of drip irrigation and fertilizer regimes on fruit quality of a pomegranate (*Punica granatum* (L.) cv. Rabab) orchard. *Agricultural Water Management* 156, 70–78. DOI: 10.1016/j.agwat.2015.04.002.

Peña, M.E., Artés-Hernández, F., Aguayo, E., Martínez-Hernández, G.B., Galindo, A. *et al.* (2013) Effect of sustained deficit irrigation on physicochemical properties, bioactive compounds and postharvest life of pomegranate fruit (cv. 'Mollar de Elche'). *Postharvest Biology and Technology* 86, 171–180. DOI: 10.1016/j.postharvbio.2013.06.034.

Peña-Estévez, M.E., Gómez, P.A., Artés, F., Aguayo, E., Martínez-Hernández, G.B. *et al.* (2015) Quality changes of fresh-cut pomegranate arils during shelf life as affected by deficit irrigation and postharvest vapour treatments. *Journal of the Science of Food and Agriculture* 95(11), 2325–2336. DOI: 10.1002/jsfa.6954.

Peña-Estévez, M.E., Artés-Hernández, F., Artés, F., Aguayo, E., Martínez-Hernández, G.B. *et al.* (2016a) Quality changes of pomegranate arils throughout shelf life affected by deficit irrigation and pre-processing storage. *Food Chemistry* 209, 302–311. DOI: 10.1016/j.foodchem.2016.04.054.

Peña-Estévez, M.E., Gómez, P.A., Artés, F., Aguayo, E., Martínez-Hernández, G.B. *et al.* (2016b) Changes in bioactive compounds and oxidative enzymes of fresh-cut pomegranate arils during storage as affected by deficit irrigation and postharvest vapor heat treatments. *Food Science and Technology International* 22(8), 665–676. DOI: 10.1177/1082013216635323.

Phene, C.J., Hoffman, G.J., Clark, T.A., Mead, R.M., Johnson, R.S. *et al.* (1991) Automated lysimeter for irrigation and drainage control. *Proceedings of Lysimeters for Evapotranspiration and Environmental Measurements. ASCE*, New York, pp. 28–36.

Prasad, R.N. and Mali, P.C. (2002) Effect of drip irrigation on physico-chemical characteristics of pomegranate fruits in arid region. *Annals of Arid Zone* 41, 65–68.

Rodríguez, P., Mellisho, C.D., Conejero, W., Cruz, Z.N., Ortuño, M.F. *et al.* (2012) Plant water relations of leaves of pomegranate trees under different irrigation conditions. *Environmental and Experimental Botany* 77, 19–24. DOI: 10.1016/j.envexpbot.2011.08.018.

Ruiz-Sánchez, M.C., Domingo, R. and Castel, J.R. (2010) Review. deficit irrigation in fruit trees and vines in Spain. *Spanish Journal of Agricultural Research* 8(S2), 5–S20. DOI: 10.5424/sjar/201008S2-1343.

Schneider, A.D., Ayars, J.E. and Phene, C.J. (1996) Combining monolithic and repacked soil tanks for lysimeter from high water table sites. *Applied Engineering in Agriculture* 12, 649–654.

Selahvarzi, Y., Zamani, Z., Fatahi, R. and Talaei, A.-R. (2017) Effect of deficit irrigation on flowering and fruit properties of pomegranate (*Punica granatum* cv. Shahvar). *Agricultural Water Management* 192, 189–197. DOI: 10.1016/j.agwat.2017.07.007.

Shulman, Y., Fainberstein, L. and Lavee, S. (1984) Pomegranate fruit development and maturation. *Journal of Horticultural Science* 59, 265–274.

Sulochanamma, B.N., Yellamanda Reddy, T. and Subbi Reddy, G. (2005) Effect of Basin and drip irrigation on growth, yield and water use efficiency in pomegranate cv. Ganesh. *Acta Horticulturae* 696, 277–279. DOI: 10.17660/ActaHortic.2005.696.48.

Tagliavini, M., Zavalloni, C., Rombolà, A.D., Quartieri, M., Malaguti, D. *et al.* (2000) Mineral nutrient partitioning to fruits of deciduous trees. *Acta Horticulturae* 512,131–140. DOI: 10.17660/ActaHortic.2000.512.13.

Thompson, R.B., Gallardo, M. and Voogt, W. (2015) Optimizing nitrogen and water inputs for greenhouse vegetable production. *Acta Horticulturae* 1107, 15–30. DOI: 10.17660/ActaHortic.2015.1107.2.

Thompson, R.B., Incrocci, L., Voogt, W., Pardossi, A. and Magán, J.J. (2017) Sustainable irrigation and nitrogen management of fertigated vegetable crops. *Acta Horticulturae* 1150, 363–378. DOI: 10.17660/ActaHortic.2017.1150.52.

Trout, T.J. and Gartung, J. (2006) Use of canopy size to estimate crop coefficient for vegetable crops. *Proceeding of 2006 World Environment and Water Resources Congress*, Omaha, Nebraska, 21–25 May.

Varasteh, F., Arzani, K., Zamani, Z. and Tabatabaei, S.Z. (2008) Physico-chemical seasonal changes of pomegranate (*Punica granatum* L.) fruit 'Malas-e-Torsh-e-Saveh' in Iran. *Acta Horticulturae* 769,255–258. DOI: 10.17660/ActaHortic.2008.769.36.
Wullschleger, S.D., Meinzer, F.C. and Vertessy, R.A. (1998) A review of whole-plant water use studies in tree. *Tree Physiology* 18(8–9), 499–512. DOI: 10.1093/treephys/18.8-9.499.
Zhang, H., Anderson, R.G. and Wang, D. (2015) Satellite based crop coefficient and regional water use estimates for Hawaiian sugarcane. *Field Crops Research* 180, 143–154. DOI: 10.1016/j.fcr.2015.05.023.
Zhang, H., Wang, D., Ayars, J.E. and Phene, C.J. (2017) Biophysical response of young pomegranate trees to surface and sub-surface drip irrigation and deficit irrigation. *Irrigation Science* 35(5), 425–435. DOI: 10.1007/s00271-017-0551-y.

11 Physiological Disorders

K.S. Shivashankara* and G.A. Geetha
Division of Plant Physiology and Biochemistry, ICAR-Indian Institute of Horticultural Research, Bengaluru, India

11.1 Introduction

Fruit tree productivity and crop quality are affected by many factors including biotic and abiotic stresses and physiological disorders. Physiological disorders are more common in horticultural crops, especially in fruit crops, and cause economic losses worldwide. Many tropical and subtropical fruit crops like apple, mango, banana, citrus, grape, papaya, pomegranate, loquat, etc. are vulnerable to different physiological disorders including browning, spongy tissue, chock throat, granulation, pink berries, bumpy fruits, fruit cracking, purple spot, etc. Many physiological disorders are genetically controlled, but some of them are influenced by adverse environmental conditions such as extreme temperatures, irregular water supply, nutritional and hormonal imbalances, and pollination and fertilization problems. The current review elaborates a few major physiological disorders of pomegranate fruit, which affect the marketable yield as well as quality, leading to enormous losses for fruit growers.

Pomegranate (*Punica granatum* L.) belongs to the family *Punicaceae* and the name is derived from the Latin phrase 'malum granatum', which means 'grainy apple'. The fruits, along with seeds, fruit peel, flowers and juice, have been used for reducing various health problems due to their nutritional quality and health-promoting substances. At present, pomegranate is grown in various Mediterranean, subtropical and tropical countries, while it is marketed in all parts of the world. Major pomegranate producer countries are India, Iran, China, Turkey, Spain and the USA (California), while Italy, Argentina, Australia, Brazil, Peru, Chile, Tunisia, Egypt, Afghanistan and South Africa are also smaller producers. According to the International Plant Genetic Resources Institute (IPGRI), about 50 pomegranate varieties are commercially cultivated in various parts of the world (IPGRI, 2001).

Pomegranate fruits are botanically classified as a berry. The pericarp of the fruit is tough and leathery, and the internal structure has many irregular segments separated by non-edible white piths and thin carpellary membrane. Each segment is closely packed with small arils, which are anchored on a soft and fleshy placental tissue. Each aril contains a seed surrounded by edible juicy pulp. Pomegranate arils are rich sources of sugars, vitamins, polyphenols and minerals (Melgarejo and Artes, 2000; Ferrara *et al.*, 2014). The fruit skins and membranes are rich in ellagitannins, which have a wide array of health-promoting activities (Seeram *et al.*, 2006) and the seeds have a considerable amount of oil (Ferrara *et al.*, 2011, 2014).

*Corresponding author: shivaiihr@yahoo.com

Pomegranate, like any other fruit crop, provides livelihood security to the farmers and nutritional benefits to the people. Insects, pests and diseases affect both production and quality of the pomegranate. Thus, it is essential to pay more attention to preventing losses owing to the broad range of constraints in its production. Physiological disorders are non-pathological problems, and one of the significant threats to pomegranate production, causing economic losses worldwide. Fruit development and maturation is controlled by various physiological and biochemical processes. Any dysfunction or malfunction of the physiological processes due to changes in temperature, humidity, water availability, incorrect application of chemicals and abnormal mineral nutrients results in physiological disorders (Ladaniya, 2008). Pomegranate cultivation and its potential quality yield are affected by various physiological disorders like browning, fruit cracking, internal breakdown, scald and chilling injury (Nath *et al.*, 2008). In this chapter, the various physiological disorders commonly occurring in pomegranate and their control measures are described in detail. Among them, aril browning or internal breakdown or aril blackening and fruit cracking are important disorders leading to substantial economic losses to farmers.

11.2 Aril Browning (Internal Breakdown)

The disintegration of arils in matured pomegranate fruits is known as the internal breakdown or aril blackening (Fig. 11.1). Internal breakdown can also be called aril browning since the arils turn brown and become unfit for consumption. Aril breakdown or browning is also characterized by soft, light creamy brown to blackish, slightly flattened arils, which are deformed and possess an unpleasant odour when the fruit is cut open. It is one of the major threats to pomegranate fruit quality, leading to 50–60% losses every year in some production areas (Shete and Waskar, 2005). This disorder cannot be identified externally. No insect or organism is responsible for this particular disorder. The symptoms of internal breakdown typically start at 90 days after anthesis and later on arils become creamy-brown to dark blackish-brown from the outside in, and the disorder is severe if matured fruits are allowed to remain on the tree. Thus, harvesting of fruits at the maturation stage is usually recommended to avoid this disorder (Sharma, 2005). It is also reported that internal breakdown is significantly correlated with the weight of the fruit: the higher the fruit weight, the higher is the chance of internal breakdown. Mango cv. 'Alphonso' fruits of more than 350 g in weight had a higher incidence of internal breakdown (60%) than the fruits of 150–200 g in weight (26.6%) (Joshi and Limaye, 1986).

Fig. 11.1. Arils of pomegranate cultivar 'Ganesh' showing browning symptoms. (Photo: K.S. Shivashankara.)

The intensity of the disorder in fully matured fruits can go up to 50%, which leads to severe loss of fruit quality. Since the fruits with aril browning are devoid of external symptoms, separation of affected fruits is not possible, and thus, it becomes a serious problem during export (Shivashankar *et al.*, 2012).

11.2.1 Physiological and biochemical changes in the arils

Higher respiration, transpiration, starch, acidity, pyruvic acid, malondialdehyde (MDA) (a lipid peroxidation product) and higher activities of antioxidative enzymes like superoxide dismutase, peroxidase and polyphenol oxidase enzymes have been observed in the affected arils (Shivashankara *et al.*, 2004) (Table 11.1). Higher sugars, and lower starch, anthocyanins and phenols have also been reported in the

Table 11.1. Various biochemical and physiological properties of the healthy and browned arils of pomegranate cultivar 'Ganesh' (Shivashankara *et al.*, 2004).

Parameters	Healthy arils	Brown arils
Respiration (mg CO_2 kg/h)	25.38	33.70
Transpiration (mM H_2O kg/h)	74.02	93.31
Moisture (%)	81.00	80.00
Starch (mg/g fresh weight (fw))	20.25	28.25
Total sugars (g of glucose/100 g fw)	9.99	8.02
Reducing sugars (g/100 g fw)	9.49	7.92
Total protein (mg/g fw)	8.32	8.12
Total amino acids (mg/g fw)	0.39	0.43
Titrable acidity (mg of citric acid/g fw)	4.19	6.31
Ascorbic acid (mg/100 g)	7.78	3.33
Anthocyanins (absorbance/g fw)	0.12	0.08
Pyruvic acid (μmol of pyruvic acid/g fw)	0.23	0.56
Malondialdehyde (MDA) (absorbance/g fw)	0.31 (0.06)[a]	0.39 (0.08)[a]
Super oxide dismutase (units/g fw)	16.20	20.00
Peroxidase (units/g fw)	0.47	3.75
Polyphenol oxidase (units/g fw)	0.33	0.71

[a]Values in parenthesis are μg of MDA/100 g fw.

affected arils (Singh *et al.*, 2013). Aril pigmentation is related to the activity of phenylalanine ammonia lyase (Gil *et al.*, 1997). On the other hand, higher activity of phenylalanine ammonia lyase (PAL) has also been observed along with high pectin methylesterase activity in the affected arils. However, ascorbic acid, anthocyanins, total phenols and antioxidant activity were lower in the affected arils along with the increased activities of antioxidant enzymes and respiration in cultivar 'Malase Saveh' (Meighani *et al.*, 2014). Affected arils show significant changes in firmness and colour parameters like L, a and b values (Darsana *et al.*, 2016).

A significant increase in the activity of antioxidant enzymes, especially peroxidase and polyphenol oxidase, and respiration rate indicate severe oxidative stress and membrane damage leading to oxidation of phenols in the arils. Therefore, identifying varietal variation in the activity of polyphenol oxidase is a viable option to identify the genotypes with less aril browning (Shete *et al.*, 2006). Variation in the activity of polyphenol oxidase enzyme has been reported in four cultivars ('G-137', 'Gulsha Red', 'Jalore Seedless' and 'Mridula'), which show differences in the extent of aril blackening (Shete *et al.*, 2006). Browning appeared early in 'G-137' and late in 'Gulsha Red'. Cultivar 'Gulsha Red' showed a very low activity of polyphenol oxidase (PPO) and peroxidase (POD) and had the lowest browning of arils, indicating that these characters are associated with lower aril browning. However, the study by Shivashankara *et al.* (2004) indicated that the increase in PPO and POD are mainly due to mineral deficiency-induced higher membrane leakage, leading to diffusion of phenols.

Apart from biochemical changes in the arils, enzyme activities in the seeds have also been found to vary between the affected and healthy arils. Lower activities of amylase and total dehydrogenase, but higher polyphenol oxidase were observed in the seeds of the affected arils compared with the seeds of healthy arils. Weak sink activity of certain seeds due to an imbalance in hormones was reported to be associated with the formation of aril browning in pomegranate cv. 'Bhagwa' (Shivashankar *et al.*, 2012). The authors of the study also mentioned that the membrane leakage of the arils leads to increased polyphenol oxidase and browning, but the study did not indicate the causes of membrane leakage. It is also not clear from the study how weaker sinks are developing within fruit and receive less water. Affected arils had lower calcium and higher free radicals with higher oxidation of membranes along with higher PPO enzyme activity. This indicated that the calcium-related membrane leakage leads to the leakage of phenols and their further oxidation resulting in aril browning (Shivashankara *et al.*, 2004; Meighani *et al.*, 2014). Furthermore, lower concentrations of Cu in the affected arils, and K, Mg and Mn in the peel of affected fruits were observed when compared with the healthy fruit in cv. 'Malase Saveh' (Meighani *et al.*, 2014). The direct oxidation of phenolic compounds by PPO and POD enzymes is a major cause of fruit tissue browning (Tomás-Barberán and Espín, 2001).

A similar increase in the activities of PPO, superoxide dismutase (SOD) and MDA content has been reported in browned peels of pomegranate (Liu *et al.*, 1998). Increased activities of peroxidase and polyphenol oxidase were also associated with browning in other fruits such as apple, lychee, rambutan and longan (Zhang *et al.*, 2005; Yingsanga *et al.*, 2008; Holderbaum *et al.*, 2010; Venkatachalam and Meenune, 2011). Franck *et al.* (2007) suggested that browning disorder of pear was caused by an imbalance between oxidative and reductive processes due to a metabolic gas gradient inside the fruit, which may be due to reduced diffusion of gases or altered membrane diffusivity.

11.2.2 Germplasm diversity for aril browning resistance

In order to develop varieties resistant to aril browning, it is necessary to evaluate a large amount of germplasm and to identify the important traits associated with it. A large number of pomegranate progenies (158) of different crosses were evaluated for severity of aril browning (Jalikop *et al.*, 2010). The severity of aril browning was related to fruit traits like skin colour, aril colour, total soluble solids (TSS) and seed mellowness. It was correlated negatively with the anthocyanin content and positively with TSS. Therefore, to develop aril browning-free pomegranate cultivars, breeders may look for higher anthocyanin types with slightly lower TSS values. Cultivars 'G-137', 'Gulsha Red', 'Jalore Seedless' and 'Mridula' showed variability in aril browning, with cv. 'Gulsha Red' showing lower aril browning (Shete *et al.*, 2006). Variation in POD and PPO was found to be associated with the aril browning in these cultivars. Therefore, selecting genotypes for lower activity of POD and PPO enzymes along with traits of higher anthocyanins and moderate TSS may be a good strategy for developing aril browning-resistant types.

a b

Fig. 11.2. Healthy (a) and browned (b) arils of pomegranate. (Photos: Hossein Meighani.)

11.2.3 Control of aril browning

Summer foliar application of salicylic acid (SA) and sodium nitroprusside (SNP) at a concentration of 0.1 mM significantly reduced the aril browning (Fig. 11.2), and increased the total anthocyanin content, ascorbic acid and the activities of ascorbate peroxidase and SOD enzymes (Khodaei *et al.*, 2015). Pruning intensity was reported to reduce aril browning in some of the susceptible cultivars (Pawar *et al.*, 1994). Light exposure and reduced humidity around the fruits played an essential role in the development of the disorder. Covering the fruits to reduce the light exposure and increase the humidity significantly increased aril browning in 'Ganesh' (Shivashankara, 2004). Higher light exposure and lower humidity around the fruits might aid better movement of calcium to fruits thereby reducing the incidence, as the affected arils were found to have significantly lower calcium in cv. 'Ganesh'. Lower calcium, boron and copper were also observed in affected arils of 'Malase Saveh' pomegranates (Meighani *et al.*, 2014). Therefore, pruning of shoots to increase the light penetration may have resulted in better transport of calcium and boron to fruits leading to the reduction in aril browning (Shivashankara, 2004). Spraying of calcium and boron directly on to the fruits in the initial stages of fruit growth, coupled with harvesting at proper maturity time may reduce aril browning. Care must be taken to ensure that mature fruits are not left on the trees for too long particularly in areas where the soil pH is a little acidic and atmospheric humidity is higher.

11.3 Fruit Cracking

Fruit cracking or splitting is another disorder that causes a severe reduction in the marketable yield of fruits (Peet, 1992). Fruit cracking is the physical failure of the fruit peel, which manifests as fractures in the peel or cuticle of fruits or splitting. In severe cases the splitting penetrates deep into the pulp (Opara *et al.*, 1997). Fruit cracking and splitting in fruits like apple, cherry, pear, grapes, banana, tomato, etc. has been reviewed extensively (Opara *et al.*, 1997). Fruit cracking is described as the small radiating cracks developed mainly on the surface/cuticle of the fruits, which usually do not extend to the deeper layers of the fruit. Wider cracks extending deep into the fruits are called fruit splitting. Opara *et al.* (1997) summarized the role of rootstocks, cultivars, nutrition, soil moisture variations, rainfall, temperature, hormones and growth regulators like paclobutrazol and promalin on fruit cracking, and different methods to control this disorder. However, not much has been said about the fruit cracking in pomegranates. In pomegranates, fruits completely spilt open exposing the arils with either horizontal or vertical splits. It can lead to a loss of up to 40% of the total yield in pomegranates. As explained in earlier reviews for other fruits, in pomegranates fruit cracking is also influenced by the wider fluctuations in soil moisture caused by irregular irrigation, erratic rainfall or overirrigation during the ripening period (Mars, 2000; Meshram *et al.*, 2010), late harvest (Shulman *et al.*, 1984; El-Khawaga, 2007; Khalil and Aly, 2013), physical injuries to peel and sunburn (Shulman *et al.*, 1984), imbalance in plant nutrition, temperature difference between day and night periods (El-Kassas *et al.*, 1992; El-Rhman, 2010), warm winds during an arid period, instant decrease of the temperature (Plamenac, 1972) and genetic factors (Josan *et al.*, 1979).

11.3.1 Causes of fruit cracking

Hot, dry weather, the genetics of the cultivar, fruit growth and cultural practices are the main factors involved in the enhancing of fruit cracking in pomegranate (Saad *et al.*, 1988). Prasad and Mali (2000) also reported erratic irrigation or excessive rainfall patterns during the maturation period as the cause of fruit cracking in pomegranate. Previously, studies on different fruit species indicated the high influence of plant nutrients and transpiration levels on fruit cracking (Aksoy and Akyuz, 1993). Kumar *et al.* (2010) also found that variabilities in soil moisture, climate and tree nutrition are the factors responsible for fruit cracking. In addition to the soil moisture fluctuations, heavy irrigation after a dry spell or dry wind is also reported to increase fruit cracking (Kumar, 1990). During a drought period, strengthened tissue develops in xylem and phloem and they lose their ability to divide and enlarge. After a dry spell (April–May) if the water supply is increased the meristematic tissue quickly resumes growth but the strengthened tissue does not, and owing to the differential growth rate, tissue ruptures appear. Dry heat accompanied by the hot, dry wind at the time of fruit ripening in pomegranate was the primary cause of cracking; during the rapid flesh growth, temperatures higher than 38°C combined with less than 60% humidity favoured cracking. Maintenance of at least 25% of available soil moisture in summer reduces fruit cracking significantly (Sheikh and Manjula, 2012). The later formed fruits have more tendency to split than the early ones (Mohamed, 2004). This may be due to the higher temperatures at the later stages and is further influenced by the lower calcium content in the late-formed fruits.

Calcium and magnesium are the constituents of the pectates which form the middle lamella that bind the cells. Also, Ca and Mg are responsible for strengthening the bonds between epidermal and other fruit cells resulting in better strength and low cracking (Poovaiah, 1986). Tree age also influences the extent of fruit cracking in pomegranate. Old trees are more susceptible to soil moisture fluctuations and erratic rainfall than young trees. The disorder in young trees is often due to boron deficiency rather than climatic variabilities (El-Rhman, 2010). Older trees usually show lower indole-3-acetic acid (IAA) and higher abscisic acid (ABA) content compared with younger trees.

11.3.2 Genotypic variability in fruit cracking

As well as the soil moisture variability-induced fruit cracking, pomegranate also exhibits genotypic variability for fruit cracking resistance. Diversity among the cultivars for fruit cracking has been reported by many researchers. Cultivars like 'Wonderful', 'Jalore Seedless', 'Khogand', 'BedanaBosec', 'PS 75K3', 'Appuli', 'Shirvan', 'Burachni', 'Apsherconskil', 'Krasnyl', 'Sur-Anar', 'Kyrmyz-Kabukh', 'Francis', 'Kadi', 'Lefon', (resistant), 'Malase Saveh' (moderately susceptible), 'Koycegiz' and 'Siyah' (susceptible) have been reported in many countries (Karp, 2006; Singh *et al.*, 2006; Salih, 2017). Varietal variations in leaf nitrogen content, transpiration rate and peel calcium content were found to be related to fruit cracking (Hepaksoy *et al.*, 2000). Leaf transpiration rate is known to affect the absorption and translocation of calcium and boron in the plant. The peel calcium content was found to be related to diversity in fruit cracking among the cultivars; however, fruit peel thickness variations were not related to cracking. Excessive calcium content was seen in the susceptible cultivar, which was reported to increase the peel toughness leading to cracking. Peel toughness increased during the ripening stage. Among hormones, ABA content was high in the cracked fruits, indicating the increased stress levels in these fruits (Yilmaz and Ozguven, 2006). There was no difference in IAA and GA_3 levels between the susceptible and resistant cultivars. Changes in ABA and IAA indicate that cell wall extension is affected under adverse conditions. Lower expression of cell wall expansion genes was observed in the affected fruits (Khadivi-Khub, 2015). The germplasm may be evaluated for phenotypic traits like leaf and fruit transpiration, fruit peel thickness, root water, and mineral acquisition properties, root volume and root length, which can contribute to better mineral uptake and lower fruit cracking. Identification of such rootstocks may help in completely avoiding fruit cracking problems in pomegranates.

11.3.3 Control of fruit cracking

Application of 5 ppm of N-(2-chloro-4-pyridyl)-N′-phenylurea (CPPU) in mid-May with *in-situ* moisture conservation decreased fruit cracking and improved fruit quality in pomegranate cv. 'Kandhari' (Sahu *et al.*, 2013). Foliar application of boron (50 ppm), zinc sulfate, calcium hydroxide and GA_3 (40 ppm) minimized the incidence of cracking in the young fruits (Sepahi, 1986; Bambal *et al.*, 1991; Singh *et al.*, 2003; Yilmaz and Ozguven, 2006; Lal *et al.*, 2011). Spraying 0.3% of boric acid also reduced fruit cracking in pomegranates (Sahu *et al.*, 2013). Application of 300 ppm of paclobutrazol also reduced fruit cracking significantly (Khalil and Aly, 2013) probably by reducing the GA_3 content in the fruit, which might have affected the overall growth of the fruit itself. Application of commercial formulations like cytozymes (4 ml/l) also reduced fruit cracking in pomegranates (Abubakar *et al.*, 2013). Fertilization with a low Ca concentration (0.50 g Ca/l) in the form of a nano-Ca formulation resulted in lower fruit cracking (Davarpanah *et al.*, 2018). Suitable orchard management that minimizes water stress, and takes into account balanced nutrition and physiological factors contributes to lower fruit cracking (Khadivi-Khub, 2015). Application of 2, 4-D and NAA at 10 ppm concentration also reduced fruit cracking significantly (Kumar *et al.*, 2017). It is suggested that spraying trees four times with a mixture containing salicylic acid at 100 ppm, magnesium sulfate at 0.5%; chelated Zn at 0.05%, boric acid at 0.05% and calcium chloride at 2% reduced fruit splitting and improved productivity of 'Manfalouty' pomegranate trees grown under Assiut region conditions (Ahmed *et al.*, 2014). Application of calcium and magnesium is known to strengthen the cell walls and also bind the cells tightly. Boron is involved in the transport of sugars,

Fig. 11.3. Pomegranate fruits showing fruit cracking patterns. (Photos: Alimohammad Yavari.)

which helps in cell division and synthesis of cell wall materials.

Cracking might be caused by stress (either soilborne or airborne)-induced hardening of the fruit peel, which later cracks due to the pressure created by the expansion of internal arils (Fig. 11.3). Cell wall thickening due to lignification and reduced cell wall extension caused by lower boron and calcium concentrations might increase fruit cracking. Therefore, the application of cell elongation hormone GA_3 has been reported to reduce fruit cracking (Patil, 2018). On the other hand, application of growth inhibitors like ABA and paclobutrazol decreased cracking by reducing the growth of internal tissues of the fruit, thereby reducing the pressure on the fruit peel.

11.4 Sunburn and Sunscald

Pomegranates are terminal-bearing plants with thin branches; therefore, being exposed to direct solar radiation continuously throughout the fruit growth period. Since the fruits are harvested at the end of summer or the beginning of autumn, they are exposed to high temperatures throughout the summer, leading to burn symptoms on the exposed side of the fruits. This symptom on the fruits due to excess solar radiation is called sunburn. The estimated loss of the harvested fruit due to sunburn is about 30% (Melgarejo and Martínez, 1992). The incidence of sunburn is more common in arid and semi-arid regions, where the tree is exposed to excessive heating. Excess solar radiation is reported to cause sunburn either directly (Barber and Sharpe, 1971; Schrader *et al.*, 2001) or indirectly by increasing the radiant heating (Thorpe, 1974; Schrader *et al.*, 2001), which increases the fruit surface temperature. Three types of sunburn are commonly noted in pomegranate, namely sunburn necrosis, sunburn browning and photo-oxidative sunburn. Excess radiant heating and/or exposure to excess sunlight are the direct factors that cause sunburn, while the effect can be influenced by relative humidity, wind velocity, acclimation of fruit and cultural management practices. Sunburn affects pomegranate fruit in many ways: it causes structural and morphological changes, alters the pigment composition, influences the adaptive mechanisms, impairs photosynthesis and consequently decreases the fruit quality.

11.4.1 Control of sunburn

The incidence of sunburn could be reduced/controlled by several approaches. Leaves can provide shade, preventing direct sunlight from falling on the fruits, hence growth of cultivars with more leaf area in places where the maximum mean temperature is high can reduce incidences of sunburn. Better fertilization and

Fig. 11.4. Pomegranate fruits showing sunburn symptoms (left), healthy fruit (right). (Photo: Alimohammad Yavari.)

irrigation improves vegetative growth, thereby protecting the fruits from direct sunlight. Erection of shades and screens could shelter the trees as well as fruits from direct exposure to sunlight. Another approach is the application of chemicals to reduce sunburn. Covering the fruits with reflective materials may also reduce the incidence (Fig. 11.4).

Pomegranate trees sprayed twice with 2% kaolin (first spray at 21 days after fruit set and second spray 1 month later) in each season recorded significantly increased fruit weight (g), yield (kg/tree) and non-edible parts (%) as compared with the control. Meanwhile, application of 4% kaolin led to a marked reduction in cracked and sunburned fruits, increased the percentage of marketable fruits as well as fruit chemical quality (TSS, TSS/TA, vitamin C, anthocyanin pigment and total sugars) and redness of the fruit (Abou El-Wafa, 2015).

A white coating of kaolin suspension (6%) on pomegranate fruits cv. 'Ardestani' resulted in a significant reduction in fruit surface temperature from 35.4 to 29.3°C and sunburn damage from 22.3 to 15.3% (Vatandoost *et al.*, 2014). Similarly, spraying of 'Terra Alba' on pomegranate fruits resulted in the development of a white coating on the surface, which significantly reduced fruit surface temperatures relative to the control by averages of 4.9 and 2.5°C, respectively, and sunburn damage was reduced from 21.9 to 9.4% (Parashar and Ansari, 2012). The kaolin-based sunscreens 'Surround' and 'Parasol' significantly reduced the severity of sunburn damage on pomegranate fruit, but treatment with 'Anti-stress-500' did not show any difference (Weerakkody *et al.*, 2010). Yazici and Kaynak (2009) reported that applications of 3% kaolin not only prevented sunburn in fruits of 'Hicaznar' pomegranate cultivar but also increased the soluble dry matter content and red colour of fruits.

It is known that in plants the surface temperature of leaves and fruits is controlled by evaporative cooling. If this process is not able to reduce the raised temperature caused by the direct solar radiation, then discolouration of the surface occurs owing to oxidation/destruction of pigments like chlorophylls, carotenoids and anthocyanins. High temperatures for prolonged periods will result in the death of cell layers and necrosis of the tissues. Therefore, regulation of water status, transpiration and surface wax status to reflect the radiation is important for reducing the sunburn of fruits. Pomegranate germplasm needs to be screened for these physiological parameters as well as for foliage cover of the canopy.

11.4.2 Sunscald

A high temperature in combination with excessive light, drought and suboptimal relative humidity causes sunscald in pomegranate fruits (Fig. 11.5). It is also known as lesion browning or pericarp necrosis (Lal and Sahu, 2017). Sunscald symptoms appear as a superficial skin browning, which initially appears at the stem end of the fruit and later on spreads towards the blossom end (Defilippi *et al.*, 2006; Kader, 2006). Sunscald-affected fruits easily succumb to various diseases and decay. The scald incidence and severity have been reported to be greater on pomegranate fruits harvested during the late season than those harvested during mid-season, indicating that this disorder is associated with longer duration of sunlight on fruit surface. The chemical compounds diphenylamine (DPA) and 1-methyl-cyclopropane (1-MCP) alone or together are used to control scald in other fruits, but do not show a similar effect on pomegranate (Defilippi *et al.*, 2006). Zhang and Zhang (2008) reported that tannins are the basic components responsible for pomegranate peel browning; they also revealed that the activities of ascorbic acid oxidase, PPO and PDA were positively

Fig. 11.5. Sunscald symptom on pomegranate fruit. (Photo: Alimohammad Yavari.)

correlated with peel browning, whereas catalase activity was negatively correlated. Protection of fruits from direct sunlight either by bagging or by covering is the ideal way to control sunscald. Heavy pruning also causes severe sunscald; thus, it is essential to maintain a good canopy to avoid sunscald. Sunscald seems to be similar to sunburn except for the severity of the exposure. Therefore, most of the methods used for control of sunburn also reduce sunscald.

11.5 Salinity Tolerance

Screening of 22 genotypes of pomegranate for salinity tolerance resulted in significant reduction in length (25%) and dry weight (32%) of new shoots in all the cultivars with large variations among cultivars (Wu *et al.*, 2015). Salinity stress was found to increase the antioxidant activity in pomegranate (Tavousi *et al.*, 2014).

Among pomegranate genotypes cv. 'Manfalouty' is considered as highly salt tolerant compared with cultivars 'Nab-Elgamal' and 'Wonderful in descending order, with respect to fruit set and fruit drop (El-Khawaga *et al.*, 2013). On the other hand, Bhantana and Lazarovitch (2010) noted that there were no differences between two cultivars of pomegranate, 'Wonderful and 'SP-2' in response to varying saline water stress. In another study with 10 Iranian pomegranate genotypes, cv. 'Voshike-e-Saravan was found to be the most salinity tolerant, followed by 'Malas-e-Yazdi' and 'Tab-o-Larz', and other cultivars were susceptible to salinity, with cvs 'Gabri', 'Malas-e-Esfahani' and 'Khafri-e-Jahrom' as the most sensitive among them (Okhovatian-Ardakani *et al.*, 2010).

Differential response of cuttings to salinity was also reported in pomegranate. Quick responses of cuttings to salinity in terms of changes in membrane permeability, MDA, proline content and soluble protein were noticed at high salinity (0.5%). The activities of SOD, catalase and POD were also found to increase initially up to 0.4% salt concentration. Pomegranate cv. 'Tunisi' was a moderately salt-tolerant cultivar (0.4% NaCl) and a growth-promoting effect was found below 0.1% salinity (Liu *et al.*, 2018).

Foliar treatments of spermidine and putrescine polyamines were applied to investigate the responses of the commercial genotype of pomegranate, 'P. Rabbab', to salinity (Amri *et al.*, 2011). Application of polyamines at 1 and 2 mM concentration as foliar spray after 72 h of salinity treatments helped in improving the growth rates of rooted cuttings up to 70 mM NaCl concentration. Results indicated polyamines could be used to reduce the effect of salinity in pomegranate.

11.6 Conclusions

There is a need for more complete knowledge of physiological disorders of pomegranate crop to help in developing suitable technologies to overcome them. This is also important for identifying the right traits for crop improvement programmes. In pomegranates, there exists diversity in resistance to aril browning and fruit cracking. There is a need to use these genotypes for transferring the resistance genes to a good horticultural background. In addition to the identification of traits for resistance, development of management practices also needs more attention. Better water and nutrient management along with canopy management are some of the methods to be adopted to overcome these disorders. Identification of rootstocks for increasing water and nutrient use efficiencies also needs more emphasis in pomegranates.

References

Abou El-Wafa, M. (2015) Effect of some treatments on reducing sunburn in wonderful pomegranate fruit trees. *Egyptian Journal of Horticulture* 42, 795–806.

Abubakar, A.R., Ashraf, N. and Ashraf, M. (2013) Effect of plant biostimulants on fruit cracking and quality attributes of pomegranate cv. *Kandharikabuli. Scientific Research and Essays* 8(44), 2171–2175.

Ahmed, F.F., Mohamed, M.M., Abou El-Khashab, A.M.A. and Aeed, S.H.A. (2014) Controlling fruit splitting and improving productivity of Manfalouty pomegranate trees by using salicylic acid and some nutrients. *World Rural Observations* 6(1), 87–93.

Aksoy, U. and Akyuz, D. (1993) Changes in K, Ca and Mg contents in different parts of the fig fruit during development. In: Fragoso, M.A.C., Van Beusichem, M.L. and Houwers, A. (eds) *Optimization of Plant Nutrition*. Kluwer Academic Publishers, Dordrecht, The Netherlands, pp. 305–309.

Amri, E., Mirzaei, M., Moradi, M. and Zare, K. (2011) The effects of spermidine and putrescine polyamines on growth of pomegranate (*Punica granatum* L. cv. Rabbab) in salinity circumstance. *International Journal of Plant Physiology and Biochemistry* 3(3), 43–49.

Bambal, S.B., Wavhal, K.N. and Nasalkar, S.D. (1991) Effect of foliar application of micro-nutrients on fruit quality and yield of pomegranate (*Punica granatum* L. cv. Ganesh). *Maharashtra Journal of Horticulture* 5(2), 32–36.

Barber, H.N. and Sharpe, P.J.H. (1971) Genetics and physiology of sunscald of fruits. *Agricultural Meteorology* 8, 175–191.

Bhantana, P. and Lazarovitch, N. (2010) Evapotranspiration, crop coefficient and growth of two young pomegranate (*Punica granatum* L.) varieties under salt stress. *Agricultural Water Management* 97(5), 715–722.

Darsana, K., Bhosale Yuvraj, K. and Sinija, V.R. (2016) Effect of aril browning on physicho-chemical properties of pomegranate. *International Journal of Science, Environment and Technology* 5, 1116–1126.

Davarpanah, S., Tehranifar, A., Abadía, J., Val, J., Davarynejad, G. *et al*. (2018) Foliar calcium fertilization reduces fruit cracking in pomegranate (*Punica granatum* cv. Ardestani). *Scientia Horticulturae* 230, 86–91.

Defilippi, B.G., Whitaker, B.D., Hess-Pierce, B.M. and Kader, A.A. (2006) Development and control of scald on wonderful pomegranates during long-term storage. *Postharvest Biology and Technology* 41(3), 234–243.

El-Kassas, S.E., Amen, K.I.A., Hussein, A.A. and Osman, S.M. (1992) Effect of certain methods of weed control and nitrogen fertilization on the yield, fruit quality and some nutrient contents of Manfalouty [local variety] pomegranate trees. 1. Flowering and fruit setting. *Assiut Journal of Agricultural Research* 23(3), 199–218.

El-Khawaga, A.S. (2007) Reduction in fruit cracking in Manfalouty pomegranate following a foliar application with paclobutrazol and zinc sulphate. *Journal of Applied Sciences Research* 3(9), 837–840.

El-Khawaga, A.S., Zaeneldeen, E.M.A. and Yossef, M.A. (2013) Response of three pomegranate cultivars (*Punica granatum* L.) to salinity stress. *Middle East Journal of Agriculture Research* 1(1), 64–75.

El-Rhman, A.I. (2010) Physiological studies on cracking phenomena of pomegranates. *Journal of Applied Sciences Research* 6(6), 696–703.

Ferrara, G., Cavoski, I., Pacifico, A., Tedone, L. and Mondelli, D. (2011) Morpho-pomological and chemical characterization of pomegranate (*Punica granatum* L.) genotypes in Apulia region, south eastern Italy. *Scientia Horticulturae* 130(3), 599–606.

Ferrara, G., Giancaspro, A., Mazzeo, A., Giove, S.L., Matarrese, A.M.S. *et al*. (2014) Characterization of pomegranate (*Punica granatum* L.) genotypes collected in Puglia region, south eastern Italy. *Scientia Horticulturae* 178, 70–78.

Franck, C., Lammertyn, J., Ho, Q.T., Verboven, P., Verlinden, B. *et al*. (2007) Browning disorders in pear fruit. *Postharvest Biology and Technology* 43(1), 1–13.

Gil, M.I., Holcroft, D.M. and Kader, A.A. (1997) Changes in strawberry anthocyanins and other polyphenols in response to carbon dioxide treatments. *Journal of Agricultural and Food Chemistry* 45(5), 1662–1667.

Hepaksoy, S., Aksoy, U., Can, H.Z. and Ui, M.A. (2000) Determination of relationship between fruit cracking and some physiological responses, leaf characteristics and nutritional status of some pomegranate varieties. *CIHEAM Options Méditerranéennes* Série A(Séminaires Méditerranéens No.42), 87–92.

Holderbaum, D.F., Kon, T., Kudo, T. and Guerra, M.P. (2010) Enzymatic browning, polyphenol oxidase activity, and polyphenols in four apple cultivars: dynamics during fruit development. *HortScience* 45(8), 1150–1154.

IPGRI (2001) *Regional report CWANA 1999–2000*. International Plant Genetic Resources Institute, Rome, Italy, pp. 20–28.

Jalikop, S.H., Venugopalan, R. and Kumar, R. (2010) Association of fruit traits and aril browning in pomegranate (*Punica granatum* L). *Euphytica* 174(1), 137–141.

Josan, J.S., Sawanda, J.S. and Uppal, D.K. (1979) Studies on the floral biology of pomegranate. III. Mode of pollination, fruit development and fruit cracking. *Punjab Horticultural Journal* 19, 134–138.

Joshi, G.D. and Limaye, V.P. (1986) Effects of tree location and fruit weight on spongy tissue occurrence in Alphonso mango. *Journal of Maharashtra Agricultural University* 11, 104–109.

Kader, A.A. (2006) Postharvest biology and technology of pomegranates. In: Seeram, N.P., Schullman, R.N. and Heber, D. (eds) *Pomegranates: Ancient Roots to Modern Medicine*. CRC Press, Taylor and Francis Group, Boca Raton, Florida, pp. 211–220.

Karp, D. (2006) The pomegranate: for one and all. *Fruit Gardener* 38(5), 8–11.

Khadivi-Khub, A. (2015) Physiological and genetic factors influencing fruit cracking. *Acta Physiologiae Plantarum* 37(1), 1718. DOI: 10.1007/s11738-014-1718-2.

Khalil, H.A. and Aly, H.S. (2013) Cracking and fruit quality of pomegranate (*Punica granatum* L.) as affected by pre-harvest sprays of some growth regulators and mineral nutrients. *Journal of Horticultural Science Ornamental Plants* 5(2), 71–76.

Khodaei, M., Nahandi, F.Z., Motallebi-Azar, A. and Dadpour, M. (2015) Effect of salycilic acid and sodium nitro proside on the pomegranate aril browning disorder. *Biological Forum* 7, 1014.

Kumar, G.N.M. (1990) Pomegranate. In: Nagy, S., Shaw, P.E. and Wardowski, W.F. (eds) *Fruits of Tropical and Subtropical Origin*. Florida Science Source, Inc, Ocala, FL, pp. 328–347.

Kumar, R., Bakshi, P. and Srivastava, J.N. (2010) Fruit cracking: a challenging problem of the fruit industry. Krishi Sandesh. Available at: www.krishisandesh.com/fruit-cracking-a-challenging-problem-of-fruit-industry (accessed 1 October 2020).

Kumar, K., Pinder, R., Dabas, S.T.K., Yadav, B. and Rana, S. (2017) Effect of growth regulators and micronutrients on fruit cracking and fruit yield in pomegranate. *Indian Journal of Agricultural Research* 51(3), 272–276.

Ladaniya, M.S. (2008) Growth, maturity, grade standards and physico-mechanical characteristics of fruit. In: Ladaniya, M.S. (ed.) *Citrus Fruit: Biology, Technology and Evaluation*. Academic Press, London, pp. 191–213.

Lal, S., Ahmed, N. and Mir, J.I. (2011) Effect of different chemicals on fruit cracking in pomegranate under karewa condition of Kashmir Valley. *Indian Journal of Plant Physiology* 16(3&4), 326–330.

Lal, N. and Sahu, N. (2017) Management strategies of sun burn in fruit crops – a review. *International Journal of Current Microbiology and Applied Sciences* 6(6), 1126–1138.

Liu, X., Kou, L., Sing, S. and Hu, Q. (1998) Biochemical characteristics studies on peel browning of pomegranate after picking. *Journal of Northwest Forestry College* 13(4), 19–22.

Liu, C., Yan, M., Huang, X. and Yuan, Z. (2018) Effects of salt stress on growth and physiological characteristics of pomegranate (*Punica granatum* L.) cuttings. *Pakistan Journal of Botany* 50(2), 457–464.

Mars, M. (2000) Pomegranate plant material: genetic resources and breeding, a review. *Options Mediterraneennes Serie A* 42, 55–62.

Meighani, H., Ghasemnezhad, M. and Bakshi, D. (2014) Evaluation of biochemical composition and enzyme activities in browned arils of pomegranate fruits. *International Journal of Horticultural Science and Technology* 1(1), 53–65.

Melgarejo, P. and Artes, F. (2000) Organic acids and sugar composition of pomegranate juice. *European Food Research and Technology* 4, 30–31.

Melgarejo, M.P. and Martínez, V.R. (1992) *El Granado*. 163. Ediciones Mundi-Prensa, Madrid.

Meshram, D.T., Gorantiwar, S.D., Teixeira da Silva, J.A., Jadhav, V.T. and Chandra, R. (2010) Water management in poimegranate (*Punica granatum* L.). *Fruit, Vegetable and Cereal Science and Biotechnology* 4(2), 106–112.

Mohamed, A.K.A. (2004) Effect of gibberellic acid (GA_3) and benzyl adenine (BA) on splitting and quality of Manfalouty pomegranate fruits. *Asian Journal of Agricultural Sciences* 35(3), 11–21.

Nath, V., Kumar, D. and Pandey, V. (2008) *Fruits for the Future*. Vol. 1. SSPH, Delhi, pp. 285–299.

Okhovatian-Ardakani, A.R., Mehrabanian, M., Dehghani, F. and Akbarzadeh, A. (2010) Salt tolerance evaluation and relative comparison in cuttings of different pomegranate cultivar. *Plant, Soil and Environment* 56(4), 176–185.

Opara, L.U., Studman, C.J. and Banks, N.H. (1997) Fruit skin splitting and cracking. *Horticultural Reviews* 19, 217–262.

Parashar, A. and Ansari, A. (2012) A therapy to protect pomegranate (*Punica granatum* L.) from sunburn. *International Journal of Comprehensive Pharmacy* 3, 1–3.

Patil, S.N. (2018) Effect of foliar application of gibberellic acid and nutrients on physiology and quality of pomegranate (*Punica granatum* L.) cv. Bhagwa under northern dry zone of Karnataka. *IJCS* 6(5), 3403–3407.

Pawar, S.K., Desai, U.T. and Choudhari, S.M. (1994) Effect of pruning and thinning on growth, yield and quality of pomegranate. *Annals of Arid Zone* 33, 45–45.

Peet, M.M. (1992) Fruit cracking in tomato. *HortTechnology* 2(2), 216–223.

Plamenac, M. (1972) A contribution to studies on the fruiting of pomegranate varieties in the bar district. *Jugoslovensko Vocarstvo* 5(17–18), 233–240.

Poovaiah, B.W. (1986) Role of calcium in prolonging storage life of fruits and vegetables. *Food Technology* 40(5), 86–89.

Prasad, R.N. and Mali, P.C. (2000) Changes in physico-chemical characteristics of pomegranate squash during storage. *Indian Journal of Horticulture* 57(1), 18–20.

Saad, F.A., Shaheen, M.A. and Tawfik, H.A. (1988) Anatomical study of cracking in pomegranate fruit [Saudi Arabia]. *Alexandria Journal of Agricultural Research* 32(2), 313–323.

Sahu, P., Sharma, N. and Sharma, D.P. (2013) Effect of in situ moisture conservation, forchlorfenuron and boron on growth, fruit cracking and yield of pomegranate cv. Kandhari under rain fed conditions of Himachal Pradesh. *Indian Journal of Horticulture* 7, 501–505.

Salih, R.F. (2017) Disease and environmental factor of cracking pomegranate fruit (*Punica granatum* L). *Eurasian Journal of Science and Engineering* 3(2), 37–46.

Schrader, L.E., Zhang, J. and Duplaga, W.K. (2001) Two types of sunburn in apple caused by high fruit surface (peel) temperature. *Plant Health Progress* 2(1), 3. DOI: 10.1094/PHP-2001-1004-01-RS.

Seeram, N.P., Henning, S.M., Zhang, Y., Suchard, M., Li, Z. *et al.* (2006) Pomegranate juice ellagitannin metabolites are present in human plasma and some persist in urine for up to 48 hours. *The Journal of Nutrition* 136(10), 2481–2485.

Sepahi, A. (1986) GA_3 concentration for controlling fruit cracking in pomegranates. *Iran Agricultural Research* 5(2), 93–99.

Sharma, R.R. (2005) *Physiological Disorders of Tropical and Subtropical Fruits – Causes and Control, Problems and Solutions*. Indian Agricultural Research Institute, New Delhi, pp. 310–311.

Shete, A.M., Wadhawa, G., Banat, I.M. and Chopade, B.A. (2006) Mapping of patents on bioemulsifier and biosurfactant: a review. *Journal of Scientific and Industrial Research* 65, 91–115.

Sheikh, M.K. and Manjula, N. (2012) Effect of chemicals on control of fruit cracking in pomegranate (*Punica granatum* L.) var. Ganesh. In: Melgarejo, P. and Valero, D. (eds) *II International Symposium on the Pomegranate*. CIHEAM / Universidad Miguel Hernández, Zaragoza, pp. 133–135.

Shete, M.B. and Waskar, D.P. (2005) Internal breakdown of pomegranate (*Punica granatum* L) fruits – a review. *Journal of Maharashtra Agricultural Universities* 30(1), 59–61.

Shivashankar, S., Hemlata, S. and Sumathi, M. (2012) Aril browning in pomegranate (*Punica granatum* L.) is caused by the seed. *Current Science* 103(1), 26–28.

Shivashankara, K.S. (2004) Physiological and nutritional basis of internal breakdown in pomegranate. Final report of the project funded by Indian Council of Agricultural Research. pp. 17.

Shivashankara, K.S., Chander, M.S., Laxman, R.H., Vijayalaxmi, G.P. and Bujjibabu, C.S. (2004) Physiological and biochemical changes associated with aril browning of pomegranate (*Punica granatum* cv. Ganesh). *Journal of Plant Biology* 31, 149–152.

Shulman, Y., Fainberstein, L. and Lavee, S. (1984) Pomegranate fruit development and maturation. *Journal of Horticultural Science* 59(2), 265–274.

Singh, D.B., Sharma, B.D. and Bhargava, R. (2003) Effect of boron and GA_3 to control fruit cracking in pomegranate (*Punica granatum*). *Current Agriculture* 27(1/2), 125–127.

Singh, D.B., Kingsly, A.R.P. and Jain, R.K. (2006) Controlling fruit cracking in pomegranate. *Indian Horticulture* 51(1), 14–22.

Singh, H., Singh, N., Marathe, A. and Ugalat, J. (2013) Influence of aril browning on biochemical properties of pomegranate (*Punica granatum* L). *Journal of Crop and Weed* 9(1), 184–187.

Tavousi, M., Kaveh, F., Alizadeh, A., Babazadeh, H. and Tehranifar, A. (2014) Integrated impact of salinity and drought stress on quantity and quality of pomegranate (*Punica granatum* L). *Bulletin of Environment, Pharmacology and Life Sciences* 4(1), 146–151.

Thorpe, M.R. (1974) Radiant heating of apples. *Journal of Applied Ecology*, 755–760.

Tomás-Barberán, F.A. and Espín, J.C. (2001) Phenolic compounds and related enzymes as determinants of quality in fruits and vegetables. *Journal of the Science of Food and Agriculture* 81(9), 853–876.

Vatandoost, S., Davarynejad, G.H. and Tehranifar, A. (2014) Would kaolin particle film avoid sunburn in 'Ardestani' pomegranate. *Advances in Environmental Biology* 8(12), 607–610.

Venkatachalam, K. and Meenune, M. (2011) Changes in physiochemical quality and browning related enzyme activity of longkong fruit during four different weeks of on-tree maturation. *Food Chemistry* 131(4), 1437–1442.

Weerakkody, P., Jobling, J., Infante, M.M.V. and Rogers, G. (2010) The effect of maturity, sunburn and the application of sunscreens on the internal and external qualities of pomegranate fruit grown in Australia. *Scientia Horticulturae* 124(1), 57–61.

Wu, S., Sun, Y., Niu, G., Altland, J. and Cabrera, R.I. (2015) Salt tolerance of 22 pomegranate cultivars. *Annual Conference of ASHS*, New Orleans, LA.

Yazici, K. and Kaynak, L. (2009) Effects of kaolin and shading treatments on sunburn on fruit of Hicaznar cultivar of pomegranate (*Punica granatum* L. cv. Hicaznar). *Acta Horticulturae* 818, 167–174. DOI: 10.17660/ActaHortic.2009.818.24.

Yilmaz, C. and Ozguven, A.I. (2006) Hormone physiology of preharvest fruit cracking in pomegranate (*Punica granatum* L.). *Acta Horticulturae* 727, 545–550. DOI: 10.17660/ActaHortic.2006.727.67.

Zhang, Y., Ni, J., Zhou, G., Yuan, J., Ren, W. *et al*. (2005) Cloning, expression and characterization of the human NOB1 gene. *Molecular Biology Reports* 32(3), 185–189.

Yingsanga, P., Srilaong, V., Kanlayanarat, S., Noichinda, S. and McGlasson, W.B. (2008) Relationship between browning and related enzymes (PAL, PPO and POD) in rambutan fruit (*Nephelium lappaceum* Linn.) cvs. Rongrien and See-Chompoo. *Postharvest Biology and Technology* 50(2–3), 164–168.

Zhang, Y.L. and Zhang, R.G. (2008) Study on the mechanism of browning of pomegranate (*Punica granatum* L. cv. Ganesh) peel in different storage conditions. *Agricultural Sciences in China* 7(1), 65–73.

12 Diseases and Management

J. Sharma[1]*, G. Manjunath[2], K.V. Xavier[3] and G.E. Vallad[3]
[1]ICAR-National Research Centre on Pomegranate, Solapur, India; [2]University of Horticultural Sciences, Bagalkot, India; [3]University of Florida, Gulf Coast Research and Education Center, Wimauma, Florida, USA

12.1 Introduction

Global awareness on the health benefits of pomegranate has increased market demand for this superfruit. Moreover, better returns from its cultivation under changing climatic conditions have led to a tremendous increase in acreage during the past decade, especially in arid and semi-arid regions of India, China, Iran, Turkey and South America. This ancient crop that once had no major biotic problems except fruit borer, today faces a multitude of biotic stresses. Among them, bacterial and fungal diseases are prominent in pomegranate production regions, leading to considerable fruit losses throughout the growing season. More than 55 pathogens have been reported to cause around 37 different diseases on pomegranate. The major ones are discussed in this chapter.

12.2 Bacterial Blight

Bacterial blight of pomegranate caused by *Xanthomonas axonopodis* pv. *punicae* was first reported in India in 1952 (Hingorani and Mehta, 1952), later it was reported in other countries (Table 12.1). In India, losses ranging between 10 and 100% have been reported under favourable conditions. Bacterial pathogens like *Pseudomonas* sp. associated with pomegranate blight and *Pseudomonas savastanoi* pv. *savastanoi* associated with knots on trunks have also been reported in Turkey (Table 12.1). However, *X. axonopodis* pv. *punicae* is the most prevalent and devastating bacterial pathogen of pomegranate.

Disease symptoms caused by *X. axonopodis* pv. *punicae* affect aboveground parts of pomegranate such as leaves, twigs and fruits (Sharma *et al.*, 2017b). Flower buds and flowers are not affected. On leaves one to several small, regular to irregular, greyish-black, water-soaked lesions are first observed on the abaxial leaf surface. The spots appear translucent yellow when observed against a source of light. These lesions increase in size and turn necrotic dark brown to black and are visible on both leaf surfaces. These necrotic lesions are surrounded by a water-soaked margin that appears as a yellow halo against the light. Infection through hydathodes results in large lesions from tip or margins. Infection can result in premature defoliation (Fig. 12.1).

Stem infections are mostly observed at nodes and sometimes on injured stems. Water-soaked lesions initiate on the nodes of twigs and branches, which extend above and below the infection sites to form brown to black sticky lesions of various sizes. Later these lesions become sunken (Fig. 12.2). Nodal blight/stem lesions

*Corresponding author: jyotisharma128@yahoo.com

Table 12.1. Bacterial diseases of pomegranate reported in different countries.

Disease	Causal organism	Country	Reference
Bacterial blight	*Xanthomonas axonopodis* pv. *punicae*	India	Hingorani and Mehta, 1952; Chand and Kishun, 1991; Sharma *et al.*, 2012
		Pakistan	Akhtar and Bhatti, 1992
		South Africa	Petersen *et al.*, 2010
		Turkey	Icoz *et al.*, 2014
	Pseudomonas sp.	India	Jagdale *et al.*, 2018
Bacterial knot disease	*Pseudomonas savastanoi* pv. *savastanoi*	Turkey	Bozkurt *et al.* (2014)

due to secondary fungal infections develop into large cankers. At advanced stages, nodal blight girdles stems and twigs, which leads to drying and breaking of branches (Sharma *et al.*, 2017a).

Fruits are most vulnerable to blight infection, especially after they reach the green lemon stage. Fruits exhibit one to many, isolated or coalesced water-soaked lesions that gradually become necrotic. Cracks of various shapes and sizes develop on lesions. Bacterial ooze may be observed on the lesions during humid conditions, later it dries to form a white shiny encrustation on the fruit lesion (Fig. 12.3). Bacterial blight symptoms are sometimes confused with those caused by fungal pathogens, such as *Cercospora*. However, several characteristics, such as the presence of water-soaked foliar lesions with yellow halos and stickiness, fruit lesions with cracking, and additional testing for bacterial streaming in the laboratory, can be used to confirm the bacterial nature of the disease.

Fig. 12.1. Bacterial blight symptoms on leaves: (a) initial water-soaked lesions on undersurface of leaves; (b) water-soaked lesions as seen against light; (c) advanced stage necrotic spots on leaf lamina; (d) necrotic spots with yellow hallo when seen against light; (e) midrib infection with chlorotic lamina; and (f) infection from leaf tip in winter season. (Photos: Jyotsana Sharma.)

Fig. 12.2. Bacterial blight symptoms on stems: (a) initial water-soaked bark around stem node; (b) advanced stage necrotic lesion on stem node; and (c) formation of canker on the lesion after secondary fungal infections. (Photos: Jyotsana Sharma.)

Bacterial blight is caused by *X. axonopodis* pv. *punicae* (Hingorani and Singh) Vauterin, Hoste, Kersters and Swings (Hingorani and Mehta, 1952; Vauterin *et al.*, 1995) with earlier nomenclature being *Xanthomonas campestris* pv. *punicae* (Hingorani and Mehta, 1952). The bacterium *Xanthomonas* belongs to the kingdom: Bacteria; phylum: Proteobacteria; class: Gammaproteobacteria; order: Xanthomonadales; family: Xanthomonadaceae. *Xanthomonas axonopodis* pv. *punicae* is a motile, gram-negative, rod-shaped bacterium measuring 0.4–0.75 × 1.0–3.0 μm, with a single polar flagellum. The optimum growth conditions for the bacterium include a temperature of 28°C and pH 7.0. The colonies on nutrient agar medium appear after 48–96 h of inoculation and are

Fig. 12.3. Bacterial blight symptoms on fruits: (a) water-soaked lesions with minute cracks; (b) necrotic black lesions with cracks; (c) fruit splitting due to bacterial blight with a shiny white encrustation of bacterial ooze on blight lesions; and (d) blight-infected fruit with secondary infections. (Photos: Jyotsana Sharma.)

smooth, circular, raised, light yellow, glistening and mucoid, with entire margins. Brown-fuscan pigmentation is observed in the growth medium after 5 days. Mondal *et al.* (2012) developed a DNA-based polymerase chain reaction (PCR) assay based on the *gyr*B housekeeping gene to differentiate *X. axonopodis* pv. *punicae* from other xanthomonads. A recent phylogenetic study based on multi-locus sequence typing (MLST) of 25 *X. axonopodis* pv. *punicae* strains collected throughout India from 2008 through 2016 (including a strain collected in the 1950s), concluded that bacterial blight of pomegranate in India is caused by a single clonal lineage (Kumar *et al.*, 2020).

Xanthomonas axonopodis pv. *punicae* can survive in dormant buds, stem cankers, infected fruits and plant debris for several months, but cannot survive for more than 30 days in soil without its host (Rani and Verma, 2002). Long-distance dissemination is primarily through propagated planting materials, with nodes and dormant buds serving as latent carriers of the pathogen (Sharma, 2017). Vegetative propagation of infected planting materials or inoculum from neighbouring infected orchards serve as the primary sources of inoculum. The pathogen enters the host in water droplets through natural openings like stomata, lenticels, hydathodes or wounds made by various agents (Sharma *et al.*, 2017b). Initial blight symptoms often appear at the nodes above soil level, 5–7 months after planting. The secondary spread of the pathogen takes place through splash dispersal from rain and spray equipment, insect activity, contaminated pruning tools, animals and handling by field workers. Disease progression is not systemic in nature. A complete disease cycle is given in Fig. 12.4.

Bacterial blight develops when the air temperature is between 20 and 35°C and relative humidity (RH) above 30% for 24 h. Air temperatures of 25–30°C and RH above 50% for 16 h coupled with intermittent rains and wind are optimum for disease development and spread. Under optimal conditions, water-soaked lesions appear between 4–7 days and the disease cycle is completed within 21 days post-infection. Hot

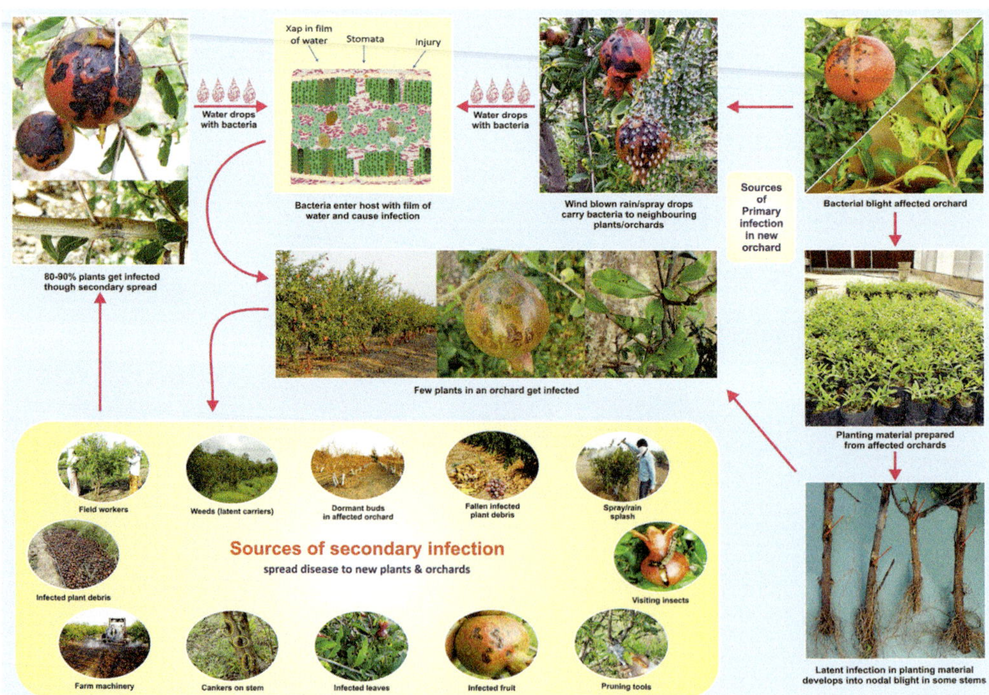

Fig. 12.4. Disease cycle of pomegranate bacterial blight. (From: Jyotsana Sharma.)

humid conditions are favourable for blight development. Nevertheless, the pathogen survives and spreads throughout the year; though, incubation periods can extend up to 10–19 days and disease cycle can be completed in 37–39 days during periods of very high/low temperatures and/or low humidity in semi-arid regions (Sharma *et al.*, 2017b). Alternate hosts like *Azadirachta indica*, *Aegle marmelos*, and weed genera *Tridax* and *Achyranthes* have been reported (Kumar *et al.*, 2006; Yenjerappa, 2009), although contradictory results were reported in subsequent cross-inoculation tests on *A. indica* and other weed hosts (NRCP, 2012; Sharma *et al.*, 2012).

12.2.1 Management of bacterial blight

Bacterial diseases are best managed with either disease-resistant or tolerant cultivars. ICAR-National Research Centre on Pomegranate (ICAR-NRCP), Solapur, India, has a collection of 345 germplasm lines including indigenous lines from India, as well as exotic lines from Afghanistan, France, Iran, Italy, Japan, Oman, Russia, Sri Lanka, Turkey, Turkmenistan and the USA. However, none of the lines was resistant to blight. Some germplasm lines were found to have low susceptibility to bacterial blight, but they were associated with negative horticultural traits like small fruit size, hard seeds and a sour taste. In a study on genetic diversity and association mapping of bacterial blight and other horticulturally important traits with microsatellite markers in pomegranate from India, Singh *et al.* (2015) identified marker PGCT001 associated with both fruit weight and bacterial blight. Hence, all commercial varieties where large fruit size is preferred are susceptible to this disease, as the locus for fruit size is tightly linked to bacterial blight susceptibility.

In the absence of bacterial blight-resistant varieties, integrated disease management is the best option, which includes the use of blight-free planting material, maintaining soil health for best nutrition, orchard sanitation, cultural practices and preventative pesticide applications.

Planting materials

As already discussed above, planting material (air layers and hardwood cuttings) are the primary sources of infection in a new orchard. Hence, propagated plant materials should always be sourced from blight-free orchards. ICAR-NRCP developed a protocol for bio-hardened tissue culture plants, which ensures propagated plant tissues are free from latent infections of the bacterial blight pathogen and are a viable option over other means of vegetative propagation (Singh *et al.*, 2016).

Orchard sanitation

Sanitation plays a significant role in blight management. Infected plant debris should be removed from the orchard and destroyed. In blight-infected orchards, soon after harvest pruning should be practised to eliminate all stems with fresh bacterial blight infections. Infected parts must be pruned about 5–7 cm below the infected area. Pruning tools should be sterilized after handling each infected tree with a 2.5–10% sodium hypochlorite solution or 70% ethanol. Bordeaux mixture (1%) should be sprayed on the plants immediately after pruning.

Cultural practices

Planted wind breaks around the orchard can act as a barrier to help limit pathogen spread by wind-driven rains. In pomegranate-growing regions like India, where options are available for regulating crops during any of the three seasons – rainy, winter or autumn – avoiding rainy season crops and regulating winter season crops is the best option to reduce losses due to this disease.

Nutrient management

Nutrient imbalance indices are closely related to bacterial blight severity in pomegranate. As per Diagnosis and Recommendation Integrated System (DRIS) analysis, the leaf nutrient range detailed in Table 12.2 reduced the severity of bacterial blight on pomegranate (Maity *et al.*, 2016).

Organic manures like well-decomposed farmyard manure, poultry manure, vermicompost,

Table 12.2. Pomegranate leaf nutrient status imparting moderate resistance to bacterial blight (Maity *et al.*, 2016).

Macronutrient		Micronutrient	
Nutrient	Status (%)	Nutrient	Status (mg/kg)
N	1.56–2.05	Fe	132.5–187.0
P	0.11–0.28	Mn	31.60–58.40
K	0.83–1.20	Zn	13.20–27.40
Ca	1.60–2.16	Cu	26.00–47.80
Mg	0.38–0.82		
S	0.09–0.16		

green manures and oil cake of *A. indica* along with balanced inorganic nutrition at proper growth stages can help manage bacterial blight and improve fruit yield and quality (Benagi and Kumar, 2009; Maity *et al.*, 2016). It has also been observed that postharvest nutrient applications helps plants recover following the production cycle and resist blight infection during the following season (A. Maity, Solapur, India, 2012, personal communication).

Biological control

Soil application of bioformulations containing *Aspergillus niger* AN27 and mycorrhizal preparation with *Rhizophagus irregularis* (Syn. *Glomus intraradices*) have been found to be beneficial for the management of bacterial blight (Sharma, 2017). In addition, bioagents, namely *Bacillus subtilis*, *Pseudomonas fluorescens* and *Pseudomonas putida*, and plant-derived oils from *Oscimum sanctum*, ginger and clove reduced disease under field conditions (Puneeth, 2015; Katwal, 2015; Mali, 2015).

Defence inducers

An array of plant defence inducers such as salicylic acid, jasmonic acid, laminarin, II-amino butyric acid, III-amino butyric acid (GABA) and eugenol were identified to activate systemic resistance in pomegranate against bacterial blight. Among them, salicylic acid and GABA at 50 ppm were effective in reducing the disease as preventative sprays (Maity *et al.*, 2018; Praveen, 2018). Three foliar applications of salicylic acid at 300 ppm starting pre-flowering and repeated at monthly intervals were effective for bacterial blight management (Maity *et al.*, 2018).

Chemical control

Options for using chemicals in the management of blight are very limited and no single chemical gives complete disease control. Copper compounds (copper oxy-chloride, copper hydroxide, copper sulfate, Bordeaux mixture), 2-bromo, 2-nitro propane-1,3-diol (Bronopol) and an antibiotic combination of streptomycin sulfate (90%) + oxytetracycline (10%) are the only chemicals that give some degree of bacterial blight control (Jadhav and Sharma, 2011; Sharma, 2017). These combinations achieved better efficacy by reducing inoculum load and arresting further disease spread under favourable weather conditions. A spray schedule consisting of streptomycin sulfate + oxytetracycline alone or in combination with copper oxychloride, copper hydroxide or carbendazim at 15-day intervals resulted in 82.2% blight control and increased marketable fruit yield (Sharma *et al.*, 2008, Sharma *et al.*, 2010a; Sharma, 2017).

Integrated disease management

Integrating previously discussed approaches is the ideal option for successful management of bacterial blight. Published studies by Sharma and Jadhav (2011) and Sharma and Jadhav (2012) demonstrated the success of integrated disease management for mitigating pomegranate bacterial blight in a network mode in three major pomegranate growing states of India. The management schedule was demonstrated during the period 2008–2012 in farmers' orchards suffering losses of between 40 and 100% due to bacterial blight disease in the states of Maharashtra, Karnataka and Andhra Pradesh in India. The integrated management schedule in participating orchards included avoiding production during seasonal rains; following proper orchard sanitation; maintaining soil and plant health through recommended applications of nutrients, organic manures and beneficial microorganisms during the rest period soon after harvest and again at flowering; making three or four foliar applications of salicylic acid, ZnSO4 and MnSO4 at 1-month intervals; and preventive bactericidal sprays at recommended doses at 10–14-day intervals along with required insecticides. In all, 57 successful demonstrations (1 ha each) were conducted in collaboration with local research organizations. Average reduction in bacterial blight incidence was 74% (max. 100%), average productivity 9.28 t/ha (max. 18.4 t/ha) and average benefit:cost ratio was 4.31:1 (max. 8.57:1) in demonstration plots. The management of bacterial blight had a positive impact on the pomegranate-growing scenario in India, with constant increases in area and productivity.

12.3 Wilt Diseases

Pomegranate wilt, also reported as decline or dieback by some authors, is prevalent in several pomegranate-growing regions across the globe. Crop losses ranging from 30 (Xu *et al.*, 2011) to 91.7% (Sharma *et al.*, 2012) have been reported. Delays in taking appropriate control measures to manage wilt diseases have compelled some farmers to uproot entire orchards. Wilt in pomegranate is reported to be associated with several organisms throughout the world (Table 12.3).

Table 12.3. Wilt pathogens/parasites reported on pomegranate from different countries.

Disease	Causal organism	Country	Reference
1. Fungal wilt	*Ceratocystis fimbriata*	India	Somasekhara, 1999; Sharma *et al.*, 2010b; Khosla *et al.*, 2011
		China	Huang *et al.*, 2003
		Iran	Banihashemi, 1998
		Greece	Tziros and Tzavella-Klonari, 2008
		Pakistan	Alam *et al.*, 2017
	Fusarium solani and *Fusarium oxysporum*	India	Kore and Mitkar, 1993; Chavan and Dake, 2001; Kumar *et al.*, 2001
	Rhizoctonia solani	India	NRCP, 2008, NRCP, 2016
	Verticillium dahliae	Greece	Tziros and Tzavella-Klonari, 2008
	Macrophomina phaseolina	India	Sharma *et al.*, 2012
2. White root rot	*Dematophora* (*Rosellinia*) *necatrix*	India	Sztejnberg and Madar, 1980
3. Parasitic nematodes	*Meloidogyne incognita*	India	Verma, 1985; Darekar *et al.*, 1989
		Jordan	Hashim, 1983
	Meloidogyne javanica	Egypt	Elazab and Elzawahry, 2016.
	Helicotylenchus spp., *Xiphinema insigne*, *Rotylenchulus reniformis*, *Helicotylenchus multicinctus*, *Pratylenchus coffeae*, *Xiphinima index* and *Aphelenchus sp.*	India	Darekar *et al.*, 1989; Ilangovan and Poornima, 2017; Sharma and Sharma, 2017
	Helicotylenchus pseudorobustus, *Tylenchorhynchus clams*, *Longidorus* sp.	Jordan	Hashim, 1983

The severity of wilt symptoms will vary depending on the causal pathogen. The fungus *Ceratocystis fimbriata* alone and/or root-knot nematode *Meloidogyne incognita* are reported as the two most important causes of wilt in India (NRCP, 2015), China, Iran, Greece and Pakistan (Table 12.3).

12.3.1 Wilt caused by *Ceratocystis fimbriata*

Wilt symptoms due to *C. fimbriata* initially appear in scattered isolated plants; later in continuous patches in affected orchards. The first symptoms are yellowing of leaves in one or two branches followed by drooping and drying of foliage. Eventually, the entire tree dries resulting in plant death. This may take a few weeks to months or sometimes even a year depending on factors like soil type, rainfall, plant age and plant stage. Fruit-bearing trees are more susceptible to vascular wilt caused by *C. fimbriata*, hence maximum mortality is often observed during fruiting periods.

The *C. fimbriata* fungus infects the vascular and adjoining cortical tissues of the root and stems. Affected roots initially display a dark yellow vascular discolouration that further develops into dark greyish-brown/black/purplish streaks in vascular and adjoining cortical tissues. Diseased roots often have an alcoholic, fruity smell when sectioned for examination. As the disease advances, discolouration progresses into vascular and cortical tissues of affected stems, sometimes several feet above ground level (Fig. 12.5).

Ceratocystis fimbriata belongs to kingdom: Fungi; phyllum: Ascomycota; class: Sordariomycetes; order: Microascales; family: Ceratocystidaceae.

Ceratocystis fimbriata is a slow-growing fungus; it initially produces greyish-white, raised mycelial growth on potato dextrose agar (PDA) with endoconidia production. Later the colony turns dark grey with the production of reproductive structures such as perithecia and aleurioconidia. The endoconidia are formed endogenously in hyphae and are unicellular, hyaline, smooth, cylindrical with flattened ends, biguttulate, measuring 10.2–42.1 × 2.4–4.6 μm and borne in chains of variable length. Perithecia are superficial or embedded in the substrate, have a globose or subglobose base measuring 137.1–287.0 × 130.4–263.4 μm and are black to dark brownish-black in colour. It has a characteristically long neck measuring 109.7–713.1 μm, ostiolar hyphae are divergent, light or hyaline in colour, smooth-walled. Ascospores are hat-shaped, measuring 3.0–5.6 × 2.6–3.6 μm and are liberated by early ascus deliquescence. The ascospores come out in cirrus or accumulate in a cream-coloured mass at the tip of the neck. Aleurioconidia are thick-walled, golden-brown, ellipsoidal, pyriform or obpyriform, truncate at the base, measuring 7.6–35.6 × 6.1–6.3 μm, borne singly or in short chains, laterally or terminally on hyphae with zero to five septa. Most isolates are self-fertile (Harrington and McNew, 1997).

Ceratocystis fimbriata is known to attack a wide variety of annual and perennial plants, some of which include species of *Prunus*, *Malus*, *Acacia*, *Manihot*, *Citrus*, *Coffea*, *Colocasia*, *Annona*, *Cassia*, *Eucalyptus*, *Ficus*, *Mangifera*, *Quercus*, *Populus*, *Hevea*, *Ipomoea*, *Cajanus* and *Crotalaria* (Halsted, 1890; Webster and Butler, 1967; Baker and Harrington, 2001; Harrington, 2004). However, *C. fimbriata* isolates from pomegranate are unable to infect plants of other hosts including *Acharas sapota*, *Carica papaya*, *Citrus aurantifolia*, *Curcuma longa*, *Ficus elastica*, *Mangifera indica*, *Piper beetle*, *Murraya koenigi*, *Psidium guajava*, *Saccahrum officinarum*, *Tectona grandis*, *Vitis vinifera*, and *Zizypus mauritiana* (Somasekhara and Gaddanakeri, 2009). There are several apparently host-specialized strains that are sometimes called 'types', 'races' or 'forms' (Harrington, 2000; Baker *et al.*, 2003). Three broad geographic clades – the North American, the Latin American and the Asian clades – have been proposed by Harrington (2000) as cryptic species within *C. fimbriata*.

Ceratocystis fimbriata survives as mycelium or reproductive structures within the infected host or in plant debris in the soil for several years. Grosclaude *et al.* (1995) could observe viable spores of *C. fimbriata* in infected wood and sawdust of *Platanus* sp. for up to 5 years. In soil, the pathogen survives as aleurioconidia (Accordi, 1989), does not require an alternate host to complete its life cycle and can complete several generations within the rhizosphere (Beaver, 1989).

Fig. 12.5. Wilt symptoms due to *Ceratocystis fimbriata:* (a) yellowing of leaves; (b) severe wilt-affected plot; (c) yellow dicolouration of inner wood of roots; (d) blue-black staining of inner wood of roots in advanced stage; and (e) discolouration of inner wood of aerial stem in advanced stage. (Photos: Jyotsana Sharma.)

Infected planting material (hardwood cuttings, unsterilized potting mixture) is the primary source of inoculum for *C. fimbriata* and other wilt pathogens. Secondary spread in the orchard is facilitated through farm operations and agricultural implements or through wind/water dispersal of the soil from infected plants in the orchard (Sharma *et al.*, 2012). The fungus is also disseminated by the scolytid beetle *Xyleborus fornicatus* (Hinds, 1972; Viegas, 1960), which are attracted by the fruity aroma produced by *C. fimbriata* (Crone and Bachelder, 1961; Iton, 1966; Hulcr *et al.*, 2011). This has been assumed to be an adaptation for dispersal by insects, which are attracted to diseased plants and can become contaminated with sticky spores. *Ceratocystis fimbriata* spores can be carried on the bodies of ambrosia beetles and can survive passage through an insect gut. Shothole borer-affected plants (*X. fornicates and X. perforans*) show pin head-sized holes on roots and aboveground stems (Somasekhara, 2002; Kulkarni and Gupta, 2007; Sharma *et al.*, 2010b). The borer most of the time is associated with trees weakened by wilt caused by *C. fimbriata* or other causes.

Root grafts in close or high-density orchard plantings may also transmit the disease to adjacent trees (Accordi, 1989). Pruning tools can serve as a source for the secondary spread of *C. fimbriata*, which can enter through fresh wounds (Teviotdale and Harper, 1991). Use of infected wood for various purposes can result in long-distance spread of *C. fimbriata* and other fungal pathogens.

Fig. 12.6. Root rot symptoms due to: (a) *Rhizoctonia solani*; and (b) *Macrophomina phaseolina*. (Photos: Jyotsana Sharma.)

12.3.2 Other agents causing wilt on pomegranate

Other important wilt pathogens include species of *Fusarium*, *Rhizoctonia*, *Sclerotium* and *Macrophomina* however, these are not typically associated with major losses. Wilt due to *Fusarium* spp., namely *F. solani* and *F. oxysporum*, results in discolouration of the central pith and xylem, collar rot and drying of roots. *Rhizoctonia solani* results in sunken collar rot symptoms on trunks near the soil surface. Symptoms can lead to girdling that can cause a sudden toppling down or wilting of green plants. Sometimes a white coating of *Thanatephorus cucumeris*, the perfect stage of *Rhizoctonia solani*, is observed on affected roots when the soil is removed. In infection of *Sclerotium rolfsii* and *Macrophomina phaseolina*, white/black fungal growth may appear on the root surface (J. Sharma, Solapur, India, 2010 and 2018, personal communication). Moreover, feeder roots rot completely. Rotted roots may appear slimy to the touch or dry depending on the associated organisms (Fig. 12.6).

Pomegranate plants affected by root-knot nematode (*Meloidogyne incognita*) exhibit knots of various sizes on primary and/or secondary roots (Fig. 12.7). Foliar symptoms resemble a nutrient deficiency. Plants continue surviving for long periods, with an increased severity of foliar symptoms and a reduction in yield. Affected plants exhibit stunted plant growth when plants are infested at the initial growth stages. It has also been observed that fully grown trees affected by nematodes exhibit a drastic reduction in flowering or no flowering after a few years even though affected trees show lush foliar growth. The root system of such trees exhibit larger and more intense root-knot galling (J. Sharma, Solapur, India, 2008 and 2012, personal communication).

The nematode *M. incognita* belongs to kingdom: Animalia; phylum: Nematoda; class: Secernentea; order: Tylenchida; family: Heteroderidae. Some other nematode species have been reported associated with pomegranate in different countries, including *Helicotylenchus pseudorobustus*, *Tylenchorhynchus clarus*, *Longidorus* sp., *Meloidogyne javanica* and *Xiphinema insigne* (Hashim, 1983; Darekar *et al.*, 1989; Siddiqui and Khan, 1986).

Pomegranate production in arid and semi-arid regions of the globe provides favourable temperatures and humidity for the various fungal wilt pathogens to survive, multiply and infect the plant throughout the year. Soil type is another factor. Sandy and sandy loam soils provide ideal conditions for the survival of all types of wilt pathogens. Pomegranate production in sandy soils under elevated temperature conditions is also ideal for nematodes (Yadav *et al.*, 1970).

Insect injuries in the root or association of root-knot nematodes predispose the plant to fungal wilt infections (Walter *et al.*, 1952; Francl and Wheeler, 1993). Temperatures in the range of 18–30°C (optimum 25–26°C) with adequate moisture as well as frequent rains are favourable for the development of wilt caused by *C. fimbriata* (Huang *et al.*, 2003). Imbalanced plant nutrition, especially boron deficiency in soil (Hu *et al.*, 1999) may result in more wilt. Root injuries during operations for weed control, fertilizer application or intercropping can also predispose the plants to severe infections. Latent wilt infection during critical growth stages such as flowering and fruit-bearing can lead to plant death.

12.3.3 Management of wilt in pomegranate orchards

Successful wilt management requires an integrated approach right from planting, among which healthy, pathogen-free planting material and use of sufficient organic fertilizers as well as efficient biological agents in the rhizosphere for bio-hardening are important prophylactic measures. Further, cultural practices, orchard sanitation, use of therapeutic fungicidal treatments at the first observance of wilt, and improving host

Fig. 12.7. Foliar and root symptoms typical of wilt caused by root-knot nematode on pomegranate: (a) foliar symptoms in a young plant; (b) foliar symptoms in an old pomegranate orchard (>4 years of age); (c) an orchard with mature trees showing lush foliage but no flowering due to root-knot infestation; (d) knots on primary roots of plants with initial nematode infestation; and (e) knots on secondary roots of plants with severe nematode infestation. (Photos: Jyotsana Sharma.)

resistance through nutrients and defence inducers are other management practices that have helped manage wilt diseases.

Pre-plant soil preparation

Root-knot nematodes and other wilt pathogens are not host specific; hence, the site selected for planting pomegranate should be treated for weeds, nematodes and fungal wilt pathogen before planting. Solarization is recommended as an effective pre-plant soil treatment (Arora *et al.*, 2006; Maity *et al.*, 2012; Singh *et al.*, 2012; Satyagopal *et al.*, 2014; Kahramanoglu and Usanmaz, 2016), especially in arid/semi-arid production regions that have at least 2–3 months of day temperatures above 35°C. Otherwise, growers will need to consider other cultural or chemical practices to reduce soilborne pest and pathogen populations prior to planting.

Plantation should be done on raised beds. The beds should be made with 60 cm wide × 30 cm deep trenches below the soil surface, with aboveground beds of 200 cm width and 45 cm height in the middle sloping down towards both ends, separated by parallel dead furrows oriented in the direction of land slope. This leads to appropriate aeration in the active root zone, avoiding waterlogging and increased nutrient uptake. These conditions facilitate reduction in soilborne diseases (Marathe *et al.*, 2017).

Planting

Orchards should be established with certified disease-free planting material or tissue culture saplings. The wilt organisms, especially fungi and nematodes, are easily transmitted in planting material propagated through hardwood cuttings. Soil used for vegetative propagation of planting material (hardwood cuttings, air layers or tissue culture plants) should be pasteurized, solarized or fumigated to ensure it is free of wilt pathogens or nematodes. Bio-hardening of planting material with promising bioagents will be advantageous (Singh *et al.*, 2016). Planting should be done at the recommended spacing for the variety planted (depending on spread) to avoid root contact between neighbouring plants. One should ensure the application of well-decomposed organic manures and beneficial microorganisms to the soil at planting and repeat at least twice a year (Sharma *et al.*, 2014).

Orchard sanitation

Orchard sanitation is one of the most important measures to prevent and reduce the impact of many diseases (Walter, 1946; Clark and Mayer, 1988). Wilt affected plants should be treated with an appropriate fungicide, soon after the first symptoms (yellowing) are observed. Wilt-infected plants, should be carefully removed and soil at the affected site treated before resetting with a new plant. Precautions should be taken to avoid the spread of infested soil and plant material during plant removal. Pomegranate orchards should be kept weed-free, as many weeds can act as reservoirs for nematodes or other pathogens of pomegranate (Somasekhara *et al.*, 2009).

Intercropping

Crops like onion, tomato, chilli, potato, capsicum, gram, pea, cucurbit as well as other legumes known to be susceptible to nematodes, should be avoided as intercrops, since they can lead to elevated nematode levels in the soil. However, green manure crops like sunhemp (*Crotalaria juncea*) and species of *Sesbania* help to improve levels of nitrogen in the soil and the abundance of beneficial microflora, making them ideal for intercropping (Dubey *et al.*, 2015). Planting African marigold (*Tagetes erecta*) varieties for a period of 6–7 months is beneficial for reducing nematode population in infested orchards (Sharma and Sharma, 2017).

Biological control

Various studies have demonstrated the potential benefit of various soil-applied biological control agents in managing wilt pathogens. Some effective biological control agents found promising in pomegranate-growing regions of India include strains of *Bacillus subtilis*, *Paecilomyces lilacinus*, *Pseudomonas fluorescens*, *Trichoderma harzianum* and *Aspergillus niger* (Somasekhara, 2002; Raghuvanshi, 2007; Mhase, 2007; Sonyal, 2010; Sharma *et al.*, 2012). Controlled pot studies conducted at ICAR-NRCP demonstrated that the combined use of the vesicular arbuscular mycorrhizal (VAM) fungus, *Rhizophagus irregularis* (=*Glomus intaradices*), along with *A. niger* strain AN 27 in soil infested with *C. fimbriata*, delayed wilt symptoms for 62 weeks (1 year and 10 weeks), in comparison with a non-treated control where initial wilt symptoms developed 12 weeks after planting, with all plants wilting within a few weeks (Sharma and Sharma, 2017). In another study at a naturally infested field site, the use of *A. niger* and *R. irregularis* at crop regulation/flowering and soon after harvest along with recommended organic fertilizers have effectively checked the wilt due to *C. fimbriata* and root-knot nematode *M. incognita*, along with increased plant resistance to other foliar diseases and improved yield and quality of the produce (Sharma and Sharma, 2017). In addition, the use of bio-formulations of *A. niger* and *R. irregularis* was effective for promoting bio-hardening of vegetative propagated pomegranate plants through tissue culture, hardwood cuttings or air layering methods (Sharma and Sharma, 2017). Bompadre *et al.* (2018) recently reported the use of *R. irregularis* to reduce transplanting stress.

Well-decomposed farmyard manure including poultry manure, vermicompost and green manure should be used twice a year to support and establish beneficial microflora (Sharma *et al.*, 2014; Singh *et al.*, 2016). De-oiled seed cakes of *A. indica*, *Pongamia pinnata*, *Bassia latifolia* (=*Madhuca longifolia*) and *Ricinus communis* along with fertilizer doses were also found to be effective in the management of root-knot nematodes (Darekar *et al.*, 1989).

Chemical control

Several chemical options are also available for the management of wilt caused by fungal pathogens. Fungicides containing carbendazim, difenoconazole, propiconazole, copper oxychloride and boric acid have reported efficacy against fungal wilt pathogens when applied to soil (Kore and Mitkar, 1993; Kumar *et al.*, 2001; Somasekhara, 2006; Somasekhara *et al.*, 2009; Sharma, 2009; Sonyal, 2010).

In a field trial conducted in 2017–2018 at ICAR-NRCP at a site with a history of wilt due to *C. fimbriata*, repeated soil applications of propiconazole + chlorpyriphos; alternated soil applications of fosetyl Al and tebuconazole; and a rotational programme that included an application of propiconazole + chlorpyriphos followed by *A. niger* AN27 and then *R. irregularis* successfully controlled wilt caused by *C. fimbriata* (J. Sharma, Solapur, India, 2017, personal communication). Application of humic acid or organic manures help new root development and the establishment of beneficial microorganisms; this further improves survival and establishment of the treated plants. Thiabendazole and pyrazophos delayed development of *Dematophora* root rots. Stem pasting up to 30–60 cm from ground level with 10% Bordeaux paste or paste prepared with red soil + chlorpyriphos + copper oxychloride + water help control shot hole borer and collar rots/cankers caused by soilborne fungi. Stem pasting twice a year, once during the rest period and once before rains start, is beneficial in checking collar infections or borer pest damage (Kulkarni and Gupta, 2007).

Chemical management of nematodes is not very promising. Preventive applications of phorate, carbofuran (Mhase, 2007) or fenamiphos provided protection to roots for 60 days against nematode invasion and development of root gall (Siddiqui and Khan, 1986); however, these chemicals are not recommended for export crops, due to the lack of registration in Europe (UTZ, 2015). A new class of nematicide, fluensulfone, was reported to be effective against root-knot nematodes (Westerdahl *et al.*, 2014), and gave promising root-knot control in field trials at ICAR-NRCP, Solapur.

12.4 Phytophthora Blight

Phytophthora spp. have been reported to cause losses in pomegranate in India, Turkey, Greece and Iran (Table 12.4).

Three different species of *Phytophthora* have been reported to cause damping-off of seedlings, leaf blight, fruit rot, crown and root rot in pomegranate, particularly during the rainy season, when humidity is high and temperatures conducive to disease development. Affected seedlings exhibit damping-off symptoms, whereas affected leaves and young twigs exhibit a blighted appearance (J. Sharma, Solapur, India, 2010, personal communication). Disease initially appears as water-soaked lesions along the leaf

Table 12.4. Phytophthora diseases reported in pomegranate from different countries.

Disease	Causal organism	Country	Reference
1. Collar rot	*Phytophthora cactorum*	Iran	Alavi and Zackii, 1985
2. Seedling blight and damping-off of propagating material	*Phytophthora nicotianae* var. *nicotianae*	India	Sharma *et al*., 2010b
3. Fruit rot	*Phytophthora* sp., *Phytophthora nicotianae*	India	Neama and Sharma, 2006
4. Crown/root rot	*Phytophthora palmivora*	Turkey	Turkolmez *et al*., 2015
		Greece	Markakis *et al*., 2017
		India	J. Sharma, India, 2012, personal communication
5. Crown rot	*Coniella granati* (syn. *Pilidiella granati*)	Greece Turkey Italy	Thomidis and Exadaktylou, 2011; Çeliker *et al*., 2012; Pollastro *et al*., 2016a

margins. The entire leaf develops light brown, rotted areas, as the infection progresses, which gradually turn dark. Finally, defoliation occurs, as the affected leaves die. The fungus also attacks flowers and fruits at all stages and can cause fruit rot (Khosla and Gupta, 2014). Light brown/tan coloured, soft (but not watery) rot appears on fruits from the stem end or fruit surface. Rot appears first as tan-coloured lesions on fruits near ground level; in later stages the affected fruits turn darker and rot spreads rapidly with white sporulation on the surface (Fig. 12.8).

Phytophthora cactorum (Alavi and Zackii, 1985) and *P. palmivora* have been reported to cause crown and root rot (Turkolmez *et al.*, 2015). Symptoms consist of brown lesions on the collar region of the roots below the soil line, bark cracking on stems just above the soil line, associated with a rotting of the stem and a lack of capillary roots. On a few occasions, trees show sudden death – where a plant that looked green and healthy on one day, suddenly topples down the next day – associated with a *Phytophthora* sp. These plants show depressed lesions above and below the stem at soil line and infected tissue under the bark (J. Sharma, Solepur, India, 2012, personal communication).

Phytophthora nicotianae (Neama and Sharma, 2006; Sharma *et al.*, 2012; Khosla and Gupta, 2014), *P. palmivora* (More *et al.*, 1989) and *P. cactorum* (Alavi and Zackii, 1985) have been reported to infect pomegranate.

Periods of persistent rains when temperatures are between 25 amd 30°C and humidity is high favour diseases caused by *Phytophthora* spp. (J. Sharma, 2010, Solapur, India, personal communication). *Phytophthora* produces sporangia that can either germinate directly or release zoospores. Sporangia are produced on infected tissues and are able to germinate directly on the plant surface or in the soil. They can also produce small zoospores. The zoospores swim in soil water or on a wet plant surface and finally enter the plant. Pomegranate leaves and fruits near soil are typically infected first due to rain-splashed propagules from the soil. With subsequent rains, pathogen sporangia and zoospores are easily spread to other parts of the tree and orchard. The pathogen is devastating under rainy seasons and can spread to large areas within 2–3 days in the absence of appropriate control measures. Collar and root rot are more common where poor drainage leads to continuous waterlogged soils surrounding the plant. *Phytophthora* spp. produces asexual resting spores called chlamydospores, which can survive for long periods in soil. These serve as a source of primary infection for the next season. Many *Phytophthora* spp. also produce oospores, which can survive long periods in the soil. For heterothallic species like *P. palmivora* and *P. nicotianae*, oospore formation requires the presence of two mating types; whereas *P. cactorum* is a homothallic species that does not require the presence of two mating

Fig. 12.8. *Phytophthora* blight symptoms on: (a) leaves, (b) buds and (c) fruit of pomegranate; (d) dropped pomegranate fruit on the ground with white sporulation of *Phytophthora* spp. on the surface. (Photos: Jyotsana Sharma.)

types for oospore production. Hence, for the homothallic *Phytophthora* species, asexual structures known as chlamydospores may serve as an important source of inoculum between production seasons.

12.4.1 Management of *Phytophthora* diseases

Orchard sanitation and soil water management are important preventive measures for managing *Phytophthora* diseases. Use of organic manures in sufficient quantities is the best way to manage any soilborne inoculum. Organics enrich the soil with beneficial microflora that are nutrient solubilizers, hence they improve plant nutrition, release biochemicals that impart resistance to the plant and antagonize the growth of pathogenic microorganisms. Organic amendments therefore are good for management of soilborne *Phytophthora* spp. (Jambhulkar *et al.*, 2015). Copper fungicides like copper hydroxide, copper oxychloride and Bordeaux mixture have been effectively used to control *Phytophthora* diseases (Pscheidt, 2019). Several studies report the use of calcium sulfate fertilizers to reduce soil inoculum levels of *Phytophthora* (von Broembsen and Deacon, 1997; Sugimoto *et al.*, 2008). Foliar-applied fungicides containing metalaxyl, mancozeb, dimethomorph or fosetyl-Al 80WP, as solo or mixed formulations, are effective for disease management (Sharma *et al.*, 2012). Preventative fungicide applications are recommended when weather conditions are conducive to rapid disease development, especially in

Table 12.5. Fungal spots reported on pomegranate from different countries.

Disease	Causal organism	Country	Reference
1. Scab	*Sphaceloma* (syn. *Elsinoe*) *punicae*	India	NRCP, 2009
		South Africa	Carstens *et al.*, 2018
		Iran	Arzanlou *et al.*, 2018
2. Cercospora spots	*Cercospora punicae*	Most pomegranate-growing regions	Alfieri Jr, 1978
3. Leaf and fruit spots/blotch	*Pseudocercospora punicae*	Mexico and Florida (USA)	Ayala-Escobar *et al.*, 2019; Wolf, 1927
4. Black spot disease	*Cladosporium cladosporioides*	China	Zhou *et al.*, 2018
5. Fruit spot	*Drechslera rostrata*	India	Utikar *et al.*, 1977; Sharma *et al.*, 2012
6. Fruit spot	*Beltraniella humicola*	India	Sherkar and Utikar, 1982
7. Powdery mildew	*Erysiphe* sp.	Italy	Pollastro *et al.*, 2016b

orchards that have a history of crop losses due to *Phytophthora* spp.

12.5 Fungal Pathogens Causing Foliar and Fruit Diseases

Pomegranate is susceptible to several fungal pathogens causing leaf and fruit spots as well as fruit rots. However, only some of them are economically important. Under favourable environmental conditions, these pathogens can lead to reductions in overall yield or fruit quality that can be financially significant for the grower. Common fungal pathogens that cause foliar and fruit diseases on pomegranate are summarized in Table 12.5 and Table 12.6. Most of the listed pathogens overwinter on infected crop residues. Rain and wind disperse spores throughout the orchard, which remain latent until favourable temperature and humidity conditions prevail for further disease development. In addition, many of the listed pathogens can survive on plant surfaces as epiphytes and cause infection during favourable weather conditions or specific stages of host development.

The most important fruit pathogens responsible for commercial losses in almost all pomegranate-growing regions include *Alternaria alternata* (heart rot), *Colletotrichum gloeosporioides* (anthracnose), *Coniella granati* (syn. *Pilidiella granati*) and *Coniella noviae* (fruit rot), *Phytophthora nicotianae* (fungal blight) and *Botrytis cinerea* (grey mould). In addition, *Phoma* sp., *Phomopsis* sp. (hard rot), *Cytospora punicae* (corky rot), *Aspergillus flavus*, *Aspergillus fumigatus*, *Aspergillus niger*, *Aspergillus variecolor*, *Penicillium frequentans*, *Penicillium implicatum* and *Rhizopus* spp. (secondary soft rot, weak pathogens) are also identified as fruit rot pathogens (Sharma *et al.*, 1982; Neama and Sharma, 2006; Pala *et al.*, 2009; Mincuzzi *et al.*, 2017), although causing fewer losses.

Some of the fungal pathogens that infect foliage can also infect fruit, causing spots and rots that reduce quality and yield.

12.5.1 Scab

Since 2007, scab has become a limiting factor in pomegranate production in arid and semi-arid regions of India (J. Sharma, Solapur, India, 2008, 2009, personal communication). In areas where the disease is prevalent, it is not uncommon to find orchards where 90–100% of fruits are affected. Losses to scab can vary depending on the timing of infection (Sharma *et al.*, 2012).

The pathogen attacks at any stage from flowering through fruit maturity (Fig. 12.9). Infections of flower buds or small fruits lead to deformation and consequently yield losses.

Table 12.6. Fungal rots reported on pomegranate from different countries.

Disease	Causal organism	Country	Reference
1. Anthracnose	*Colletotrichum gloeosporioides* *Colletotrichum fioriniae* *Colletotrichum nymphaeae* *Colletotrichum siamense* *Colletotrichum simmondsii* *Colletotrichum theobromicola* *Glomerella cingulata*	USA	Xavier *et al*., 2019a
		India	Singh and Chohan, 1972; Sataraddi *et al*., 2011; Jayalakshmi *et al*., 2015
		Venezuela	Mazzani, 1994
		Greece	Thomidis, 2014
		Cyprus	Nattrass, 1932
2. Heart rot + leaf spots	*Alternaria alternata*	Greece	Tziros *et al*., 2008
		USA	Zhang and McCarthy, 2012
		India	Sharma and Sharma, 2017
		Israel	Ezra *et al*., 2010
		Spain	Berbegal *et al*., 2014
3. Leaf spot and fruit rot	*Dwiroopa punicae*	USA	Xavier *et al*., 2019b
4. Fruit spots and rots	*Aspergillus niger*	USA, India, Italy, Spain, Saudi Arabia	Munhuweyi *et al*., 2016
	Aspergillus variecolor	India	Sharma *et al*., 1982
	Chaetomella raphigera	India	Gajbhiye *et al*., 2016
	Coniella granati (syn. *Pilidiella granati*)	China, Greece, India, Israel, Italy, Spain, Tunisia, Turkey, USA	Çeliker *et al*., 2012; Hebert and Clayton, 1963; Jabnoun-Khiareddine *et al*., 2018; Levy *et al*., 2011; Mincuzzi *et al*., 2016; Palou *et al*., 2010; Pollastro *et al*., 2016a; Sharma and Jain (1978); Sharma and Tegta, 2011; Thomidis and Exadaktylou, 2011
5. Grey mould rot	*Botrytis cinerea*	USA	Day and Wilkins, 2011
		Armenia, Azerbaijan, Georgia	Kechakmadze *et al*., 1990 in Sharma *et al*., 2012
		Greece	Bardas *et al*., 2009
		Spain	Palou *et al*., 2013
6. Blue/green mould	*Penicillium* spp.	India, Pakistan, Slovak Republic, Spain	Palou *et al*., 2013; Munhuweyi *et al*., 2016
7. Fruit rot	*Phomopsis* sp.	India	Jamadar *et al*., 2011
8. Fruit rot and fruit spot	*Pestalotiopsis versicolor*	Kenya, India	Utikar *et al*., 1980; Siboe *et al*., 1982
9. Nematode-associated fruit decay	*Sheraphelenchus sucus* *Panagrellus* sp.	Italy	Fanelli *et al*., 2017

Fig. 12.9. Different types of scab symptoms (a–f) caused by *Sphaceloma* (syn. *Elsinoe*) *punicae* on flowers and fruits of different pomegranate germplasm. (Photos: Jyotsana Sharma.)

Infection of mature fruits affects the appearance and exterior quality of fruits. The lesions affect only the outer rind surface, without affecting aril or juice quality. Fruit symptoms can vary among varieties and wild germplasm (NRCP, 2016). The lesions on fruits may be small or large, brown, rough, raised, covering small areas to entire fruit surface, giving a russet scab appearance to the rind. The spots at times may enlarge to form larger spots with a light centre and darker edge and are rough to the touch (Sharma *et al.*, 2012). Spots on leaves are reported, but not commonly observed (Jamadar *et al.*, 2011).

Sphaceloma (syn. *Elsinoe*) *punicae* Bitanc & Jenkins is the pathogen associated with pomegranate scab (Table 12.5). *Sphaceloma punicae* is extremely slow-growing in culture; hence, improper sterilization of affected host tissue during isolation usually results in the growth of other rapid-growing secondary fungi, like *Alternaria*. Visible growth of the pathogen on PDA after isolation is observed after 7–10 days at 25°C. The colony measures 10–15 mm in 3–4 weeks, is hard and purplish/reddish-brown in colour, mostly without sporulation. Mycelium is immersed in the substrate, branched, septate and hyaline to pale brown. Fruiting structures, mostly formed on affected tissue, are conidiomata, acervular, foliicolous or caulicolous, initially separate but often coalescing. Conidiophores are sparse, small, one- to two-septate, cylindrical, unbranched, pale brown or hyaline. Conidia generally are produced on the host and are extremely minute, hyaline, aseptate, ellipsoid, smooth and eguttulate.

12.5.2 Pseudocercospora and cercospora spots

Pseudocercospora punicae (Henn.) Deighton or *Cercospora punicae* (Henn.) have been reported in the USA, Africa, Mauritius, India, Taiwan, Iran, Japan and almost all pomegranate-growing regions worldwide causing leaf and fruit spots (Alfieri Jr, 1978).

Initial symptoms appear on mature leaves as a few or many small subcircular to irregular brown spots, which later develop into large spots measuring 1–5 mm with reddish-brown to black colouration. A diffuse yellow or greenish halo may be observed around lesions. Leaf infection often leads to premature leaf drop (Wolf, 1927).

Lesions on infected fruits first appear as small, irregular light brown spots that may

Fig. 12.10. Symptoms caused by *Cercospora punicae*: (a) reddish-brown necrotic lesions on leaves; (b) black necrotic spots surrounded by a green halo; (c) black necrotic spots without halo; and (d) brown necrotic spots on fruits with secondary pathogens. (Photos: Jyotsana Sharma.)

coalesce into bigger blotches. Fruit lesions enlarge becoming brownish-black or dark black, irregular, discrete, of various sizes, resembling symptoms of bacterial blight, but without the cracks or stickiness that are characteristic of bacterial blight (Wolf, 1927; Phengsintham *et al*., 2011; Sharma *et al*., 2012). The pathogen remains superficial on the fruit surface, and does not colonize the inner parts of the fruits like rind and arils. Lesions may vary in colour and size among different cultivars (Fig. 12.10). Under high humidity, fruit or leaf lesions may have a grey-coloured centre that corresponds to pathogen sporulation.

The pathogen was first described as *Cercospora punicae* in Japan by Hennings (1906), and later reclassified by Deighton (1976) as *Pseudocercospora punicae*. As per some reports, *Pseudocercospora* is considered synonymous with *Cercospora* (Rawla, 1971; Nakashima *et al*., 2016). As symptoms produced by both the fungi are similar and morphologically it is difficult to distinguish between *C. punicae* and *P. punicae*, reports on both *C. punicae* and *P. punicae* spots are considered here. Colonies on PDA are erumpent, grey with irregular patches of white or smoke-grey, the reverse iron-grey, slow-growing, with sparse aerial mycelium. Sporulation is not typically observed in culture on PDA; however, incubating sections of symptomatic tissues at room temperature in a moist chamber can induce conidia production. The conidiophores are formed in aggregates and measure 5.5–40 × 2–4 μm with zero to five septa. The conidia are 20–80 μm long cylindrical, straight to slightly curved, truncate at base, obclavate, 2.5–4.0 μm wide, with three to nine transverse septa, rarely catenulate and subhyaline to pale olivaceous in colour (Nakashima *et al.*, 2016).

12.5.3 Alternaria black spot and heart rot

Alternaria alternata has been reported to cause black spot disease on leaves and fruits and an internal fruit rot in pomegranate (Pantidou, 1973; Madhukar and Reddy, 1976; Ezra *et al.*, 2010). Lesions on leaves are irregular to round, 1–4 mm, brownish-black to dark black, with concentric rings, although not always distinct. The affected leaves may turn chlorotic, dry and fall off. *Alternaria* leaf spots are generally observed when the plant is under any type of stress (Sharma *et al.*, 2012). Black spot symptoms on fruits may appear as isolated spots covering up to 50% fruit surface. The spots are typically small (1–3 mm), round, with a green-yellow halo and limited to the fruit surface (Ezra *et al.*, 2010).

Alternaria fruit rot also known as 'heart rot' or 'black heart' has been reported to cause heavy losses in many pomegranate-growing regions of the world (Sonawane *et al.*, 1986; Tziros *et al.*, 2008; Holland *et al.*, 2009; Pala *et al.*, 2009; Day and Wilkins, 2011; Ezra *et al.*, 2015). No symptoms of the pathogen are observed on the rind surface. However, disintegration of arils inside in the advanced stage of fruit development leads to the appearance of a shrunken or softened rind that can reduce fruit weight. The pathogen is reported to enter fruits at the time of full bloom through pistils; however, arils develop rot depending on the environmental and physiological conditions of the plant. Rot is only revealed when the fruit is opened. Affected fruits show rotting of some or all the arils along with fungal mycelium. (Fig. 12.11) Some of the arils exhibit brown (soft) or black (dry) rot, presumably due

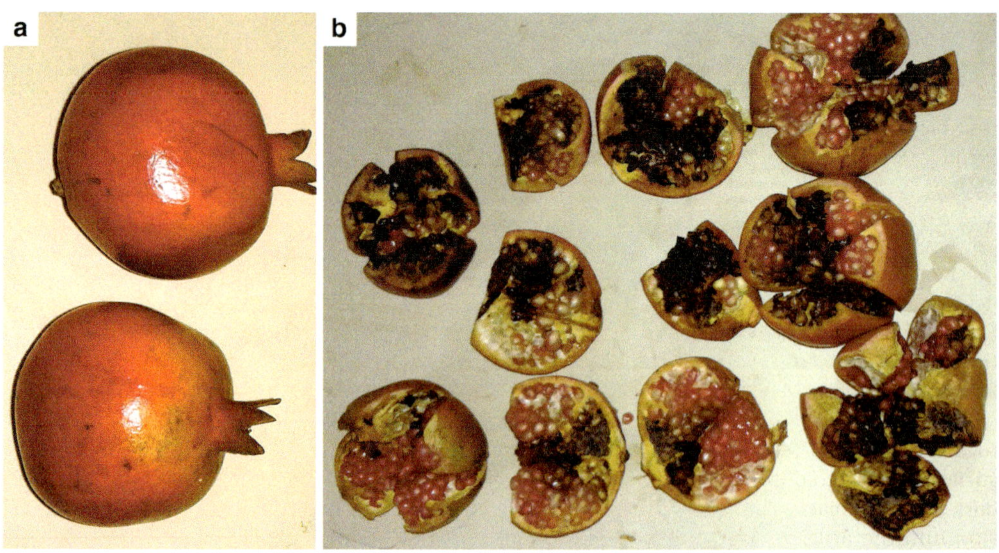

Fig. 12.11. Symptoms of heart rot of fruits: (a) *Alternaria*-affected apparently healthy fruits; and (b) rotten arils with black fungal sporulation. (Photos: Jyotsana Sharma.)

to differences in rot development stage. No rotting of the mesocarp is observed in heart rot-affected fruits (Ezra *et al.*, 2015). Discolouration of rind surface may be seen when secondary rot fungi including *Aspergillus and Penicillium* gain entry through the crown region or by mechanical injury.

The *Alternaria* strains affecting pomegranate are generally considered weak pathogens and mostly attack mature or stressed leaves and fruits. Different pathogenic strains of *A. alternata* cause black leaf and fruit spot as well as fruit rot (Ezra *et al.*, 2010). The pathogen *A. alternata* is a fast-growing fungus in culture with whitish mycelial growth turning dark grey with the development of conidia. The conidiophores are short, septate, branched or unbranched and greenish-brown in colour. The conidia are obpyriform with conical or cylindrical beak and often produced in branched chains. Average conidial size is 17 × 6 µm, ranging from 10–21 µm in length and 4–10 µm in width at the broadest point (Simmons, 1967).

Fig. 12.12. Symptoms of anthracnose: (a) discoloured rind from calyx end; (b) numerous isolated black-brown spots on the fruit surface; and (c) discoloured rind in advanced stage. (Photos: Jyotsana Sharma.)

12.5.4 Anthracnose

Anthracnose caused by *Colletotrichum* spp. is becoming a major problem in semi-arid conditions. The disease is common in tropical and subtropical regions (Prashanth and Sataraddi, 2011; Nargund *et al.*, 2012) with high humidity (62–95%) and temperatures (20–30°C) that favour disease development (Arauz, 2000; Whitelaw-Weckert *et al.*, 2007; Cannon *et al.*, 2012).

Symptoms of *Colletotrichum* develop on flowers, fruits, leaves or twigs, but fruits are most susceptible. On leaves, small circular spots develop with yellowish halos; later infected leaves turn yellow leading to premature defoliation. On fruits, the first symptoms are observed as discolouration of fruit rind from calyx end or numerous isolated spots that later coalesce together forming irregular blotches on the fruit surface. The discoloured areas become reddish/dark brown to black. The rot extends beyond the rind into the arils, which disintegrate and are dark grey/brown-black coloured but not watery. The affected fruits may drop at a later stage. The fruits are most vulnerable to this disease at all stages from setting. Symptoms are dry hard rot (Fig. 12.12) covering half to full fruit within a week (Sharma *et al.*, 2012). High humidity, rainfall and warm temperatures during the summer months increase disease severity. Under these wet conditions, pathogen signs become visible as orange spore masses on the lesion. Acervuli develop on the lesions from which emerge oblong, hyaline, non-septate conidia, which are capable of causing fresh infections.

Similar to other fruit crops, *Colletotrichum* can infect flowers during spring and remain latent till later in the season when environmental conditions become favourable to disease development and the fruits begin to ripen (Jeffries *et al.*, 1990; Wharton and Diéguez-Uribeondo, 2004; Peres *et al.*, 2005).

Several species of *Colletotrichum* are reported to be associated with anthracnose fruit rot and leaf blight on pomegranate. The genus *Colletotrichum* was first described infecting pomegranate in India in 1924 (McRae, 1924). Globally, *C. gloeosporioides* (Penz.) Penz. and Sacc. and *C. theobromicola* Delacr. are the most commonly reported species from the USA

(Alfieri Jr *et al.*, 1984), Greece (Thomidis, 2014), Brazil (Mendes *et al.*, 1998), India (Sataraddi *et al.*, 2011; Sharma *et al.*, 2015; Munhuweyi *et al.*, 2016), Iran (Rahimlou *et al.*, 2014) and Australia (Shivas *et al.*, 2016), followed by other reported species of *Colletotrichum*, like *C. acutatum* (Mincuzzi *et al.*, 2017; Bellé *et al.*, 2018) and *C. fructigenum* (Gómez and Pallás, 2001). *Colletotrichum gloeosporioides* species was reported infecting foliage in Turkey (Uysal and Kurt, 2018) and *C. acutatum* infecting flowers in Brazil (Bellé *et al.*, 2018). Recently, six new species of *Colletotrichum* have been characterized on pomegranate across the south-eastern USA (Xavier *et al.*, 2019a).

On PDA, colonies of *C. gloeosporioides* are fluffy, white changing to velvety grey, with a salmon-greyish colour later turning peachy-pink. The colony later is covered by salmon-coloured conidial masses. The fungus produces conidia in acervuli. The conidia are oblong/cylindrical or slightly dumbbell-shaped, hyaline, aseptate with rounded ends and one to two oil globules. They are rarely found in aggregates. Masses of conidia appear pink or salmon-coloured. The average conidia size in culture on PDA is 15.7 × 5.4 µm, whereas on host fruit the range is slightly larger, 16.8 µm × 6.4 µm.

Colletotrichum spp. overwinter on crop residue as conidia within fungal structures, called acervuli, that erupt through the epidermis of the host tissues. Conidia are deposited on to pomegranate plant surfaces such as flowers, buds, leaves or shoots, and under wet conditions geminate to cause infection (Arauz, 2000). As a result of infection, infected plant cells die and disease symptoms become visible. In the last stages of the disease cycle, the pathogen produces acervuli, containing conidiospores and setae. These conidia serve as a source of secondary infections in the same season or overwinter in crop debris and serve as the source of primary inoculum in the next season (Arauz, 2000).

12.5.5 New fungal spot and rot pathogen: *Dwiroopa punicae*

Dwiroopa punicae is a novel fungal species associated with pomegranate fruit rot and leaf spot and has been identified during disease surveys between 2014 and 2017 across the south-eastern USA. *Dwiroopa punicae*, together with *Colletotrichum* spp., was capable of causing almost 100% fruit loss where pomegranate trees were left unmanaged in the surveyed areas (Xavier *et al.*, 2019b).

Foliar symptoms consist of oval, brown spots, ranging from 0.1–1.5 cm in diameter, scattered on the leaf surface. Fruit symptoms start as small, brown lesions on the fruit calyx at different stages of fruit development. Lesions on leaves and fruit become black due to the formation of black, erumpent, globose pycnidia on the host surface. Arils of the infected fruits rot leading to premature fruit drop from the trees.

The pathogen associated with the disease is *D. punicae*. Multilocus phylogenetic analyses and morphological characterization of the isolates revealed that the novel species belongs to the genus *Dwiroopa* (Farr and Rossman, 2003); the genus is placed in the order Diaporthales. The proposed name of the new species is *Dwiroopa punicae* sp. nov. and a new family has been also described in this work as Dwiroopaceae (Xavier *et al.*, 2019b). This was the first report for *D. punicae* associated with any plant species globally, although the type specimen for the genus, *Dwiroopa ramya* (Farr and Rossman, 2003), was originally isolated as a potential biocontrol agent from purple loosestrife (*Lythrum salicaria* L.) in North America (Farr and Rossman, 2001).

The pathogen produces black globose pycnidia, measuring 150–300 µm in diameter. Each pycnidium has a central ostiole from which emerges a slimy, black conidial mass. Conidia are solitary, dark brown, single-celled, broadly ellipsoid, with an obtuse apex and truncate base, 2–2.5 µm in diameter. Conidia have longitudinal slits running along the entire length. The conidial size varies from 13 to 16 × 12–14 µm.

12.5.6 Minor spots and fruit rots

Drechslera rostrata is reported to cause minor spotting on fruits in India (Lande and Utikar, 1979). Fruit symptoms consist of small, irregular, brown spots surrounded by a greenish-yellow border. Lesions may expand to form blotches and in severe cases may result in discolouration of inner tissue or even the seeds, where arils become brown. Different pathogens may produce similar leaf spot symptoms, but most of them are controlled by common fungicide schedules, hence, these are considered of little economic importance.

Aspergillus niger and A. variecolor cause *Aspergillus* rot (Table 12.6). *Aspergillus niger* is reported to cause black spots on fruit, but is more often associated with an internal soft rot of fruit. Internal rotted tissues are tan-coloured, with a depressed centre in later stages where blackish sporulation of the fungus can be seen (Kumar and Chahal, 2016). The pathogen generally gains entry through the calyx or from injury to the fruit surface associated with insect feeding, fruit cracking or sunburn. Internal symptoms can develop with few external symptoms, but external symptoms of decay are typically close to the calyx. The rind of infected fruit is often off-colour developing a yellow to brownish-red discolouration as the symptoms progress. In the absence of sporulation, the rotted rind can have a papery translucent appearance, like that of a boiled potato (J. Sharma, 2010, Solapur, India, personal communication). The underlying pulp becomes soft, but fruits retain their shape.

Penicillium expansum is another soft rot pathogen that generally initiates at fruit wounds and produces a light-coloured (not much different from rind colour) mushy soft rot, with a bluish-green mould growth in later stages (Thomidis, 2014). Other secondary fruit rot pathogens, like *Rhizopus* spp., produce irregular, brown, watery rot with rapid tissue disintegration, occasionally causing water to ooze out from infected fruits. Species of *Phoma* and *Phomopsis* are reported to affect flowers and young fruits. Affected fruits may rot and drop. On bigger fruits, rapidly enlarging yellowish-tan turning brown to black spots appear all over the fruit (Jamadar *et al.*, 2011).

12.5.7 Management of fungal spots and fruit rots

The fungal spots on fruits reduce fruit quality and market value, whereas fruit rots result in complete loss of produce. Hence, if neglected, both the diseases result in substantial economic losses. Integrated management strategies including orchard sanitation, cultural practices and fungicide sprays are necessary to reduce potential losses.

Orchard sanitation and other cultural practices

Orchard sanitation is critical to minimize seasonal carryover of pathogen inoculum. Proper plant nutrition and good cultural practices can improve plant resistance to diseases and insect damage, which can provide entry sites for weak secondary pathogens. Diseased plant parts or debris, such as infected twigs and fruits, where the pathogen can sporulate, should never be dumped in or near the orchard. They should be removed periodically from the orchard and destroyed, to reduce inoculum load. Suckers, low-hanging and crowded branches should be pruned to improve air circulation and microclimate of the orchard. Tools used to prune the trees must be disinfested with 10% bleaching solution or 70% ethanol to reduce the risk of pathogen spread from tree to tree. The orchard should be kept weed-free. Nutrition during the rest period, soon after harvest and at different growth stages plays an important role in disease resistance. Silicon plays an important role by enhancing physical as well as chemical resistance of plants to the entry of several pathogens. It increases production of defence enzymes, and antifungal compounds like phenolic metabolism products, phytoalexins and pathogenesis-related proteins (Weerahewa and Somapala, 2016). It is a well-established fact that excessive nitrogen or nitrogen deficiency can increase the incidence and severity of many diseases; hence, appropriate fertility is essential to orchard health (Thind, 2017).

Chemical control

In addition to sanitation and other cultural practices, the use of effective pesticides is recommended for the management of pomegranate diseases. Timing initial pesticide applications and maintaining appropriate application intervals is essential during and after the production season. Orchards neglected after harvest, in terms of nutrition and disease management, result in the survival and build-up of pathogen inoculum within the orchard, which contributes to outbreaks in the following production cycle. Fungicide applications during the flowering period are most important for the management of fruit rots and start at pre-flowering/fruit set and continue at 10–15-day intervals depending on weather conditions and the nature of the fungicide. Numerous studies have reported that fungicide applications at bloom stage are critical to effectively manage disease in many *Colletotrichum*–host systems (Arauz, 2000; Peres *et al.*, 2002). Therefore, it is essential to identify effective fungicides for pomegranate, as well as determine the application timing necessary to minimize fruit losses in tropical and subtropical production areas. Fungicide programmes should avoid sequential applications of fungicides with the same mode of action, and limit repeated applications of the same fungicide throughout the season, with the exception of contact fungicides that have multiple modes of action, such as copper. Fungicide applications should be scheduled based on the recommended pre-harvest interval to avoid fungicide residue issues.

Most foliar and fruit diseases caused by fungal pathogens are effectively managed by preventative fungicide applications. Effective fungicides include carbendazim, propiconazole, thiophanate methyl, mancozeb, difenoconazole, captan, benomyl, ziram, Bordeaux mixture and various fixed-copper compounds. Mixtures of carbendazim + mancozeb and benomyl + mancozeb also improved fruit quality and yield. In India, anthracnose and calyx end rot of pomegranate caused by *C. gloeosporioides* was effectively managed by applications of tricyclazole + mancozeb and hexaconazole + zineb (Jamadar *et al.*, 2011; Jadhav and Sharma, 2011; Nargund *et al.*, 2012; Sharma *et al.*, 2014).

Fungicides thiophanate methyl and tebuconazole are effective in managing fruit rot caused by pycnidia-forming pathogens, namely *Coniella granati (syn. P. granati)*, *Botryosphaeria* (Ma *et al.*, 2002), *Diaporthe* (Thomidis and Michailides, 2009) and *Phoma* (Schmitz *et al.*, 2006). Fungicides containing cymoxanil, dimethomorph, fosetyl aluminium, metalaxyl, mancozeb or combinations thereof are recommended for the management of diseases caused by *Phytophthora*. These fungicides reduced losses when applied preventatively in orchards with a history of *Phytophthora* blight or fruit rot (Jadhav and Sharma, 2011).

12.6 Shoot Blight and Canker Disease

Shoot blight and canker disease of pomegranate caused by various fungi have been reported from different countries (Table 12.7). However, the relative frequency and economic importance of these pathogens is not commonly reported. Hence a brief description of the most common causes of shoot blight and canker, *Coniella granati* Saccardo (syn. *P. granati* Saccardo), is given below.

Major disease symptoms caused by *C. granati* include dieback, shoot blight, stem cankers and

Table 12.7. Shoot and stem diseases of pomegranate reported from different countries.

Disease	Causal organism	Country	Reference
1. Shoot blight and stem canker	*Coniella granati*	Greece	Thomidis, 2015
2. Collar rot	*Coniella granati*	Iran	Mirtalebi *et al.*, 2015
3. Stem scab	*Botryosphaeria dothidea*	China	Liu *et al.*, 2009
4. Shoot blight	*Neofusicoccum parvum*	Greece	Palavouzis *et al.*, 2015a
5. Dieback	*Lasiodiplodia gilanensis*	California	Urbez-Torres *et al.*, 2017
6. Wood canker and branch dieback	*Cytospora punicae*	USA	Peduto Hand *et al.*, 2014
		Greece	Palavouzis *et al.*, 2015b
		Cyprus	Samouel and Kanetis, 2016
		Iran	Mahdikhani and Davoodi, 2017
	Ceuthospora phyllosticta	India	Sohi *et al.*, 1965

Coniella granati (syn. *Pilidiella granati*).

collar rot. Several other pathogens can also cause shoot blight and stem cankers of pomegranate trees, as discussed in previous sections; including stem cankers formed due to secondary fungal infections in nodal bacterial blight symptoms caused by *Xanthomonas axonopodis* pv. *punicae* (see Section 12.2).

On artificial media, colonies of *C. granati* are light yellow, leathery mycelia with abundant black, solitary pycnidia of various sizes. Hyphae are septate and conidia are hyaline, one-celled and ellipsoid to fusiform, ranging from 10.1 to 20.2 × 3.2–4.3 µm in size (Thomidis, 2015). *Coniella granati* grows between 2 and 35°C and optimum temperatures are between 25 and 30°C. Pycnidia of the pathogen have been reported on plant debris left in the orchards including, mummified fruits, pruned shoots, and blighted dead shoots and crown of trees, which serve as a source of inoculum (Thomidis, 2015). Irrigation water and rains can spread spores from overwintered pycnidia present on the bark of the trees and the surface of young fruit and cause infections.

12.6.1 Management

Management strategies include pruning and removal of dead branches, safe disposal of plant debris, sanitization of pruning tools, and applications of Bordeaux mixture throughout the growing season and offseason, and shortly after pruning. In India, the application of a 10% Bordeaux paste on the main stems up to 0.45–0.60 m (1.5–2 ft) above soil level during crop production and following harvest is a common practice to limit the incidence of stem cankers. Branches with visible cankers should be pruned and safely disposed. Fungicide applications for fungal spots and rot pathogens during the season can also help in the management of dieback and blight pathogens.

12.7 Viral, Viroid and Phytoplasmal Diseases

Economic losses in pomegranate have rarely been reported due to diseases caused by viruses, viroids or phytoplasmas, though there are reports of their association with some symptoms on pomegranate (Table 12.8).

Juretic and Horvath (1984) first reported virus infection on pomegranate in the former Yugoslavia. They observed pomegranate showing leaf deformities, variegation and yellowing, as well as a reduction in flowering. The pathogen was identified as an isolate of CMV (CMV-Pg), by test plant reactions, stability in sap, virus particle size and serological studies. In 2000, hop stunt viroid (HSVd) was detected in 10 pomegranate cultivars in Turkey (Onelge, 2000). Gazel *et al.* (2016) detected viroids or viroid-like RNAs from leaves of six different cultivars of pomegranate. Pomegranate was confirmed as an alternate host for grapevine leafroll-associated virus 1

Table 12.8. Viruses, viroids and phytoplasmas reported in pomegranate.

Viruses, viroids and phytoplasmas	Country	Reference
1. Cucumber mosaic virus CMV-Pg	Former Yugoslavia	Juretic and Horvath, 1984
2. Hop stunt viroid (HSVd)	Turkey	Onelge, 2000
3. Viroid-like RNAs	Spain	Gazel *et al.*, 2016
4. Tomato ring spot virus	Italy	EPPO Global Database, 2015
5. Cherry leaf roll virus	Italy	Bichcheri *et al.*, 2015
6. *Candidatus* Phytoplasma *pruni*	Iran	Karimi *et al.*, 2015
7. 16SrII phytoplasma	Iran	Salehi *et al.*, 2016
8. Grapevine leafroll-associated virus 1	Turkey	Caglayan *et al.*, 2016
9. 16SrI-B and 16SrXII-A phytoplasma	Turkey	Gazel *et al.*, 2016
10. Pomegranate fasciation (PoF) phytoplasma	China	Rui *et al.*, 2018

(GLRaV-1) causing foliar symptoms of yellowing, chlorotic spots, oak-leaf and vein clearing (Caglayan *et al.*, 2016).

Phytoplasma infection has been reported in pomegranate plants from different countries. Symptoms like yellowing of top leaves with rolled-up margins, reddening and thickening of veins, dieback, reduced plant height and vigour resulting in slow decline over the years and fasciation symptoms, namely stunting and shortened internodes, have been reported associated with different phytoplasma infections (Caglayan *et al.*, 2014; Karimi *et al.*, 2015; Gazel *et al.*, 2016; Rui *et al.*, 2018). In Turkey phytoplasmas were detected in pomegranate using universal phytoplasma primers P1/P7 followed by R16F2n/R2 (Caglayan *et al.*, 2014). Phytoplasmas related to '*Candidatus* Phytoplasma *pruni*' were found associated with 'Khazar' pomegranate in Iran (Karimi *et al.*, 2015). In Turkey, Gazel *et al.* (2016) reported 16SrI-B and 16SrXII-A phytoplasmas on pomegranate trees. In China, symptoms of fasciation were observed in pomegranate infected with phytoplasma belonging to the 'Candidatus Phytoplasma asteris' group (16SrI) (Rui *et al.*, 2018).

References

Accordi, S.M. (1989) The survival of *Ceratocystis fimbriata* f.sp. *platani* in the soil. *Informatore Fitopatologica* 39, 57–62.

Akhtar, M.A. and Bhatti, M.H.R. (1992) Occurrence of bacterial leaf spot of pomegranate in Pakistan. *Pakistan Journal of Agricultural Research* 13, 95–97.

Alam, M.W., Gleason, M.L., Mehboob, S., Riaz, K. and Rehman, A. (2017) First report of *Ceratocystis fimbriata* causing pomegranate wilt in Pakistan. *Plant Disease* 101(1), 251. DOI: 10.1094/PDIS-06-16-0835-PDN.

Alavi, A. and Zackii, Z. (1985) Crown and root rot of pomegranate. *Iranian Journal of Plant Pathology* 21, 22–70.

Alfieri Jr, S.A. (1978) *Cercospora leaf spot of pomegranate*. Plant Pathology Circular No. 194, Florida Department of Agriculture and Consumer Services, Tallahassee, Florida. Available at: https://www.freshfromflorida.com/content/download/11201/143497/pp194.pdf (accessed 19 November 2019).

Alfieri Jr, S.A., Langdon, K.R., Wehlburg, C. and Kimbrough, J.W. (1984) Index of plant diseases in Florida (revised). *Florida Department of Agriculture and Consumer Services, Division of Plant Industries Bulletin* 11, 389.

Arauz, L.F. (2000) Mango anthracnose: economic impact and current options for integrated management. *Plant Disease* 84(6), 600–611. DOI: 10.1094/PDIS.2000.84.6.600.

Arora, R.K., Sharma, J. and Singh, R.K. (2006) Soil solarization. Technologies in Aid of Healthy Potato. Technical Bulletin 35. Potato Research Institute, Shimla, India.

Arzanlou, M., Dalili, A.R., Aghajaninasab, M.A. and Rabbaninasab, H. (2018) Morphological and molecular characterization of *Elsinoë punicae*, the causal agent of pomegranate scab disease in Golestan and Mazandaran provinces. *Iranian Journal of Plant Pathology* 54(2), 147–157.

Ayala-Escobar, V., Pérez-López, A., Suaste-Dzul, A.P., Leyva-Mir, S.G. and Tovar-Pedraza, J.M. (2019) First report of *Pseudocercospora punicae* causing black spot of pomegranate fruit in Mexico. *Journal of Plant Pathology* 101(2), 403. DOI: 10.1007/s42161-018-0181-0.

Baker, C.J. and Harrington, T.C. (2001) *Ceratocystis fimbriata*. In: *Crop Protection Compendium*. CAB International, Wallingford, UK. Available at: www.cabi.org/cpc (accessed 6 October 2020).

Baker, C.J., Harrington, T.C., Krauss, U. and Alfenas, A.C. (2003) Genetic variability and host specialization in the Latin American clade of *Ceratocystis fimbriata*. *Phytopathology* 93(10), 1274–1284. DOI: 10.1094/PHYTO.2003.93.10.1274.

Banihashemi, Z. (1998) Etiology of pomegranate decline in Fars Province of Iran. *Phytopathologia Mediterranea* 37, 127–132.

Bardas, G.A., Tzelepis, G.D., Lotos, L. and Karaoglanidis, G.S. (2009) First report of *Botrytis cinerea* causing gray mold of pomegranate (*Punica granatum*) in Greece. *Plant Disease* 93(12), 1346. DOI: 10.1094/PDIS-93-12-1346C.

Beaver, R.A. (1989) Insect-fungus relationships in the bark and ambrosia beetles. In: Wilding, N., Collins, N.M., Hammond, P. and Webber, J.F. (eds) *Insect–Fungus Interactions*. Elsevier Ltd, London, pp. 121–137.

Bellé, C., Moccellin, R., Souza-Júnior, I.T., Maich, S.L.P., Neves, C.G. *et al*. (2018) First report of *Colletotrichum acutatum* causing flower anthracnose on pomegranate (*Punica granatum*) in southern Brazil. *Plant Disease* 102(11), 2373–2374. DOI: 10.1094/PDIS-04-18-0599-PDN.

Benagi, V.I. and Kumar, M.R. (2009) Present status of pomegranate bacterial blight and its management. *Acta Horticulturae* 890, 475–480.

Berbegal, M., López-Cortés, I., Salazar, D., Gramaje, D., Pérez-Sierra, A. *et al*. (2014) First report of *Alternaria* black spot of pomegranate caused by *Alternaria alternata* in Spain. *Plant Disease* 98(5), 689–689. DOI: 10.1094/PDIS-07-13-0717-PDN.

Bichcheri, R., Mirott, A., Babini, A.R. and Poggi Pollini, C. *et al*. (2015) Viral infections in one collection field of pomegranate (*Punica granatum*) in Italy. AGRIS FAO. Available at: http://agris.fao.org/agrissearch/search.do?recordID=US201700109294 (accessed 19 November 2019).

Bompadre, M.J., Colombo, R.P., Silvani, V.A., Fernández Bidondo, L., Pardo, A.G. *et al*. (2018) Pomegranate transplant stress can be ameliorated by *Rhizophagus intraradices* under nursery management. *Journal of Soil Science and Plant Nutrition* 18(ahead), 772–789. DOI: 10.4067/S0718-95162018005002203.

Bozkurt, I.A., Soylu, S., Mirik, M., Ulubas Serce, C. and Baysal, Ö. (2014) Characterization of bacterial knot disease caused by *Pseudomonas savastanoi* pv. *savastanoi* on pomegranate (*Punica granatum* L.) trees: a new host of the pathogen. *Letters in Applied Microbiology* 59(5), 520–527. DOI: 10.1111/lam.12309.

Caglayan, K., Gazel, M., Serce, C.U., Kaya, K. and Cengiz, F.C. (2014) Fruit tree phytoplasmas and their possible insect vectors in Turkey. In: Bertaccini, A. (ed.) *Phytoplasmas and Phytoplasma Disease Management: How to Reduce Their Economic Impact*. IPWG – International Phytoplasmologist Working Group, Bologna, Italy, pp. 130–136.

Caglayan, K., Elçi, E. and Gazel, M. (2016) Detection and partial characterization of grapevine leafroll-associated virus 1 in pomegranate trees in Turkey. *European Journal of Plant Pathology* 145(1), 199–202. DOI: 10.1007/s10658-015-0807-4.

Cannon, P.F., Damm, U., Johnston, P.R. and Weir, B.S. (2012) *Colletotrichum* – current status and future directions. *Studies in Mycology* 73(1), 181–213. DOI: 10.3114/sim0014.

Carstens, E., Langenhoven, S.D., Pierron, R., Laubscher, W., Serfontein, J.J. *et al*. (2018) *Elsinoë punicae* causing scab of pomegranates in South Africa does not cause disease on citrus. *Australasian Plant Pathology* 47(4), 405–411. DOI: 10.1007/s13313-018-0572-x.

Çeliker, N.M., Uysal, A., Çetinel, B. and Poyraz, D. (2012) Crown rot on pomegranate caused by *Coniella granati* in Turkey. *Australasian Plant Disease Notes* 7(1), 161–162. DOI: 10.1007/s13314-012-0074-6.

Chand, R. and Kishun, R. (1991) Studies on bacterial blight (*Xanthomonas campestris* pv. *punicae*) of pomegranate. *Indian Phytopathology* 44, 370–372.

Chavan, S. and Dake, G. (2001) In vitro inhibition of *Fusarium* associated with wilt of pomegranate by rhizobacteria. *Journal of Maharashtra Agricultural Universities* 26, 257–259.

Clark, C.A. and Mayer, J.W. (1988) *Compendium of Sweet Potato Diseases*. 74. American Phytopathological Society, Saint Paul, Minnesota.

Crone, L.J. and Bachelder, S. (1961) Insect transmission of canker stain fungus, *Ceratocystis fimbriata* f.sp. platani. *Phytopathology* 77, 576–582.

Darekar, K.S., Mhase, N.L. and Shelke, S.S. (1989) Management of nematodes infesting pomegranate. *International Nematology Network Newsletter* 6, 15–17.

Day, K.R. and Wilkins, E.D. (2011) Commercial pomegranate (*Punica granatum* L.) production in California. *Acta Horticulturae* 890, 275–286.

Deighton, F.C. (1976) Studies on *Cercospora* and allied genera. *VI. Pseudocercospora Speg., Pantospora Cif., and Cercoseptoria Petr. Mycological Papers* 140, 168.

Dubey, L., Dubey, M. and Jain, P. (2015) Role of green manuring in organic farming. *Plant Archives* 15(1), 23–26.

Elazab, D.S. and Elzawahry, A.M. (2016) Effect of salicylic scid on enhancing growth, some nutrients concentration and carbohydrates metabolism on pomegranate seedlings infected with *Meloidogyne javanica*. *Assiut Journal of Agricultural Sciences* 47, 61–77.

EPPO Global Database (2015) Tomato ringspot virus detected in *Punica granatum* in Italy. Available at: https://gd.eppo.int/reporting/article-5141 (accessed 19 November 2019).

Ezra, D., Gat, T., Skovorodnikova, Y., Vardi, Y. and Kosto, I. (2010) First report of *Alternaria* black spot of pomegranate caused by *Alternaria alternata* in Israel. *Australasian Plant Disease Notes* 5(1), 1–2. DOI: 10.1071/DN10001.

Ezra, D., Kirshner, B., Hershcovich, M., Shtienberg, D. and Kosto, I. (2015) Heart rot of pomegranate: disease etiology and the events leading to development of symptoms. *Plant Disease* 99(4), 496–501. DOI: 10.1094/PDIS-07-14-0707-RE.

Fanelli, E., Troccoli, A., Vovlas, N., Scarcia, G., Mincuzzi, A. *et al.* (2017) Occurrence of *Sheraphelenchus sucus* (Nematoda: Aphelenchoidinae) and *Panagrellus* sp. (Rhabditida: Panagrolaimidae) associated with decaying pomegranate fruit in Italy. *Journal of Nematology* 49(4), 418–426. DOI: 10.21307/jofnem-2017-091.

Farr, D.F. and Rossman, A.Y. (2001) *Harknessia lythri*, a new species on purple loosestrife. *Mycologia* 93(5), 997–1001. DOI: 10.1080/00275514.2001.12063231.

Farr, D.F. and Rossman, A.Y. (2003) *Dwiroopa*, a coelomycetous genus with two species. *Mycoscience* 44(6), 443–446. DOI: 10.1007/S10267-003-0141-0.

Francl, L.J. and Wheeler, T.A. (1993) Interaction of plant-parasitic nematodes with wilt-inducing fungi. In: Khan, M.W. (ed.) *Nematode Interactions*. Chapman & Hall, London, UK, pp. 79–103.

Gajbhiye, M., Sathe, S., Shinde, V. and Kapadnis, B. (2016) Morphological and molecular characterization of pomegranate fruit rot pathogen, *Chaetomella raphigera*, and its virulence factors. *Indian Journal of Microbiology* 56(1), 99–102. DOI: 10.1007/s12088-015-0554-4.

Gazel, M., Çağlayan, K., Başpınar, H., Mejia, J.F., Paltrinieri, S. *et al.* (2016) Detection and identification of phytoplasmas in pomegranate trees with yellows symptoms. *Journal of Phytopathology* 164, 136–140.

Gómez, G. and Pallás, V. (2001) Detection of viroid-like RNAs in pomegranate (*Punica granatum* L.). *Acta Horticulturae* 550, 321–326. DOI: 10.17660/ActaHortic.2001.550.46.

Grosclaude, C., Olivier, R. and Romiti, C. (1995) Chancre colore du platane. Comment l'agent responsable peut survivre dans le sol. *Phytoma* 479, 41–42.

Halsted, B.D. (1890) Some fungous diseases of the sweet potato. *New Jersey Agricultural College Experiment Station Bulletin* 76, 7–14.

Harrington, T.C. (2000) Host specialization and speciation in the American wilt pathogen *Ceratocystis fimbriata*. *Fitologia Brasileira* 25S, 262–263.

Harrington, T.C. (2004) *Ceratocystis fimbriata*. Crop Protection Compendium CD-ROM, CABI.

Harrington, T.C. and McNew, D.L. (1997) Self-fertility and uni-directional mating-type switching in *Ceratocystis coerulescens*, a filamentous ascomycete. *Current Genetics* 32(1), 52–59. DOI: 10.1007/s002940050247.

Hashim, Z. (1983) Plant parasitic nematodes associated with pomegranate (*Punica granatum* L.) in Jordan and an attempt to chemical control. *Nematologia Mediterranea* 11, 199–200.

Hebert, T.T. and Clayton, C.N. (1963) Pomegranate fruit rot caused by *Coniella granati*. *Plant Disease Reporter* 47, 222–223.

Hennings, P. (1906) *Cercospora punicae* Henn. *Botanische Jahrbücher für systematik Pflanzengeschichte und Pflanzengeographie* 37, 165.

Hinds, T.E. (1972) Insect transmission of *Ceratocystis* species associated with aspen cankers. *Phytopathology* 62(2), 221–225. DOI: 10.1094/Phyto-62-221.

Hingorani, M.K. and Mehta, P.P. (1952) Bacterial leaf spot of pomegranate. *Indian Phytopathology* 5, 55–56.

Holland, D., Hatib, K. and Bar-Ya'akov, I. (2009) Pomegranate: botany, horticulture, breeding. *Horticultural Reviews* 35, 127–191.

Hu, F.P., Ke, O.H., Tu, R. and Huang, Y.Y. (1999) Causes of black rot and bitter rot of sweet potato in Liancheng, Fujian. *Journal of Fujian Agriculture and Forestry University* 28, 441–444.

Huang, Q., Zhu, Y.Y., Chen, H.R., Wang, Y.Y., Liu, Y.L. *et al.* (2003) First report of pomegranate wilt caused by *Ceratocystis fimbriata* in Yunnan, China. *Plant Disease* 87(9), 1150–1161. DOI: 10.1094/PDIS.2003.87.9.1150B.

Hulcr, J., Mann, R. and Stelinski, L.L. (2011) The scent of a partner: ambrosia beetles are attracted to volatiles from their fungal symbionts. *Journal of Chemical Ecology* 37(12), 1374–1377. DOI: 10.1007/s10886-011-0046-x.

Icoz, S.M., Polat, I., Sulu, G., Yilmaz, M., Unlu, A. *et al.* (2014) First report of bacterial blight of pomegranate caused by *Xanthomonas axonopodis* pv. *punicae* in Turkey. *Plant Disease* 98(10), 1427. DOI: 10.1094/PDIS-06-14-0656-PDN.

Ilangovan, M. and Poornima, K. (2017) Occurrence and distribution of plant parasitic nematodes in pomegranate growing areas of Tamil Nadu. *International Journal of Biological and Pharmaceutical Research* 8, 106–111.

Iton, E.F. (1966) *Ceratocystis wilt*. University of the West Indies, St. Augustine, Trinidad, pp. 44–56.

Jabnoun-Khiareddine, H., Ibrahim, N., Ben Abdallah, R.A., Mars, M., Kthiri, Z. *et al*. (2018) *Coniella granati* (saccardo) a new potential threat to pomegranate (*Punica granatum* L.) in Tunisia causing twig dieback and fruit rot. *Journal of Plant Pathology & Microbiology* 09(09), 450. DOI: 10.4172/2157-7471.1000450.

Jadhav, V.T. and Sharma, K.K. (2011) Integrated management of diseases in pomegranate. *Acta Horticulturae* 890, 467–474.

Jagdale, S.B., Sonawane, M.S. and Kapadnis, B.P. (2018) A new bacterial blight of pomegranate caused by *Pseudomonas* sp. in Maharashtra, India. *Australasian Plant Disease Notes* 13(1), 27. DOI: 10.1007/s13314-018-0311-8.

Jamadar, M.M, Patil, P.V., Jawadagi, R.S. and Patil, D.R. (2011) Status of pomegranate diseases of northern Karnataka in India. *Acta Horticulturae* 890,501–507. DOI: 10.17660/ActaHortic.2011.890.70.

Jambhulkar, P.P., Sharma, M., Lakshman, D. and Sharma, P. (2015) Natural mechanisms of soil suppressiveness against diseases caused by *Fusarium*, *Rhizoctonia*, *Pythium*, and *Phytophthora*. In: Meghvansi, M. and Varma, A. (eds) *Organic Amendments and Soil Suppressiveness in Plant Disease Management: Soil Biology*. 46. Springer, Cham, Switzerland, pp. 95–123.

Jayalakshmi, K., Nargund, V.B., Raju, J., Benagi, V.I., Ram, R. *et al*. (2015) Pomegranate anthracnose caused by *Colletotrichum gloeosporioides*: A menance in quality fruit production. *Journal of Pure and Applied Microbiology* 9, 3093–3097.

Jeffries, P., Dodd, J.C., Jeger, M.J. and Plumbley, R.A. (1990) The biology and control of *Colletotrichum* species on tropical fruit crops. *Plant Pathology* 39(3), 343–366. DOI: 10.1111/j.1365-3059.1990.tb02512.x.

Juretic, N. and Horvath, J. (1984) Isolation of cucumber mosaic virus from pomegranate (*Punica granatum* L) in Yugoslavia. *Acta Phytopathologica Academiae Scientiarum Hungaricae* 19, 309–313.

Kahramanoglu, I. and Usanmaz, S. (2016) *Pomegranate Production and Marketing*. CRC Press, Taylor & Francis Group, Boca Raton, FL.

Karimi, M.R., Paltrinieri, S., Contaldo, N., Kamali, H., Sajadinejad, M. *et al*. (2015) Phytoplasma detection and identification in declining pomegranate in Iran. *Phytopathogenic Mollicutes* 5(2), 95–99. DOI: 10.5958/2249-4677.2015.00067.5.

Katwal, V.S. (2015) Studies on bacterial blight of pomegranate caused by *Xanthomonas axonopodis* pv. *punicae* (Hingorani and Singh). Vauterin, et al. PhD Thesis. University of Horticulture & Forestry, Solan, India.

Kechakmadze, L.A., Beradze, L.A. and Kikvadze, I.V. (1990) Grey rot of pomegranate, persimmon and fig. *Subtrop.Kul't* 5, 116–118.

Khosla, K., Gupta, A.K. and Bhardwaj, S.S. (2011) Occurrence of pomegranate wilt caused by *Ceratocystis fimbriata* in Himachal Pradesh. *Journal of Mycology and Plant Pathology* 41(3), 117–118.

Khosla, K. and Gupta, A.K. (2014) *Phytophthora* fruit rot of pomegranate – a new report from Himachal Pradesh. *Plant Disease Research* 29, 105–107.

Kore, S.S. and Mitkar, P.L. (1993) Dry root rot disease of pomegranate incited by *Fusarium solani*. *Journal of Maharashtra Agricultural University* 18, 256–258.

Kulkarni, S.R. and Gupta, N.S. (2007) Integrated pest management of pomegranate. In: *Advances in Arid Zone Fruit Culture*. Centre of Advanced Studies in Horticulture (Fruits) MPKVV, Rahuri, India, pp. 180–187.

Kumar, M.R., Shamarao, J., Srinivaschary. and Ryagi, Y.H. (2001) Management of clump rot of pomegranate caused by *Fusarium* spp. *Agricultural Science Digest* 21, 210.

Kumar, M.R., Shamarao, J., Yenjereapp, S.T. and Patil, H.B. (2006) Epidemiology and management of bacterial blight of pomegranate caused by *Xanthomonas axonopodis* pv. *punicae*. *Acta Horticulturae* 818, 291–296.

Kumar, A. and Chahal, T.S. (2016) Studies of *Aspergillus* black spot disease of pomegranate caused by *Aspergillus niger* in Punjab. *The Bioscan* 11(4), 2775–2781.

Kumar, A., Sharma, J., Munjal, V., Sakthivel, K., Thalor, S.K. *et al*. (2020) Polyphasic phenotypic and genetic analysis reveals clonal nature of *Xanthomonas axonopodis* pv. *punicae* causing pomegranate bacterial blight. *Plant Pathology* 69(2), 347–359.

Lande, P.S. and Utikar, P.G. (1979) Studies on fruit spot of pomegranate caused by *Drechslera rostrata* (India). *Indian Journal of Mycology and Plant Pathology* 8, 205.

Levy, E., Elkind, G., Ben-Arie, R. and Ben-Ze'ev, I.S. (2011) First report of *Coniella granati* causing pomegranate fruit rot in Israel. *Phytoparasitica* 39(4), 403–405. DOI: 10.1007/s12600-011-0171-7.

Liu, H.X., Li, X.D., Zhu, X.P. and Liu, A.X. (2009) First report of pomegranate stem scab caused by *Botryosphaeria dothidea* in China. *Plant Pathology* 58(2), 400. DOI: 10.1111/j.1365-3059.2008.01964.x.

Ma, Z., Morgan, D.P., Felts, D. and Michailides, T.J. (2002) Sensitivity of *Botryosphaeria dothidea* from California pistachio to tebuconazole. *Crop Protection* 21(9), 829–835. DOI: 10.1016/S0261-2194(02)00046-7.

Madhukar, J. and Reddy, S.M. (1976) Some new leaf spot diseases of pomegranate. *Indian Journal of Mycology and Plant Pathology* 18, 171–172.

Mahdikhani, M. and Davoodi, A. (2017) First report of wood canker of pomegranate caused by *Cytospora punicae* in western Iran. *New Disease Reports* 35, 1. DOI: 10.5197/j.2044-0588.2017.035.001.

Maity, A., Sharma, J., Jadhav, V.T., Babu, D. and Chandra, R. (2012) Effect of solarization on nutrient availability, enzyme activity and growth of pomegranate (*Punica granatum*) air-layered on various potting mixtures. *Indian Journal of Agricultural Sciences* 82, 775–782.

Maity, A., Sharma, J., Sarkar, A., More, A.K. and Pal, R.K. (2016) Nutrient imbalance indices are closely related with susceptibility of pomegranate to bacterial blight disease. *Scientia Horticulturae* 211, 79–86. DOI: 10.1016/j.scienta.2016.08.012.

Maity, A., Sharma, J., Sarkar, A., More, A.K., Pal, R.K. *et al.* (2018) Salicylic acid mediated multi-pronged strategy to combat bacterial blight disease in pomegranate caused by *Xanthomonas axonopodis* pv. *punicae*. *European Journal of Plant Pathology* 150, 923–937.

Mali, R.B. (2015) Investigation on bacterial blight of pomegranate incited by *Xanthomonbas axonopodis* pv. *punicae* (Hingorani and Singh). MSc Thesis. Vasantrao Naik Marathwada Agricultural University, Parbhani, India.

Marathe, R.A., Sharma, J., Babu, K.D. and Murkute, A.A. (2017) Bedding system: a unique plantation method of pomegranate in arid and semi-arid region. *National Academy Science Letters* 40(4), 249–251. DOI: 10.1007/s40009-017-0567-0.

Markakis, E.A., Tzima, A.K., Palavouzis, S.C., Antoniou, P.P., Paplomatas, E.J. *et al.* (2017) First report of *Phytophthora palmivora* causing fruit rot on pomegranate in Greece. *Plant Disease* 101(6), 1060. DOI: 10.1094/PDIS-11-16-1691-PDN.

Mazzani, C. (1994) *Colletotrichum gloeosporioides* causing a severe spotting and rot of pomegranate fruits in Venezuela. *Fitopatologia-Venezolana* 7, 28.

McRae, W. (1924) *Economic Botany. Part III. Mycology*. 1924. Annual Report of Board of Science and Advisory Council, India, pp. 31–35.

Mendes, M.A.S., da Silva, V.L., Dianese, J.C., Ferreira, M.A.S.V., dos Santos, C.E.N. *et al.* (1998) *Fungos em plantas no Brasil*. 555. EMBRAPA-SPI, Brasilia.

Mhase, N.L. (2007) Management of plant parasitic nematodes of dryland fruit crops. In: Patil, R.S. (ed.) *Advances in Arid Zone Fruit Culture*. Centre of Advanced Studies in Horticulture, MPKV, Rahuri, India, pp. 173–177.

Mincuzzi, A., Garganese, F., Ippolito, A. and Sanzani, S.M. (2016) First report of Pilidiella granati causing postharvest fruit rot on pomegranate in southern Italy. *Journal of Plant Pathology* 98, 377.

Mincuzzi, A., Sanzani, S.M., Garganese, F., Ligorio, A. and Ippolito, A. (2017) First report of *Cytospora punicae* causing fruit rot on pomegranate in Italy. *Journal of Plant Pathology* 99, 302.

Mirtalebi, M., Banihashemi, Z. and Sabahi, F. (2015) First report of *Pilidiella granati* on pomegranate with symptoms of crown rot in Fars Province. *Iranian Journal of Plant Pathology* 51, 111–113.

Mondal, K.K., Rajendran, T.P., Phaneendra, C., Mani, C. and Shrama, J. (2012) The reliable and rapid polymerase chain reaction (PCR) diagnosis for *Xanthomonas axonopodis* pv. *punicae* in pomegranate. *African Journal of Microbiology Research* 6, 5950–5596.

More, W.D., Banger, S.G. and Khetmalab, M.D. (1989) A new fruit rot disease of pomegranate in Maharashtra. *Journal of Maharashtra Agricultural Universities* 14, 386.

Munhuweyi, K., Lennox, C.L., Meitz-Hopkins, J.C., Caleb, O.J. and Opara, U.L. (2016) Major diseases of pomegranate (*Punica granatum* L.), their causes and management – a review. *Scientia Horticulturae* 211, 126–139.

Nakashima, C., Motohashi, K., Chen, C., Groenewald, J.Z. and Crous, P.W. (2016) Species diversity of *Pseudocercospora* from Far East Asia. *Mycological Progress* 15, 1093–1117.

Nargund, V.B., Jayalakshmi, K., Benagi, V.I., Byadgi, A.S. and Patil, R.V. (2012) Status and management of anthracnose of pomegranate in Karnataka state of India. In: Melgarejo, P. and Valero, D. (eds) *II International Symposium on the Pomegranate, Options Méditerranéennes, Series A: Mediterranean Seminars, No. 103*. Madrid, Spain, pp. 117–120.

Nattrass, R.M. (1932) *Annual Report of the Mycologist*. Department of Agriculture, Cyprus, pp. 44–49.

Neama, S. and Sharma, N.D. (2006) Occurrence of *Phytophthora nicotianae* on pomegranate in Madhya Pradesh. *Indian Phytopathology* 59, 128.

NRCP (2008) *ICAR–NRCP Annual Report 2007–08*. National Research Centre on Pomegranate, Solapur, India, pp. 38–41.

NRCP (2009) *ICAR–NRCP Annual Report 2008–09*. National Research Centre on Pomegranate, Solapur, India, pp. 47–48.

NRCP (2012) *ICAR–NRCP Annual Report 2011–12*. National Research Centre on Pomegranate, Solapur, India, pp. 92–93.

NRCP (2015) *ICAR–NRCP Annual Report 2014–15*. National Research Centre on Pomegranate, Solapur, India, pp. 58–59.

NRCP (2016) *ICAR–NRCP Annual Report 2015–16*. National Research Centre on Pomegranate, Solapur, India, p. 77.

Onelge, N. (2000) Occurrence of hop stunt viroid (HSVd) on pomegranate (*Punica granatum*) trees in Turkey. *Journal of Turkish Phytopatholgy* 29, 49–52.

Pala, H., Tatli, A., Yilmaz, C. and Ozguven, A.I. (2009) Important diseases of pomegranate fruit and control possibilities in Turkey. *Acta Horticulturae* 818, 285–290.

Palavouzis, S.C., Tzamos, S., Paplomatas, E. and Thomidis, T. (2015a) First report of *Neofusicoccum parvum* causing shoot blight of pomegranate in northern Greece. *New Disease Reports* 32, 10.

Palavouzis, S.C., Tzamos, S., Paplomatas, E. and Thomidis, T. (2015b) First report of *Cytospora punicae* isolated from pomegranate plants with symptom of collar rot in northern Greece. *Journal Plant Pathology* 97, 209–220.

Palou, L., Montesinos-Herrero, C. and Guardado, A. (2010) A first report of *Penicillium* spp. and *Pilidiella granati* causing postharvest fruit rot of pomegranate in Spain. *New Disease Reports* 22, 21.

Palou, L., Taberner, V., Guardado, A., Del Río, M. A. and Montesinos-Herrero, C. (2013) Incidence and etiology of postharvest fungal diseases of pomegranate (*Punica granatum* cv. Mollar de Elche) in Spain. *Phytopathologia Mediterranea* 52, 478–489.

Pantidou, M.E. (1973) *Fungus – Host Index for Greece*. 382. Benaki Phytopathology Institute, Kiphissia, Athens.

Peduto Hand, F., Choudhury, R.A. and Gubler, W.D. (2014) First report of *Cytospora punicae* causing wood canker and branch dieback of pomegranate (*Punica granatum*) in the United States. *Plant Disease* 98(6), 853–853. DOI: 10.1094/PDIS-11-13-1133-PDN.

Peres, N.A.R., Kuramae, E.E., Dias, M.S.C. and De Souza, N.L. (2002) Identification and characterization of *Colletotrichum* spp. affecting fruit after harvest in Brazil. *Phytopathology* 150, 128–134.

Peres, N., Timmer, L., Adaskaveg, J. and Correll, J. (2005) Lifestyles of *Colletotrichum acutatum*. *Plant Disease* 89, 784–796.

Petersen, Y., Mansvelt, E.L., Venter, E. and Langenhoven, W.E. (2010) Detection of *Xanthomonas axonopodis* pv. *punicae* causing bacterial blight on pomegranate in South Africa. *Australasian Plant Pathology* 39, 544–546.

Phengsintham, P., Chukeatirote, E., McKenzie, E.H.C., Hyde, K.D. and Braun, U. (2011) Tropical phythopathogens 1: *Pseudocercospora punicae*. *Plant Pathology and Quarantine* 1(1), 1–6.

Pollastro, S., Dongiovanni, C., Gerin, D., Pollastro, P., Fumarola, G. *et al*. (2016a) First report of *Coniella granati* as a causal agent of pomegranate crown rot in southern Italy. *Plant Disease* 100, 1498.

Pollastro, S., Gerin, D., Marullo, S., De Miccolis Angelini, R.M. and Faretra, F. (2016b) First report of *Erysiphe* sp. as a causal agent of powdery mildew on *Punica granatum* in Italy. *Plant Disease* 100, 1949.

Prashanth, A. and Sataraddi, A.R. (2011) Variability in different isolates of anthracnose of pomegranate caused by *Colletotrichum gloeosporioides*. *Acta Horticulturae* 890, 533–538.

Praveen, Y. (2018) Effect of elicitors in induction of resistance against bacterial blight and phenylepropanoid pathway in pomegranate. MSc Thesis. University of Horticultural Sciences, Bagalkot, India.

Pscheidt, J.W. (2019) Diagnosis and control of Phytophthora diseases. In: Pscheidt, J.W. and Ocamb, C.M. (eds) *Pacific Northwest Plant Disease Management Handbook*. Oregon State University, Corvallis, Oregon.

Puneeth, M.E. (2015) *Biocontrol of bacterial blight of pomegranate caused by Xanthomonas axonopodis pv. punicae (Hingorani and Singh) Vauterin, et al. MSc Thesis*. 112. Department of Plant Pathology University of Agricultural Sciences, Bengaluru, Karnataka, India.

Raghuvanshi, K.S. (2007) Management of pomegranate wilt. In: Patil, R.S. (ed.) *Advances in Arid Zone Fruit Culture*. Centre of Advanced Studies in Horticulture (Fruits), MPKVV Rahuri, Maharashtra, India, pp. 178–179.

Rahimlou, S., Babaeizad, V. and Sayari, M. (2014) First report of fruit spot of pomegranate caused by *Colletotrichum gloeosporioides* in Iran. *Journal of Plant Pathology* 96(3), 603–611.

Rani, U. and Verma, K.S. (2002) Perpetuation and spread of *Xanthomonas axonopodis* pv. *punicae* causing black spot of pomegranate. *Plant Disease Research* 17, 46–50.

Rawla, G.S. (1971) *Cercospora granati* sp. *nov*. on *Punica*. *Transactions of the British Mycological Society* 56(3), 483–484.

Rui, G., Jie, W., Tiansheng, Z., Xi, J. and Xiangdong, L. (2018) Identification and molecular characterization of a phytoplasma associated with pomegranate fasciation disease. *Horticultural Plant Journal* 4, 30–34.

Salehi, M., Hosseini, S.A.E., Rasoulpour, R., Salehi, E. and Bertaccini, A. (2016) Identification of a phytoplasma associated with pomegranate little leaf disease in Iran. *Crop Protection* 87, 50–54.

Samouel, S. and Kanetis, L. (2016) First report of *Cytospora punicae* causing trunk canker of pomegranate (*Punica granatum*) in Cyprus. *Plant Disease* 100, 222.

Sataraddi, A.R., Prashanth, A., Prabhu, H.V., Jamadar, M.M. and Aski, S. (2011) Role of bio-agents and botanicals in the management of anthracnose of pomegranate. *Acta Horticulturae* 890, 539–544.

Satyagopal, K., Sushil, S.N., Jeyakumar, P., Shankar, G., Sharma, O.P. *et al*. (2014) *AESA Based IPM Package for Pomegranate*. National Institute of Plant Health Management, Hyderabad, India. Available at: https://farmer.gov.in/imagedefault/ipm/pomegranate.pdf (accessed 15 November 2019).

Schmitz, S., Zini, J., Etienne, M., Moreau, J.M., Chandelier, A. *et al*. (2006) Effectiveness of thiophanate-methyl. trifloxystrobin and vinclozolin on canker caused by *Phoma exigua* Desm. on ash tree seedlings. *Biotechnology, Agronomy, Society and Environment* 10(1), 25–31.

Sharma, R., Roy, A. and Singh, G. (1982) A new fruit rot of pomegranate caused by *Aspergillus variecolor*. *Current Science* 51, 378.

Sharma, K.K., Sharma, J., Jadhav, V.T. and Chandra, R. (2008) Bacterial blight of pomegranate and its management. *Indian Phytopathology* 61, 380–381.

Sharma, K.K., Sharma, J. and Jadhav, V.T. (2010a) Status of bacterial blight of pomegranate in India. In: Chandra, R. (ed.) *Pomegranate. Fruit, Vegetable and Cereal Science and Biotechnology*, Special Issue 2. Vol. 4. Global Science Books, Japan, pp. 102–105.

Sharma, K.K., Sharma, J. and Jadhav, V.T. (2010b) Etiology of pomegranate wilt and its management. In: Chandra, R. (ed.) *Pomegranate. Fruit, Vegetable and Cereal Science and Biotechnology*, Special Issue 2. Vol. 4. Global Science Books, Japan, pp. 96–101.

Sharma, J., Sharma, K.K. and Jadhav, V.T. (2012) Diseases of pomegranate. In: Misra, A.K., Chowdappa, P., Sharma, P. and Khetrapal, R.K. (eds) *Diseases of Fruit Crops*. Indian Phytopathological Society, New Delhi, pp. 181–224.

Sharma, J., Chandra, R., Sharma, K.K., Babu, K.D. and Meshram, D.T. (2014) Pomegranate cultivation, marketing and utilization.Technical Bulletin. No. NRCP/2014/1. ICAR-National Research Centre on Pomegranate, Solapur, India.

Sharma, G., Pinnaka, A.K. and Shenoy, B.D. (2015) Resolving the *Colletotrichum siamense* species complex using ApMat marker. *Fungal Diversity* 71, 247–264.

Sharma, K.K. (2009) Vascular wilt of pomegranate caused by *Ceratocystis fimbriata* Ellis and Halsted and its control. *5th International Conference on Plant Pathology in the Globalized Era*, IARI, New Delhi, 10–13 November.

Sharma, J. (2017) Pomegranate bacterial blight: present status and research developments. *2nd National Seminar-cum-Farmers' Fair*, ICAR-National Research Centre on Pomegranate, Solapur, India, 28–30 April, pp. 148–155.

Sharma, J. and Jadhav, V.T. (2011) Network project on mitigating the bacterial blight disease of pomegranate in Maharashtra, Karnataka and Andhra Pradesh: abridged progress report, 2008–09 to 2010–11. National Research Centre on Pomegranate, Solapur, India.

Sharma, J. and Jadhav, V.T. (2012) Network project on mitigating bacterial blight of pomegranate in Maharashtra, Karnataka and Andhra Pradesh: progress report April 2011 to October 2012. National Research Centre on Pomegranate, Solapur, India.
Sharma, N.D. and Jain, A.C. (1978) Two new fruit rot diseases of pomegranate (*Punica granatum* L.) caused by *Coniella* spp. *Current Science* 47(23), 908–909.
Sharma, J. and Sharma, K.K. (2017) *Pomegranate Diseases and their Management*. In: Pal, R.K. and Singh, N.V. (eds) *Pomegranate for Nutrition, Livelihood Security and Entrepreneurship Development*. Daya Publishing House, New Delhi, India, pp. 169–176.
Sharma, R.L. and Tegta, R.K. (2011) Incidence of dry rot of pomegranate in Himachal Pradesh and its management. *Acta Horticulturae* 890, 491–499.
Sharma, J., Gharate, R. and Chinchure, S. (2017a) Fungal flora associated with bacterial blight stem cankers in pomegranate. *2nd National Seminar-cum-Farmer's Fair on Pomegranate for Health, Growth and Prosperity*, Jointly organized by ICAR- National Research Centre on Pomegranate, Solapur and Society for Advancement of Research on Pomegranate, Solapur, India, 28–30 April.
Sharma, J., Sharma, K.K., Kumar, A., Kalyan, K. and Thalor, S. (2017b) Pomegranate bacterial blight: symptomatology and rapid inoculation technique for *Xanthomonas axonopodis* pv. *punicae*. *Journal of Plant Pathology* 99, 109–119.
Sherkar, B.V. and Utikar, P.G. (1982) *Beltraniella humicola* – a new fruit spotting fungus on pomegranate. *Indian Journal of Mycology and Plant Pathology* 12, 50.
Shivas, R.G., Tan, Y.P., Edwards, J., Dinh, Q., Maxwell, A. *et al*. (2016) *Colletotrichum* species in Australia. *Australasian Plant Pathology* 45, 447–464.
Siboe, G.M., Birgen, J.K. and Subramaniam, V. (1982) Leaf blotch and fruit rot of pomegranate. *FAO Plant Protection Bulletin* 30, 161–162.
Siddiqui, Z.A. and Khan, M.W. (1986) A survey of nematodes associated with pomegranate in Libya and evaluation of some systemic nematicides for their control. *Pakistan Journal of Nematology* 4, 83–90.
Simmons, E.G. (1967) Typification of *Alternaria*, *Stemphylium*, and *Ulcoladium*. *Mycologia* 59(1), 67–92.
Singh, R.K., Sharma, J., Jha, S. and Singh, A. (2012) Solarization technique: Its use in the multiplication of *in vitro* planting materials. *Current Science* 102(10), 1433–1436.
Singh, N.V., Abburi, V.L., Ramajayam, D., Kumar, R., Chandra, R. *et al*. (2015) Genetic diversity and association mapping of bacterial blight and other horticulturally important traits with microsatellite markers in pomegranate from India. *Molecular Genetics and Genomics: MGG* 290(4), 1393–1402. DOI: 10.1007/s00438-015-1003-0.
Singh, N.V., Sharma, J., Chandra, R., Babu, K.D., Shinde, Y.R. *et al*. (2016) Bio-hardening of *in-vitro* raised plants of Bhagwa pomegranate (*Punica granatum*). *Indian Journal of Agricultural Sciences* 86(1), 132–136.
Singh, R.S. and Chohan, J.S. (1972) A new fruit spot disease of pomegranate. *Current Science* 41, 651.
Sohi, H.S., Sharma, S.L. and Gupta, G.K. (1965) New diseases of pomegranate. *Plant Protection Bulletin, FAO* 13, 113.
Somasekhara, Y.M. (1999) New record of *Ceratocystis fimbriata* causing wilt of pomegranate in India. *Plant Disease* 83(4), 400–408. DOI: 10.1094/PDIS.1999.83.4.400B.
Somasekhara, Y.M. (2002) Application of Bacillus subtilis in the management of pomegranate (*Punica granatum* Linn.) wilt (*Ceratocystis fimbriata* Elli. and Halst). *Disease Research Crops* 3, 202–203.
Somasekhara, Y.M. (2006) Spacious distribution of wilt (*Ceratocystis fimbriata*) of pomegranate (*Punica granatum* L) in India. *Research on Crops* 7, 844–853.
Somasekhara, Y.M. and Gaddanakeri, M.A. (2009) Host specificity of pomegranate (*Punica granatum* L.) wilt pathogen, *Ceratocystis fimbriata*. *2nd International Symposium on Pomegranate and Minor including Mediterranean Fruits*, UAS Dharwad, India, 23–27 June.
Somasekhara, Y.M., Wali, S.Y. and Shaik, M.K. (2009) Studies and the management of pomegranate (*Punica granatum* L.) wilt (*Ceratocystis fimbriata*, Latin American group). *2nd International Symposium on Pomegranate and Minor including Mediterranean Fruits*, University of Agricultural Sciences, Dharwad, India, 23–27 June, pp. 132–133.
Sonawane, C.S., Utikar, P.G. and Shinde, P.A. (1986) Postharvest fungal flora of pomegranate. *Journal of the Agricultural University Maharashtra* 11, 107–110.
Sonyal, S. (2010) Studies on pomegranate wilt complex. MSc Thesis. University of Agricultural Sciences, Dharwad, India.

Sugimoto, T., Watanabe, K., Yoshida, S., Aino, M., Irie, K. *et al.* (2008) Select calcium compounds reduce the severity of *Phytophthora* stem rot of soybean. *Plant Disease* 92(11), 1559–1565. DOI: 10.1094/PDIS-92-11-1559.
Sztejnberg, A. and Madar, Z. (1980) Host range of *Dematophora necatrix*, the cause of white root rot disease in fruit trees. *Plant Disease* 64, 662–664.
Teviotdale, B.L. and Harper, D.H. (1991) Infection of pruning and small bark wounds in almond by *Ceratocystis fimbriata*. *Plant Disease* 75, 1026–1030.
Thind, S.K. (2017) Principles of disease management in fruit crops. *International Clinical Pathology Journal* 4(5), 123–137.
Thomidis, T. (2014) Fruit rots of pomegranate (cv. Wonderful) in Greece. *Australasian Plant Pathology* 43, 583–588.
Thomidis, T. (2015) Pathogenicity and characterization of *Pilidiella granati* causing pomegranate diseases in Greece. *European Journal of Plant Pathology* 141, 45–50.
Thomidis, T. and Exadaktylou, E. (2011) First report of *Pilidiella granati* on pomegranate with symptoms of crown rot in the Prefecture of Xanthi, Greece. *Plant Disease* 95(1), 79. DOI: 10.1094/PDIS-07-10-0514.
Thomidis, T. and Michailides, T.J. (2009) Studies on *Diaportheeres* as a new pathogen of peach trees in Greece. *Plant Disease* 93(12), 1293–1297. DOI: 10.1094/PDIS-93-12-1293.
Turkolmez, S., Ciftci, O., Serce, C.U. and Dervis, S. (2015) First report of *Phytophthora palmivora* causing crown and root rot on pomegranate (*Punica granatum* L.) in Turkey. *Plant Disease* 100, 227.
Tziros, G.T., Lagopodi, A.L. and Tzavella-Klonari, K. (2008) *Alternaria alternata* fruit rot of pomegranate (*Punica granatum*) in Greece. *Plant Pathology* 57, 379.
Tziros, G.T. and Tzavella-Klonari, K. (2008) First report of *Verticillium* wilt of pomegranate caused by *Verticillium* dahliae in Greece. *Journal of Plant Pathology* 90, 589–595.
Urbez-Torres, J.R., Hand, F.P. and Trouillas, F.P. *et al.* (2017) Pomegranate dieback caused by *Lasiodiplodia gilanensis* in California. *European Journal of Plant Pathology* 148, 223–228.
Utikar, P.G., Lande, P.S. and More, B.B. (1977) *Drechslera rostrata* – a new pathogen of pomegranate. *Indian Phytopathology* 29, 189.
Utikar, P.G., Sherkar, B.V., More, B.B. and Shinde, P.A. (1980) *Pestalotiopsis versicolor* – a new fruit spot pathogen on pomegranate from India (*Punica granatum* L). *Indian Phytopathology* 33, 343–344.
UTZ (2015) List of banned pesticides and pesticides watchlist, version 1.0, 17pp. Available at: https://utz.org/wp-content/uploads/2015/12/EN_UTZ_List-of-Banned-PesticidesWatchlist_v1.0_2015.pdf (accessed 20 November, 2019).
Uysal, A. and Kurt, S. (2018) *Colletotrichum gloeosporiodes* causing anthracnose on pomegranate in Turkey. *Australasian Plant Disease Notes* 13, 19.
Vauterin, L., Hoste, B., Kersters, K. and Swings, J. (1995) Reclassification of *Xanthomonas*. *International Journal of Systematic Bacteriology* 45, 472–489.
Verma, R.R. (1985) Susceptibility of some pomegranate varieties to root-knot nematode. *Indian Journal of Nematology* 15, 247–247.
Viegas, A.P. (1960) Mango blight. *Review of Applied Mycology* 19, 163–182.
von Broembsen, S.L. and Deacon, J.W. (1997) Calcium interference with zoospore biology and infectivity of *Phytophthora parasitica* in nutrient irrigation solutions. *Phytopathology* 87(5), 522–528. DOI: 10.1094/PHYTO.1997.87.5.522.
Walter, J.M. (1946) *Canker Stain Plane Trees*. No. 742. 12. USDA Circular.
Walter, J.M., Rex, E.G. and Schreiber, R. (1952) The rate of progress and destructiveness of canker stain of plane trees. *Phytopathology* 42, 236–239.
Webster, R.K. and Butler, E.E. (1967) A morphological and biological concept of the species *Ceratocystis fimbriata*. *Canadian Journal of Botany* 45, 1457–1468.
Weerahewa, D. and Somapala, K. (2016) Role of silicon on enhancing disease resistance in tropical fruits and vegetables: a review. *OUSL Journal* 11, 135–162.
Westerdahl, B., Long, D. and Schiller, C.T. (2014) Nimitz (mcw-2) for management of root-knot nematode on annual crops. *Acta Horticulturae* 1044, 353–358.
Wharton, P.S. and Diéguez-Uribeondo, J. (2004) The biology of *Colletotrichum acutatum*. *Anales del Jardín Botánico de Madrid* 61, 3–22.
Whitelaw-Weckert, M.A., Curtin, S.J., Huang, R., Steel, C.C., Blanchard, C.L. *et al.* (2007) Phylogenetic relationships and pathogenicity of *Colletotrichum acutatumisolates* from grape in subtropical. *Australasian Plant Pathology* 56, 448–463.
Wolf, F.A. (1927) Pomegranate blotch. *Journal of Agricultural Research* 35(5), 465–469.

Xavier, K.V., Kc, A.N., Peres, N.A., Deng, Z., Castle, W. *et al*. (2019a) Characterization of *Colletotrichum* species causing anthracnose of pomegranate in the southeastern United States. *Plant Disease* 103(11), 2771–2780. DOI: 10.1094/PDIS-03-19-0598-RE.

Xavier, K.V., Kc, A.N., Crous, P.W., Groenewald, J.Z. and Vallad, G.E. (2019b) *Dwiroopa punicae* sp. nov. (Dwiroopaceae fam. nov., Diaporthales), associated with leaf spot and fruit rot of pomegranate (*Punica granatum*). *Fungal Systematics and Evolution* 4(1), 33–41. DOI: 10.3114/fuse.2019.04.04.

Xu, B., Zheng, X.H., Guo, W.X., Zhou, X.P. and He, P. (2011) First report of pomegranate wilt caused by *Ceratocystis fimbriata* in Sichuan Province. *Plant Disease* 95(6), 776. DOI: 10.1094/PDIS-02-11-0146.

Yadav, B.S., Varma, M.K. and Naik, S.M. (1970) A note on the prevalence of *Meloidogyne incogni*ta in various plants of Rajasthan. *Current Science* 39(20), 470–471.

Yenjerappa, S.T. (2009) Epidemiology and management of bacterial blight of pomegranate caused by *Xanthomonas axonopodis* pv. *punicae* (Hingorani and Singh). PhD Thesis. University of Agricultural Sciences, Dharwad, India.

Zhang, L. and McCarthy, M.J. (2012) Black heart characterization and detection in pomegranate using NMR relaxometry and MR imaging. *Postharvest Biology and Technology* 67, 96–101. DOI: 10.1016/j.postharvbio.2011.12.018.

Zhou, Y., Yang, Y., Guo, J., Bai, J. and Hu, X. (2018) *Black spot disease of pomegranate caused by Cladosporium cladosporioides in China*. 170. Earth and Environmental Science, IOP Conference Series, pp. 1–4.

13 Arthropod Pests and Their Management

G.E. Massimino Cocuzza[1*], S.H. Goldansaz[2] and M. Harsur[3]
[1]*Department of Agriculture, Food and Environment, University of Catania, Catania, Italy;* [2]*Department of Plant Protection, University of Teheran, Karaj, Iran;* [3]*ICAR-National Research Centre on Pomegranate, Solapur, India*

13.1 Introduction

The pomegranate is traditionally and optimistically considered a rustic crop, easily adaptable to a multiplicity of environments and with very few pests. Although it is a plant that has been well known since antiquity, in recent years it has undergone an unexpected revaluation, which has led to a considerable increase in cultivated areas. India and Iran are still the largest producer countries in which 90% of production is concentrated, but other countries have specialized plantations, among which Turkey, Spain, the USA and China stand out. In most of these countries, for decades pomegranate has been restricted to agricultural plots for family use or public gardens as an ornamental plant, and only recently is it emerging as an income crop to be planted as an alternative to other less profitable crops or on marginal land with limited irrigation availability (La Malfa *et al.*, 2009; Pal *et al.*, 2014). Even the traditional producers of India and Iran have seen an increasing demand for the fruit to be exported, but fruit must have high quality standards and low or no pesticide residues.

The global change of perspective on pomegranate has been accompanied by the introduction of new varieties, more appreciated in terms of both quality and quantity. These varieties have led to an increase in interest in the phytosanitary problems of pomegranate, which if not correctly managed can seriously compromise production. The control of pests is not easy, owing to the lack of information available and restrictions on the use of pesticides imposed in many countries. Worldwide, more than 80 phytophagous pests have been reported on pomegranate, and less than a dozen are to be considered as potential serious pests. Except for *Aphis punicae* Passerini and some mites, all the phytophagous pests are polyphagous and not strictly correlated with pomegranate. Typically, their infestations are to be considered as abnormal events, often a consequence of incorrect culture management. In pomegranate, the pest control should be above all preventive, applying integrated methods that do not favour the population of phytophagous pests and that maximize the activity of their numerous natural enemies. The use of pesticides should be minimized and justified by the lack of valid alternatives.

In the present chapter, the first part deals with pomegranate integrated pest management (IPM), followed by a description of the main

*Corresponding author: cocuzza@unict.it

pests in relation to their potential economic importance worldwide and grouped according to the part of the plant attacked (foliage, flowers, buds, fruits, branches and trunk, and the root system).

13.2 Integrated Pest Management

As already mentioned, the pomegranate in the common imaginary of farmers and consumers has been frequently considered a rustic and easy plant to grow. This was true when the cultivation had no economic interest, as traditional varieties required relatively little care compared with other crops; however the new varieties need more productive input (e.g. irrigation, fertilization) due to the higher income expectation reserved for specialized crops. Equally, the specialized crop has led to a series of phytosanitary problems whose management is not easy for several reasons. For instance, in countries where intensive cultivation of pomegranate is recent, specific knowledge on the pests and pathogens is rather superficial. Moreover, the culture is still considered a 'minor crop' and, especially in countries with strict pesticide regulations, there are few chemicals available. A pomegranate orchard, if the environmental and agronomic conditions are optimal for its growth, can be considered a rather stable agro-ecosystem, in which the strategy of control adopted may potentially modify the ecology of the environment. This stability is important as it allows IPM to be carried out more easily than for other crops (e.g. stone or pome fruits). IPM foresees that all methods useful to pest control should be considered, combining strategies that prevent, mitigate and reduce the eventual pest populations under the economic threshold of damage, and requires an overall tactic that prevents pest outbreaks by not creating the right conditions for their establishment (Kogan, 1998; Letourneau *et al.*, 2017; Daniel *et al.*, 2018). Before establishing a pomegranate orchard it is necessary to verify in advance the suitability of the pedo-climatic and irrigation conditions. These checks are part of the control strategies because plants in good health can cope better with pest infestations.

Constant monitoring of plants is another important aspect of pests control strategies. Plant sampling and monitoring allow early detection of infestation outbreaks and allows their early control (Letourneau *et al.*, 2017). Moreover, monitoring makes it easier to understand what led to the infestation. In fact, infestations often result from environmental disequilibrium or changes in ecological stability that may have been caused by agronomic or pesticide errors (Vacante and Bonsignore, 2018). For example, excessive fertilization induces an increase in new vegetation that favours aphid and mealybug populations (Massimino Cocuzza and Rapisarda, 2017). Again, the use of broad-spectrum pesticides depresses the populations of beneficial, non-target arthropods, increases the risk of occurrence of resistant populations and disrupts natural biological control of non-pest organisms (Bass *et al.*, 2015; Letourneau *et al.*, 2017). IPM in the pomegranate should have as its main objective the reduction of pest damages and the preservation of natural antagonists. This goal can be achieved by applying all the methods (cultural, physical and biological) useful to maintain pests under the economic threshold of damages. The use of pesticides is possible but only when it is strictly necessary and selective for beneficial communities. The the IPM methods for the main pomegranate pests are reported in Table 13.1.

The basis for correct pest management is the knowledge of their main morphological and biological characteristics, harmfulness and the environmental factors that can favour their outbreaks. Luckily, we are far away from the times when pest control was carried out exclusively through the massive use of pesticides according to a calendar scheme, without awareness of the ecological context in which agricultural activity took place. Furthermore, consumers demand no pesticide residues on fruits, and this is a key point when exporting the fresh product. The farmer today is an entrepreneur, an ecologist and an environmentalist, who is not only delegated the responsibility to provide consumers with safe products, but also to be the custodian of the territory and the environment to protect and maintain it for future generations.

Table 13.1. Chemical management of pomegranate pests.

Pest species	Management, active ingredients, natural antagonist, biotechnical methods	Notes
Dialeurodes citri, *Siphoninus phillyreae*, *Aleurocanthus spiniferus*	*Agronomic* Air the foliage of the plants with pruning and reduce nitrogen fertilization *Monitoring* Visual control of vegetation and fruits (10% plants/ha) *Chemical* - IPM authorized: Spirotetramat, buprofezin, imidacloprid, methomil[a] - Organic authorized: Mineral oil, neem oil, azadirachtin	Before applying chemicals, verify the activity of natural enemies [a] Not authorized in Europe
Aphis punicae, *A. gossypii*	*Agronomic* Avoid drastic pruning and excessive fertilization, soil milling to disturb the nests of the ants *Chemical* Normally, for trees up to 4 years, chemical control is not necessary. For young trees, intervene upon reaching the threshold of 25% of infested shoots in spring and in absence of flowers - IPM authorized: Flupyradifurone, spirotetramat, sulfoxaflor, imidacloprid, clothianidin[a], acetamiprid[b], flonicamid, methomyl[c], - Organic authorized: Pyrethrins, potassium salts[d], mineral oil, azadirachtin, neem oil, rosemary oil, peppermint oil	• Never use insecticides during flowering • Before applying chemicals, verify the activity of natural enemies (coccinellids, *Aphidoletes aphidimyza*, *Lysiphlebus* spp., *Aphidius*, etc.) [a] Not authorized in Europe [b] Not authorized in Europe [c] To use only in case of outbreaks; destructive for natural enemies; not authorized in Europe [d] Potassium salts applied at the beginning of colony formation

Continued

Table 13.1. Continued

Pest species	Management, active ingredients, natural antagonist, biotechnical methods	Notes
Planococcus citri, *Maconellicoccus hirsutus*, *Ferrisia virgata*	*Agronomic* Air the foliage of the plants with pruning and reduce nitrogen fertilization. Light ploughing of the ground to control the ants' nests *Monitoring* Visual control of vegetation and fruits (10% plants/ha); pheromone traps for capturing males *Biological* For *P. citri* use *Cryptolaemus montrouzieri* (two throws of 800–1000 insects/ha), *Leptomastix dactylopii* and *Anagyrus pseudococci* (two or three throws of 1000–2000 insects/ha) when temperatures are stable at 18°C. We recommend the use of pheromone traps (one trap/ha) to verify the consistency of the populations *Chemical* Intervene only at reaching of the threshold of 2–3% (fruits at summer maturation) or 5–10% (fruits at autumn maturation) of infested fruits, in conditions of no or low parasitization - IPM authorized: Spirotetramat, buprofezin, sulfoxaflor, imidacloprid, clothianidin[a], methomil[b], thiametoxam[c] - Organic authorized: Mineral oil, azadirachtin, rosemary oil, peppermint oil	[a] Not authorized in Europe. [b] To use only in case of outbreaks; destructive for natural enemies; not authorized in Europe [c] Not authorized in Europe
Saissetia oleae, *Coccus pseudomagnoliarum*, *Ceroplastes spp.*	*Agronomic* Air the foliage of the plants with pruning and reduce nitrogen fertilization *Monitoring* Visual control of vegetation *Chemical* - IPM authorized: Spirotetramat, buprofezin, flupyradifurone, imidacloprid, methomil[a] - Organic authorized: Mineral oil, neem oil, rosemary oil, peppermint oil	[a] Not authorized in Europe

Continued

Table 13.1. Continued

Pest species	Management, active ingredients, natural antagonist, biotechnical methods	Notes
Aonidiella aurantii, *Parlatoria oleae*, *Lepidosaphes granati*, *Pinnaspis buxi*, *Chrysomphalus aonidium*	*Agronomic* Avoid drastic pruning and the presence of dust on the vegetation, soil milling to disturb the nests of the ants *Monitoring* Check four fruits per plant on four different arrangements on 10% of plants/ha. Place two yellow pheromone traps/ha and chemically intervene after 2–4 weeks only when the males' catch peak is reached and at the threshold of 20% of infested fruits in conditions of no or low parasitiation *Biological* For *A. aurantii* use *Aphytis melinus* (different throws every 15–20 days of 10,000–20,000 insects/ha for a total of 100,000) *Chemical* - IPM authorized: Spirotetramat, buprofezin, sulfoxaflor, pyriproxyfen, acetamiprid[a], cypermethrin, chlorpyrifos[b], chlorpyrifos-metile[b], fosmet - Organic authorized: Mineral oil, pyrethrin	[a] To use only in case of outbreaks; destructive for natural enemies; not authorized in Europe [b] Verify the authorization for use
Zeuzera pyrina	*Monitoring* In spring, inspect the young twigs to identify the attached ones and eliminate them. In larger branches, use a wire to be introduced into the galleries to kill the larvae. Use one trap/ha to monitor the population and five traps/ha for mass capture *Chemical* - IPM authorized: Triflumuron - Organic authorized: *Bacillus thuringiensis* kurstaki	
Deudorix isocrates, *Dudorix livia*		

Continued

Table 13.1. Continued

Pest species	Management, active ingredients, natural antagonist, biotechnical methods	Notes
Ceratitis capitata	*Monitoring* Use monitoring traps active with para-pheromone *Chemical* Intervene at reaching of the threshold of 20 adults caught per trap Use attract & kill device Use bait poisoned with spinosad Poisoned protein baits - IPM authorized Acetamiprid[a], etofenprox, fosmet, malation[b] - Organic authorized: Pyrethrin	[a] Not authorized in Europe [b] Verify the authorization for use
Ectomyelois ceratoniae	*Monitoring* Visual control of vegetation and fruits - Organic authorized: *Bacillus thuringiensis* *Trichogramma* spp.[a]	[a] Not authorized in Europe

13.3 Foliage, Twig and Shoot Pests

On pomegranate, leaves, twigs and shoots are infested mainly by Hemiptera (flatid, psyllids, aphids, aleurodids, mealybugs and scales), and by a few other moths and mites (Table 13.2). The Hemiptera are sap-feeding insects and their activity can affect metabolism and photosynthetic capacity of the plants with potential consequences for their development and productivity. Of course, the damage is commensurate with the number of insect populations that infest plants. These insects produce honeydew on which saprophytic fungi develop causing the appearance of sooty mould. Damage to leaves caused by lepidopterans and mites is normally of secondary importance. The sap-feeding insect populations develop mainly during the spring period in the presence of the new developing vegetation, and then decline with the hardening of the foliage (Cocuzza *et al.*, 2016; Elango and Sridharan, 2017a).

13.3.1 Citrus flatid planthopper

Metcalfa pruinosa (Say) (Hemiptera, Flatidae) is a species that spread quickly in the 1990s in Europe, causing considerable problems for its control. However, in a few years, the species has entered the ecosystem and, being well controlled by indigenous and introduced natural antagonists, is currently no longer an important entomological problem.

Identification

The species is easily recognizable by the filamentous white wax that covers the body of the nymphs, while the adults (7–8 mm in length with wings) are whitish-greyish in colour (Mead, 1969).

Distribution

Probably of North American origin, today the flatid is present also in Central America and widespread in the European continent.

Biology and ecology

Metcalfa pruinosa develops one generation per year, with eggs wintering in the vegetation.

Hosts

It is a polyphagous pest recorded on more than 200 botanical species. Only occasionally infests pomegranate.

Table 13.2. Pomegranate pests listed in this chapter.

Pest species	Order	Family	Tree part infested	Distribution	Pest importance
Metcalfa pruinosa	Hemiptera	Flatidae	Leaves, twigs	N, C America, Europe	Minor
Aleurocanthus spiniferus	Hemiptera	Aleyrodidae	Leaves, twigs	Asia, Africa, Europe	Minor
Dialeurodes citri	Hemiptera	Aleyrodidae	Leaves, twigs	Cosmopolitan	Minor
Siphoninus phillyreae	Hemiptera	Aleyrodidae	Leaves, twigs	Cosmopolitan	Minor
Aphis punicae	Hemiptera	Aphididae	Leaves, shoots, fruits	Cosmopolitan	Moderate/high
Aphis gossypii	Hemiptera	Aphididae	Leaves, shoots, fruits	Cosmopolitan	Moderate/high
Planococcus citri	Hemiptera	Pseudococcidae	Leaves, twigs, fruits	Cosmopolitan	Moderate
Ferrisia virgata	Hemiptera	Pseudococcidae	Leaves, twigs, fruits	Cosmopolitan	Moderate
Ceroplastes spp.	Hemiptera	Coccidae	Twigs, branches	Cosmopolitan	Low
Saissetia oleae	Hemiptera	Coccidae	Twigs, branches	Cosmopolitan	Low
Coccus pseudomagnoliarum	Hemiptera	Coccidae	Twigs, branches	Cosmopolitan	Low
Aonidiella aurantii	Hemiptera	Diaspididae	Twigs, branches, fruits	Cosmopolitan	Low
Parlatoria oleae	Hemiptera	Diaspididae	Twigs, branches, fruits	Cosmopolitan	Low
Lepidosaphes granati	Hemiptera	Diaspididae	Twigs, branches, fruits	Europe, Asia Minor	Low
Dysgonia spp.	Lepidoptera	Erebidae	Leaves	Asia, Europe	Low
Tenuipalpus granati	Prostigmata	Tenuipalpidae	Leaves, shoots	Asia, Europe	Low
Aceria granati	Acarina	Eryophidae	Leaves, shoots	Cosmopolitan	Low
Zeuzera pyrina	Lepidoptera	Cossidae	Trunk, branches, twigs	Cosmopolitan	High
Inderbela spp.	Lepidoptera	Cossidae	Trunk, branches, twigs	Indian subcontinent	Moderate
Euzophera bigella	Lepidoptera	Pyralidae	Trunk, branches	Indian subcontinent	Low/moderate
Celosterna scabrator	Coleoptera	Cerambicidae	Trunk	Indian subcontinent	Low/moderate
Apate monachus	Coleoptera	Bostrichidae	Trunk	Cosmopolitan	Low/moderate
Xyleborus spp.	Coleoptera	Scclytidae	Trunk, branches	Cosmopolitan	Moderate
Scirtothrips dorsalis	Thysanoptera	Thripidae	Leaves, fruits	Asia, Africa, America, Oceania	Minor

Continued

Table 13.2. Continued

Pest species	Order	Family	Tree part infested	Distribution	Pest importance
Rhiphiphorothrips cruentatus	Thysanoptera	Thripidae	Leaves, fruits	Asia	Minor
Leptoglossus clypealis	Hemiptera	Coreidae	Fruits	America	Minor
Leptoglossus zonatus	Hemiptera	Coreidae	Fruits	America	Minor
Leptoglossus gonagra	Hemiptera	Coreidae	Fruits	Africa, N, C America	Minor
Ceratitis capitata	Diptera	Tephritidae	Fruits	Cosmopolitan	Moderate
Apomyelois ceratoniae	Lepidoptera	Pyralidae	Fruits	Cosmopolitan	Moderate
Deudorix isocrates	Lepidoptera	Lycaenidae	Fruits	Indian subcontinent	High
Deudorix livia	Lepidoptera	Lycaenidae	Fruits	Africa, Asia Minor	High
Deudorix epijarbas	Lepidoptera	Lycaenidae	Fruits	Indian subcontinent	Moderate
Cryptoblades gnidiella	Lepidoptera	Totricidae	Fruits	Cosmopolitan	Moderate/high
Thaumatotibia leucotreta	Lepidoptera	Totricidae	Fruits	Africa	Moderate
Carpophilus spp.	Coleoptera	Nitidulidae	Fruits	Cosmopolitan	Minor
Polyphylla olivieri	Coleoptera	Melolonthidae	Roots	S-E Europe, Asia Minor	Minor
Meloidogyne spp.	Tylenchida	Heteroderidae	Roots	Cosmopolitan	Minor

Damage

In addition to the removal of sap and abundant production of honeydew on which develops sooty mould, *M. pruinosa* heavily smears the vegetation with the wax produced by nymphs. Fruits can be heavily depreciated by the presence of honeydew.

Survey methods

It is necessary to ascertain the presence of white wax produced by nymphs on vegetation, especially on young shoots and stems.

Management

The flatid is well controlled by the parasitoid of Nearctic origin *Neodrynus typhlocybae* (Ashmead) (Hemiptera, Driinidae) (Lucchi, 2000), introduced in Europe and produced by several bio factories. Frequently, outbreaks are the consequence of ecological disequilibrium (sometimes caused by the inappropriate use of broad-spectrum insecticides) and, the population returns below the damage threshold, restoring the standard conditions that favour the activity of natural enemies. In case of need, it is advisable to perform only one treatment in the late spring–early summer to eliminate juvenile stages and the first adults of the new generation. Treatments with potassium salts applied at the beginning of the development of the new annual generation can be useful in organic farming.

13.3.2 Whiteflies (Homoptera, Aleyrodidae)

Several whiteflies can occasionally infest pomegranate. The citrus whitefly *Dialeurodes citri* (Ashmead), the ash whitefly *Siphoninus phillyreae* (Holiday) (Fig. 13.1a) and the orange spiny whitefly *Aleurocanthus spiniferus* (Quaintance) (Fig. 13.1b) are the most frequent on pomegranate (Cocuzza *et al.*, 2016).

Fig. 13.1. Some of the foliage, twig and shoot pests of pomegranate plant. (a) Shoot infested by *Siphoninus phillyreae*. (b) *Aleurocanthus spiniferus*. (c) Adult of *Clitostethus arcuatus* (d) *Aphis punicae* (e) *Aphis gossypii* (f) *Planococcus citri*. (Photos: (a) Antonios Tsagkarakis; (b, e) Giuseppe Eros Massimino Cocuzza; (c) Gaetano Siscaro; (f, d) Alimohammad Yavari.)

Identification

Whiteflies are tiny, sap-sucking insects that may become abundant in vegetable and ornamental plantings, especially during warm weather. The adults of these insects (about 1 mm long) have the body and wings covered with powdery white wax (with the exception of *A. spiniferus*, whose colour is bluish). The nymphs are flattened, ovoid and fixed on the vegetation and without the body covered with wax.

Distribution

Dialeurodes citri and *S. phillyreae* are cosmopolitan, while *A. spiniferus*, widespread in Asia and Africa, is rapidly spreading also in the Mediterranean basin (EPPO, 2019a).

Biology and ecology

Generally, females may live up to 30–60 days, while males live an average of 9 days. A winged female lays eggs on the underside of the leaves. When the nymphs emerge, they do not move far and feed on the plant sap by digging their mouthparts into the leaf tissue to suck sap until pupation (Gillespie, 2000). On first observation, the pupal case of *D. citri* appears similar to the white encrustation of snow scale, while *A. spiniferus* can easily be mistaken for a Diaspididae scale. The pupal case is 0.8–1.0 mm long and 0.55–0.7 mm wide. The entire life cycle from egg to adult usually takes place under the same leaf.

Damage

Infestations of whiteflies on pomegranate are not frequent. They mainly infest leaves and buds. Serious damage is caused by the excretion of honeydew by whitefly. Under moist conditions, sooty mould develops on honeydew reducing photosynthesis and respiration of plants. Curling of leaves and growth of black sooty mould can be seen on the tender leaves. Heavy infestations cause leaf yellowing, early leaf drop and smaller fruits (Bellows *et al.*, 1990). Massive attacks can cause defoliation and qualitative decay of the fruit, mainly due to the sooty mould that develops on the abundant honeydew produced by these insects.

Hosts

All the abovementioned species are polyphagous (reported on more than 50 botanical species of various families).

Survey methods

The infestation is recognized by the presence of honeydew and the sooty mould on all the vegetation. In addition, it is easy to observe the presence of the adult and juvenile forms on the undersides of the leaves.

Management

On pomegranate, whiteflies are considered as secondary pests. Their infestations are mainly the consequence of incorrect agricultural practices (among these, the lack of the pruning of the plants), or the use of broad-spectrum insecticides that adversely affect the population of their natural enemies (Balika *et al.*, 1999; Tsagkarakis, 2012). All the abovementioned species have numerous antagonists, among which the most efficient seem to be *Clitostethus arcuatus* (Rossi) (Fig. 13.1c) and *Delphastus catalinae* (Horn) (Coleoptera, Coccinellidae). Also effective are the parasitoids of the genus *Encarsia* (Hymenoptera) such as *E. lahorensis* (Howard) for *D. citri* and *E. smithi* (Silvestri) for *A. spiniferus* (Van Den Berg *et al.*, 2000; Abd-Rabou and Simmons, 2014). With severe infestations, their control can be reached by restoring the natural equilibrium of the agro-ecosystem, promoting or safeguarding the natural enemies of the whiteflies and non-target arthropods, and limiting or eliminating broad-spectrum insecticides (Gyeltshen *et al.*, 2014).

13.3.3 Aphids (Hemiptera, Aphididae)

In spring, pomegranate sprouts are usually infested by aphids. The presence of *Aphis punicae* Passerini (pomegranate aphid) and *Aphis gossypii* (cotton aphid) is not always considered serious, as infestations are in most cases not harmful to production. However, infestations concerning the fruits are to be controlled.

Identification

The apterous females of *A. punicae* (Fig. 13.1d) are entirely greenish-yellow in colour, slightly glossy, while the winged forms have blackish head and thorax. The siphunculus is green with the proximal part tending to brown. The apterous females of *A. gossypii* (Fig. 13.1e) take on a more or less dark greenish colour, with the siphons entirely brown. The winged form differs having brown head and thorax. The two species are morphologically similar and, especially in the field, the distinction is rather difficult. Discrimination through morphological analysis on a slide or DNA barcode is quite easy (Cocuzza *et al.*, 2009).

Distribution

Both species are cosmopolitan.

Biology and ecology

The presence of *A. punicae* and *A. gossypii* is quite usual on pomegranate, which in spring produces conspicuous colonies on sprouts. *Aphis punicae* lives almost exclusively on pomegranate on which it develops for the whole life cycle (Blackman and Eastop, 2006). The life cycle of *A. punicae* begins early in the spring, with the fundatrix born from the eggs laid in the previous autumn, near the buds. In contrast, the colonies of *A. gossypii* develop later, thanks to the winged forms coming from other plants, which settle on pomegranate when the buds are already developed. With favourable climatic conditions, the colonies can infest new vegetation in a very short time, affecting all the new vegetation, including the flowers and the growing fruits. The decline of populations occurs with the hardening of the shoots, and the numerous natural enemies of the two pests also contribute to this decline. Moreover, in this period, the winged forms of *A. gossypii* leave the pomegranates to move to other plants on which to spend the summer. In contrast, *A. punicae* strongly reduce their populations and remain on the same plant by slowing down the metabolic and reproductive activity, finding refuge in the late sprouting shoots. In September, with the decreasing temperatures, the colonies of *A. punicae* resume vigour, occupying younger leaves or fruits. Then, in late autumn the sexual morphs mate and produce eggs to overcome the winter period.

Damage

Infestations of *A. gossypii* and *A. punicae* should be kept under careful observation, as sometimes the production losses can be of some importance. In fact, although a high amount of extracted sap is well tolerated in irrigated pomegranates, abundant honeydew is produced, on which sooty mould develops, which in sunny areas can trigger a 'lens effect', causing damage to the vegetation (Cocuzza *et al.*, 2016). The fruits can be covered by sooty mould and the epicarp discoloured as a consequence of the feeding activity of the aphids.

Hosts

The presence of *A. punicae* Passerini is quite usual on pomegranate, whereas *A. gossypii* is a polyphagous species, recorded on more than 600 botanical species (Holman, 2009). On pomegranate they frequently develop mixed colonies (Lee *et al.*, 2015).

Survey methods

In spring, on the tender shoots and then on the newly attached fruits, the colonies of both species can be observed. The populations grow in a few weeks and then regress. In late summer populations of *A. punicae* can infest the fruits.

Management

Numerous predators (Coleoptera: Coccinellidae, Diptera: Syrphidae and Cecidomiidae, Neuroptera: Chrysopidae) and parasitoids (mainly Hymenoptera of the genera *Lysiphlebus* and *Aphidius*) control their populations, although their effectiveness is often rather late (Cocuzza *et al.*, 2016). Minimizing the use of broad-spectrum insecticides together with appropriate agronomic practices (irrigation, pruning and balanced fertilization) are the most effective strategies to containing infestations in mature plants (over 4–5 years). It is useful also to control the ants that contribute to spreading the aphids among shoots. Normally, chemical control is not necessary in spring, because the colonies are exhausted with the hardening of the shoots. In some environments, in the case of heavy infestation of growing fruits, it may be necessary to use insecticides with specific aphicide action. In the case of young plants, severe attacks could cause a delay in their development and fruit production, and consequently it may be necessary to use selective insecticides (Cocuzza *et al.*, 2016).

13.3.4 Mealybugs (Hemiptera, Pseudococcidae)

The female mealybug is easy to recognize by its oval body, flattened, soft and covered with floury wax. On the marginal parts of the body, waxy filaments of variable length are visible. Being polyphagous, they can occasionally infest even the pomegranate.

Identification

The females of the citrus mealybug, *Planococcus citri* (Risso) are oval and 3–4 mm long. It is entirely covered by a powdery wax and on the marginal parts are 18 pairs of stout waxy filaments, with the anal one and those that surround it, slightly longer than the rest. Nymphs of the first instar (crawlers) are yellow, oval-shaped and covered in white wax (Gill *et al.*, 2013). Females of the striped mealybug *Ferrisia virgata* (Cockerell) are recognizable by the presence of two longitudinal dark stripes on the dorsum and two long caudal wax filaments (Kaydan and Gullan, 2012). Other mealybugs occasionally reported on pomegranate are the cocoa or oriental mealybug, *Planococcus lilacinus* (Cockerell) and the pink hibiscus mealybug, *Maconellicoccus hirsutus* (Green) (Mani and Krishnamoorthy, 1990, 1991, 2001; Williams, 1986; Ben-Dov and German, 2003; Halima-Kamel *et al.*, 2015). In view of the expansion of the pomegranate fruit and the polyphagy of these last insects, a careful monitoring of these species is necessary.

Distribution

Planococcus citri is present in all warm-temperate regions, especially in citrus fruit areas. *Ferrisia virgata* and *Maconellicoccus hirsutum*, native to Central America and southern Asia, respectively, have become widespread throughout the world in the past 10 years, probably due to the increase in the international plant material trade (CABI, 2019a).

Biology and ecology

The citrus mealybug spends the winter in the various stages under the bark and other repaired parts of the plant. When vegetative growth begins again, the female nymphs reach maturity and, after having been fertilized, they begin to lay eggs (300–600/female) inside a cottony white ovisac. *Planococcus citri* can develop three to six generations/year depending on the climatic conditions. The highest population densities are reached in late summer–autumn. Similarly, *F. virgata* can develop three to five generations/year and each female can lay about 700 eggs inside a cottony ovisac (Awadallah *et al.*, 1979).

Damage

The main damage caused by mealybugs is due to the wax and honeydew produced, which smears the fruit, and causes wilting and discolouration of the leaves and stems. Nymphs and adults of both types of mealybug tend to remain clustered in the crown area of small fruits or the contact points between them (Fig. 13.1f). In the field, its spread is favoured by the ants that carry and distribute the various forms among the vegetation. Damage caused by *P. citri* has been reported in Spain (Bartual *et al.*, 2012), Israel (Holland *et al.*, 2009), Turkey (Özturk and Ulusoy, 2009), Cyprus (Kahramanoglu and Usanmaz, 2013) and India (Mani and Krishnamoorthy, 1989).

Hosts

All the species that can infest the pomegranate are polyphagous.

Survey methods

Checking for mealybugs is carried out on the youngest fruits and twigs, especially observing the most sheltered and internal parts of the plants. On these parts the insect colonies and the surface smeared with honeydew can easily be seen. The infestation density can be measured on a sample of 10 fruits/twigs on 5–10% of the plants. The intervention threshold is when more than 10–15% of fruits are infested.

Management

Several ladybirds and parasitoid Hymenopterans are reported for both species (Noyes, 2016), and usually their control activity is rather successful. Among them, the most effective are *Cryptolaemus montrouzieri* and *Scymnus* spp. (Coleoptera, Coccinellidae), *Leptomastix dactylopii* and *Anagyrus* spp. (Hymenoptera, Encyrtidae) (Williams, 1986; Mani and Krishnamoorthy, 1989, 2001; Mani *et al.*, 2011; García *et al.*, 2016). However, some experiences indicate that natural enemies are able to reduce the population density of *P. citri*, but not to prevent damage to fruits (Bartual *et al.*, 2012). Sudden infestations are often a consequence of ecological imbalances caused by incorrect control strategies (e.g. use of broad-spectrum insecticides). In case of severe infestations (the intervention threshold can be considered as between 2–3% and 5–10% of infested fruits), authorized selective insecticides could be used only if there are no other options. An alternative is to apply light mineral oil with caution to avoid phytotoxicity phenomena, directing the jet towards the points of the greatest presence of the mealybugs. Moreover, it would also be advisable to check the role of ants in the spread of populations. These insects defend mealybugs from natural enemies and can weaken their control activity. A winter pruning on the most heavily populated branches and on which the winter forms are found helps to lighten future infestations.

13.3.5 Soft scale (Hemiptera, Pseudococcidae), scale (Hemiptera, Coccidae) and armoured scale insects (Hemiptera, Diaspididae)

Armoured scale (Hemiptera, Diaspididae)

About 20 scale insect species are reported on pomegranate around the world. The body of these insects is more or less flattened, ovoid and covered with wax or silk distributed in various ways.

Identification

The females of the soft scale *Ceroplastes sinensis* Del Guercio (Chinese wax scale), *Ceroplastes japonicus* Green (Japanese wax scale) and *Ceroplastes floridensis* Comstock (Florida wax scale) have bodies that are ovoid in form and covered with convex waxy plates in variable numbers (often specific for species, although not always clearly distinguishable), each of which has a central depression from which the wax comes out. The females of the scales *Coccus pseudomagnoliarum* (Kuwana) (grey citrus scale) and *Saissetia oleae* Olivier (black or olive scale) have a sub-oval, convex and rigid wax-covered body ((Holland *et al.*, 2009; Özturk and Ulusoy, 2009; Ma and Bai, 2004). The armoured scales have a subcircular or elongated body, covered with a robust, dorsal, sericeous layer. The main species reported on pomegranate are *Aonidiella aurantii* (Maskell) (California red scale), *Lepidosaphes granati* Koroneus (pomegranate scale), *Parlatoria oleae* (Colvée) (olive scale) and *Pinnaspis buxi* (Bouché) (coconut scale) (Juan *et al.*, 2000; Moghaddam, 2013; Pollini, 2013). Populations of these species rarely cause economic damage to the crop.

Distribution

All the abovementioned species are cosmopolitan, with the exception of *L. granati*, which is reported only in Europe and Asia Minor (García *et al.*, 2016).

Biology and ecology

Almost all the species complete one or two generations/year, except for *A. aurantii*, which in optimal conditions can develop up to six generations. The number of eggs that each female can produce varies from 60 to 150 (*A. aurantii*) to 2500 (*S. oleae*) or 4000 (*Ceroplastes* spp.). Some species are amphigonic and dimorphic, with winged males, while others are parthenogenetic. The nymphs, upon escaping from the female follicle, are dispersed in the plant substrate in search of a suitable point for fixing. This period lasts about 24–36 h, and during this phase, the young nymphs are particularly exposed to adverse weather events that can drastically reduce their number. The time needed to complete nymphal development varies depending on the weather conditions.

Damage

The damage caused by the soft scale and scale insects includes the removal of sap and the production of abundant honeydew on which sooty mould develops. The latter can alter the physiological activities of the plant and cause aesthetic damage to the fruits. In cases of severe and prolonged infestations, the attacked branches can decay. Armoured scales do not produce honeydew but cause foliar and fruit yellowing. The infested fruits can be rejected by the market or undergo a commercial depreciation.

Hosts

With the exception of *L. granati*, which is specific to pomegranate, all the other species are polyphagous.

Survey methods

The presence of scale insects is verified by checking 2–3-year-old twigs, the leaves and the fruits (in the case of the armoured scales).

Management

Infestations of these insects are usually considered rare events and are often the result of incorrect management of the agro-ecosystem (use of non-selective insecticides, poor or absent plant pruning). For each species, different antagonists are known – both predatory (*Chilocorus* spp., *Exochomus* spp., etc.) and specific parasitoids (*Aphytis melinus* DeBach and *Encarsia perniciosi* Tower for *A. aurantii*) – which, in the case of numerically limited populations, manage to control them quite effectively. Moreover, the nymphs, during the wandering period, are particularly vulnerable to atmospheric events (rain and strong wind) that drastically reduce the population (a mortality of 90–95% can be reached). Equally effective in this period are any chemical treatments, so when these are necessary, it is always advisable to position them after having ascertained the presence of the wandering nymphs on the leaves. There are no specific intervention thresholds for pomegranate scales; however, it is advisable to intervene with insecticidal treatments when reaching one sample per centimetre of branch and/or four specimens per fruit (armoured scales) or leaf.

13.3.6 Passenger

The larvae of *Dysgonia algira* (L.) and *Dysgonia torrida* (Gaenée) (Lepidoptera, Erebidae) are two very similar species distributed mainly in Asia and the Mediterranean basin. These polyphagous moths are also reported on pomegranate. In adults, the anterior wings have a dark brown background colour with a median and marginal greyish band. The larvae feed on the leaves, causing typical semi-circular erosions from the margin, without attacking flowers or fruit (Sannino *et al*., 1986). The occasional infestation on pomegranate is rather negligible.

13.3.7 Mites

Among the mites, only *Tenuipalpus granati* Sayed (pomegranate mite), *Tenuipalpus punicae* Pirchard & Baker (false spider mite) (Prostigmata, Tenuipalpidae) and *Aceria granati* Canestrini & Massalongo (pomegranate leaf curl mite) (Acarina, Eryophyidae), are rarely harmful to pomegranate (Jeppson *et al*., 1975; Döker *et al*., 2013). *Tenuipalpus* spp. is common in the Mediterranean area and Asia, whereas *A. granati* is cosmopolitan. Their infestations occur exclusively on the leaves or shoots (Al-Gboory and El-Haidani, 1989; Vacante, 2016). All the species are normally kept below the threshold of harmfulness by different predatory mites, and no particular control interventions are required.

13.4 Trunk, Branch and Stem Pests

13.4.1 Leopard moth

The leopard moth, *Zeuzera pyrina* (L.) (Lepidoptera, Cossidae) (Fig. 13.2a) is one of the most frequent and harmful species on pomegranate, especially in young plants (3–4 years).

Fig. 13.2. Some trunk, branch and stem pests of pomegranate plant. (a) Pomegranate branch dug internally by *Zeuzera pyrina*. (b) Pomegranate branch dug internally by *Indarbela quadrinotata*. (c) *Celosterna scabrator*. (d) Adults of *Apate monachus* on pomegranate branch. (e) Exit hole caused by *Celosterna scabrator*. (f) Pomegranate branch dug internally by *Apate monachus*. (Photo: (a, e) Giuseppe Eros Massimino Cocuzza; (b, d, f) Alimohammad Yavari; and (c) Mallikarjun Harsur.)

Identification

The larvae (about 5–6 cm long at maturity) are easily recognizable by the yellowish background colour and for the blackish tubercles present longitudinally along the body. The adults (about 2.5–3 cm long with a wingspan of 3.5–6 cm) are whitish in colour with black-bluish spots on the front wings and six large bluish spots on the thorax. Females are larger than males and are also recognizable for the filiform antennae, which are bipectinate in males.

Distribution

The species is present on all continents.

Biology and ecology

The eggs are laid in groups from May to September, depending to the climate, in the most sheltered parts of the trunks of 2–3-year-old trees, of about 100 per female. However, only a few larvae will emerge, as most are damaged by high summer temperatures, and preyed on or parasitized by natural enemies (Guario *et al.*, 2001). The biological cycle is completed in 1–2 years, depending if the eggs hatch in early or late summer, respectively (Kutinkowa *et al.*, 2006; Hegazi *et al.*, 2015). The larvae, after hatching, attack the bases of the leaves or buds and begin to penetrate the woody tissues, digging tunnels that develop in a centripetal way and parallel to the axis of the twig. With growth, the larvae switch to other larger branches.

Damage

The attack of the larvae causes the desiccation of the shoots, while on branches leads to desiccation of the vegetation and diminishes mechanical resistance, which, in strong winds, can result in their breakage (Hegazi *et al.*, 2015). Younger plants (1 year) can perish due to the attack of larvae, while the larger ones (3 years old) can lose entire branches (Cocuzza *et al.*, 2016).

Hosts

Zeuzera pyrina is a polyphagous moth that attacks numerous cultivated arboreal plants, such as *Prunus*, *Pyrus*, *Olea*, *Malus*, *Acer*, *Tilia*, *Fagus* and *Salix* (Gatwick, 1992).

Survey methods

The presence of the larvae is not easy to notice and it is necessary to observe the base of 2–3-year-old branches to see the presence of the entry holes made by the xylophagous. It is useful also to observe the base of the plants or around the outlet holes on twigs, trunk or branches to verify the presence of excrement or sawdust (Hegazi *et al.*, 2015).

Management

The control of *Z. pyrina* is overall preventive, especially in areas where its presence is well known, such as for pomegranate adjacent to olive orchards, normally subject to their attacks (Hegazi *et al.*, 2015). The most harmful infestation occurs on young trees, so regular inspection of the orchard is advisable for early detection of the damage. Pheromone traps are very useful to verify and quantify the presence of the lepidopterous species, to follow the trend of flights, which could suggest carrying out an insecticide intervention against the larvae at an early age. Also, mass trapping can be effective in suppressing populations of *Z. pyrina*, particularly if used in isolated areas (Hegazi *et al.*, 2009). Moreover, when possible, it is always advisable to perform a careful observation plant by plant, to search for any symptoms that indicate the presence of the moth (Kutinkowa *et al.*, 2006). Once identified, the larvae can be killed by introducing a wire inside the tunnels, remembering to clean and close the holes with healing mastic for plants to prevent stagnant water or the entrance of fungi. Control activity performed by the natural xylophagous antagonists is not very effective. The entomopathogenic fungus *Aspergillus candidus* is reported to cause mortality of bark-eating caterpillars on pomegranate in Haryana (Ramsingh *et al.*, 1982). Similarly, natural incidence in the field and 100% pathogenicity of *Beauveria bassiana* have been recorded in the laboratory on bark-eating caterpillars (Fasih and Srivastava, 1988), but no further studies have been done in the field. Despite the numerous parasitoids found on *Z. pyrina*, the percentage of larvae attacked is very low (Campadelli, 1995; Hegazi *et al.*, 2015). The parasitoid *Podagrionella indarbelae* (Chalcidoidea: Torymidae) has been collected and reared from the eggs of *Indarbella tetraonis* collected from cashew in Kerala (Narendran and Sureshan, 1988). Authorized insecticide used at early larval stages, when they are still present on the superficial layers of the wood, can be effective to control the Cossidae.

13.4.2 Bark-eating caterpillars

The female of *Indarbela quadrinotata* (Walker) (Fig. 13.2b) and *I. tetraonis* (Moore) (Lepidoptera, Cossidae) (wingspan of 26–29 mm) are pale brownish with forewings having a row of dark rusty red spots. Adult males (22–25 mm) are whitish-grey, with brownish streaks on pale whitish forewings and pale hind wings. Eggs are laid in clusters of 15–20 directly on the bark of branches. The larvae are pinkish-white with brown spots along the body. The species are primarily distributed on the Indian subcontinent, and are polyphagous species that can occasionally be a severe problem to pomegranate, mostly in neglected and unmanaged orchards (Varma *et al.*, 1974; Methews and Rugmini, 1998; Senguttuvam, 2000; CABI, 2019a). Damage is caused by larvae that bore tunnels downwards into the wood, usually at the junction of branches and feed on the bark of the tree at night for 9–11 months. Frass is visible in the form of a web around the affected portion (Chandra *et al.*, 2011; Sharma *et al.*, 2012). The tunnels cause weakening of the tree and potential breakage. Both species develop one generation per year.

Another species that occasionally can infest pomegranate is the quince or woodborer moth, *Euzophera bigella* (=*punicaella*) (Zeller) (Lepidoptera, Pyralidae). Larvae attacking the cortical tissue of the trunk or branches can cause direct or indirect damage (introduction of secondary parasites) to the plants (Mehrnejad and Ebrahimi, 1993; Atay and Öztürk, 2010). Sometimes, also the fruits can be infested. Adult forewings are brownish-grey in colour, with two lighter transverse strips. Very rarely *E. bigella* causes economic damage (Simoglou *et al.*, 2012).

13.4.3 Stem borer beetles

Celosterna scabrator (F.) (=*Coelosterna spinator* Fletcher) (Coleoptera, Cerambicidae) and the black borer *Apate monachus* (F.) (Coleoptera, Bostrichidae) are polyphagous insects that occasionally attack and seriously damage pomegranate plants. However, both species are considered minor pests of pomegranate (Nair, 2007; Bonsignore, 2012).

Identification

Adults of *C. scabrator* are about 4 cm long, dull yellowish-brown in colour, with tiny spots distributed on all surfaces of the body (Fig. 13.2c). Larvae and pupae are whitish. The adults of *A. monachus* (1–2 cm long) are elongated and cylindrical in shape, dark brown-black in colour (Fig. 13.2d) with evident hairs on the front of females (Walker, 2008; Bonsignore, 2012; Braham and Gahbiche, 2016).

Distribution

Celosterna scabrator is widespread on the Indian subcontinent, whereas *A. monachus* is largely distributed worldwide (CABI, 2019a).

Biology and ecology

Eggs of *C. scabrator* are laid singly under the bark of the stem and the newly emerged larvae dig tunnels down to reach the roots. The adult emerges from July to September through a round hole made by chewing wood (Fig. 13.2e). The larvae carry the frass outside the hole and it is accumulated at the base of the trunk. During the day, the adults feed by gnawing the green bark of shoots. It usually completes one generation per year (Ahuja *et al.*, 2013). The various development stages of *A. monachus* feed and lay their eggs inside the holes that they themselves produce on branches and trunks. However, there is very little information on the biological cycle of this insect. Both species also attack the wood of dead plants (Ahuja *et al.*, 2013; Braham and Gahbiche, 2016).

Damage

Celosterna scabrator and *A. monachus* frequently attack stems, branches and trunks of pomegranates and the infestation can be noticed by the yellowing of the leaves of the attached branches. The holes dug by *A. monachus* reach a diameter of 4–10 mm, depending on the size of the twigs, branches or trunks (Fig. 13.2f) (Braham and Gahbiche, 2016). These tunnels lead to a mechanical weakening of branches that can evolve into ruptures as a result of violent meteorological events.

The percentage of pomegranate plants up to 6 years attacked by *A. monachus* could be up to 23% (Braham and Gahbiche, 2016), whereas in nurseries the numbers affected can reach 100% (Bonsignore, 2012). Both species frequently infest weakened, suffering (due to fungal diseases or water stress), neglected plants and dead wood, but also healthy pomegranate plants can be of interest (Sharma *et al.*, 2012; Braham and Gahbiche, 2016). The latter can resist the attack of the beetles, albeit the yield and quality of production can be affected (Sharma *et al.*, 2012). In contrast, young trees, especially if weakened by water stress or lack of nutrients, can die.

Hosts

Celosterna scabrator and *A. monachus* are polyphagous species that occasionally attack pomegranate (Öztop *et al.*, 2010; Bonsignore, 2012).

Survey methods

The infestation can be intuited by the leaves of attacked branches that turn yellow and subsequently by the weakening of the tree. The presence of sawdust-like pellets under the tree indicates the grubs of borers are causing the damage.

Management

The preventive control of a xylophagous insect involves keeping the plants in a good phytosanitary state and carrying out careful inspections, especially in the spring and early summer, to identify the signs of the presence of the beetle (afflicted branches and holes). Inserting a wire into the holes that contain them can kill the larvae. The parts of the attached plants must be immediately cut and eliminated (Bonsignore, 2012). Chemical control can be performed by injecting the insecticide directly on the stems or into the holes made by the beetles. Another method consists of the smearing of long-lasting insecticide (imidacloprid or chlorpyriphos) from the collar up to about 60 cm, to discourage the oviposition and the excavation of new holes (Jagginavar *et al.*, 2008; Naik *et al.*, 2011; Bonsignore, 2012; Braham and Gahbiche, 2016). There are no reports of specific natural enemies of both beetles. The use of *Metarhizium anisopliae* Matschnikoff and *Beauveria bassiana* (Bals. & Criv.) can be quite effective (Bonsignore, 2012).

In Turkey and Greece attacks on pomegranate have been reported from another Bostrychidae, the dog grape borer *Amphicerus* (=*Schistoceros*) *bimaculatus* (Olivier), whose larvae and adults bore holes in the branches and trunk, and in cases of severe infestation, cause the death of the plant (Tezcan, 2008; Andreadis *et al.*, 2016).

13.4.4 Shot hole borer

Xyleborus perforans (Wollaston) and *Xyleborus* (=*Euwellacea*) *fornicatus* (Eichhoff) (Coleoptera, Scolitydae), commonly known as shot hole borers, are becoming serious pests of pomegranate in some regions of India.

Identification

The adult beetles of both species are brown or dark brown in colour, with females 2.5 mm long (males 1.5 cm) and cylindrical in shape.

Distribution

Both species are cosmopolitan (CABI, 2019b).

Biology and ecology

Xyleborus perforans and *X. fornicatus* remain active throughout the year with higher activity during the post-monsoon period (Jagginavar and Naik, 2005).

Damage

Like all the scolytids, these species also attack plants weakened by water stress or diseases, on which they dig galleries in the trunk and branches. Moreover, the insect introduces inside the holes the spores of *Ambrosia* fungus to feed the larvae. The attack of the shot hole borer, and the development of *Ambrosia* inside the holes that interfere with vascular transport and the activity of secondary pathogens can lead to the death of the pomegranate plant.

Hosts

Both beetles are polyphagous. *Xyleborus fornicatus* is also a key pest of tea (Hill, 2008; Karunaratne *et al.*, 2008).

Survey methods

The plants, especially the afflicted ones, should be continuously checked to identify the holes excavated by the bark beetles.

Management

The control of the shot hole borer on the infested plant is complicated owing to its well-concealed habitats (Walgama and Pallemulla, 2005). It should be primarily preventive, keeping the plants in good irrigated and nutritive condition, because a healthy tree is more likely to fend off stem borer attack. As for other crops (e.g. tea or grape), where these insects are particularly harmful, on pomegranate also it is advisable to apply some cultural methods of control such as pruning the dried branches during the pre-monsoon season, burning or burial of pruned debris and, in the summer, eliminating eggs and larvae with mechanical instruments able to reach them (e.g. by inserting hooked wires in the holes) (Thirugnanasundaran, 1989). It is important to not stack dead infested wood within or near pomegranate orchards. Adults of *C. scabrator* can be attracted and killed by installing light traps with 200 W sodium bulbs immediately after the first rains by keeping kerosene/insecticide-mixed water below the trap. For scolytids various types of traps and pheromone lures are available to monitor their flight period (James *et al.*, 2007; Hill, 2008). The mass capture of these can be carried out by placing partly dried stems of *Montanoa bipinnatifida* (www.tnau.ac.in). The use of *Beauveria bassiana* and *Bacillus thuringiensis* for control of *X. fornicatus* in the laboratory is reported, but needs further investigation in the field (Walgama, 2012). Spraying pesticides in the trunk or branches is ineffective, whereas injection of authorized insecticide inside holes with larvae can control *C. scabrator* (Biradar *et al.*, 2005; Jagginavar *et al.*, 2008; Naik *et al.*, 2011). In India, both cerambycids and scolytids are managed with a paste obtained from red soil (4 kg), chlorpyriphos (20 ml) and copper oxychloride (25 g) mixed in water (10 l) (Mote and Tambe, 2000), applied with a brush from the plant base up to 50–60 cm aboveground to deter oviposition. Applications of tar oil to the tree trunks in the spring drive away adults from the galleries (Hill, 2008).

13.5 Fruit Pests

13.5.1 Thrips (Thysanoptera, Thripidae)

Thrips (Thysanoptera, Thripidae) are tiny insects (less than 2 mm long) that can superficially damage leaves, shoots and fruits (Table 13.2). The main species that occasionally can affect the pomegranate are the chilli or yellow tea thrips, *Scirtothrips dorsalis* Hood and the grapevine thrips, *Rhiphiphorothrips cruentatus* Hood (Bagle, 1993; Satyagopal *et al.*, 2014).

Identification

The adults of both species are elongated and slightly flattened. Adults of *S. dorsalis* are yellowish and 2 mm long, while those of *R. cruentatus* are blackish-brown in colour and 1.4 mm long (Palmer and Mound, 1983; Chandra and Verma, 2010). Thanks to the two pairs of wings, narrow and fringed, they can make flights several metres long.

Distribution

Scirtothrips dorsalis is widely distributed in Asia, Africa, Oceania and America, while *R. cruentatus* is present only in Asia (CABI, 2019c, CABI, 2019d).

Biology and ecology

Females of both species lay about 40–70 eggs on the undersurface of leaves and floral buds. The newly hatched nymphs of *R. cruentatus* are reddish to yellowish-brown, while those of *S. dorsalis* are yellowish. In both species, the first two nymphal instars are active feeders, while the last two are non-feeding, quiescent prepupal and pupal instars. The life cycle (egg to adult) is completed in about 14–15 days. *Scirtothrips dorsalis* can complete eight generations in a year (Kumar *et al.*, 2012).

Damage

On pomegranate, both species infest leaves and flowers; on the latter, feeding on developing fruits by pricking and tearing the epidermal cells and the underlying parenchyma, and consequently the tissues appear scarified and browned. The main damage occurs on fruits, whose peel appears scarred (Fig. 13.3a. and Fig. 13.3b), causing a strong economic depreciation, especially for export production. On leaves, thrips can cause distortions and silvering of leaves, discolouration of buds and flowers; however, normally this damage is negligible. Often the scarification caused by thrips is attributed to other causes (rubbing of the fruits on the branches, phytotoxicity of insecticides, phytophagous mites) and these errors occur because the damage is only evident when thrips are absent from the plants as they migrate to other plants.

Fig. 13.3. Some pomegranate fruit pests (a) Fruit exterior damage caused by thrips. (b) Fruit interior damage caused by thrips. (c) Pomegranate fruit infested by *Leptoglossus* sp. (d) Adult of *Ceratitis capitata* on pomegranate fruit. (e) Larvae of *Ceratitis capitata* on pomegranate fruit. (f) Dig feeding tunnels caused by larvae of *Ceratitis capitata* on pomegranate fruit. (g–h) Damage caused by larvae of *Ceratitis capitata* on pomegranate fruit. (i) *Apomyelois ceratoniae* larvae entering inside the fruit from the crown. (j–k) Larvae of *Apomyelois ceratoniae*. (l) Behavioural feeding of the *Apomyelois ceratoniae* larvae facilitates some saprophytic fungi such as *Aspergillus* spp. and/or *Penicillium* spp. (m) Rotten and unmarketable fruit caused by larvae of *Apomyelois ceratoniae*. (n) Damage of larvae of *Deudorix isocrates*. (o) *Drosophila melanogaster* L. (p) Beetle *Carpophilus*. (Photos: (a, b, e, f, g, h, I, j, k, l, m, p) Alimohammad Yavari; (c) Kathy Keatley Garvey; (d) Santi Longo; (n–o) Mallikarjun Harsur.)

Hosts

Both thrips are polyphagous and particularly injurious to many vegetable and fruit crops.

Survey methods

Measuring population density can be useful for assessing its consistency and deciding on any control actions. Thrips can be caught with water or sticky traps, but these methods do not allow one to know the plant of origin. The direct collection of thrips can be done by lightly beating the vegetable parts where there is possible infestation (flowers or leaves), making them fall on a flat surface and collecting them in a jar containing ethyl alcohol (60%). The investigation should be performed during fruit set, which represents the vegetative phase in which thrips cause the greatest damage.

Management

The control of thrips should begin by monitoring the populations present in the pomegranate orchard, through yellow sticky traps (six to eight traps per hectare) (Elango *et al.*, 2017b). There are numerous natural enemies of thrips, which can be very effective in their control. For *S. dorsalis*, the predators *Orius* spp. (Hemiptera, Anthocoridae), *Neoseiulis cucumeris* and *Amblyseius swirskii* (Acari, Phytoseiidae), and various zoophagous thrips have proven effective on vegetable crops (Arthurs *et al.*, 2009; Dogramaci *et al.*, 2011; Kumar *et al.*, 2012). Unfortunately, there are no specific studies on the natural enemies of the chilli thrips on pomegranate. There is little information concerning the natural enemies of *R. cruentatus*, and it is limited to the parasitoid *Ceranisus* sp. (Hymenoptera, Eulophidae) reported by Chiu (1984). The various insecticides available should be considered with caution to favour the activity of beneficials. Bioinsecticides (spinosad, azadirachtin, neem oil) and other products such as neonicotinoids or phosphoric esters have shown good efficacy, but they must not be used during the flowering period to avoid damaging insect pollinator populations (Dongarjal *et al.*, 2018; Elango *et al.*, 2018). Often the thrips move onto the cultivated plants, especially the arboreal ones, because the spontaneous herbaceous plants on which they normally live have been eliminated by weeding. Promoting flora diversity with a controlled grassing around and inside the crop can prevent the thrips migration, significantly reducing the damage on cultivated plants and encouraging the maintenance of their natural enemies (Nicholls *et al.*, 2000; Gurr *et al.*, 2003; Altieri, 2012).

13.5.2 Leaf-footed plant bugs

On the American continent, occasional damage is reported to the pomegranate fruits by *Leptoglossus clypealis* Heidemann, *Leptoglossus zonatus* Dallas and *Leptoglossus gonagra* (F.) (=*membranaceous*) (Hemiptera, Coreoidea) (Fig. 13.3c) commonly called leaf-footed plant bugs.

Identification

These bugs (19–25 mm long) are red-brown in colour and are easily recognizable by the characteristic expansion on the hind tibiae and by the presence of a white sinuous strip across the centre of the forewings. Typically, the eggs are laid in chains of 50 or more on branches and leaves of host plants.

Distribution

Leptoglossus clypealis and *L. zonatus* are distributed on the American continent, while *L. gonagra* is reported in Africa, Central and North America, Oceania and southern Asia (ITIS, 2019).

Biology and ecology

Leptoglossus spp. overwinter as adults under the tree bark or herbaceous plants. The development from egg to adult is completed in about 45–50 days (at 25°C) and, in the case of *L. zonatus*, develops about three generations per year (Tepole-García *et al.*, 2016). Both adults and nymphs emit an alarm pheromone (Panizzi, 2004).

Damage

On pomegranate fruits, adults and nymphs insert the rostrum through the peel of ripe fruits or the ones already split, to suck out the juice of the arils that wither (Haviland *et al.*, 2013). However, damaged fruits are difficult to detect from the outside. Moreover, the opening caused by the bugs can allow the entry of opportunistic fungi or bacteria that cause fruit rot. A detailed study carried out in San Joaquin Valley stated that the dominant species on pomegranate is *L. zonatus* (Joyce *et al.*, 2017). A few years ago, in the Maharashtra region of India, severe damage of pomegranate was reported caused by the congeneric *L. gonagra* (F.) (=*membranaceous*) (Jadhav *et al.*, 1976).

Hosts

All species are polyphagous and recently there has been an increase in reports of damage, especially in San Joaquin Valley (California, USA) on almond, pistachio and pomegranate orchards (Xiao and Fademiro, 2011; Joyce *et al.*, 2017).

Survey methods

The best way to prevent *Leptoglossus* activity is by careful inspection of fruit, especially in those areas with pistachio, almond and peach cultivations, because these insects are skilled flyers and move from one crop to another in search of new sources of food.

Management

In California *Gryon pennsylanicum* (Ashmead) (Hymenoptera, Platygastridae), *Ooencyrtus johnsoni* (Howard) (Hymenoptera, Encyrtidae) and *Anastatus pearsalli* Ashmead (Hymenoptera, Eupelmidae) are reported as egg parasites of the large-sized species of Coreidae, effective in controlling the populations of the leaf-footed plant bugs (Mitchell *et al.*, 1999; Maltese *et al.*, 2012). Any outbreaks, especially with ripening fruits, should be controlled quickly with chemical treatments, using products that act by contact (Haviland *et al.*, 2013).

13.5.3 Mediterranean fruit fly

The Mediterranean fruit fly (medfly), *Ceratitis capitata* (Wiedemann) (Fig. 13.3d) (Diptera, Tephritidae) is considered one of the most destructive fruit pests of the world. Meanwhile, it has a great ability to adapt to even the freshest climates and it is likely that, as a result of the ongoing global warming of the earth, it may spread further.

Identification

The adults of *C. capitata* are easily recognizable by their yellowish-brown colour due to the typical combination of orange stripes and brown spots crossing the wings (Fig. 13.3d). The larvae are whitish, cylindrical and elongated, typically narrower in the apical part and broader in the posterior part (Fig. 13.3e).

Distribution

The species is native to sub-Saharan Africa, and is widespread throughout the subtropical and warm-temperate areas of the world.

Biology and ecology

Ceratitis capitata is a very prolific species, with females that can lay 300–800 eggs, in groups of 4–10 per fruit. The larvae, as soon as they emerge, begin to dig feeding tunnels that lead to the decay of the pulp in a short time (Fig. 13.3f). The larval development is completed in 6–15 days, depending on the climatic conditions and the host fruit. The species develops two to four (cold climates) or six to seven (warm climates) annual generations. The overwintering occurs as pupa in the soil.

Damage

Occasionally, even on the ripening fruits of the pomegranate, especially when a superficial crack is present on the epicarp, females may lay their eggs (Özturk and Ulusoy, 2009; Braham, 2015). The larvae that emerge from them develop by feeding inside the fruit, which will become unmarketable (Fig. 13.3g and and Fig. 13.3h). In some years with favourable climatic trends, the lack of alternative fruits increases its appetite for pomegranate. The susceptibility of some varieties or the availability of fruits with a small crack on the epicarp can cause an abnormal increase in infestations.

Hosts

The medfly is an extraordinarily polyphagous species, reported on about 260 host plants, preferring above all the fruit ones (Thomas *et al.*, 2010).

Survey methods

The intensity of infestations may depend on many factors (susceptibility of the pomegranate variety, climatic conditions, presence of other fruit plants around the pomegranates) and may vary over the years. The use of attractive traps can inform about the dynamics and consistency of medfly populations. In addition, the direct observation of ripening fruits may give some indication of the need for a control intervention.

Management

The control of *C. capitata* can be performed using the 'attract & kill' method, or products composed of attractive substances and insecticides. These substances can be distributed directly on the foliage of plants or released by traps. Authorized insecticides can be applied if the infestations are consistent, following carefully the indications on the label and considering that pomegranate is susceptible to phytotoxicity (Cocuzza *et al.*, 2016).

13.5.4 Carob moth

The carob moth, *Apomyelois* (=*Ectomyelois*) *ceratoniae* Zeller (Lepidoptera, Pyralidae) (Fig. 13.3i) has been reported in many countries as causing damage to pomegranate fruits.

Identification

The adults (0.8–1 cm long with forewings of 2–2.4 cm) are variable in colour from creamy white to grey or brownish, on which are distinguishable two darker transverse bands. The larvae are pinkish with a brown head.

Distribution

The species is cosmopolitan.

Biology and ecology

The moth develops four to five generations per year, with the overwintering period spent as larvae. Acacia pods, pomegranate fruits, fig and almonds are used by larvae to overwinter.

Damage

Severe infestations on pomegranate have been repeatedly reported in Morocco and Israel (Blumberg *et al.*, 2001), Iran (Norouzi *et al.*, 2008), Iraq (Al-Izzi *et al.*, 1985), Turkey (Yıldırım and Başpınar, 2015), Saudi Arabia (Moawad *et al.*, 2011) and Tunisia (Braham, 2015). The moth can attack also fruits in storage (Yıldırım and Başpınar, 2015). In addition cracked fruits increase the pest population during the season (Hosseini *et al.*, 2017) because the females lay more eggs on the cracked part of the fruit than into the calyx of the flowers or on the developing fruits of pomegranate (Talaee, 2009). The eggs are laid in the calyx of developing fruits through which larvae penetrate (Fig. 13.3j and Fig. 13.3k). This behavioural feeding of the larvae facilitates some saprophytic fungi such as *Aspergillus* spp. and/or *Penicillium* spp. (Fig. 13.3l) to penetrate the fruit, and then the fruit will partially be rotten and unmarketable (Shakeri, 2004; Fig. 13.3m). The infested fruits are not always easily identifiable, even after the harvest, because externally there may not be evident signs of the presence of the moth larvae. This creates serious problems especially for the export product. The use of visible/near infrared spectroscopy has been successfully experimented with in Iran to recognize preventively infested fruits (Khodabakhshian *et al.*, 2016).

Hosts

The moth is polyphagous and considered a serious pest of fruits such as pomegranate, pistachio, date, almond, fig, walnut and dried fruits (Norouzi *et al.*, 2008).

Survey methods

The detection on pomegranate of *A. ceratoniae* is particularly difficult and often the attached fruits can escape even a careful external examination. In areas where the presence of the moth is known, pheromone traps may monitor the flight of adults during the year.

Management

Several parasitoids have been reported for *A. ceratoniae*, such as *Phanerotoma ocuralis* Kohl, *Bracon hebetor* Say, *Apanteles myeloenta* Wilkinson (Hymenoptera, Braconidae) and *Brachymeria minuta* (L.) (Chalcididae), and different species of *Trichogramma* (Trichogrammatidae) (Hassan *et al.*, 2001; Ksentini *et al.*, 2010; Poorjavad *et al.*, 2011; Nobakht *et al.*, 2015). Twelve parasitoid species from three regions of Iran are reported by Kishani *et al.* (2011) and Kishani *et al.* (2012). In addition, resistant cultivars (Hashemi *et al.*, 2011; Sobhani *et al.*, 2015), sterile insect technique (SIT) (Dhouibi and Abderahmane, 2002; Soufbaf *et al.*, 2018), stamen elimination to disrupt oviposition behaviour of the females (Sabahi and Shakeri, 2009), essential oil of *Ferula assafoetida* as a fumigant to disrupt reproductive behaviour of the pest (Kamelshahi, 2010) or the egg parasitoid *Trichogramma turkistanica* Meyer (Sayed *et al.*, 2015) may be used to suppress the pest population.

13.5.5 Common guava blue or anar butterfly

The common guava blue or anar butterfly, *Deudorix* (=*Virachola*) *isocrates* (F.), the pomegranate butterfly, *Deudorix* (=Virachola) *livia* (Klug) and the cornelian butterfly, *Deudorix epijarbas* (Moore) (Lepidoptera, Lycaenide) (Fig. 13.3n) are considered serious pests in some of the most important areas of pomegranate cultivation.

Identification

Adult females of *D. isocrates* are brownish with an orange patch on the forewings and a black and orange spot on the hindwings, and males are bluish-brown (Bhakare, 2019). Whereas, the females of *D. livia* are bluish-brown (1.2 cm long with a forewing of 3 cm), while males are brown-orange (Beladis *et al.*, 2018). Larvae of *D. isocrates* are creamy-whitish after hatching, turning from greenish-brown or light brown to dark brown in the following stages, with yellowish-orange patches on the dorsal and lateral part of the body (Kumar *et al.*, 2017).

Distribution

Deudorix isocrates is widespread on the Indian subcontinent, *D. epijarbas* is present in Southeast Asia, India and Oceania, whereas the pomegranate butterfly, *D.* (=*V.*) *livia* (Klug) is distributed from Middle Eastern Asia to the sub-Saharan areas and coastal areas of the African continent (Vattakaven *et al.*, 2016).

Biology and ecology

The three species have very similar behaviour. Eggs are laid singly in the crown of the developing fruits, from which the newly hatched larvae penetrate inside to feed on pulp and seeds, until the completion of the juvenile development (Gharbi, 2010; Kumar *et al.*, 2017). Several studies have reported a total development of 43–75 days for *D. isocrates* reared on pomegranate under laboratory conditions, and a fecundity of 20–30 eggs per female (Kumar *et al.*, 2017; Khandare *et al.*, 2018; Mallikarjun and Pal, 2018). The pupation takes place on the soil.

Damage

Both species are considered key pests of pomegranate, as they can cause significant production losses (Obeidat and Akkawi, 2002; Bagle, 2009; Ksentini *et al.*, 2011; Moawad *et al.*, 2011; Kahramanoglu and Usanmaz, 2013; Braham, 2015; Mallikarjun and Pal, 2018). Balikai *et al.* (2011) and Ksentini *et al.* (2011) reported that pomegranate yield losses can reach 50%. The infested attached fruits by *D. livia* show on the epicarp a hole 1–5 mm wide, which, as a result of the fermentation processes following the feeding activity of the larvae, attracts adults of *Drosophila melanogaster* L. and the invasion of opportunistic fungi and bacteria. The infestation can also cause falling of the fruits (Beladis *et al.*, 2018) (Fig. 13.3o). In any case, the infested fruits are unmarketable. In India, the damage of *D. isocrates* has increased in recent years after the introduction of the pomegranate cultivars and the specialization of the culture (Mallikarjun and Pal, 2018).

Hosts

All the species are polyphagous. *Deudorix isocrates* is reported on citrus, guava, lychee, aonla, woodapple, mulberry, peach, plum and other fruits (Khan, 2016); *D. livia* on date and acacia (Beladis *et al.*, 2018); while *D. epijarbas* is reported on lychee, longan, soapnut, Indian horse-chestnut and tulipwood (Vattakaven *et al.*, 2016).

Survey methods

In areas where the presence of the butterflies is recognized, it is essential to carry out periodic inspection of the fruits especially in the early stages of enlargement after setting, in search of the entry holes caused by the larvae.

Management

Control of both species is rather difficult. A treatment with neem oil after fruit set could be useful to prevent the penetration of larvae inside the fruits. An effective control method is closing the fruits (30–50 days after setting) inside paper bags, after having treated them with neem oil or *B. thuringiensis* (Kumar and Kamala Jayanthi, 2018). Some trials carried out in Tunisia and Cyprus showed the efficacy of *B. thuringiensis* and spinosad to control *V. livia* (Kahramanoglu and Usanmaz, 2013; Sayed *et al.*, 2015). It is also important to remove and eliminate the infested fruits from the field in order to prevent population increase (Kumar and Kamala Jayanthi, 2018). Several studies have shown the effectiveness of various insecticides to control the infestations of *Deudorix* spp., such as pyrethroids, avermectins and diamides (Khan *et al.*, 2017; Kumar and Gupta, 2018).

13.5.6 Honeydew moth and false codling moth

Cryptoblades gnidiella Millière, also known as Christmas berry moth or citrus pyralid, and *Thaumatotibia leucotreta* (Meyrick) (Lepidoptera, Tortricidae) are species of secondary importance that can sometimes attack the fruits of the pomegranate.

Identification

The larvae of *C. gnidiella* are variable in colour from yellowish to olive or reddish-brown. Forewings of adults (12–18 mm wingspan) can vary from greyish to brown, with whitish cross-lines and reddish-purple scales (Lucchi *et al.*, 2011, 2019). Larvae of the last instar of *T. leucotreta* are about 1.5 cm long, reddish-pink in colour with a yellow-brown head. It is a dimorphic species with females bigger than males (wingspan of 1.5 and 2 cm, respectively). Forewings are basically grey-brown, with numerous minute black, orange and dark brown markings and a whitish dot in the central area (EPPO, 2019b).

Distribution

Cryptoblades gnidiella is a cosmopolitan species of probable Mediterranean origin, while *T. leucotreta* is widespread in Africa and occasionally intercepted in Europe and North America.

Biology and ecology

The females of both species lay 200–300 eggs directly on the leaves or close to the fruits, so that the larvae penetrate inside to feed on it. There is no diapause, with *C. gnidiella* and *T. leucotreta* developing three and five generations per year, respectively.

Damage

The larvae penetrate the fruits, digging deep galleries that make them unmarketable. Severe losses of pomegranate fruits caused by *C. gnidiella* have been reported in Spain and Turkey during harvesting and storage (Juan *et al.*, 2000; Öztürk and Ulusoy, 2012; Demirel, 2016; García-Martínez *et al.*, 2017). *T. leucotreta* in South Africa is considered a species particularly harmful to citrus and pomegranate. In several Mediterranean countries, larvae of the moth have been repeatedly found on imported plant material (Kirkman and Moore, 2007; EPPO, 2019b; Mazza *et al.*, 2014).

Hosts

Both species are polyphagous species, recorded on more than 80 host plants (Mazza *et al.*, 2014; Lucchi *et al.*, 2019).

Survey methods

Several studies conducted on pomegranate and grapevine for *C. gnidiella* have shown a good correlation between the catches of males and the intensity of infestations, and can be informative to set up any insect control action (Vidart *et al.*, 2013; Demirel, 2016; Lucchi *et al.*, 2019). Pheromone traps to identify flights and for mating disruption of *T. leucotreta* are available, but their use has not always given satisfactory results, especially with regard to the optimal blend for male capture (Levi-Zada *et al.*, 2020). For both *C. gnidiella* and *T. leucotreta*, visual inspection of fruits is always necessary to prevent massive infestations.

Management

Chemical control of the larvae is rather hard as they are located within the fruit and are difficult to reach. Infestations can be controlled with *B. thuringiensis* var. *kurstaki* (Ben Yehuda *et al.*, 1993; Harari *et al.*, 2007; Sellanes and González, 2014). The possibilities of biological control are limited as there are very few reports of antagonistic species of the insect (Lucchi *et al.*, 2011). Recently, the precise mixture of pheromone components has been identified and the chances of developing an effective product are high (Levi-Zada *et al.*, 2020). Particular attention must be paid to the control of imported plant material, as, for example, *T. leucotreta* has been intercepted several times in checks carried out in various European and US ports (Mazza *et al.*, 2014).

13.5.7 Dried fruit beetles or sap beetle

On the mature fruit, beetles of the genus *Carpophilus* (Coleoptera, Nitidulidae) are frequently found (Fig. 13.3p). They are small detrimental insects (adults are about 1.5 mm long), recognizable by the elytra shorter than the abdomen. *Carpophilus* spp. normally infest decaying fruits, in which the pericarp is cracked and attacked by opportunistic fungi. These species are widespread worldwide. Severe infestations have been reported in Israel (Zvi Mendel, personal communication) and south Italy (Nuzzaci, 1968). The eggs are laid inside the calix, and the emerged larvae penetrate inside the fruit. Their food activity causes rotting and fall of the fruits. In areas where *Carpophilus* spp. are a recurrent problem, control occurs by eliminating all the infested fruits to prevent the numerical development of the populations, or with insecticidal treatments to prevent oviposition on the fruit. Aggregative pheromones are available to monitor the populations (Hossain *et al.*, 2013).

13.6 Root Pests

13.6.1 White grub

Occasionally, pomegranates are attacked by grubs of *Polyphylla olivieri* Laporte de Castelnau (Coleoptera, Melolonthidae), a soil-dwelling pest whose larvae feed on the roots, sometimes causing severe damage especially in young trees.

Identification

Mature larvae or grubs of *P. olivieri* are 2.5–5.5 cm long, greyish-white, reddish head and typically C-shaped. The body of adults (2.5–3 cm long) is robust, convex and reddish-brown in colour with fine whitish pubescence that gives a marbled appearance. At the apex, antennae of males present a typical enlarged fan.

Distribution

Polyphylla olivieri is a species widespread in south-eastern Europe and Asia Minor.

Biology and ecology

In summer, the larvae emerge from the eggs and head towards the roots of the host plants where they winter. In spring, they resume feeding. The life cycle is completed in 3–4 years, with the adults that emerge in late spring.

Damage

The grubs or larvae that feed on root tissues cause the damage. The initial symptoms of the infestation are similar to those of drought stress, with yellowing of vegetation and plants that appear wilted. Strong infestations can cause the death of young plants. The damage occurs 3–4 years after first detection of the adult in the field. In Iran, *P. olivieri* is considered a serious pest of fruit trees (Kharazi-Pakdel and Karimi, 2008), but there is no specific report in pomegranate orchards.

Hosts

Polyphylla olivieri is polyphagous.

Survey methods

The presence of adults on pomegranate flowers indicates that careful control is needed in the underground parts to check if there are grubs, which can be easily pulled from the ground.

Management

Young grubs of *P. olivieri* are very vulnerable to long periods of drought that significantly lower populations. Late spring or early autumn brings the larvae to the surface exposing them to various predators (birds, moles, skunks, etc.). The maintenance of groundcover can offer alternative food sources to grubs, which can lighten the infestations on the pomegranates. Chemical control can be performed by applying registered insecticides in the root area via soil incorporation. In dry periods, it is recommended to carry out the treatments before the forecasted rain or irrigation, since the humidity pushes the larvae to move towards the more superficial layers of the soil, resulting in them being more vulnerable to pesticides. Among natural entomopathogens, effective ones are nematodes (*Steinernema feltiae* and *Heterorhabditis megidis*) and fungi (*Metarhizium anisopliae* and *Beauveria bassiana*) (Kharazi-Pakdel and Karimi, 2008).

13.6.2 Root-knot nematodes

Meloidogyne spp. (Heteroderidae, Meloidogyne). The root-knot parasitic nematodes are very small and slender worms living in soil, which infest occasionally pomegranate causing yield losses and quality decaying of production.

Identification

Among root-knot nematodes, five species, *Meloidogyne incognita*, *Meloidogyne javanica*, *Meloidogyne arenaria*, *Meloidogyne hapla* and *Meloidogyne acrita* are reported on pomegranate (Carneiro *et al.*, 2007; Somasekhara *et al.*, 2012). All *Meloidogyne* spp. are dimorphic, with females whitish in colour, about 400–600 μm long, thickened and pyriform due to the abnormal development of the ovary when they remain fixed to the roots of the host plant. Males, when present, are wormlike.

Distribution

Meloidogyne spp. are widespread in all continents, with prevalence in tropical and subtropical countries.

Biology and ecology

The root-knot nematodes are sedentary endoparasites associated with the root systems of the host plant. Reproduction is parthenogenetic, although males are present. In tropical countries, the nematodes reproduce year-round, managing to complete over eight generations each year. The females, as soon as they reach the adult stage, attach themselves to the radicle tissue through the stylet and assume the typical spherical shape. Because of the alimentary activity, galls are formed, of variable dimensions, which represent the first symptom of nematode infestation. Each female can produce over 1000 eggs (Ambrogion *et al.*, 2014).

Damage

Meloidogyne incognita is the most common, causing substantial yield losses and reducing the quality of pomegranates (Mhase and Kadam, 2000). The damage is the consequence of their feeding activity on roots and the formation of galls that decrease significantly the absorption of water and nutrients. Infested plants show pale green or yellowish leaves and reduced plant growth that sometimes can evolve to wilting and death of the plants (Perry and Ploeg, 2010).

Hosts

Meloidogyne spp. are polyphagous, and are able to attack over 700 plant species (Ambrogion *et al.*, 2014).

Survey methods

In the field, single or close groups of pomegranate plants infested show a stunted development, chlorotic and/or early falling leaves, with fruits of reduced dimensions that result in lower fruit production.

Management

A deep ploughing of the land is advisable before hosting the new plant, and soil solarization for 30–45 days in the summer period to reduce nematode populations in the top 35 cm. However, soil solarization does not provide long-term protection for fruit crops under orchards (Rao and Krishnappa, 1995). Seedlings must be free of nematodes before planting in the orchard. Inspect for the presence of root galls by uprooting a few pomegranate seedlings from the polythene bag in the nursery. Organic amendments (peat, manure, composts and green manure crops) increase the water- and nutrient-holding capacity of the soil, especially the sandy ones. Mustard and castor oil cakes are effective in reducing the nematode populations in pomegranate (Khan *et al.*, 2011), whereas non-edible oil cakes enhance the activity of predacious fungi that feed on nematodes (Singh *et al.*, 2002). The control efficacy of the application of farmyard manure compost or oil cakes has been observed in the pomegranate nursery as well as in the main field (Khan *et al.*, 2014). Of course, maintaining the plants in good water and nutritive conditions also reduces nematode damage (Perry and Ploeg, 2010). Some of the marigold species produce phytochemical exudates that reduce populations of root-knot nematodes (Motsinger *et al.*, 1977; Singh and Gupta, 1993; Hooks *et al.*, 2010). The trap effects of cucumber and tomato on pomegranate root-knot nematode were found to be significantly better than aubergine, spinach and pepper in pomegranate orchards (YanFen *et al.*, 2013). The application of authorized fumigant chemical products with nematocidal action to the soil can be useful in particularly infested soils. However, it is necessary to calculate the costs and assess whether the economic damage due to the loss of production exceeds the cost of the treatment. Several biocontrol agents have been tested against root-knot nematodes, and the most promising was the fungus *Paecilomices lilacinus* (Khan and Goswami, 2002; Kiewnick and Sikora, 2006). *Paecilomices lilacinus* penetrates the eggs of *M. incognita* and eats the contents leaving only shell (Bhatt *et al.*, 2002).

References

Abd-Rabou, S. and Simmons, A.M. (2014) Survey of natural enemies of whiteflies (Hemiptera: Aleyrodidae) in Egypt with new local and world records. *Entomological News* 124(1), 38–56. DOI: 10.3157/021.124.0106.

Ahuja, D.B., Sharma, J. and Suroshe, S.S. (2013) Pests of fruits: banana, mango and pomegranate. In: Ahuja, D.B. (ed.) *Pest Surveillance and Pest Management Advisory*. National Centre for Integrated Pest Management, Dept. of Horticulture, New Delhi, India

Al-Gboory, I. and El-Haidani, H. (1989) Some ecological aspects of the pomegranate false spider mite, *Tenuipalpis punicae* (Acari: Tenuipalpidae) in Iraq. In: Channabasavanna, G.P. and Viraktamath, C.A. (eds) *Progress in Acarology*. 2. Leiden Brill, Leiden, the Netherlands, pp. 73–79.

Al-Izzi, M.A.J., Al-Maliky, S.K., Younis, M.A. and Jabbo, N.F. (1985) Bionomics of *Ectomyelois ceratoniae* (Lepidoptera: Pyralidae) on pomegranates in Iraq. *Environmental Entomology* 14(2), 149–153. DOI: 10.1093/ee/14.2.149.

Altieri, M.A. (2012) Insect pest management in the agroecosystems of the future. *Atti Accademia NazionaleItaliana di Entomologia* LX, 137–144.

Ambrogion, L., Carletti, B. and Roversi, P.F. (2014) Nematodi galligeni. In: Ambrogioni, L., d'Errico, F.P., Greco, N., Marinari Palmisano, A. and Roversi, P.F. (eds) *Nematologia Agraria Generale e Applicata*. Societa Italiana Entomologia, Tip, Coppini, Italy, pp. 337–365.

Andreadis, S.S., Navrozidis, E. I. and Katerinis, S. (2016) First record of the grape cane borer, *Amphicerus bimaculatus* (Olivier, 1790) (Coleoptera: Bostrichidae), on pomegranate in Greece. *Turkish Journal of Zoology* 40, 286–289. DOI: 10.3906/zoo-1505-41.

Arthurs, S., McKenzie, C.L., Chen, J., Dogramaci, M., Brennan, M. *et al.* (2009) Evaluation of *Neoseiulus cucumeris* and *Amblyseius swirskii* (Acari: Phytoseiidae) as biological control agents of chilli thrips, *Scirtothrips dorsalis* (Thysanoptera: Thripidae), on pepper. *Biological Control* 49(1), 91–96. DOI: 10.1016/j.biocontrol.2009.01.002.

Atay, E. and Öztürk, N. (2010) *Euzophera semifuneralis* (Walker, 1863) (Lepidoptera, Pyralidae) detected in pomegranate orchards in Adana and Osmaniye and its type of damage. *Ziraat Fakultesi Dergisi, Mustafa Kemal Universitesi* 15, 51–58.

Awadallah, K.T., Ammar, E.D., Tawfik, M.F.S. and Rashad, A. (1979) Life history of the white mealybug *Ferrisia virgata* (Ckll.) (Homoptera: Pseudococcidae). *Deutsche Entomologische Zeitschrift* 26(1–3), 101–110. DOI: 10.1002/mmnd.19790260111.

Bagle, B.G. (1993) Seasonal incidence and control of thrips *Scirtothrips dorsalis* Hood in pomegranate. *The Indian Journal of Entomology* 55, 148–153.

Bagle, S. (2009) Studies on varietal reaction, extent of damage and management of anar butterfly, *Deudorix* (=Virachola) *isocrates* Fab., in pomegranate. *Acta Horticulturae* 890, 557–560.

Balika, R.A., Biradar, A.P. and Teggelli, R.G. (1999) Severe incidence of pomegranate whitefly, *Siphoninus phillyreae* in Northern Karnataka. *Insect Environment* 5, 76.

Balikai, R.A., Kotikal, Y.K. and Prassana, P.M. (2011) Status of pomegranate pests and their management strategies in India. *Acta Horticulturae* 890, 569–584.

Bartual, J., Loyoza, A., García, J. and Valdés, G. (2012) Efficacy and residues of selected insecticides for control of cotton aphid *(Aphis gossypii)* and citrus mealybug (*Planococcus citri*) in pomegranate. *Options Mediterranean, Series A* 103, 108–111.

Bass, C., Denholm, I., Williamson, M.S. and Nauen, R. (2015) The global status of insect resistance to neonicotinoid insecticides. *Pesticide Biochemistry and Physiology* 121, 78–87. DOI: 10.1016/j.pestbp.2015.04.004.

Beladis, B., Verheggen, F., Baba Aissa, N., Boukraa, S., Salah Ou Elhadj, B. *et al.* (2018) Premier signalement de *Deudorix livia* (Lepidoptera: Lycanidae) en Algérie: un ravageur important du grenadier et du palmier dattier. *EPPO Bulletin* 48(2), 281–286. DOI: 10.1111/epp.12478.

Bellows, T., Paine, T., Arakawa, K.Y., Meisenbacher, C., Leddy, P. *et al.* (1990) Biological control sought for ash whitefly. *California Agriculture* 44(1), 4–6.

Ben Yehuda, S., Wysoki, M. and Rosen, D. (1993) Laboratory evaluation of microbial pesticides against the honeydew moth. *Insect Science and its Application* 14, 627–630.

Ben-Dov, Y. and German, V. (2003) ScaleNet, *Maconellicoccus hirsutus*. Available at: http://scalenet.info/catalogue/Maconellicoccus%20hirsutus/ (accessed 1 October 2020).

Bhakare, M. (2019) *Virachola isocrates* (Fabricius, 1793) Common guava blue. In: Kunte, K., Sondhi, S. and Roy, P. (eds) *Butterflies of India, v. 2.60*. Indian Foundation for Butterflies. Available at: www.ifoundbutterflies.org/sp/635/Virachola-isocrates

Bhatt, J., Chaurasia, R.K. and Sengupta, S.K. (2002) Management of *Meloidogyne incognita* by *Paecilomyces lilacinus* and influence of different inoculum levels of *Rotylenchulus reniformis* on betelvine. *Indian phytopathology* 5, 348–350.

Biradar, A.P., Jagginavar, S.B. and Sunitha, N.D. (2005) Management of stem borer, *Coelosterna scrabrator* Fabr. (Coleoptera: Cerambycidae) in pomegranate. *International Journal of Agricultural Sciences* 1(1), 16–17.

Blackman, R.L. and Eastop, V.F. (2006) *Aphids on the World's Herbaceous Plants and Shrubs*. Vol. 2. John Wiley and Sons and The Natural History Museum, London. Available at: www.aphidsonworldsplants.info

Blumberg, D., Navon, A., Kehat, M., Eliahu, M., Levski, S. *et al.* (2001) Date palm pests in Israel at the beginning of the third millennium. *Alon Hanotea* 55, 42–48.

Bonsignore, C.P. (2012) *Apate monachus* (Fabricius, 1775), a bostrichid pest of pomegranate and carob trees in nurseries – Short Communication. *Plant Protection Science* 48(2), 94–97. DOI: 10.17221/53/2011-PPS.

Braham, M. (2015) Insect larvae associated with dropped pomegranate fruits in an organic orchard in Tunisia. *Journal on Entomology and Nematology* 7, 5–10.

Braham, M. and Gahbiche, H. (2016) Occurrence of *Apate monachus* Fabriccius 1775 (Coleoptera: Bostrichidae), a black borer attacking pomegranate trees in the central-east region of Tunisia. *International Journal of Entomology and Nematology* 2, 27–41.

CABI (2019a) *Farrisia virgata*. In: *Invasive Species Compendium*. CAB International, Wallingford, UK. Available at: www.cabi.org/isc

CABI (2019b) *Xyleborus perforans*. In: *Invasive Species Compendium*. CAB International, Wallingford, UK. Available at: www.cabi.org/isc

CABI (2019c) *Scirtothrips dorsalis*. In: *Invasive Species Compendium*. CAB International, Wallingford, UK. Available at: www.cabi.org/isc

CABI (2019d) *Rhiphiphorothrips cruentatus*. In: *Invasive Species Compendium*. CAB International, Wallingford, UK. Available at: www.cabi.org/isc

Campadelli, C. (1995) Su alcuni parassitoidi di *Zeuzera pyrina* L. (Lep. Cossidae) in Emilia Romagna. *Bollettino dell'Istituto di Entomologia "G. Grandi"* 50, 127–131.

Carneiro, R.M.D.G., Almeida, M.R.A., Cofcewicz, E.T., Magunacelaya, J.C. and Aballay, E. (2007) *Meloidogyne ethiopica*, a major root-knot nematode parasitizing *Vitis vinifera* and other crops in Chile. *Nematology* 9(5), 635–641.

Chandra, M. and Verma, R.K. (2010) Morphological variation in adult male and female *Rhiphiphorothrips cruentatus* Hood (Thysanoptera: Thripidae). *World Applied Sciences Journal* 9, 1421–1423.

Chandra, R., Suroshe, S.S., Sharma, J., Marathe, R.A. and Meshram, D.T. (2011) *Pomegranate Growing Manual*. NRC on Pomegranate, Solapur, India

Chiu, H.T. (1984) The ecology and chemical control of grapevine thrip (*Rhiphiphorothrips cruentatus* Hood) on wax apple. *Plant Protection Bulletin, Taiwan* 26, 365–377.

Cocuzza, G.E., Cavalieri, V., Zappal, L. and Barbagallo, S. (2009) Genetic relationships inside *Aphis frangulae/gossypii* group based on mitochondrial DNA sequences. *Redia* XCII, 65–68.

Cocuzza, G.E.M., Mazzeo, G., Russo, A., Giudice, V.L. and Bella, S. (2016) Pomegranate arthropod pests and their management in the Mediterranean area. *Phytoparasitica* 44(3), 393–409. DOI: 10.1007/s12600-016-0529-y.

Daniel, C., Barloggio, G., Stoeckli, S., Luka, H. and Niggli, U. (2018) Management of crop to prevent pest outbreaks. In: Vacante, V. and Kreiter, S. (eds) *Handbook of Pest Management in Organic Farming*. CAB International, Wallingford, UK, pp. 1–23.

Demirel, N. (2016) Seasonal flight patterns of the honeydew moth, *Cryptoblades gnidiella* Millère (Lepidoptera: Pyralidae) in pomegranate orchards as observed using pheromone traps. *Entomology and Applied Science Letters* 3, 1–5.

Dhouibi, M.H. and Abderahmane, C.T. (2002) *The effect of substerilizing doses of gamma radiation on the pupae of the carob moth Ectomyelois ceratoniae (Lepidoptera: Pyralidae). Evaluation of Lepidoptera population suppression by irradiation*. IAEA, Insect Pest Control Section, Vienna, Austria.

Doğramaci, M., Arthurs, S.P., Chen, J., McKenzie, C., Irrizary, F. *et al.* (2011) Management of chilli thrips *Scirtothrips dorsalis* (Thysanoptera: Thripidae) on peppers by *Amblyseius swirskii* (Acari: Phytoseiidae) and *Orius insidiosus* (Hemiptera: Anthocoridae). *Biological Control* 59(3), 340–347. DOI: 10.1016/j.biocontrol.2011.09.008.

Dongarjal, R.P., Ilyas, M. and Shendge, S.A. (2018) Bioefficacy of newer insecticides on thrips of pomegranate. *Journal of Entomology and Zoology Studies* 6, 1034–1036.

Döker, I., Kazak, C. and Karut, K. (2013) A new pomegranate pest for Turkey, pomegranate false spidermite, *Tenuipalpus punicae* Pirchard and Baker (Acari: Tenuipalpidae). *Turkish Bulletin of Entomology* 3, 113–117.

Elango, K. and Sridharan, S. (2017a) Population dynamics of pomegranate sucking pests under high density planting in Tamil Nadu. *Journal of Entomology and Zoology Studies* 5, 377–380.

Elango, K., Sridharan, S., Saravanan, P.A. and Balakrishnan, S. (2017b) Relative performance of different colour laden sticky traps on the attraction of sucking pests in pomegranate. *International Journal of Current Microbiology and Applied Sciences* 6(11), 2997–3004. DOI: 10.20546/ijcmas.2017.611.350.

Elango, K., Sridharan, S., Saravanan, P.A. and Balakrishnan, S. (2018) Managing of thrips *Scirtothrips dorsalis* Hood (Thysanoptera: Thripidae) in pomegranate under high density planting. *Advances in Life Sciences* 7, 82–85.

EPPO (2019a) *Aleurocathus spiniferus*. Available at: https://gd.eppo.int/taxon/ALECSN/ (accessed 5 March 2019).

EPPO (2019b) PM 7/137 (1) *Thaumatotibia leucotreta*. *EPPO Bulletin* 49(2), 248–258. DOI: 10.1111/epp.12580.

Fasih, M. and Srivastava, R.P. (1988) Natural occurrence of *Beauveria bassiana* an entomogenous fungus on bark eating caterpillar, *Indarbela* spp. *Indian Journal of Plant Pathology* 6(1), 11–16.

García-Martínez, O., Pérez-Hedo, M., Urbaneja, A., Beltrán, V., Bartual, J. *et al.* (2017) Importance of moth species as pests on pomegranate and persimmon crops in the Valencian Community (Spain). *IV International Symposium on Pomegranate and Minor Mediterranean Fruits*, Elche, Spain, 22 September.

García, M., Denno, B., Miller, D.R., Miller, G.L., Ben-Dov, Y. *et al.* (2016) ScaleNet: a literature-based model of scale insect biology and systematics. Available at: http://scalenet.info (accessed 8 June 2020).

Gatwick, J. (1992) *Crop pests in the UK. Collected edition of MAFF leaflets*. Chapman and Hall, London, UK.

Gharbi, N. (2010) Laboratory rearing of the pomegranate fruit butterfly *Virachola livia* on two host plants in Tunisia. *Tunisian Journal of Plant Protection* 5, 195–199.

Gill, H.K., Goyal, G. and Gillett-Kaufman, J. (2013) Citrus mealybug *Planococcus citri* (Risso) (Insecta: Hemiptera: Pseudococcidae). *University of Florida, Institute of Food and Agricultural Sciences (IFAS) EENY* 537, 1–4.

Gillespie, P.S. (2000) A new whitefly for NSW: the ash whitefly. NSW Agriculture. Available at: www.agric.nsw.gov.au/Hort/ascu/insects/ashwf.htm (accessed 1 August 2020).

Guario, A., Bari, G., Marinuzzi, V., Alfarano, L., Falco, R. *et al.* (2001) Biologia della *Zeuzera pyrina* e controllo con la confusion sessuale. *L'Informatore Agrario* 44, 57–61.

Gurr, G.M., Wratten, S.D. and Luna, J.M. (2003) Multi-function agricultural biodiversity: pest management and other benefits. *Basic and Applied Ecology* 4(2), 107–116. DOI: 10.1078/1439-1791-00122.

Gyeltshen, J., Hodges, A. and Hodges, G.S. (2014) Orange spiny whitefly, *Aleurocanthus spiniferus* Quaintance (Insecta: Hemiptera: Aleyrodidae), University of Florida, Institute of Food and Agricultural Sciences (IFAS). Available at: http://entnemdept. ufl.edu/creatures/citrus/orange_spiny_white fly.htm (accessed 5 March 2019).

Halima-Kamel, M.B., Germain, J.F. and Mdellel, L. (2015) First records of two mealybugs, *Maconellicoccus hirsutus* (Green) and *Phenacoccus peruvianus* Granara de Willink, in Tunisia and the North of Africa. *EPPO Bulletin* 45(1), 139–143. DOI: 10.1111/epp.12186.

Harari, A.R., Zahavi, T., Gordon, D., Anshelevich, L., Harel, M. *et al.* (2007) Pest management programmes in vineyards using male mating disruption. *Pest Management Science* 63(8), 769–775. DOI: 10.1002/ps.1365.

Hashemi, S., Karimizadeh, J., Jalalzand, A.R., Besharatnejad, M.H. and Modaresi, M. (2011) Studying on damage of carob moth in three pomegranate cultivars of Isfahan (Iran). *Procedia Environmental Sciences* 8, 257–261.

Hassan, B., Chemseddine, M., Abbassi, M. and Brun, J. (2001) The date moth in the area of Tafilalet in the Southeast of Morocco. Available at: http://webspirs.ziur.co.il/webspirs/doLS.ws?ss=Fruits:189-196 (accessed 8 June 2020).

Haviland, D.R., Carroll, D., Bentley, W.J. and Walton, W. (2013) *UC IPM Pest Management Guidelines: Pomegranate*. UC ANR Publication 3474.

Hegazi, E., Khafagi, W.E., Konstantopoulou, M., Raptopoulos, D., Tawfik, H. *et al.* (2009) Efficient mass-trapping method as an alternative tactic for suppressing populations of leopard moth (Lepidoptera: Cossidae). *Annals of the Entomological Society of America* 102(5), 809–818. DOI: 10.1603/008.102.0507.

Hegazi, E., Schlyter, F., Khafagi, W., Atwa, A., Agamy, E. *et al.* (2015) Population dynamics and economic losses caused by *Zeuzera pyrina*, a cryptic wood-borer moth, in an olive orchard in Egypt. *Agricultural and Forest Entomology* 17(1), 9–19. DOI: 10.1111/afe.12075.

Hill, D.S. (2008) *Pests of Crops in Warmer Climates and their Control*. Springer, Berlin.

Holland, D., Hatip, K. and Bar-Ya'akov, I. (2009) Pomegranate: botany, horticulture and breeding. In: Janick, J. (ed.) *Horticultural Reviews*. Wiley, pp. 119–127.

Holman, J. (2009) *Host Plant Catalogue of Aphids: Palaearctic Region*. Springer, Berlin

Hooks, C.R.R., Wang, K.-H., Ploeg, A. and McSorley, R. (2010) Using marigold (Tagetes spp.) as a cover crop to protect crops from plant-parasitic nematodes. *Applied Soil Ecology* 46(3), 307–320. DOI: 10.1016/j.apsoil.2010.09.005.

Hossain, M.S., Hossain, M.A.B.M., Williams, D.G. and Chandra, S. (2013) Management of *Carpophilus* spp. beetles (Nitidulidae) in stone fruit orchards by reducing the number of attract-and-kill traps in neighbouring areas. *International Journal of Pest Management* 59(2), 135–140. DOI: 10.1080/09670874.2013.782077.

Hosseini, S.A., Goldansaz, S.H., Menken, S.B.J., van Wijk, M., Roessingh, P. *et al.* (2017) Field attraction of carob moth to host plants and conspecific females. *Journal of Economic Entomology* 110(5), 2076–2083. DOI: 10.1093/jee/tox218.

ITIS (2019) Integrated taxonomic information system on-line database. Available at: www.itis.gov (accessed 31 July 2019).

Jadhav, L.D., Ajri, D.S., Kadam, M.V. and Dorge, S.K. (1976) Leaf-footed plant bug on pomegranate in Maharashtra. *Entomologists' Newsletter* 6, 57.

Jagginavar, S.B. and Naik, L.K. (2005) Management of shothole borer, *Xyleborus perforans* (Wollostan) (Coleoptera: Scolytidae) in pomegranate. *Indian journal of agricultural research* 39, 133–137.

Jagginavar, S.B., Sunitha, N.D. and Paitl, D.R. (2008) Management strategies for grape stem borer *Celosterna scrabrator* Fabr (Coleoptera: Cerambycidae). *Indian Journal of Agricultural Research* 42(4), 307–309.

James, S.P., Babu, A., Selvasundaram, R. and Muraleedharan, N. (2007) Field evaluation of traps for attracting shot hole borer. *Newsletter – UPASI Tea Research Foundation* 17(1), 3.

Jeppson, L.R., Keifer, H.H. and Baker, E.W. (1975) *Mites Injurious to Economic Plants*. University of California Press, Berkeley, California.

Joyce, A.L., Higbee, B.S., Haviland, D.R. and Brailovsky, H. (2017) Genetic variability of two leaffooted bugs, *Leptoglossus clypealis* and *Leptoglossus zonatus* (Hemiptera: Coreidae) in the central valley of California. *Journal of Economic Entomology* 110(6), 2576–2589. DOI: 10.1093/jee/tox222.

Juan, P., Martìnez, J., Martìnez, J.J., Oltra, M.A. and Fernandez, M. (2000) Current situation of pomegranate growing (*Punica granatum* L.) in southern Alicante. Chemical control of pests and diseases and financial cost. In: Malgarejo, P., Martìnez-Nicol S.J.J. and Martìnez-Tomé, J. (eds) *Production, Processing and Marketing of Pomegranate in the Mediterranean Region: Advances in Research and Technology*. CHIEAM, Zaragoza, Spain, pp. 157–161.

Kahramanoglu, I. and Usanmaz, S. (2013) Management strategies of fruit damaging pests of pomegranate *Planococcus citri*, *Ceratitis capitata* and *Deudorix* (*Virachola*) *livia*. *African Journal of Agricultural Research* 8, 6563–6568.

Kamelshahi, G. (2010) Effect of essential oil from *Ferula assafoetida* on some reproductive behaviour and some life parameters of *Ectomyelois ceratoniae* under field and laboratory conditions. MSc Thesis. Department of Plant Protection, University of Tehran, Iran.

Karunaratne, W.S., Kumar, V., Pettersson, J. and Kumar, N.S. (2008) Response of the shot-hole borer of tea, *Xyleborus fornicatus* (Coleoptera: Scolytidae) to conspecifics and plant semiochemicals. *Acta Agriculturae Scandinavica, Section B – Plant Soil Science* 58(4), 345–351. DOI: 10.1080/09064710701788802.

Kaydan, M.B. and Gullan, P.J. (2012) A taxonomic revision of the mealybug genus *Ferrisia* Fullaway (Hemiptera: Pseudococcidae), with descriptions of eight new species and a new genus. *Zootaxa* 3543(1), 1–65. DOI: 10.11646/zootaxa.3543.1.1.

Khan, A., Shaukat, S.S. and Sayed, M. (2011) Management of plant nematodes associated with pomegranate (*Punica granatum* L.) using oil-cakes in Balochistan, Pakistan. *Indian Journal of Nematology* 41(1), 1–3.

Khan, M.R., Jain, R.K., Ghule, T.M. and Pal, S. (2014) *Root Knot Nematodes in India – a Comprehensive Monograph*. Indian Agricultural Research Institute, New Delhi.

Khan, I., Khan, S.A., Hussain, S., Maula, F., Shah, H.A. *et al*. (2017) To study the infestation level and effective chemical control of pomegranate fruit borer (*Virachola isocrates*). *Journal of Entomology and Zoology Studies* 5, 282–284.

Khan, M.M.H. (2016) Biology and management of fruit borer, *Virachola isocrates* (Fab.) infesting guava. *Bangladesh Journal of Agricultural Research* 41(1), 41–51. DOI: 10.3329/bjar.v41i1.27666.

Khan, M.R. and Goswami, B.K. (2002) Evaluation of *Paecilomyces lilacinus* isolate 2 against *Meloidogyne incognita* infecting tomato. *International Journal of Nematology* 12, 111–113.

Khandare, R.Y., Kadam, D.R. and Jayewar, N.E. (2018) Biology of pomegranate fruit borer, *Deudorix isocrates* (Fab.) (Lycaenidae: Lepidoptera) on pomegranate, *Punica granatum* L. *Journal of Pharmacognosy and Phytochemistry* 7, 328–330.

Kharazi-Pakdel, A. and Karimi, J. (2008) New insights into management of the white grub Polyphylla olivieri in fruit orchards in Iran. *7th IOBC Conference on Integrated Fruit Production*, Avignon, France, 27-30 October, p. 90.

Khodabakhshian, R., Emadi, B., Khojastehpour, M. and Golzarian, M.R. (2016) Carob moth, *Ectomyelois ceratoniae*, detection in pomegranate using visible/near infrared spectroscopy. *Computers and Electronics in Agriculture* 129, 9–14. DOI: 10.1016/j.compag.2016.09.006.

Kiewnick, S. and Sikora, R.A. (2006) Biological control of the root-knot nematode *Meloidogyne incognita* by *Paecilomyces lilacinus* strain 251. *Biological Control* 38(2), 179–187. DOI: 10.1016/j.biocontrol.2005.12.006.

Kishani, H., Goldansaz, S.H., Sabahi, Q. and Shakeri, M. (2011) A study on the larval parasitoids of carob moth, *Ectomyelois ceratoniae* in Varamin, Qom and Saveh. *Iranian Journal of Plant Protection Sciences* 41, 337–344.

Kirkman, W. and Moore, S. (2007) A study of alternative hosts for the false codling moth, *Thaumatotibia* (=*Cryptophlebia*) *leucotreta* in the eastern Cape. *South Africa Fruit Journal* 6, 33–38.

Kishani-Farahani, H., Goldansaz, S.H. and Sabahi, Q. (2012) A survey on the overwintering larval parasitoids of *Ectomyelois ceratoniae* in three regions in Iran. *Crop Protection* 36, 52–57. DOI: 10.1016/j.cropro.2012.01.018.

Kogan, M. (1998) Integrated pest management: historical perspectives and contemporary developments. *Annual Review of Entomology* 43(1), 243–270. DOI: 10.1146/annurev.ento.43.1.243.

Ksentini, I., Monje, J.C., Jardak, T. and Zeghal, N. (2010) Naturally occurring egg parasitoids of the genus *Trichogramma* (Hymenoptera: Trichogrammatidae) in a pomegranate orchard in Tunisia. *Entomological Science* 13(1), 99–106. DOI: 10.1111/j.1479-8298.2009.00356.x.

Ksentini, I., Jardak, T. and Zeghal, N. (2011) First report on *Virachola livia* Klug. (Lepidoptera: Lycaenidae) and its effects on different pomegranate varieties in Tunisia. *EPPO Bulletin* 41(2), 178–182. DOI: 10.1111/j.1365-2338.2011.02451.x.

Kumar, S. and Gupta, D. (2018) Management of pomegranate fruit borer, *Deudorix epijarbas* (Moore) using new group insecticides and some biopesticides. *Journal of Entomology and Zoology Studies* 6, 1060–1063.

Kumar, K.P. and Kamala Jayanthi, P.D. (2018) Management strategies for the pomegranate fruit borer, *Deudorix isocrates* (Fab.) (Lycaedidae: Lepidoptera): a sustainable solution for small scale farmers. *Agrotechnology* 7, 41.

Kumar, V., Kakkar, G., McKenzie, C.L., Seal, D.R. and Osborne, L.S. (2012) *An overview of chilli thrips, Scirtothrips dorsalis (Thysanoptera: Thripidae), biology, distribution and management*. Intrachem.

Kumar, K.P., Kamala Jayanthi, P.D., Onkara Naik, P.D., Verghese, A., Chakravarthy, A.K. *et al.* (2017) Biology of anar butterfly, *Deudorix isocrates* (Fab.) (Lycaenidae: Lepidoptera) on pomegranate, *Punica granatum* L. *International Journal of Pure & Applied Bioscience* 5(1), 498–503. DOI: 10.18782/2320-7051.2564.

Kutinkowa, H., Andreev, R. and Arnaoudov, V. (2006) The leopard moth borer, *Zeuzera pyrina* L., (Lepidoptera: Cossidae) – important pest in Bulgaria. *Journal of Plant Protection Research* 46, 111–115.

La Malfa, S., Gentile, A., Domina, F. and Tribulato, E. (2009) PRIMOSOLE: a new selection from Sicilian pomegranate germplasm. *Acta Horticulturae* 818,125–132. DOI: 10.17660/ActaHortic.2009.818.17.

Lee, Y., Lee, W., Kim, H. and Lee, S. (2015) A new record of *Aphis punicae* Passerini, 1863 (Hemiptera: Aphididae) from Korea. *Journal of Asia-Pacific Entomology* 18(2), 157–163.

Letourneau, D.K., Fitzsimmons, M.I. and Mieto, D.J. (2017) Approaches in plant protection: science, technology, environment and society. In: Coll, M. and Wajnberg, E. (eds) *Environmental Pest Management: Challenges for Agronomists, Ecologists, Economists and Policymakers*. Wiley & Sons Ltd, Oxford, UK, pp. 21–53.

Levi-Zada, A., Fefer, D., Madar, R., Steiner, S. and Kaspi, R. (2020) Evaluation of pheromone of false codling moth *Thaumatotibia leucotreta* in Israel by sequential SPME/GCMS analysis and field trials. *Journal of Pest Science* 93(1), 519–529. DOI: 10.1007/s10340-019-01138-0.

Lucchi, A. (2000) *Metcalfa pruinosa (Say) (Homoptera: Flatidae): Biologia, morfologia, dannosit controllo*. ARSIA, Regione Toscana, Italy.

Lucchi, A., Botton, M. and Bagnoli, B. (2011) Tignola rigata SU vite dA tenere sotto controllo. *L'Informatore Agrario* 31, 65–69.

Lucchi, A., Ricciardi, R., Benelli, G. and Bagnoli, B. (2019) What do we really know on the harmfulness of *Cryptoblabes gnidiella* (Millière) to grapevine? From ecology to pest management. *Phytoparasitica* 47(1), 1–15. DOI: 10.1007/s12600-018-0705-3.

Ma, J.L. and Bai, H.Y. (2004) The main pests in the pomegranate producing areas of Sichuan Province and their integrated control. *South China Fruits* 33, 70–71.

Mallikarjun, M.H. and Pal, R.K. (2018) Laboratory rearing protocol for pomegranate fruit borer (*Deudorix isocrates*). *International Journal Current Microbiology Applied Science* 6, 883–888.

Maltese, M., Caleca, V., Guerrieri, E. and Strong, W.B. (2012) Parasitoids of *Leptoglossus occidentalis* Heidemann (Heteroptera: Coreidae) recovered in western North America and first record of its egg parasitoid *Gryon pennsylvanicum* (Ashmead) (Hymenoptera: Platygastridae) in California. *The Pan-Pacific Entomologist* 88(3), 347–355. DOI: 10.3956/2012-23.1.

Mani, M. and Krishnamoorthy, A. (1989) Occurrence of mealybugs and their natural enemies on custard apple around Bangalore, south India. *Journal of Biological Control* 3, 77.

Mani, M. and Krishnamoorthy, A. (1990) Outbreak of mealybugs and record of their natural enemies on pomegranate. *Journal of Biological Control* 4, 61–62.

Mani, M. and Krishnamoorthy, A. (1991) *Maconellicoccus hirsutus* (Green) on pomegranate. *Entomon* 16, 103.

Mani, M. and Krishnamoorthy, A. (2001) Biological suppression of mealybugs *Planococcus citri* (Risso) and *Planocccus lilacinus* (Ckll) on pomegranate in India. *Indian Journal of Plant Protection* 28, 187–189.

Mani, M., Krishnamoorthy, A. and Shivaraj, U. (2011) Biological suppression of major mealybug species on horticultural crops in India. *Journal Horticulural Science* 6, 85–100.

Massimino Cocuzza, G.E. and Rapisarda, C. (2017) Integrated pest management in citrus. In: Rapisarda, C. and Massimino Cocuzza, G.E. (eds) *Integrated Pest Management in the Tropics*. CAB International, Wallingford, UK, pp. 246–269.

Mazza, G., Strangi, A., Marianelli, L., Del Nista, D. and Roversi, P.F. (2014) *Thaumatotibia leucotreta* (Meyrick) (Lepidoptera Tortricidae) intercepted for the first time in Italy. *Redia* XCVII, 147–149.

Mead, F.W. (1969) *Citrus flatid planthopper, Metcalfa pruinosa (Say) (Homoptera: Flatidae)*. Florida Department of Agriculture, Division of Plant Industry, Entomology Circular no. 85.

Mehrnejad, M.R. and Ebrahimi, S.J. (1993) Damage of quince moth [Euzophera bigella (Zell.)] to trunk and branch of pomegranate and fig trees. *Proceedings of the 11th Plant Protection Congress of Iran*, Rasht, Iran, 28 August – 2 September.

Methews, G. and Rugmini, P. (1998) Control of bark eating caterpillar in forest plantations of *Paraserienthes falcataria*. *Indian Journal of Environment and Toxicology* 8(1), 37–40.

Mhase, N.L. and Kadam, D.B. (2000) Management of root-knot nematode, M. incognita infesting pomegranate. *National Symposium on Integrated Nematode Management*, O.U.A.T., Bhubaneswar, India, 23–24 November.

Mitchell, P.L., Paysen, E.S., Muckenfuss, A.E., Schaffer, M. and Shepard, B.M. (1999) Natural mortality of leaf footed bug (Hemiptera: Heteroptera: Coreidae) eggs in cowpea. *Journal of Agricultural and Urban Entomology* 16, 25–36.

Moawad, S.S., Hassan, S.A. and Al Barty, A.M. (2011) Enumeration and estimation of insect attack fruits of some cultivars of *Punica granatum*. *African Journal of Biotechnology* 10, 3880–3887.

Moghaddam, M. (2013) An annotated checklist of the scale insects of Iran (Hemiptera, Sternorrhyncha, Coccoidea) with new records and distribution data. *ZooKeys* 334, 1–92.

Mote, U.N. and Tambe, A.B. (2000) Effective and economic management of shot-hole borer on pomegranate. *Journal of Maharashtra Agricultural Universities* 25(2), 155–157.

Motsinger, R.E., Moody, E.H. and Gay, C.M. (1977) Reaction of certain French marigold (*Tagetes patula*) cultivars to three *Meloidogyne* spp. *Journal of Nematology* 9, 278.

Naik, L.K., Jagginavar, S.B. and Biradar, A.P. (2011) Beetle enemies of pomegranate and their management. *Acta Horticulturae* 890, 565–568.

Nair, K.S.S. (2007) *Tropical Forest Insect Pests, Ecology Impact and Management*. Cambridge University Press, Cambridge, UK.

Narendran, T.C. and Sureshan, P.M. (1988) A contribution to our knowledge of *Torymidae* of India (Hymenoptera: Chalcidoideae). *Bollettino-del-Laboratorio-di-Entomologia-Agraria-Filippo-Silvestri* 45, 37–47.

Nicholls, C.I., Parrella, M.P. and Altieri, M.A. (2000) Reducing the abundance of leafhoppers and thrips in a northern California organic vineyard through maintenance of full season floral diversity with summer cover crops. *Agricultural and Forest Entomology* 2(2), 107–113. DOI: 10.1046/j.1461-9563.2000.00054.x.

Nobakht, Z., Karimzadeh, J., Shakaram, J. and Jafari, S. (2015) Identification of parasitoids of *Apomyelois ceratoniae* (Zeller) (Lepidoptera, Pyralidae) on pomegranate in Isfahan province. *Journal of Entomology and Zoology Studies* 3, 287–289.

Norouzi, A., Talebi, A.A. and Fathipour, Y. (2008) Development and demographic parameters of the carob moth *Apomyelois ceratoniae* on four diet regimes. *Bulletin of Insectology* 61, 291–297.

Noyes, J. (2016) Universal Chalcidoidea database. Natural history museum, London. Available at: www.nhm.ac.uk/our-science/data/chalcidoids/database/ (accessed 8 June 2020).

Nuzzaci, G. (1968) Danni da *Carpophilus mutilatus* Erich. a frutti di melograno. *Entomologica* 4, 167–173.

Obeidat, W.M. and Akkawi, M. (2002) Bionomics and control of pomegranate butterfly *Virachola* (*Deudorix*) *livia* (Klug) (Lepidoptera: Lycaenidae) in northern Jordan. *Dirasat-Agricultural Sciences* 29, 1–12.

Öztop, A., Keçec, M. and Kivradim, M. (2010) Investigation on pomegranate pests in Antalya Province: stem and branch pests. *Batı Akdeniz Tarımsal Araştırma Enstitüsü Derim Dergisi* 27, 12–17(in Turkish).

Özturk, N. and Ulusoy, M.R. (2009) Pests and natural enemies determined in pomegranate orchards in Turkey. *Acta Horticulturae* 818,277–284.

Öztürk, N. and Ulusoy, M.R. (2012) Determination of adult population dynamics and generation number of honeydew moth (*Cryptoblabes gnidiella* Milliére, 1867) (Lepidoptera: Pyralidae) in pomegranate orchards in the Eastern Mediterranean Region. *Turkish Journal of Entomology* 36, 101–112(in Turkish).

Pal, R.K., Babu, K.D., Singh, N.V., Maity, A. and Gaikwad, N. (2014) Pomegranate research in India – status and future challenges. *Progressive Horticulture* 46, 184–201.

Palmer, J.M. and Mound, L.A. (1983) The *Scirtothrips* species of Australia and New Zealand (Thysanoptera: Thripidae). *Journal of Natural History* 17(4), 507–518. DOI: 10.1080/00222938300770441.

Panizzi, R.A. (2004) A possible territorial or recognition behaviour of *Leptoglossus zonatus* (Dallas) (Heteroptera, Coreidae). *Centro Nacional de Pesquisa de Soja. Revista Brasileira de Entomologia* 48, 577–579.

Perry, E.J. and Ploeg, A.T. (2010) Pest notes. Publication no. 7489. University of California.

Pollini, A. (2013) *Entomologia Applicata*. Edagricole – New Business Media, Milan, Italy.

Poorjavad, N., Goldansaz, S.H., Hosseininaveh, V., Nozari, J., Dehghaniy, H. *et al.* (2011) Fertility life table parameters of different strains of *Trichogramma* spp. collected from eggs of the carob moth *Ectomyelois ceratoniae*. *Entomological Science* 14(3), 245–253. DOI: 10.1111/j.1479-8298.2011.00443.x.

Ramsingh, J., Singh, J., Singh, R. and Singh, J. (1982) New record of *Aspergillus candidus* Link – a potential entomogenous fungus on *Indarbela* spp. *Science and Culture* 48, 282–283.

Rao, V.K. and Krishnappa, K. (1995) Integrated management of *Meloidogyne incognita-Fusarium oxysporum* f. sp. *ciceri* wilt disease complex in chickpea. *International Journal of Pest Management* 41(4), 234–237. DOI: 10.1080/09670879509371956.

Sabahi, Q. and Shakeri, M. (2009) Effect of stamen removal on pomegranate infestation to carob moth *Ectomyelois ceratoniae*. *Iranian Journal of Plant Protection Science* 39, 55–65.

Sannino, L., Balbiani, A. and Parenzan, P. (1986) *Dysgonia algira* L. (Lepidoptera: Noctuidae – Catocolinae) dannosa a melograno (*Punica granatum* L). *Entomologica* 21, 127–139.

Satyagopal, K., Sushil, S.N., Jeyakumar, P., Shankar, G., Sharma, O.P. *et al.* (2014) AESA based IPM package for pomegranate. 38 P. Available at: https://farmer.gov.in/imagedefault/ipm/pomegranate.pdf (accessed 8 June 2020).

Sayed, S.M., Elsayed, G., Mahmoud, S.F. and Elzahrany, O.M. (2015) Efficacy of *Bacillus thuringiensis* and indigenous *Trichogramma turkistanica* for controlling lepidopterous pests on Taify pomegranate fruits. *African Entomology* 23(2), 443–450. DOI: 10.4001/003.023.0229.

Sellanes, C. and González, A. (2014) The potential of sex pheromones analogues for the control of *Cryptoblabes gnidiella* (Lepidoptera: Pyralidae), an exotic pest in South America. *IOBC-WPRS Bulletin* 99, 55–60.

Senguttuvam, T. (2000) Bark borer – a polyphagous pest on agroforestry trees. *Insect Environment* 6(1), 28.

Shakeri, M. (2004) A review on investigations on pomegranate neck worm in Iran. Proceedings on Evaluation of Finding and Current Problems Associated with *Spectrobates ceratoniae* Management in Pomegranate. Ministry of Jihad-e-Agriculture, Tehran, Iran.

Sharma, J., Suroshe, S.S. and Shinde, Y. (2012) Diagnosis and integrated management of diseases and insect pests of pomegranate (in Marathi). *Extention Bulletin* (no. 7), 58pp.

Simoglou, K.B., Karataraki, A., Roditakis, N.E. and Roditakis, E. (2012) *Euzophera bigella* (Zeller) (Lepidoptera: Pyralidae) and *Dasineura oleae* (FLow) (Diptera: Cecidomyiidae): emerging olive crop pests in the Mediterranean. *Journal of Pest Science* 85, 169–177.

Singh, D. and Gupta, D.C. (1993) Evaluation of marigold cultivars/hybrids for resistance against *Meloidogyne javanica*. *Haryana Agricultural University Journal of Research* 23, 156–159.

Singh, K.P., Bandopadhyay, P., Vaish, S.S., Makeshkumar, T. and Gupta, R.C. (2002) Growth and population dynamics of *Catenaria anguilluiae* in relation to oilcakes. *Indian Phytopathology* 55(3), 286–289.

Sobhani, M., Goldansaz, S.H., Hatami, B. and Hosseini, S.A. (2015) A field screening of pomegranate cultivars for resistance to the carob moth, *Ectomyelois ceratoniae*, and compatibility with its larval parasitoids. *International Journal of Pest Management* 61(4), 346–352. DOI: 10.1080/09670874.2015.1069418.

Somasekhara, Y.M., Ravichandra, N.G. and Jain, R.K. (2012) Bio-management of root knot (*Meloidogyne incognita*) infecting pomegranate (*Punica granatum* L.) with combination of organic amendments. *Research on Crops* 13(2), 647–651.

Soufbaf, M., Salehi, B., Kalantarian, N., Zanganeh, A.H., Fathollahi, H. *et al.* (2018) Is sterile insect technique's efficiency affected by pomegranate variety in mixed cultivars? New insights from a case study on the carob moth, *Apomyelois ceratoniae* Zeller (Lepidoptera: Pyralidae) in Iran. *Oriental Insects* 52(2), 210–220. DOI: 10.1080/00305316.2017.1397064.

Talaee, L. (2009) Some biological characters and dynamic population of *E. ceratoniae* and its larval parasitoids in Esfahan. MSc Thesis. Department of Plant Protection, University of Tehran, Karaj, Iran.

Tepole-García, R.E., Rámirez-Rojas, S., Bartolo-Reyes, J.C. and Castrejón-Gómez, V.R. (2016) Life cycle and climate risk analysis of *Leptoglossus zonatus* Dallas (Hemiptera: Coreidae) for sorghum producing areas in the state of Morelos, Mexico. *Acta Zoologca Mexicana* 32, 300–309.

Tezcan, S. (2008) Grape cane borer (*Schistoceros bimaculatus*) (Coleoptera: Bostrichidae): an insect species gaining importance in pomegranate orchards in western Turkey. *Hasad* 24, 80–84.

Thirugnanasundaran, K. (1989) A review on control methods of shot-hole borer in tea in Sri Lanka. *Sri Lanka Journal of Tea Science* (Conference Issue. Proceedings of the Regional Tea (Scientific) Conference).

Thomas, M.C., Heppner, J.B., Woodruff, R.E., Weems, H.U. and Steck, G.J. (2010) *Ceratitis capitata* (Wiedemann) (Insecta: Diptera: Tephritidae). *DPI Entomology Circulars* 4, 230, 273, Dept. of

Agriculture and Consumer Services, Division of Plant Industry, University of Florida. Available at: http://entnemdept.ufl.edu/creatures/fruit/mediterranean_fruit_fly.htm (accessed 8 June 2020).

Tsagkarakis, A.E. (2012) First record of *Siphoninus phillyreae* on pomegranate in Greece. *Entomologia Hellenica* 21(1), 39–43. DOI: 10.12681/eh.11516.

Vacante, V. (2016) *The Handbook of Mites of Economic Plants: Identification, Bio-ecology and Control.* CAB International, Wallingford, UK.

Vacante, V. and Bonsignore, C.P. (2018) Natural enemies and pest control. In: Vacante, V. and Kreiter, S. (eds) *Handbook of Pest Management in Organic Farming. CAB International*. Wallingford, UK, pp. 60–77.

Van Den Berg, M.A., Hoppner, G. and Greenland, J. (2000) An economic study of the biological control of the spiny blackfly, *Aleurocanthus spiniferus* (Hemiptera: Aleyrodidae), in a citrus orchard in Swaziland. *Biocontrol Science and Technology* 10(1), 27–32. DOI: 10.1080/09583150029350.

Varma, A.N., Khurana, A.D. and Singh, R. (1974) Chemical control of the bark eating caterpillar, *Indarbela quadrinotata* (Walker) infesting pomegranate. *Indian Journal of Entomology* 36, 297–301.

Vattakaven, T., George, R.M., Balasubramanian, D., Réjou-Méchain, M., Muthusankar, G. *et al.* (2016) India biodiversity portal: an integrated, interactive and participatory biodiversity informatics platform. *Biodiversity Data Journal* 4(4), e10279. DOI: 10.3897/BDJ.4.e10279.

Vidart, M.V., Mujica, M.V., Calvo, M.V., Duarte, F., Bentancourt, C.M. *et al.* (2013) Relationship between male moths of *Cryptoblabes gnidiella* (Millière) (Lepidoptera: Pyralidae) caught in sex pheromone traps and cumulative degree-days in vineyards in southern Uruguay. *SpringerPlus* 2(1), 258. DOI: 10.1186/2193-1801-2-258.

Walgama, R.S. (2012) Ecology and integrated pest management of *Xyleborus fornicatus* (Coleoptera: Scolytidae) in Sri Lanka. *Journal of Integrated Pest Management* 3(4), 1–8. DOI: 10.1603/IPM11031.

Walgama, R.S. and Pallemulla, R.M.D.T. (2005) The distribution of shot-hole borer *Xyleborus fornicatus* Eichh. (Coleoptera: Scolytidae) across tea growing areas in Sri Lanka – A reassessment. *Sri Lanka Journal of Tea Science* 70, 105–120.

Walker, K. (2008) Black borer (*Apate monachus*). Pest and diseases image library. Available at: www.padil.gov.au (accessed 30 June 2011).

Williams, D.J. (1986) The identity and distribution of the genus *Maconellicoccus* Ezzat (Hemiptera: Pseudococcidae) in Africa. *Bulletin of Entomological Research* 76(2), 351–357. DOI: 10.1017/S0007485300014814.

Xiao, Y. and Fademiro, H. (2011) Evaluation of damage to satsuma mandarin (*Citrus unshiu*) by the leaf-footed bug, *Leptoglossus zonatus* (Hemiptera: Coreidae). *Journal Applied Entomology* 134, 964–703.

YanFen, M., ShengYong, Y., Chun, X. and XianQi, H. (2013) Screening of trap plants of pathogenic root-knot nematode in pomegranate orchard. *Journal of Henan Agricultural Sciences* 42, 99–102(in Chinese).

Yıldırım, E.M. and Başpınar, H. (2015) The population fluctuations of carob moth, Apomyelois ceratoniae (Zell.) (Lepidoptera: Pyralidae) and honeydew moth Cryptoblabes gnidiella Mill. (Lepidoptera: Pyralidae), and investigation on their damage and natural enemies on pomegranate in West Aegean region of Turkey. *Agriculture and Food* 3, 186–192.

14 Fruit Maturity, Harvest Methods and Technologies

F. Varasteh[1]*, M. Alizadeh[1] and A. Sarkhosh[2]
[1]Horticulture Department, Faculty of Plant Production, Gorgan University of Agricultural Sciences and Natural Resources, Gorgan, Iran; [2]Department of Horticultural Sciences, University of Florida, Gainesville, Florida, USA

14.1 Introduction

In recent years the horticulture industry has funded research and development for novel fruit harvesting systems. The motivation for the research is to decrease harvesting costs and increase the value of their products to the consumers (Li *et al.*, 2011). The conventional harvesting method is highly labour-intensive and inefficient in terms of both economy and time. Machine harvesting systems are a partial solution to overcome these issues by removing fruits from the trees efficiently thus reducing the harvesting costs to about 35–45% of total production costs (Sanders, 2005; Li *et al.*, 2011).

On the other hand, postharvest technology, most popularly known as postharvest handling, is a series of methods and techniques used in the preservation of agricultural products after harvest. It is a science applied to agricultural commodities for preservation, conservation, quality control and enhancement, processing, packaging, storage, distribution, marketing and utilization to meet the food and nutritional requirements of consumers. Governed by these objectives, postharvest technology is applied to minimize rough handling, sorting for grading and removing damaged and diseased produce, and strategically controlling the storage environment (for more information about postharvest of pomegranate, see Chapter 15).

Although fruit harvesting research is reasonably active in some countries, it is scarce or even totally lacking or without defined objectives in most developing countries. Damage resulting from harvest operations increases the storage losses of pomegranate. The present chapter describes fruit harvesting methods, technologies and related topics in pomegranate.

14.2 Maturity Definition

The nutritional value, freshness and flavour of fresh produce (which includes all fruits and vegetables marketed fresh) will depend on the stage of maturity and the time of day when they are harvested. If the harvested produce is overly mature, it will be stringy and coarse. Produce picked too soon may be too tender and will lack quality and flavour (Sudheer and Indira, 2007). In order to seek a standardized definition, it can be stated that fruit maturity is the stage of full development of fruit tissues only after which it will ripen normally. This is the period when a fruit has reached optimum colour, flavour,

*Corresponding author: feryalvarasteh@gmail.com

aroma, firmness and texture for consumption. Depending on the plant species or type of fruit, this period can occur directly after removal from the tree (e.g. peach, cherry, citrus) or after a period of storage or conditioning (e.g. avocado, banana, etc.). During the process of maturation, the fruit receives a regular supply of food materials from the mother plant. When mature, the abscission or corky layer that forms at the stem end stops this inflow. Afterward, the fruit depends on its reserves, carbohydrates are dehydrated and sugars accumulated until the sugar–acid ratio forms. In addition to this, typical flavour and characteristic colour also develop (FAO, 2003).

It has been determined that the stage of maturity at the time of picking influences the storage life and quality of fruit (Kader, 2002). When picked green (immature), the fruit lacks the normal °Brix–acid ratio or sugar–acid ratio, taste, flavour and colour. On the other hand, if the fruits are harvested when they are overmature or fully ripe they are easily susceptible to microbial and physiological spoilage and their storage life is considerably reduced (NAIP, 2011a). Such fruits encounter numerous problems during handling, storage and transportation. Therefore, it is necessary or essential to pick the fruits at the correct stage of maturity to facilitate proper ripening, distant transportation and maximum storage life (FAO, 1997; Dhatt *et al.*, 2007; NAIP, 2011b).

In the literature, fruit maturity has been divided into two categories: physiological maturity and harvest maturity. Physiological maturity is the stage when fruit is capable of further development or ripening when it is harvested, that is, ready for eating or processing. Harvest maturity refers to the stage of development when plant and plant parts possess the prerequisites for use by consumers for a particular purpose, that is, ready for harvest (Thompson, 1996; Reid, 2002; Babu *et al.*, 2017). The relationship between physiological maturity and fruit ripening is shown in Fig. 14.1.

In pomegranate, flowering and the subsequent fruit set last for about 1 month. The fruit ripens 5–8 months after fruit set, depending on the cultivar and climatic conditions. Early maturing varieties are harvested in August–September and late mature ones may be harvested by October–November in most countries of the northern hemisphere. On the other hand, in most countries of the southern hemisphere early maturing varieties are harvested in February and March and late maturing ones may be harvested by April and May .

14.3 Fruit Growth, Development and Maturation

The pomegranate fruit is a type of berry, commonly known as 'Balausta', with a leathery rind enclosing many tightly packed juicy arils,

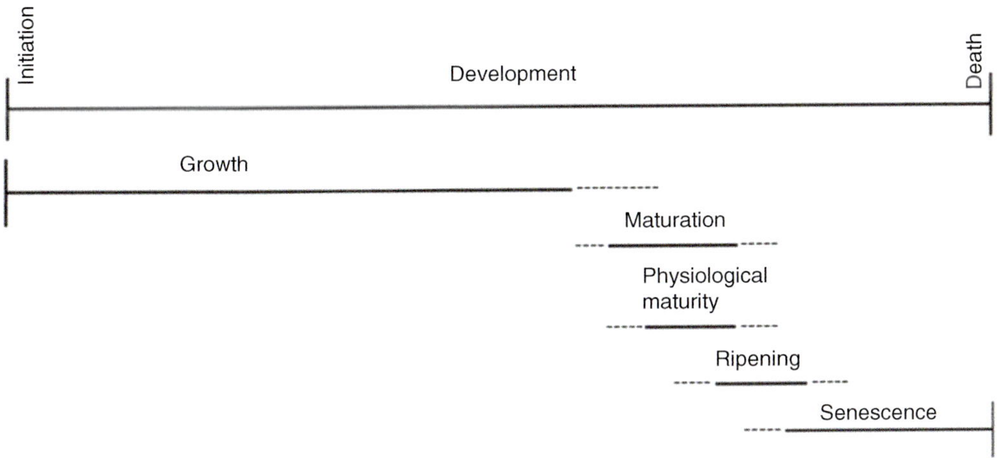

Fig. 14.1. Different stages of fruit maturity. (From: Reid, 2002.)

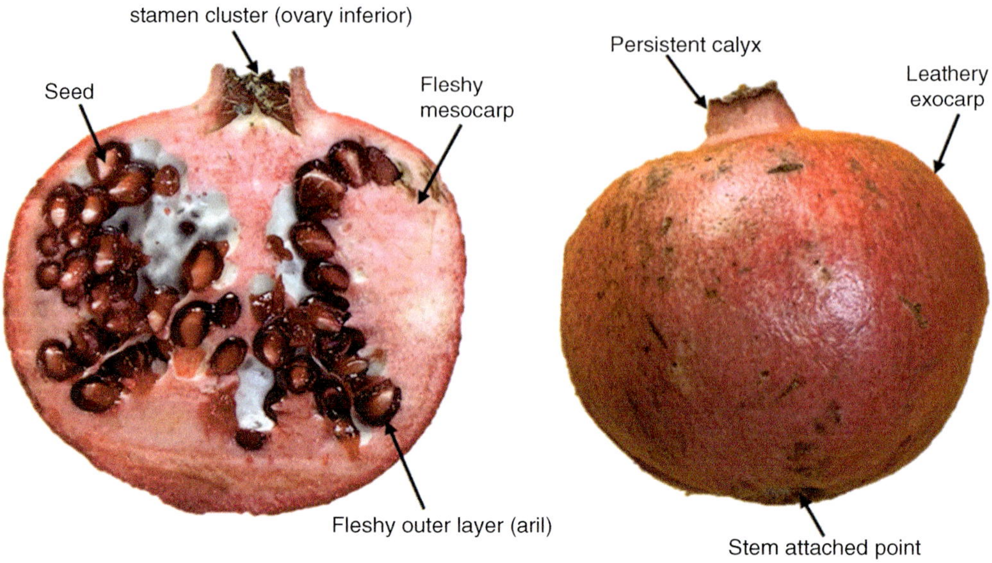

Fig. 14.2. Pomegranate fruit showing arils (edible parts), seeds, persistent crown at the top and stem attached point at the bottom. (Photo: Ali Sarkhosh.)

which comprise the edible portion of the fruit (Fig. 14.2). Depending on the cultivar, the arils constitute approximately half of the fresh fruit weight (Kulkarni and Aradhya, 2005; Stover and Mercure, 2007). Therefore, the pomegranate fruit can be divided into three parts: the outer skin, mesocarp plus endocarp, and arils, which also include the interior network of membranes (Fawole and Opara, 2013b). The edible portion of pomegranate can be about 60% of the total fruit weight and contains 80% juice and 20% seeds (these numbers vary depending on cultivars and cultural practices). The fresh juice contains about 85% water and 15% sugars, pectins, acids (ascorbic acid), polyphenols, flavonoids, anthocyanins and amino acids (Mirdehghan and Rahemi, 2007). Pomegranate is exploited for the nutritional value of its fruit, the medicinal properties of different parts of the tree and its use as an ornamental (Naovi et al., 1991; Jayesh and Kumar, 2004; Johanningsmeier and Harris, 2011).

The pomegranate fruit grows continuously from fruit set until the commercial harvest time. The pomegranate fruit growth pattern has been characterized as a single sigmoid curve (Fig. 14.3) from the beginning of fruit set till maturity (Gozlekci̦ and Kaynak, 2000; Varasteh et al., 2008). According to Kumar and Purohit (1989), there are periods of fast fruit growth rates, which alternate with periods of slow growth rates. The initial rapid increment in fruit growth occurs during cell division, which is characterized by growing kernel tissue and the increment in testa hardness (Shulman et al., 1984), after which a slowdown in fruit growth occurs (Gozlekci̦ and Kaynak, 2000). However, while the kernel stops growing, the aril continues to grow steadily as the fruit increases to its final size through cell enlargement during maturation (Ben-Arie et al., 1984; Shulman et al., 1984; Melgarejo et al., 1997).

Upon maturity, the sepal colour changes to orange-red or deep red. The colour of the sepal is directly associated with the colour of fruit skin, with darker red flowers producing deep red fruit skin (Holland et al., 2009). The sepals remain on the fruit as it matures, forming a prominent calyx (Pande and Akoh, 2016). The crown portion is considered important in marketability of table pomegranates and, while harvesting, the fruits must be picked with care to avoid breaking or deformation of the crown. The fruit ripens 5–8 months after fruit set, depending on the cultivar (Fig. 14.4).

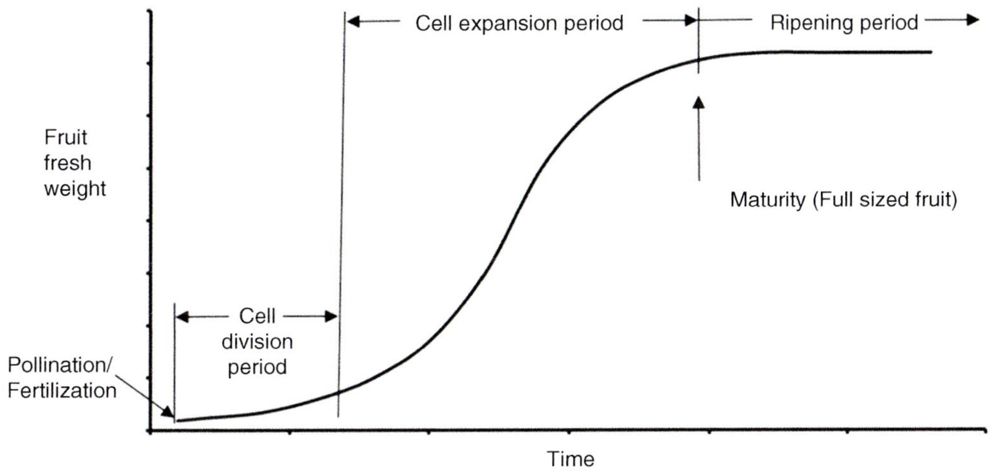

Fig. 14.3. Single sigmoid fruit growth pattern in pomegranate. (Drawing: Ali Sarkhosh.)

The seeds range in hardness from very hard to soft. The colour of the edible juicy layer or aril can vary from white to deep red, depending on the cultivar (Holland *et al.*, 2009). The fruit size can vary from 6–12 cm in diameter and the fruit has a tough, leathery, yellow to brown/ red, bitter skin (peel or pericarp) (Pande and Akoh, 2016). It has been reported that CO_2 evolution from the fruits is low, and no climacteric peak or measurable ethylene are detected during fruit maturation, thus suggesting that the pomegranate is a non-climacteric fruit (Babu *et al.*, 2017).

14.4 Biochemical Changes During Maturity and Ripening

For most fruits, advancing maturity corresponds to a number of coordinated physiological, biochemical and structural processes that result in changes in size, firmness, colour and flavour, making the fruit desirable for consumption (Moing *et al.*, 2001; Nunes *et al.*, 2009; Wilson and Downs, 2012; Mphahlele *et al.*, 2014) (Table 14.1).The composition of pomegranate varies primarily based on the cultivar, agronomic practices, geographical location, storage and processing. Some studies have been carried out on the compositional analysis of different pomegranate varieties and also different parts of pomegranate such as peel, seed, juice, leaves and flowers. The composition of the edible fraction of pomegranate determines its nutritional value and food applications (Pande and Akoh, 2016). During fruit development and maturation, significant changes have been found in the physical parameters (fruit weight and volume) and chemical profile of pomegranate arils and peels (Shwartz *et al.*, 2009). In general, aril colour changes to dark red (according to cultivar), juice content increases, total soluble solids (TSS) content either remains constant or increases, juice colour changes from white to pink and then intense red, sugars reach maximum at ripening, titratable acid (TA) decreases, pH increases, phenols decrease with advancement of ripening, anthocyanins increase rapidly at ripening and ascorbic acid decreases with ripening (Pareek *et al.*, 2015).

Further details of the biochemical changes of pomegranate fruits during maturity are described in the following sections.

14.4.1 Phenolics

Phenols, sometimes called phenolics, are a class of aromatic organic compounds consisting of one or more hydroxyl groups attached to an aromatic hydrocarbon group. Phenol is a benzene derivative and is the simplest member of the phenolic chemicals (Pande and Akoh, 2016). Physiologically relevant phenolics found in pomegranate are gallic acid, ellagic acid,

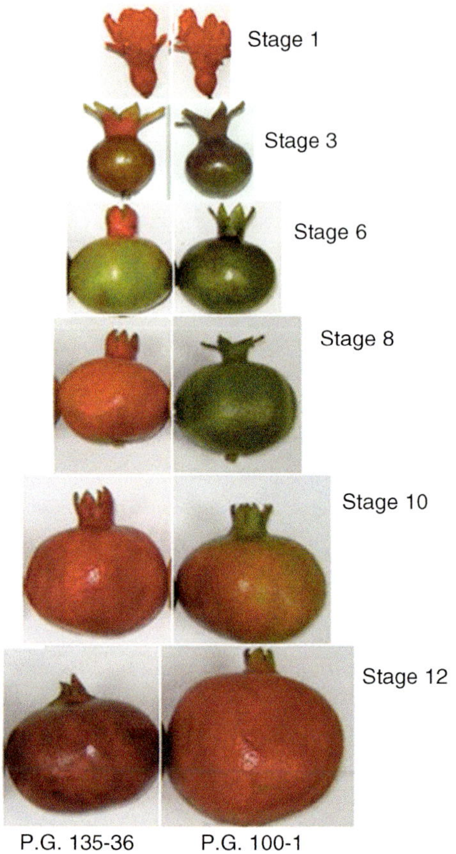

Fig. 14.4. Flower and fruit of two pomegranate accessions, P.G. 135–36 and P.G. 100–1 ('Wonderful' landraces), at six different developmental stages, including: flower (stage 1), young fruit (stages 3 and 6), nearly mature fruit (stage 8), ripened fruit (stage 10) and overripened fruit according to commercial practice (stage 12). P.G. 135–36 is an early anthocyanin-accumulating accession and P.G. 100–1 is a late anthocyanin-accumulating accession (see stage 8). (Photos: Ben-Simhon *et al.*, 2011, Springer.)

catechin, epicatechin, punicalagin and anthocyanins (Pande and Akoh, 2016). A reduction in total phenolic compounds in the aril juice during ripening and maturation was previously reported for the 'Ganesh' pomegranate (Kulkarni and Aradhya, 2005). A decrease in phenolic compounds during ripening has also been reported for other fruits, such as pear (Amiot *et al.*, 1995) and guava (Bashir *et al.*, 2003).

Mirdehghan and Rahemi (2007) demonstrated that total phenolics levels increased at the early stage of growth both in peel and arils of fruit, but thereafter generally decreased during maturation and reached 3.70 and 50.22 mg/g dry weight in arils and peel, respectively, at harvest. They also found that the amount of total phenolics in peel was markedly higher than arils of pomegranate fruit. Shwartz *et al.* (2009) evaluated the changes in the major chemical composition in arils and peels during fruit maturation in two commercial accessions, 'Wonderful' and 'Rosh-Hapered'. In both accessions, the levels of total phenolics, antioxidant activity and hydrolysable tannins were reduced in the peel during maturation, while the anthocyanin level was increased.

14.4.2 Sugars and total soluble solids

The edible part of the fruit contains a considerable amount of sugars. In one raw pomegranate fruit (10 cm diameter/282 g weight), the amount of total sugars was about 38.55 g (Pande and Akoh, 2016). One of the processes occurring in fruit during ripening is the hydrolysis of the starch that accumulates in the early stages of fruit development and degrades into simple sugars at ripening. As a result, the fruit gets its sweetness. Starch and sucrose will change into glucose during pomegranate fruit ripening (Zarei *et al.*, 2011). The concentration of total soluble solids (TSS) and total sugars increase significantly during pomegranate fruit ripening (Zarei *et al.*, 2011). The increment of TSS and sugars during ripening has already been reported in various pomegranate cultivars (Du *et al.*, 1975; Gil *et al.*, 1995a; Al-Maiman and Ahmad, 2002). The ratio of TSS/TA has been reported as one of the most reliable indicators of fruit maturity in pomegranate cultivars, although it is mostly dependent on cultivar type (whether the fruit is sweet, sweet-sour or sour type), as well as the agro-climatic conditions (Ben-Arie *et al.*, 1984; Al-Maiman and Ahmad, 2002; Kulkarni and Aradhya, 2005; Shwartz *et al.*, 2009). Melgarejo *et al.* (2000) and Carbonell-Barrachina *et al.* (2012) reported higher levels of fructose than glucose in pomegranate cultivars grown in Spain. However, literature studies indicate that there are variations

Table 14.1. Cultivar differences in bioactive compounds in pomegranate fruit (Mphahlele *et al.*, 2014).

Factor levels	Country	Fruit part	Bioactive compounds	Key findings	References
'Bhagwa', 'Arakta', 'Ruby'	South Africa	Juice	TP, total anthocyanin, total flavonoids, gallic acid	'Bhagwa' had the highest TP, total flavonoid, anthocyanin than 'Arkata' and 'Ruby'	Fawole *et al.*, 2012
13 sour vs. sweet cultivars	Tunisia	Juice	TP, total anthocyanin	Higher TP and AA, less delphinidin-3,5-diglucoside in sour cultivars, higher TP and anthocyanin (delphinidin-3,5-diglucoside) were recorded in sweet cultivars	Zaouaya *et al.*, 2012
Eight sour vs. sweet cultivars	Italy	Juice	Polyphenols, vitamin C	Sour cvs exhibited higher polyphenols and vitamin C than sweet cvs	Ferrara *et al.*, 2011
10 cultivars	Morocco	Juice	TP	'Hamde' gave a higher TP content than 'Mesri'	Legua *et al.*, 2012
32 accessions	Egypt	Juice	Vitamin C, total anthocyanin, ellagic acid	Vitamin C content ranged between 2.77 and 9.48 mg/100 ml; total anthocyanin (0.045–1.37 mg/ml); ellagic acid ranged between 0.84 and 10 mg/l	Hassan *et al.*, 2012
Six cultivars	Turkey	Arils	TP, total monomeric anthocyanin	TP varied between 1245 and 2076 mg gallic acid equivalents (GAE/l) while total monomeric anthocyanin ranged between 6.1 and 219 mg Cy3-gluc/l	Ozgen *et al.*, 2008
20 cultivars	Iran	Juice	TP, total anthocyanin, AA	Total anthocyanin varied between (5.56 mg/100 g and 30.11 mg/100 g); TP (295.79–985.37 mg/100 g); AA (9.91–20.92 mg/100 g)	Tehranifar *et al.*, 2010
Six cultivars	Iran	Juice	Total anthocyanin, AA, TP, TTs, CTs	Higher TP, total anthocyanin, TTs, CTs were recorded in 'Aghaye' but it recorded the lowest AA; 'Shahvar' showed the lowest TP; 'Shirin-e-Bihaste' had the lowest CTs but higher AA content	Zarei *et al.*, 2010

Continued

Table 14.1. Continued

Factor levels	Country	Fruit part	Bioactive compounds	Key findings	References
Four cultivars	China	Seed oil	TP, total flavonoid content, proanthocynadins	'Suanshiliu' had the highest TP, total flavinoid content and proanthocynidins followed by 'Tianhongdan', 'Sanbaitian' and 'Jingpitian'	Jing *et al.*, 2012

TP, total phenolics;AA, ascorbic acid;TT, total tannins;CT, condensed tannins;

with little consistency concerning the relative concentrations of glucose and fructose in pomegranate cultivars. For example, other studies have shown that glucose was higher than fructose in some other pomegranate cultivars (Al-Maiman and Ahmad, 2002; Miguel *et al.*, 2004; Ozgen *et al.*, 2008).

Pomegranate aril juice shows a significant increase in content of glucose and fructose during fruit maturation. The rise in total soluble sugars also affects TSS content, which is one of the most widely used parameters measured during fruit ripening as a quality characteristic. The TSS value indeed increases significantly during the maturation stage and a strong correlation has been found between TSS and glucose and fructose levels during fruit development. A similar increase in TSS level was also reported for other pomegranate accessions during fruit development (Gil *et al.*, 1995a; Al-Maiman and Ahmad, 2002; Kulkarni and Aradhya, 2005). According to Ben-Arie *et al.* (1984), pomegranate fruit of 'Wonderful' grown in Israel attained optimum quality when the TSS of the fruit reached 15%. The increase in TSS could be attributed to the hydrolysis of starch to sugars as the pomegranate fruit advanced in maturity (Kulkarni and Aradhya, 2005).

14.4.3 TA and pH

The TA and TSS measurements are used to identify the fruit and juice qualities. Organic acids found in the pomegranate include citric, malic, acetic, fumaric, tartaric and lactic acids. However, the major acid accounting for TA in pomegranate arils is citric acid (Melgarejo *et al.*, 2000). During the immature stage, the fruit has a high acid content, which could be attributed to the organic acids and their composition at this stage of fruit development (Zarei *et al.*, 2011). In general, as the ripening period progresses, significant decreases occur in TA. According to Fawole and Opara (2013b) the decrease in acids coincided with an increase in pH. The pH of the juice decreased from early immature (3.57) to early half-ripe stage (3.18), but did not differ significantly until the full-ripe stage. They also observed a significant increase in TSS/TA ratio, which plays a significant role in juice taste and flavour, that peaked at 140 days after full bloom (DAFB).

14.4.4 Organic acids

Acid content is an important maturity parameter in pomegranate fruit because it plays a major role in the development of juice flavour (Fawole and Opara, 2013b). Although several organic acids have been found in pomegranate aril juice, the major acid accounting for titrable acidity is citric acid (Melgarejo *et al.*, 2000). Tartaric, citric and malic acids contents are decreased gradually during fruit maturation, followed by a significant decrease as the fruit reaches the final stages of maturation. This pattern of acid evolution corroborates the general phenomenon that organic acids accumulate during fruit growth and are used as respiratory substrates in mature fruit (Diakou *et al.*, 2000; Moing *et al.*, 2001). There are contradictory reports concerning ascorbic acid changes during fruit ripening. While the ascorbic acid content was decreased significantly with ongoing maturity in the 'Ganesh' and 'Taifi' pomegranate accessions (Al-Maiman and Ahmad, 2002; Kulkarni and Aradhya, 2005), Shwartz *et al.* (2009) found that its level increased in two pomegranate accessions. Hence, further studies are required to explain these differences in ascorbic acid accumulation among pomegranate fruit cultivars.

14.4.5 Pigments

Pomegranate fruits are recognized for their high anthocyanin content accumulation in skin and arils (Holton and Cornish, 1995; Holland *et al.*, 2009) resulting in vibrant colours and high antioxidant capacity (Fawole and Opara, 2013a; Khaksar *et al.*, 2015). Anthocyanins are water-soluble pigments primarily responsible for the attractive purple to red colour of many fruits, including pomegranate, and they are well known for their antioxidant activity (Shwartz *et al.*, 2009). Anthocyanins are the major pigments responsible for the pomegranate fruit skin colour, and the high variability in fruit external colour in pomegranate cultivars reflects variations in anthocyanin composition (Ben-Simhon *et al.*, 2011). The changes in pomegranate colour may be considered in both peel and juice pigmentation. The skin colour of a pomegranate fruit ranges from yellow, green or pink overlaid with pink to deep red or indigo to entirely red, or deep purple cover, depending on the variety and stage of ripening (Holland *et al.*, 2009). Upon maturity, the sepal colour changes to orange-red or deep red, and the fruit peel becomes usually orange-red or pink and, rarely, white. The colour of the sepal is directly associated with the colour of the fruit skin, also, darker red flowers produce deep red fruit skin (Holland *et al.*, 2009). The anthocyanins levels in the aril juice are significantly increased during maturation in cultivars with red aril colour (Shwartz *et al.*, 2009). However, in cultivars that have a low content of these pigments, the level does not significantly change. A rapid increase in anthocyanin pigment concentration during ripening was also reported for the 'Mollar' and 'Ganesh' pomegranate accessions (Gil *et al.*, 1995a; Kulkarni and Aradhya, 2005). For further details on changes in anthocyanin pigments during fruit maturity please refer to Section 14.6.

14.4.6 Minerals

Pomegranate is an important source of mineral nutrients. The chemical composition of pomegranate fruit differs depending on cultivar, growing region, climate, maturity and cultural practices (Mirdehghan and Rahemi, 2007). The dry weight of peel and arils increase regularly throughout the season. Al-Maiman and Ahmad (2002) showed that the composition of minerals varied markedly among the three ripening stages. The amounts of potassium, calcium and sodium were highest in both juice and seeds followed by magnesium, phosphorus, zinc, iron and copper. They demonstrated that pomegranate fruit can be a good source of nutrients and variations could originate from the pomegranate cultivars and the agro-climatic conditions.

The changes in levels of macronutrients (N, P, K, Ca, Mg and Na) and micronutrients (Zn, Cu, Mn, Fe and B) in arils and peel of pomegranate fruit were recorded from 10 DAFB until harvest (Mirdehghan and Rahemi, 2007). The concentration of most elements in arils and peel decreased during fruit growth and development. At harvest the relative order of concentration of macronutrients both in arils and peel was K > N > Ca > P > Mg > Na. The concentration of most micronutrients was greater in the arils than in the peel, especially in the early season. The relative order of concentration of micronutrients in arils was B > Fe > Zn > Cu > Mn. The accumulation of all the macro- and microelements within the fruit also increased during fruit growth and development.

14.5 Genetic Differences for Ripening Among Cultivars

In a report released by the International Plant Genetic Resources Institute (IPGRI, 2001) more than 500 cultivars of pomegranate have been named, but such ancient and widespread fruits often have considerable synonymy, so that the same basic genotype may be named differently in different regions. The most important traits are fruit size, husk colour (ranging from yellow to purple, with pink and red most common), aril colour (ranging from white to red), seed hardness, maturity time, juice content,

Fig. 14.5. Fruit peel and arils of various varieties displaying a wide range of colours. (Photo: Doron Holland.)

acidity, sweetness and astringency (Fig. 14.5). Pareek *et al.* (2015) have also listed some important pomegranate cultivars in major growing countries.

Genetic differences for ripening among cultivars are precisely apparent even in a single orchard in the same region. While minor genetic differences are present in pomegranate fruit ripening, the fruit growth pattern is similar in most varieties. Hence, many researchers have described a single sigmoidal pattern (Fig. 14.3) for the growth of pomegranate fruit, with rapid early growth followed by a slowing down. The first phase of growth may be related to the determination of fruit size, and the second phase to the increase in dry weight (Worrell *et al.*, 1998). Working with 'Malas-e-Torsh-e-Saveh', a cultivar grown in Iran, the average fruit weight and volume increased rapidly until 45 days after fruit set and then continued more slowly until harvest time (Varasteh *et al.*, 2008). Similarly, 'Mule's Head' cultivar studied by Shulman *et al.* (1984) also followed a simple sigmoid curve whereas the growth pattern of 'Wonderful' was linear. However, linear fruit growth patterns for Omani cultivars grown in the Al-Jabal Al-Akhadar area (Al-Yahyai *et al.*, 2009) and 'Wonderful' cultivar grown in Australia (Weerakkody *et al.*, 2010) have been reported. Differences in ripening time among pomegranate cultivars are not dependent on the differences in flowering dates and could be calculated from anthesis (Holland *et al.*, 2009).

14.6 Maturity and Quality Components and Indices

Maturity at harvest is the most important determinant of storage life and final fruit quality. Maturity indices are determined for many fruit, vegetable and floral crops. Harvesting crops at proper maturity allows handlers to begin their work with the best possible quality produce (Sudheer and Indira, 2007). The principles dictating at which stage of maturity a fruit or vegetable should be harvested are crucial to its subsequent storage and marketable life and quality. The term quality implies the degree of excellence of a product or its suitability for particular use. The quality of produce includes sensory properties (appearance, texture, taste and aroma), nutritional values, chemical and biochemical constitutions, mechanical properties, functional properties and defects (Abbott, 1999; Dhatt *et al.*, 2007; Wills and Golding, 2018). Postharvest physiologists distinguish three stages in the later lifespan of most fruits and vegetables: maturation, ripening and senescence. Maturation is indicative of the produce being ready for harvest. At this point, the edible part of the fruit or vegetable is fully developed in size, although it may not be ready for immediate consumption. Ripening follows or overlaps the maturation, rendering the produce edible, as indicated by taste. Senescence is the last stage, characterized by natural degradation of the fruit or vegetable, such as the loss of texture, flavour,

Table 14.2. Maturity indices for some fruit crops (Reid, 2002).

Index	Example
Elapsed days from full bloom to harvest	Apples, pear
Mean heat units during development	Apples, pear
Development of abscission layer	Apples, feijoa
Surface morphology and structure	Cuticle formation on grape
Size	All fruits
Specific gravity	Cherries
Shape	Angularity of banana fingers Full cheeks of mangoes
Textural properties	
Firmness	Apples, pears, stone fruits
Tenderness	Pears
External colour	All fruits
Internal colour and structure	Flesh colour of some fruits
Compositional factors	
Starch content	Apples, pears
Sugar content	Apples, pears, stone fruits, grapes
Acid content, sugar/acid ratio	Pomegranates, citrus, papaya, kiwifruit
Juice content	Citrus fruits
Oil content	Avocadoes
Astringency (tannin content)	Persimmons, dates
Internal ethylene concentration	Apples, pears

etc. (senescence ends at the death of the tissue of the fruit, NAIP, 2011b; Sudheer and Indira, 2007). Determination of maturity indices help ensure the sensory quality (flavour, colour, aroma and texture) and nutritional quality, and also ensures an adequate shelf-life, facilitates scheduling harvest and packing operations besides facilitating marketing over the phone or through the internet (Babu *et al.*, 2017). Maturity indices for some selected fruit crops are shown in Table 14.2.

Fruits harvested too early may lack flavour and may not ripen properly. Being non-climacteric, the pomegranate fruits should be harvested once they attain proper maturity on the plant itself (Gaikwad *et al.*, 2014). The non-climacteric nature of pomegranate fruit during development and ripening was first reported by Lee *et al.* (1974), who observed a decline in respiration rate during fruit development. The pomegranate fruit reaches full maturity within 4.5–6 months after full bloom, depending on cultivar and climatic conditions (Ben-Arie *et al.*, 1984; Kader, 2006). At the same time, late harvesting should be avoided as it limits the market life of fruits and increases the incidence of physiological disorder and internal breakdown. So, it is essential to harvest the fruits of pomegranate at the right stage of maturity to ensure better quality and optimum market life of harvested fruits. As the cultivars differ in their maturity period, it becomes indispensable to determine the maturity indices for different cultivars (Babu *et al.*, 2017). However, specific maturity indices may apply to some commercial cultivars of pomegranate.

Pomegranate fruit quality depends on the following characteristics (Pareek *et al.*, 2015):

- Fruits should be free from pre-harvest defects such as fruit cracking, sunburn, surface blackening, scald, etc.
- Careful harvesting is required, and fruits should not be injured during harvesting and transportation. Fruits must be free from surface abrasions, impact bruising, vibration injury, cuts, wounds, etc.
- Skin colour should be characteristic of the cultivar. Dark red or pink-red colour is preferred.
- Large fruit size and arils are preferred.
- Aril colour should be dark red and intense, with soft seeds (called seedless).
- Flavour in pomegranate depends mostly on sugar/acid ratio, which varies among cultivars and the stage of maturation.
- Soluble solids content above 15% and total phenolics content (TPC) below 0.25% are desirable for optimal levels of sweetness and astringency, respectively (Kader *et al.*, 1984; Crisosto *et al.*, 2000).

Pomegranates can be harvested when they reach a certain size and skin colour. Other maturity indices are TA and TSS content. Each pomegranate type requires a certain acid/soluble solids ratio at harvest. The TA of pomegranates varies between 0.13 and 4.98% at harvest. It is <1% in sweet cultivars, 1–2% in sweet-sour cultivars and >2% in sour cultivars (Onur and Kaska, 1985). The TSS content of pomegranates varies between 8.3 and 20.5% at harvest. Thus maturity indices depend on the cultivar. A juice tannins content below 0.25% is preferred, and red juice colour equal to or darker than Munsell colour chart 5R-5/12 is desirable (Crisosto *et al.*, 2000). The fruits should be harvested before they become overripe and crack (split) open, especially under rainy conditions (Pareek *et al.*, 2015).

14.7 Fruit Colour Development and Indices

Fruit skin colour is commonly applied for fruit harvesting, since skin colour changes as fruit ripens or matures. Some fruits exhibit no perceptible colour change during maturation, depending on the type of fruit. Assessment of harvest maturity by skin colour depends on the judgement of the harvester, but colour charts are available for cultivars, such as apples, peaches, etc. (NAIP, 2011b). Pomegranate fruits are a rich source of anthocyanins, which accumulate in the skin and in the arils (Gil *et al.*, 1995a; Hernández *et al.*, 1999). Six anthocyanin pigments were identified in pomegranate fruit, including mono- and diglucosides of cyanidin (red pigments), delphinidin (purple pigment) and pelargonidin (orange pigments) (Du *et al.*, 1975; Gil *et al.*, 1995a). All six anthocyanin pigments were detected in Spanish, Californian, Tunisian and Italian pomegranates (Gil *et al.*, 1995a, b), with differences in their relative amounts, depending on cultivar and climatic and cultural variables (Gil *et al.*, 1995b). Another study found that among these pigments, the pelargonidin derivatives were always present in small amounts (Hernández *et al.*, 1999).

Working with 'Mollar' cultivar (Gil *et al.*, 1995a), it was shown that the amount of 3,5-diglucosides was higher than that of 3-glucosides during the first fruit development stages, and in early maturation stages the amount of delphinidin glucosides was higher than that of the cyanidin glycosides, while in the later maturity stages the cyanidin glycosides were the main pigments of the fruit juice, and the 3-glucosides reached similar or higher concentrations than the 3,5-diglucosides. A similar trend was later confirmed in other cultivars (Hernández *et al.*, 1999). In the pomegranate fruit, the skin and aril colour development are independent of one another, differing with respect to both timing and intensity during fruit development (Holland *et al.*, 2009). Fruit from different pomegranate accessions exhibit a high variability in skin and aril colour. The pomegranate skin colour range is particularly wide, from yellow, through orange, pink, and red to deep purple (Holland and Bar-Ya'akov, 2008; Shwartz *et al.*, 2009; Dafny-Yalin *et al.*, 2010). The colour variability is also reflected in the pattern of skin colour accumulation during fruit development. While some pomegranate cultivars constantly accumulate anthocyanins, others lose their colour at early stages and develop it again during the final stages of fruit development (Holland *et al.*, 2009).

The relative amounts of the various anthocyanin components dramatically change during fruit development (Fig. 14.4). According to Ben-Simhon *et al.* (2011) the most prominent anthocyanin components in pomegranate flower are pelargonidin derivatives (orange pigments). During the early stages of fruit development, when the colour of the skin is mostly green, low levels of all anthocyanins were detected, whereas cyanidin derivatives (red pigments) prevailed in the skin during the late stages of fruit development, when the colour of the skin became red. Thus, the colour in the fruit skin was mostly due to accumulation of cyanidin derivatives. The dominant presence of cyanidin derivatives and deficiency of delphinidin derivatives (purple pigments) in the pomegranate fruit skin were also reported for Spanish pomegranate cultivars (Gil *et al.*, 1995a). The data presented by Ben-Simhon *et al.* (2011) as well as Gil *et al.* (1995a) indicate that anthocyanin biosynthesis in pomegranate skin is highly regulated during fruit development with respect to both anthocyanin quantity and composition.

The colour development in pomegranate fruit skin is mainly dependent on cultivar and certain climatic and cultural variables

(Figs. 14.4 and 14.5). Anthocyanin accumulation changed inversely to the season's temperatures. Cyanidins were generally more abundant, but delphinidin accumulation was enhanced in cooler seasons. Monoglucosylated anthocyanins prevailed at cooler temperatures and subsided during seasonal warming with a concomitant increase in diglucoside proportion (Borochov-Neori *et al.*, 2011). Fruit that matured and ripened under extremely hot temperatures had lower external and internal colour and accumulated fewer anthocyanins compared with moderate climate conditions. Anthocyanin concentration was very low in the summer, somewhat higher in autumn and substantially higher in winter fruit arils. Delphinidins and cyanidins were present throughout the whole study period; small levels of pelargonidins were detected only in winter fruit. Cyanidins were always the most abundant anthocyanins; however, during winter the accumulation rate of delphinidins was higher than that of cyanidins. Summer arils' anthocyanins were all diglucosylated. The content of monoglucosides increased with climate cooling, reaching over 60% by midwinter (Borochov-Neori *et al.*, 2011). Fruits that matured and ripened under extremely hot temperatures had lower external and internal colour and accumulated fewer anthocyanins compared with moderate climate conditions. The effects were particularly high in the arils (Borochov-Neori *et al.*, 2011). Furthermore, the fruits located on the north side of the trees showed an earlier increase in anthocyanin pigmentation and showed a significant increase in juice pigments during September. This could be explained by the lower temperatures, specially reached at night in those fruits facing the cold north wind (Hernández *et al.*, 1999). Manera *et al.* (2012) also ascertained the significant effect of air temperature on rind colour development. They found that the evaluated colourimetric parameters of pomegranate rind were highly correlated with the air temperature during fruit development and ripening. All correlation coefficients were higher than 0.9, which indicated the significant contribution of air temperature to rind colour development in pomegranates. The development of anthocyanins was also tested in fruits produced in two different orchards, and fruits from different locations in the tree. Red-coloured fruits, located in the outer parts of the tree, and yellow fruits on the inner branches, were analysed. The juices obtained from both fruit types and from the two selected orchards showed the same anthocyanin profile. However, the total amount of pigments in the juice was generally smaller in those fruits with reddish skins (outer located fruits) than in those with yellow skins (inner located fruits) (Gil *et al.*, 1995a).

14.8 Non-Destructive Methods of Maturity Indices

The determination of optimum ripening stage is one of the most fundamental aspects that influences the quality evaluation and depends on a number of internal attributes such as firmness, TSS and pH (Moing *et al.*, 1998; Opara, 2000; Nunes *et al.*, 2009; Moghimi *et al.*, 2010). Moreover, scientific techniques for determining the maturation state after harvesting are needed in order to decide the best uses and storage duration of this kind of fruit. There exist numerous instrumental techniques to carry out these determinations, but these techniques require samples from internal fruit tissues and, therefore, are destructive tests (Castro-Giráldez *et al.*, 2013). So, to ensure the minimum acceptability of the quality to consumers, developing efficient and non-destructive methods to measure internal attributes of fruit is essential.

The advantages of non-destructive methods of pomegranate fruit ripening are as follows:

- Various methods to measure biochemical attributes are destructive in nature.
- Destructive methods are tedious and inapplicable to grading and sorting.
- Pomegranates are non-climacteric fruits, so it is very important that they are harvested at their proper ripening stage.
- Measurement is almost accurate and applicable to different cultivars.
- Intact and uncut fruits are used.
- Non-destructive methods can be automated. Non-destructive methods may be applied as complementary to human inspection in automatic fruit sorting lines.
- A high rate of sample processing and fast data collection are provided as compared with human operators.

Table 14.3. Non-destructive methods for determination of pomegranate fruit ripening.

Method	Aim	Reference
Colourimetric index	To assess the optimal time for harvesting	Manera *et al.*, 2013
Dielectric spectroscopy	To study the ripeness	Castro-Giráldez *et al.*, 2013
Lustre sensor	To measure the glossiness of the rind	Czieczor *et al.*, 2018
X-rays	To quantify the volume of different parts of the fruit	Salmanizadeh *et al.*, 2015; Arendse *et al.*, 2016a
X-rays	To detect blackheart disease and fruit moth	Arendse *et al.*, 2016b
NMR[a]	To assess physiological changes induced by *Alternaria* spp. and *Aspergillus* spp.	Zhang and McCarthy, 2012
NMR	To detect blackheart infection	Zhang and McCarthy, 2012
NMR	To detect fruit internal decay	Khoshroo *et al.*, 2009
NIR[b]	To predict the quality attributes	Khodabakhshian *et al.*, 2017c
NIR	To predict rind scald	Arendse *et al.*, 2018
Image processing	The appearance of the pomegranate with the colours of the arils	Fashi *et al.*, 2019
Multispectral imaging	To determine texture and TSS of intact pomegranate fruit	Khodabakhshian *et al.* (2017a)
Machine vision system	To assess the quality of pomegranate fruits	Kumar *et al.*, 2018

[a] Nuclear magnetic resonance; [b] near-infrared spectroscopy.

Many research studies have been conducted worldwide to develop non-destructive methods to determine the overall quality of pomegranate fruit. Machine vision, nuclear magnetic resonance (NMR), dielectric spectroscopy, and X-ray computed tomography are some of the most recent non-destructive techniques used for quality evaluation of pomegranate (Blasco *et al.*, 2009; Zhang and McCarthy, 2013; Castro-Giráldez *et al.*, 2013; Magwaza and Opara, 2014). Some attempts to develop cheap, reliable and non-destructive methods of pomegranate ripening are shown in Table 14.3.

14.8.1 Multispectral imaging

Multispectral imaging is believed to be a useful new technique for fruit internal quality evaluation and assessment of postharvest storability. It is of interest to growers, breeders and postharvest technologists, particularly when implemented non-destructively. Khodabakhshian *et al.* (2016) investigated the use of multispectral imaging techniques to quantify pomegranate fruit quality. Three quality factors, including TSS, pH and firmness, were studied at four different maturity stages of 88, 109, 124 and 143 DAFB and were correlated with the spectral information extracted from images taken at four wavelength spectra. In this research, partial least squares regression (PLSR) models were developed to relate reflectance spectra obtained from a multispectral imaging system with four wavelengths to the quality parameters of pomegranates during maturity. The results showed that these models could achieve good predictions of TSS, pH and firmness of samples. Therefore, multispectral imaging could be used as a non-destructive method to distinguish pomegranate fruit ripening/maturity stages. The results demonstrated the capability of multispectral imaging and chemometrics as useful techniques to non-destructively monitor the main quality attributes of pomegranate (Khodabakhshian *et al.*, 2016). In additional reports published by the same group, a prototype multispectral imaging system for online quality assessment of pomegranate fruit was developed (Khodabakhshian *et al.*, 2017b). Initially, a visible (VIS)/near infrared (NIR) spectroscopy (400–1100 nm) was tested for non-destructive determination of TSS, TA and pH. The spectral data were analysed using PLS analysis. Then to establish a consistent multispectral imaging system, the highest absolute values of Π-coefficients corresponding to wavelengths from the best PLS calibration model were selected and used for identifying the optimal wavelengths. Consequently, a multispectral imaging system was developed based on the effective wavelengths 700, 800, 900 and 1000 nm. The performance of the developed multispectral imaging system was evaluated by multiple linear regression (MLR) models. The MLR model predicted TSS with $r=0.97$, root mean square error of calibration (RMSEC) = 0.21°Brix and ratio performance deviation (RPD) = 6.7°Brix. Also, the results showed that the model had good predictive ability for pH and TA. Their results showed that the developed multispectral imaging system based on the optimal wavelengths could be used for online quality assessment of pomegranate fruit (Khodabakhshian *et al.*, 2017b).

As mentioned earlier, most instrumental techniques to measure quality attributes (TSS, TA) of pomegranate are destructive in nature, time-consuming, and inapplicable to grading and sorting. Any technology such as spectroscopic and hyperspectral imaging systems that can classify the pomegranates non-destructively based on these quality attributes will be very useful for producers, processors and distributors to ascertain fast evaluation (Khodabakhshian *et al.*, 2017b).

14.8.2 Computer vision

A computer vision system includes the application of techniques in which computers are employed to examine and extract image contents in solving specific problems concerning the fruit surface. Quality assessment of agricultural produce offers definite challenges as the 'appearance' is inconsistent and vague (Deepa and Geethalakshmi, 2011). An efficient machine vision system was designed and implemented in order to assess the quality of pomegranate fruits by Kumar *et al.* (2018). The sample images of pomegranate fruits were captured using a custom-made image acquisition system. Two sets of features, namely spatial domain feature set and wavelet feature set, were extracted for all of the sample images. Experiments were conducted by training both artificial neural networks (ANNs) and support vector machines (SVMs) using both sets of features. The results of the experiments illustrated that ANNs outperformed SVMs with a difference in the accuracy of 12.65%. Further, the selection of wavelet feature set for training yielded more accurate results than spatial domain feature set (Kumar *et al.*, 2018).

The potential of colour and hyperspectral imaging has been evaluated to monitor the quality of 'Mollar de Elche' intact pomegranate fruit and arils during maturity. Pomegranate fruits were collected at seven different harvest times. Colour and hyperspectral images of the intact fruit and arils were acquired at each harvest. Physicochemical properties were measured in the juice of each fruit. Relationships between colour (L*, a*, b*) and spectral (720–1050 nm) data obtained from the images of the intact fruit and the arils were investigated. Discrimination of the different maturity stages was also carried out using PLS discriminant analysis models. Similar results were obtained in the prediction of the physicochemical properties using the colour and hyperspectral images of the intact fruit. However, the predictions achieved for the information about the arils were better using hyperspectral imaging. In the discrimination of maturity stage, the highest accuracies were obtained using hyperspectral imaging, where 95% of intact fruit and 100% of arils were correctly classified. These results indicate the great potential of machine vision techniques, especially hyperspectral imaging, for monitoring the quality of intact 'Mollar de Elche' pomegranate fruit and arils (Munera *et al.*, 2019).

14.8.3 Colourimetric index

Pomegranate acceptability by consumers and processors depends basically on a combination of several quality attributes, among them rind colour, sugar content, TA and flavour are more significant (Manera et al., 2013). Some researchers have studied the correlation between rind colour parameters (L*, a*, b*, C* and hue*) and TA, TSS, citric acid and anthocyanin content (Dafny-Yalin et al., 2010). Manera et al. (2013) evaluated the external colour of the pomegranate fruit from the early state to harvest, in order to develop a maturity index for the varietal group 'Mollar de Elche'. Their results provided growers with an objective criterion, a cheap, rapid and non-destructive way of assessing the optimal time for harvesting. Previously, Shwartz et al. (2009) also used a colour index to evaluate changes in the juice colour during the 10 weeks preceding harvest. However, it seems that their colour index was not sensitive to changes in the external fruit colour.

14.8.4 Dielectrics spectroscopy

This method was presented as an interesting technique to monitor online the fruit quality standards and ripening changes. Castro-Giráldez et al. (2013) analysed the effect of the major components of pomegranate and its structure on the dielectric spectrum between 500 MHz and 20 GHz. They measured the dielectric properties of pomegranate fruit in the arils, in the spongy white tissues of the locular septa and in the peel. They followed the evolution of pomegranate ripening after harvest by monitoring the citric acid content and the respiration rate. Hence, they developed a sensor system to predict these variables to follow the ripening in a rapid and non-destructive way. A dielectric factor (df) based on loss factor at 1.2 and 2.4 GHz for evaluating the citric acid content of standard solutions was defined. This factor was applied in pomegranate fruit, demonstrating its utility for determining the days from harvest, as well as the fruit's physiological activity. Moreover, a maturity index based on dielectric properties in the microwave frequency range was developed for predicting the maturity of apple fruits (Castro-Giraldez et al., 2010a, b). The feasibility of using VIS/NIR spectroscopy along with chemometrics was also investigated to predict quality parameters (pH, TSS and firmness) of pomegranate fruit in a non-destructive manner (Khodabakhshian et al., 2016). They found that VIS/NIR spectroscopy and chemometrics combined with different preprocessing techniques could be an accurate and fast method for non-destructive prediction of key pomegranate quality attributes (Khodabakhshian et al., 2016).

14.8.5 NIR spectroscopy

The potential of visible and NIR has already been investigated to classify the maturity stage and to predict the quality attributes of fruits. NIR spectroscopy has been used to evaluate the internal quality parameters non-destructively for different agricultural produce such as apples (Lu et al., 2000), cherries (Lu, 2001), citrus (Lee et al., 2004), grapes (Herrera et al., 2003) and kiwifruit (McGlone et al., 2002). The usefulness of this technique was recently evaluated in 'Ashraf' pomegranate as well (Khodabakhshian et al., 2017c).

14.9 Pomegranate Harvest Time

Commercial orchards of different pomegranate cultivars are grown in different countries (Holland et al., 2009), with about 90% of the world pomegranate production occurring in the northern hemisphere (Holland et al., 2009; Citrogold, 2011) (Table 14.4). The timing of harvest is of greatest importance, either for immediate fresh market or for storage, if fruit are to reach the customer in prime condition. Studies have shown the effects of cultivar differences, growing region and maturity status on pomegranate fruit maturity indices (Al-Maiman and Ahmad, 2002; Shwartz et al., 2009).

Harvesting pomegranate at an early stage of maturity may result in fruit that have good appearance and can withstand postharvest handling, but with poor aril colour intensity and unacceptable flavour. On the other hand, fruit harvested at late maturity are more susceptible to spoilage and have short storage potential (Fawole, 2013). The time taken for fruit to reach harvest maturity varies among cultivars, growing locations and seasons (Shulman et al., 1984; Gil et al., 1995a). The study by Shulman et al. (1984) showed that in the hot valley region in

Table 14.4. Harvesting time of pomegranate varieties in some selected countries.

Country	Cultivar	Ripening time	Reference
Northern hemisphere			
Oman	All	July–September (Peak in August)	Al-Yahyai *et al.*, 2009
Turkmenistan	Early varieties	September	
Iran	All	The pomegranates that ripen as of September are called early types, the ones that ripen as of October are called late types	Mohseni, 2009
Iran	'Malas-e-Yazd'	160–180 days from flowering	Kahramanoğlu and Usanmaz, 2016
Iran	'Ashraf'	Between September and October	Khodabakhshian *et al.*, 2017a
Italy	'Primosole'	29 October	Aquino *et al.*, 2009
USA	'Wonderful'	11 October	Wetzstein *et al.*, 2011
USA	'Granada'	August	Wetzstein *et al.*, 2011
USA	'Foothill Early'	1–2 weeks before Wonderful	Wetzstein *et al.*, 2011
Spain	'Mollar'	150–180 days from flowering	Manera *et al.*, 2013
Turkey	'Çekirdeksiz'	19 September and 11 October	Polat *et al.*, 2012
Turkey	'Hicaznar'	170–190 days from flowering	Kahramanoğlu and Usanmaz, 2016
India	'Bhagwa'	150–180 days from flowering	Kahramanoğlu and Usanmaz, 2016
Israel	'Emek'	Mid- to late August	Holland *et al.*, 2014
Israel	'Acco'	130–150 days from flowering	Kahramanoğlu and Usanmaz, 2016
Southern hemisphere			
South Africa	'Herschkovitz'	Early April to early May	Ferreira, 2013
South Africa	'Wonderful'	End of February to the end of March	Ferreira, 2013
Peru	'Wonderful'	February to March	Personal communication
Chile	'Wonderful'	March to April	Personal communication
Australia	'Wonderful'	March to April	Personal communication

Israel, fruit matured more rapidly than in the coastal plain. Fruit ripens 5–8 months from fruit set, involving a sequence of changes in fruit characteristics from flowering to maturity and senescence. These changes include physical and structural, biochemical, physiological and elemental changes, reflecting differences in fruit appearance during maturation and ripening and maturation time among cultivars (Ben-Arie *et al.*, 1984; Shulman *et al.*, 1984; Al-Maiman and Ahmad, 2002; Holland *et al.*, 2009; Shwartz *et al.*, 2009).

Harvest time is different in the northern and southern hemispheres (Table 14.4). India is one of the largest producers of pomegranate, followed by Iran, Turkey, the USA, Spain and Israel. Less than 1.25% of world pomegranate production is in the southern hemisphere (predominantly South America and South Africa). Australia only produces 2% of that 1.25%. There are fewer than 500 ha of pomegranates planted in Australia; however, much of this area has not reached its full potential or full production, owing to poorly understood tree health problems in all regions. There are a number of smaller orchards growing pomegranates for the whole fresh and ready-to-eat arils market.

Harvesting time of pomegranate cultivars in some selected countries of both hemispheres is shown in Table 14.4. It is noticeable that the fruit ripening and consequently harvesting in South Africa takes place from April to May. Hence, South Africa hits the European markets in this period, when South Africa is the only supplier of pomegranates to the northern hemisphere and prices are high. The South American pomegranates hit the European markets a month later, so prices dip a little but are still higher than during the northern hemisphere production season (Ferreira, 2013). The differences in harvesting time in the two hemispheres are important for the pomegranate fresh market. During harvest months in South Africa, late February to early May, in the northern hemisphere there is a shortage of pomegranates, so prices on the fresh food market can go up by about 250% during this period.

14.10 Harvesting Methods and Techniques

Harvesting is an important operation in horticultural crop production and any insufficiency during this time may lead to the loss of a whole year's work (Prasad *et al.*, 2017). The goals of harvesting are to gather a commodity from the field at the proper level of maturity with a

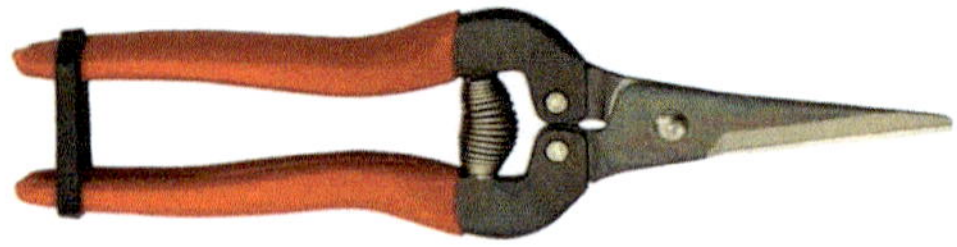

Fig. 14.7. Curved blade fruit shears. (Photo: Ali Sarkhosh.)

minimum of damage and loss, as rapidly as possible, and at a minimum cost (Kader, 2002). Harvesting practices should cause as little mechanical damage to produce as possible. Gentle picking and handling will help reducing the crop losses (Kitinoja and Kader, 2003).

Depending on cultivars and growing conditions, pomegranate fruits become ready for harvesting about 5–8 months after the appearance of blossoms (Fig. 14.6). Fruit are harvested with special secateurs called fruit shears or fruit snipper. To eliminate mechanical damage to the fruit during harvesting, a curved blade pruner/fruit snipper is recommended (Fig. 14.7). Outer appearance, total soluble content (°Brix) of the fruit, juice content of the arils and inner colour are the important characteristics that need to be considered before harvesting. Hand-held refractometers are proper tools to measure °Brix in the pomegranate fruit in the orchard during harvesting (Fig. 14.8). It is very important to pick the fruits at the right time, as delay in harvesting may cause fruit cracking. Owing to flowering in three distinct flushes, fruits can also mature and ripen at different times, therefore will be picked

Fig. 14.6. Pomegranate fruit 'Wonderful' 3 weeks before harvesting (left) and ready for harvesting (right). (Photo: Ali Sarkhosh.)

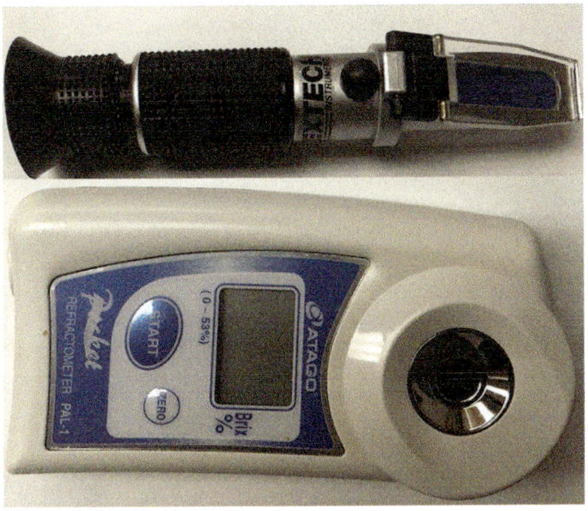

Fig. 14.8. Hand-held refractometers, digital (top) and manual (bottom). (Photos: Ali Sarkhosh.)

Fig. 14.9. Correct method of harvesting and handling of pomegranate fruits in the field. (Photos: www.cairodoutcher.com)

in three periods. In the case of desirable fruit set and fruit thinning keeping only the fruits of the first flowers, all the fruits can be harvested at once. However, in other situations, fruits must be harvested when they mature; first of all, the apical fruits, which come from the first flowers, then the axillar fruits, which come from the second and third flowers, and last the remaining fruits (Kahramanoğlu and Usanmaz, 2016).

Most horticultural crops intended for fresh consumption are picked by hand. This method reduces mechanical damage of the commodity; however, it is slow and can be expensive when labour is short and/or expensive. Hand harvesting is selective and can be done several times (Yahia *et al.*, 2008). Compared with other similar crops, pomegranates are easy to harvest, and if trees are properly trained, minimal ladder work is required. Fruit are harvested by clipping them from the stem with shears (Fig. 14.9), as close to the fruit (Fig. 14.10) as possible to prevent a sharp point of wood from piercing and rubbing against other fruit in the bin. Fruit can be placed directly into the picking bag, picking trays or bins in the orchard. Pomegranate fruit is quite sensitive and should be handled with care in order to minimize bruising (Fig. 14.11). The main benefit of hand over mechanized harvesting is that humans are able to select the produce at its correct stage of ripening and handle it carefully. The result is a higher quality product with minimum damage. Harvest labourers need to be adequately trained to give them the necessary skills to select produce at the correct stage of ripeness or degree of maturity as well as sorting techniques to minimize damage (Simson and Straus, 2010).

During harvesting, factors like the delicacy of crop, maturity criteria, time, method of harvesting, mode of packaging and transportation, the economy of the operations and the need for the harvesting method to fulfil the market requirement should be taken into consideration (Rathore *et al.*, 2012). Properly trained workers can pick and handle the product with a minimum of damage (Kader, 2002). Harvesting with improper methods results in the damage of crop by bruising, which can be caused by compression (due to overfilling of boxes or in bulky stores), impact (due

Fig. 14.10. Correct method in harvesting of pomegranate fruits by fruit shears, picking the fruits without leftover wood residue. The sharp point of wood remaining on the fruit may penetrate other fruits and damage them inside the bin. (Photos: Ali Sarkhosh.)

Fig. 14.11. Correct handling of pomegranate fruit in the orchard during harvesting. (Photos: www.pomegranate.az)

Fig. 14.12. Different types of picking bins and boxes can be used for harvesting pomegranate fruit in the field. The figure shows a 40 kg crate and 300 kg bin used for carrying pomegranate fruit in the field. (Photos: www.cairodoutcher.com and Alimohammad Yavari.)

to dropping of crop or from something hitting the crop), or vibration (due to loose packing during transportation) (Prasad *et al.*, 2017).

As previously mentioned, pomegranate fruits for fresh consumption must be picked by hand (by shears) and carefully handled. Pomegranates are easy to harvest and if pruning and trellising are done well, the use of ladders will be minimal. A size grading ring is used to ensure large enough fruit are picked. A bruise on the skin may cause a dark blemish on the shiny rind, but not actually damage the inside of the fruit. However, skin damage causes the inner quality to decrease in time during storage. Mechanical damage also increases moisture losses and reduces weight during storage. Also, the external appearance alone is enough to lead to a reduction in commercial value. Shears should be used to cut the fruit off. The fruit should be protected from sharp twigs (Fig. 14.10) (Kahramanoğlu and Usanmaz, 2016). The fruits are sensitive and should be placed into buckets or boxes carefully. Containers and harvesting tools should be clean and free from rough edges. Stackable and nestable plastic crates could be used as field containers during harvest (Fig. 14.12). Plastic crates are durable, reusable and can easily be cleaned.

14.11 Field Handling (Fruit Hauling)

Before harvesting, workers should wait until the dew dries on fruits. Otherwise, blemishes occur on fruits, which causes damage during storage. The second important point is the protection of harvested fruits from direct sunlight. Immediately after harvest, fruits begin to lose weight through transpiration. The main cause of transpiration is the temperature and fruits must be kept at cooler temperatures and protected from sunlight (Kahramanoğlu and Usanmaz, 2016). Harvested fruits may be stored under a tree or shaded areas until they are transported to the place for packing or should be immediately packed at the field and transported to markets or storage. Providing shade in the field by utilizing shady areas under plants/trees is possible but it should be remembered that shady areas change during the day. Covers on containers may be necessary when there is a risk of bird droppings from the shade trees.

In some cases, frequent transport to the final destination is preferred to reduce weight loss. If possible, it is highly recommended to transport in controlled conditions, that is, temperature and humidity (Kahramanoğlu and Usanmaz, 2016). Fruits are very susceptible to bruising and other forms of mechanical damages, and therefore should not be handled more than necessary. Fruits are normally transported and stored in bulk boxes (bins) kept in the orchard. Bins should not be allowed to sit for extended periods in direct sunlight and should be handled with care to avoid injury and damage and reduce bruising.

Harvesting involves several other activities undertaken in the field. These include those of commercial interests. Examples of operations to facilitate preparation for the market include pre-sorting and removal of foliage and other non-edible parts. In some cases, the product is completely prepared for the market in the field. However, the standard practice is to unload the harvest containers into larger ones for transportation to the packing house (Simson and Straus, 2010).

Various systems and packing materials have been developed for this purpose. The selection of type and design is related to the protection of the produce, convenience of handling and cost-effectiveness. Sometimes the crop is harvested into one type of container and is then transferred to another type for transport from the field. Picking boxes can be used, and fruits may be transferred from these to a pallet box to be taken to and stacked in the storage chamber. Types of packages and the materials from which they are made vary considerably.

Harvest containers should be cushioned, smooth and free of sharp edges. Field containers should not be overfilled and should be moved carefully. Drop heights when transferring produce to other containers should be minimized (Simson and Straus, 2010). Trained harvest labourers should be employed to handle produce gently and identify correct maturity for harvest. Gloves should be worn during harvest and handling to avoid damage to fruits. For short trips, covered trucks can be used to transport produce from the field.

14.12 Mechanical Injury (Physical Damage)

Mechanical (physical) injury is the major problem causing losses during harvesting. Pomegranate peel is sensitive to mechanical damages. Mechanical damages may be caused on trees because of hard winds and incorrect pruning. Wind causes fruits to lurch on trees and touch the spiny parts of the branches (Fig. 14.13). Thus, mechanical damage may occur on fruits, which may then enhance the growth of pathogens on or inside the fruits. Therefore, protection of the orchard from hard winds by wind breakers is very important to

Fig. 14.13. A view of damaged pomegranate fruits, damage caused by detached stem, sunburn, leftover wood residue and mechanical. (Photo: Ali Sarkhosh.)

prevent mechanical damage. Also, branches that may touch each other and cause crowding on trees should be removed during pruning to prevent mechanical damage (Kahramanoğlu and Usanmaz, 2016).

Postharvest and economic losses suffered by the horticultural industry annually due to mechanical damage of fresh produce from harvest to postharvest handling are considerable (Montero *et al.*, 2009; Ahmadi *et al.*, 2010; Ghaffari *et al.*, 2015).

Containers should be clean and separate containers should be used for organically grown produce to avoid any possibility of chemical contamination. Containers may be washed at high pressure, rinsed and sanitized prior to use. Clean containers should be covered to avoid contamination after cleaning. Containers should not have rough surfaces that can damage produce. Damaged or spoiled produce should be separated and left in the field to reduce contamination by decay organisms. Containers must be appropriate for the product: shallow enough to avoid compression damage. Overfilling of containers can further damage the product, and damage increases if containers are stacked.

Harvested fruits should be transported carefully to avoid damage. In addition to impact damage that can occur by allowing fruits to bounce around, friction damage can occur as a result of products moving against each other and against container walls. Strategies to make the transport smoother include grading rough farm roads (usually before the season) and reducing tyre pressure on transport vehicles.

Where harvesting and handling operations are not carried out with sufficient care and attention, damage can occur to the crop that may have repercussions during subsequent marketing and storage operations. These include: increasing water loss, shortening the potential maximum crop storage life due to increased respiration or ethylene biosynthesis, increased levels of microorganism infection through damaged areas and increases in some physiological disorders. The types of injury that can be inflicted on the crop include cuts, scuffs and bruises. Compression, impact or vibration can cause bruises.

14.12.1 Scuffing

This occurs when fruits are caused to move across a hard, usually rough surface, so that the cuticle and layers of cells are scraped away by abrasion.

14.12.2 Compression bruising

Where the downward force on the crop is above a threshold level it can be bruised. This damage may also be a function of time, especially where the pressure is close to the threshold level. It may be as a result of overfilling boxes and then stacking the boxes so that the crop in the lower boxes supports the weight.

14.12.3 Impact bruising

This results either from the crop being dropped or from something hitting it. The damage might be obvious on the surface of the crop or it might be internal.

14.12.4 Vibration bruising

This occurs when crops are being transported, especially in lorries. It is common when the crop is packed loosely in the lorry or even in boxes and is largely the result of the fruit moving and impacting on each other or the walls of the lorry or box. It can result in an increase in the respiration rate of the crop in addition to surface bruising. To minimize the effect, the crop needs to be packed tightly to reduce its movement. Equipment is available in packhouses to 'tight fill' boxes. Internal dividers can be used on accurately size-graded produce to reduce mutual impacts. Like any other fresh produce, pomegranate fruits are subjected to mechanical damage during postharvest handling due to the action of static and dynamic forces (Shafie *et al.*, 2015), which may occur due to fruit-to-fruit contact or contact between fruit and hard surfaces. Bruising is the most common type of mechanical damage, which results from excessive impact and compression forces due to improper handling, poorly designed equipment or improper packaging (Ahmadi *et al.*, 2010; Tabatabaekoloor, 2013; Opara and Pathare, 2014).

Bruise damage susceptibility of three commercially grown pomegranate fruit cultivars ('Acco', 'Herskawitz' and 'Wonderful') during postharvest handling were reported by Hussein *et al.* (2017). They found that bruise damage size and bruise susceptibility were cultivar dependent. 'Wonderful' pomegranate was found to be more susceptible to bruise damage, followed by 'Herskawitz'. Furthermore, an increase in drop height (or impact energy) increased the potential for bruise damage to occur on fruit. Also, they demonstrated that equivalent fruit drop height for bruise damage to occur was lowest for 'Wonderful' (3.13 cm) and highest for 'Acco' (11.13 cm). Therefore, to reduce bruise damage incidence, impacts should be minimized during fruit harvesting and postharvest. 'Wonderful' pomegranate fruit requires additional care during handling due to its critically lower bruise threshold (Hussein *et al.*, 2017). Low temperature (5°C) influenced bruising more, whereas higher fruit temperature (20°C) reduced the bruise damage of all fruit cultivars (Hussein *et al.*, 2016).

14.13 Fruit Harvesting

In recent years, research and developments in novel fruit harvesting systems have been sponsored to overcome problems related to harvest issues in horticultural industry. These techniques may decrease harvesting costs and increase the value of products to the consumers. The conventional harvesting method is highly labour-intensive and inefficient in terms of both economy and time. Machine harvesting systems are partial solutions to overcome these issues by removing fruits from the trees efficiently thus reducing the harvesting costs to about 35–45% of total production costs (Li *et al.*, 2011). The main benefit of mechanical harvest is that machines can often harvest at high speed rates. Machine harvesting also reduces management problems associated with workers. Effective use of mechanical harvesters requires operation by dependable, well-trained people (Kader, 2002).

Harvesting of pomegranate fruit is still done traditionally by hand. Manual harvesting of the fruits is hard work, which has many adverse physical and physiological effects on farmers' health, also it has the disadvantage of low capacity and high labour costs. Mechanization of harvesting is a way to keep up with the competition as labour costs rise and the supply of workers reduces every year. Therefore, it seems very important to develop an automated machine to harvest fruits (Jafari and Bakhshipour, 2011). Mechanized harvesting of pomegranate is still a challenge due to the tangled shape of the tree. Robotic harvesting seems to be the final solution in this case. The first step towards this aim is to recognize the pomegranate fruit and locate its spatial position on the tree.

Jafari and Bakhshipour (2011) developed an intelligent stereoscopic vision-based algorithm for a pomegranate fruit harvester robot. A typical fruit harvesting robot utilizes machine vision systems to recognize and locate the fruits on the tree. The pomegranate recognition algorithm developed in this study used colour differences components (YCrCb colour space) as criteria for discrimination of the pomegranate fruits from the leaves and branches. None of the other objects in the images was considered by the algorithm as pomegranate, while 3.7% of the fruit samples were removed from the images. However, this is not a problem for the harvester because the robot is assumed to turn around the tree and look for the fruits. So the omitted fruits in one image will be detected in other closer views at different positions. Two cameras with the same resolution and characteristics were used to take images. The cameras were in the same elevation and same distance from the fruits and there was a defined distance between them. At first, red colour difference was used to separate fruits from other subjects in the images. The centre of the fruit area was calculated by the algorithm as the 2D location of the fruit in each image. Then the third dimension was calculated using geometric equations. The resulting coordinate was the spatial location of fruit on the tree that must be detected by the harvesting machine. Locating the fruits was carried out by means of a stereoscopic vision system. As there was not a large distance between the two cameras to affect the distortion of the image, and pomegranate fruits are almost spherical, the centroid of the fruit was accurate enough to be considered as the target point for a picking arm (Jafari and Bakhshipour, 2011). A maximum distance estimation error (DEE) of

2.4 cm was determined for the stereoscopic vision system, which shows its consistency to be used as a recognition and locating tool for a pomegranate harvesting robot (Bakhshipour *et al.*, 2012).

14.14 Aril Extraction

The manual extraction of arils is a laborious and tedious process (Singh *et al.*, 2007). Pomegranate arils are highly susceptible to mechanical damage incurred by the use of inappropriate methods for extracting, packaging and transporting arils. This can lead to tissue wounds, abrasion, breakage and squashing of the arils. Hess-Pierce and Kader (1997) reported that mechanical damage reduces the commercial value of the arils and it may increase their susceptibility to decay and the growth of microorganisms. In the past few decades both small and commercial-scale aril extractors have been developed. For more information about aril extraction processes, refer to Chapter 16 'Processing and Industrialization'.

References

Abbott, J.A. (1999) Quality measurement of fruits and vegetables. *Postharvest Biology and Technology* 15(3), 207–225. DOI: 10.1016/S0925-5214(98)00086-6.

Ahmadi, E., Ghasemzadeh, H.R., Sadeghi, M., Moghadam, M. and Zarifneshat, S. (2010) The effect of impact and fruit properties on the bruising peach. *Journal of Food Engineering* 97, 110–117.

Al-Maiman, S.A. and Ahmad, D. (2002) Changes in physical and chemical properties during pomegranate (*Punica granatum* L.) fruit maturation. *Food Chemistry* 76(4), 437–441. DOI: 10.1016/S0308-8146(01)00301-6.

Al-Yahyai, R., Al-Said, F. and Opara, L. (2009) Fruit growth characteristics of four pomegranate cultivars from northern Oman. *Fruits* 64, 335–341.

Amiot, J.M., Tacchini, M., Aubert, S.Y. and Oleszek, W. (1995) Influence of cultivar, maturity stage and storage conditions on phenolic composition and enzymatic browning of pear fruit. *Journal of Agriculture and Food Chemistry* 43, 1132–1137.

Aquino, S.D., Schirra, M., Palma, A., Angioni, A., Cabras, P. *et al.* (2009) Effectıveness of fludioxonil in control storage decay on pomegranate fruit. *Acta Horticulturae* 818, 313–318.

Arendse, E., Fawole, O.A., Magwaza, L.S. and Opara, U.L. (2016a) Non-destructive characterization and volume estimation of pomegranate fruit external and internal morphological fractions using X-ray computed tomography. *Journal of Food Engineering* 186, 42–49.

Arendse, E., Fawole, O.A., Magwaza, L.S. and Opara, U.L. (2016b) Estimation of the density of pomegranate fruit and their fractions using X-ray computed tomography calibrated with polymeric materials. *Biosystems Engineering* 148, 148–156.

Arendse, E., Fawole, O.A., Magwaza, L.S., Nieuwoudt, H. and Opara, U.L. (2018) Evaluation of biochemical markers associated with the development of husk scald and the use of diffuse reflectance NIR spectroscopy to predict husk scald in pomegranate fruit. *Scientia Horticulturae* 232, 240–249.

Babu, K.D., Singh, N.V., Gaikwad, N., Maity, A., Suryavanshi, S.K. *et al.* (2017) Determination of maturity indices for harvesting of pomegranate (*Punica granatum*). *Indian Journal of Agricultural Sciences* 87, 1225–1230.

Bakhshipour, A., Jafari, A. and Hosseini, S.M. (2012) Recognition of pomegranate on tree and stereoscopic locating of the fruit. *American-Eurasian Journal of Agricultural & Environmental Sciences* 12, 1288–1294.

Bashir, H.A., Abu-Bakr, A. and Abu-Goukh, A. (2003) Compositional changes during guava fruit ripening. *Food Chemistry* 80, 557–563.

Ben-Arie, R., Segal, N. and Guelfat-Reich, S. (1984) The maturation and ripening of the 'Wonderful' pomegranate. *Journal of the American Society for Horticultural Science* 109, 898–902.

Ben-Simhon, Z., Judeinstein, S., Nadler-Hassar, T., Trainin, T., Bar-Ya'akov, I. *et al.* (2011) A pomegranate (*Punica granatum* L.) WD40-repeat gene is a functional homologue of *Arabidopsis* TTG1 and is involved in the regulation of anthocyanin biosynthesis during pomegranate fruit development. *Planta* 234(5), 865–881. DOI: 10.1007/s00425-011-1438-4.

Blasco, J., Cubero, S., Gomez-Sanchis, J., Mira, P. and Molto, E. (2009) Development of a machine for the automatic sorting of pomegranate (*Punica granatum*) arils based on computer vision. *Journal of Food Engineering* 90(1), 27–34. DOI: 10.1016/j.jfoodeng.2008.05.035.

Borochov-Neori, H., Judeinstein, S., Harari, M., Bar-Ya'akov, I., Patil, B.S. *et al.* (2011) Climate effects on anthocyanin accumulation and composition in the pomegranate (*Punica granatum* L.) fruit arils. *Journal of Agricultural and Food Chemistry* 59, 5325–5334.

Carbonell-Barrachina, A.A., Calín-Sánchez, A., Bagatar, B., Hernández, F., Legua, P. *et al.* (2012) Potential of Spanish sour-sweet pomegranates (cultivar C25) for the juice industry. *Food Science and Technology International* 18, 129–138.

Castro-Giraldez, M., Fito, P.J., Chenoll, C. and Fito, P. (2010a) Development of a dielectric spectroscopy technique for determining key chemical components of apple maturity. *Journal of Agricultural and Food Chemistry* 58, 3761–3766.

Castro-Giraldez, M., Fito, P.J., Chenoll, C. and Fito, P. (2010b) Development of a dielectric spectroscopy technique for the determination of apple ('Granny Smith') maturity. *Innovative Food Science and Emerging Technologies* 11, 749–754.

Castro-Giráldez, M., Fito, P.J., Ortolá, M.D. and Balaguer, N. (2013) Study of pomegranate ripening by dielectric spectroscopy. *Postharvest Biology and Technology* 86, 346–353.

Citrogold (2011) Producing pomegranates in South Africa. Available at: www.citrogold.co.za/Producing%20Pomegranates%20in%20South%20Africa%20Citrogold%202011.pdf (accessed 24 April 2013).

Crisosto, C.H., Mitcham, E.J. and Kader, A.A. (2000) Pomegranates. Produce facts [online]. Available at: http://postharvest.ucdavis.edu/Produce/ProduceFacts/Fruit/Pomegranate.html (accessed 4 February 2015).

Czieczor, L., Bentkamp, C., Damerow, L. and Blanke, M. (2018) Non-invasive determination of the quality of pomegranate fruit. *Postharvest Biology and Technology* 136, 74–79.

Dafny-Yalin, M., Glazer, I., Bar-Ilan, I., Kerem, Z., Holland, D. *et al.* (2010) Colour, sugars and organic acids composition in aril juices and peel homogenates prepared from different pomegranate accessions. *Journal of Agricultural and Food Chemistry* 58, 4342–4352.

Deepa, P. and Geethalakshmi, S.N. (2011) Improved water-shed segmentation for apple fruit grading. *Proceedings of the International Conference on Process Automation*, Control and Computing, IEEE, p. 5.

Dhatt, A.S., Mahajan, B.V.C., Sandhu, K.S., Garg, A. and Sharma, S.R. (2007) *Handbook on Post Harvest Handling of Fruits and Vegetables*, 3rd edition. PHPTC, PAU, Ludhiana, India.

Diakou, P., Svanella, L., Raymond, P., Gaudillère, J.-P. and Moing, A. (2000) Phosphoenolpyruvate carboxylase during grape berry development: protein level, enzyme activity and regulation. *Australian Journal of Plant Physiology* 27, 221–229.

Du, C.T., Wang, P.L. and Francis, F.J. (1975) Anthocyanins of pomegranate, *Punica granatum*. *Journal of Food Science* 40, 417–418.

FAO (1997) Guidelines for small-scale fruit and vegetable processors. Agricultural Service Bulletin 127. FAO, Rome.

FAO (2003) Handling and preservation of fruits and vegetables by combined methods for rural areas. Technical manual FAO Agricultural Services Bulletin 149. Available at: www.fao.org/3/y4358e/y4358e05.htm (accessed 21 October 2019).

Fashi, M., Naderloo, L. and Javadikia, H. (2019) The relationship between the appearance of pomegranate fruit and color and size of arils based on image processing. *Postharvest Biology and Technology* 154, 52–57.

Fawole, O.A. (2013) Maturity indexing, pharmacological properties and postharvest performance of pomegranate fruit grown in South Africa. PhD Thesis. Stellenbosch, South Africa, Stellenbosch University.

Fawole, O.A. and Opara, U.L. (2013a) Developmental changes in maturity indices of pomegranate fruit: a descriptive review. *Scientia Horticulturae* 159, 152–161. DOI: 10.1016/j.scienta.2013.05.016.

Fawole, O.A. and Opara, U.L. (2013b) Effects of maturity status on biochemical content, polyphenol composition and antioxidant capacity of pomegranate fruit arils (cv Bhagwa). *South African Journal of Botany* 85, 23–31. DOI: 10.1016/j.sajb.2012.11.010.

Fawole, O.A., Opara, U.L. and Theron, K.I. (2012) Chemical and phytochemical properties and antioxidant activities of three pomegranate cultivars grown in South Africa. *Food and Bioprocess Technology* 5(7), 2934–2940. DOI: 10.1007/s11947-011-0533-7.

Ferrara, G., Cavoski, I., Pacifico, A., Tedone, L. and Mondelli, D. (2011) Morpho-pomological and chemical characterization of pomegranate (*Punica granatum* L.) genotypes in Apulia region, southeastern Italy. *Scientia Horticulturae* 130(3), 599–606. DOI: 10.1016/j.scienta.2011.08.016.

Ferreira, J. (2013) A look at South Africa's pomegranate production. Available at: https://www.farmersweekly.co.za/crops/field-crops/project-pomegranate-production/ (accessed 21 October 2019).

Gaikwad, N.N., Pal, R.K. and Babu, K.D. (2014) Entrepreneurship development in pomegranate through value addition. *National Seminar-cum-Exhibition on Pomegranate for Nutrition, Livelihood Security and Entrepreneurship Development*, Solepur, India, 5–7 December, pp. 264–271.

Ghaffari, H., Ghassemzadeh, H.R., Sadeghi, M. and Alijani, S. (2015) Some physical, mechanical and chemical properties of tomato fruit related to mechanical damage and bruising models. *Biological Forum – An International Journal* 7, 712–718.

Gil, M.I., Garcia-Viguera, C., Artes, F. and Tomas- Barberan, F.A. (1995a) Changes in pomegranate juice pigmentation during ripening. *Journal of the Sciences of Food and Agriculture* 68, 77–81.

Gil, M.I., Cherif, J., Ayed, N., Artes, F. and Tomfis-Barberfin, F.A. (1995b) Influence of cultivar, maturity stage and geographical location on the juice pigmentation of Tunisian pomegranates. *Z Lebensm Unters Forsch* 201, 361–364.

Gozlekci,, I.S. and Kaynak, L. (2000) Physical and chemical changes during fruit development and flowering in pomegranate (*Punica granatum* L.) cultivar Hicaznar grown in Antalya region. *Iamz-Ciheam* 42, 79–85.

Hassan, N.A., El-Halwagi, A.A. and Sayed, H.A. (2012) Phytochemicals, antioxidant and chemical properties of 32 pomegranate accessions growing in Egypt. *World Applied Sciences Journal* 16, 1065–1073.

Hernández, F., Melgarejo, P., Tomás-Barberán, F.A. and Artés, F. (1999) Evolution of juice anthocyanins during ripening of new selected pomegranate (*Punica granatum*) clones. *European Food Research and Technology* 210, 39–42.

Herrera, J., Guesalaga, A. and Agosin, E. (2003) Shortwave near infrared spectroscopy for non-destructive determination of maturity of wine grapes. *Measurement Science and Technology* 14, 689–697.

Hess-Pierce, B. and Kader, A.A. (1997) Carbon dioxide enriched atmospheres extend post-harvest life of pomegranate arils. *Proceedings of the Seventh International Controlled Atmosphere Research Conference, CA'97, Proceedings Vol. 5, Fresh-Cut Fruits and Vegetables and MAP (Gorny, J.R., ed.)*, Davis, California, p. 122.

Holland, D. and Bar-Ya'akov, I. (2008) The pomegranate: new interest in an ancient fruit. *Chronicle Horticultural* 48, 12–14.

Holland, D., Hatib, K. and Bar-Ya'akov, I. (2009) Pomegranate: botany, horticulture, breeding. *Horticultural Reviews* 35, 127–191.

Holland, D., Bar-Ya'akov, I. and Hatib, K. (2014) 'Emek', a red and very early-ripening new pomegranate cultivar. *American Society for Horticultural Science* 49(7), 968–970. DOI: 10.21273/HORTSCI.49.7.968.

Holton, T.A. and Cornish, E.C. (1995) Genetics and biochemistry of anthocyanin biosynthesis. *The Plant Cell* 7, 1071.

Hussein, Z., Fawole, O.A. and Opara, U.L. (2016) Reducing susceptibility of fresh produce to physical damage during postharvest handling: the case of pomegranate fruit. *Fifth RUFORUM Biennial Regional Conference*, Cape Town, South Africa, 17–21 October, pp. 904–909.

Hussein, Z., Fawole, O.A. and Opara, U.L. (2017) Investigating bruise susceptibility of pomegranate cultivars during postharvest handling. *African Journal of Rural Development* 2, 33–39.

IPGRI (2001) *Rome regional report CWANA 1999–2000*. IPGRI, Rome, pp. 20–28.

Jafari, A. and Bakhshipour, A. (2011) A novel algorithm to recognize and locate pomegranate on tree for a harvesting robot using a stereo vision system. *8th European Conference on Precision Agriculture, ECPA*.

Jayesh, K.C. and Kumar, R. (2004) Crossability in pomegranate (*Punica granatum* L). *Indian Journal of Horticulture* 61, 209–210.

Jing, P., Ye, T., Shi, H., Sheng, Y., Slavin, M. *et al.* (2012) Antioxidant properties and phytochemical composition of China-grown pomegranate seeds. *Food Chemistry* 132, 1457–1464.

Johanningsmeier, S.D. and Harris, G.K. (2011) Pomegranate as a functional food and nutraceutical source. *Annual Review of Food Science and Technology* 2, 181–201.

Kader, A.A. (2002) *Postharvest Technology of Horticultural Crops*, 3rd Edition. University of California, Agricultural and Natural Resources, Davis, California.

Kader, A.A. (2006) Postharvest biology and technology of pomegranates. In: Seeram, N.P., Schulman, R.N. and Heber, D. (eds) *Pomegranates: Ancient Roots to Modern Medicine*. CRC Press, Boca Raton, Florida, pp. 211–220.

Kader, A.A., Chordas, A. and Elyatem, S. (1984) Response of pomegranates to ethylene treatment and storage temperature. *California Agriculture* 38, 14–14.

Kahramanoğlu, I. and Usanmaz, S. (2016) *Pomegranate Production and Marketing*. CRC Press, Taylor & Francis Group, Boca Raton, Florida, p. 148.

Khaksar, G., Sayed Tabatabaei, B.E., Arzani, A., Ghobadi, C. and Ebrahimie, E. (2015) Functional analysis of a pomegranate (*Punica granatum* L.) MYB transcription factor involved in the regulation of anthocyanin biosynthesis. *Iranian Journal of Biotechnology* 13, 17–25.

Khodabakhshian, R., Emadi, B., Khojastehpour, M. and Golzarian, M.R. (2016) Visible-NIR inferared spectroscopy for pomegranate fruit quality assessment: chemometrics and common preprocessing methods. *Annals Food Science and Technology* 17, 224–238.

Khodabakhshian, R., Emadi, B., Khojastehpour, M. and Golzarian, M.R. (2017a) Determining quality and maturity of pomegranates using multispectral imaging. *Journal of the Saudi Society of Agricultural Sciences* 16, 322–331.

Khodabakhshian, R., Emadi, B., Khojastehpour, M., Golzarian, M.R. and Sazgarnia, A. (2017b) Development of a multispectral imaging system for online quality assessment of pomegranate fruit. *International Journal of Food Properties* 20, 107–118.

Khodabakhshian, R., Emadi, B., Khojastehpour, M., Golzarian, M.R. and Sazgarnia, A. (2017c) Non-destructive evaluation of maturity and quality parameters of pomegranate fruit by visible/near infrared spectroscopy. *International Journal of Food Properties* 20, 41–52.

Khoshroo, A., Keyhani, A., Zoroofi, R.A., Rafiee, S., Zamani, Z. *et al*. (2009) Classification of pomegranate fruit using texture analysis of Mr images. *CIGR* XI, 1182.

Kitinoja, L. and Kader, A.A. (2003) *Small-Scale Postharvest Handling Practices: A Manual for Horticultural Crops*, 4th edition. University of California, Davis, Postharvest Technology Research and Information Center, Davis, California, p. 260.

Kulkarni, A.P. and Aradhya, S.M. (2005) Chemical changes and antioxidant activity in pomegranate arils during fruit development. *Food Chemistry* 93, 319–324.

Kumar, B.P. and Purohit, G. (1989) Studies on fruit growth and development in pomegranate. *Horticultural Abstract* 62, 521.

Kumar, R.A., Rajpurohit, V.S. and Jirage, B.J. (2018) Pomegranate fruit quality assessment using machine intelligence and wavelet features. *Journal of Horticultural Research* 26, 53–60.

Lee, S.W., Kim, K.S. and Kim, S.D. (1974) Studies on the compositional changes of pomegranate fruit during maturation. I. changes in sugars, organic acids, amino acids, and the respiration rate. *Journal of the Korean Society for Horticultural Science* 15, 57–63.

Lee, K., Kim, G., Kang, S., Son, J., Choi, D. *et al*. (2004) Measurement of sugar content in citrus using near infrared transmittance. *Key Engineering Materials* 270, 1014–1019. DOI: 10.4028/www.scientific.net/KEM.270-273.1014.

Legua, P., Melgarejo, P., Abdelmajid, H., Martinez, J.J., Martinez, R. *et al*. (2012) Total phenols and antioxidant capacity in 10 Moroccan pomegranate varieties. *Journal of Food Science* 77(1), 115–120. DOI: 10.1111/j.1750-3841.2011.02516.x.

Li, P., Lee, S. and Hsu, H.Y. (2011) Review on fruit harvesting method for potential use of automatic fruit harvesting systems. *Procedia Engineering* 23, 351–366.

Lu, R. (2001) Predicting firmness and sugar content of sweet cherries using near-infrared diffuse reflectance spectroscopy. *Transaction of the ASAE* 44, 1265–1271.

Lu, R., Guyer, D. and Beaudry, R.M. (2000) Determination of firmness and sugar content of apple using NIR diffuse reflectance. *Journal of Texture Studies* 31, 615–630.

Magwaza, L.S. and Opara, U.L. (2014) Investigating non-destructive quantification and characterization of pomegranate fruit internal structure using X-ray computed tomography. *Postharvest Biology and Technology* 95, 1–6.

Manera, F.J., Legua, P., Melgarejo, P., Martínez, R., Martínez, J.J. *et al*. (2012) Effect of air temperature on rind colour development in pomegranates. *Scientia Horticulturae* 134, 245–247. DOI: 10.1016/j.scienta.2011.11.016.

Manera, F.J., Legua, P., Melgarejo, P., Brotons, J.M., Hernández, F. *et al*. (2013) Determination of a colour index for fruit of pomegranate varietal group 'Mollar de Elche'. *Scientia Horticulturae* 150, 360–364.

McGlone, V.A., Jordan, R.B., Seelye, R. and Martinsen, P.J. (2002) Comparing density and NIR methods for measurement of kiwifruit dry matter and soluble solids content. *Postharvest Biology and Technology* 26, 191–198.

Melgarejo, P., Martinez-Valero, R., Guillamon, J.M., Miro, M. and Amoros, A. (1997) Phenological stages of the pomegranate tree (*Punica granatum* L). *Annals of Applied Biology* 130, 135–140.

Melgarejo, P., Salaza, D.M. and Artes, F. (2000) Organic acids and sugars composition of harvested pomegranate fruits. *European Food Research and Technology* 211, 185–190.
Miguel, G., Fontes, C., Antunes, D., Neves, A. and Martins, D. (2004) Anthocyanin concentration of 'Assaria' pomegranate fruits during different cold storage conditions. *Journal of Biomedicine and Biotechnology* 5, 338–342.
Mirdehghan, S.H. and Rahemi, M. (2007) Seasonal changes of mineral nutrients and phenolics in pomegranate (*Punica granatum* L.) fruit. *Scientia Horticulturae* 111, 120–127.
Moghimi, A., Aghkhani, M.H., Sazgarnia, A. and Sarmad, M. (2010) Vis/NIR spectroscopy and chemometrics for the prediction of soluble solids content and acidity (pH) of kiwifruit. *Biosystems Engineering* 106(2010), 295–302.
Mohseni, A. (2009) The situation of pomegranate orchards in Iran. *Acta Horticulturae* 818, 35–42.
Moing, A., Svanella, L., Rolin, D., Gaudillère, M., Gaudillère, J.P. *et al*. (1998) Compositional changes during the fruit development of two peach cultivars differing in juice acidity. *Journal of the American Society for Horticultural Science* 123, 770–775.
Moing, A., Renaud, C., Gaudillère, M., Raymond, P., Roudeillac, P. *et al*. (2001) Biochemical changes during fruit development of four strawberry cultivars. *Journal of the American Society for Horticultural Science* 126, 394–403.
Montero, C.R.S., Schwarz, L.L., Dos Santos, L.C., Andreazza, C.S., Kechinski, C.P. *et al*. (2009) Postharvest mechanical damage affects fruit quality of 'Montenegrina' and 'Rainha' tangerines. *Pesquisa Agropecuária Brasileira, Brasília* 44, 1636–1640.
Mphahlele, R.R., Fawole, O.A., Stander, M.A. and Opara, U.L. (2014) Preharvest and postharvest factors influencing bioactive compounds in pomegranate (*Punica granatum* L.)—a review. *Scientia Horticulturae* 178, 114–123.
Munera, S., Hernández, F., Aleixos, N., Cubero, S. and Blasco, J. (2019) Maturity monitoring of intact fruit and arils of pomegranate cv. Mollar de Elche using machine vision and chemometrics. *Postharvest Biology and Technology* 156. Available at: https://doi.org/10.1016/j.postharvbio.2019.110936.
NAIP (2011a) Maturity and ripening process. Tamil Nadu agricultural university. Available at: http://eagri.org/eagri50/HORT381/lec03.html (accessed 21 October 2019).
NAIP (2011b) Maturity indices, harvesting and post harvest handling of fruits and vegetables. Tamil Nadu agricultural university. Available at: http://eagri.org/eagri50/HORT381/lec02.html (accessed 21 October 2019).
Naovi, S.A.H., Khan, M.S.Y. and Vohora, S.B. (1991) Antibacterial, antifungal and antihelmintic investigations on Indian medicinal plants. *Fitoterapia* 62, 221–228.
Nunes, C., Rato, A.E., Barros, A.S., Saraiva, J.A. and Coimbra, M.A. (2009) Search for suitable maturation parameters to define the harvest maturity of plums (*Prunus domestica* L.): a case study of candied plums. *Food Chemistry* 112, 570–574.
Onur, C. and Kaska, N. (1985) Akdeniz bolgesi narlarının (*Punica granatum* L.) seleksiyonu [Selection of pomegranate of Mediterranean region]. *Turkish Journal of Agriculture and Forestry* 9, 25–33.
Opara, L.U. (2000) Fruit growth measurement and analysis. *Horticultural Reviews* 24, 373–431.
Opara, U.L. and Pathare, P.B. (2014) Bruise damage measurement and analysis of fresh horticultural produce – a review. *Postharvest Biology and Technology* 91, 9–24.
Ozgen, M., Durgac, C., Serce, S. and Kaya, C. (2008) Chemical and antioxidant properties of pomegranate cultivars grown in the Mediterranean region of Turkey. *Food Chemistry* 111, 703–706.
Pande, G. and Akoh, C.C. (2016) Pomegranate cultivars (Punica granatum L). In: Simmonds, M.S.J. and Preedy, V.R. (eds) *Nutritional Composition of Fruit Cultivars*. Elsevier, USA, pp. 667–689.
Pareek, S., Valero, D. and Serrano, M. (2015) Postharvest biology and technology of pomegranate. *Journal of the Science of Food and Agriculture* 95, 2360–2379.
Polat, A., Çaliskan, O. and Kamiloglu, O. (2012) Determination of pomological characteristics of some pomegranate cultivars in Dörtyol (Turkey) conditions. *Acta Horticulturae* 940, 401–405.
Prasad, K., Jacob, S. and Siddiqui, M.W. (2017) Fruit maturity, harvesting, and quality standards. In: Siddiqui, M.W. (ed.) *Preharvest Modulation of Postharvest Fruit and Vegetable Quality*. Academic Press, London, pp. 41–69.
Rathore, N.S., Mathur, G.K. and Chasta, S.S. (2012) *Postharvest Management and Processing of Fruits and Vegetables*. ICAR, New Delhi.
Reid, M.S. (2002) Maturation and maturity indices. In: Kader, A.A. (ed.) *Postharvest Technology of Horticultural Crops*. University of California Agriculture & Natural Resources, Davis, California, pp. 55–62.

Salmanizadeh, F., Nassiri, S.M., Jafari, A. and Bagheri, M.H. (2015) Volume estimation of two local pomegranate fruit (*Punica granatum* L.) cultivars and their components using non-destructive X-ray computed tomography technique. *International Journal of Food Properties* 18(2), 439–455. DOI: 10.1080/10942912.2013.833521.

Sanders, K.F. (2005) Orange harvesting systems review. *Biosystems Engineering* 90, 115–125.

Shafie, M.M., Rajabipour, A., Castro-García, S., Jiménez-Jiménez, F. and Mobli, H. (2015) Effect of fruit properties on pomegranate bruising. *International Journal of Food Properties* 18, 1837–1846.

Shulman, Y., Fainberstein, L. and Lavee, S. (1984) Pomegranate fruit development and maturation. *The Journal of Horticultural Science and Biotechnology* 59, 265–274.

Shwartz, E., Glazer, I., Bar-Ya'akov, I., Matityahu, I., Bar-Ilan, I. *et al.* (2009) Changes in chemical constituents during the maturation and ripening of two commercially important pomegranate accessions. *Food Chemistry* 115, 965–973.

Simson, S.P. and Straus, M.C. (2010) *Post-harvest Technology of Horticultural Crops*. Oxford Book Company, New Delhi, p. 360.

Singh, D.B., Kingly, A.R.P. and Jain, R.K. (2007) Studies on separation techniques of pomegranate arils and their effect on quality of anardana. *Journal of Food Engineering* 79, 671–674.

Stover, E. and Mercure, E.W. (2007) The pomegranate: a new look at the fruit of paradise. *HortScience* 42, 1088–1092.

Sudheer, K.P. and Indira, V. (2007) *Postharvest Technology of Horticultural Crops*. New India Publishing Agency, Pitampura, New Delhi.

Tabatabaekoloor, R. (2013) Engineering properties and bruise susceptibility of peach fruits (*Prunus persica). Agricultural Engineering International: CIGR Journal* 15, 244–252.

Tehranifar, A., Zarei, M., Nemati, Z., Esfandiyari, B. and Vazifeshenas, M.R. (2010) Investigation of physico-chemical properties and antioxidant activity of twenty Iranian pomegranate (*Punica granatum* L.) cultivars. *Scientia Horticulturae* 126, 180–185.

Thompson, A.K. (1996) *Postharvest Technology of Fruits and Vegetables*, 1st edition. Blackwell Science, Inc., Boston, Massachusetts.

Varasteh, F., Arzani, K., Zamani, Z. and Tabatabaei, S.Z. (2008) Physicochemical seasonal changes of pomegranate (*Punica granatum* L.) fruit 'Malas-e-Torsh-e-Saveh' in Iran. *Acta Horticulturae* 769, 255–258.

Weerakkody, P., Jobling, J.I., María, M.V. and Rogers, G. (2010) The effect of maturity, sunburn and the application of sunscreens on the internal and external qualities of pomegranate fruit grown in Australia. *Scientia Horticulturae* 124, 57–61.

Wetzstein, H.Y., Zhang, Z., Ravid, N. and Wetzstein, M.E. (2011) Characterization of attributes related to fruit size in pomegranate. *HortScience* 46, 908–912.

Wills, R.B.H. and Golding, J. (2018) *Advances in Postharvest Fruit and Vegetable Technology*, 1st edition. CRC Press, Taylor & Francis, Boca Raton, Florida.

Wilson, L. and Downs, C.T. (2012) Fruit nutritional composition and non-nutritive traits of indigenous South African tree species. *South African Journal of Botany* 78, 30–36.

Worrell, D.B., Sean Carrington, C.M. and Huber, D.J. (1998) Growth, maturation and ripening of breadfruit, *Artocarpus altilis* (Park.) Fosb. *Scientia Horticulturae* 76(1–2), 17–28. DOI: 10.1016/S0304-4238(98)00134-4.

Yahia, E.M., El Tamzini, M.I., El Saied, A.A.F. and Al Yateem, S.E.D. (2008) *Training Manual on Postharvest Handling and Marketing of Horticultural Commodities*. Food and Agriculture Organization of the United Nations Regional Office for the Near East, Cairo, p. 276.

Zaouaya, F., Mena, P., Garcia-Viguera, C. and Mars, M. (2012) Antioxidant activity and physico-chemical properties of Tunisian grown pomegranate (*Punica granatum* L.) cultivars. *Industrial Crops and Products* 40, 81–89.

Zarei, M., Azizi, M. and Bashiri-Sadr, Z. (2010) Studies on physico-chemical properties and bioactive compounds of six pomegranate cultivars grown in Iran. *Journal of Food Technology* 8, 112–117.

Zarei, M., Azizi, M. and Bashir-Sadr, Z. (2011) Evaluation of physicochemical characteristics of pomegranate (*Punica granatum* L.) fruit during ripening. *Fruits* 66(2), 121–129. DOI: 10.1051/fruits/2011021.

Zhang, L. and McCarthy, M.J. (2012) Black heart characterization and detection in pomegranate using NMR relaxometry and MR imaging. *Postharvest Biology and Technology* 67, 96–101.

Zhang, L. and McCarthy, M.J. (2013) Assessment of pomegranate postharvest quality using nuclear magnetic resonance. *Postharvest Biology and Technology* 77, 59–66.

15 Postharvest Biology and Storage

D. Valero[1], M.E. García-Pastor[1], S.A. Sargent[2], J. Medina[1], F. Guillén[1], S. Castillo[1], D. Martínez-Romero[1], P.J. Zapata[1] and M. Serrano[3*]

[1]Department of Food Technology, EPSO, University Miguel Hernández, Orihuela, Alicante, Spain; [2]Horticultural Sciences Department, University of Florida, Gainesville, Florida, USA; [3]Department of Applied Biology, EPSO, University Miguel Hernández, Orihuela, Alicante, Spain

15.1 Introduction

Pomegranate fruit (*Punica granatum* L.) is one of the oldest edible fruits. It originated in the area now occupied by Iran and Afghanistan and spread to India, China and westwards to Turkey, Egypt, Tunisia, Morocco and Spain, where it is extensively cultivated nowadays (Ward, 2003). Chandra *et al.* (2010) provided a detailed description of the history of pomegranate, showing that it was one of the earliest domesticated fruit crops, first planted between 4000 and 3000 BCE and even mentioned in the Bible and the Koran. The fruit is developed from an inferior ovary and contains the arils or seeds, which are the edible part, contributing to 50–60% of the whole fruit. The arils contain 75–85% of the juice, coming from the external cover of the seeds and the remaining 15–25% is from the seeds themselves, which can be more or less hard, depending on cultivar. The aril juice is composed of 85% water and 15% sugars, pectins, ascorbic acid, polyphenols, flavonoids, anthocyanins and amino acids (Kader, 2006), which confers both nutritional and antioxidant properties on this edible fruit (Cerdá *et al.*, 2003; Faria *et al.*, 2011; Johanningsmeier and Harris, 2011). In fact, pomegranate has been used in the traditional medicine of many countries from ancient times, and its beneficial effects against several diseases, such as atherosclerosis, inflammatory and infective-mediated diseases and cancer have been reported recently and attributed to phenolic compounds and anthocyanins (Mertens-Talcott *et al.*, 2006; Lansky and Newman, 2007; Stover and Mercure, 2007; Faria *et al.*, 2011; Ismail *et al.*, 2012; Panth *et al.*, 2017).

To date, more than 1000 pomegranate cultivars have been reported in the world (Chandra *et al.*, 2010), destined for fresh consumption and the preparation of different products, such as juices, jellies, jams, liqueurs, energy drinks, etc., the most known worldwide being 'Wonderful', which was discovered in Florida and brought to California in 1896. In Spain, 'Mollar de Elche' is the most cultivated pomegranate cultivar, and is much appreciated by consumers due to its high concentration of sugars and low acidity and its barely discernible seeds, since they are very small and soft and can be easily eaten (Melgarejo *et al.*, 2000; Nuncio-Jáuregui *et al.*, 2014).

In general, it is considered that maximum quality attributes and full flavour are reached at harvest (Kader *et al.*, 1984; Pareek *et al.*, 2015) and thereafter, pomegranate quality decreases during storage, transport and retail processes.

*Corresponding author: m.serrano@umh.es

Thus, this chapter provides an overview of the changes in the parameters responsible for organoleptic, nutritional and antioxidant properties of pomegranate during ripening on the tree and during postharvest storage, as well as the recent technologies and treatments focused on preserving these quality parameters.

15.2 Pomegranate Growth and Changes in Quality Parameters During Ripening on the Tree

The chemical composition of pomegranate arils changes as the fruit matures and ripens on the tree. Thus, it is important to know the most appropriated harvest date to have fully ripe fruits with high-quality attributes, which lasts from 4.5–6 months after full bloom, depending on cultivar and agronomic and environmental conditions (Ben-Arie *et al.*, 1984; Kader, 2006; Fawole and Opara, 2013b). Pomegranate fruit quality depends on external and internal attributes. External quality is largely dependent on fruit size, skin colour, which ranges from light yellow to deep red depending on the cultivar, and absence of visual defects such as sunburn, cracks, cuts, bruises and decay. On the other hand, internal quality depends on sugar and acid content and on the presence of small and soft seeds as well as on the aril colour, which varies from light pink to deep red, depending on the cultivar; although, in general, no correlation exists between skin and aril colour. In addition, the pomegranate fruit contains a wide range of bioactive compounds (vitamins and other non-nutrient phytochemicals) with antioxidant properties and beneficial health effects.

15.2.1 Fruit growth

The pomegranate fruit growth pattern, in terms of changes in fruit diameter and length, follows a single sigmoid growth curve. A rapid increase in the weight and volume occurs early in the season and then continues slowly until harvest. This increase in size and weight is due mainly to enhancement in aril weight, since skin thickness decreases during the growing season leading to significant increases in arils/skin ratio (Varasteh *et al.*, 2008). During the last phases of fruit development on the tree, important changes occur in aril chemical composition, the major changes being a decrease in total acids, and increases in sugar content, anthocyanins and aroma compounds (Pareek *et al.*, 2015).

15.2.2 Sugars and organic acids

During ripening of pomegranate fruit, an accumulation of sugars and a decrease in total acids occurs in arils. The major sugars are fructose and glucose, with concentrations at harvest between 3 and 8%, depending on the cultivar, leading to a total soluble solids (TSS) concentration in the range of 10–18% (Melgarejo *et al.*, 2000; Al-Maiman and Ahmad, 2002; Poyrazoğlu *et al.*, 2002; Fadavi *et al.*, 2005; Kulkarni and Aradhya, 2005; Mirdehghan *et al.*, 2006; Ozgen *et al.*, 2008). The organic acid composition is different depending on the type of cultivar since they are clarified as acidic, semi-acidic and sweet cultivars. Thus, acidic group cultivars have a titrable acidity (TA) value of 2–2.5%, and citric acid is the major one, while sweet cultivars have a TA of 0.2–0.4% and similar amounts of citric acid and malic acid or a higher concentration of the latter (Hernández *et al.*, 1999; Melgarejo *et al.*, 2000; Poyrazoğlu *et al.*, 2002; Mirdehghan *et al.*, 2006; Ozgen *et al.*, 2008). Special attention should be paid to ascorbic acid, for its role as vitamin C and its antioxidant properties: its concentration decreases during the early stages of fruit development and remains more or less stable in the final stages of maturation, with values between 10 and 36 mg/100 g aril, depending on the cultivar (Kulkarni and Aradhya, 2005; Sayyari *et al.*, 2010).

15.2.3 Colour and anthocyanins

Colour evolves during ripening through cream to pink to red in the skin and pink to dark red in the arils and is due to anthocyanin pigments, although important differences are found among cultivars. The major anthocyanin in acid cultivars is cyanidin 3,5-diglucoside, followed by cyanidin 3-glucoside and delphinidin 3,5-diglucoside, whereas in sweet ones the major anthocyanin is cyanidin 3-glucoside, while delphinidin 3-glucoside and pelargonidin 3-glucoside are found in minor concentrations (Miguel *et al.*, 2004; Kulkarni and Aradhya, 2005; Alighourchi *et al.*, 2008; D'Aquino *et al.*, 2010). In addition, the total concentration of anthocyanins in the arils of the mature fruit is also dependent on cultivar, with values between 10 and ~500 mg/100 g, leading to arils varying in colour from light pink to dark red (Mirdehghan *et al.*, 2006; Tzulker *et al.*, 2007; Ozgen *et al.*, 2008; Sayyari *et al.*, 2010, 2011a; Mphahlele *et al.*, 2014).

15.2.4 Phenolic compounds and antioxidant activity

Phenolics, including anthocyanins, are important antioxidant compounds in pomegranate skin and arils, with profile and concentration depending on cultivar, developmental stage, environmental conditions and agronomic practices (Mphahlele *et al.*, 2014). During fruit development the concentration of total phenols in the arils decreases sharply in the early stages, this decrease being slow at the end of development, reaching concentrations in the ripe fruit ranging from 90 to ~1000 mg/100 g, depending on cultivar (Al-Maiman and Ahmad, 2002; Kulkarni and Aradhya, 2005; Mirdehghan *et al.*, 2006; Ozgen *et al.*, 2008; Sayyari *et al.*, 2011a; Mphahlele *et al.*, 2014). Flavonoids, including flavonols, anthocyanins and phenolic acids, are mainly found in the peel and juice of pomegranate, while hydrolysable tannins including gallotannins and ellagitannins are found in the peel and carpelar membranes. In addition, condensed tannins are mainly located in the peel and juice (Aviram *et al.*, 2008; Mphahlele *et al.*, 2014). The main phenolic compounds in pomegranate arils are the phenolic acids, gallic, chlorogenic, caffeic, ferulic and *o*-and *p*-coumaric acids, as well as catechin and quercetin (Poyrazoğlu *et al.*, 2002). In addition, it is interesting to note that in the skin the phenol content is much higher than in the arils, and in turn, pomegranate skin could be an important source of antioxidants (Li *et al.*, 2006). Fischer *et al.* (2011) identified 48 phenolic compounds in the arils, mesocarp and skin of an unknown Peruvian pomegranate cultivar and in all tissues the main phenolic compound was found to be punicalagin.

Antioxidant activity also decreases during the early stages of fruit development, but then increases again, reaching the highest levels in the state of commercial maturity (Kulkarni and Aradhya, 2005). In a study carried out with different genotypes of pomegranate from the Germplasm Bank of the Higher Polytechnic School of Orihuela, it was shown that pomegranate juice has a high antioxidant activity, superior to other fruits, characteristic of the Mediterranean diet (Valero *et al.*, 2015). It was found that the total antioxidant activity of the hydrophilic fraction (AAT-H) in these varieties correlated with its phenolic compound content, in accordance with Gil *et al.* (2000); although in other works it has been found that ascorbic acid is a water-soluble compound that also contributes significantly to the antioxidant capacity of this fruit (Mirdehghan *et al.*, 2006; Mirdehghan *et al.*, 2007b). On the other hand, the AAT of the lipophilic-soluble fraction (AAT-L) correlates with the content of carotenoids, which contribute 20% to the AAT of the arils (Valero and Serrano, 2010; Valero *et al.*, 2015).

15.2.5 Aroma compounds

The synthesis of aroma compounds is an integral part of the ripening process of pomegranate fruit, and the volatile profiles change with the advance of fruit growth and maturation, although the composition and relative proportions of the aroma compounds are dependent on cultivar (Fawole and Opara, 2013b; Hussein *et al.*, 2015). For instance, in 'Bhagwa' and 'Ruby' cultivars, hexanol and limonene were the two aroma compounds at S1 maturity stage, with hexanol constituting 52.4% in 'Bhagwa' and 95.2% in 'Ruby', while limonene constituted 47.6 and 4.8%, respectively. At S2 the alcohol group had the highest relative proportion, with an increase observed in 'Bhagwa' and a decrease in 'Ruby', when compared with S1 maturity stage. Ketone groups, comprising heptanone (7.8%) and 2-octanone (27.7%) were more prominent in 'Bhagwa' at maturity stage S3, while hexylacetate, 2-ethyl acetate and butyl acetate predominated in 'Ruby' and continued to dominate the rest of the maturity stages (S4 = 49%;

S5 = 33.4%). On the other hand, when fruit were harvested at maturity stage S4, 'Bhagwa' was characterized by the dominance of esters, while in S5 it was dominated by the ethanol group with a total proportion of 36.4 and 44.5%, respectively (Fawole and Opara, 2013b).

In nine different Spanish pomegranate cultivars, differences in the composition and relative proportions of the aroma volatile compounds have also been reported, with a total concentration of volatiles ranging from 1.7–10.9 g/kg. Eighteen compounds were found in pomegranate aroma profiles, including monoterpenes, aldehydes, alcohols, monoterpenoids and linear hydrocarbons. The most abundant compounds were trans-2-hexenal, 3-carene, α-terpinene and α-terpineol. The overall consumer liking of pomegranate juices was associated with the presence of monoterpenes such as α-pinene, β-pinene, β-myrcene, limonene and γ-terpinene, while the presence of aldehydes such as hexanol, hexanal and cis-3-hexenol was correlated with poor overall consumer liking (Calín-Sánchez *et al.*, 2011).

Given the high organoleptic, nutritive and functional properties of pomegranate fruits, it can be concluded that pomegranate has a high potential for commercialization in new markets. However, it should be harvested at full maturity stage, since if harvesting is performed too early fruits will have low-quality attributes because they have not fully developed their colour, aroma and flavour. However, if harvesting is done too late, high-quality fruits could be achieved, but they deteriorate faster. In this case, fruits should be harvested at full ripe stage, with high-quality attributes, and stored in appropriate conditions. Also, specific treatments are applied to the fruits to maintain quality properties as long as possible by delaying the postharvest ripening and senescence processes, as well as the appearance of decay, chilling injuries and other disorders that depreciate fruit quality, as will be discussed in the next sections.

15.3 Harvest, Packing and Cooling Operations

Harvest and postharvest handling of pomegranates involve many steps, and these must be carefully integrated in order to minimize losses. Precautions must also be taken to minimize build-up of microbial populations in the field via adequate spray programmes. Additionally, strict standard sanitation operating procedures (SSOPs) must be in place to minimize the potential for cross-contamination of the fruit with human pathogens from field through harvest, handling, storage and shipping (Lepper *et al.*, 2017, 2019). Workers must be trained in the areas of personal hygiene and appropriate handling of fresh produce such as pomegranates. Portable restroom facilities with potable water, soap and disposable hand towels must be available at all time for field crews.

15.3.1 Harvest maturity and fruit quality

Pomegranates require up to 7 months for growth and ripening, and are harvested from two to four times, depending upon the cultivar (Morton, 1987). As a non-climacteric fruit, however, pomegranates cannot be ripened postharvest, therefore growers must harvest when minimal aril quality parameters are met (Kader, 2006; Kader *et al.*, 2006). Since skin colour does not correlate well with aril ripeness, the most common commercial harvest index is based on established minimums for juice SSC and/or maximums for total TA concentration for the widely grown commercial cultivars. In the state of California, USA, for example, >90% of pomegranate juice samples must have <1.8% total TA in order to be harvested (Barclays Official California Code of Regulations, 2019); juice colour standards for redness are also set. Fresh market fruit commands the highest prices, and hence, has the most stringent cosmetic standards, which include freedom from mechanical injuries, cuts, bruises, sunburn and cracks in the skin. Fruit with minor defects are sent to processing for arils while those with severe defects are sent for juice extraction (Haug and Zalom, undated). Quality standards for major cultivars are also established by other production areas and by importing regions such as the European Union; these were thoroughly reviewed by Pareek *et al.* (2015).

15.3.2 Harvest and transport operations

Prior to harvest, fruit samples should be taken to confirm that aril quality conforms with prevailing standards. Experienced harvest crews also estimate fruit quality by tapping on individual fruit and listening for a metallic-like sound (Morton, 1987). The harvest crew must be properly dressed for protection against tree thorns. Fruit stems should be clipped close to fruit, placed in picking bags, then carefully transferred into field bins on low trailers or field lugs (crates) on pallets or trucks. Care must be taken to avoid scratches from thorns at harvest and excessive drop heights during transfers. Use of field bins minimizes the number of transfers the fruit undergo, reducing mechanical injuries. Harvest during cooler times of the day minimizes heat accumulation and saves on cooling times and costs later at the packing facility. Fruit can be presorted in the field as they are harvested, discarding those that are out-of-grade (inedible). Serious defects include cracking and latent diseases caused by *Alternaria* sp. (Ezra *et al.*, 2015) and *Aspergillus niger*, among others (Pala *et al.*, 2009).

Harvested pomegranates should not be exposed to direct sunlight in the field while waiting for timely transport to the packing facility. Trailers can be temporarily parked under shade to minimize the accumulation of field heat. Orchard lanes should be graded as necessary to minimize fruit jostling during transport to the main road. SSOPs during harvest operations include personal hygiene of the harvest crew, cleaning and disinfecting field containers prior to use, diverting dropped fruit to juice processing, sanitizing cutting implements regularly while harvesting, and keeping harvest containers and pallets off the ground (Lepper *et al.*, 2019).

15.3.3 Packing operations

At the packing house, bins and palletized lugs are unloaded from the transport vehicle with fork lifts and temporarily held under shade prior to being moved directly into the facility (Fig. 15.1). Pomegranates are ideally run over the packing line on the day of harvest and sorted for three purposes: (i) immediate packing into shipping containers; (ii) packing in bulk bins for

Fig. 15.1. Temporary storage under ambient conditions prior to packing. (Photo: Steven Sargent.)

short-, medium- or long-term storage; and (iii) diversion to aril or juice operations, or culling.

Although pomegranates have a leathery skin, they must be handled carefully during packing as they are very susceptible to physical blemishes such as bruises, scratches and abrasions. Typical packing line operations include: unloading, presorting washing, application of coatings or fungicides (if applied), drying, sorting/grading, sizing and packing (see Section 15.5). Unloading fruit in bins and lugs into a padded receiving hopper or directly on to the packing line with an inverted unloading unit (Fig. 15.2) is beneficial for transfer to the packing line. Drop heights at transfer points between packing line components must be minimized to reduce mechanical injuries (Sargent *et al.*, 1991). A quality assurance programme should be in-place to ensure that minimum quality standards are met for each packed lot of fruit. SSOPs during packing operations include use of potable water in washing operations, personal hygiene for workers touching the fruit and regular cleaning of all fruit contact surfaces (Lepper *et al.*, 2017). Cold room walls, floors and evaporator coils should be regularly cleaned and sanitized.

15.3.4 Cooling and storage operations

Proper temperature management is essential for successful storage and marketing of high-quality pomegranates, whether for the fresh market

Fig. 15.2. Gentle transfer on to the packing line using a mass air flow (MAF) unloading unit. (Photo: Steven Sargent.)

(whole fruit or arils) or for juice. For maximum storage (shelf) life, the fruit should be cooled to near ideal pulp temperature within a few hours of harvest. This means removing seven-eighths of the field heat (the temperature difference between the incoming pulp temperature and the cooling air). Room cooling can be employed for fruit destined for short-term storage. However, forced-air cooling (FAC) should be considered for longer storage periods; it is a much more rapid means to uniformly remove field heat (within 1–2 h) compared with room cooling (>12 h). The faster the cooling rate, the less moisture that is lost during the cooling process. (Watson *et al.*, 2015). Following cooling, fruit is cold-stored until shipment or repacking. Since pomegranates are susceptible to chilling injury, fruit should not be stored at <5°C; recommended storage temperatures vary by cultivar (Pareek *et al.*, 2015). Humidification of the cold room to 85–90% relative humidity will minimize moisture loss during storage. SSOPs during cooling and storage include regular cleaning and sanitation of the evaporator coils in the forced-air cooler unit and in the cold room and of cold room walls and floors.

15.4 Postharvest Physiology and Fruit Deterioration

Pomegranate fruits exhibit important quality losses during postharvest storage, due to several physiological and enzymatic disorders, the most important being weight loss, browning symptoms in both peel and arils, and decay occurrence. The internal fruit quality is also depreciated mainly due to reduced firmness, changes in aril colour, and decreases in total acids content and bioactive compounds concentration (Pareek *et al.*, 2015; Valero *et al.*, 2015).

15.4.1 Weight loss

One of the major problems associated with losses in pomegranate fruit quality during storage is excessive weight loss due mainly to water loss by transpiration through the fruit surface, which may result in hardening of the husk and browning of the rind and ultimately of the arils, resulting in loss of visual appearance. Thus, the storage potential of pomegranate fruit at 21°C and 82% RH may not be extended for more than 15 days due to excessive weight loss. However, under refrigerated conditions and high RH, most cultivars can be stored for longer periods (Caleb *et al.*, 2012a; Pareek *et al.*, 2015). For instance, Al-Mughrabi *et al.* (1995) observed that weight loss increased with storage temperature and time, reaching values of ~18, 22 and 33% after 8 weeks of storage at 5, 10 and 20°C, respectively, for 'Taeifi', 'Manfaloti' and 'Ganati' pomegranates. Accordingly, Fawole and Opara (2013a) reported a weight loss of 3.85% in 'Halow' pomegranates stored at 7°C and 95% RH for 6 weeks, whereas at 21°C and 65% RH the weight loss reached values significantly higher, 15.42%.

15.4.2 Respiration rate and ethylene production

Pomegranate fruit behaves as a non-climacteric fruit, showing a decline in respiration rate during fruit development on the tree and producing trace amounts of ethylene (Pareek *et al.*, 2015). The respiration rate in pomegranate fruit declines with time after harvest although higher values are found with increased storage temperature (Ben-Arie *et al.*, 1984; Kader *et al.*, 1984; Caleb *et al.*, 2012b). The Q10 values for respiration were 3.4 between 0 and 10°C, 3.0

between 10 and 20°C, and 2.3 between 20 and 30°C (Elyatem and Kader, 1984).

On the other hand, pomegranate fruit has low sensitivity to ethylene treatments. Thus, treatment of pomegranates with 10, 100 or 1000 ppm of ethylene for 48 h at 20°C had little or no effect on their skin colour, juice colour and composition in TSS or TA when evaluated immediately following treatment or after an additional 7 days at 20°C in air (Ben-Arie *et al.*, 1984; Elyatem and Kader, 1984). In addition, it was found that exposure of pomegranates to 1, 10 or 100 ppm of ethylene for up to 13 days at 20°C stimulated their respiration rate, the degree of stimulation being proportional to the ethylene concentration, although this increase in respiration rate was temporary and declined to near the levels of control fruits after 3 days of treatments (Kader *et al.*, 1984). This response occurred again when the fruits were exposed to a second ethylene treatment for 2 days after 7 days of storage. However, none of the ethylene treatments had a significant effect on skin colour, juice colour, TSS, pH or acids of the fruits. Therefore, these treatments confirm the non-climacteric ripening pattern of pomegranates, showing that they do not ripen once removed from the tree and that they should be picked when fully ripe to ensure the best eating quality for the consumer.

15.4.3 Changes in sugars and organic acids

Different results have been published with respect to the evolution of sugar content during storage of pomegranate fruits. Thus, reduction in sugar content has been found in 'Wonderful' cultivar, due to their utilization in respiration (Shaarawi and Nagy, 2017). However, some previous studies have reported increases in TSS contents in pomegranate arils throughout storage, showing that sugar content increased, which was attributed to moisture loss (Köksal, 1989; Ghafir *et al.*, 2010). On the other hand, with respect to organic acid content, a decrease of TA in pomegranate arils during storage has been reported (Mirdehghan *et al.*, 2007a; Selcuk and Erkan, 2015), which was attributed to the use of citric acid in the respiratory process of fruit during storage, although it has also been reported that the ability of fruit to synthesize acids decreases with postharvest maturity (Shaarawi and Nagy, 2017).

Fig. 15.3. Husk scald in 'Mollar de Elche' pomegranate manifested as skin browning developing from the stem end of the fruit. (Photo: Salvador Castillo.)

15.4.4 Physiological disorders and diseases

Pomegranate fruits suffer from different physiological disorders during postharvest storage, such as high CO_2 injury, husk scald and chilling injury as well as fungal diseases leading to important quality losses with high economic impact. Husk scald, as can be observed in Fig. 15.3, is manifested as skin browning, similar to superficial scald of apples, which generally develops from the stem end of the fruit spreading toward the blossom end as the severity increases (Defilippi *et al.*, 2006). This disorder is due to the oxidation of phenolic compounds on the fruit husk when stored at a temperature higher than 5°C (Pareek *et al.*, 2015). Pomegranate is also very sensitive to storage at temperatures below 5°C, since it develops a wide range of alterations known as chilling injury symptoms, manifested as surface pitting and skin browning (Fig. 15.4), as well as internal skin browning, which increase with storage time at chilling

Fig. 15.4. 'Mollar de Elche' pomegranates showing external chilling injury symptoms, manifested as surface pitting and skin browning. (Photo: Salvador Castillo.)

temperatures (Fig. 15.5). Pomegranates also have high susceptibility to decay, which increases with the duration of storage and may even reach the arils, depreciating the external and internal quality of the fruit (Mirdehghan *et al.*, 2006; Sayyari *et al.*, 2010; Valero and Serrano, 2010). Figure 15.6 and Fig. 15.7 show external decay caused by *Botrytis cinerea* and internal

Fig. 15.6. 'Mollar de Elche' pomegranate showing decay by *Botrytis cinerea* after 2 months of storage at 4°C. (Photo: Salvador Castillo.)

Fig. 15.5. 'Mollar de Elche' pomegranates showing internal skin chilling injury symptoms after storage for 2, 4, 6, 8 and 10 weeks at 4°C plus 3 days at 20°C. (Photo: Salvador Castillo.)

Fig. 15.7. 'Mollar de 'Elche' pomegranate showing decay in the arils caused by *Alternaria* sp. (Photo: Salvador Castillo.)

decay due to *Alternaria* sp in 'Mollar de Elche' pomegranates after long-term storage at 4°C.

15.4.5 Changes in bioactive compounds during storage

In addition, some evidence exists regarding the changes in bioactive compounds and antioxidant activity during cold storage of pomegranate fruits. Thus, losses of ascorbic acid have been generally reported in several pomegranate cultivars, such as 'Malas Saveh', 'Mollar de Elche' and 'Mridula' leading to diminution of their health-beneficial effects (Sayyari *et al.*, 2009, 2010, 2011b; Barman *et al.*, 2014). However, the reduction of ascorbic acid levels during the storage of whole fruit is lower than those found in ready-to-eat arils or juice (Mphahlele *et al.*, 2014). With respect to anthocyanins, increases have been found in 'Wonderful' and 'Mollar de Elche' among others, which has been correlated with the antioxidant capacity of the pomegranate arils and is attributed to the advancement of the postharvest ripening process (Holcroft *et al.*, 1998; Sayyari *et al.*, 2010; Valero *et al.*, 2015). Accordingly, higher levels of phenolic content were found in 'Hicaznar' pomegranates after 180 days of cold storage, as compared with those at harvest, although during this period some fluctuations were detected (Selcuk and Erkan, 2015). However, in 'Mollar de Elche' pomegranate arils, total phenolic content after 80 days of cold storage was lower than that at harvest (Sayyari *et al.*, 2010).

15.5 Postharvest Treatments to Maintain Fruit Quality

To preserve pomegranate quality and avoid losses in bioactive compounds with antioxidant properties as well as chilling injuries, several postharvest treatments have been carried out with satisfactory results, which are discussed below.

15.5.1 Polyamine treatments

Polyamines are polycationic molecules at physiological pH, which can bind strongly to phospholipids and other anionic components of the cell membranes as well as to cell wall components, leading to stabilization of both the bilayer surface and the cell wall structure (Valero *et al.*, 2002). Thus, it has been reported that postharvest treatments with polyamines have beneficial effects on maintaining fruit quality attributes during postharvest storage (Valero *et al.*, 2002; Serrano *et al.*, 2016). Specifically, in pomegranate, pre-storage treatments of 'Mollar de Elche' with 1 mM putrescine (Put) or spermidine (Spd), applied either by immersion or by pressure infiltration, led to maintenance of quality parameters and a significant reduction in chilling injury occurrence after storage at chilling temperatures. This was related to increases in Put and Spd concentrations in the skin, which were 3- and 2-fold higher, respectively, in treated than in control fruits (Mirdehghan *et al.*, 2007a). These results suggest that part of the exogenous Put can be used to be transformed into Spd, while an up-regulation of arginine decarboxylase (ADC), a key enzyme of one of the routes for Put biosynthesis, could account for the increased concentration of Put after Spd treatment.

Furthermore, Put and Spd treatments were also effective in maintaining higher concentrations of ascorbic acid, total phenolics, total anthocyanins and antioxidant capacity in arils of treated pomegranates than in controls (Mirdehghan *et al.*, 2007c). Accordingly, 2 mM Put treatment of 'Mridula', by dipping for 8 min, led to higher retention of anthocyanins, ascorbic acid, tannin and sensory qualities as compared with control fruits during 60 days of storage at 3 and 5°C, and, in turn, antioxidant capacity of pomegranate arils was significantly higher in treated than in control fruits (Barman *et al.*, 2014). CI was also reduced and other quality parameters maintained as a consequence of Put treatment, leading to enhancement of shelf-life (Barman *et al.*, 2011).

15.5.2 Heat treatment

Heat treatment as a postharvest tool to partially control green and blue moulds (*Penicillium digitatum* and *Penicillium italicum*) was used for the first time in the USA. Later, Ben-Yehoshua and his colleagues in Israel introduced the concept of 'curing' as the heat treatment of citrus fruits at 36°C for 3 days before cold storage in order to reduce decay (Ben-Yehoshua and Porat, 2005). Nowadays, the use of heat treatments is actually considered as an environmentally friendly method of decay control, when it is used either alone or in combination with other methods. Moreover, this treatment has also been shown to have a beneficial effect on delaying the evolution of postharvest fruit ripening and, hence, preserving fruit quality and increasing storage time, with interesting possibilities for commercial application in the horticultural industry (Biswas *et al.*, 2016).

In addition, heat treatments, applied as hot water dip, hot forced air or vapour heat, at proper temperatures and duration, have been shown to alleviate chilling injuries of a wide range of fruits (such as avocados, citrus fruits, cucumbers, mangos, sweet peppers, persimmons, tomatoes and zucchini squash among others) including pomegranate (Valero and Serrano, 2010; Aghdam and Bodbodak, 2013; Lurie, 2016). Chilling mitigation in heat-treated fruits and vegetables has been attributed to the enhancement of membrane integrity by increasing the unsaturated/saturated fatty acid ratio, heat shock proteins gene expression and accumulation, antioxidant system activity and sugar metabolism (Aghdam and Bodbodak, 2014). Moreover, these effects in many of the studied commodities have been related to increases in polyamine concentration. For instance, heat treatment of pomegranate fruit by hot water dip at 45°C for 4 min before storage at chilling temperatures led to a decrease in chilling injury symptoms, which were manifested in control fruit as skin browning and electrolyte leakage, the severity of damages being related to softening and loss of fatty acids, with a concomitant reduction in the ratio of unsaturated/saturated fatty acids during storage. These chilling injury symptoms were reduced in heat-treated pomegranates and were related to an increase in Put and Spd concentrations in the skin during cold storage (Mirdehghan *et al.*, 2007b). In addition, arils from heat-treated pomegranates exhibited higher total antioxidant activity than controls, which was correlated with the high levels of total phenolics and to a lesser extent with ascorbic acid and anthocyanin contents. Moreover, concentration of sugars (glucose and fructose), organic acids (malic, citric and oxalic acids), total phenolics, ascorbic acid and anthocyanin were higher in arils of treated fruits than in controls (Mirdehghan *et al.*, 2006). This shows that this simple and non-contaminant technology increased the functional and nutritive properties of pomegranates.

Intermittent warming at 20°C every 6 days of storage at 2 or 5°C has also been tested with satisfactory results in maintaining pomegranate quality during storage, in terms of retention of anthocyanin and TA, reduction of decay and alleviation of chilling injuries (Artés *et al.*, 2000a; Nanda *et al.*, 2001). On the other hand, a preconditioning treatment, at moderate temperatures (30–40°C) and high RH (90–95%) for a short period of time (1–4 days), a technique also known as curing, applied before cold storage, reduced considerably the pitting and husk superficial scald (produced by the enzyme polyphenol oxidase) compared with control fruits, the effects being more evident when the conservation was made at 2°C than at 5°C, and particularly after an additional period of 1 week at 15–20°C and 70–75% RH, to simulate the retail sale period (Pareek *et al.*, 2015).

However, it is worth noting that the use of inappropriate heat treatments causes both external and internal fruit damage, and finding the optimum treatment temperature for each commodity is absolutely necessary (Valero and Serrano, 2010; Aghdam and Bodbodak, 2014).

15.5.3 Treatments with salicylates and jasmonates

Salicylic acid (SA), acetyl salicylic acid (ASA), methyl salicylate (MeSa), jasmonic acid (JA) and methyl jasmonate (MeJa) are natural hormonal compounds with important roles in a wide range of physiological processes, such as inducing systemic acquired resistance, modulation of opening and closing of stomata aperture, flowering, seedling germination and providing plant tolerance to different kinds of stress (Creelman and Mullet, 1997; Kumar, 2014). These natural compounds have been recently reported as easy treatments to apply in order to alleviate chilling injuries in fruits and vegetables (Asghari and Aghdam, 2010; Aghdam and Bodbodak, 2013; Glowacz and Rees, 2016), with additional effects on improving or maintaining other quality properties, such as appearance, texture, nutritional and antioxidant compounds, during fruit storage (Zapata *et al.*, 2014).

In this sense, SA treatments on pomegranate fruit (at 0.7, 1.4 or 2 mM), applied by dipping for 10 min, were highly effective in delaying vitamin C losses in a concentration-dependent manner in the sweet-sour pomegranate 'Malas Saveh', while losses of 25% occurred in control fruits, with additional effects on decreasing chilling injuries (Sayyari *et al.*, 2009). Similar effects on maintaining ascorbic acid concentration during pomegranate storage were obtained in the sweet pomegranate 'Mollar de Elche' after ASA or SA treatments (Sayyari *et al.*, 2011b). Accordingly, a 2 mM SA dipping treatment (for 10 min) of 'Mollar de Elche' pomegranate before storage at 2°C or 10°C for 90 days plus 3 days at 20°C was effective in maintaining higher antioxidant activity (both in hydrophilic and lipophilic fractions) in the arils during storage at both temperatures (Sayyari *et al.*, 2017). In addition, the chilling injury index, which was manifested when pomegranates were stored at 2°C, was reduced in SA-treated pomegranates as well as ion leakage and respiration rate. These beneficial effects were related to the ability of the SA treatment to reduce the loss of fatty acids and the concomitant reduction in the ratio of unsaturated/saturated fatty acids occurring in chilling injury-damaged pomegranates. Similar results for SA dipping treatments were found in 'Mallas Saveh' (Sayyari *et al.*, 2016) and 'Taify' (Awad *et al.*, 2013) cultivars.

Accordingly, MeJa or MeSa treatments, at 0.01 or 0.1 mM, before storage under chilling temperature reduced ascorbic acid losses during storage. In addition, these treatments significantly increased total phenolics, total anthocyanins and antioxidant capacity in the arils compared with controls, leading to improvements in the health beneficial effects of pomegranate fruit consumption (Sayyari *et al.*, 2011a). Moreover, other parameters related to fruit quality, such as fruit firmness, TSS and TA, were also maintained in MeJa- and MeSa-treated pomegranates, while significant losses occurred in control pomegranates (Sayyari *et al.*, 2011a). However, further studies are necessary to understand the mechanism of action by which MeJA and MeSA enhance the phytochemicals in pomegranate.

15.5.4 Oxalic acid treatment

Oxalic acid (OA) is a final metabolite product in plants and has many physiological functions, the most important ones being related to the induction of systemic resistance against diseases caused by fungi, bacteria and viruses, by increasing defence-related enzyme activities and biosynthesis of secondary metabolites, such as phenolics (Zheng *et al.*, 2012). In addition, it has been reported that OA treatments have beneficial effects on fruit quality properties. Particularly in 'Mollar de Elche' pomegranates, OA dipping treatments at 2, 4 and 6 mM led to lower losses of total phenolics, and significant increases in both ascorbic acid content and antioxidant activity, also having important effects on reducing chilling injury symptoms (Sayyari *et al.*, 2010). Accordingly, a dipping treatment in OA at 3, 5 and 7 mM for 10 min reduced chilling injuries during cold storage at 2°C in 'Taify' pomegranates in a dose-dependent manner (Awad *et al.*, 2013). In addition, OA at 7 mM maintained membrane stability index, although no significant effects were observed on fruit firmness, TSS, TA or vitamin C concentration.

On the other hand, OA treatments induced maintenance of TAA derived from lipophilic compounds (L-TAA), such as carotenoids and tocopherols, after 84 days of cold storage, while decreases were observed in control arils (Sayyari *et al.*, 2010). The mechanism by which OA retains L-TAA is still unclear, but there is evidence supporting the fact that OA is a natural antioxidant compound reducing lipid peroxidation *in vitro* (Kayashima and Katayama, 2002), which could account for avoiding lipophilic antioxidant deterioration.

15.5.5 Storage under modified atmosphere packaging (MAP) and controlled atmosphere (CA) conditions

MAP consists of packaging a certain amount of fruit or vegetables inside plastic films with selective permeability to CO_2, O_2 and water vapour diffusion. The respiration rate of the commodities inside the package increases CO_2 partial pressure and decreases O_2 partial pressure inside the package, which will diffuse through the film surface according to Fick's law. Thus, a steady state will be reached when O_2 uptake level by the product is equal to that permeating into the package, and CO_2 production by the product equals CO_2 escape from the package; meanwhile transpiration rate increases water pressure inside the package (Rai and Paul, 2007; Valero and Serrano, 2010). These atmosphere modifications lead to a delay in the postharvest ripening and senescence processes, and to the maintenance of fruit and vegetable quality for longer periods. In addition, there are a few reports showing a beneficial effect of MAP storage on the content of bioactive compounds with antioxidant activity in fruit and vegetables (Artés *et al.*, 2006; Valero and Serrano, 2010; Díaz-Mula *et al.*, 2011; Serrano *et al.*, 2011).

Specifically, in pomegranates, it has been demonstrated that storage in MAP reduces water loss, visible shrivelling symptoms, husk scald and decay, and extends storage life as compared with control fruits under control air conditions in 'Mollar de Elche' (Artés *et al.*, 2000a), 'Ganesh' (Nanda *et al.*, 2001), 'Primosole' (D'Aquino *et al.*, 2010), 'Wonderful' (Porat *et al.*, 2016), 'Hicranar' and 'Hicaznar' (Selcuk and Erkan, 2015) pomegranates. In addition, beneficial effects of MAP were reported in terms of maintaining quality properties of the arils, such as TA and colour. However, the best steady-state atmosphere composition to preserve pomegranate quality depends on the cultivar. For instance, 13.5–17.60 kPa O_2 and 4.40–8.1 kPa CO_2 were suggested to minimize weight loss and decay, and maintain overall visual quality of 'Hicranar' and 'Hicaznar' pomegranate cultivars for 120 and 210 days of cold storage 6°C, respectively (Selcuk and Erkan, 2015). For pomegranate cv. 'Mollar de Elche' stored in unperforated polypropylene MAP bags the best results after 12 weeks at 2°C were obtained with 8 kPa of O_2 and 10 kPa of CO_2 (Artes *et al.*, 2000b). Accordingly, Sudhakar Rao and Shivashankara (2018) have reported that pomegranate fruits cv. 'Bhagwa' can be stored safely for 3 weeks at ambient temperature and 3 months at 8°C under MAP storage with 3–8 KPa O_2 and 7–11 KPa CO_2 in which conditions lower weight loss and maintenance of organoleptic, nutritional and functional quality occurred.

In addition, it has been reported in 'Wonderful' pomegranate that fruit can be directly harvested into large Xtend® MAP bags (80 kg or even 320 kg in bulk) and then immediately transferred from the field to commercial cold storage rooms. Under these conditions, pomegranate fruits maintain excellent quality with minimal losses for at least 12 weeks after harvest (Porat *et al.*, 2016). This new procedure saves time as well as extra packaging materials and labour costs, and better maintains the cold storage chain, since it allows direct transfer of large volumes of fruit from the field to cold storage rooms without intermediate transferring to the packing house. Accordingly, MAP by using Xtend™ film delayed chilling injury in 'Wonderful' pomegranates, which was attributed to the low O_2 concentration in the packages, and reduced fungi growth due to the higher water vapour transmission and a better potential to reduce humidity in the package made from Xtend™ films compared with that made of polyethylene (Valdenegro *et al.*, 2018). Most of the research papers regarding the use of MAP to preserve pomegranate quality are focused on using pomegranate arils and not whole fruits, which are not considered in this chapter.

On the other hand, CA is a storage technique based on the use of cold storage chambers with low O_2 and/or high CO_2 atmospheres to preserve quality and freshness of fruit and vegetables, avoiding the use of any chemicals. Under these conditions, similar effects as those for MAP conditions can be achieved, such as reduced respiration and ethylene production rates, as well as reduced losses of acids, starch to sugar conversion, sugar inter-conversions and biosynthesis of flavour volatiles. Usually, the level of O_2 is reduced below 8% and CO_2 is increased above 1%, although the optimum atmospheric composition is dependent on the commodity (Singh and Goswami, 2006; Shahbaz *et al.*, 2014; Nunes, 2008; Valero and Serrano, 2010; Zhang and McCarthy, 2013).

Several studies have shown that storage of pomegranates in different CA conditions significantly extended their postharvest life, not only by delaying fruit senescence but also by inhibiting fruit decay. Thus, Küpper *et al.* (1995) found that 'Hicaz' pomegranate could be successfully stored for 6 months at 6°C in CA containing 6 kPa CO_2 and 3 kPa O_2. In addition, CA storage in an atmosphere containing 5 kPa O_2 and 0 or 5 kPa CO_2 reduced weight loss, decay development and husk scald in 'Mollar de Elche' pomegranates stored at 5°C for 2 months (Artes *et al.*, 1996). Similarly, Hess-Pierce and Kader (2003) suggested storage at 7.5°C in CA with 5 kPa O_2 + 15 kPa CO_2 was the optimum condition to maintain the initial quality of 'Wonderful' pomegranates for up to 20 weeks. In this cultivar, it was also reported that moderate CO_2 atmospheres (10 kPa) prolonged the storage life and maintained the quality of pomegranates, including adequate red colour intensity of the arils (Holcroft *et al.*, 1998).

In addition, during cold storage at 7°C of three pomegranate cultivars ('PG100-1', 'EVE' and 'PG116-17'), a CA with 2 kPa O_2 + 5 kPa CO_2 has been suggested as a better storage regime than regular cold air in view of apparent fruit quality, since husk scald and decay development were decreased under CA storage compared with control fruit. However, storage in normal air was more beneficial for maintaining the anthocyanin level and preventing off-flavour development in aril juice (Matityahu *et al.*, 2016). Thus, it is evident from the published work that pomegranate is highly sensitive to low-oxygen atmospheres (<5 kPa) and, in turn, the beneficial effect of CA can be achieved only if the optimal atmosphere conditions are worked out for each cultivar.

15.5.6 Short-term CO_2 treatment

Short-term high CO_2 treatment before storage has been proved to be useful to maintain the quality of various fruits such as blueberry (Jiang *et al.*, 2011), citrus (Montesinos-Herrero *et al.*, 2012) and persimmon (Besada *et al.*, 2015). Specifically, in 'Mollar de Elche' pomegranate, the growth of artificially inoculated *B. cinerea* was significantly inhibited by exposure to high CO_2 (15, 50 or 95 kPa) for 48 h at 20°C, in a concentration-dependent manner (Palou *et al.*, 2016). Accordingly, short-term high CO_2 treatment (CO_2 85% for 12 h at 18°C), of 'Shishe-Kab' pomegranate fruits delayed decay, chilling injury and weight loss during storage at 5°C, leading to a two-fold reduction in unmarketable fruits after 12 weeks of storage (Moradinezhad *et al.*, 2018). In addition, no significant defect attributed to treatment was observed in organoleptic aspects of pomegranate arils. Thus, high CO_2 treatment before cold storage could be a useful tool to control postharvest decay and maintain the postharvest quality of the pomegranate.

15.5.7 1-Methylcyclopropene (MCP) treatments

1-MCP is an ethylene action inhibitor that is used to delay the ripening process and to extend the storage- as well as the shelf-life of fruits and vegetables (Watkins, 2006; Valero *et al.*, 2016). Nowadays 1-MCP is commercially used in cut flowers and fruits such as apples, bananas, melons, plums and tomatoes, among others, and it is registered under the trade name of Smartfresh®, EthylBloc®, SmartFresh™, SmartTabs™ and EthylBloc™ by Agrofresh Inc., a subsidiary of Rohm & Haas, Spring House, Pennsylvania. 1-MCP treatment is also effective in reducing chilling injury symptoms in a wide range of fruits, including pomegranate (Valero *et al.*, 2016). For instance, it has been reported that 1-MCP reduced shrivelling in Iranian 'Malas Saveh' pomegranate fruits after 45 days of storage at 13°C followed by 7 days at 20°C, as well as superficial scald incidence and peel browning in 'Wonderful' and 'Dahongpao' pomegranates, with additional effects on maintaining the internal quality (Defilippi *et al.*, 2006; Gamrasni *et al.*, 2015; Li *et al.*, 2016).

15.5.8 Edible coatings

Chitosan is a high-molecular-weight cationic polysaccharide, obtained by the deacetylation of chitin and has been shown to be an ideal edible coating because of its selective permeability to O_2 and CO_2, good mechanical properties and antimicrobial effects against several pathogens. In addition, it is considered as safe for the consumer and environmentally friendly and, in turn, as a suitable alternative treatment to replace the use of synthetic fungicides (Romanazzi *et al.*, 2017). Thus, it has been reported that coatings based on chitosan are effective in controlling fungal decay (which was mostly caused by *B. cinerea*, *Penicillium* sp., *Alternaria* sp. and *Aspergillus* sp.), reducing weight loss and respiration rate, and maintaining postharvest quality in several fruits, including pomegranates (Varasteh *et al.*, 2012, Varasteh *et al.*, 2017; Meighani *et al.*, 2015; Romanazzi *et al.*, 2017).

Chitosan coating (1%-dipping for 1 min) was effective in controlling fungal decay during cold storage of 'Hicaznar' pomegranate fruit and its effect continued during the shelf-life period (Candir *et al.*, 2018). In addition, the arils of chitosan-coated fruit were deep red and had higher antioxidant activity, total monomeric anthocyanin and total phenolic content compared with control fruits. Munhuweyi *et al.* (2016) reported that chitosan coating was a safe antimicrobial alternative to control *Botrytis* sp., *Penicillium* sp. and *Pilidiella granati* on 'Herskawitz' and 'Wonderful' pomegranates during postharvest storage. These antifungal effects were most efficient when chitosan was applied as a preventative treatment regardless of cultivar, and could be attributed to both its fungicidal activity and its ability to form a film on the fruit surface, acting as a mechanical barrier to protect the fruit from pathogen infection (Bautista-Baños *et al.*, 2006; Ghasemnezhad *et al.*, 2013).

The application of a 1–2% chitosan coating prevented colour deterioration in the arils and delayed changes in anthocyanin and total phenolic contents of 'Rabbab-eNeyriz' (Varasteh *et al.*, 2012) and 'Malase Torshe Saveh' (Meighani *et al.*, 2015) pomegranates during cold storage. Accordingly, a 0.5% chitosan treatment of 'Mallas Saveh' significantly reduced chilling injury during storage, and induced higher increases in total phenolic and anthocyanin concentrations as well as antioxidant activity during storage compared with control fruits (Sayyari *et al.*, 2016).

15.5.9 Gamma irradiation

Gamma irradiation is a quarantine procedure against different pests, which was approved by the US Department of Agriculture in February 2012 in imported fresh pomegranates, providing that the fruits have a minimum absorbed dose of 0.4 kGy. Shahbaz *et al.* (2014) evaluated the effects of the application of different gamma-irradiation doses (0.4, 1 and 2 kGy) on the chemical and sensory characteristics of pomegranate, showing that TSS, TA and pH values remained unaffected up to 1 kGy treatment. However, irradiation caused a significant decrease in the total anthocyanin and phenolic content, although a stronger preference was shown by sensory panellists for the juice from irradiated fruits.

15.5.10 Melatonin

Recently, exogenous melatonin treatments have been reported to serve as a beneficial strategy to confer chilling and fungal decay tolerance, to delay senescence, and to preserve sensory and nutritional quality in fruit and vegetables, all of which have great economic benefits for the horticultural industry (Gao *et al.*, 2016; Aghdam and Fard, 2017). Melatonin treatment at 100 μM reduced chilling injury in 'Malas Saveh' pomegranate fruit manifested by lower husk browning, electrolyte leakage and malondialdehyde (MDA) accumulation, showing an effect of melatonin treatment on maintaining membrane integrity during storage. This higher membrane integrity may arise from lower H_2O_2 accumulation owing to higher activity of the antioxidant enzymes catalase (CAT), superoxide dismutase (SOD), ascorbate peroxidase (APX) and glutathione reductase (GR), as well as to lower activity of the membrane degrading enzymes, such as phospholipase D (PLD) and lipoxygenase (LOX). In addition, higher accumulation of phenolic compounds has been reported in arils from melatonin-treated fruits as a consequence of higher phenylalanine ammonialyase (PAL) enzyme activity concomitant with lower polyphenol oxidase (PPO) enzyme activity (Jannatizadeh, 2019).

15.5.11 Propolis and black seed oil

Propolis extract and black seed oil have been recently investigated for their positive influence on fruit postharvest life. Thus, in 'Wonderful' pomegranate, dipping treatment in propolis extract (0.01 and 0.1%) or in black seed oil (0.1%) significantly affected the maintenance of fruit weight, juice content and visual quality during storage at 6.5 ± 1°C and 90 ± 95% for 150 days (Kahramanoğlu *et al.*, 2018). Both treatments reduced grey mould development in a similar way to the commercial fungicide fludioxonil and delayed the ascorbic acid losses that usually occur during pomegranate storage.

15.5.12 Arginine treatments

Arginine is a metabolically versatile amino acid since apart from being a structural component of proteins, it is also involved in biosynthesizing signalling molecules, such as PAs, nitric oxide (NO), III-aminobutyric acid (GABA) and proline, leading to enhanced tolerance of horticultural commodities to chilling stress (Aghdam and Bodbodak, 2013). Specifically, in pomegranate, it has been reported that preharvest treatments with arginine at 0.5, 1 and 2 mM applied three times at 20-day intervals before commercial harvest combined with postharvest dipping treatment with arginine at 0.5, 1 and 2 mM for 15 min at 20°C decreased chilling injury symptoms (manifested by external husk browning), although the highest effect was obtained at 1 mM concentration. This 1 mM combined treatment also reduced H_2O_2 and MDA accumulation (partially due to higher activity of the antioxidant enzymes SOD, CAT and APX) and electrolyte leakage in skin and arils, showing higher membrane integrity in arginine-treated fruits compared with the control (Babalar *et al.*, 2018). In addition, arginine treatments led to arils with higher total phenolics, anthocyanins and ascorbic acid levels, showing that arginine treatment could be a promising technology not only for reducing chilling injury but also for maintaining nutraceutical properties of pomegranate fruit by promoting antioxidant system activity. Nevertheless, it is necessary to point out that they combined pre- and postharvest arginine treatments, so it is not possible to know the effects of pre- or postharvest treatment separately.

Taking into account that arginine is a precursor of polyamines, the reduced chilling injury in arginine-treated pomegranates could be due to higher endogenous polyamine accumulation. However, this issue deserves further research.

15.5.13 Combined treatments

Recently, it has been reported that combined treatments have higher effects than single ones on maintaining pomegranate quality properties during long storage periods. For example, dipping treatments of 'Hicaznar' pomegranates with SA, OA or Put at 6, 2 and 2 mM concentrations, respectively, for 10 min before storage in CA (5% O_2 and 15% CO_2 concentrations) for 6 months led to better results in terms of reducing weight loss and maintaining ascorbic acid, TA, total phenolic contents and antioxidant activity at higher levels compared with the control (Koyuncu *et al.*, 2018). The most effective combination in reducing weight loss was Put plus CA, with this effect being attributed to the strong delaying effect of Put on cell integrity and the senescence process of fruit. CA and Put combined treatment was also the best for reducing TA, glucose, fructose and ascorbic acid losses, skin colour changes and maintaining higher levels of phenolics and antioxidant activity. Thus, the combination of postharvest Put treatments with CA storage can be a promising tool to delay quality loss and maintain or enhance some bioactive compounds and antioxidant activity of pomegranate during cold storage.

On the other hand, combined treatment of 'Mridula' pomegranates with 2 mM Put at 25°C for 8 min followed by dipping in carnauba wax emulsion (1:10) at 40°C for 2 min was more effective than Put treatment alone in maintaining external and internal pomegranate quality properties during cold storage. Moreover, after 60 days of storage, Put + carnauba wax-treated fruits retained about 25% higher antioxidant activity than control fruits, both at 3 and 5°C storage temperatures, mainly due to their higher anthocyanin and ascorbic acid concentrations (Barman *et al.*, 2014). Similarly, the combination of chitosan with SA has been proved to be a more effective treatment in maintaining quality properties and bioactive compounds during storage of pomegranates than chitosan or SA applied independently (Sayyari *et al.*, 2016). In 'Wonderful' pomegranate, treatment with *Aloe vera* extract at 17% followed by 0.5 g/l SA treatment has been recommended to maintain quality of fruits during cold storage at 5°C for 60–90 days, in comparison with each of them applied separately (Kamel *et al.*, 2016). As mentioned in the previous section, chitosan has also been proved to be effective in maintaining pomegranate quality. However, better results in terms of controlling husk scald, decay and weight loss, and maintaining visual quality and initial red aril colour intensity for 6 months of cold storage plus shelf-life were obtained when chitosan treatment was combined with MAP storage (Candir *et al.*, 2018).

On the other hand, it has been reported that combined treatment of hot water (50°C) for 3 min, SA (1 or 2 mM) or calcium chloride (1 or 2%) for 3 min at 20°C and storage under MAP conditions had a greater effect on controlling fruit decay, reducing chilling injury and extending 'Shishe-Kab' pomegranate shelf-life than individual applications of each treatment. Shelf-life roughly doubled in fruit treated with a combination of hot water and SA (1 mM) and stored in MAP compared with the control fruit with no packaging, although treatments had no significant effect on TSS or sensory assessments of arils (Moradinezhad *et al.*, 2013). Accordingly, a combination of CA storage (5 kPa O_2 + 15 kPa CO_2) with pre-storage antifungal treatment, such as dipping in 3% potassium sorbate for 3 min at 21°C, has been found to be very effective in controlling pomegranate decay and maintaining internal and external fruit quality during 15 weeks of storage at 7.2°C (Palou *et al.*, 2007).

The combination of 1-MCP treatment and MAP has been reported to be more effective than individual treatment in delaying the appearance of chilling injury symptoms (Valdenegro *et al.*, 2018). Accordingly, the effects of 0.5% black seed oil and 0.1% propolis treatments were increased when they were combined with MAP, since they protected the quality and marketability of pomegranate fruits for 150 days at 6.5 ± 1°C and 90–95% RH, by controlling grey mould development (as an alternative to fungicide application) and slowing the occurrence of chilling injury (Kahramanoğlu *et al.*, 2018). Finally, the combined application of dipping in nitric oxide (300 μM) for 2 min and further cellophane wrapping led to significantly increased antioxidant activity and total anthocyanin content, and reduced chilling injury symptoms and electrolyte leakage during cold storage compared with nitric oxide or wrapping individual treatments (Ranjbari *et al.*, 2018).

15.6 Effects of Preharvest Treatments on Pomegranate Fruit Quality

The effects of preharvest treatments on pomegranate quality at harvest and during postharvest storage have been studied in a limited number of papers. Thus, 1 and 2 mM MeJa spray treatment of 'Malas' pomegranate trees 15 days before harvest has been recently reported to improve aril colour at harvest and reduce the postharvest chilling injury index, manifested as a reduction in electrolyte leakage. In addition, MeJa treatment significantly increased flavonoids, total antioxidant activity, total phenolics and total anthocyanins, and delayed TA and vitamin C losses in pomegranate arils in comparison with untreated controls (Koushesh Saba and Zarei, 2019).

15.7 Ready-to-Eat Pomegranate Arils

The hard skin of pomegranate fruits often makes it unpopular as a table fruit due to the difficulty in taking out the arils manually, which limits its market possibilities. As a result, minimally processed or ready to-eat pomegranate arils have become popular due to their attractive aspect, ease of consumption and desirable sensory characteristics. However, minimally processed arils deteriorate quickly, mainly due to changes in texture, colour, microbial spoilage and release of undesirable aroma compounds, leading to overall quality reduction in a short period of time. Thus, maintaining the nutritional and microbial quality of pomegranate arils is a major challenge that can be achieved by proper temperature management (as described in Section 15.3) and using MAP and CA technologies (Caleb *et al.*, 2012b). For instance, quality and convenience of minimally processed 'Hicaznar' pomegranate arils packed in polypropylene (PP) trays sealed with bioriented polypropylene (BOPP) film were maintained for 18 days of storage at 5°C, with aerobic mesophilic bacteria in the range of 2.30–4.51 log CFU/g (Ayhan and Eştürk, 2009). Accordingly, the use of perforated polyethylene films led to maintenance of sensory and microbial quality parameters in 'Acco' pomegranate arils as well as their cascorbic acid and anthocyanin concentration levels (Opara *et al.*, 2017). In a recent study, by using arils from 19 Iranian pomegranate cultivars packaged in rigid polyethylene boxes (15 × 10 × 5 cm) with air-tight screw caps stored at 5 °C, Ghasemi Soloklui *et al.* (2019) have shown that the onset of decay and shelf-life varied among cultivars, ranging from 7–21 days. In general, cultivars exhibiting a very slow process of decay contained the highest content of TA, gallic acid equivalent and TSS content.

However, the shelf-life of ready-to-eat pomegranate arils can be extended by using combined treatments. Thus, Martínez-Romero *et al.* (2013) performed several treatments (water as control, ascorbic + citric acids (at 0.5 or 1%), *A. vera* gel (at 50 or 100%), 50% *A. vera* gel + 0.5% ascorbic and 0.5% citric acid, and 100% *A. vera* gel + 1% ascorbic and 1% citric acid) on 'Mollar de Elche' pomegranate arils prior to storage in rigid PP boxes for 12 days at 3°C. Results showed that *A. vera* (alone or in combination with acids) led to firmness retention, increased levels of total anthocyanins and total phenolics, and reduction of counts for both mesophilic aerobics and yeast and moulds. In addition, sensory analysis scores for flavour, texture, aroma, colour and purchase decision were higher in arils treated with *A. vera*, especially in those combined with 1% ascorbic and citric acids, which had no off-flavours as a consequence of *A. vera* gel treatment.

15.8 Conclusions and Future Trends

In this chapter, it has been pointed out that several postharvest technologies, such as PAs, heat treatments, SA, ASA, OA, and vapour treatments with MeJA and MeSA, are effective tools to alleviate physiological disorders and maintain fruit quality during cold storage. Interestingly, the application of these postharvest treatments, considered as natural compounds and to be environmentally friendly, had an additional benefit in terms of enhanced content of bioactive compounds and antioxidant activity. In addition, ready-to-eat pomegranate arils is an interesting method to increase pomegranate consumption due to convenience and aril quality attributes. Further research should be focused on new preharvest treatments with an impact on pomegranate quality properties during postharvest storage.

Acknowledgements

This work has been funded by the Spanish Ministry of Economy and Competitiveness through Projects AGL2015-63986R and RTI2018-099664-B-I00 and the European Commission with FEDER funds. The authors thank Ciruelo SA for granting permission to use their plots and for the technical support received, and University Miguel Hernández for the PhD scholarship of M.E. García-Pastor.

References

Aghdam, M.S. and Bodbodak, S. (2013) Physiological and biochemical mechanisms regulating chilling tolerance in fruits and vegetables under postharvest salicylates and jasmonates treatments. *Scientia Horticulturae* 156, 73–85. DOI: 10.1016/j.scienta.2013.03.028.

Aghdam, M.S. and Bodbodak, S. (2014) Postharvest heat treatment for mitigation of chilling injury in fruits and vegetables. *Food and Bioprocess Technology* 7(1), 37–53. DOI: 10.1007/s11947-013-1207-4.

Aghdam, M.S. and Fard, J.R. (2017) Melatonin treatment attenuates postharvest decay and maintains nutritional quality of strawberry fruits (*Fragaria×anannasa* cv. Selva) by enhancing GABA shunt activity. *Food Chemistry* 221, 1650–1657. DOI: 10.1016/j.foodchem.2016.10.123.

Al-Maiman, S.A. and Ahmad, D. (2002) Changes in physical and chemical properties during pomegranate (*Punica granatum* L.) fruit maturation. *Food Chemistry* 76(4), 437–441. DOI: 10.1016/S0308-8146(01)00301-6.

Al-Mughrabi, M.A., Bacha, M.A. and Abdelrahman, A.O. (1995) Effects of storage temperature and duration on fruit quality of three pomegranate cultivars. *Journal of King Saud University* 7, 239–248.

Alighourchi, H., Barzegar, M. and Abbasi, S. (2008) Anthocyanins characterization of 15 Iranian pomegranate (*Punica granatum* L.) varieties and their variation after cold storage and pasteurization. *European Food Research and Technology* 227(3), 881–887. DOI: 10.1007/s00217-007-0799-1.

Artes, F., Marin, J.G. and Martinez, J.A. (1996) Controlled atmosphere storage of pomegranate. *European Food Research and Technology* 203(1), 33–37.

Artes, F., Villaescusa, R. and Tudela, J.A. (2000b) Modified atmosphere packaging of pomegranate. *Journal of Food Science* 65(7), 1112–1116. DOI: 10.1111/j.1365-2621.2000.tb10248.x.

Artés, F., Tudela, J.A. and Villaescusa, R. (2000a) Thermal postharvest treatments for improving pomegranate quality and shelf life. *Postharvest Biology and Technology* 18(3), 245–251. DOI: 10.1016/S0925-5214(00)00066-1.

Artés, F., Gómez, P.A. and Artés-Hernández, F. (2006) Modified atmosphere packaging of fruits and vegetables. *Stewart Postharvest Review* 5, 2.

Asghari, M. and Aghdam, M.S. (2010) Impact of salicylic acid on post-harvest physiology of horticultural crops. *Trends in Food Science & Technology* 21(10), 502–509. DOI: 10.1016/j.tifs.2010.07.009.

Aviram, M., Volkova, N., Coleman, R., Dreher, M., Reddy, M.K. *et al*. (2008) Pomegranate phenolics from the peels, arils, and flowers are antiatherogenic: studies in vivo in atherosclerotic apolipoprotein E-deficient (E ^{0}) mice and *in* vitro in cultured macrophages and lipoproteins. *Journal of Agricultural and Food Chemistry* 56(3), 1148–1157. DOI: 10.1021/jf071811q.

Awad, M.A., Al-Qurashi, A.D. and Elsayed, M.I. (2013) Effect of pre-storage salicylic acid and oxalic acid dipping on chilling injury and quality of 'Taify' pomegranates during cold storage. *Journal of Food, Agriculture and Environment* 11(2), 117–122.

Ayhan, Z. and Eştürk, O. (2009) Overall quality and shelf life of minimally processed and modified atmosphere packaged 'ready-to-eat' pomegranate arils. *Journal of Food Science* 74(5), C399–C405. DOI: 10.1111/j.1750-3841.2009.01184.x.

Babalar, M., Pirzad, F., Sarcheshmeh, M.A.A., Talaei, A. and Lessani, H. (2018) Arginine treatment attenuates chilling injury of pomegranate fruit during cold storage by enhancing antioxidant system activity. *Postharvest Biology and Technology* 137, 31–37. DOI: 10.1016/j.postharvbio.2017.11.012.

Barclays Official California Code of Regulations (2019) Title 3. food and agriculture division 3. economics, chapter 1. fruit and vegetable standardization. Subchapter 4. fresh fruits, nuts and vegetables. article 39. § 1464. pomegranates, standards. Available at: https://govt.westlaw.com/calregs/Document/I2E983B40D45911DEB97CF67CD0B99467?originationContext=document&transitionType=StatuteNavigator&needToInjectTerms=False&viewType=FullText&t_querytext=pomegranate&contextData=%28sc.Default%29 (accessed 10 June 2020).

Barman, K., Asrey, R. and Pal, R.K. (2011) Putrescine and carnauba wax pretreatments alleviate chilling injury, enhance shelf life and preserve pomegranate fruit quality during cold storage. *Scientia Horticulturae* 130(4), 795–800. DOI: 10.1016/j.scienta.2011.09.005.

Barman, K., Asrey, R., Pal, R.K., Kaur, C. and Jha, S.K. (2014) Influence of putrescine and carnauba wax on functional and sensory quality of pomegranate (*Punica granatum* L.) fruits during storage. *Journal of Food Science and Technology* 51(1), 111–117. DOI: 10.1007/s13197-011-0483-0.

Bautista-Baños, S., Hernández-Lauzardo, A.N., Velázquez-del Valle, M.G., Hernández-López, M., Ait Barka, E. *et al.* (2006) Chitosan as a potential natural compound to control pre and postharvest diseases of horticultural commodities. *Crop Protection* 25(2), 108–118. DOI: 10.1016/j.cropro.2005.03.010.

Ben-Arie, R., Segal, N. and Guelfat-Reich, S. (1984) The maturation and ripening of the 'Wonderful' pomegranate. *Journal of the American Society for Horticultural Science* 109, 898–902.

Ben-Yehoshua, S. and Porat, R. (2005) Heat treatments to reduce decay. In: Ben-Yehoshua, S. (ed.) *Environmentally Friendly Technologies for Agricultural Produce Quality*. Taylor & Francis, Boca Raton, Florida, pp. 11–42.

Besada, C., Llorca, E., Novillo, P., Hernando, I. and Salvador, A. (2015) Short-term high CO_2 treatment alleviates chilling injury of persimmon cv. Fuyu by preserving the parenchyma structure. *Food Control* 51, 163–170. DOI: 10.1016/j.foodcont.2014.11.013.

Biswas, P., East, A.R., Hewett, E.W. and Heyes, J.A. (2016) Intermittent warming in alleviating chilling injury—a potential technique with commercial constraint. *Food and Bioprocess Technology* 9(1), 1–15. DOI: 10.1007/s11947-015-1588-7.

Caleb, O.J., Mahajan, P.V., Opara, U.L. and Witthuhn, C.R. (2012a) Modelling the respiration rates of pomegranate fruit and arils. *Postharvest Biology and Technology* 64(1), 49–54. DOI: 10.1016/j.postharvbio.2011.09.013.

Caleb, O.J., Opara, U.L. and Witthuhn, C.R. (2012b) Modified atmosphere packaging of pomegranate fruit and arils: a review. *Food and Bioprocess Technology* 5(1), 15–30. DOI: 10.1007/s11947-011-0525-7.

Calín-Sánchez, A., Martínez, J.J., Vázquez-Araújo, L., Burló, F., Melgarejo, P. *et al.* (2011) Volatile composition and sensory quality of Spanish pomegranates (*Punica granatum* L.). *Journal of the Science of Food and Agriculture* 91(3), 586–592. DOI: 10.1002/jsfa.4230.

Candir, E., Ozdemir, A.E. and Aksoy, M.C. (2018) Effects of chitosan coating and modified atmosphere packaging on postharvest quality and bioactive compounds of pomegranate fruit cv. 'Hicaznar'. *Scientia Horticulturae* 235, 235–243. DOI: 10.1016/j.scienta.2018.03.017.

Cerdá, B., Llorach, R., Cerón, J.J., Espín, J.C. and Tomás-Barberán, F.A. (2003) Evaluation of the bioavailability and metabolism in the rat of punicalagin, an antioxidant polyphenol from pomegranate juice. *European Journal of Nutrition* 42(1), 18–28. DOI: 10.1007/s00394-003-0396-4.

Chandra, R., Babu, K.D., Jadhav, V.T. and Teixeira da Silva, J.A. (2010) Origin, history and domestication of pomegranate. *Fruit, Vegetable and Cereal Science and Biotechnology* 4(Special issue 2), 1–6.

Creelman, R.A. and Mullet, J.E. (1997) Biosynthesis and action of jasmonates in plants. *Annual Review of Plant Physiology and Plant Molecular Biology* 48(1), 355–381. DOI: 10.1146/annurev.arplant.48.1.355.

Defilippi, B.G., Whitaker, B.D., Hess-Pierce, B.M. and Kader, A.A. (2006) Development and control of scald on wonderful pomegranates during long-term storage. *Postharvest Biology and Technology* 41(3), 234–243. DOI: 10.1016/j.postharvbio.2006.04.006.

Díaz-Mula, H.M., Martínez-Romero, D., Castillo, S., Serrano, M. and Valero, D. (2011) Modified atmosphere packaging of yellow and purple plum cultivars. 1. Effect on organoleptic quality. *Postharvest Biology and Technology* 61(2-3), 103–109. DOI: 10.1016/j.postharvbio.2011.02.010.

D'Aquino, S., Palma, A., Schirra, M., Continella, A., Tribulato, E. *et al.* (2010) Influence of film wrapping and fludioxonil application on quality of pomegranate fruit. *Postharvest Biology and Technology* 55(2), 121–128. DOI: 10.1016/j.postharvbio.2009.08.006.

Elyatem, S.M. and Kader, A.A. (1984) Post-harvest physiology and storage behaviour of pomegranate fruits. *Scientia Horticulturae* 24(3–4), 287–298. DOI: 10.1016/0304-4238(84)90113-4.

Ezra, D., Kirshner, B., Hershcovich, M., Shtienberg, D. and Kosto, I. (2015) Heart rot of pomegranate: disease etiology and the events leading to development of symptoms. *Plant Disease* 99(4), 496–501. Available at: https://doi.org/ DOI: 10.1094/PDIS-07-14-0707-RE.

Fadavi, A., Barzegar, M., Azizi, M.H. and Bayat, M. (2005) Note. physicochemical composition of ten pomegranate cultivars (*Punica granatum* L.) grown in Iran. *Food Science and Technology International* 11(2), 113–119. DOI: 10.1177/1082013205052765.

Faria, A., Calhau, C. and Conceição, C. (2011) The bioactivity of pomegranate: impact on health and disease. *Critical Reviews in Food Science and Nutrition* 51(7), 626–634. DOI: 10.1080/10408391003748100.

Fawole, O.A. and Opara, U.L. (2013a) Effects of storage temperature and duration on physiological responses of pomegranate fruit. *Industrial Crops and Products* 47, 300–309. DOI: 10.1016/j.indcrop.2013.03.028.

Fawole, O.A. and Opara, U.L. (2013b) Seasonal variation in chemical composition, aroma volatiles and antioxidant capacity of pomegranate during fruit development. *African Journal of Biotechnology* 12, 4006–4019.

Fischer, U.A., Carle, R. and Kammerer, D.R. (2011) Identification and quantification of phenolic compounds from pomegranate (*Punica granatum* L.) peel, mesocarp, aril and differently produced juices by HPLC-DAD–ESI/MSn. *Food Chemistry* 127(2), 807–821. DOI: 10.1016/j.foodchem.2010.12.156.

Gamrasni, D., Gadban, H., Tsvilling, A., Goldberg, T., Neria, O. *et al.* (2015) 1-MCP improves the quality of stored 'Wonderful' pomegranates. *Acta Horticulturae* 1079,229–234. DOI: 10.17660/ActaHortic.2015.1079.26.

Gao, H., Zhang, Z.K., Chai, H.K., Cheng, N., Yang, Y. *et al.* (2016) Melatonin treatment delays postharvest senescence and regulates reactive oxygen species metabolism in peach fruit. *Postharvest Biology and Technology* 118, 103–110. DOI: 10.1016/j.postharvbio.2016.03.006.

Ghafir, S.A.M., Ibrahim, I.Z., Zaied, S.A. and Abusrewel, G.S. (2010) Response of local variety 'Shlefy' pomegranate fruits to packaging and cold storage. *Acta Horticulturae* 877,427–431. DOI: 10.17660/ActaHortic.2010.877.55.

Ghasemi Soloklui, A.A., Gharaghani, A., Oraguzie, N. and Ramezanian, A. (2019) Shelf life and biochemical changes of ready-to-eat arils among nineteen Iranian pomegranate cultivars (*Punica granatum* L.) during storage. *Journal of Food Science and Technology* 56(3), 1416–1426. DOI: 10.1007/s13197-019-03620-0.

Ghasemnezhad, M., Zareh, S., Rassa, M. and Sajedi, R.H. (2013) Effect of chitosan coating on maintenance of aril quality, microbial population and PPO activity of pomegranate (*Punica granatum* L. cv. Tarom) at cold storage temperature. *Journal of the Science of Food and Agriculture* 93(2), 368–374. DOI: 10.1002/jsfa.5770.

Gil, M.I., Tomás-Barberán, F.A., Hess-Pierce, B., Holcroft, D.M. and Kader, A.A. (2000) Antioxidant activity of pomegranate juice and its relationship with phenolic composition and processing. *Journal of Agricultural and Food Chemistry* 48(10), 4581–4589. DOI: 10.1021/jf000404a.

Glowacz, M. and Rees, D. (2016) Using jasmonates and salicylates to reduce losses within the fruit supply chain. *European Food Research and Technology* 242(2), 143–156. DOI: 10.1007/s00217-015-2527-6.

Haug, M. and Zalom, J. (undated) Pomegranate: harvest & postharvest. Fruit & Nut Research & Information Center, Dept. of Plant Sciences, UC Davis. Available at: http://fruitandnuteducation.ucdavis.edu/fruitnutproduction/Pomegranate/PomeHarvestPost/ (accessed 10 June 2020).

Hernández, F., Melgarejo, P., Tomás-Barberán, F.A. and Artés, F. (1999) Evolution of juice anthocyanins during ripening of new selected pomegranate (*Punica granatum*) clones. *European Food Research and Technology* 210(1), 39–42. DOI: 10.1007/s002170050529.

Hess-Pierce, B. and Kader, A.A. (2003) Responses of 'Wonderful' pomegranates to controlled atmospheres. *Acta Horticulturae* 600, 751–757.

Holcroft, D.M., Gil, M.I. and Kader, A.A. (1998) Effect of carbon dioxide on anthocyanins, phenylalanine ammonia lyase and glucosyltransferase in the arils of stored pomegranates. *Journal of the American Society for Horticultural Science* 123(1), 136–140. DOI: 10.21273/JASHS.123.1.136.

Hussein, Z., Caleb, O.J. and Opara, U.L. (2015) Perforation-mediated modified atmosphere packaging of fresh and minimally processed produce—a review. *Food Packaging and Shelf Life* 6, 7–20. DOI: 10.1016/j.fpsl.2015.08.003.

Ismail, T., Sestili, P. and Akhtar, S. (2012) Pomegranate peel and fruit extracts: a review of potential anti-inflammatory and anti-infective effects. *Journal of Ethnopharmacology* 143(2), 397–405. DOI: 10.1016/j.jep.2012.07.004.

Jannatizadeh, A. (2019) Exogenous melatonin applying confers chilling tolerance in pomegranate fruit during cold storage. *Scientia Horticulturae* 246, 544–549. DOI: 10.1016/j.scienta.2018.11.027.

Jiang, A., Meng, X., Hu, W., Tian, M. and Wang, Y. (2011) Effects of high CO_2 shock treatment on physiological metabolism and quality of postharvest blueberry fruits. *Transactions from the Chinese Society of Agricultural Engineering* 27, 362–368.

Johanningsmeier, S.D. and Harris, G.K. (2011) Pomegranate as a functional food and nutraceutical source. *Annual Review of Food Science and Technology* 2(1), 181–201. DOI: 10.1146/annurev-food-030810-153709.

Kader, A.A., Chordas, A. and Elyatem, S. (1984) Response of pomegranates to ethylene treatment and storage temperature. *California Agriculture* 38(7), 14–15.

Kader, A.A. (2006) Postharvest biology and technology of pomegranates. In: Seeram, N.P., Schulman, R.N. and D, Heber. (eds) *Pomegranates. Ancient Roots to Modern Medicine*. CRC Press-Taylor & Francis, Boca Raton, Florida, pp. 211–218.

Kader, A.A., Chordas, A. and Elyatem, S. (2006) Responses of pomegranates to ethylene treatment and storage temperature. *California Agriculture* 38, 14–15.

Kahramanoğlu, İ., Aktaş, M. and Gündüz, Ş. (2018) Effects of fludioxonil, propolis and black seed oil application on the postharvest quality of "Wonderful" pomegranate. *PloS one* 13(5), e0198411. DOI: 10.1371/journal.pone.0198411.

Kamel, H.M., Zaki, Z.A. and Abd El-Moneim, E.A.A. (2016) Influence of treatment with aloe vera extract, honey solution and salicylic acid on quality maintenance of 'Wonderful' pomegranate fruits during cold storage. *International Journal of ChemTech Research* 9(3), 0–15.

Koyuncu, M.A., Erbas, D., Onursal, C.E., Secmen, T., Guneyli, A. *et al.* (2018) Postharvest treatments of salicylic acid, oxalic acid and putrescine influences bioactive compounds and quality of pomegranate during controlled atmosphere storage. *Journal of Food Science and Technology* 56(1), 350–359. DOI: 10.1007/s13197-018-3495-1.

Kulkarni, A.P. and Aradhya, S.M. (2005) Chemical changes and antioxidant activity in pomegranate arils during fruit development. *Food Chemistry* 93(2), 319–324. DOI: 10.1016/j.foodchem.2004.09.029.

Kumar, D. (2014) Salicylic acid signalling in disease resistance. *Plant Science* 22, 127–134.

Küpper, W., Pekmezci, M. and Henze, J. (1995) Studies on CA-STORAGE of pomegranate (*Punica granatum* L., cv. HICAZ). *Acta Horticulturae* 398,101–108. DOI: 10.17660/ActaHortic.1995.398.10.

Kayashima, T. and Katayama, T. (2002) Oxalic acid is available as a natural antioxidant in some systems. *Biochimica et Biophysica Acta (BBA) – General Subjects* 1573(1), 1–3. DOI: 10.1016/S0304-4165(02)00338-0.

Köksal, A.I. (1989) Research on the storage of pomegranate (cv. Gök Bahçe) under different conditions. *Acta Horticulturae* 258, 295–302.

Koushesh Saba, M. and Zarei, L. (2019) Preharvest methyl jasmonate's impact on postharvest chilling sensitivity, antioxidant activity, and pomegranate fruit quality. *Journal of Food Biochemistry* 43(3), e12763. DOI: 10.1111/jfbc.12763.

Lansky, E.P. and Newman, R.A. (2007) *Punica granatum* (pomegranate) and its potential for prevention and treatment of inflammation and cancer. *Journal of Ethnopharmacology* 109(2), 177–206. DOI: 10.1016/j.jep.2006.09.006.

Lepper, J.A., Sreedharan, A., Goodrich-Schneider, R.M. and Schneider, K.R. (2017) Food safety on the farm: good agricultural practices and good handling practices – packing operation sanitation. Publ. FSHN12=05. University of Florida/IFAS Extension Service, Gainesville, Florida.

Lepper, J.A., De, J., Pabst, C.R., Sreedharan, A. and Goodrich-Schneider, R.M. (2019) Food safety on the farm: good agricultural practices and good handling practices – field sanitation. Publ. FSHS10-12. University of Florida/IFAS Extension Service, Gainesville, Florida.

Li, Y., Guo, C., Yang, J., Wei, J., Xu, J. *et al.* (2006) Evaluation of antioxidant properties of pomegranate peel extract in comparison with pomegranate pulp extract. *Food Chemistry* 96(2), 254–260. DOI: 10.1016/j.foodchem.2005.02.033.

Li, L., Lichter, A., Chalupowicz, D., Gamrasni, D., Goldberg, T. *et al.* (2016) Effects of the ethylene-action inhibitor 1-methylcyclopropene on postharvest quality of non-climacteric fruit crops. *Postharvest Biology and Technology* 111, 322–329. DOI: 10.1016/j.postharvbio.2015.09.031.

Lurie, S. (2016) Prestorage heat stress to improve storability of fresh produce: a review. *Israel Journal of Plant Sciences* 63(1), 17–21. DOI: 10.1080/07929978.2016.1159411.

Martínez-Romero, D., Castillo, S., Guillén, F., Díaz-Mula, H.M., Zapata, P.J. *et al.* (2013) *Aloe vera* gel coating maintains quality and safety of ready-to-eat pomegranate arils. *Postharvest Biology and Technology* 86, 107–112. DOI: 10.1016/j.postharvbio.2013.06.022.

Matityahu, I., Marciano, P., Holland, D., Ben-Arie, R. and Amir, R. (2016) Differential effects of regular and controlled atmosphere storage on the quality of three cultivars of pomegranate (*Punica granatum* L.). *Postharvest Biology and Technology* 115, 132–141. DOI: 10.1016/j.postharvbio.2015.12.018.

Meighani, H., Ghasemnezhad, M. and Bakhshi, D. (2015) Effect of different coatings on post-harvest quality and bioactive compounds of pomegranate (*Punica granatum* L.) fruits. *Journal of Food Science and Technology* 52(7), 4507–4514. DOI: 10.1007/s13197-014-1484-6.

Melgarejo, P., Salazar, D.M. and Artés, F. (2000) Organic acids and sugars composition of harvested pomegranate fruits. *European Food Research and Technology* 211(3), 185–190. DOI: 10.1007/s002170050021.

Mertens-Talcott, S.U., Jilma-Stohlawetz, P., Ríos, J., Hingorani, L. and Derendorf, H. (2006) Absorption, metabolism, and antioxidant effects of pomegranate (*Punica granatum* L.) polyphenols after ingestion. *Journal of Agricultural and Food Chemistry* 5, 8956–8961.

Miguel, G., Fontes, C., Antunes, D., Neves, A. and Martins, D. (2004) Anthocyanin concentration of 'Assaria' pomegranate fruits during different cold storage conditions. *Journal of Biomedicine and Biotechnology* 2004(5), 338–342. DOI: 10.1155/S1110724304403076.

Mirdehghan, S.H., Rahemi, M., Serrano, M., Guillén, F., Martínez-Romero, D. *et al.* (2006) Prestorage heat treatment to maintain nutritive and functional properties during postharvest cold storage of pomegranate. *Journal of Agricultural and Food Chemistry* 54(22), 8495–8500. DOI: 10.1021/jf0615146.

Mirdehghan, S.H., Rahemi, M., Castillo, S., Martínez-Romero, D., Serrano, M. *et al.* (2007a) Pre-storage application of polyamines by pressure or immersion improves shelf-life of pomegranate stored at chilling temperature by increasing endogenous polyamine levels. *Postharvest Biology and Technology* 44(1), 26–33. DOI: 10.1016/j.postharvbio.2006.11.010.

Mirdehghan, S.H., Rahemi, M., Martínez-Romero, D., Guillén, F., Valverde, J.M. *et al.* (2007b) Reduction of pomegranate chilling injury during storage after heat treatment: role of polyamines. *Postharvest Biology and Technology* 44(1), 19–25. DOI: 10.1016/j.postharvbio.2006.11.001.

Mirdehghan, S.H., Rahemi, M., Serrano, M., Guillén, F., Martínez-Romero, D. *et al.* (2007c) The application of polyamines by pressure or immersion as a tool to maintain functional properties in stored pomegranate arils. *Journal of Agricultural and Food Chemistry* 55(3), 755–760. DOI: 10.1021/jf062985v.

Montesinos-Herrero, C., del Río, M.Á., Rojas-Argudo, C. and Palou, L. (2012) Short exposure to high CO2 and O2 at curing temperature to control postharvest diseases of citrus fruit. *Plant Disease* 96(3), 423–430. DOI: 10.1094/PDIS-07-11-0595.

Moradinezhad, F., Khayyat, M. and Saeb, H. (2013) Combination effects of postharvest treatments and modified atmosphere packaging on shelf life and quality of Iranian pomegranate fruit cv. Sheshikab. *International Journal of Postharvest Technology and Innovation* 3(3), 244–256. DOI: 10.1504/IJPTI.2013.059286.

Moradinezhad, F., Khayyat, M., Ranjbari, F. and Maraki, Z. (2018) Physiological and quality responses of Shishe-Kab pomegranates to short-term high CO2 treatment and modified atmosphere packaging. *International Journal of Fruit Science* 18(3), 287–299. DOI: 10.1080/15538362.2017.1419399.

Morton, J.F. (1987) Pomegranate. In *Fruits of Warm Climates*. Creative Resource Systems, Inc., Winterville, North Carolina.

Mphahlele, R.R., Stander, M.A., Fawole, O.A. and Opara, U.L. (2014) Effect of fruit maturity and growing location on the postharvest contents of flavonoids, phenolic acids, vitamin C and antioxidant activity of pomegranate juice (cv. Wonderful). *Scientia Horticulturae* 179, 36–45. DOI: 10.1016/j.scienta.2014.09.007.

Munhuweyi, K., Lennox, C.L., Meitz-Hopkins, J.C., Caleb, O.J., Sigge, G.O. *et al.* (2016) *In vitro* effects of crab shell chitosan against mycelial growth of *Botrytis* sp., *Penicillium* sp. and *Pilidiella granati*. *Acta Horticulturae* 1144,403–408.10.17660/ActaHortic.2016.1144.60 DOI: 10.17660/ActaHortic.2016.1144.60.

Nanda, S., Sudhakar Rao, D.V. and Krishnamurthy, S. (2001) Effects of shrink film wrapping and storage temperature on the shelf life and quality of pomegranate fruits cv. Ganesh. *Postharvest Biology and Technology* 22(1), 61–69. DOI: 10.1016/S0925-5214(00)00181-2.

Nuncio-Jáuregui, N., Calín-Sánchez, A., Carbonell-Barrachina, A. and Hernández, F. (2014) Changes in quality parameters, proline, antioxidant activity and color of pomegranate (*Punica granatum* L.) as affected by fruit position within tree, cultivar and ripening stage. *Scientia Horticulturae* 165, 181–189. DOI: 10.1016/j.scienta.2013.11.021.

Nunes, M.C.D.N. (2008) Impact of environmental conditions on fruit and vegetable quality. *Stewart Postharvest Review* 4(4), 4.

Opara, U.L., Hussein, Z. and Caleb, O.J. (2017) Phytochemical properties and antioxidant activities of minimally processed 'Acco' pomegranate arils as affected by perforation-mediated modified atmosphere packaging. *Journal of Food Processing and Preservation* 41(3), e12948. DOI: 10.1111/jfpp.12948.

Ozgen, M., Durgaç, C., Serçe, S. and Kaya, C. (2008) Chemical and antioxidant properties of pomegranate cultivars grown in the Mediterranean region of Turkey. *Food Chemistry* 111(3), 703–706. DOI: 10.1016/j.foodchem.2008.04.043.

Pala, H., Tatli, A., Yilmaz, C. and Özgüven, A.I. (2009) Important diseases of pomegranate fruit and control possibilities in Turkey. *Acta Horticulturae* 818(Special Issue I International Symposium on Pomegranate and Minor Mediterranean Fruits), 285–290.

Palou, L., Crisosto, C.H. and Garner, D. (2007) Combination of postharvest antifungal chemical treatments and controlled atmosphere storage to control gray mold and improve storability of

'Wonderful' pomegranates. *Postharvest Biology and Technology* 43(1), 133–142. DOI: 10.1016/j.postharvbio.2006.08.013.

Palou, L., Rosales, R., Montesinos-Herrero, C. and Taberner, V. (2016) Short-term exposure to high CO2 and O2 atmospheres to inhibit postharvest gray mold of pomegranate fruit. *Plant Disease* 100(2), 424–430. DOI: 10.1094/PDIS-06-15-0637-RE.

Panth, N., Manandhar, B. and Paudel, K.R. (2017) Anticancer activity of *Punica granatum* (pomegranate): a review. *Phytotherapy Research* 31(4), 568–578. DOI: 10.1002/ptr.5784.

Pareek, S., Valero, D. and Serrano, M. (2015) Postharvest biology and technology of pomegranate. *Journal of the Science of Food and Agriculture* 95(12), 2360–2379. DOI: 10.1002/jsfa.7069.

Porat, R., Kosto, I. and Daus, A. (2016) Bulk storage of 'Wonderful' pomegranate fruit using modified atmosphere bags. *Israel Journal of Plant Sciences* 63(1), 45–50. DOI: 10.1080/07929978.2016.1152839.

Poyrazoğlu, E., Gökmen, V. and Artik, N. (2002) Organic acids and phenolic compounds in pomegranates (*Punica granatum* L.) grown in Turkey. *Journal of Food Composition and Analysis* 15(5), 567–575. DOI: 10.1016/S0889-1575(02)91071-9.

Rai, D.R. and Paul, S. (2007) Packaging requirements of highly respiring produce under modified atmosphere: a review. *Journal of Science and Technology* 44, 10–15.

Ranjbari, F., Moradinezhad, F. and Khayyat, M. (2018) Effect of nitric oxide and film wrapping on quality maintenance and alleviation of chilling injury on pomegranate fruit. *Journal of Agricultural Science and Technology* 20, 1025–1036.

Romanazzi, G., Feliziani, E., Baños, S.B. and Sivakumar, D. (2017) Shelf life extension of fresh fruit and vegetables by chitosan treatment. *Critical Reviews in Food Science and Nutrition* 57(3), 579–601. DOI: 10.1080/10408398.2014.900474.

Sargent, S.A., Brecht, J.K. and Zoellner, J.J. (1991) Analyses of tomato and bell pepper packing lines using the instrumented sphere. *Applied Engineering in Agriculture* 8(1), 76–83.

Sayyari, M., Babalar, M., Kalantari, S., Serrano, M. and Valero, D. (2009) Effect of salicylic acid treatment on reducing chilling injury in stored pomegranates. *Postharvest Biology and Technology* 53(3), 152–154. DOI: 10.1016/j.postharvbio.2009.03.005.

Sayyari, M., Valero, D., Babalar, M., Kalantari, S., Zapata, P.J. *et al.* (2010) Prestorage oxalic acid treatment maintained visual quality, bioactive compounds, and antioxidant potential of pomegranate after long-term storage at 2 degrees C. *Journal of Agricultural and Food Chemistry* 58(11), 6804–6808. DOI: 10.1021/jf100196h.

Sayyari, M., Babalar, M., Kalantari, S., Martínez-Romero, D., Guillén, F. *et al.* (2011a) Vapour treatments with methyl salicylate or methyl jasmonate alleviated chilling injury and enhanced antioxidant potential during postharvest storage of pomegranates. *Food Chemistry* 124(3), 964–970. DOI: 10.1016/j.foodchem.2010.07.036.

Sayyari, M., Castillo, S., Valero, D., Díaz-Mula, H.M. and Serrano, M. (2011b) Acetyl salicylic acid alleviates chilling injury and maintains nutritive and bioactive compounds and antioxidant activity during postharvest storage of pomegranates. *Postharvest Biology and Technology* 60(2), 136–142. DOI: 10.1016/j.postharvbio.2010.12.012.

Sayyari, M., Aghdam, M.S., Salehi, F. and Ghanbari, F. (2016) Salicyloyl chitosan alleviates chilling injury and maintains antioxidant capacity of pomegranate fruits during cold storage. *Scientia Horticulturae* 211, 110–117. DOI: 10.1016/j.scienta.2016.08.015.

Sayyari, M., Valero, D. and Serrano, M. (2017) Prestorage salicylic acid treatment affects functional properties, unsaturated/saturated fatty acids ratio and chilling resistance of pomegranate during cold storage. *International Food Research Journal* 24, 637–642.

Selcuk, N. and Erkan, M. (2015) Changes in phenolic compounds and antioxidant activity of sour–sweet pomegranates cv. 'Hicaznar' during long-term storage under modified atmosphere packaging. *Postharvest Biology and Technology* 109, 30–39. DOI: 10.1016/j.postharvbio.2015.05.018.

Serrano, M., Díaz-Mula, H.D. and Valero, D. (2011) Antioxidant compounds in fruits and vegetables and changes during postharvest storage and processing. *Stewart Postharvest Review* 7(1), 1.

Serrano, M., Zapata, P.J., Martínez-Romero, D., Díaz-Mula, H.M. and Valero, D. (2016) Polyamines as an eco-friendly postharvest tool to maintain fruit quality. In: Siddiqui, M.W. (ed.) *Eco-Friendly Technology for Postharvest Produce Quality*. Elsevier, Amsterdam, pp. 219–242.

Shaarawi, S.A. and Nagy, K.S. (2017) Effect of modified atmosphere packaging on fruit quality of 'Wonderful' pomegranate under cold storage conditions. *Middle East Journal of Agriculture Research* 6(2), 495–505.

Shahbaz, H.M., Ahn, J., Akram, K., Kim, H., Park, E. *et al.* (2014) Chemical and sensory quality of fresh pomegranate fruits exposed to gamma radiation as quarantine treatment. *Food Chemistry* 145, 312–318.

Singh, A.K. and Goswami, T.K. (2006) Controlled atmosphere storage of fruits and vegetables: a review. *Journal of Food Science and Technology* 34, 1–7.

Stover, E. and Mercure, E.W. (2007) The pomegranate: a new look at the fruit of paradise. *HortScience* 42(5), 1088–1092. DOI: 10.21273/HORTSCI.42.5.1088.

Sudhakar Rao, D.V. and Shivashankara, K.S. (2018) Effect of modified atmosphere packaging on the extension of storage life and quality maintenance of pomegranate (cv. 'Bhagwa') at ambient and low temperatures. *Journal of Food Science and Technology* 55(6), 2103–2113. DOI: 10.1007/s13197-018-3125-y.

Tzulker, R., Glazer, I., Bar-Ilan, I., Holland, D., Aviram, M. *et al.* (2007) Antioxidant activity, polyphenol content, and related compounds in different fruit juices and homogenates prepared from 29 different pomegranate accessions. *Journal of Agricultural and Food Chemistry* 55(23), 9559–9570. DOI: 10.1021/jf071413n.

Valdenegro, M., Huidobro, C., Monsalve, L., Bernales, M., Fuentes, L. *et al.* (2018) Effects of ethrel, 1-MCP and modified atmosphere packaging on the quality of 'Wonderful' pomegranates during cold storage. *Journal of the Science of Food and Agriculture* 98(13), 4854–4865. DOI: 10.1002/jsfa.9015.

Valero, D. and Serrano, M. (2010) *Postharvest Biology and Technology for Preserving Fruit Quality*. Taylor and Francis, Boca Raton, FL.

Valero, D., Mirdehghan, S.H., Sayyari, M. and Serrano, M. (2015) Vapor treatments, chilling, storage, and antioxidants in pomegranates. In: Preedy, V. (ed.) *Processing and Impact on Active Components in Food*. Academic Press-Elsevier, USA, pp. 189–196.

Valero, D., Guillén, F., Valverde, J.M., Castillo, S. and Serrano, M. (2016) Recent development of 1-methylcyclopropene (1-MCP) treatments on fruit quality attributes. In: Siddiqui, M.W. (ed.) *Eco-Friendly Technology for Postharvest Produce Quality*. Elsevier, Amsterdam, pp. 185–202.

Varasteh, F., Arzani, K., Zamani, Z. and Tabatabaei, S.Z. (2008) Physico-chemical seasonal changes of pomegranate (*Punica granatum* L.) fruit 'Malas-e-Torsh-e-Saveh' in Iran. *Acta Horticulturae* 769,255–258. DOI: 10.17660/ActaHortic.2008.769.36.

Valero, D., Martínez-Romero, D. and Serrano, María. (2002) The role of polyamines in the improvement of the shelf life of fruit. *Trends in Food Science & Technology* 13(6-7), 228–234. DOI: 10.1016/S0924-2244(02)00134-6.

Varasteh, F., Arzani, K., Barzegar, M. and Zamani, Z. (2012) Changes in anthocyanins in arils of chitosan-coated pomegranate (*Punica granatum* L. cv. Rabbab-e-Neyriz) fruit during cold storage. *Food Chemistry* 130(2), 267–272. DOI: 10.1016/j.foodchem.2011.07.031.

Varasteh, F., Arzani, K., Barzegar, M. and Zamani, Z. (2017) Pomegranate (*Punica granatum* L.) fruit storability improvement using pre-storage chitosan coating technique. *Journal of Agricultural Science and Technology* 19, 389–400.

Ward, C. (2003) Pomegranates in eastern Mediterranean contexts during the late bronze age. *World Archaeology* 34(3), 529–541. DOI: 10.1080/0043824021000026495.

Watkins, C.B. (2006) The use of 1-methylcyclopropene (1-MCP) on fruits and vegetables. *Biotechnology Advances* 24(4), 389–409. DOI: 10.1016/j.biotechadv.2006.01.005.

Watson, J.A., Treadwell, D.D., Sargent, S.A., Brecht, J.K. and Pelletier, W. (2015) Postharvest storage, packaging and handling of specialty crops: a guide for Florida small farm producers. HS 1270. EDIS Publ., Gainesville, Florida.

Zapata, P.J., Martínez-Esplá, A., Guillén, F., Díaz-Mula, H.M., Martínez-Romero, D. *et al.* (2014) Preharvest application of methyl jasmonate (MeJA) in two plum cultivars. 2. Improvement of fruit quality and antioxidant systems during postharvest storage. *Postharvest Biology and Technology* 98, 115–122. DOI: 10.1016/j.postharvbio.2014.07.012.

Zhang, L. and McCarthy, M.J. (2013) Effect of controlled atmosphere storage on pomegranate quality investigated by two dimensional NMR correlation spectroscopy. *LWT – Food Science and Technology* 54(1), 302–306. DOI: 10.1016/j.lwt.2013.04.015.

Zheng, X., Jing, G., Liu, Y., Jiang, T., Jiang, Y. *et al.* (2012) Expression of expansin gene, MiExpA1, and activity of galactosidase and polygalacturonase in mango fruit as affected by oxalic acid during storage at room temperature. *Food Chemistry* 132(2), 849–854. DOI: 10.1016/j.foodchem.2011.11.049.

16 Processing and Industrialization

N.N. Gaikwad[1]*, P.G. Patil[1], N.B. Patil[1], M. Cano-Lamadrid[2] and Á.A. Carbonell-Barrachina[2]

[1]ICAR-National Research Centre on Pomegranate, Solapur, India; [2]Departamento Tecnología Agroalimentaria, Universidad Miguel Hernández, Orihuela, Spain

16.1 Introduction

Pomegranate is an important fruit crop of the tropical and subtropical regions of the world. It is grown in central and south Asia, the Middle East, the Caucasus region, the Mediterranean region, North and tropical Africa and North America. Pomegranate is also called a wonder crop due to its nutritional and medicinal properties. The fruit has gained wide acceptability among consumers due to its juicy arils with attractive colour and sweet and/or sour taste. Due to its adaptability to a wide range of climatic conditions, the pomegranate crop has become an important crop among the farming community. The crop has become popular among farmers in arid and semi-arid regions due to high returns on unit investment, its role in the alleviation of poverty and because of its high nutritional value.

Globally, the growing area and production of pomegranate have increased tremendously. The expected rise in production of pomegranate has compelled stakeholders to think about future marketing and utilization strategies for this high-value produce. In this context, postharvest management of pomegranate to improve its storage and shelf-life by adopting modern handling, storage, packaging and transportation practices is of high importance in order to market this fruit to distant market destinations. A high percentage of pomegranate fruits are consumed in fresh form. In addition, the fruits are also used for the preparation of various value-added processed products such as minimally processed arils, juice and juice-based beverages, juice concentrate, carbonated drinks, seed oil, dehydrated arils ('anardana'), wine, molasses, jam, jelly, antioxidant extracts and biocolours, among others. The fruit are known to be difficult to eat due to the presence of tough peel over the edible arils that is hard to remove. Thus, minimal processing and packaging of pomegranate arils offer convenience in handling and consumption. A high demand for minimally processed arils is evident in the international market.

The biochemical and nutritional composition of processed products such as minimally processed arils is affected by various factors including those related to the fruit itself (cultivar, season, region where it is grown, cultivation practices, maturity level at harvest, etc.) and those related to industrial processing (e.g. heat treatment, initial moisture content, storage, etc.). The presence of abundant bioactive compounds has been reported in many parts of this plant such as leaves, bark, roots and, especially, fruit peel, and this is linked to its high antimicrobial and antioxidant activity (Seeram *et al.*,

*Corresponding author: nileshgaikwad98@gmail.com

2006; Orgil *et al.*, 2014). Fruits have potential value for the development of pharmaceutical and cosmetic products on a commercial scale because of their high content of bioactive compounds and antioxidant activity. The edible portion, pomegranate arils, contains low but interesting amounts of flavonoids, anthocyanins, punicic acid, ellagitannins, alkaloids, fructose, sucrose, glucose, simple organic acids, vitamins, minerals, polyphenols and other components (Zarfeshany *et al.*, 2014).

The pomegranate fruit is composed of approximately 50% edible parts consisting of arils, and 50% non-edible peel on a fresh weight basis. Within the arils, seeds are found and represent between 4 and 10% of the total fruit weight (Fadavi *et al.*, 2006). The pomegranate juice made from arils contains 85% water, 10% total sugars (mainly fructose and glucose), and 1.5% pectin, organic acids (ascorbic, citric and malic) and bioactive compounds such as phenolic compounds and flavonoids, mainly anthocyanins (Aviram *et al.*, 2000; Viuda-Martos *et al.*, 2010). Arils also provide 12% of the daily value (DV) for vitamin C and 16% for vitamin K per 100 g serving. The red colour of the arils and juice can be attributed to anthocyanins, such as delphinidin, cyanidin and pelargonidin glycosides. Pomegranate peel contains higher bioactive compounds than arils and seeds. Therefore, the peel is an important source of phenolics, ellagitannins (ETs), proanthocyanidins, flavonoids, minerals (mainly K, Ca, P, Mg, Na) and complex polysaccharides (Li *et al.*, 2006). Pomegranate seeds are an important co-product of the pomegranate juice processing industry. Seeds contain a significant amount of lipid, protein, sugars, dietary fibre and essential minerals (El-Nemr *et al.*, 1990). In addition to this, the pomegranate seeds contain nutraceuticals, such as sterols, γ-tocopherol, hydroxyl benzoic, punicic acid and phenyl aliphatic glycosides such as phenethyl rutinoside; in addition, the contents of most minerals excluding K are higher in the seeds than in the arils (El-Nemr *et al.*, 1992). The soft-seeded varieties of pomegranate contain seed oil up to 25–26% (v/w), which is rich in conjugated linolenic acid (70%). Seeds contain punicic, palmitic, stearic, oleic and linoleic acids. Seed oil is also a rich source of a steroidal oestrone, and contains γ-tocopherol, a rare and potent form of vitamin E and the phytosterols: β-sitosterol, stigmasterol and campesterol. Pomegranate flowers contain a variety of secondary metabolites with strong antioxidant activity, such as polyphenols. It contains mainly ellagic acid, anthocyanins, triterpenes, oleanolic acid, ursolic acid and gallic acid (Table 16.1) (Li *et al.*, 2008).

Approximately 60–70% of pomegranate orchard produce is high-quality fruits that fetches a high price in the market for fresh consumption. However, the remaining 30–40% of produce does not obtain acceptable prices owing to peel defects, bruising, sunscald, fungal spots, etc. However, these fruits can be used for industrial processing to prepare pomegranate-based products with high value, producing extra benefits for farmers and the industries. Therefore, postharvest processing for value-added food, pharmaceutical and nutraceutical products

Table 16.1. Important constituents and bioactive compounds in pomegranate plant.

Juice	Ellagic acid, quinic acid, flavonols, catechin, quercetin, rutin, caffeic acid, amino acids, minerals (K, Ca, P, Mg and Na), ascorbic acid, simple sugars (fructose, glucose), organic acids (citric, malic, oxalic, acetic)
Seeds	Punicic acid, palmitic acid, stearic acid, oleic acid and linoleic acid, eleostearic acid, catalpic acid, sterols, tocopherols, fibre
Peel	Punicalin, ellagic acid, punicalagin, caffeic acid, ellagitannins, pelletierine alkaloids, luteolin, kaempferol, pedunculagin, gallic acid, quercetin
Flower	Polyphenols, gallic acid, triterpenoids, fatty acids, punicalagin, ursolic acid, punicalin, ellagic acid
Leaves	Carbohydrates, reducing sugars, sterols, saponins, flavanoids, tannins, piperidine, alkaloids, flavone, glycoside, ellagitannins
Root and bark	Ellagitannins, piperidine alkaloids, pyrrolidine alkaloid, pelletierine alkaloids

is important. This will generate additional demand for the fruits and thereby overcome situations where there is a glut of fruit or a depressed market. The pomegranate-processing industries can be set up in production catchments at appropriate scales depending on the availability of raw material, marketing potential of the processed products and finance. The establishment of integrated pomegranate-processing industries for total utilization of fruits contributes to the reduction of postharvest losses, creates new employment opportunities and also brings nutritional security to the masses.

In this chapter, information on the following points will be provided: various pomegranate value-added products, industrial processes thereof and recent scientific developments in pomegranate processing that will benefit the processing industry and researchers.

16.2 Basic Industrial Processes

The richness of bioactive compounds and unique sensory attributes of the pomegranate fruit come with a short shelf-life, necessiting processing and the development of new products with high added value and extended shelf-life. Although a large proportion of pomegranate fruits are consumed in fresh form for table purposes worldwide, they are also processed into various value-added products on a commercial scale such as pomegranate juice, juice-based beverages (e.g. nectars), concentrated juice, carbonated beverages, dried arils ('anardana'), anar-rub, jelly and molasses, among others. Pomegranate products have also been used as toppings, and flavouring and colouring agents. In addition to the demand for fresh fruits and traditional processed products, innovative non-traditional high-value processed products, such as minimally processed arils, pomegranate wine, pomegranate seed oil and pomegranate tea, are also gaining importance in the world trade. Figure 16.1 depicts quantitative details of the products processed from 100 kg of pomegranate fruits.

16.2.1 Quality guidelines for procurement of fruits for processing

Quality guidelines for fresh pomegranate marketing and export are currently available, since a major portion of pomegranate fruits are used for table consumption. However, guidelines for processable grade pomegranates are not available at the national or international level. For fresh marketing, pomegranate fruits are classified into extra class, class I and class II by Codex (CODEX STAN 310-2013); however, fruits for processing have been excluded from this classification, though, the fruits meant for processing may include fruits of the above-mentioned grades. However, for fresh fruit marketing high-quality fruits are always preferred owing to higher returns (Fig. 16.2). Mature, wholesome and sound fruits are preferred for processing. In addition to this, fruits with surface bruising, peel affected by frost or sunburn can also be acceptable; however, fruits affected by rotting or deterioration rendering them unfit for consumption should be excluded.

16.2.2 Reception and cleaning

Fruits procured from farms or markets are received and weighed before processing. Preliminary sorting is applied to remove rotten fruits before cleaning. Subsequently, the pomegranate crown should be removed before washing, because the crown is the place where the highest microbial load is usually observed. The removal of the crown is achieved either manually with a knife or by using a mechanical rotary crown cutting machine, which can trim the crowns effectively without exposing arils. Then, fruits are washed to remove soil, dust, dirt, surface microflora, pesticides and other chemical residues, as well as physical and microbial contaminants. The washing of the fruits before entering into the processing line is a mandatory unit operation. Fruit washing is generally carried out in two stages using fruit washers with a conveyor belt system. In the first wash, all the fruits are treated with sodium hypochlorite at 200 ppm for 1–2 min. This concentration is sufficient for reducing microbial load and surface disinfection. During this process agitation, turbulence and bubbles are created with pressurized air. This is followed by removal of the surface dirt and microflora, achieved by soft roller bristles. In the second stage, all the fruits are immersed in the clean water. The washing

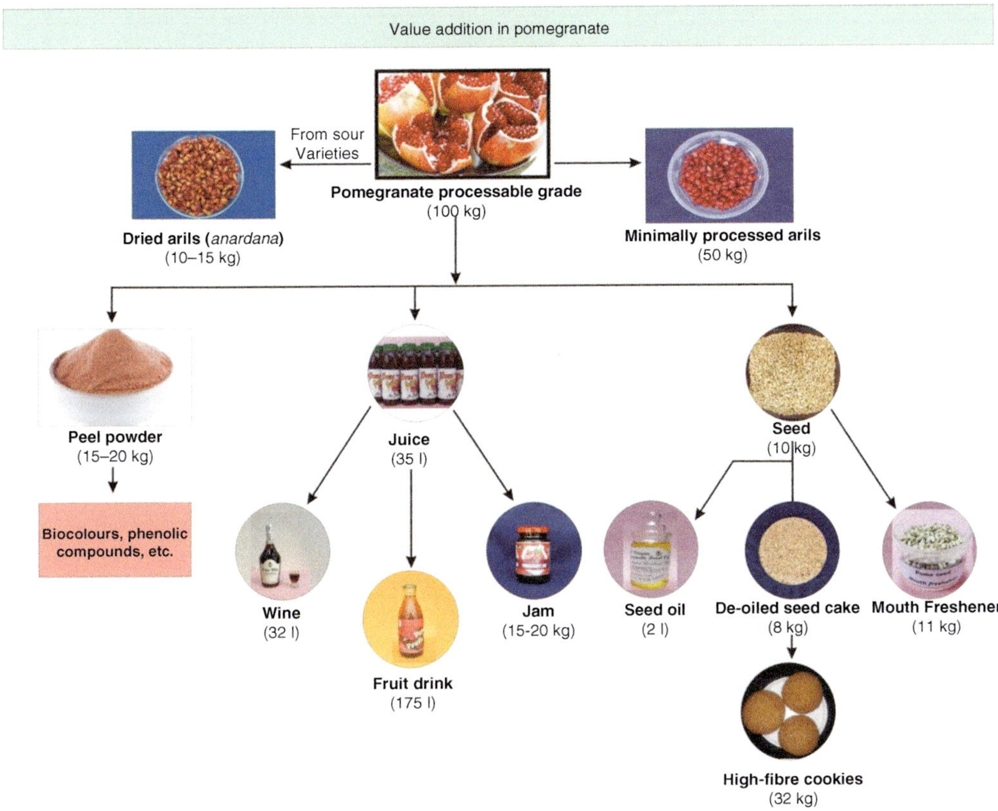

Fig. 16.1. Different important processed products from pomegranate. (From: Nilesh N. Gaikwad.)

operation is followed by allowing the surface water to drain onto the conveyor belt and mopping the surface of the fruits using the sponge rollers. Subsequently, hot air is blown over the fruits to remove any leftover surface moisture.

Fig. 16.2. Mature pomegranate fruits cv. 'Bhagawa' ready for harvest. (Photo: Nilesh N. Gaikwad.)

16.2.3 Sorting

The washed and surface-dried fruits are sorted for removal of rotten, damaged, immature or unripe fruits, which are left out of preliminary sorting. The sorting is generally carried out either manually by trained labourers, or by passing fruits over roller conveyers followed by hand removal and separating fruits based on predefined criteria for the intended processing operation.

16.2.4 Aril extraction

Aril extraction is a very important preliminary step in processing, where pomegranate fruits are opened, and arils are separated either manually or by using machines. The separated arils are then used for further processing into various value-added products. Aril separation or extraction is a very tedious and time-consuming process when performed manually. The soft arils are too delicate to extract in intact form and are a challenging proposition. Therefore, machines and hand tools have been developed for aril extraction and have been found to be successful to varying degrees. Various machines have been developed worldwide for aril extraction by various researchers. Khazaei *et al.* (2008) tested an impingement method, where pomegranate fruits are first cut into two halves and arils are extracted with the help of pressurized air jets. Safa *et al.* (2006) also used the pneumatic method, where fruits are opened (with a cut) into two halves and then arils are dislodged and extracted with the help of a scanning air jet. Juran Technologies, an Israel-based company has come up with an integrated mechanized solution for pomegranate aril extraction and packaging. This machine comes in different variants with varied capacities, that is, 300–1650 kg fruits/h. The machine has various units such as the reception section, where fruits are first washed with hot water for removal of dirt and the the aril extraction section, where the peel of the fruit is scored, gently opened and arils are extracted by subjecting the opened fruits to cold water from a water jet. Finally, separated arils are passed over the conveyor and dried with air jets.

The Central Institute of Post-Harvest Engineering and Technology, the constituent organization of the Indian Council of Agriculture Research (ICAR-CIPHET), has developed two types of aril extractors, one is a hand-operated tool for easy opening of pomegranate fruits and separation of arils from the peel, and another is an automated high-throughput aril extractor of capacity 500 kg/h. The hand tool consists of fruit holders with a blade arranged in such a way that it penetrates only into the peel. The pomegranate fruit has to be held between holders, and the holders are turned by hand in the opposite direction. Due to the rotating action of the holders, the fruit opens up into two irregular halves as a simultaneous effect of tension with a twist on the peel. During this action the whole fruit experiences a shearing effect, and due to this effect, the inside arils become loosened, which provides an opportunity for easy separation. About 20–25% of arils are successfully separated in this process of irregular breaking due to shearing action on the inner sheath and outer peel. The motorized unit is a unique design, in which an action similar to the hand tool is achieved during aril extraction. It consists of two concentric cones that are fitted with blades and clearance between the cones reduces as fruits pass through them. A vibratory screen is provided for separation of arils and peel. This machine causes less than 2% of arils to be damaged during processing.

16.2.5 Minimal processing

The pomegranate has a non-climacteric ripening pattern and is consumed mainly in fresh form. The difficulty encountered in separating edible arils from fruit has several limitations for its direct consumption, unlike other fruits such as oranges, bananas and grapes. Pomegranate arils are highly perishable and have <24 h shelf-life. Hence, minimally processed 'ready-to-eat' pomegranate arils have become very popular recently due to their convenience, high value, health benefits and unique sensory characteristics. Pomegranate fruits affected by thrips, peel bruising and fungal scab can be utilized for minimal processing. Although surface defects are acceptable, fruits with rot and/or brown arils are not acceptable in minimal processing. Fruits with the characteristic 'intense' dark-red colored arils are desirable for minimal processing (Fig. 16.3).

The minimally processed arils show browning caused by the oxidation of phenolic compounds during storage, indicating that stabilization of anthocyanin pigments is essential in order to achieve good quality of arils during storage. The minimally processed arils easily deteriorate in colour, texture and overall quality with a consequent reduction in shelf-life. Thus, maintaining the microbial and nutritional quality of pomegranate arils is a major challenge.

The minimal processing of pomegranate mainly involves washing them with sanitizing agents to reduce the initial microbial load, pH modifications, use of antioxidants, modified atmosphere packaging (MAP) and temperature control (Sepulveda *et al.*, 2000). Minimal processing of pomegranate arils requires separation of the processing unit into low-care and

Fig. 16.3. Minimally processed and packaged arils of cv. 'Bhagawa'. (Photo: Nilesh N. Gaikwad.)

high-care areas. The low-care area includes reception, weighing, washing and sorting. The high-care area includes aril extraction, aril weighing, filling of punnets, MAP and temporary storage before despatch. Aril extraction, punnet filling and packaging are carried out at low temperatures of 16–18°C. Manual aril extraction is preferred to machine extraction in India for minimal processing owing to the high cost of sophisticated aril extraction machines, unavoidable damage to arils during extraction and availability of labourers. The maintenance of a cold chain from the processing unit to the market destination is absolutely essential in minimal processing of pomegranate arils.

16.2.6 Pretreatments

In order to maintain quality and extend shelf-life of the minimally processed arils during storage, pretreatment has emerged as an easy alternative. Natural and chemical pretreatment with antimicrobial agents and antioxidants play an important role in extending the shelf-life of pomegranate arils. The use of aloe vera gel, ascorbic acid, citric acid, chlorinated water, honey, salicylic acid, potassium sorbate, 4-hexyl resorcinol and ozone have been found to extend the shelf-life and maintain the fresh quality of minimally processed arils by delaying microbial development, pigment changes and quality loss (Gil *et al.*, 1996a). Recently, radiation processing was found by researchers to extend shelf-life of arils.

16.2.7 Packaging and storability

Nutritional, microbial and sensory quality are the most important criteria to determine acceptability limits and the shelf-life of minimally processed pomegranate arils. The microbial standards, in particular, are crucial to ensure that processed produce is free of pathogenic microorganisms and is within the limits of the acceptable microbial count. Pomegranate aril storage at optimal modified atmospheric conditions is used to reduce the risk of mesophilic, psychrotrophic, enterobacteria and lactic acid bacteria as well as moulds and yeasts (Sepulveda *et al.*, 2000). In the Spanish regulations, the acceptable maximum aerobic bacterial count is 7 log colony forming units (CFU)/g.

Packaging plays an important role in maintaining the nutritional and microbial quality of minimally processed fresh pomegranate arils. Either active or passive MAP can be aused for extension of the shelf-life of arils. The hygienically extracted arils are weighed and packaged in polyethylene terephthalate (PET) punnets. Various types of selectively permeable polymeric films are used for packaging of minimally processed pomegranate arils to create a microenvironment that reduces respiratory activity by maintaining conditions unfavourable for the action of many contaminating microorganisms. MAP and low-temperature storage have been reported to be excellent methods to extend shelf-life by reducing microbial hazards yet maintaining the sensory and nutritional properties of minimally processed pomegranate arils. Active or passive MAP is used for packaging of pomegranate arils on a commercial scale by many suppliers. Research is ongoing with the aim of improving the shelf-life of minimally processed arils using different storage temperatures and packaging conditions (Table 16.2). MAP and low storage temperatures (1–5°C) have been shown to prolong the shelf-life of pomegranate arils for up to 18 days.

Table 16.2. Optimum conditions reported for minimal processing and storage to extend shelf-life of arils.

	Modified atmosphere packaging (MAP) composition		Storage temp		Storability	
Cultivars	CO_2	O_2	°C	Packaging film	Days	Reference
'Malas-e-Saveh'	10%	15%	4	3-Layer coating comprising polyethylene (LDPE), polyrpamide and polyethylene	15	Tavasoli Talarposhti *et al.*, 2016
'Wonderful'	40%	30%	5	High barrier polymeric film (polylid)	12	Banda *et al.*, 2015
	Passive			High barrier polymeric film (polylid)	9	
	Passive			Clamshell	6	
	1%	22%	4	BB4 (cryovac based on ethyl vinyl acetate)	14	Sepulveda *et al.*, 2000
	20%	-	5	-	16	Hess-Pierce and Kader, 2003
Acco	Mediated MAP		5	Perforation mediated	15	Hussein *et al.*, 2015
Acco and Herskawitz	Passive MAP		5	Polymeric film (polylid)	10	Caleb *et al.*, 2013
Primosole	6.5%	11.4%	5	Polypropylene	10	Palma *et al.*, 2009
'Mollar de Elche'	20.1–21.6 kPa	2–5 kPa	5	Polypropylene basket sealed with bi-axial-oriented polypropylene (BOPP)	15	López-Rubira *et al.*, 2005
	1%	30%	4	Semi-permeable plastic bag	10	Garcia *et al.*, 2000
	188 ml/l	22 ml/l	1	Oriented polyproylene (OPP)	7	Gil *et al.*, 1996a
	12.5%	8.5%	8	OPP	7	Gil *et al.*, 1996b

Ayhan and Eştürk (2009) reported yeast and mould growth below the limit of detection and aerobic mesophilic bacteria in the range of 2.30–4.51 log CFU/g in four different MAP conditions, namely air; 100% N 5% O_2 + 10% CO_2 + 85% N_2, and 70% O_2 + 10% CO_2 + 20% N_2 for 18 days' storage, without the sensory qualities being affected. The lowest count was observed at high oxygen application as 2.3 log CFU/g and the highest count was observed at 100% nitrogen application as 4.51 log CFU/g. High levels of O_2 are effectively used in preventing anaerobic fermentation reactions, inhibiting enzymatic discolouration, and inhibiting aerobic and anaerobic microbial growth. Among the MAP gases (N_2, CO_2 and O_2), the CO_2 has direct and significant antimicrobial activity due to alteration of cell membrane function including effects on nutrient uptake and absorption, direct inhibition of enzymes or decreases in the rate of enzyme reactions, and penetration of bacterial membranes leading to intracellular pH changes, and changes to the physicochemical properties of proteins (Farber, 1991). The arils from overmatured fruits have higher metabolic activity leading to lower shelf-life. Treatment of arils with UV-C can extend shelf-life by 10–12 days due to reduced microbial growth (López-Rubira *et al.*, 2005).

Various types of packaging materials have been used for MAP, which include polypropylene (PP), high-density polyethylene (HDPE), low-density polyethylene (LDPE), heat seal trays with oriented polypropylene film (OPPF), metalized polyester (MP) bags, polypropylene modular mates (PPMMs), rigid polystyrene vessels (RPVs), polyethylene standing pouches (PESPs), polyethylene terephthalate packs (PETPs) and perforated polypropylene trays (PPTs).

Minimal processing of pomegranate arils is a growing industry. The POM Wonderful US-based enterprise is a pioneer in minimal processing of arils and has popularized pomegranate and pomegranate-based products such as arils in North America and Europe. India is the largest producer of pomegranates in the world, with few major pomegranate processing enterprises who have international footprints. The major pomegranate aril-processing industries in India include Sam Agri Tech, INI farms, Kay Bee exports and Santosh exports. Figure 16.4 depicts the aril extraction and packaging process at one of the modern and high-tech minimal processing facilities.

16.2.8 Juice extraction

Pomegranate juice can be extracted by pressing arils in a screw, hydraulic or basket press, pressing halved fruits or by crushing whole fruits. The juice recovery from pomegranate fruits is affected by cultivar, the season of growing, production practices and the juice extraction method. Juice yield is generally found to be around 35–40% on a fruit weight basis. Phadnis (1974) reported a juice yield of about 42% on a fruit weight basis and 70% on aril weight basis. Similarly, Herrera-Hernandez *et al.* (2013) reported 61.12–73.26% of juice yield on aril basis. Saxena *et al.* (1987) recovered 36.41% juice by cutting the pomegranate fruits into quarters and pressing them in a rack and cloth hydraulic press under moderate pressure. The juice yield is generally higher in screw presses, followed by hydraulic presses. The juice from crushed whole fruits contains excess tannin from the rind (as much as 0.175%), which may need to be precipitated out by a gelation process. The hydraulic extraction of juice should be at a pressure less than 690 kPa (100 PSI) to avoid undue yield of tannins from the rind.

Different pomegranate juice extraction methods influence the physicochemical constituents, sensorial attributes and microbial population. Fruit juice pressed with the rind, membrane and non-edible parts contains more total phenols than that obtained from pressing only the arils. The increase in total anthocyanin content, total phenols, ascorbic acid and total antioxidant capacity was 33.96, 20.70, 22.49 and 65.54%, respectively (Gaikwad *et al.*, 2017a), when the juice was extracted from halved fruits compared with pressing only arils due to contribution from the peel. The peel is a rich source of phenolic compounds, anthocyanin and ascorbic acid (Li *et al.*, 2006; Shiban *et al.*, 2012; Pimenta Barros *et al.*, 2014; Janbi and Al-Said, 2014). The phenolic constituents contributed from the peel during juice extraction give colour, bitterness and astringency to the pomegranate juice.

Fig. 16.4. Automated aril extraction and packaging. (Photos: Ali Sarkhosh.)

16.2.9 Juice clarification and filtration

The pomegranate juice extracted by any means contains suspended particulate, looks turbid and hence has low storability. The polymerization of phenolic compounds and their interaction with other components (e.g. proteins) could cause a haze complex and turbidity in fruit juices (Bayindirli *et al.*, 1994). Phenolic compounds present in the juice also affect its taste and colour. Therefore, clarification of pomegranate juice is necessary to prevent the formation of a cloudy appearance and to improve taste. Fining agents and filtration processes are used to remove cloud from juices to enhance colour, flavour and storage stability. Fining agents such as gelatin, bentonite, albumin, activated carbon, casein, clay, ion-exchange waxes and polyvinyl poly pyrrolidone (PVPP) are used to enhance settling of the formed haze. Pomegranate juice can be clarified by 1 g/l gelatin addition before heat treatment (Vardin and Fenercioglu, 2003). Improved organoleptic score, retention of anthocyanin and colour density, and reduction in phenolic substances are also observed. Settled solids and other suspended matter are removed by filtration. The membrane filtration used more recently for juice clarification is more efficient, requires shorter time and less manpower. The operational cost for this membrane-based filtration method is quite low. The pressure-driven different membrane filtration processes used in the juice industry are based on the size of particulates, such as ultrafiltration (0.01–0.1 μm) and microfiltration (0.1–10 μm) (Girard and Fukumoto, 2000). The juices clarified via microfiltration have a greater loss of anthocyanins than those treated with gelatine. Ultrafiltration and microfiltration are the most commonly used methods to clarify and stabilize pomegranate juice. Ultrafiltration is a single-unit operation that can replace both conventional clarification and fining of fruit juice, and is becoming more common practice in all of the juice industry. The ultrafiltration removes suspended solids as well as haze-inducing and turbidity-causing substances to obtain a clear juice. Mirsaeedghazi *et al.* (2012) studied the use of ultrafiltration and microfiltration for pomegranate juice. Both of these processes showed similarities in turbidity reduction and removal of phenolic compounds. However, microfiltration retained a greater level of anthocyanins than ultrafiltration. Bagci (2014) studied different clarification pretreatments before ultrafiltration and reported that sequentially treatment with 0.4 g/l of PVPP and 0.5 g/l of bentonite at 50°C for 1 h for each treatment followed by ultrafiltration was the best, and reduced haze active phenolics (such as catechins) and proteins by the combined adsorptive effects of PVPP and bentonite during the pre-clarification step. These contributed to preventing oxidation and polymerization reactions and therefore precipitation, turbidity, browning and astringent flavour formation during further storage. The enzymatic treatment of juice with hemicellulase, cellulase, xylanase, glucanase, pectinase, carbohydrate or arabinose is also used prior to clarification to obtain clear juice.

16.2.10 Thermal and non-thermal juice processing

Thermal processing alone or in combination with biochemical or chemical preservative techniques is known to be the most effective method for inactivation of enzymes and microorganisms to increase shelf-life in the food industry. Pasteurization treatment is an important step in pomegranate juice processing. The pomegranate juice is susceptible to microbial contamination with acid-tolerant bacteria, fungi (yeasts and moulds) and pathogenic bacteria, which leads to deterioration of nutritional and sensorial properties such as functional ingredients, colour, flavour and odour, as well as incidence of food-borne diseases due to pathogenic bacteria or toxigenic fungi. The temperature and time combinations for pasteurization and storage conditions are critical in degradation of the organoleptic and microbiological stability of pomegranate juice. Thermal processing at 80°C for 5 min is found to be the most suitable condition for maintaining sensory quality and reducing the microbial load to below detectable limits (Gaikwad *et al.*, 2017a).

The bioactive compounds are affected to a great extent by exogenic factors such as oxygen, light and, especially, pH and temperature. To avoid losses of bioactive compounds alternative treatments such as pulsed electric field (PEF), ohmic heating, ultrasound, irradiation, high-pressure processing, active packaging and ozone treatment are being studied. Guo *et al.* (2014) reported feasibility of a commercial scale PEF processing system at 35 and 38 kV/cm for 281 μs at 55°C with a flow rate of 100 l/h for treatment of pomegranate juice. They found PEF treatment significantly inhibited the growth of total aerobic bacteria, which remained at <2.5 log CFU/ml during the 12-week storage period. No yeast and mould were detected (<0.69 log CFU/ml) in the PEF-treated juices during storage up to 10 and 12 weeks, which was similar to the thermally processed juice. PEF processing did not alter the contents of total phenolic and anthocyanin as compared with unprocessed juice. PEF processing had less impact on the colour of pomegranate juice than thermal processing. Alighourchi *et al.* (2013) studied the effect of sonication treatment on the quality of pomegranate juice and found no significant differences ($P < 0.01$) in pH, acidity and solid soluble content. Further, the degradation percentage of total and individual anthocyanins, total phenols, the antioxidant activity of juice and colour parameters were not considerable. These results indicated the non-destructive effects of ultrasound on the visual and chemical properties of pomegranate juice.

16.3 Pomegranate Products

16.3.1 Juice and juice-based beverages

The popularity of consuming pomegranate juice and juice-based beverages such as smoothies or nectars has increased during the past few years, along with the consumption of healthy beverages worldwide (Fig. 16.5). Pomegranate juice is consumed frequently for its bioactive compounds. It is worth mentioning that the antioxidant content in pomegranate juice is even higher than in black, white and green tea and red wines (Aslam *et al.*, 2006). The antioxidant activity of pomegranate juice is due to the presence of phenolic compounds and anthocyanins. The composition of pomegranate juice depends on cultivar type, postharvest and environmental

Fig. 16.5. Pomegranate juice of a local pomegranate cultivar grown in Yazad-Iran. (Photo: Alimohammad Yavari.)

factors, processing and storage factors. The colour of the juice is one of the most important attributes for consumers, who prefer bright and reddish pomegranate juice, avoiding brown and pale colours (Fig. 16.6). The final colour of the juice depends mainly on pomegranate cultivar, heat treatment and storage conditions. Pomegranate juice can be utilized for preparation of fruit drinks by using a varied proportion of juice, adjusting total soluble solids (TSS) and acidity to an acceptable level. The TSS is generally adjusted to 15°Brix and 0.25% acids by addition of cane sugar and citric acid (Vaidya *et al.*, 1998). Blended pomegranate fruit drinks are also being prepared commercially with the addition of various fruit juices or fruit purée (smoothies); for example, blueberry, cranberry, sweet orange, guava, grape juice and aloe vera (Cano-Lamadrid *et al.*, 2019). The quality of these pomegranate-based drinks (colour, bioactive compounds and organoleptic attributes) depends on the maturity of fruits, cultivar, processing and storage.

Commercial pomegranate juice or juice-based beverage processing at an industrial scale has important unit operations such as reception, washing, sorting, juice extraction, juice filtration, clarification, pasteurization, filling, sealing and bottle cooling. However, many industrial units also prefer to go for reconstitution of juices from juice concentrates. The reception, washing and sorting processes for the production of concentrated juice are the same as discussed in Section 16.2.

16.3.2 Concentrated juice

Concentrated pomegranate juice at a commercial scale is made by using evaporators without adding sugars or preservatives. The evaporated juice is generally concentrated five fold to get a TSS value of around 65–70% and pH 2.7–3.1. Dhumal *et al.* (2013) reported concentrated pomegranate juice of 65°Brix achieved in 37, 78 and 106 min by using atmospheric heating, microwave and rotary vacuum evaporation techniques, respectively. Following the pomegranate juice concentration process, Orak (2009) reported an increase in reducing sugars, glucose and fructose level to 46.5, 23.9 and 22.5%, respectively. Concentrated pomegranate juice may have two- to threefold higher minerals, antioxidant activity and other compounds than fresh pomegranate juice, but heat-sensitive bioactive compounds such as anthocyanin and punicalagin decrease owing to heat treatment. Juice concentrates are aseptically packaged in commercial processing units by using a bag in barrel type packaging. The improved storability of juice concentrates is due to high sugars and low moisture content and aseptic packaging. Due to longer storability and reduction in volume, it can be exported. This product is the preferred choice for many juice beverage industries worldwide for reconstitution into juice and juice-based beverages.

16.3.3 Carbonated drinks

As the taste of carbonated drinks is much liked by many consumers, especially the younger generation, soft drinks based on pomegranate can be very popular. Chemically, adding CO_2 to water creates carbonic acid, which is tasted by the sour-sensing taste cells on the tongue. Research

Fig. 16.6. Colour difference for pomegranate juices extracted from different cultivars. (Photos: Ali Sarkhosh.)

has suggested that a certain enzyme called carbonic anhydrase, which sits on those cells and reacts with the acid, causes carbonated water's familiar popping sensation. This popping sensation leads to the typical taste of carbonated beverages.

Pomegranate carbonated drinks are a healthy and tasty alternative to non-fruit-based carbonated beverages. Carbonated juice drinks can be prepared with different juice levels maintaining the TSS at 15°Brix and 0.3% acid levels after carbonation. Carbonated pomegranate beverages can be prepared by blending with various fruit juices such as grape, ginger, lemon, etc.

16.3.4 Seed and seed oil

Pomegranate seeds are the by-product of the juice-processing industry. The pomegranate aril contains a seed, which is around 4–10% of fresh fruit weight. Pomegranate seeds are an excellent source of dietary fibre, a rich source of various minerals and micronutrients (K, P, Mg, Ca, Zn, Mn and Fe) and a unique profile of fatty acids. The industrial importance of pomegranate seed is down to the oil, which is a unique natural product (the majority of the fatty acids are polyunsaturated ones, PUFAs) as compared with other plant oils. As an example of the quantity of oil in pomegranate fruit, around 25–26% (v/w) of its fatty acids are PUFAs in soft-seed cultivars (Gaikwad *et al.*, 2017b). Pomegranate seed oil (Fig. 16.7) consists of 65–80% conjugated fatty acids, the most important of which is punicic acid (9-*trans*-11-*cis*-13-*trans* octadecatrienoic acid). In a recent study the fatty acid profiles of 20 cultivars were classified depending on the type of pomegranate taste (sour, sour-sweet and sweet). Punicic acid is the most prevalent fatty acid in all of the studied cultivars (Table 16.3) (Alcaraz-Mármol *et al.*, 2015). Conjugated fatty acids are important because they inhibit eicosanoid metabolism at several points by the synthesis of prostaglandins from arachidonic acid. This makes them significant natural

Fig. 16.7. Pomegranate seed oil. (Photo: Nilesh N. Gaikwad.)

anti-inflammatory agents. Pomegranate seed oil contains other significant bioactive compounds; for example, it is the richest known plant source of a steroidal oestrogen and oestrone. Studies have shown pomegranate seed oil contains over 3 mg/g of 17-α-oestradiol, the mildest and safest steroidal oestrogen, which is an exceptionally potent antioxidant and brain-preserving compound. Other important compounds found in pomegranate seed oil include gamma-tocopherol, a rare and potent form of vitamin E, and phytosterols, that is, beta-sitosterol, stigmasterol, and campesterol, etc. Seed oil has been linked to improving heart health and also may protect against cancer (Lansky and Newman, 2007) and atherosclerosis (Boussetta *et al.*, 2009).

The pomegranate seed oil can be extracted by various methods such as soxhlet, supercritical CO_2, subcritical propane, superheated hexane, cold pressing, sonication-assisted microwaves and ultrasound-assisted extraction. Among these methods, pomegranate seed oil is mainly extracted through cold pressing due to its high value and tendency for oxidation. There are large amounts of conjugated unsaturated fatty acids in its triglyceride composition, which are very sensitive to heat and easily undergo cis/trans isomerization. Abbasi *et al.* (2008) reported the extraction of pomegranate seed oil using hexane and petroleum benzene by applying four extraction methods. Different methods of extraction with organic solvents (normal stirring, soxhlet, microwave irradiation and ultrasonic irradiation) showed a significant difference in the extraction yield. However, no differences were found when a single method was applied using different organic solvents. On the other hand, different extraction conditions from the various runs of supercritical CO_2 extraction resulted in different extraction yields, all of which were lower than those of the other extraction methods using organic solvents.

The industrial cold press technique of pomegranate seed oil extraction involves cleaning and separation of pulp from the seeds, drying and size reduction of seeds. Pomegranate seeds are then pressed in the cold press for extraction of seed oil. The seed cake remaining after oil extraction in the cold press can be utilized for the development of fibre-rich cookies. The effect of microwave pretreatment on pomegranate seed oil extraction yield and quality was studied (Gaikwad *et al.*, 2017b). The results suggested oil extraction yield was increased with an increase in microwave power, pretreatment time and extraction time. The microwave pretreatment was also found to be effective in reducing the extraction time. The optimum pretreatment conditions of 720 W, for 60 s and extraction time of 4 h was recommended for

Table 16.3. Fatty acid composition of pomegranate type as % of total fatty acid profile.

Fruit type	Palmitic	Linoleic	Oleic	Stearic	Punicic	Linolenic	Arachidic
Sour	2.73	4.42	4.82	0.47	67.2	19.30	0.50
Sour-sweet	3.82	4.76	5.34	0.59	67.9	16.04	0.48
Sweet	3.79	4.65	4.76	0.60	66.7	18.71	0.61

microwave-assisted soxhlet extraction of pomegranate seed oil. A light microscopic image of microwave-treated seeds shows loosened cell wall and expanded lipid bodies. Microwave pretreatment in oil seeds helps in increasing oil extraction yield due to an increase in mass transfer coefficient as the cell membrane gets ruptured. In addition, permanent pores are generated as a result of microwave pretreatment, enabling the oil to move through the permeable cell walls.

16.3.5 Dried products (dehydrated arils, spray-dried powder, leather skin)

Dried pomegranate-based products are good options for the pomegranate industry to increase its product catalogue and to avoid losses. Consumers are interested in these types of products due to their ease of consumption and the stability of their healthy and sensory properties, including aroma and taste attributes. Dried products are quite stable at room temperature with retention of flavour and taste due to their low water activity. Dried arils have low weight and, thus, their transport is cheap and they have high protection against enzymatic and oxidative spoilage. One example of dehydrated pomegranate arils is 'anardana', which is found around the world. Anardana (Fig. 16.8) consists of dried arils of sour-type pomegranates with hard seeds that are highly acidic in nature. It is popular in India, Pakistan, Iran and other southern Asian countries, and is widely used in traditional cuisine and especially for chutneys and curry. Pomegranate arils contain citric acid as the major acid besides malic acid, oxalic acid, succinic acid and tartaric acid (Saxena *et al.*, 1987). Anardana is used in ayurveda medicine as digestive and stomachic agent. Ground anardana powder is also used to provide more intense flavour to food preparations. It adds a peculiar taste to some famous north Indian delicacies. The dehydrated seeds are acidic (7.8–15.4% acids) with great mouth-feel and are good for digestion. Anardana is rich in vitamin C and minerals (Ca, Zn, Mn). Traditional healers use a number of formulations of anardana as ayurvedic medicine in treatment of dysentery, diarrhoea, stomach ache, inflammations, hymenoletidosis, dyspepsia, bronchitis and cardiac problems. Pomegranate varieties with high natural acid content are suitable for preparation of anardana. 'Solapur anardana' and 'Amalidana' are the two pomegranate hybrid varieties developed by ICAR-National Research Centre on Pomegranate and ICAR-Indian Institute of Horticulture Research, respectively, in India with high acids for anardana purposes. Anardana has good export potential in east Asian countries. Anardana is not enjoyed by some European consumers such as Spaniards due to the sourness and the hard wooden part. Dehydrated arils made from sweet or sour-sweet pomegranate cultivars with soft seeds, such as 'Mollar de Elche', can be produced and are a completely different product to Anardana, although the concept behind them is the same, dehydrating pomegranate arils.

Fig. 16.8. Anardana. (Photo: Nilesh N. Gaikwad.)

Various drying techniques are used currently for preparation of dehydrated arils around the world, such as sun drying, tray drying, greenhouse drying, osmotic drying, freeze drying, vacuum drying and vacuum freeze drying (Fig. 16.9). Various pretreatments are also used before drying of arils, such as steam blanching, chemical dips in sodium benzoate, citric acid and KMS (potassium metabisulfite) solution and osmotic dehydration with other juices. Steam blanching of arils for 30 s followed by sulfur fumigation at 0.3% for 60 min was found to be the best method as, of the methods tested, this took the least time to dry a given tray load, and had

Fig. 16.9. Pomegranate arils dehydrated by different techniques. (Photo: Calidad y Seguridad Alimentaria, UMH.)

the lowest non-enzymatic browning, furfural, hydroxyl methyl furfural and moisture contents (Thakur *et al.*, 2010).

It is worth mentioning other dehydrated pomegranate products. One of the newest and most interesting products is pomegranate juice powder, which can be incorporated into water or in other vegetable-based drinks (Fig. 16.10). The method used for its preparation is spray drying, which is an established drying technique for transforming liquid products into dry powders in a one-step processing operation, creating new products. This method is widely used in the food industry to obtain powders with positive aspects such as low moisture content and high- quality (with flavour resembling that of the fresh product), and increasing its shelf-life. However, some disadvantageous aspects are also produced with this type of drying – mainly stickiness and hygroscopicity of products, influenced by the presence of low molecular weight sugars and acids. Although spray drying is mainly considered a technique to create solid particles from a liquid matrix through a drying method, another interesting application is the encapsulation of bioactive compounds by this technique. Spray drying can be used to microencapsulate the bioactive pomegranate compounds, enriching the initial matrix by a concentration step leading to a highly valuable dried powder. The cost of production is lower than for other products (about 50 times less than for freeze drying), and the required equipment is widely available.

Fig. 16.10. 'Wonderful' (left) and 'Mollar de Elche' (right) pomegranate juice powder made by spray drying. (Photos: Calidad y Seguridad Alimentaria, UMH.)

Additionally, dehydrated fruit edible films have been developed recently, such as wraps and leathers. Preparation of fruit leather is not only a preservation method but it also has the advantage of improving value and diversifying fruit products. Fruit leather has been considered as a healthy choice and a big opportunity to improve fruit consumption. Apart from fruit pulp and fruit juice, it is also common to add starch, maltodextrins, gums or/and pectin to reduce the adhesion to the drying layers. Dried fruit-based leather is very common in Iran, Azerbaijan and Persian countries, and it is known as 'lavashak', especially the one based on sour pomegranate. In other countries, there are some developments using different pomegranate cultivars, which are more acceptable for their local consumers and mixing with other local fruits (Fig. 16.11).

16.3.6 Wine

Pomegranate wine has high antioxidant activity due to the presence of polyphenols and other bioactive compounds. It is associated with many health benefits coupled with excellent taste and aroma, which has led to increased interest in the recent past. It is an alcoholic beverage resulting from anaerobic fermentation of pomegranate fruit juice by yeast, in which sugars are converted into alcohol and carbon dioxide. Pomegranate wine (Fig. 16.12) has higher antioxidant capacity than juice (Sezer *et al.*, 2007) and a greater protective effect on LDL oxidation than red wine. Melatonin (N-acetyl-5-methoxytryptamine) is a neurohormone related to a broad array of physiological functions and has proven therapeutic properties. Melatonin was observed to be absent in pomegranate juice, but it is detected in prominent amounts with respect to other food

Fig. 16.11. Leather based on 'Wonderful' pomegranate juice and quince purée. (Photo: Calidad y Seguridad Alimentaria, UMH.)

Fig. 16.12. Pomegranate wine. (Photo: Nilesh N. Gaikwad.)

matrixes (0.54–5.50 ng/ml) in pomegranate wine (Mena *et al.*, 2012).

The industrial process of pomegranate winemaking involves the pressing of whole fruits without crushing to avoid excessive astringency in the wine. Alternatively, separated arils can be pressed in a hydraulic press to get fresh juice. The juice is placed in jacketed fermentation tanks of capacity 10,000 or 20,000 l for fermentation. The jacketed fermentation tanks are made of food-grade stainless steel and fitted with a motorized agitator for agitation of juice for facilitating fermentation. The fermentation tank has an arrangement for cooling by recirculation of chilled water through the jackets of the tank. A preservative, potassium metabisulfite, is added to the juice to an extent that should not affect the yeast. The juice is pasteurized to avoid unwanted microbial growth. Pomegranate juice has a TSS of around 12–16°Brix and acids of 0.3–0.75 g/100 g, and for winemaking the juice needs to be adjusted to higher levels of about 22–26°Brix using sucrose. pH also needs to be adjusted to pH 4 by using sodium bicarbonate or calcium carbonate. Activated yeast culture of a strain of *Saccharomyces cerevisiae* is used for preparing a starter culture, which will be later used for fermentation. However, at the commercial scale encapsulated yeast cultures are used for fermentation. Usually, the pomegranate fermentation process takes longer and is carried out at a lower temperature than grape juice (Cohen *et al.*, 2012). In order to prepare sweet table wine, sugar is added to bring its TSS to 8–10°Brix after ageing. The wine is clarified using bentonite, flash pasteurized and hot bottles are sealed. Pomegranate wine is prepared on a commercial scale in Turkey and India.

An innovative, functional beverage based on a dealcoholized red wine supplemented with a pomegranate extract enriched in ellagitannins with beneficial cardiovascular health properties has also been developed (Tárrega *et al.*, 2014). The addition of 0.16% (w/v) pomegranate juice extract resulted in a product enriched with 16.3 mg of specific ellagitannins per 100 ml. The use of immobilized cells instead of free cells for fermentation is gaining ground due to enhanced fermentation productivity, the feasibility of continuous processing of cells and downstream processing (Stewart and Russell, 1986).

Fig. 16.13. Pomegranate molasses in a local market in Yazad-Iran. (Photo: Alimohammad Yavari.)

Fig. 16.14. Pomegranate jam in a local market in Yazad-Iran. (Photo: Alimohammad Yavari.)

16.3.7 Molasses

Pomegranate molasses (pomegranate paste, called 'Robb' in Iran) is a traditional condiment commonly used in Middle Eastern, African and Mediterranean cooking (Fig. 16.13). It is a thick pomegranate syrup, slightly astringent, sweet-sour, deep and dark red in colour used in cooking, in salads and in many dishes to improve the taste and aroma . It is also used to marinate meat as it contains proteolytic enzymes, which act as a meat tenderizer (Hobani and Elansari, 2004). It is a concentrated product produced by boiling of pomegranate juice without the further addition of sugar or other additives. Traditional methods are still being used to produce pomegranate molasses, a thick, dark red liquid (70°Brix) formed after evaporation in an open vessel or under a vacuum of pomegranate juice. Commercial production of pomegranate molasses typically includes cleaning and crushing of pomegranates, extraction, filtration, clarification and concentration of pomegranate juice.

16.3.8 Jam

Pomegranate jam can be prepared by concentrating pomegranate juice and heating the mixture on a low heat for a long period with the addition of sugar (Fig. 16.14). The finished product contains 70–75%TSS with a thick consistency. Maestre *et al.* (2000) reported on pomegranate jam and preserves made from frozen 'Mollar' pomegranate juice by adding citric acid, pectin and sucrose. It was observed that, during processing treatment, 25% of the colour pigments are destroyed. High methoxypectins yielded better pomegranate jams. Jam stored at 5°C and without light exposure had an extended shelf-life.

16.3.9 Jelly

In the preparation of pomegranate jellies, the percentage of sugar added is less than that used in marmalades and jams. The addition of jellying agents (pectin, agar, etc.) is necessary for preparation of jellies, which generally maintain very well the colour characteristics of the original juice. Better colour and pigment stability was observed in jellies prepared from the juice of sour pomegranates than those produced with sweet pomegranate cultivar 'Mollar'. Also, the acidification of juice produced a noteworthy improvement in colour of the jellies, both initially and during storage.

In sweeter pomegranate varieties jelly is prepared from a combination of juice and sugar 1:1, and citric acid as acidulate to achieve better quality, colour, flavour and acceptance.

16.3.10 Antioxidants from pomegranate marc

Recently, natural antioxidants have become very popular in preference to synthetic antioxidants such as BHA (butylated hydroxyanisole) and BHT (butylated hydroxytoluene) for medical and food applications among consumers (Wang *et al.*, 2011). In the commercial juice processing industry, the marc is left over after juice extraction. The marc consists of approximately 73% peel and 27% seeds. It has a high potential for value addition as a source of phenolics, proanthocyanidins and flavonoids, which are basically antioxidants. Pomegranate peel, in particular, possesses relatively higher antioxidant activity than seed and pulp. This is because of polyphenols in peel composed of condensed tannins and anthocyanins. In general, the solvent used, temperature, solid–liquid ratio and particle size are influential parameters for extraction of antioxidants from pomegranate marc (Bucić-Kojić *et al.*, 2007). Commercial-scale antioxidant extraction will increase in the future and marc has great potential for use as a food additive.

16.3.11 Bio-colours

Natural dyes obtained from plant sources with minimal chemical processing are recognized to be eco-friendly, cost-effective, renewable and non-carcinogenic in nature and have no allergic reaction on the skin. The colourants (from plant origin) used in dyeing various fabrics are mainly flavonoids, along with indigo and anthraquinones. The extracted natural dye from the brown dry rind of pomegranate fruit has been used as a natural colourant for textiles from ancient times. The principal colouring components of pomegranate rind include tannins and flavonols. The flavonoids give a variety of yellow, brown and green colour shades. Various techniques are used to extract natural colourants from pomegranate rind, such as ultrasonic-assisted, enzyme-assisted and enzyme-mediated ultrasonic-assisted extraction (Tiwari *et al.*, 2010), solvent extraction (Kulkarni *et al.*, 2011), microwave-assisted (Sinha *et al.*, 2012) and supercritical fluid. The rind of pomegranate contains a considerable amount of tannin, about 19% with pelletierine (Tiwari *et al.*, 2010). The main colouring agent in pomegranate peel is granatonine, which is present in the alkaloid form N-methyl granatonine (Goodarzian and Ekrami, 2010). Natural dyes from the peel of pomegranate are used in colouration of lipsticks and other cosmetics. Anthocyanins are considered as potential replacements for synthetic colouring agents because of their bright, attractive colours and water solubility (Kong *et al.*, 2003). The incorporation of anthocyanin in food systems not only increases colour intensity but also increases the medicinal and therapeutic values of food products.

16.3.12 Protected brands

The European Union scheme of geographical indications known as Protected Designation of Origin (PDO) promotes and protects high-quality agricultural products and foodstuffs by marking them with specific logos and thus helping in their identification. Currently, only a pomegranate cultivar in Spain is under PDO generating a novel quality product in the market (DOP 'Mollar de Elche') (R2016/83, www.granadaselche.com). 'Mollar de Elche' pomegranate has characteristics and qualities that differentiate it from the rest of the pomegranates (Fig. 16.15). Noted for its particular sweetness, it has a colour that can range from cream to deep red and its nugget (woody part) is soft. About 30–40% of Spanish production, around 20,000 t, is destined for the domestic market and of this, 10% is for the juice industry. The remaining 60–70% production of 'Mollar de Elche' pomegranates is exported.

On the other hand, the 'HydroSOStainable' brand is recognized due to the environmental friendly agronomic practices applied in different Mediterranean orchards, such as olive, almond, pistachio and pomegranate, which simultaneously reduce farmers' expenses and improve quality of fruits (bioactive compounds, sensory attributes and consumer acceptance). To enhance these fruit parameters, pomegranate trees are grown under regulated deficit irrigation (RDI) and the obtained fruits are called 'hydroSOStainable' (Cano-Lamadrid *et al.*, 2018; Corell *et al.*, 2019) (Fig. 16.16).

Fig. 16.15. 'Mollar de Elche' DOP pomegranate fruits. (Photo: Calidad y Seguridad Alimentaria, UMH.)

Fig. 16.16. HydroSOStainable logo for fruits such as pomegranate fruits and pomegranate-based products. (From: Calidad y Seguridad Alimentaria, UMH.)

References

Abbasi, H., Rezaei, K. and Rashidi, L. (2008) Extraction of essential oils from the seeds of pomegranate using organic solvents and supercritical CO2. *Journal of the American Oil Chemists' Society* 85(1), 83–89. DOI: 10.1007/s11746-007-1158-x.

Alcaraz-Mármol, F., Nuncio-Jáuregui, N., Calín-Sánchez, Á., Carbonell-Barrachina, Á.A., Martínez, J.J. *et al.* (2015) Determination of fatty acid composition in arils of 20 pomegranates cultivars grown in Spain. *Scientia Horticulturae* 197, 712–718. DOI: 10.1016/j.scienta.2015.11.004.

Alighourchi, H.R., Barzegar, M., Sahari, M.A. and Abbasi, S. (2013) Effect of sonication on anthocyanins, total phenolic content, and antioxidant capacity of pomegranate juices. *International Food Research Journal* 20, 1703–1709.

Aslam, M.N., Lansky, E.P. and Varani, J. (2006) Pomegranate as a cosmeceutical source: pomegranate fractions promote proliferation and procollagen synthesis and inhibit matrix metalloproteinase-1 production in human skin cells. *Journal of Ethnopharmacology* 103(3), 311–318. DOI: 10.1016/j.jep.2005.07.027.

Aviram, M., Dornfeld, L., Rosenblat, M., Volkova, N., Kaplan, M. *et al.* (2000) Pomegranate juice consumption reduces oxidative stress, atherogenic modifications to LDL, and platelet aggregation: studies in humans and in atherosclerotic apolipoprotein E-deficient mice. *The American Journal of Clinical Nutrition* 71(5), 1062–1076. DOI: 10.1093/ajcn/71.5.1062.

Ayhan, Z. and Eştürk, O. (2009) Overall quality and shelf-life of minimally processed and modified atmosphere packaged 'ready-to-eat' pomegranate arils. *Journal of Food Science* 74(5), C399–C405. DOI: 10.1111/j.1750-3841.2009.01184.x.

Bagci, P.O. (2014) Effective clarification of pomegranate juice: a comparative study of pretreatment methods and their influence on ultrafiltration flux. *Journal of Food Engineering* 141, 58–64. DOI: 10.1016/j.jfoodeng.2014.05.009.

Banda, K., Caleb, O.J., Jacobs, K. and Opara, U.L. (2015) Effect of active-modified atmosphere packaging on the respiration rate and quality of pomegranate arils (cv. Wonderful). *Postharvest Biology and Technology* 109, 97–105. DOI: 10.1016/j.postharvbio.2015.06.002.

Bayindirli, L., Sahin, S. and Artik, N. (1994) The effects of clarification methods on pomegranate juice quality. *Fruit Processing* 9, 264–270.

Boussetta, T., Raad, H., Lettéron, P., Gougerot-Pocidalo, M.-A., Marie, J.-C. *et al.* (2009) Punicic acid a conjugated linolenic acid inhibits TNF α-induced neutrophil hyperactivation and protects from experimental colon inflammation in rats. *PLoS ONE* 4(7), e6458. DOI: 10.1371/journal.pone.0006458.

Bucić-Kojić, A., Planinić, M., Tomas, S., Bilić, M. and Velić, D. (2007) Study of solid–liquid extraction kinetics of total polyphenols from grape seeds. *Journal of Food Engineering* 81(1), 236–242. DOI: 10.1016/j.jfoodeng.2006.10.027.

Caleb, O.J., Opara, U.L., Mahajan, P.V., Manley, M., Mokwena, L. *et al.* (2013) Effect of modified atmosphere packaging and storage temperature on volatile composition and postharvest life of minimally-processed pomegranate arils (cvs. 'Acco' and 'Herskawitz'). *Postharvest Biology and Technology* 79, 54–61. DOI: 10.1016/j.postharvbio.2013.01.006.

Cano-Lamadrid, M., Galindo, A., Collado-González, J., Rodríguez, P., Cruz, Z.N. *et al.* (2018) Influence of deficit irrigation and crop load on the yield and fruit quality in Wonderful and Mollar de Elche pomegranates. *Journal of the Science of Food and Agriculture* 98(8), 3098–3108. DOI: 10.1002/jsfa.8810.

Cano-Lamadrid, M., Turkiewicz, I.P., Tkacz, K., Sánchez-Rodríguez, L., López-Lluch, D. *et al.* (2019) A critical overview of labeling information of pomegranate juice-based drinks: phytochemicals content and health claims. *Journal of Food Science* 84(4), 886–894. DOI: 10.1111/1750-3841.14497.

Cohen, S.D., Tarara, J.M., Gambetta, G.A., Matthews, M.A. and Kennedy, J.A. (2012) Impact of diurnal temperature variation on grape berry development, proanthocyanidin accumulation, and the expression of flavonoid pathway genes. *Journal of Experimental Botany* 63(7), 2655–2665. DOI: 10.1093/jxb/err449.

Corell, M., Martín-Palomo, M.J., Sánchez-Bravo, P., Carrillo, T., Collado, J. *et al.* (2019) Evaluation of growers' efforts to improve the sustainability of olive orchards: Development of the hydroSOStainable index. *Scientia Horticulturae* 257, 108661. DOI: 10.1016/j.scienta.2019.108661.

Dhumal, S.S., Karale, A.R., More, T.A., Nimbalkar, C.A., Chavan, U.D. *et al.* (2013) Preparation of pomegranate juice concentrate by various heating methods and appraisal of its physicochemical characteristics. *Acta Horticulturae* 1089, 473–484.

El-Nemr, S.E., Ismail, I.A. and Ragab, M. (1990) Chemical composition of juice and seeds of pomegranate fruit. *Food / Nahrung* 34(7), 601–606. DOI: 10.1002/food.19900340706.

El-Nemr, S.E., Ismail, I.A. and Ragab, M. (1992) The chemical composition of the juice and seeds of pomegranate fruits. *Fruit Processing* 211, 162–164.

Fadavi, A., Barzegar, M. and Azizi, M.H. (2006) Determination of fatty acids and total lipid content in oilseed of 25 pomegranates varieties grown in Iran. *Journal of Food Composition and Analysis* 19(6-7), 676–680. DOI: 10.1016/j.jfca.2004.09.002.

Farber, J.M. (1991) Microbiological aspects of modified-atmosphere packaging technology a review. *Journal of Food Protection* 54(1), 58–70. DOI: 10.4315/0362-028X-54.1.58.

Gaikwad, N.N., Pal, R.K., Suryawanshi, S., Babu, K.D., Maity, A. *et al.* (2017a) Effect of extraction method and thermal processing on retention of bioacvtive compounds of pomegrante (*Punica granatum*, cv. *Bhagwa) juice. Indian Journal of Agricultural Sciences* 87, 1445–1452.

Gaikwad, N.N., Yedle, V.H., Yenge, G., Suryavanshi, S., Babu, K.D. *et al.* (2017b) Effect of microwave pretreatment on extraction yield of pomegranate seed (cv. Bhagwa) oil. *International Journal of Chemical Studies* 5, 1291–1294.

Garcia, E., Salazar, D.M., Melgarejo, P. and Coret, A. (2000) Determination of the respiration index and of the modified atmosphere inside the packaging of minimally processed products. *Options Mediterranéennes. Serie A, Seminaires Mediterraneens* 42, 247–251.

Gil, M., Martinez, J. and Artes, F. (1996a) Minimally processed pomegranate seeds. *Lebensmittel-Wissenschaft und Technologie* 29(8), 708–713. DOI: 10.1006/fstl.1996.0110.

Gil, M.I., Artes, F. and Tomas-Barberan, F.A. (1996b) Minimal processing and modified atmosphere packaging effects on pigmentation of pomegranate seeds. *Journal of Food Science* 61(1), 161–164. DOI: 10.1111/j.1365-2621.1996.tb14749.x.

Girard, B. and Fukumoto, L.R. (2000) Membrane processing of fruit juices and beverages: a review. *Critical Reviews in Food Science and Nutrition* 40(2), 91–157. DOI: 10.1080/10408690091189293.

Goodarzian, H. and Ekrami, E. (2010) Wool dyeing with extracted dye from pomegranate (*Punica granatum* L.) peel. *World Applied Sciences Journal* 8, 1387–1389.

Guo, M., Jin, T.Z., Geveke, D.J., Fan, X., Sites, J.E. *et al.* (2014) Evaluation of microbial stability, bioactive compounds, physicochemical properties, and consumer acceptance of pomegranate juice processed in a commercial scale pulsed electric field system. *Food and Bioprocess Technology* 7(7), 2112–2120. DOI: 10.1007/s11947-013-1185-6.

Herrera-Hernandez, M.G., Mondragon-Jacobo, C., Soria-Lara, D.M. and Maldonado, S.H.G. (2013) Comparative study of physicochemical and functional characteristics in juices from new Mexican pomegranate cultivars (*Punica granatum* L.) and Wonderful variety. *Biochemistry and Biophysics* 1, 35–42.

Hess-Pierce, B. and Kader, A. (2003) Responses of 'Wonderful' pomegranates to controlled atmosphere. *Acta Horticulturae* 600, 751–757.

Hobani, A.I. and Elansari, A.M. (2004) Thermal transitions of pomegranate extracts using modulated differential scanning calorimeter (MDSC). *International Journal of Food Properties* 7(3), 671–681. DOI: 10.1081/JFP-200033086.

Hussein, Z., Caleb, O.J., Jacobs, K., Manley, M. and Opara, U.L. (2015) Effect of perforation-mediated modified atmosphere packaging and storage duration on physicochemical properties and microbial quality of fresh minimally processed 'Acco' pomegranate arils. *LWT – Food Science and Technology* 64(2), 911–918. DOI: 10.1016/j.lwt.2015.06.040.

Janbi, A.H.A. and Al-Said, S.A. (2014) The effect of technological treatments of the pasteurized juice processing on the antioxidant compounds content and the antioxidant activity of pomegranate (*Punica granatum* L) fruits juice. *Middle East Journal of Applied Sciences* 4, 232–242.

Khazaei, J., Ekrami-Rad, N., Safa, M. and Nosrati, S.-Z. (2008) Effect of air-jet impingement parameters on the extraction of pomegranate arils. *Biosystems Engineering* 100(2), 214–226. DOI: 10.1016/j.biosystemseng.2008.02.010.

Kong, J.-M., Chia, L.-S., Goh, N.-K., Chia, T.-F. and Brouillard, R. (2003) Analysis and biological activities of anthocyanins. *Phytochemistry* 64(5), 923–933. DOI: 10.1016/S0031-9422(03)00438-2.

Kulkarni, S.S., Gokhale, A.V., Bodake, U.M. and Pathade, G.R. (2011) Cotton dyeing with natural dye extracted from pomegranate (*punica granatum*) peel. *Universal Journal of Environmental Research and Technology* 1, 135–139.

Pimenta Barros, Z.M., Salgado, J.M., Melo, P.S. and Biazotto, F.O. (2014) Enrichment of commercially-prepared juice with pomegranate (*Punica granatum* L.) peel extract as a source of antioxidants. *Journal of Food Research* 3(6), 179–187. DOI: 10.5539/jfr.v3n6p179.
Wang, Z., Pan, Z., Ma, H. and Atungulu,, G.G. (2011) Extract of phenolics from pomegranate peels. *The Open Food Science Journal* 5(1), 17–25.
Lansky, E.P. and Newman, R.A. (2007) *Punica granatum* (pomegranate) and its potential for prevention and treatment of inflammation and cancer. *Journal of Ethnopharmacology* 109(2), 177–206. DOI: 10.1016/j.jep.2006.09.006.
Li, Y., Guo, C., Yang, J., Wei, J., Xu, J. *et al.* (2006) Evaluation of antioxidant properties of pomegranate peel extract in comparison with pomegranate pulp extract. *Food Chemistry* 96(2), 254–260. DOI: 10.1016/j.foodchem.2005.02.033.
Li, Y., Qi, Y., Huang, T.H., Yamahara, J. and Roufogalis, B.D. (2008) Pomegranate flower: a unique traditional antidiabetic medicine with dual PPAR-α/-γ activator properties. *Diabetes. Obesity and Metabolism* 10, 10–17.
López-Rubira, V., Conesa, A., Allende, A. and Artés, F. (2005) Shelf life and overall quality of minimally processed pomegranate arils modified atmosphere packaged and treated with UV-C. *Postharvest Biology and Technology* 37(2), 174–185. DOI: 10.1016/j.postharvbio.2005.04.003.
Maestre, J., Melgarejo, P., Tomas-Barberan, F.A. and Garcia-Viguera, C. (2000) New food products derived from pomegranate. *Symposium on Production Processing and Marketing of Pomegranate in the Mediterranean Region. Advances in Research and Technology*, Orihuela, Spain, pp. 243–245.
Mena, P., Gil-Izquierdo, Á., Moreno, D.A., Martí, N. and García-Viguera, C. (2012) Assessment of the melatonin production in pomegranate wines. *LWT – Food Science and Technology* 47(1), 13–18. DOI: 10.1016/j.lwt.2012.01.009.
Mirsaeedghazi, H., Mousavi, S.M., Emam-Djomeh, Z., Rezaei, K., Aroujalian, A. *et al.* (2012) Comparison between ultrafiltration and microfiltration in the clarification of pomegranate juice. *Journal of Food Process Engineering* 35(3), 424–436. DOI: 10.1111/j.1745-4530.2010.00598.x.
Orak, H.H. (2009) Evaluation of antioxidant activity, colour and some nutritional characteristics of pomegranate (*Punica granatum* L.) juice and its sour concentrate processed by conventional evaporation. *International Journal of Food Sciences and Nutrition* 60(1), 1–11. DOI: 10.1080/09637480701523306.
Orgil, O., Schwartz, E., Baruch, L., Matityahu, I., Mahajna, J. *et al.* (2014) The antioxidative and antiproliferative potential of non-edible organs of the pomegranate fruit and tree. *LWT – Food Science and Technology* 58(2), 571–577. DOI: 10.1016/j.lwt.2014.03.030.
Palma, A., Schirra, M., D'Aquino, S., La Malfa, S. and Continella, G. (2009) Chemical properties changes in pomegranate seeds packaged in polypropylene trays. *Acta Horticulturae* 818,323–330. DOI: 10.17660/ActaHortic.2009.818.48.
Phadnis, N.A. (1974) Pomegranate for dessert and juice. *Indian Horticulture* 19, 9.
Safa, M., Khazaei, J. and Kianmehr, M.H. (2006) Extracting the pomegranate arils (seeds) using a pneumatic method. *Acta Horticulturae* 818, 353–362.
Saxena, A.K., Manan, J.K. and Berry, S.K. (1987) Pomegranates: post-harvest technology, chemistry and processing. *Indian Food Packer* 41, 43–60.
Seeram, N.P., Zhang, Y., Reed, J.D., Krueger, C.G. and Vaya, J. (2006) Pomegranate phytochemicals. In: *Pomegranates: Ancient Roots to Modern Medicine*. CRC Press, Taylor and Francis, Boca Raton, Florida, pp. 3–29.
Sepulveda, E., Galletti, L., Sáenz, C. and Tapia, M. (2000) Minimal processing of pomegranate var Wonderful. *CIHEAM-Options Mediterraneennes* 42, 237–242.
Sezer, E.D., Akçay, Y.D., İlanbey, B., Yıldırım, H.K. and Sözmen, E.Y. (2007) Pomegranate wine has greater protection capacity than red wine on low-density lipoprotein oxidation. *Journal of Medicinal Food* 10(2), 371–374. DOI: 10.1089/jmf.2006.210.
Shiban, M.S., Al-Otaibi, M.M. and Al-Zoreky, N.S. (2012) Antioxidant activity of pomegranate (*Punica granatum* L.) fruit peels. *Food and Nutrition Sciences* 3, 991–996.
Sinha, K., Saha, P.D. and Datta, S. (2012) Response surface optimization and artificial neural network modeling of microwave assisted natural dye extraction from pomegranate rind. *Industrial Crops and Products* 37(1), 408–414. DOI: 10.1016/j.indcrop.2011.12.032.
Stewart, G.G. and Russell, I. (1986) One hundred years of yeast research and development in the brewing industry. *Journal of the Institute of Brewing* 92(6), 537–558. DOI: 10.1002/j.2050-0416.1986.tb04453.x.

Tavasoli Talarposhti, S., Barzegar, M., Hamidi-Esfahani, Z. and Faculty of Agriculture, Tarbiat Modares University, Tehran, Iran., (2016) Effect of modified atmosphere packaging on aril physico-chemical and microbial properties of two pomegranate cultivars (*Punica granatum* L.) grown in Iran. *Nutrition and Food Sciences Research* 3(4), 29–40. DOI: 10.18869/acadpub.nfsr.3.4.29.

Thakur, N.S., Bhat, M.M., Rana, N. and Joshi, V.K. (2010) Standardization of pre-treatments for the preparation of dried arils from wild pomegranate. *Journal of Food Science and Technology* 47(6), 620–625. DOI: 10.1007/s13197-010-0091-4.

Tiwari, H.C., Singh, P., Mishra, P.K. and Srivastava, P. (2010) Evaluation of various techniques for extraction of natural colorants from pomegranate rind – ultrasound and enzyme assisted extraction. *Indian Journal of Fiber and Textile Research* 35, 272–276.

Tárrega, M.A., Varela, P., Fromentin, E., Feuillère, N., Issaly, N. *et al.* (2014) Specific phenolic compounds and sensory properties of a new dealcoholized red wine with pomegranate (*Punica granatum* L.) extract. *Food Science and Technology International* 20(6), 421–429. DOI: 10.1177/1082013213489128.

Vaidya, R.N., Kotecha, P.M. and Kadam, S.S. (1998) Studies on mixed fruit juice beverages based on beer, pomegranate and guava. *Beverage and Food World* 25, 41–47.

Vardin, H. and Fenercioglu, H. (2003) Study on the development of pomegranate juice processing technology: clarification of pomegranate juice. *Nahrung/Food* 47(5), 300–303. DOI: 10.1002/food.200390070.

Viuda-Martos, M., Fernández-López, J. and Pérez-Álvarez, J.A. (2010) Pomegranate and its many functional components as related to human health: a review. *Comprehensive Reviews in Food Science and Food Safety* 9(6), 635–654. DOI: 10.1111/j.1541-4337.2010.00131.x.

Zarfeshany, A., Asgary, S. and Javanmard, S.H. (2014) Potent health effects of pomegranate. *Advanced Biomedical Research* 3, 100. DOI: 10.4103/2277-9175.129371.

Lansky, E.P. and Newman, R.A. (2007) *Punica granatum* (pomegranate) and its potential for prevention and treatment of inflammation and cancer. *Journal of Ethnopharmacology* 109(2), 177–206. DOI: 10.1016/j.jep.2006.09.006.

Li, Y., Guo, C., Yang, J., Wei, J., Xu, J. *et al.* (2006) Evaluation of antioxidant properties of pomegranate peel extract in comparison with pomegranate pulp extract. *Food Chemistry* 96(2), 254–260. DOI: 10.1016/j.foodchem.2005.02.033.

Li, Y., Qi, Y., Huang, T.H., Yamahara, J. and Roufogalis, B.D. (2008) Pomegranate flower: a unique traditional antidiabetic medicine with dual PPAR-α/-γ activator properties. *Diabetes. Obesity and Metabolism* 10, 10–17.

López-Rubira, V., Conesa, A., Allende, A. and Artés, F. (2005) Shelf life and overall quality of minimally processed pomegranate arils modified atmosphere packaged and treated with UV-C. *Postharvest Biology and Technology* 37(2), 174–185. DOI: 10.1016/j.postharvbio.2005.04.003.

Maestre, J., Melgarejo, P., Tomas-Barberan, F.A. and Garcia-Viguera, C. (2000) New food products derived from pomegranate. *Symposium on Production Processing and Marketing of Pomegranate in the Mediterranean Region. Advances in Research and Technology*, Orihuela, Spain, pp. 243–245.

Mena, P., Gil-Izquierdo, Á., Moreno, D.A., Martí, N. and García-Viguera, C. (2012) Assessment of the melatonin production in pomegranate wines. *LWT – Food Science and Technology* 47(1), 13–18. DOI: 10.1016/j.lwt.2012.01.009.

Mirsaeedghazi, H., Mousavi, S.M., Emam-Djomeh, Z., Rezaei, K., Aroujalian, A. *et al.* (2012) Comparison between ultrafiltration and microfiltration in the clarification of pomegranate juice. *Journal of Food Process Engineering* 35(3), 424–436. DOI: 10.1111/j.1745-4530.2010.00598.x.

Orak, H.H. (2009) Evaluation of antioxidant activity, colour and some nutritional characteristics of pomegranate (*Punica granatum* L.) juice and its sour concentrate processed by conventional evaporation. *International Journal of Food Sciences and Nutrition* 60(1), 1–11. DOI: 10.1080/09637480701523306.

Orgil, O., Schwartz, E., Baruch, L., Matityahu, I., Mahajna, J. *et al.* (2014) The antioxidative and anti-proliferative potential of non-edible organs of the pomegranate fruit and tree. *LWT – Food Science and Technology* 58(2), 571–577. DOI: 10.1016/j.lwt.2014.03.030.

Palma, A., Schirra, M., D'Aquino, S., La Malfa, S. and Continella, G. (2009) Chemical properties changes in pomegranate seeds packaged in polypropylene trays. *Acta Horticulturae* 818,323–330. DOI: 10.17660/ActaHortic.2009.818.48.

Phadnis, N.A. (1974) Pomegranate for dessert and juice. *Indian Horticulture* 19, 9.

Safa, M., Khazaei, J. and Kianmehr, M.H. (2006) Extracting the pomegranate arils (seeds) using a pneumatic method. *Acta Horticulturae* 818, 353–362.

Saxena, A.K., Manan, J.K. and Berry, S.K. (1987) Pomegranates: post-harvest technology, chemistry and processing. *Indian Food Packer* 41, 43–60.

Seeram, N.P., Zhang, Y., Reed, J.D., Krueger, C.G. and Vaya, J. (2006) Pomegranate phytochemicals. In: *Pomegranates: Ancient Roots to Modern Medicine*. CRC Press, Taylor and Francis, Boca Raton, Florida, pp. 3–29.

Sepulveda, E., Galletti, L., Sáenz, C. and Tapia, M. (2000) Minimal processing of pomegranate var Wonderful. *CIHEAM-Options Mediterraneennes* 42, 237–242.

Sezer, E.D., Akçay, Y.D., İlanbey, B., Yıldırım, H.K. and Sözmen, E.Y. (2007) Pomegranate wine has greater protection capacity than red wine on low-density lipoprotein oxidation. *Journal of Medicinal Food* 10(2), 371–374. DOI: 10.1089/jmf.2006.210.

Shiban, M.S., Al-Otaibi, M.M. and Al-Zoreky, N.S. (2012) Antioxidant activity of pomegranate (*Punica granatum* L.) fruit peels. *Food and Nutrition Sciences* 3, 991–996.

Sinha, K., Saha, P.D. and Datta, S. (2012) Response surface optimization and artificial neural network modeling of microwave assisted natural dye extraction from pomegranate rind. *Industrial Crops and Products* 37(1), 408–414. DOI: 10.1016/j.indcrop.2011.12.032.

Stewart, G.G. and Russell, I. (1986) One hundred years of yeast research and development in the brewing industry. *Journal of the Institute of Brewing* 92(6), 537–558. DOI: 10.1002/j.2050-0416.1986.tb04453.x.

Tavasoli Talarposhti, S., Barzegar, M., Hamidi-Esfahani, Z. and Faculty of Agriculture, Tarbiat Modares University, Tehran, Iran., (2016) Effect of modified atmosphere packaging on aril physico-chemical and microbial properties of two pomegranate cultivars (*Punica granatum* L.) grown in Iran. *Nutrition and Food Sciences Research* 3(4), 29–40. DOI: 10.18869/acadpub.nfsr.3.4.29.

Thakur, N.S., Bhat, M.M., Rana, N. and Joshi, V.K. (2010) Standardization of pre-treatments for the preparation of dried arils from wild pomegranate. *Journal of Food Science and Technology* 47(6), 620–625. DOI: 10.1007/s13197-010-0091-4.

Tiwari, H.C., Singh, P., Mishra, P.K. and Srivastava, P. (2010) Evaluation of various techniques for extraction of natural colorants from pomegranate rind – ultrasound and enzyme assisted extraction. *Indian Journal of Fiber and Textile Research* 35, 272–276.

Tárrega, M.A., Varela, P., Fromentin, E., Feuillère, N., Issaly, N. *et al.* (2014) Specific phenolic compounds and sensory properties of a new dealcoholized red wine with pomegranate (*Punica granatum* L.) extract. *Food Science and Technology International* 20(6), 421–429. DOI: 10.1177/1082013213489128.

Vaidya, R.N., Kotecha, P.M. and Kadam, S.S. (1998) Studies on mixed fruit juice beverages based on beer, pomegranate and guava. *Beverage and Food World* 25, 41–47.

Vardin, H. and Fenercioglu, H. (2003) Study on the development of pomegranate juice processing technology: clarification of pomegranate juice. *Nahrung/Food* 47(5), 300–303. DOI: 10.1002/food.200390070.

Viuda-Martos, M., Fernández-López, J. and Pérez-Álvarez, J.A. (2010) Pomegranate and its many functional components as related to human health: a review. *Comprehensive Reviews in Food Science and Food Safety* 9(6), 635–654. DOI: 10.1111/j.1541-4337.2010.00131.x.

Zarfeshany, A., Asgary, S. and Javanmard, S.H. (2014) Potent health effects of pomegranate. *Advanced Biomedical Research* 3, 100. DOI: 10.4103/2277-9175.129371.

17 Pomegranate Bioactive Compounds and Health

S. Asgary[1]*, Z. Azadeh[2], S. Dodman[2], A. Soleymani[3], M.H. Mirjalili[4], M.H. Farzaei[5], F. Hoseinzadeh-chahkandak[6], S. Hozeifi[7], E. Jasemi[8], M. Keshvari[1], Z. Lorigooini[2], S. Momtaz[8,9], A.M. Pourbagher-Shahri[7], R. Rahimi[10] and M. Sedaghat[7]

[1]Isfahan Cardiovascular Research Center, Cardiovascular Research Institute, Isfahan University of Medical Sciences, Isfahan, Iran; [2]Medical Plants Research Center, Basic Health Sciences Institute, Shahrekord University of Medical Sciences, Shahrekord, Iran; [3]Department of Chemical Engineering, University of Florida, Gainesville, Florida, USA; [4]Department of Agriculture, Medicinal Plants and Drugs Research Institute, Shahid Beheshti University, Tehran, Iran; [5]Pharmaceutical Sciences Research Center, Kermanshah University of Medical Sciences, Kermanshah, Iran; [6]Social Determinants of Health Research Center, Birjand University of Medical Sciences, Birjand, Iran; [7]Faculty of Medicine, Birjand University of Medical Sciences, Birjand, Iran; [8]Medicinal Plants Research Center, Institute of Medicinal Plants, ACECR, Karaj, Iran; [9]Toxicology and Diseases Group, the Institute of Pharmaceutical Sciences (TIPS), Tehran University of Medical Sciences, Tehran, Iran; [10]Department of Traditional Pharmacy, School of Persian Medicine, Tehran University of Medical Sciences, Tehran, Iran

17.1 Introduction

Cultivation and consumption of pomegranate (*Punica granatum* L.) dates back to at least 3000 BCE (Wu and Tian, 2017). Pomegranate originated in the region extending from Iran, Afganistan to northern India (Holland *et al.*, 2009), and has been cultivated since ancient times throughout the Mediterranean region. It was introduced into Spanish America in the late 16th century and into California by Spanish settlers in 1769. The name pomegranate derives from medieval Latin – pōmum 'apple' and grānātum 'seeded.' Possibly stemming from the old French word for the fruit, pomme-grenade, the pomegranate was known in early English as 'apple of Grenada' – a term that today survives only in heraldic blazons. This is a folk etymology, confusing the Latin granatus with the name of the Spanish city of Granada, which derives from Arabic.

17.2 Ethnobotany and Traditional Medicine

The pomegranate is a symbol of life, longevity, health, femininity, fecundity, knowledge,

*Corresponding author: sedighehasgary@gmail.com

morality, immortality and spirituality, if not divinity (Mahdihassan, 1984). It has appeared in Greek mythology (Newman and Lansky, 2007). Pomegranate appears in the coats of arms of several British medical societies (Langley, 2000). For thousands of years, many cultures have believed in various benefits of pomegranate for health, fertility, longevity and rebirth. Pomegranates feature prominently in Judaism, Christianity, Islam, Buddhism and Zoroastrianism.

Pomegranate with the common Persian name of 'Anaar' was much used in Zoroastrian ritual and domestic observances. In Buddhist art, the fruit represents the essence of favourable influences, and along with the citrus and the peach, the pomegranate is one of the three blessed fruits. In Judaism, pomegranate seeds are said to number 613 – one for each of the Bible's 613 commandments. The pomegranate was revered for the beauty of its shrub, flowers and fruit, symbolizing sanctity, fertility and abundance. A symbol of resurrection and life everlasting in Christian art, the pomegranate is often found in devotional statues and paintings of the Virgin and Child. The heavenly paradise of the Koran describes four gardens with shade, springs and fruits – including the pomegranate. Legend holds that each pomegranate contains one seed that has come down from paradise. Pomegranates have had a special role as a fertility symbol in weddings among the Bedouins of the Middle East. Abundant seeds ensure that the couple who eat it will have many children (Langley, 2000; Newman and Lansky, 2007).

Preparations of different parts of the plant, including flower, fruit juice, peel and root bark, have been used for a wide variety of disorders. Dioscorides in his book *Materia Medica* written in the 1st century describes some of them:

> All sorts of pomegranates are of a pleasant taste and good for ye stomach. The juice of the kernells prest out, being sod and mixed with Hony, are good for the ulcers that are in ye mouth and in ye Genitalls and in the seate, as also for the Pterygia in digitis and for the Nomae and ye excrescencies in ulcers, and for ye paines of ye eares, and for the griefs in ye nosthrills... The decoction of ye flowers is a collution of moist flagging gummes and of loose teeth ... ye rinde having a binding faculty ... but ye decoction of ye roots doth expell and kill the Latas tineas ventris. (Langley, 2000)

In Algeria, pomegranate has been claimed to be haemostatic and antiseptic, especially in the intestine, and is used for the treatment of various aliments including anaemia, gum and teeth diseases, peptic ulcers and eczema (Merzouki *et al.*, 2000; Ouelbani *et al.*, 2016). In China, the pomegranate is widely represented in ceramic art symbolizing fertility, abundance, posterity, numerous and virtuous offspring, and a blessed future (Langley, 2000). A popular wedding present in Chinese culture is a picture of a split pomegranate (Newman and Lansky, 2007). Pomegranate is known as shí liú in China and is traditionally used as an antihelmintic and vermifuge, to eliminate parasites, to treat mouth ulcers, diarrhoea, acidosis, dysentery, hemorrhage, microbial infections and respiratory pathologies, and as an antipyretic (Lee *et al.*, 2012). Among the 216 prescriptions of Tibetan medicine in the treatment of spleen and stomach diseases, pomegranate seed was used at the highest frequency (Gongbao *et al.*, 2018). Ancient Egyptians buried their rulers with pomegranate (Newman and Lansky, 2007). In Egyptian culture, several common ailments such as inflammation, diarrhoea, intestinal worms, cough and infertility have been treated by exploiting pomegranate peel extract (PoPx) (Ismail *et al.*, 2012). In Guatemala, *P. granatum*, with the common name of Granada, is used topically for treatment of various dermatomucosal disorders. Its flower is used for management of eye irritation, leucorrhea, mouth lesions and stomatitis, as well as wounds and ulcers (Cáceres *et al.*, 1987). Pomegranate fruit has been used by siddha practitioners with the vernacular name of Mātuḷai for the treatment of various disorders, including bromhirdosis, hyperacidity, constipation, gastric ulcers, anorexia, wheezing, rheumatalgia and hypothyroidism (Esakkimuthu *et al.*, 2018). In ayurvedic medicine the pomegranate is considered 'a pharmacy unto itself' : the bark and roots are believed to have anthelmintic and vermifuge properties (Naqvi *et al.*, 1991), the peel acts as a powerful astringent and cure for diarrhoea and oral aphthae, and the juice as a 'refrigerant' (Arseculeratne *et al.*, 1985) and 'blood tonic' (Frawley and Lad, 1986). In the ancient ayurvedic system of medicine, the rind of the fruit and the bark of the pomegranate tree are used as a traditional remedy against diarrhoea, dysentery and intestinal parasites. The seeds and juice are considered a tonic for the heart, throat, eyes and for a variety of purposes, such as stopping nose bleeds and bleeding gums, toning skin, firming up sagging breasts and treating haemorrhoids (Bhowmik *et al.*, 2013). In the Rayalaseema region of India, immature fruit and stem bark of pomegranate are administered as a powder or decoction for treatment of peptic ulcers (Nagaraju and Rao, 1990). In Iran and Afghanistan it was much used in Zoroastrian rituals and domestic observances. In Persian mythology Isfandiyar eats a pomegranate and becomes invincible. In *The Persian Wars* Herodotus mentions golden pomegranates adorning the spears of warriors in the Persian phalanx (Langley, 2000). From the view of traditional Iranian medicine, the pharmacological activities and indications of sweet, sweet-sour, and sour pomegranate are different. Sweet pomegranate is a laxative, diuretic, liver tonic and aphrodisiac for people with a hot temperament and is useful for hepatic disorders, cough and pruritus. Sweet-sour pomegranate is useful for exacerbation of yellow bile, and other indications are similar to sweet pomegranate. Sour pomegranate is useful for temperatures of the stomach and liver, nausea and vomiting and as a mouthwash it is a perfect remedy for aphthous. Pomegranate flower is useful for aphthous and gingivitis (Avicenna, 1983). Pomegranate juice has been recommended in traditional Iranian medicine for improving vision. Local administration of pomegranate flower is suitable for treating inflammatory reactions of the eye (Namdar *et al.*, 2015). Pomegranate fruit is prescribed in traditional Iranian medicine for the treatment of depression (Tavakkoli-Kakhki *et al.*, 2014). *Punica granatum* L., with the Tunisian (Arabian) name of Romman, is traditionally used for the treatment of various ailments. Its fruit peel is used as a decoction for treatment of gastric ulcers and diarrhoea, and as mouthwash for treatment of gingivitis. The fruit itself is claimed to be hypoglycaemic and hypotensive (Boukef *et al.*, 1982). A decoction of root bark, stem and peel is an anthelminthic agent, and mouthwash made from the aforementioned parts, as well as its juice, is used for the treatment of gingivitis and pyorrhoea. A peel decoction is used for management of diarrhoea and its fruit is claimed to be effective for arthritic pain (Boukef *et al.*, 1982). The use of pomegranate was reported for the treatment of gastrointestinal ailments including diarrhoea and dysentery in Mexico (Lozoya *et al.*, 1987). Pomegranate ('Anar') is used among the rural and urban communities of Swat valley as a wild edible fruit for management of gastrointestinal disorders. Powdered fruit peels are given to children to relieve stomach pain, colic, control dysentery and improve digestion (Khan and Ahmad, 2015).

17.3 Chemical Composition

Different parts of the pomegranate, including leaves, flowers, fruits, seeds, roots and bark of the tree (Lansky and Newman, 2007), are the sources of various compounds of high chemical value. Phenolics, flavonoids, tannins, alkaloids, terpenoids, phytosterols, lignins, glycosides, saponins, fatty acids, amino acids, carbohydrates and vitamins have been identified in the different parts of the plant (McMahon *et al.*, 1995; Sharma and Maity, 2010; Cheng *et al.*, 2012). Study of the composition of this plant could result in the development of important chemical drugs.

17.3.1 Phenolic compounds

The presence of phenolics in the fruit peel, whole fruit, root and bark of the tree (Akkiraju *et al.*, 2016; Sharma *et al.*, 2018), seeds (Syed *et al.*, 2007), pericarp or rind (Moorthy *et al.*, 2013), mesocarp, exocarp, arils (Jaiswal *et al.*, 2010), juice (Jurenka, 2008), flowers (Jurenka, 2008; Tripathi and Kohli, 2011) and leaves (Sreedevi *et al.*, 2017) of the pomegranate has been confirmed. Total phenolic content (TPC) of pomegranate peel has been reported to be higher than that of its flowers, leaves and seeds (Elfalleh *et al.*, 2012b), and much higher than the phenolic content reported for its juice (Elfalleh *et al.*, 2009), therefore it is two- to three-fold higher than that of the whole fruit (Masci *et al.*, 2016). In addition, pomegranate peel extract has been reported to have approximately 10-fold higher total phenolic compounds than its pulp extract (Li *et al.*, 2006). TPC varies among pomegranate cultivars depending on the geographical region (Elfalleh *et al.*, 2011) and the peel colour (Table 17.1). The cultivars with dark red peel have higher TPC levels than the light-coloured ones (Gözlekçi *et al.*, 2011). Caffeic acid and its derivatives, caffeic acid phenethylester, caffeoylquinic acid, *p*-coumaric acid glucuronide (Al-Rawahi *et al.*, 2014), chlorogenic acid, *o*-coumaric acid, *p*-coumaric acid and *cis-p*-coumaric acid have been reported in the pomegranate (Wu and Tian, 2017). Caffeic acid has been found in different parts of the pomegranate (Jurenka, 2008). The amounts of phenolic acids such as caffeic, ferulic and *p*-coumaric acids were reported to be 18.9–20.7, 17.1–18.8 and 3.8–5.2 mg/100 g of fresh weight basis, respectively, in pomegranate peel of six Georgian cultivars (Pande and Akoh, 2009).

Caffeic acid (3.88–75.19 μg/g), *p*-coumaric acid (0.12–14.87 μg/g), ferulic acid (0.15–8.84 μg/g), sinapic acid (2.13–3.58 μg/g), syringic acid (15.17–88.24 μg/g) and vanillic acid (65.87–108.36 μg/g) have been reported to be the main phenolic acids of pomegranate peel of Pakistani cultivars (Mushtaq *et al.*, 2015). The amounts of caffeic, chlorogenic, ferulic, *p*-coumaric and 4-hydroxybenzoic acids have been reported as 21.4, 18.5, 9.8, 5.6 and 22.7 mg/g, respectively, in methanolic peel extract of Turkish pomegranate (Dikmen *et al.*, 2011). The mean values for content of caffeic, *p*-coumaric, gallic and vanillic acids in 21 Iranian pomegranate accessions have been reported to be 14.7, 3.9, 124.1 and 1.0 mg/100g dry weight, respectively (Mansour *et al.*, 2013). The active fraction of pomegranate peel from India ('Ruby' cultivar) was found to contain 0.707, 0.651 and 2.75 mg/g dry weight *p*-coumaric, cinnamic and gallic acids, respectively (Arun *et al.*, 2017). Chlorogenic, caffeic and gallic acids have been reported from Chinese pomegranate peel extract (0.37, 0.03 and 2.53 mg/100 mg, respectively) (Song *et al.*, 2016). The amount of gallic acid has been reported to be 8.91 mg/g in the peel of a Chinese pomegranate cultivar (Ma *et al.*, 2015) and 30.4 mg/g in the methanolic peel extract of Turkish pomegranate (Dikmen *et al.*, 2011). Gallic acid (5.52 mg/g) is the main phenolic acid identified in the peel extract of a Serbian pomegranate cultivar (Stojanović *et al.*, 2017). Malic acid, malic acid glucoside derivative and fragments of malic acid (Al-Rawahi *et al.*, 2014) as hydroxybutanedioic acids have also been reported in the pomegranate. Quinic acid, quinic acid methyl ester and acetyl glucoside derivative are some of the phenolic acids of hydroxy cyclohexane carboxylic acids that have also been found in pomegranate (Al-Rawahi *et al.*, 2014).

17.3.2 Tannins

There are several reports regarding the presence of tannins in different parts of the pomegranate that hydrolyse during the ripening of the fruits (Tzulker *et al.*, 2007). The peel of pomegranate contains approximately 12.1- and 16.6-fold higher tannin content than the seed and juice fractions, respectively. Total tannin content of pomegranate peel fractions was higher than their total flavonoid content and total anthocyanin content (Pande and Akoh, 2009). Hydrolysable tannins (HTs) are found in different parts of the pomegranate, such as fruit juice, whole fruits, seeds, leaves and bark of the tree (Tanaka *et al.*, 1985, 1986; El-Toumy and Rauwald, 2002). HTs (ellagitannins and gallotannins) are the predominant phenolic compounds in pomegranate peel (Çam and Hişil, 2010). Ellagitannins and gallotannins are structurally different (Schofield *et al.*, 2001). Hydrolysable tannin content (HTC) of Turkish pomegranate peel derived by the pressurized water extraction method in optimized conditions was reported to be 262.7 mg tannic acid equivalents (TAE)/g (Çam and Hişil, 2010). HTC obtained for methanolic and aqueous extracts of pomegranate peel was 139.63 and 62.71 mgTAE/g, respectively (Elfalleh *et al.*, 2012b). Upon analysis of aqueous, methanol and ethanol extracts of peel of four Turkish pomegranates extracts, the highest concentration of tannins was obtained for methanol extracts (124.10–183.18 μg TAE/mg) (Orak *et al.*, 2012). HTC has been reported to be 4792.3–6894.8 mg/100 g fresh weight in peel of six pomegranate cultivars grown in

Table 17.1. Total phenolic content of different parts of the pomegranate.

Plant part	Total phenolic	Reference
Peel	40.8–230.4 mg GAE/mg DW (methanolic and aqueous extracts) in 21 Iranian pomegranates	Mansour *et al.*, 2013
	841.5 mg GAE/g (methanolic extracts) in Indian pomegranate	Arun *et al.*, 2017
	258.2 mg TAE/g was reported in pomegranates from Turkey using a conventional solid–liquid methanol method	Çam and Hişil, 2010
	244.6 mg GAE/g (water extract) in Indian pomegranate	Arun *et al.*, 2017
	1.227 mmol GAE/g (aqueous extracts)	Masci *et al.*, 2016
	264.3 mg TAE/g in pomegranates from Turkey (pressurized water method)	Çam and Hişil, 2010
	166.83 mg GAE/100g (compounds of 70% in hot water)	Nuamsetti *et al.*, 2012
	1.924 mmol GAE/g (ethyl acetate extracts)	Masci *et al.*, 2016
	259.3 mg GAE/g (ethyl acetate) in Indian pomegranate	Arun *et al.*, 2017
	152.65 mg GAE/100 g (ethanol extracts)	Nuamsetti *et al.*, 2012
	1.503 mmol GAE/g (ethanol extracts)	Masci *et al.*, 2016
	85.48 mg GAE/100 g (acetone extracts)	Nuamsetti *et al.*, 2012
	295.5 mg/g DW in 'Ganesh cultivar' and 179.3 mg/g DW in 'Mollar de Elche' cultivar	Fawole *et al.*, 2012
	134.3–181.0 mg GAE/g DW in four Tunisian pomegranate cultivars	Saad *et al.*, 2012
	10.01 mg GAE/g in extract of the 'Wonderful' cultivar from Chile under optimal conditions of extraction	Bustamante *et al.*, 2017
	In extract of 'Badana', 'Desi' and 'Kandhari' varieties was reported to be 255.35, 273.30 and 289.40 mg GAE/g, respectively	Khalil *et al.*, 2017
	During harvest was 2386.18 and after 4-month storage period inside a polyliner bag (at 7 ± 0.5°C and 92 ± 2% relative humidity) was reported to be 1537.50g GAE/kg	Mphahlele *et al.*, 2017
	In four Turkish pomegranate cultivars ranged from 1775.4–3547.8 mg GAE/l extract	Gözlekçi *et al.*, 2011
	In the extract of nine Persian cultivars ranged from 98.24–226.56 mg GAE/g	Ardekani *et al.*, 2011
	84.89–109.79 mg GAE/g DW in pomegranate peel extract of six Tunisian pomegranate ecotypes	Elfalleh *et al.*, 2009
	85.60 mg GAE/g DW	Elfalleh *et al.*, 2012b
	249.4 mg TAE/g in a Chinese pomegranate cultivar	Li *et al.*, 2006
	218.74 and 301.53 mg GAE/g in enzyme-assisted solvent and enzyme-assisted supercritical fluid extraction methods	Mushtaq *et al.*, 2015
Whole fruit	1.924 mmol GAE/g (ethyl acetate extract)	Masci *et al.*, 2016
	0.479 mmol GAE/g (ethanol extract)	Masci *et al.*, 2016
	0.406 mmol GAE/g (aqueous extract)	Masci *et al.*, 2016

Continued

Table 17.1. Continued

Plant part	Total phenolic	Reference
Juice	784.4–1551.5 mg GAE/l in extract of four Turkish pomegranate cultivars	Gözlekçi *et al.*, 2011
	6.89–13.70 mg GAE/g DW in six Tunisian pomegranate ecotypes	Elfalleh *et al.*, 2009
Seed	117.0–177.4 mg GAE/l in extract of four Turkish cultivars	Gözlekçi *et al.*, 2011
	11.84 mg GAE/g DW in Tunisian pomegranate cultivar	Elfalleh *et al.*, 2012b
Pulp	11.62–21.03 mg GAE/g in extract of nine Persian cultivars	Ardekani *et al.*, 2011
	24.4 mg TAE/g in the extract of a Chinese cultivar	Li *et al.*, 2006
Flowers	66.29 mg GAE/g DW in Tunisian cultivar	Elfalleh *et al.*, 2012a
Leaves	14.78 mg GAE/g DW in Tunisian cultivar	Elfalleh *et al.*, 2012a
	0.65 ± 274.1 mg ChAE/g of dry extract	Sreedevi *et al.*, 2017
	117.6 ± 0.761 mg ChAE/g (hydroalcoholic extract)	Sreedevi *et al.*, 2017
	240.8 ± 0.28 mg ChAE (ethyl acetate extract)	Sreedevi *et al.*, 2017
	365 mg GAE/g of FW basis	Pande and Akoh, 2009
	378.32 ± 0.92 mg GAE/g of FW	Hossain *et al.*, 2017

GAE, gallic acid equivalent; TAE, tannic acid equivalent; DW, dry weight; FW, fresh weight; ChAE, chlorogenic acid equivalent.

Georgia (Pande and Akoh, 2009). HTC of the peel of four Tunisian pomegranate cultivars ranged from 470.7–504.8 mg TAE/g (Saad *et al.*, 2012). Ellagitannins such as punicalin and puninalagin are the predominant polyphenolic compounds, with a wide variety in pomegranate peel (Table 17.2). Ellagitannins have also been identified in pomegranate roots and bark (Tanaka *et al.*, 1986; Jurenka, 2008). In addition, ellagitannins are identified with hexahydroxydiphenoyl and/or galloyl (Tanaka *et al.*, 1985; Hussein *et al.*, 1997). Fischer *et al.* (2011a) reported total ellagitannin and gallotannin content in the peel of Peruvian pomegranate fruits of an unknown cultivar to be 44 g/kg and 4.3 mg/kg, respectively. The identified ellagitannins were pedunculagin I, punicalin, punicalagin, ellagic acid pentoside, ellagic acid-hexoside, hexahydroxydiphenoyl-hexoside, ellagic acid-deoxyhexoside, galloyl-hexahydroxydiphenoyl-hexoside, pedunculagin II, casuarinin, valoneic acid bilactone, granatin B, lagerstannin B, punigluconin, lagerstannin C, castalagin derivative and brevifolin carboxylic acid. Digalloylhexoside has been identified as gallotannin, and pomegranate fruit has also been reported to contain valoneic acid bilactone and brevifolin carboxylic acid. Moreover, some ellagitannins that are found in abundance in Peruvian pomegranate peel are punicalagin (10.5 g/kg), granatin B (5.9 g/kg), lagerstannin C (3.9 g/kg), punigluconin (3.8 g/kg) and pedunculagin I (3.5 g/kg). Punicalagin content in the extract of pomegranate peel of 'Badana', 'Desi' and 'Kandhari' cultivars from Pakistan has been reported to be 88.70, 110 and 118.60 mg/g dry weight, respectively (Khalil *et al.*, 2017). Punicalagin content in the peel of Tunisian soft-seed (Li *et al.*, 2016b) and a Chinese (Ma *et al.*, 2015) pomegranate cultivar has been reported to be 83.9 and 407 mg/g, respectively. Ellagitannins (C-glycosides, punicacorteins A–D [ellagitannins with a gallagic acid component]) and punigluconin (an ellagitannin with a gluconic acid core) have been identified in pomegranate stems (Tanaka *et al.*, 1986). Ellagitannins, including punicalin and puninalagin, have also been identified in pomegranate roots and bark (Tanaka *et al.*, 1986; Jurenka, 2008). Punicalagin is the major and most studied bioactive polyphenol of pomegranate and has many biological properties. The punicalagin content of pomegranate peel has been obtained 116.6 mg/g by the pressurized water extraction method under optimized

Table 17.2. The chemical structure of important phenolic acids and tannins in pomegranate.

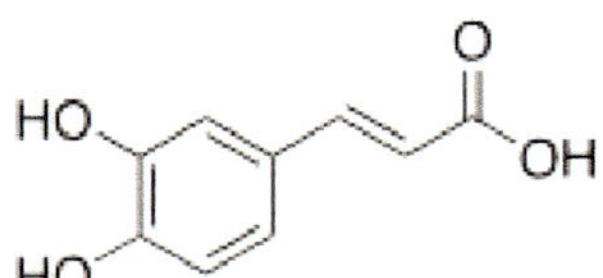 Caffeic acid: $C_9H_8O_4$	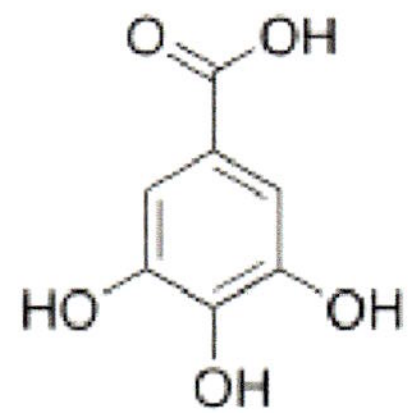 Gallic acid: $C_7H_6O_5$	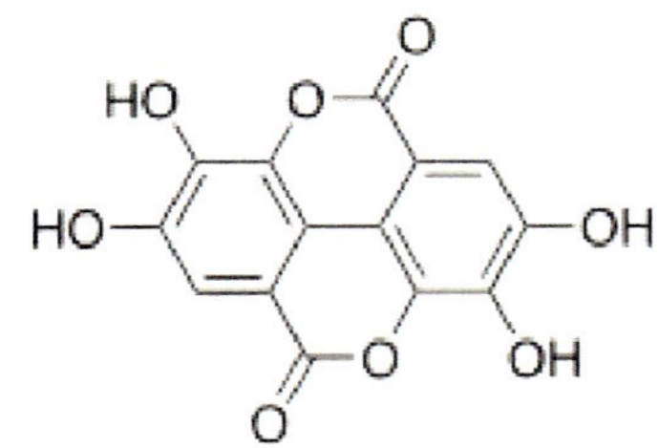 Ellagic acid:$C_{14}H_6O_8$
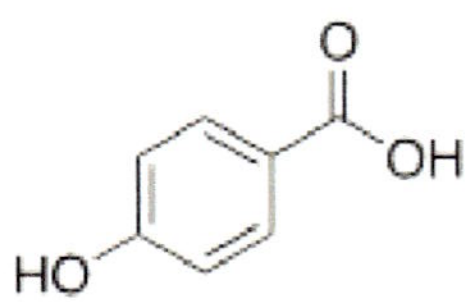 Hydroxybenzoic acid: $C_7H_7NO_7$	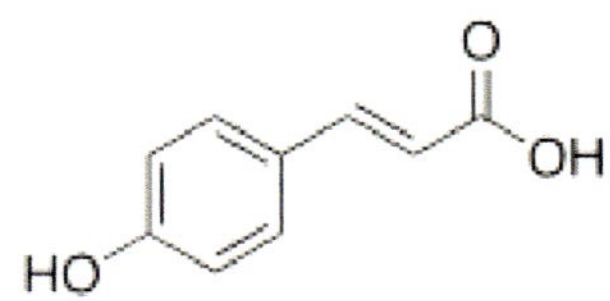 *p*-Coumaric acid: $C_9H_8O_3$	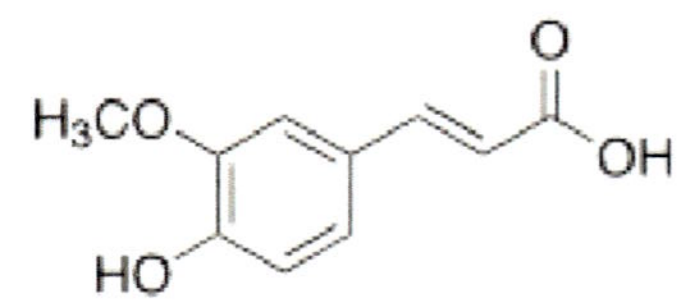 Ferulic acid: $C_{10}H_{10}O_4$

Continued

Table 17.2. Continued

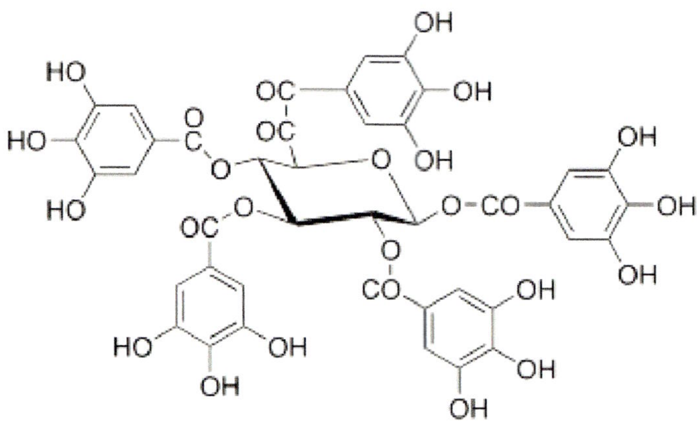

Gallotannin

Continued

Table 17.2. Continued

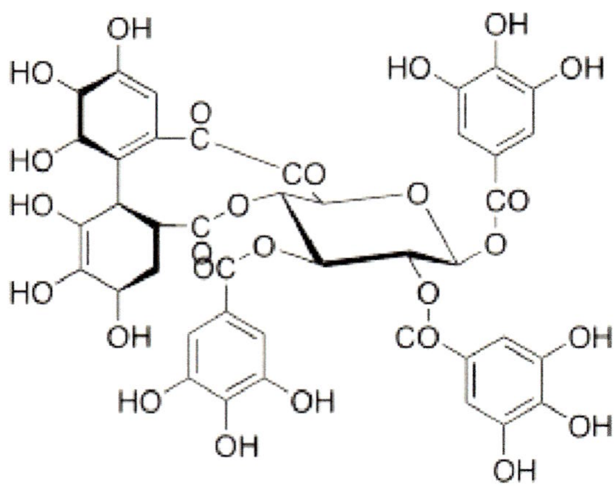

Ellagitannin

Continued

Table 17.2. Continued

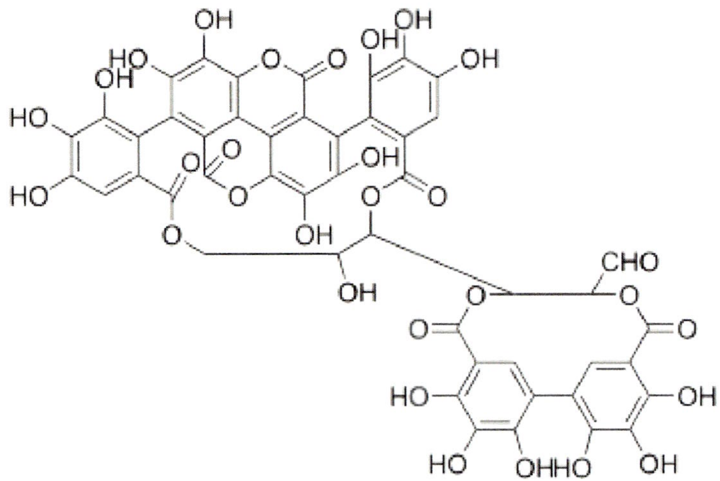

Punicalagin: $C_{48}H_{28}O_{30}$

conditions (Çam and Hişil, 2010). In ethanol, aqueous and ethyl acetate extracts of pomegranate peel of Italian origin pomegranate fruits, punicalagin content was reported to be 28.45, 41.36 and 32.68 mg/g, respectively, while in those of the whole fruit, it was reported to be 8.96, 13.03 and 21.66 mg/g, respectively (Masci *et al.*, 2016). The concentrations of α-punicalagin and β-punicalagin in peel of six Spanish pomegranate cultivars were reported to be 106–148 and 74–107 mg/g, respectively (Rosas-Burgos *et al.*, 2017). Punicalagin, with a content of 65.38 mg/100 mg, has been identified as the main phenolic compound in Chinese pomegranate peel (Song *et al.*, 2016). The concentrations of punicalagin A and B in the peel of a Chinese pomegranate (cv. 'Wonderful') were 40.09 and 43.18 mg/l, respectively (Qu *et al.*, 2012). Punicalagin, with a content of 13.74 mg/g, has been reported to be the main tannin in the peel of a Brazilian pomegranate cultivar (Morzelle *et al.*, 2016). Punicalagin (56.78 mg/g) and punicalin (200.02 mg/g) are the main tannins identified in Serbian pomegranate peel (Stojanović *et al.*, 2017). Punicalagin isomers were found to exist in lower amounts in pomegranate roots than its peel (Ono *et al.*, 2012). Punicalin content was reported to be 706.04 mg/kg in the peel of freshly harvested pomegranates from South Africa (Mphahlele *et al.*, 2017). Tannins of pomegranate leaves and peel are different. In the leaves, the most abundant tannins are grutin A and B, while punicalagin and punicillins are present in very small amounts (Tanaka *et al.*, 1985). Ellagic acid has been reported in various pomegranate species (Amakura *et al.*, 2000; Jurenka, 2008). Ellagic acid and elgiantan (e.g. urolithin M-5, brevifolin and brevifolin carboxylic acid) have been isolated from pomegranate leaves (Nawwar *et al.*, 1994). It has been found to be the most abundant phenolic compound identified in pomegranate peel extract of seven South African cultivars, with a concentration of 46.87–209.44 µg/ml (Fawole *et al.*, 2012). Ellagic acid concentration was 9.8–16.5 mg/g in a methanolic extract of peel from six Spanish pomegranate cultivars (Rosas-Burgos *et al.*, 2017). In soft seed of 'Tunisia' pomegranate and pomegranate peel, the amount of ellagic acid has been reported to be 7.3 mg/g (Li *et al.*, 2016b). Ellagic acid content in ethanol, aqueous and ethyl acetate extracts of pomegranate peel was reported to be 28.45, 41.36 and 32.68 mg/g, respectively, and in whole fruit extracts 11.85, 1.37 and 48.55 mg/g, respectively (Masci *et al.*, 2016). Ellagic acid and its glycoside derivatives have also been found in other pomegranate tissues in addition to fruit peel (Ono *et al.*, 2012). In flowers, ellagic acid and two of its derivatives, and phyllanthusiin E have been identified (Wang *et al.*, 2006). Studies on pomegranate seeds have shown that gallotannins are present only in the methanolic extract of seeds, but not in extracts of mesocarp and exocarp (Akkiraju *et al.*, 2016).

Condensed tannins (CTs) have a more complex structure than HTs. The main CTs are oligomers and polymers (e.g. monomers, dimers and trimers) of flavan-3-diols (catechin or epicatechin derivatives) known as proanthocyanidins (Schofield *et al.*, 2001). Proanthocyanidin extraction efficiency and quality are determined by different factors including solvent–solid ratio, temperature and the type of solvent used in the process (Aviram *et al.*, 2000; Lo Scalzo *et al.*, 2004). The level of CTs declines significantly during fruit ripening. The decline in condensed tannin content (CTC) may be due to increased activity of certain enzymes such as anthocyanin synthase and 3-glycosyl transferase during the formation of anthocyanins (Robbins *et al.*, 1998). CTC in pomegranate peel of four Tunisian cultivars was obtained at a level of 3.2–7.7 mg CE/g dry weight (Saad *et al.*, 2012).

17.3.3 Flavonoids

Different flavonoids have been reported in different parts of the pomegranate (Mirdehghan and Rahemi, 2007; Liu *et al.*, 2009; Sharma *et al.*, 2018) (Table 17.3). The concentrations of flavonoids in fruit peel are higher compared with those in fruit pulp (Li *et al.*, 2006). Studies have shown that the total flavonoid content (TFC) in pomegranate peel was 12.4-fold higher than that of TFC in the juice extracts of four Turkish genotypes (Elfalleh *et al.*, 2009). Masci *et al.* (2016) reported higher total flavonoids compounds in ethyl acetate, ethanol and aqueous

Table 17.3. Total flavonoid content of different parts of the pomegranate.

Plant part	Total flavonoids	Reference
Peel	TFC in extract of nine Iranian cultivars was reported as 18.61–36.40 mg catechin equivalents /g	Ardekani *et al.*, 2011
	TFC of seven South African pomegranate cultivars was reported as 97.8–121.1 mg CE /g DW (methanol extracts)	Fawole *et al.*, 2012
	In 'Badana', 'Desi' and 'Kandhari' cultivars of Pakistan, TFC was reported as 50.86, 55.21 and 58.63 mg RE/g DW, respectively	Khalil *et al.*, 2017
	TFC in Tunisian cultivars was 51.52 mg RE/g (methanol extracts)	Elfalleh *et al.*, 2012a
	TFC in Tunisian cultivars was 21.03 mg RE/g (water extracts)	Elfalleh *et al.*, 2012a
	TFC in four genotypes of Turkish pomegranates was 14.37–20.52 µg QE/mg (methanolic extract)	Orak *et al.*, 2012
	TFC ranged from 44.83–56.46 mg RE/g DW in six Tunisian ecotypes	Elfalleh *et al.*, 2009
	TFC 59.1 mg RE/g in extract	Li *et al.*, 2006
	TFC in Italian pomegranate was 0.387 mmol RE/g (aqueous extract)	Masci *et al.*, 2016
	TFC in Italian pomegranate was 0.881 mmol RE/g (ethyl acetate extract)	Masci *et al.*, 2016
	TFC in Italian pomegranate was 0.471 mmol RE/g (ethanol extract)	Masci *et al.*, 2016
	TFC was 5.83 mg CE/g in extract of Serbian pomegranate	Stojanović *et al.*, 2017
	Flavonoid level decreased with storage and was 94.48g CE/kg at harvest and 89.24g CE/kg at the end of 4 months of storage	Mphahlele *et al.*, 2017
Whole fruit	0.561 mmol RE/g (ethyl acetate extract)	Masci *et al.*, 2016
	0.122 mmol RE/g (aqueous extract)	Masci *et al.*, 2016
	0.134 mmol RE/g (ethanolic extract)	Masci *et al.*, 2016
Juice	TFC ranged from 4.11–5.75 mg RE/ml in extracts from six Tunisian ecotypes	Elfalleh *et al.*, 2009
Pulp	TFC in extract of nine uercIranian cultivars was reported in the range of 0.84–2.14 mg CE/g	Ardekani *et al.*, 2011
	TFC in pomegranate pulp extract was 17.2 mg RE/g	Li *et al.*, 2006

TFC, total flavonoid content; DW, dry weight; CE, conventional extraction; RE, rutin equivalents; QE, quercetin equivalents.

extracts of Italian pomegranate peel compared with whole fruit extracts.

Pomegranate peel is a potential source of flavonoids such as epicatechin, catechin, quercetin, anthocyanins and procyanidins (Masci *et al.*, 2016). Flavonoid composition changes in the fruit developmental stages, and the concentrations of flavonols and flavones also vary in the peel of diferent pomegranate cultivars (Zhao *et al.*, 2014). Catechin and epicatechin have been identified in the peel of seven South African pomegranate cultivars and their content varied from 570.76–61.30 mg/kg after harvest (Mphahlele *et al.*, 2017).

Catechin (110.7–125.6 mg/100g fresh weight), epicatechin (110.7–125.6 mg/100 g fresh weight) and quercetin (92.1–99.2 mg/100 g fresh weight) have been reported in the peel of six Georgian pomegranate cultivars (Pande and Akoh, 2009). The content of catechin in Indian pomegranate peel has been reported as being 76.5 mg/100 g (Singh *et al.*, 2016). The amounts of catechin, epicatechin and rutin in Chinese pomegranate peel extract have been reported to be 12.66, 0.99 and 0.34 mg/100 mg, respectively (Song *et al.*, 2016). Quercetin, myricetin, kaempferol, luteolin and apigenin have also been found in the pomegranate peel of four Chinese cultivars (Zhao *et al.*, 2014) (Table 17.4). Catechin, gallocatechin and procyanidin B have been detected in the peel of a Tunisian pomegranate cultivar by liquid chromatography–mass spectrometry (LC–MS) analysis (Wafa *et al.*, 2017). Rutin has been identified as the most abundant flavonoid in the fresh peel of pomegranate (3446.24 mg/kg) but its level significantly decreases (1191.39 mg/kg) during prolonged cold storage (Mphahlele *et al.*, 2017). In 21 Iranian pomegranate accessions, the mean value of quercetin content has been reported as 1.9 mg/100g dry weight (Mansour *et al.*, 2013). Punicaflavanol (a misnomer or flavanone), granatumflavanyl xyloside, phlorizin (a glycoside of the dihydrochalone phloretin), hovetrichoside C (a glycoside of an auronol), luteolin and trystyn have been detected in the flower of the pomegranate (Xie *et al.*, 2008; Yuan *et al.*, 2013). Eriodictyol-7-O-α-L-arabinofuranosyl(1-6)-β-D-glucoside, naringenin-4-O-methyl ether 7-O-α-L-arabinof uranosyl(1-6)-β-D-glucoside and quercetin-3,4-O-dimethyl ether 7-O-α-L-arabinofuranosyl(1-6)-β-D-glucoside are flavanone and flavonol diglycosides that have been isolated from pomegranate stem bark (Chauhan and Chauhan, 2001; Srivastava *et al.*, 2001). Pomegranate leaves contain flavons luteolin and epigenyl glycosides (Nawwar *et al.*, 1994; Jurenka, 2008). Two isoflavones (genistein and daidzein) and a flavonol (quercetin) have been reported in the seeds of pomegranate (Moneam *et al.*, 1988; Pande and Akoh, 2009). The presence of flavonoids has also been confirmed in methanolic extracts of exocarp and mesocarp of pomegranate but not in seeds (Akkiraju *et al.*, 2016). Quercetin, rutin and other flavonols (Artik, 1998), flavones and flavonones (Nawwar *et al.*, 1994) have all been identified in pomegranate pericarp. Kaempferol, luteolin and narigenin have been found in glycoside forms in some studies (Aviram *et al.*, 2000; Polagruto *et al.*, 2003). Kaempferol 3-orutinoside, kaempferol derivative, isorhamnetin and hexahydroxydiphenoyl-glucoside-acetyl glucoside derivatives (Al-Rawahi *et al.*, 2014) have been found in pomegranate's various parts such as flowers (Huang *et al.*, 2005a) and pericarp (De Pascual-Teresa *et al.*, 2000; Jurenka, 2008).

17.3.4 Anthocyanins

Anthocyanins belong to the family of flavonoids and are the most important pigments in pomegranate fruit (Gil *et al.*, 1995). Colour variation of different tissues and different species of pomegranate is due to various anthocyanins and ratios of their derivatives (Gil *et al.*, 1995; Fischer *et al.*, 2011b). Other flavonoids present in pomegranate fruit are responsible for yellow colouration (Tanaka *et al.*, 2008). Anthocyanins affect the marketability of pomegranate fruit, protect fruit tissue against UV rays (Li *et al.*, 1993; Hou *et al.*, 2004) and also increase the shelf-life and reduce the sensitivity of pomegranate fruit to grey mould (Bassolino *et al.*, 2013). Anthocyanins of peel contain about 30% of the total anthocyanins of pomegranate fruit. Anthocyanin composition in pomegranate peel varies depending on the cultivars, fruit developmental stages and colouration time (Zhao *et al.*, 2013). A rapid increase in total anthocyanin concentration during ripening has been reported for the 'Mollar' and 'Ganesh' pomegranate cultivars (Du *et al.*, 1975; Al-Maiman and Ahmad, 2002). Cyanidin 3,5-diglucoside (157.8 mg/kg), pelargonidin 3,5-diglucoside (145.8 mg/kg), pelargonidin 3-glucoside (56.7 mg/kg), cyanidin 3-glucoside (41.2 mg/kg), cyanidin 3-rutinoside (18.4 mg/kg), delphinidin 3-glucoside (13.3 mg/kg), delphinidin 3,5-diglucoside (10.8 mg/kg), cyanidin hexoside (1.7 mg/kg) and cyanidin-pentoside (1.4 mg/kg) are the main anthocyanins identified and quantified in the peel of Peruvian pomegranate (unknown cultivars) (Fischer *et al.*, 2011a). The above compounds have also been observed in the extract of 'Nana' peel grown in Tunisia (Wafa *et al.*, 2017). In the red cultivars of pomegranate, cyanidin 3,5-diglucoside in leaves; cyanidin 3,5-diglucoside, pelargonidin 3,5-diglucoside, cyanidin 3-glucoside and pelargonidin 3-glucoside in flowers (sepals only); delphinidin 3,5-diglucoside, cyanidin 3,5-diglucoside, delphinidin 3-glucoside, cyanidin 3-glucoside and

Table 17.4. The chemical structure of important flavonoids in pomegranate.

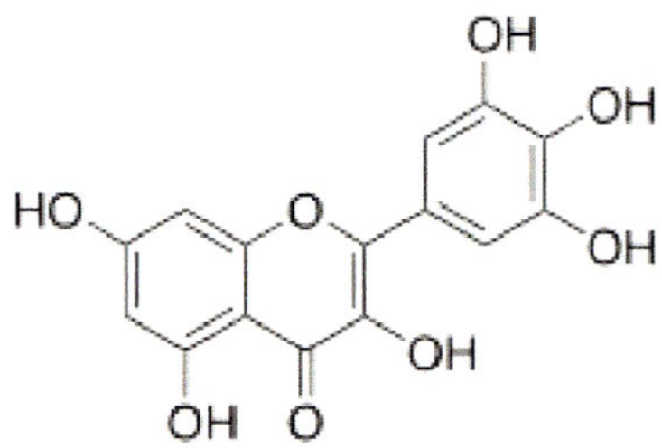 Myricitin: $C_{15}H_{10}O_8$	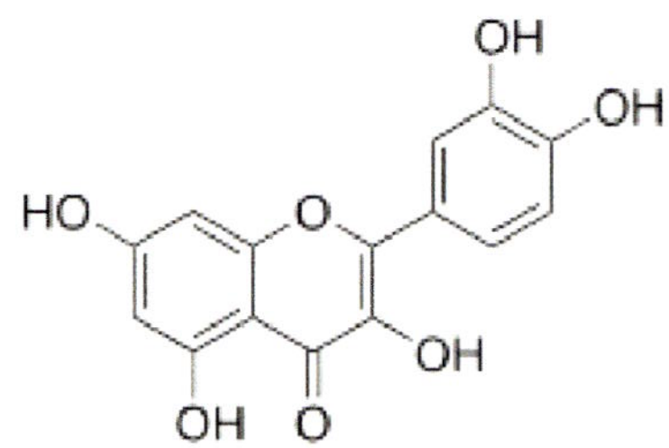 Quercetin: $C_{15}H_{10}O_7$	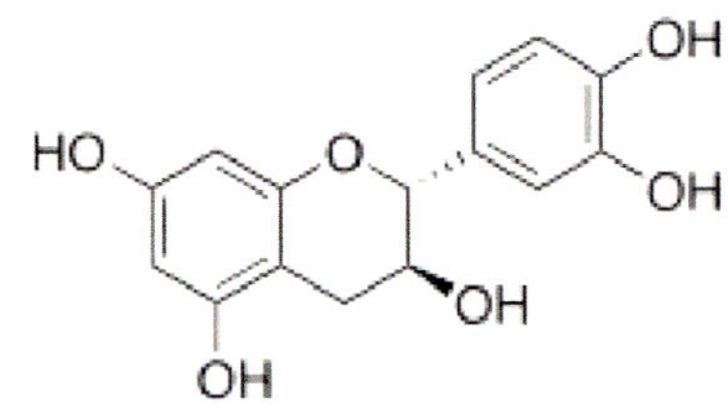 Catechin: $C_{15}H_{14}O_6$
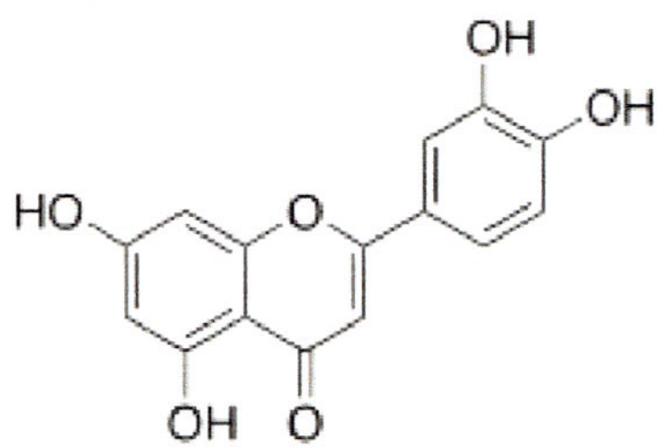 Luteolin: $C_{15}H_{10}O_6$		

Table 17.5. The chemical structure of important anthocyanins in pomegranate.

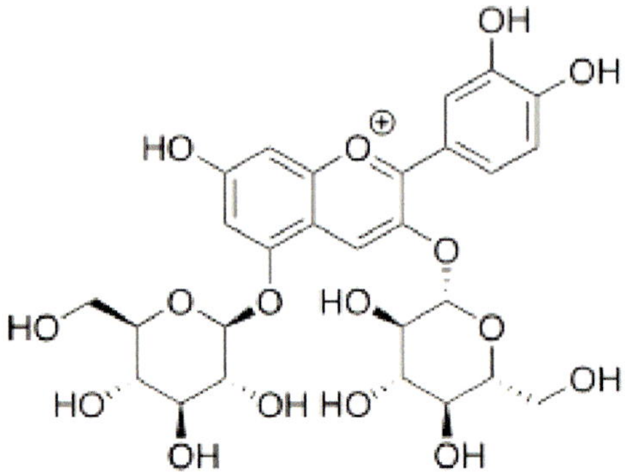

Cyanidin-3,5-O-diglucoside: $C_{27}H_{31}O_{16}$

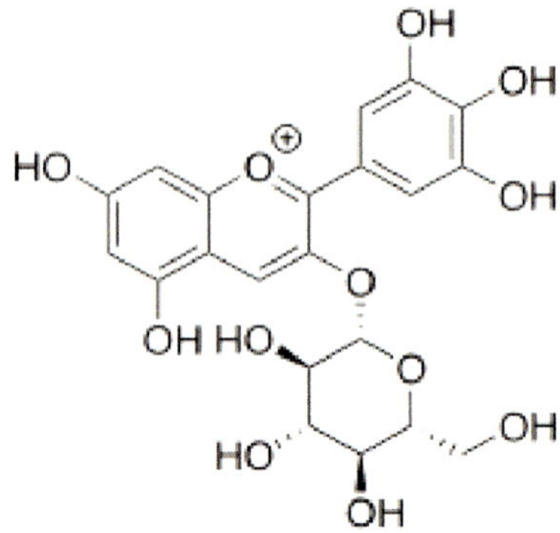

Delphinidin 3-glucoside: $C_{21}H_{21}O_{12}$

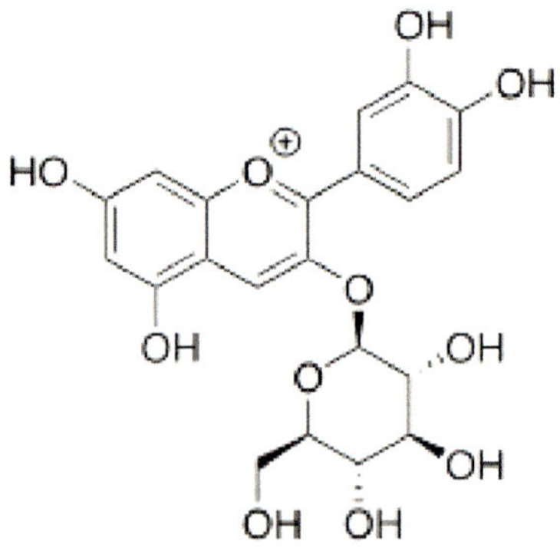

Cyanidin 3-O-glucoside: $C_{21}H_{21}O_{11}$

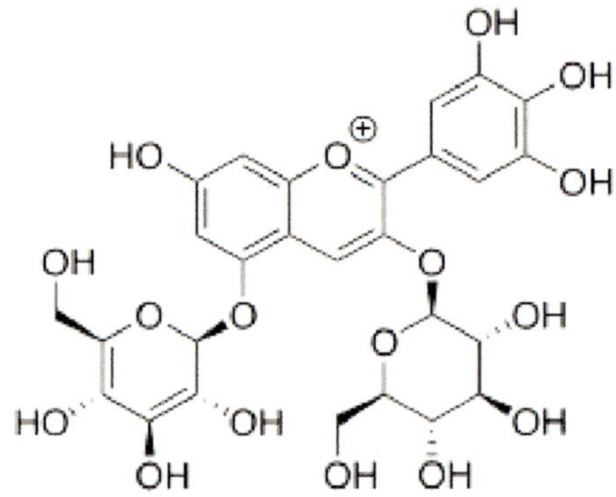

Delphinidin-3,5-O-diglucoside: $C_{27}H_{30}O_{17}$

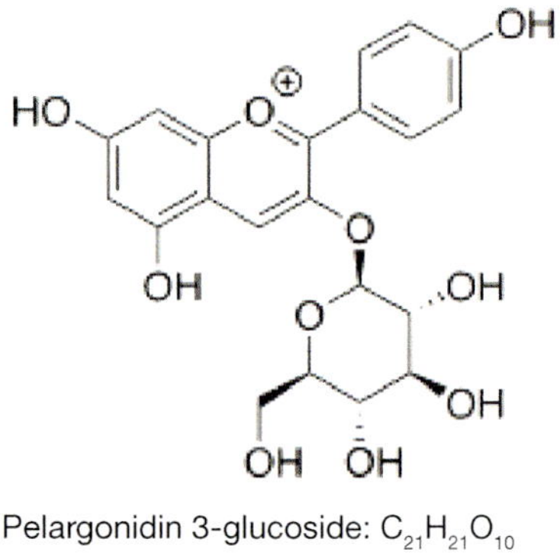

Pelargonidin 3-glucoside: $C_{21}H_{21}O_{10}$

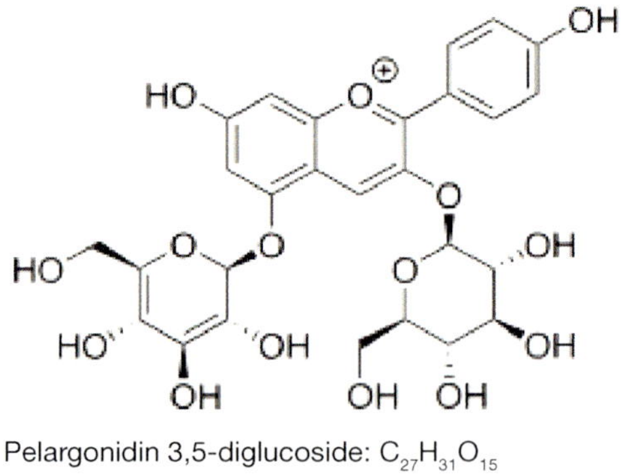

Pelargonidin 3,5-diglucoside: $C_{27}H_{31}O_{15}$

pelargonidin 3-glucoside in arils; and cyanidin 3,5-diglucoside, pelargonidin 3.5-diglucoside, cyanidin 3-glucoside and pelargonidin 3-glucoside in peel have been reported so far. Delphinidin and pelargonidin have been identified in the red cultivars as purple and orange pigments, respectively (Table 17.5).

Anthocyanins are absent from the fruit arils and peel of green and immature fruits. Sreekumar *et al.* (2014) identified delphinidin 3-glucoside, cyanidin 3-glucoside, delphinidin 3,5-diglycoside, cyanidin 3,5-diglucoside, pelargonidine 3,5-diglucoside and pelargonidine 3-glucoside in pomegranate seeds (Sreekumar *et al.*, 2014).

17.3.5 Fatty acids

Pomegranate seeds are rich source of oils (Fadavi *et al.*, 2006). The total weight of the seed oil has been reported as between 12 and 20% (Kohno *et al.*, 2004). Melgarejo and Artes (2000) reported that the seeds contain 63–272 g/kg oil dry weight. The oil yield of pomegranate seeds varies significantly among pomegranate genotypes, ranging from 10.7–26.8% in sweet genotypes and from 4.9–17.4% in sour genotypes (Ferrara *et al.*, 2014). Oil content obtained is significantly affected by extraction methods (Abbasi *et al.*, 2008; Eikani *et al.*, 2012). The saponification value of pomegranate oil is 188.9 (El-Nemr *et al.*, 1992). Peroxide number is one of the important qualitative indexes of edible oils, which, if it exceeds 9 mEq/g of peroxide per 1000 g of oil, indicates an oxidative degradation of oil. Peroxide number in pomegranate oil has been calculated at the standard level of the American Oil Chemists' Society (AOCS), with the lowest microbial corruption (Samad and Barzegar, 2006). The breakdown rates of pomegranate oils of different cultivars are higher than those of common edible oils. The high breakdown rate of pomegranate oil can be attributed to the high trans-fatty acid content (El-Nemr *et al.*, 1992).

Fatty acids are present in the pericarp (Jurenka, 2008; Moorthy *et al.*, 2013), leaf (1.7 ± 0.96%) (Sreedevi *et al.*, 2017), fruit peel (1.2%), seeds (4.8%), whole fruit (1.4%) (Sharma *et al.*, 2018) and juice (Liu *et al.*, 2009) of pomegranates at different concentrations. Pomegranate oil contains palmitic acid, stearic acid, oleic acid, linoleic acid and four isomers of linolenic acid (Table 17.6). Linolenic acid isomers (31–86%), linoleic acid isomers (4.2–4.4%) and oleic acid isomers (4.1–4.4%) are the main fatty acids of the oil. Punicic acid is the dominant form of linolenic acid in pomegranate seed oil (Özgül-Yücel, 2005; Fadavi *et al.*, 2006). Schubert *et al.* (1999) reported the amount of punicic acid in pomegranate seed oil to be 65.3% (Schubert *et al.*, 1999). Linolenic acid ranging from 74–88% and linoleic acid ranging from 5–16% have also been reported in the seed oil (Nemr *et al.*, 2001). Triacylglycerols (TAGs) containing 9E, 11Z, 13E-octadecatrieneic acid, 3-O-octadec-2-enoic acid, 9Z, 11E, 13Z-octadecatrieneic acid and 8Z, 11Z, 13E-octadecatrieneic acid have also been found in pomegranate seeds (Matthäus *et al.*, 2010). In addition, glycosphingolipid N-palmitoyl cerebroside has been identified in pomegranate seeds (Wilson and Baietto, 2009). Pomegranate oil contains phyto-oestrogens, which are similar to the natural oestrogens that are produced in the human body (Abbasi *et al.*, 2008). Pomegranate seed oil also contains steroidal oestrogens (g-tocopherol, 17-a-oestradiol, stigmasterol, β-oestriol sitosterol and testosterone) and non-steroidal compounds (compestrol, coumestrol) (van Elswijk *et al.*, 2004). Caproic, caprylic, lauric, myristic, capric, myristoleic, palmitic, palmitoleic, punicic, linoleic, α-linolenic, γ-linolenic, oleic, stearic, α-eleostearic, β-eleostearic, catalpic, arachidic, gadoleic, behenic and nervonic acids have been reported from pomegranate seed oil (Wu and Tian, 2017) (Table 17.6). Certain fatty acids including palmitic, oleic, palmitoleic, stearic, arachidonic, caprylic and lauric acids have been identified in different variants of *P. granatum*. Studies have shown that unsaturated fatty acids are the main lipid compounds of sweet cultivars of pomegranate fruit (Melgarejo and Artes, 2000). Pomegranate oil has high levels of unsaturated fatty acids of the omega-3 type (Melgarejo *et al.*, 2000).

17.3.6 Alkaloids

The presence of alkaloids in the different parts of pomegranate (fruit peel and seeds, and whole fruit powder) has been reported (Hagir *et al.*, 2016; Dixit *et al.*, 2017; Sharma *et al.*, 2018). Total alkaloid content has been reported to be 3.5 ± 1.5% in ethanolic extract of pomegranate leaves (Sreedevi *et al.*, 2017). In the research of Tripathi and Kohli (2011), no alkaloid was found in any of the extracts (chloroform, petroleum ether, ethyl acetate, ethanol, chloroform) of pomegranate flowers (Tripathi and Kohli, 2011). Piperidine alkaloids have been found in various types of pomegranate roots and bark extracts (Neuhöfer *et al.*, 1993; Jurenka, 2008). Furthermore, pelletierine $C_8H_{15}NO$ is a liquid alkaloid extracted from the root bark of *P. granatum* (El-Sakka, 2010). Pelletierine, pseudopelletierine, isopelletierine and methylisopelletierine (Table 17.7) are the most important alkaloids detected in the pomegranate (Haque *et al.*, 2015). These

Table 17.6. The chemical structure of important fatty acids in pomegranate.

 Palmitic acid: $C_{16}H_{32}O_2$	 Stearic acid: $C_{18}H_{36}O_2$
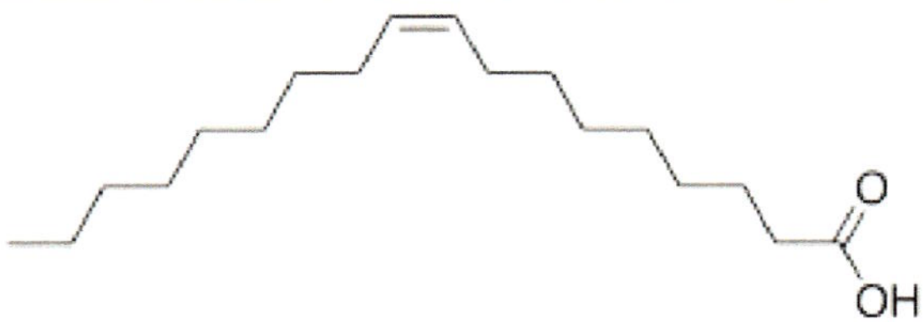 Oleic acid: $C_{18}H_{34}O_2$	
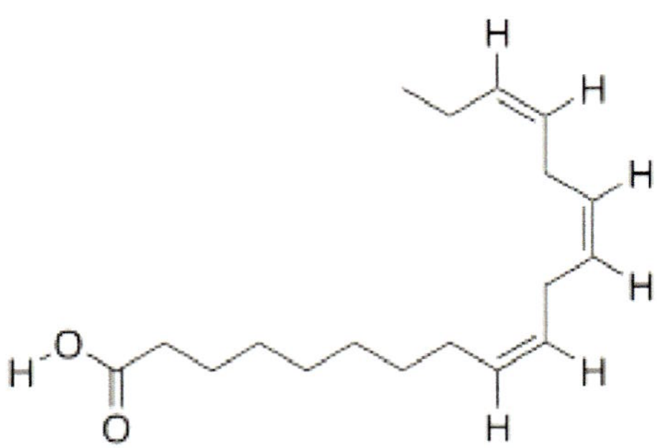 Linolenic acid: $C_{18}H_{30}O_2$	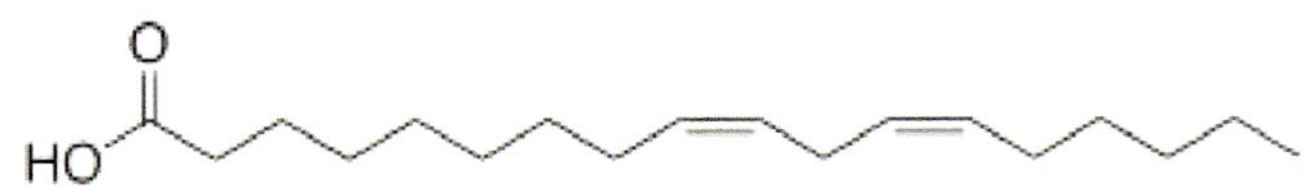 Linoleic acid: $C_{18}H_{32}O_2$

Table 17.7. The chemical structure of important alkaloids in pomegranate.

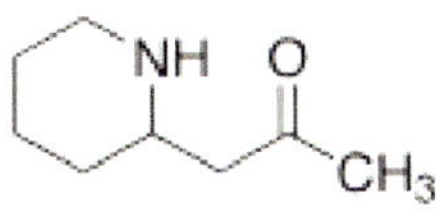 Pelletierine: $C_8H_{15}NO$	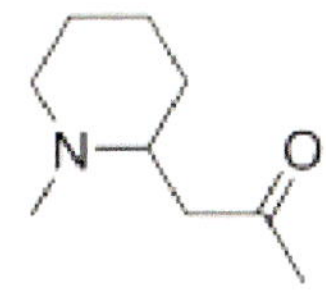 N-Methylisopelletierine: $C_9H_{17}NO$
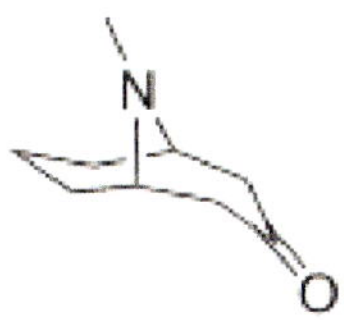 Pseudopelletierine: $C_9H_{15}NO$	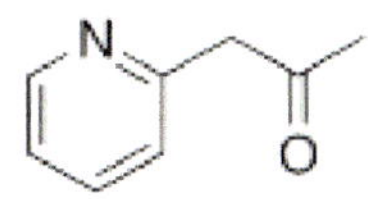 Isopelletierine: C_8H_9NO

four compounds were reported in fruit pomace peel (Gil *et al.*, 1995; Aviram *et al.*, 2000). Also, pyrrolidine-type alkaloid punigratane was recently isolated from pomegranate peel (Rafiq *et al.*, 2016). Sedridine, 2-(20-hydroxypropyl)-Δone piperideine, 2- (20-propenyl)-Δ1piperideine, norpseudopelletierine and the pyrrolidine alkaloids in small amounts have been found in the root of the pomegranate (Neuhöfer *et al.*, 1993). N-(20,50-dihydroxyphenyl) pyridinium chloride has also been identified in pomegranate leaves. Low levels of amine derivatives of indole, including tryptamine, melatonin and serotonin, have also been identified in pomegranate fruit extract (Badria, 2002).

17.3.7 Other compounds

Terpenoids, saponins, phytosterols, lignins, organic acids, amino acids, proteins and vitamins are the other chemical constituents that have been found in the pomegranate. Terpenoids have been observed in peel extract (Jayaprakash and Sangeetha, 2015; Hagir *et al.*, 2016). Triterpenoids such as maslinic and asiatic acid have been reported in the flowers of pomegranate (Jurenka, 2008; Tripathi and Kohli, 2011). These compounds are also found in ethanolic extract of pericarp (Moorthy *et al.*, 2013), stem, seeds and leaves (Mohajer *et al.*, 2016). The levels of triterpenoids were reported to be 44.9 ± 1.55 mg of ursolic acid equivalent (UE)/g dry weight in aqueous extract, 71.77 ± 1.67 mg of UE/g dry weight in hydroalcoholic extract, 112.2 ± 2.36 mg of UE/g dry weight in ethanol extract and 37.3 ± 0.92 mg of UE/g dry weight extract in ethyl acetate extract (Sreedevi *et al.*, 2017). Certain tropenoids identified in the pomegranate include asiatic acid, betulinic acid (betulic acid), friedooleanan-3-one (friedelin), maslinic acid, oleanolic acid, punicanolic acid and ursolic acid (Wu and Tian, 2017).

Phytosterols including campesterol, cholesterol, daucosterol, β-sitosterol, β-sitosterol laurate, β-sitosterol myristate and stigmasterol have been found in pomegranate leaf, flower, peel and peel tissue (Heftmann *et al.*, 1966; Abdel Wahab *et al.*, 1998; Wu and Tian, 2017) (Table 17.8). Also the seeds contain phytooestrogen coumestrol oestrone (Singh and Sethi, 2003; Syed *et al.*, 2007). Animal steroid hormones (oestrone, oestriol, oestradiol and testosterone) have already been reported in pomegranate seeds (Heftmann *et al.*, 1966; Abdel Wahab *et al.*, 1998). Steroids were found to be present in all pomegranate leaf extracts (hydroalcoholic, ethanolic, ethyl acetate and n-hexane extracts) except for the aqueous extract (Sreedevi *et al.*, 2017). Lignins have also been identified in different parts of pomegranate tissues. Isolariciresinol is the most abundant lignin in the peel of pomegranate (Fischer *et al.*, 2011c). Pomegralignan, a dihydrobenzofuran-type neolignan glycoside, was also discovered in the peel and aril (Ito *et al.*, 2014). Conidendrin, isohydroxymatairesinol, isolariciresinol, matairesinol, medioresinol, phylligenin, pinoresinol, secoisolariciresinol,

Table 17.8. Reported amounts of organic acid compounds in different parts of the pomegranate.

Organic acid	The amount of compounds	Reference
Total organic acid	The most significant variation was determined in titratable acidities with a cv. of 35.16%. Titratable acidity of the samples varied from 4.58–17.3 g/l expressed as citric acid	Cemeroglu *et al.*, 1992
	Titrable acidity and total sugar content of former Soviet Union-originated pomegranate juices reported as 0.52–1.6 and 15.2–20.5%, respectively	Gabbasova and Abdurazakova, 1969
	The titratable acidity and total sugar content of Macedonian-originated pomegranate juices were reported as 0.37–2.80 and 8.4–13.2%, respectively	Veres, 1976
	In pomegranates grown in Turkey, titrable acidity of 1.47% and total sugars of 13.9% have been reported	Cemeroglu *et al.*, 1992
Oxalic acid	Values of sweet 0.037 g/100 g, sourish 0.015 g/100 g and sour 0.017 g/100 g were observed	Melgarejo *et al.*, 2000
	0.031–1.016 g/l	Gundogdu and Yilmaz, 2012
	0.02–6.72 g/l	Poyrazoğlu *et al.*, 2002
Citric acid	Values of sweet 0.142 g/100 g, sourish 0.566 g/100 g and sour 2.317 g/100 g were observed	Melgarejo *et al.*, 2000
	0.33–8.96 g/l	Poyrazoğlu *et al.*, 2002
	0.20–3.20 g/100 ml	Ozgen *et al.*, 2008
	0.613–2.182 g/l	Gundogdu and Yilmaz, 2012
Malic acid	0.117–2.230 g/l	Gundogdu and Yilmaz, 2012
	Values of sweet 0.135 g/100 g, sourish 0.160 g/100 g and sour 0.17 g/100 g were observed	Melgarejo *et al.*, 2000
	0.56–6.86 g/l	Poyrazoğlu *et al.*, 2002
	0.09–0.15 g/100 ml	Ozgen *et al.*, 2008
Tartaric acid	0.28–2.83 g/l	Poyrazoğlu *et al.*, 2002
	0.033–0.126 g/l	Gundogdu and Yilmaz, 2012
Quinic acid	0.00–0.82 g/l	Poyrazoğlu *et al.*, 2002
	0.00–1.54 g/l	Poyrazoğlu *et al.*, 2002
	0.039–0.32 g/l	Gundogdu and Yilmaz, 2012
Lactic acid	4.516–33.115 mg/l	Gundogdu and Yilmaz, 2012
Fumaric acid	0.011–0.299 mg/l	Gundogdu and Yilmaz, 2012

syringaresinol, pomegralignan and punicatannin C are the other lignins that have been identified in the pomegranate (Wu and Tian, 2017). Pomegranate juice contains feruloyl coniferin, cyclolariciresinol hexoside, secoisolariciresinol hexoside and guaiacyl (8-5) ferulic acid hexosid (Mena *et al.*, 2012).

Studies have shown the presence of saponins in different parts of the pomegranate. Sharma *et al.* (2018) reported saponins in all extracts of the fruit and seed by using the foam test (Sharma *et al.*, 2018). These compounds were also found in methanolic extracts of seed, mesocarp and exocarp of pomegranate (Akkiraju *et al.*, 2016). Studies on pomegranate flowers have also shown that chloroform-aqueous and ethanolic extracts of pomegranate flower contain saponin, while petroleum ether, chloroform and ethyl acetate extracts do not have saponin (Tripathi and Kohli, 2011).

Organic acids affect the taste of pomegranate fruits depending on their ratio and along with sugars (Cemeroglu *et al.*, 1992). Organic acid concentration declines significantly during ripening of the fruit (Viuda-Martos *et al.*, 2010). Citric, malic, tartaric, oxalic, succinic, quinic and fumaric acids are the major organic acids in pomegranate (Aviram *et al.*, 2000; Gil *et al.*, 2000; Tezcan *et al.*, 2009; Cheng *et al.*, 2012; Wu and Tian, 2017). It has been claimed that pomegranate fruits do not contain acetic acid (Melgarejo *et al.*, 2000).

Carbohydrates have also been identified in the different parts of pomegranate such as fruit, seeds and whole fruit (Sharma and Akansha, 2018). Carbohydrate content (%) has been reported to be 64.84 ± 0.53 in fruit peel, 64.85 ± 0.07 in seed and 78.58 ± 0.23 in whole fruit (Sharma *et al.*, 2018), and also to exist in all leaf extracts (Sreedevi *et al.*, 2017). The amount of sugar in the ethanolic extract of leaves has been found to be 19.6 ± 2.3% dry weight (Sreedevi *et al.*, 2017). Reduction of sugar in leaves and peel has been reported in both *in vitro* and *in vivo* conditions but not in seeds and stem (Mohajer *et al.*, 2016). In addition, carbohydrates were not found in any of the extracts of pomegranate flowers (Tripathi and Kohli, 2011). Studies have shown that these compounds are also found in arils (Jaiswal *et al.*, 2010) and seeds (Syed *et al.*, 2007). Carbohydrates have been reported to exist in the methanolic extracts of exocarp, mesocarp and seeds (Akkiraju *et al.*, 2016), and also in ethanolic extract of pomegranate pericarp (Moorthy *et al.*, 2013). Carbohydrates have been reported to exist in juice (Jurenka, 2008), with a content of 14.5% (Bhowmik *et al.*, 2013). The main compounds of sugars in pomegranate juice include glucose, sucrose, fructose (Melgarejo *et al.*, 2000), glucose and maltose (Cheng *et al.*, 2012). Also, 5% pectin has been detected in the juice obtained from arils (Tezcan *et al.*, 2009).

Phytochemical experiments have shown the presence of amino acids and proteins in different parts of the pomegranate (Sharma *et al.*, 2018). The free amino acids of the protein reserves were investigated in the pomegranate seeds of two Tunisian and Chinese cultivars. Results showed that pomegranate seeds contained 18 common free amino acids including all essential amino acids, amino acid containing sulfuric and aromatic amino acids. The percentage of essential amino acids in total amino acids, amino acid containing sulfuric and aromatic amino acids in pomegranate seeds were approximately 30%, over 15% and less than 10%, respectively. Glutamic acid, arginine and aspartate acid were found to be the main amino acids, followed by glycine, leucine, serine and proline (Elfalleh *et al.*, 2012b). Both glutamine and asparagine, which contain glutamic acid and aspartate, respectively, are important reservoirs of amino acid groups for the body (Mahon and Escott-Stump, 1996). A team of researchers compared the Tunisian and Chinese cultivars, and reported that protein content varied significantly between cultivars, ranging from 20.70 (GR1, Tunisian cultivar) to 28.54% (SCH2, Chinese cultivar) (Elfalleh *et al.*, 2011). The pomegranate juice protein content is about 7.95 g/l. The protein content of the pomegranate seed is about 16.87% dry weight, of which globulins (62.4 mg/g dry weight) and albumins (54.12 mg/g dry weight) are the most abundant amino acids, followed by glutelins (33.2 mg/g dry weight) and prolamins (18.08 mg/g dry weight). The total amino acid content is 14.45 g/100g dry weight. In addition, the pomegranate seed is rich in amino acids glutamate, arginine, aspartate, leucine and glycine. This essential amino acid contains 33.54% total amino acid (Elfalleh *et al.*, 2011). In another study, the protein content of the pomegranate seed was found to be 13.2% (Nagy *et al.*, 1990). Sharma and Akansha (2018) also reported the highest amount of protein in pomegranate seed in a comparison of three parts of the fruit: peel, seed and whole fruit, 7.8 ± 0.16, 9.2 ± 0.35 and 6.4 ± 0.37% respectively (Sharma *et al.*, 2018). Protein content in the ethanolic extract of pomegranate leaves has been found to be 11.86 ± 0.9% (Sreedevi *et al.*, 2017). In another study, amino acids were not found in any of the aqueous, hydroalcoholic, ethanolic, ethyl acetate and n-hexane extracts of pomegranate leaf (Sreedevi *et al.*, 2017). These compounds were also reported to exist in pomegranate juice (Lansky and Newman, 2007; Jurenka, 2008) and hydroalcoholic extract of its peel (Dixit *et al.*, 2017). However, in a study by Akkiraju *et al.* (2016), amino acids were not found in methanolic extract of exocarp, mesocarp and seeds. In the flowers, the presence of proteins was investigated and the results showed that proteins were found in only two chloroform-aqueous and ethyl acetate extracts of pomegranate flower, but not in the petroleum ether, chloroform and ethanolic extracts (Tripathi and Kohli, 2011).

Different parts of the pomegranate contain certain vitamins; for example, arils (Jaiswal *et al.*, 2010), seeds (Singh and Sethi, 2003; Syed *et al.*, 2007) and fruit peel (Mirdehghan and Rahemi, 2007), which contain various vitamins including vitamins C (Hassan *et al.*, 2012), B5, B9, K (Abbasi *et al.*, 2008), B2, B1, β-carotene (Gil *et al.*, 2000; Aviram *et al.*, 2002; Cheng *et al.*, 2012) and α-tocopherol (vitamin E) (Liu *et al.*, 2009). Experiments have shown that vitamin C is found in methanolic extract of all three parts: exocarp, mesocarp and seeds (Akkiraju *et al.*, 2016). The amount (%) of this vitamin was reported to be 21.5 ± 0.06 in fruit, 20.6 ± 0.01 in seed and 19 ± 0.06 in whole fruit powder (Sharma *et al.*, 2018). Vitamin C is one of the most important vitamins found in pomegranate juice (Jurenka, 2008). High-performance liquid chromatography (HPLC) analysis showed that the amount of vitamin C was 53.7–5917.3 μg/ml, 4–3957.3 μg/ml and 497.4–2797.7 μg/mg in exocarp, mesocarp and seed extracts, respectively. The contents of tailing factor (TF) ranged from 0.65–1.6 μg/mg, 0.5–1.9 μg/mg and 0.60–1.61 μg/mg for exocarp, mesocarp and seed extracts, respectively (Akkiraju *et al.*, 2016). The amount of vitamin B (B complex fibre) has also been reported to be about 5.1% (Bhowmik *et al.*, 2013).

17.4 Functional/Health Benefit

Various biological effects of pomegranate including anti-inflammatory, antioxidant, antidiabetic, anticancer, immunomodulatory, antimicrobial and cardiovascular protective activities have been so far characterized. Pomegranate is a polyphenol-rich fruit. Virtually all the extracts from all parts of the fruit have been found to be safe and, so far, no harmful effects have been reported in humans. Extracts from different parts of the pomegranate have been associated with reduced risk for cardiovascular diseases (CVDs) such as hypertension, hypercholesterolaemia, oxidative stress, hyperglycaemia, atherosclerosis and inflammation; coronary heart diseases (CHDs); as well as other health complications (Aviram and Rosenblat, 2012). Various short-term randomized controlled trials and a limited number of prolonged clinical interventions have been conducted to elucidate the underlying mechanisms through which pomegranate exerts cardioprotective effects by means of lipid and lipoprotein profile, antioxidant status, oxidative stress, chronic inflammation and endothelial dysfunctions. Pomegranate causes a decrease in energy intake and the absorption of fats in the diet due to its inhibitory effect on pancreatic lipase. It also diminishes oxidative stress and inflammation, making it more prominent in the context of anti-obesity (Lei *et al.*, 2007). So far, much research has been accomplished to find more evidence for the role of pomegranate in the modulation of lipid profiles. By mechanisms such as reducing the cellular uptake of oxidized LDL (OxLDL) and inhibiting the cellular cholesterol biosynthesis, pomegranate juice has a straight effect on macrophage cholesterol metabolism. This action leads to a reduction in the macrophage cholesterol store and prevents the progression of atherosclerosis or delays the onset of the disease (Fuhrman *et al.*, 2005). Also, intake of pomegranate juice by LDL receptor-deficient mice, fed with a high-cholesterol diet, lessened the progression of atherogenesis at different stages of the disease (De Nigris *et al.*, 2005).

In diabetic fatty rats, oral administration of pomegranate flower extracts showed a protective effect on the serum lipid profile accompanied by antioxidant activity, lowered fatty acids, triglycerides and total cholesterol plasma levels. The extracts also reduced cardiac triglyceride content, suggesting that pomegranate supplementation can be profitable in coping with chronic atherosclerotic diseases (Huang *et al.*, 2005b). Regarding the antiobesity properties of pomegranate, Lei *et al.* (2007) have shown that in high-fat diet induced-obesity mice, treatment with pomegranate leaf extract decreased the body weight, energy intake and total cholesterol, triglyceride and glucose levels (Lei *et al.*, 2007). Therefore, it can be stated that, along with its inhibitory function on the pancreatic lipase enzyme, pomegranate tissue extracts are capable of suppressing energy uptake, thus can help to reduce obesity as an appetite suppressant or as a probiotic, especially in a fat-rich diet (Lei *et al.*,

2007). Pomegranate seed oil was found to be rich in fatty acids and mainly contains punicic acid. Oral administration of pomegranate seed oil diminished the malondialdehyde content in heart and kidney tissue homogenates, and reduced triglyceride levels in treated rats (Mollazadeh *et al.*, 2016). In high- and low-exercise-lifestyle-mimicking rats (high- and low-capacity runners), pomegranate juice consumption reduced cellular oxidation and triglyceride content, and also increased paraoxonase activity (an HDL-related esterase able to protect lipids from peroxidation) (Rosenblat *et al.*, 2015).

Studies conducted in humans indicated that pomegranate has beneficial effects on modulating harmful fats, which are connected to cardiovascular complications. Pomegranate intake reduces the accumulation of LDL and the susceptibility to LDLs by increasing the serum paraoxigenase activity (Aviram *et al.*, 2000; Aviram and Dornfeld, 2001; Chistiakov *et al.*, 2017). In type 2 diabetic patients with hyperlipidaemia, concentrated pomegranate juice caused a drop off in cholesterol absorption, increased faecal cholesterol l excretion, showed a desirable effect on cholesterol metabolism, vigorously reduced LDL cholesterol, LDL/HDL cholesterol and improved total HDL ratios (Esmaillzadeh *et al.*, 2006).

Given that free radicals play a major role in the advancement of CVDs, consideration of antioxidants to prevent these diseases is essential. Over the past few decades, extensive studies have been carried out, all emphasizing the antioxidative potency of pomegranate and its components due to the presence of high polyphenol concentrations, such as punicalagins, punicalins, anthocyanins, unique fatty acids, gallagic acid and ellagic acid. Pomegranate polyphenols effectively reduce the oxidative stress of macrophages, eliminate free radicals and reduce the peroxidation of lipids (Gil *et al.*, 2000; Les *et al.*, 2015). Following metabolism in the biological system, a few of these polyphenols turn into urolithins, which are responsible for the antioxidant properties of the plant *in vivo* (Johanningsmeier and Harris, 2011). Urolithin C and D were found to be more potent antioxidants than the parental substance ellagic acid and punicalagin (Bialonska *et al.*, 2009). In addition, the antioxidant activities of pomegranate anthocyanins, such as delphinidin, cyanidin and pelargonidin, are greater than the other natural antioxidants, vitamins E, C and A (Noda *et al.*, 2002; Youdim *et al.*, 2002).

Pharmacological evidence has showed that ellagic acid helps to prevent atherosclerosis development. Reactive oxygen species (ROS) generation was markedly suppressed in human umbilical vein endothelial cells (HUVECs) pretreated with ellagic acid. In addition, ellagic acid inhibited the endothelial lectin-like oxLDL receptor-1 (LOX-1)-induced endothelial dysfunction by hindering the NADPH oxidase-induced overproduction of superoxide, suppressing the release of NO by down-regulating inducible nitric oxide synthase (iNOS), enhancing cellular antioxidant defences and attenuating oxLDL-induced LOX-1 up-regulation and endothelial nitric oxide synthase (eNOS) down-regulation. Expression of LOX-1 was connected with the pathobiological effects of oxLDL in endothelial cells such as ROS generation and suppression of eNOS activity (Lee *et al.*, 2010). An increased production of ROS also plays a major role in ischaemic heart disease, congestive heart failure, cardiomyopathy and arrhythmias (Das and Maulik, 2000). Pretreatment with pomegranate juice and its butanolic fraction caused restoration of heart rate, reduction in vascular reactivity to various agonists, increase in the levels of antioxidant enzymes (i.e. superoxidase dismutase and catalase) and a significant decrease in the levels of cardiac marker enzymes such as creatine kinase and lactate dehydrogenase in isoproterenol-induced myocardial infarction rats (Mohan *et al.*, 2010). In the 2000s, basic research revealed that the antioxidant properties of pomegranate juice were related to tannin punicalagin, anthocyanins, ellagic acid derivatives, as well as other phenolic substances. It has been claimed that the antioxidant capacity of pomegranate is three times more than that of green tea and red wine (Gil *et al.*, 2000). Pomegranate seed oil, leaf extract, fruit juice and fruit peel extract inhibited the activity of cholinesterase in the brain of high-fat-high fructose-fed overweight rats. This diet increased the brain stress, body weight and lipid profile of the blood; the extracts modulated the lipid profile in the blood and prevented the accumulation of lipids in the brain and the body. Pomegranate fruit extracts may act as free radical scavengers and conserve the endogenous antioxidant system in rats; thus inhibiting the peroxidation of membrane lipids, consequent leakage of soluble enzymes and oxidation of -SH groups of enzymatic proteins. Protective effects of pomegranate fruit extracts were studied in cardiac toxicity induced by drugs or smoking.

For inflammation treatments, the use natural products is very effective. Several studies have been conducted, all proposing that pomegranate reduces inflammation by altering the biological activity of the cell. The main anti-inflammatory agent of the standardized extract of pomegranate is known to be ellagic acid (Bagri *et al.*, 2010; Mo *et al.*, 2013). The activity of cyclooxygenase, a key enzyme in the conversion of arachidonic acid to prostaglandins, was inhibited by pomegranate seed oil extract. Lipoxygenase, which catalyses the conversion of arachidonic acid to leukotrienes, other mediators of inflammation, was also repressed by pomegranate seed oil extract. In comparison with seed oil, fermented pomegranate juice inhibited the lipoxygenase to a lesser extent *in vitro* (Schubert *et al.*, 1999). Nitric oxide (NO) acts as both an antioxidant and anti-inflammatory agent in endothelial cells; therefore, this molecule is an obstacle to the development of atherosclerosis (Napoli and Ignarro, 2001). Pomegranate juice prevented the oxidative destruction effect of NO and improved its anti-inflammatory and antioxidant properties in bovine pulmonary artery endothelial cells (Ignarro *et al.*, 2006).

Punicalagin and ellagitannins (purified from fruit husk) reduced the inflammatory cell signalling in the HT-29 colon cancer cell line (Adams *et al.*, 2006). A cell-based experiment aimed at unravelling the contribution of ellagitannin-containing foods in CVD prevention. The study revealed that ellagic acid and some urolithins are very potent to prevent atherogenesis. Collectively, pomegranate and its active phytochemicals diminish the inflammatory reactions related to CVD in different ways.

Anticancer activity of pomegranate against breast, prostate, leukaemia, bladder, lung, pancreatic, and skin cancer cells has been extensively reported. Pomegranate extract exhibited anti-inflammatory and anticytotoxic activities on breast cancer cells both *in vitro* and *in vivo*. Pomegranate seed oil has also been reported to reduce mammary carcinogenesis in the mouse mammary organ and to prevent proliferation of the various cancer cell types *in vitro* (Kim *et al.*, 2002; Shirode *et al.*, 2015). Pomegranate extracts inhibit MCF-7 cell line proliferation by inducing apoptosis (Jeune *et al.*, 2005). In a study, linolenic acid isomers of pomegranate seeds were measured as selective oestrogen modulators (SERMs) *in vitro*. Punicic acid, a pomegranate-derived compound, inhibited ER-β and ER-α with inhibitory concentration (IC_{50}) of 8.8 and 7.22 μM, respectively. Six compounds obtained from ellagitannin-derived extract of pomegranate have been reported to exhibit antiaromatase activity, out of which urolithin B was the most effective in live cell assay (Adams *et al.*, 2010). Moreover, pomegranate-derived ellagitannin prevents oestrogen-responsive tumours in the breast, as confirmed in an experiment on cell proliferation (Adams *et al.*, 2010). Greater antiaromatase activity against MDA-MB-435 and MCF-7 breast cancer cell lines was observed in a study of the polyphenols, obtained from pomegranate fermented juice, seed oil and pericarp extracts. These polyphenols suppressed 17-beta-hydroxysteroid dehydrogenase type 1 up to 34–79% and inhibited aromatase activity up to 60–80% (Kim *et al.*, 2002). Pomegranate extracts inhibit MCF-7 cell line proliferation by inducing apoptosis (Jeune *et al.*, 2005).

Pomegranate metabolites can be also used to prevent the recurrence of prostate cancer, indicating the effect of flora-derived metabolites in the gut to prevent cancer (Vicinanza *et al.*, 2013). Pomegranate extract effects also lead to an increase in NF-κB (nuclear factor kappa-light-chain-enhancer of activated B cells) activity during the transition to androgen independence from androgen dependence in the LAPC4 (a cell line of human prostate cancer) xenograft model (Rettig *et al.*, 2008). Pomegranate polyphenols reduce gene expression for androgen-synthesizing enzymes and androgen receptor (AR) in the LNCaP cell line. Therefore, a decrease in gene expression can be helpful in androgen-independent prostate cancer cells and human prostate cancer cells where AR is upregulated (Hong *et al.*, 2008). Pomegranate juice increases prostate-specific antigen (PSA) and doubling time in prostate cancer (CAP) patients. Ellagitannins present in juice are helpful for their biological activities.

As soon as ellagitannins are consumed, they are hydrolysed, which releases ellagic acid. Pomegranate is effective for prostate cancer patients owing to increasing prostate-specific antigen doubling time. The polyphenols of fermented juice and pericarp demonstrated anticancer potential against human prostate cancer cell lines (PC3, DU145 and LNCaP) at 20–100 μg/ml (Albrecht *et al.*, 2004). Pomegranate peel extract showed inhibition of migration and invasion of the prostate cancer cells via a reduced level of MMP2/MMP9 (matrix metalloproteinase; MMP) and increased TIMP2 (tissue inhibitor of metalloproteinases 2) expression (Deng *et al.*, 2017). Moreover, total pomegranate seed ethanolic extract contains punicic, α-linoleic and α-linolenic acids, which increase the inhibition of cell growth in the hormone-dependent LNCaP prostate cancer cell line (Lucci *et al.*, 2015). Overall, different compounds of pomegranate were found to exert antiproliferative activity and antiangiogenic effects, and induce apoptosis in prostate cancer. Therefore, pomegranate can be regarded as an effective chemotherapeutic agent to treat prostate cancer.

Pomegranate seed oil exhibits chemopreventive activity against experimental colon carcinogenesis (Adams *et al.*, 2006). Adams *et al.* (2006) reported pomegranate juice exhibits an anti-inflammatory activity in the signalling proteins in HT-29 human colon cancer cell line. Pomegranate seed oil treatment remarkably prevents colonic adeno-carcinoma development and enhances peroxisome proliferator-activated receptor (PPAR) gamma protein expression in the non-tumour mucosa (Dana *et al.*, 2016). Epidemiological studies have indicated that the use of pomegranate is inversely correlated with development of colon cancer (Núñez-Sánchez *et al.*, 2015). Pomegranate has drawn attention for its antitumourigenic properties in the colon (Khan *et al.*, 2009). Compounds such as punicalagin, ellagic acid, total pomegranate tannin (TPT) and pomegranate juice were tested against colon cancer cell lines (HT-29, HCT116, SW620). Punicalagin, ellagic acid, TPT and pomegranate juice at 100 μg/ml induced apoptosis in HT-29 colon cells.

Pomegranate has been found to exert anticancer properties against leukaemia. The polysaccharide PSP001, obtained from the pomegranate fruit, produces an antioxidant effect and growth suppression response on leukaemia cell lines. 3-(4,5-imethylthiazol-2-yl)–2,5-diphenyltetrazolium bromide (MTT) assay showed an IC_{50} value of 52.8 ± 0.9 μg/ml after 72-h incubation of K562 (chronic myeloid leukaemia) cells (Joseph *et al.*, 2012). The effect of pomegranate peel extract (PPE) has been studied in chronic myeloid leukaemia (Asmaa *et al.*, 2015). Certain phyto-constituents including saponins, flavonoids and polyphenols have also been isolated from the PPE. This extract was observed to suppress the K562 cell line via cell cycle arrest at G2/M phase and by activating apoptosis. Ellagitannins can be an important bioactive, anticancer agent (Dahlawi *et al.*, 2012). Pomegranate juice extracts caused apoptosis in eight leukaemia cell lines; four lymphoid cell lines, Jurkat (peripheral blood T cell leukaemia), MOLT-3, SUP-B15 (acute lymphoplastic leukaemia) and CCRF-CEM; and four myeloid leukaemia cell lines, THP-1, K562, HL-60 and KG1a (acute myelogenous leukaemia); and no tumour hematopoietic stem cells (control cells). The effectiveness of flavonoid-rich fractions obtained from fresh pomegranate juice, fermented juice and an aqueous extract of pomegranate pericarps on differentiation of HL-60 human leukaemia cells has been examined. Both pomegranate juice and fermented juice extracts promoted strong cellular differentiation and prohibited proliferation of HL-60 cells while fresh pomegranate juice showed only a relatively mild differentiation-promoting effect. The efficacy of pomegranate juice was comparatively milder than that of fermented juice and pomegranate extract. This demonstrates that pomegranate extract activity is influenced by the high amount of flavonoids and partly because of ellagitannin compounds (Kawaii and Lansky, 2004).

The important phyto-constituents of pomegranate rind extract (PRE), isolated by HPLC-MS, are ellagitannins, namely, punicalins, punicalagin A, punicalagin B and EA. PRE specifically inhibits the viability of human bladder cancer cells (EJ cell) in comparison with PRE-insensitive rat urinary bladder epithelial cells. The anticancer activity of Taiwanese pomegranate fruit ethanol extract (PEE) on urinary bladder urothelial carcinoma (UBUC), and its mechanism has been investigated. PEE reduced UBUC T24 and J82 cell proliferation in a dose-dependent manner. Therefore, PEE treatment could lead to extreme stimulation of endoplasmic reticulum, which could be another apoptotic mechanism of PEE-induced inhibition of bladder cancer cell (Lee *et al.*, 2013).

Pomegranate is considered a highly important functional food because of its numerous beneficial effects in human health, and has been observed to play an essential role in treating bladder cancer (Lee *et al.*, 2013; Masci *et al.*, 2016). Lee *et al.* (2013) reported that PEE suppressed the proliferation of UBUC tumour cells through cell cycle arrest at S phase because of down-regulation of cdk-1 and up-regulation of cyclin A (Lee *et al.*, 2013). Furthermore, it has been observed that PEE triggers pro-caspase-3,-8 and -9, and increases the Bax/Bcl-2 ratio. More importantly, it has also been found that PEE stimulates the expression of procaspase-12, with increased expression of CHOP and Bip, which are endoplasmic reticulum stress markers (Lee *et al.*, 2013).

The effects of pomegranate extract on lung tumourigenesis have also been studied both *in vitro* and *in vivo*. These studies have shown that pomegranate extract could be used as a chemo-preventive and chemo-therapeutic agent to treat human lung cancer (Khan *et al.*, 2007; Pantuck *et al.*, 2015). Li *et al.* (2016b) have shown that pomegranate leaf extract suppresses cell proliferation in a non-small lung carcinoma cell line dose- and time-dependently (Li *et al.*, 2016a). Pomegranate leaf extract also influences H1299 cell line survival by arresting cell cycle progression in the G2/M phase. It has been suggested that pomegranate leaf extract could serve as a safe and effective chemotherapy agent in treating non-small cell lung carcinoma by inducing apoptosis, inhibiting proliferation, arresting cell cycle, and impairing cell migration and invasion. Pomegranate extract is also able to inhibit tumour growth in nude mice and suppress pro-survival pathways in human A549 lung carcinoma cells (Khan *et al.*, 2007). Khan *et al.* (2009) showed that pomegranate extract remarkably inhibited lung carcinogenesis in the mouse model and therefore could be studied for chemo-preventive effects in lung cancer in humans (Afaq *et al.*, 2008). Studies have demonstrated that pomegranate peel and seed extracts *in vitro* exert antioxidant properties and inhibit A549 cell line proliferation (Modaeinama *et al.*, 2015; Seidi *et al.*, 2016). The treatment of a non-small lung carcinoma cell line with pomegranate leaf extract reduced the expression of matrix metalloproteinase, inhibited cell invasion and migration through cell cycle arrest at the G2/M phase, and lowered the mitochondrial membrane potential and ROS (Khan *et al.*, 2007; Khan and Ahmad, 2015; Li *et al.*, 2016a). Both *in vitro* and *in vivo* studies have shown that pomegranate fruit extract treatment suppresses the degradation and phosphorylation of IκBα kinase and leads to the down-regulation of NF-κB expression. Taken together, these studies suggest that pomegranate extracts have potent chemotherapeutic properties with anticancer effects on lung carcinoma by inhibiting cell growth and proliferation, inducing apoptosis, and thereby preventing the migration and progression of lung cancer.

Hepatocellular carcinoma is a common and fatal cancer extremely stimulated by oxidative stress. Pomegranate peel has hepato-protective activity (Bhatia *et al.*, 2013). The effect of pomegranate extract on diethylnitrosamine (DENA)-induced hepato cell carcinogenesis has been studied *in vivo*. A significant chemo-preventive potential has been reported due to a decrease in the incidence, size, volume and multiplicity of the hepatic nodules. Pomegranate extract also caused a decrease in the liver lipid peroxidation and oxidation of proteins. Bishayee *et al.* (2011b) suggested and supported administration of pomegranate-derived compounds to treat and prevent hepatocellular carcinoma in humans (Bishayee *et al.*, 2011b). Bishayee *et al.* (2011a) observed that pomegranate bioactive compounds produce a chemo-preventive effect against diethyl nitrosamine-induced liver carcinogenesis by inhibiting hepatic oxidative stress. The chemo-preventive effect of pomegranate acts against hepatic carcinoma possibly via antioxidant signalling mechanisms without any toxic effect. Pomegranate bioactive properties have been reported to inhibit cell growth, induce apoptosis and regulate cell cycle progression in an animal model (Kaplan *et al.*, 2001).

In pancreatic cancer cells, pomegranate extract causes cell cycle arrest and controls cell proliferation pancreatic cancer cells in humans (PANC-1 cells). Pomegranate extract treatment increases the number of cells that do not have CD44 and CD24 expressions, which are associated with increased tumour-initiating ability. This indicates that pomegranate extract changes cell phenotype. Pomegranate extract's effectiveness in controlling the proliferation of human pancreatic cancer cells PANC-1 and AsPC-1 was greater than that of the standard drug paclitaxel. However, identified ingredients of pomegranate exhibited a modest activity. As a result, it is necessary to identify more compounds in pomegranate extract involved in the mechanism of its anticancer activity in pancreatic cancer. Nair *et al.* (2011) observed that pomegranate extract is a strong inhibitor of the human pancreatic epithelial cancer *in vitro* by targeting cell cycle progression.

Pomegranate-derived compounds have also been examined for chemo-prevention of different skin cancer types (Syed *et al.*, 2006). Pomegranate fruit extract and diallyl sulfide (DAS) slowed down the initiation and decreased the likelihood of tumour by 55 and 45%, respectively. A combination of pomegranate fruit extract and DAS at low dose synergistically decreased the development of tumour by 84%. A combination of pomegranate fruit extract and DAS caused a decrease in the cell proliferation and stimulation of apoptosis in comparison with separate components, which was confirmed by observations in the histological analysis and the analysis of cell death (George *et al.*, 2011). Pomegranate fruit extract reduced UVB-stimulated nuclear factor kappa B (NF-XB) activation and mitogen-stimulated protein kinase pathways. Treatment with pomegranate fruit extract before treatment with tetradecanoyl phorbol 13-acetate (TPA) in skin led to a decrease in latency period from 9 weeks to 14 weeks and prevention of tumour multiplicity and tumour incidence. This observation shows that pomegranate fruit extract is a strong antitumour agent because it is able to inhibit different biomarkers of tumour progression by TPA in an animal model (Afaq *et al.*, 2005).

Pomegranate extract is a rich source of ellagitannins, anthocyanins and tannins, and hence exerts a strong antioxidant activity. Tannins have a remarkable cancer-preventing activity (Li *et al.*, 2003). Tannin has been reported to stimulate release of interleukin 1 (IL-1) and IL-1 beta from mice and human macrophages *in vitro*. These results suggest that tannin exerts its antitumour effect by activating macrophages (Miyamoto *et al.*, 1993). Some *in vitro* and *in vivo* studies have shown that polyphenol rich fractions derived from the pomegranate fruit are a safe and effective chemopreventive agent against skin cancer. The potential chemopreventive and/or cancer therapeutic effects of pomegranate derivatives are revealed as potential effects for photocarcinogenesis and chemical skin carcinogenesis in several animal models (Hora *et al.*, 2003; Syed *et al.*, 2006). Pomegranate seed oil has been found, in experimental studies, to inhibit proliferation of different tumour cell types (Kim *et al.*, 2002; Lansky *et al.*, 2005) and to reduce skin carcinogenesis in mice (Viuda-Martos *et al.*, 2010). This indicates that pomegranate seed oil can serve as an effective and safe agent to fight skin cancer (Hora *et al.*, 2003). Overall, pomegranate has been found to have potent chemo-preventive properties against skin cancer without adverse side effects.

Pomegranate has also showed remarkable effects on the central nervous system (CNS). Treatment with PPE improved performance on various tests with more marked effects on spatial learning tendency and long-term memory than on retention capacity (Adiga *et al.*, 2010). A study by Braidy *et al.* (2016) showed supplementation with pomegranate in Alzheimer's disease mice model improved the decreased phosphorylation of mTOR through activation of the PI3K-Akt-mTOR pathway. Furthermore, pomegranate improved synaptic function alleviating the up-regulation of important inflammatory transcripts, TNF-α, il-1β, iNOS, ccl2 and il-1 (Braidy *et al.*, 2016). Bcl1 is an essential protein that enhances microglial phagocytosis and autophagy, which is crucial for the catabolism of abnormal Aβ aggregates by activated microglial cells in the brain (Pickford *et al.*, 2008; Jaeger *et al.*, 2010; Lucin *et al.*, 2013). In an experiment by Braidy *et al.*, pomegranate increased the protein expression of Bcl1 and LC-3 type II. Brains of Alzheimer's patients are under extensive oxidative stress and overproduction of Aβ leads to Aβ-associated free radical oxidative stress (Miranda *et al.*, 2000; Praticò and Delanty, 2000). This oxidative stress is manifested by the formation of ROS, H_2O_2, lipid peroxidation and the subsequent modification of proteins by reactive lipid peroxidation products (Butterfield *et al.*, 2002; Uttara *et al.*, 2009). In an *in vitro* study, *P. granatum* ethanol extract (containing 2,4-di-tert-butylphenol) in PC12 cells decreased

H_2O_2-induced cytotoxicity and protected PC12 cells from oxidative stress-induced cell death by protecting neuronal cell mitochondrial dysfunction. Furthermore, pomegranate extract showed an antiamnestic effect and inhibited learning and memory deficits (Choi *et al.*, 2011).

Neuroinflammation is significant in the pathogenesis and development of Alzheimer's disease. In a study by Kim *et al.* (2002), lipopolysaccharide (LPS)-treated cultured astrocytes and microglial BV-2 cells along with an Alzheimer's disease model mouse were treated with punicalagin (PUN), a component of pomegranate. Results showed PUN (1.5 mg/kg) improved LPS-induced memory impairment and prevented the LPS-induced expression of inflammatory proteins in mice. On the other hand, an *in vitro* experiment showed that PUN (1 mg/ml) inhibited the LPS-induced expression of iNOS and Cox-2, as well as the production of ROS, NO, TNF-α and IL-1b. PUN also suppressed activation of NF-κB via inhibition of IκB degradation. This study showed that PUN inhibits LPS-induced memory impairment via anti-inflammatory and antiamylogenic mechanisms through inhibition of NF-kB activation (Kim *et al.*, 2017). In another experiment, short-term treatment with a standardized pomegranate extract in an aged Alzheimer's disease animal model did not improve cognitive performance, but altered levels and ratio of the Aβ42 and Aβ40 peptides, which would favour a diminution in Alzheimer's disease pathogenesis and was supportive of a specific antiamyloidogenic mechanism of a pomegranate extract in this aged Alzheimer's disease animal model (Ahmed *et al.*, 2014). In another study, freeze-dried pomegranate water extract (PWE) treatment on IL-1β-stimulated SK-N-SH cells showed a dose-dependent reduction of COX-2-dependent PGE2 production, as well as inhibition of phosphorylation of IκB, IKK, NF-κB transactivation. Next to Alzheimer's dementia, Parkinson's disease is the second-leading neurodegenerative disorder that affects about 1% of people beyond 65 years of age, with a higher prevalence in men (Recchia *et al.*, 2004). In one study, selected pomegranate juice extract protected against 1-methyl-4-phenyl-1,2,3,6-tetrahydropyridine (MPTP)-induced neurotoxicity in primary human neurons in a dose-dependent manner via attenuating increase in extracellular LDH activity (Braidy *et al.*, 2014). Pomegranate seed oil (PSO), as food or as a water-soluble nano-emulsion, was used in TgMHu2ME199K mice (an animal model for genetic prion disease). Nano-PSO delayed disease presentation and postponed disease aggravation in already sick mice. Nano-PSO treatment did not decrease PrPSc accumulation, but reduced lipid oxidation and neuronal loss, indicating a strong neuroprotective effect (Mizrahi *et al.*, 2014). Treatment with PSO on an experimental autoimmune encephalomyelitis (EAE) mouse model of multiple sclerosis, decreased disease burden. Furthermore, PSO reduced demyelination and oxidation of lipids in the animal's brains. This study showed that lipid oxidation is an important factor in demyelinating diseases (Binyamin *et al.*, 2015). Oral administration of PSE alleviated impairment in memory and motor coordination. PSE improved active avoidance learning and motor activity in ischaemic groups (Hajipour *et al.*, 2014). It was suggested that pomegranate juice neuroprotective effects might be secondary to the suppression of both the maternal inflammatory response and inhibition of fetal brain apoptosis, neuronal nitric oxide synthase and nuclear factor-κB activation (Ginsberg *et al.*, 2018). A series of pomegranate juice extracts (i.e. helow, malasi, qusum and hamedh) on quinolinic acid (QUIN)-induced excitotoxicity on primary cultures of human neurons showed that all extracts reduced the oxidative effects of increased NO production thereby reducing the formation of 3-nitrotyrosine and poly [ADP-ribose] polymerase (PARP) activity, and hence preventing NAD+ depletion and cell death. Observed inhibitory effects of some of these compounds on specific excitotoxic processes such as calcium influx were suggestive of beneficial effects of pomegranate juice extracts in excitable tissue, particularly within the CNS (Essa *et al.*, 2013). Pomegranate juice diminished caspase-3 activation in brain regions (Loren *et al.*, 2005). In another experiment, pomegranate polyphenol extract in the neonatal HI mouse model resulted in significantly decreased H-I induced caspase-3 activation (West *et al.*, 2007). In one study, PSO, leaves, juice and peel (PP) were used to investigate their effects on cholinesterase activity, brain oxidative stress and lipid profile in high-fat-high fructose diet (HFD) induced-obese rat (Ballabh *et al.*, 2004). PSO, leaves, peel and juice inhibited cholinesterase activity in a dose-dependent manner, modulated lipid profiles in blood, and prevented accumulation of lipid in brain and body as evidenced by a decrease of their weights as compared with untreated rats. Also, these extracts protected the brain from stress oxidants, evidenced by the decrease of malondialdehyde (MDA) and protein carbonylation (PC) levels and the increase in superoxide dismutase (SOD) and glutathione peroxidase (GPx) levels and reduced AChE activity (Amri *et al.*, 2017).

Ethanol crude extracts of *P. granatum* and *Eugenia uniflora* showed potential antimicrobial activity against skin microbiota (*Staphylococcus epidermidis* and *Staphylococcus aureus*) in an *in vitro* study (Bernardo *et al.*, 2015). Bacterial biofilms facilitate bacterial adhesion to biotic and abiotic surfaces. Biofilms protect bacteria from host immune responses and protect cells from antimicrobial substances (Ando *et al.*, 2004; Kokare *et al.*, 2009). *Staphylococcus aureus* is a powerful pathogen due to its resistance to traditional antibiotics and its ability to form biofilms. Methicillin-resistant *S. aureus* (MRSA) is a common cause of hospital-acquired infections commonly associated with high morbidity and mortality. In one study, *P. granatum*, along with *Rosmarinus officinalis* and *Tetradenia riparia*, showed anti-MRSA effects *in vitro* and also showed synergistic interactions with penicillin (Endo *et al.*, 2018). Standardized pomegranate rind extract (SPRE) containing 13% w/w ellagic acid showed a bacteriostatic effect against *Propionibacterium acnes*, a Gram-positive anaerobe, although this antibacterial activity had a narrow spectrum (Panichayupakaranant *et al.*, 2010). *Helicobacter pylori* can cause chronic gastritis, which can lead to a peptic ulcer, mucosa-associated lymphoid tissue (MALT) lymphoma and gastric cancer. In an *in vitro* experiment, extracts of *P. granatum* showed significant antibacterial activity against clinical isolates of *H. pylori* as demonstrated by the mean of inhibition zone diameter ranging from 16–40 mm at 50 mg (Hajimahmoodi *et al.*, 2011). Food-borne pathogens are responsible for diseases that are environmental hazards to the food supply and human health. Pomegranate fruit peel extracts were shown to have antimicrobial activity against some food-borne pathogens such as *Listeria monocytogenes* both *in vitro* and *in situ* (Al-Zoreky, 2009). Pomegranate aril and peel extracts have been showed to inhibit bacterial growth of two essential pathogens in food-borne diseases including *Staphylococcus aureus* and *Escherichia coli*, in an *in vitro* study (Pagliarulo *et al.*, 2016). Dental diseases are caused by pathogenic bacteria in dental plaque. In one experiment, one of the wild forms *of P. granatum* named 'Daru' showed antimicrobial activity against four dental bacteria: *Streptococcus* species, *Lactobacillus* species, *Staphylococcus* species and *Proteus* species (Devi *et al.*, 2011). In another study, *P. granatum* water extracts showed antibacterial properties against five oral bacteria and prevented orthodontic wire bacterial biofilm formation (Dastjerdi *et al.*, 2014). In another study, glycolic extract of pomegranate was used against *Porphyromonas gingivalis* infection by using a *Galleria mellonella* experimental model. It was observed that pomegranate extract treatment showed antimicrobial effects via inducing higher survival rates in treated subjects (Aparecida Procópio Gomes *et al.*, 2016).

Herpes simplex virus–2 (HSV-2) can cause a lifelong infection in the immunocompromised host and intermittent in healthy persons. HSV-2 also increases the transmission of HIV. Acyclovir as the therapeutic drug for HSV-2 infection cannot control viral latency and recurrent infection (Arunkumar and Rajarajan, 2018). In one study, extracts from fruit peel of *P. granatum* treatment in epithelial cells (HEp-2) of human tissue, showed antiviral activity against HSV-2 infection (Arunkumar and Rajarajan, 2018). Hepatitis C virus (HCV) is one of the main causes of chronic liver disease. One *in vitro* experiment showed that punicalagin, punicalin and ellagic acid isolated from the crude extract of pomegranate blocked the HCV life cycle at different stages. Also, these compounds were non-toxic and safe (Reddy *et al.*, 2015). So far, there is no definite cure for AIDS. AIDS is due to infection with human immunodeficiency virus type 1 (HIV-1). Also, no anti-HIV-1 vaccines that can be applied to global immunization programmes are expected to be available for many years to come (Neurath *et al.*, 2004). Among the available prevention strategies is the application of mechanical and chemical barrier methods. One of these barriers is microbicides, which are topical formulations designed to block HIV-1 infection when applied vaginally and rectally before intercourse (Stone, 2002; Neurath *et al.*, 2004; Shattock and Solomon, 2004). In one study, pomegranate extracts were used as microbicides in an *in vitro* study and were shown to be a great HIV-1 entry inhibitor and a potential candidate as a topical microbicide (Neurath *et al.*, 2004).

The influenza virus has been responsible for several epidemics and pandemics over several centuries. The virus can not be eradicated from the human population because it has several zoonite hosts. In one study, PPE was used against influenza A virus. It was observed that PPE inhibited the replication of human influenza A/ Hong Kong (H3N2) *in vitro* (Haidari *et al.*, 2009).

Detection of multiple drug-resistant malaria parasites has led to numerous studies focused on finding alternative therapies. One experiment investigated the effects of pomegranate on murine malaria-induced splenic injury and oxidative stress. Pomegranate treatment was shown to exhibit antimalarial activity in the host by attenuating inflammatory and oxidative stress responses (Mubaraki *et al.*, 2016). Coccidiosis and helminthiasis cause worldwide economic losses. In one experiment *P. granatum* was shown to have significant anticoccidial properties *in vivo* and anthelmintic activity *in vitro*. Also, *P. granatum* improved histopathological pictures of jejunum, induced antioxidant effects and protected the host tissue from injuries by parasites (Dkhil, 2013). Another experiment showed that different solvent extracts of the fruit peel of *P. granatum* had anticoccidial activity against experimentally induced coccidial infection in broiler chicken (Ahad *et al.*, 2018). *Cryptosporidium parvum* is a parasite that causes cryptosporidiosis. *Cryptosporidium parvum* is highly resistant to drug treatments. In one study, aqueous *P. granatum* peel extract in an experimental murine model of cryptosporidiosis showed promising therapeutic effects demonstrated as continuous weight gain, improvement in intestinal histopathology and the cessation of shedding of faecal oocysts in treated animals, and did not cause any side effects (Al-Mathal and Alsalem, 2012). In one study, pomegranate peel extract treatment was used in mice against giardiasis. Results showed that pomegranate was effective in the prevention and treatment of *Giardia lamblia* infection (Al-Megrin, 2017). Taken together, pomegranate and its parts such as seeds, peel, etc. have demonstrated a variety of antimicrobial effects. Pomegranate has showed protective effects against infection with bacteria responsible for hospital infections; dental plaques; food-borne bacteria; viruses such as HSV, hepatitis, and HIV; helminths and parasites, etc.

17.5 Drug Interaction

Medications may be ingested with common fruit juices by patients. Whether pomegranate juice could show rapid drug interactions is an interesting and practical issue. Braga *et al.* (2005) indicated synergic interaction between pomegranate extract and antibiotics against *S. aureus*. The interaction between *P. granatum* (pomegranate) methanolic extract (PGME) and antibiotics against 30 clinical isolates of MRSA and methicillin-sensitive *S. aureus* has been observed. PGME increased the post-antibiotic effect of ampicillin from 3 to 7 h. In addition, PGME demonstrated the potential to either inhibit the efflux pump NorA or to enhance the influx of the drug. The detection of *in vitro* variant colonies of *S. aureus* resistant to PGME was low and they did not survive (Braga *et al.*, 2005).

In vitro and *in vivo* laboratory studies indicate that pomegranate juice inhibits intestinal cytochrome P450 (CYP) 2C9 and CYP3A4 enzymes, which leads to increased bioavailability, peak drug concentration and increased overall exposure of drugs that are metabolized by these enzymes. The enzyme inhibition wears off after 1–3 days; this means that ingestion of medication several hours after ingestion of the juice will not prevent the interaction (Srinivas, 2013).

A large number of drugs in psychiatry are substrates of cytochrome P450 (CYP) 2C9 and CYP3A4 enzymes (Srinivas, 2013). Quetiapine is an example of CYP3A4 substrate. However, there is no literature, as yet, on interactions between pomegranate juice and quetiapine (Spina and de Leon, 2007). Human studies show that pomegranate juice has no effect on the bioavailability or pharmacokinetics of representative CYP2C9 and CYP3A4 substrates (midazolam, fluconazole and flurbiprofen are examples of CYP3A4 and CYP2C9 substrates) (Farkas *et al.*, 2007; Misaka *et al.*, 2011; Hanley *et al.*, 2012). Pomegranate juice may therefore be safely consumed by patients receiving drugs that are substrates of CYP2C9 and CYP3A4 (Hanley *et al.*, 2012; Srinivas, 2013). Preclinical data show that pomegranate juice also inhibits intestinal CYP3A2 and P-glycoprotein, as well as inhibiting hepatic sulfoconjugation (Srinivas, 2013). Animal studies suggest that, although carbamezapine concentrations were increased, half-life elimination was not affected; perhaps pomegranate extract only inhibited gastrointestinal CYP3A4 and not hepatic CYP3A4. Species differences in drug metabolism prevent us from ruling out a drug interaction based on these data, as drugs may be metabolized by different CYP sub-families in rats compared with humans (Farkas *et al.*, 2007). However, the clinical significance of these effects also remains to be established.

Komperda (2009) published a report that described an interaction between pomegranate juice and warfarin. This case report describes a 64-year-old Caucasian woman who was treated with warfarin for recurrent deep vein thrombosis. She had been receiving a relatively stable dosage of warfarin 4 mg/day for several months, with stable international normalized ratios (INRs). During that time, the patient was consuming pomegranate juice two or three times a week. She stopped drinking the juice, and her INRs became subtherapeutic. Her dosage of warfarin was increased to maintain therapeutic anticoagulation. No rechallenge with pomegranate juice was performed. Use of the Drug Interaction Probability Scale indicated a possible relationship between the patient's subtherapeutic INR and the pomegranate juice (Komperda, 2009).

Jarvis *et al.* (2010) reported another case of a potential interaction between pomegranate juice and warfarin. They reported a strong temporal association between high levels of pomegranate juice consumption and uncontrolled anticoagulation (Jarvis *et al.*, 2010). Laboratory studies have shown that pomegranate juice inhibits key cytochrome P450 enzymes involved in warfarin metabolism, which provides a mechanistic explanation for this potential interaction (Nagata *et al.*, 2006). Interactions between warfarin and food/drugs occur with a wide range of drugs metabolized by these P450s; however, the anticoagulant efficacy of warfarin is affected mainly when metabolism of S-warfarin via the CYP2C9 is altered. CYP2C9 is found predominantly in the intestinal epithelium and recent studies have shown that pomegranate juice is a potent inhibitor of this enzyme (Kaminsky and Zhang, 1997). Other studies have also shown that pomegranate juice inhibits the intestinal CYP3A enzyme and may inhibit P-glycoprotein, thereby enhancing warfarin absorption (Hidaka *et al.*, 2005).

A review of potential warfarin–fruit interactions based on 23 citations (15 case reports and seven controlled clinical trials) in 2014 revealed that the majority of cases involved cranberry products; however, pomegranate juice, avocado, grapefruit juice, mango and papain were also implicated in reports of suspected warfarin–fruit interactions. Cranberry juice was also the most frequently studied fruit product. Other fruit products evaluated with warfarin in controlled clinical trials were cranberry concentrate and grapefruit juice (Norwood *et al.*, 2015).

A study reported a case of elevated tacrolimus (an immunosuppressive drug) concentrations in a heart transplant after massive myocardial infarction recipient consuming concentrated pomegranate juice popsicles. His post-operative course was uneventful, and he was discharged on post-transplant day (PTD) 14, on a stable tacrolimus dose. On PTD 35, the patient's tacrolimus concentration dropped suddenly by 50%, without an antecedent dose change (Fig. 17.1). Investigation revealed the patient had been eating fruit popsicles made from a base of pomegranate concentrate – one or two popsicles (each 51 g) (Breyers, Englewood Cliffs, New Jersey). The patient stopped eating the popsicles immediately (PTD 70) and was further instructed to avoid pomegranate along with grapefruit or any other herbal supplements. The patient showed a drop in ImmunoKnow assay (Cylex, Inc., Columbia, Maryland) results during this period, but did not develop infection. He continued to be well and without rejection or major infection at 1 year post-transplantation (Khuu *et al.*, 2013).

The specific compound(s) in pomegranate involved in the drug interactions have yet to be clearly defined. Pomegranate juice does not contain furanocoumarins like grapefruit juice. However, it does contain two known CYP inhibitors: the flavinoid quercetin and the antioxidant ellagic acid. Quercetin is a known inhibitor of CYP3A4 (Li *et al.*, 2006; Bhagwat *et al.*, 2014). Thus, although it is possible that other fruit in the popsicle formulation contributed to the interaction with tacrolimus, it is more likely largely related to the pomegranate content. Ellagic acid is found in high concentrations in pomegranate juice and is a known inhibitor of CYP2A2, -3A1 and -2C6 in rats. The effects of ellagic acid on human isozymes have yet to be described. None of the other listed ingredients has published drug interactions with CYP isozymes, P-glycoprotein or immunosuppressant drug (Khuu *et al.*, 2013).

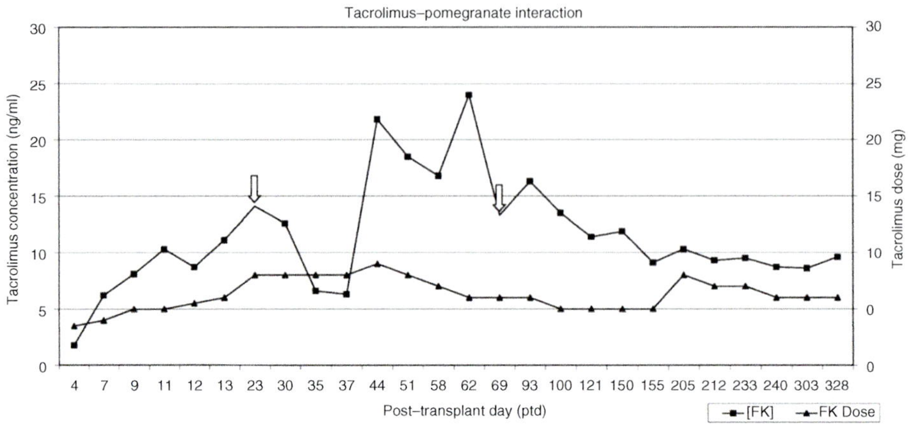

Fig. 17.1. Tacrolimus concentrations and total daily dose over time. First arrow: pomegranate intake initiated; second arrow: pomegranate intake discontinued. (From: Khuu *et al.*, 2013.)

Pomegranate juice intake could attenuate the increase in systemic oxidative stress and inflammation induced by intravenous iron during haemodialysis. Such beneficial effects are probably due to the pomegranate juice's potent antioxidant contents such as polyphenols (Shema-Didi *et al.*, 2013).

Drinking pomegranate juice while taking statin drugs may increase the risk of rhabdomyolysis, a serious condition characterized by the breakdown of muscle fibres that enter the bloodstream and are harmful to the kidneys, often resulting in kidney damage. Scientists at Hartford Hospital in Connecticut report a case of a patient being treated for high cholesterol with medications, who developed rhabdomyolysis 3 weeks after drinking pomegranate juice, according to research published in the *American Journal of Cardiology* (Sorokin *et al.*, 2006). Physicians had successfully treated him for 17 months with rosuvastatin, a statin medication for treatment of high cholesterol and ezetimibe that prevents absorption of cholesterol, and he had no signs of rhabdomyolysis prior to drinking pomegranate juice. However, 3 weeks after drinking pomegranate juice while taking these medications, the patient had symptoms of the condition. The scientists conclude pomegranate juice interacts with certain liver enzymes that increase concentration of statins, which stay in the body and in turn cause rhabdomyolysis. Symptoms of rhabdomyolysis include dark red urine, fatigue, muscle aches and joint pain. Acute kidney failure occurs in many patients with rhabdomyolysis. Early treatment of rhabdomyolysis reduces the risk of chronic kidney failure (Sorokin *et al.*, 2006).

Herbal reactions towards different types of statins are varied so grapefruit or pomegranate interacts with only some types of statins, but not with all statin types. In this context, administration of herbal materials can lead to decreased absorption of statins or decrease the plasma concentration of these drugs (Rouhi-Boroujeni *et al.*, 2015). According to Rosenblat *et al.* (2002), although simvastatin with a dose of 15 μg/ml could decrease macrophage cholesterol biosynthesis rate by 42% as compared with control cells, the combination of pomegranate and simvastatin resulted in an inhibitory effect up to 59% that was significant. Moreover, simvastatin with the same dosage modestly decreased macrophage ROS formation by 11% alone and by up to 63% concurrently with pomegranate (Rosenblat *et al.*, 2013). Simvastatin, pravastatin and lovastatin are inhibitors of HMG-CoA reductase, the rate-limiting step in cholesterol synthesis (Izzo, 2005). Thus, any herbs involved in activation or inhibition of this enzymatic pathway can induce changes in drug absorption or catalysis. Finally, it should be noted that natural products are not always necessarily safe and dosages can be important. One should be sure to follow relevant directions on product labels and consult your pharmacist or physician or other healthcare professional before usage.

References

Abbasi, H., Rezaei, K. and Rashidi, L. (2008) Extraction of essential oils from the seeds of pomegranate using organic solvents and supercritical CO2. *Journal of the American Oil Chemists' Society* 85, 83–89.

Abdel Wahab, S., El Fiki, N., Mostafa, S. and Hassan, A. (1998) Characterization of certain steroid hormones in *Punica granatum* L. seeds. *Bulletin of the Faculty of Pharmacy* 36, 11–15.

Adams, L.S., Seeram, N.P., Aggarwal, B.B., Takada, Y., Sand, D. *et al.* (2006) Pomegranate juice, total pomegranate ellagitannins, and punicalagin suppress inflammatory cell signaling in colon cancer cells. *Journal of Agricultural and Food Chemistry* 54, 980–985.

Adams, L.S., Zhang, Y., Seeram, N.P., Heber, D. and Chen, S. (2010) Pomegranate ellagitannin–derived compounds exhibit antiproliferative and antiaromatase activity in breast cancer cells in vitro. *Cancer Prevention Research* 3, 108–113.

Adiga, S., Trivedi, P., Ravichandra, V., Deb, D. and Mehta, F. (2010) Effect of *Punica granatum* peel extract on learning and memory in rats. *Asian Pacific Journal of Tropical Medicine* 3, 687–690.

Afaq, F., Saleem, M., Krueger, C.G., Reed, J.D. and Mukhtar, H. (2005) Anthocyanin-and hydrolyzable tannin-rich pomegranate fruit extract modulates MAPK and NF-κB pathways and inhibits skin tumorigenesis in CD-1 mice. *International Journal of Cancer* 113, 423–433.

Afaq, F., Zaid, M., Khan, N., Syed, D., Yun, J.-M. *et al.* (2008) Inhibitory effect of oral feeding of pomegranate fruit extract on UVB-induced skin carcinogenesis in SKH-1 hairless mice. *American Associate for Cancer Research* 68(9), 1318–1326.

Ahad, S., Tanveer, S., Malik, T.A. and Nawchoo, I.A. (2018) Anticoccidial activity of fruit peel of *Punica granatum* L. *Microbial Pathogenesis* 116, 78–83.

Ahmed, A.H., Subaiea, G.M., Eid, A., Li, L., Seeram, N.P. *et al.* (2014) Pomegranate extract modulates processing of amyloid-beta precursor protein in an aged Alzheimer's disease animal model. *Current Alzheimer Research* 11, 834–843.

Akkiraju, P.C., Suryawanshi, D.D., Jawakekar, A.J., Tambe, H.S. and Mamillapalli, S. (2016) Phytochemical analysis and HPLC study of vitamin-C from *Punica granatum* L. Aarakta variety of India. *Journal of Medicinal Plants Research* 4, 09–12.

Al-Maiman, S.A. and Ahmad, D. (2002) Changes in physical and chemical properties during pomegranate (*Punica granatum* L.) fruit maturation. *Food Chemistry* 76, 437–441.

Al-Mathal, E.M. and Alsalem, A.M. (2012) Pomegranate (*Punica granatum*) peel is effective in a murine model of experimental *Cryptosporidium parvum*. *Experimental Parasitology* 131, 350–357.

Al-Megrin, W.A. (2017) In vivo study of pomegranate (*Punica granatum*) peel extract efficacy against *Giardia lamblia* in infected experimental mice. *Asian Pacific Journal of Tropical Biomedicine* 7, 59–63.

Al-Rawahi, A.S., Edwards, G., Al-Sibani, M., Al-Thani, G., Al-Harrasi, A.S. *et al.* (2014) Phenolic constituents of pomegranate peels (*Punica granatum* L.) cultivated in Oman. *European Journal of Medicinal Plants* 4, 315.

Al-Zoreky, N.S. (2009) Antimicrobial activity of pomegranate (*Punica granatum* L.) fruit peels. *International Journal of Food Microbiology* 134, 244–248.

Albrecht, M., Jiang, W., Kumi-Diaka, J., Lansky, E.P., Gommersall, L.M. *et al.* (2004) Pomegranate extracts potently suppress proliferation, xenograft growth, and invasion of human prostate cancer cells. *Journal of Medicinal Food* 7, 274–283.

Amakura, Y., Okada, M., Tsuji, S. and Tonogai, Y. (2000) Determination of phenolic acids in fruit juices by isocratic column liquid chromatography. *Journal of Chromatography A* 891, 183–188.

Amri, Z., Ghorbel, A., Turki, M., Akrout, F.M., Ayadi, F. *et al.* (2017) Effect of pomegranate extracts on brain antioxidant markers and cholinesterase activity in high fat-high fructose diet induced obesity in rat model. *BMC Complement Alternative Medicine* 17, 339.

Ando, E., Monden, K., Mitsuhata, R., Kariyama, R. and Kumon, H.J. (2004) Biofilm formation among methicillin-resistant *Staphylococcus aureus* isolates from patients with urinary tract infection. *Acta Medica Okayama* 58, 207–214.

Aparecida Procópio Gomes, L., Alves Figueiredo, L.M., Luiza do Rosário Palma, A., Corrêa Geraldo, B.M., Isler Castro, K.C, Geraldo, C., Maria, B. *et al.* (2016) *Punica granatum* L. (pomegranate) extract: *In vivo* study of antimicrobial activity against *Porphyromonas gingivalis* in *Galleria mellonella* model. *The Scientific World Journal* 2016, 1–5. DOI: 10.1155/2016/8626987.

Ardekani, M.R.S., Hajimahmoodi, M., Oveisi, M.R., Sadeghi, N., Jannat, B. *et al.* (2011) Comparative antioxidant activity and total flavonoid content of Persian pomegranate (*Punica granatum* L.) cultivars. *Iranian Journal of Pharmaceutical Research* 10, 519.

Arseculeratne, S.N., Gunatilaka, A.L. and Panabokke, R.G. (1985) Studies on medicinal plants of Sri Lanka. Part 14: toxicity of some traditional medicinal herbs. *Journal of Ethnopharmacology* 13, 323–335.

Artik, N. (1998) Determination of phenolic compounds in pomegranate juice by using HPLC. *Fruit Processing* 8, 492–499.

Arun, K., Jayamurthy, P., Anusha, C., Mahesh, S. and Nisha, P. (2017) Studies on activity guided fractionation of pomegranate peel extracts and its effect on antidiabetic and cardiovascular protection properties. *Journal of Food Processing and Preservation* 41, e13108.

Arunkumar, J. and Rajarajan, S.J. (2018) Study on antiviral activities, drug-likeness and molecular docking of bioactive compounds of *Punica granatum* L. to Herpes simplex virus-2 (HSV-2). *Microbial Pathogenesis* 118, 301–309.

Asmaa, M.J.S., Ali, A.J.H., Farid, J.M. and Azman, S. (2015) Growth inhibitory effects of crude pomegranate peel extract on chronic myeloid leukemia, K562 cells. *International Journal of Applied and Basic Medical Research* 5, 100.

Avicenna, A. (1983) *The Cannon of Medicine*. Soroush Publications, Tehran, Iran.

Aviram, M. and Rosenblat, M. (2012) Pomegranate protection against cardiovascular diseases. *Evidence-Based Complementary and Alternative Medicine* 2012(Article ID 382763), 1–20. DOI: 10.1155/2012/382763.

Aviram, M., Dornfeld, L., Rosenblat, M., Volkova, N., Kaplan, M. *et al.* (2000) Pomegranate juice consumption reduces oxidative stress, atherogenic modifications to LDL, and platelet aggregation: studies in humans and in atherosclerotic apolipoprotein E–deficient mice. *The American Journal of Clinical Nutrition* 71, 1062–1076.

Aviram, M., Dornfeld, L., Kaplan, M., Coleman, R., Gaitini, D. *et al.* (2002) Pomegranate juice flavonoids inhibit low-density lipoprotein oxidation and cardiovascular diseases: studies in atherosclerotic mice and in humans. *Drugs Under Experimental and Clinical Research* 28, 49–62.

Aviram, M. and Dornfeld, L. (2001) Pomegranate juice consumption inhibits serum angiotensin converting enzyme activity and reduces systolic blood pressure. *Atherosclerosis* 158, 195–198.

Badria, F.A. (2002) Melatonin, serotonin, and tryptamine in some Egyptian food and medicinal plants. *Journal of Medicinal Food* 5, 153–157.

Bagri, P., Ali, M., Aeri, V., Sultana, S. and Bhowmik, M. (2010) Evalution of anti-inflammatory and analgesic activity of *Punica granatum* linn. *International Journal of Drug Devlopment Research* 2, 698–702.

Ballabh, P., Braun, A. and Nedergaard, M.J. (2004) The blood–brain barrier: an overview: structure, regulation, and clinical implications. *Neurobiology of Disease* 16, 1–13.

Bassolino, L., Zhang, Y., Schoonbeek, H.J., Kiferle, C., Perata, P. *et al.* (2013) Accumulation of anthocyanins in tomato skin extends shelf life. *New Phytologist* 200, 650–655.

Bernardo, T.H.L., Sales Santos Veríssimo, R.C., Alvino, V., Silva Araujo, M.G., Evangelista Pires Dos Santos, R.F. *et al.* (2015) Antimicrobial analysis of an antiseptic made from ethanol crude extracts of *P. granatum* and *E. uniflora* in Wistar rats against *Staphylococcus aureus* and *Staphylococcus epidermidis*. *The Scientific World Journal* 7, 1–7.

Bhagwat, S., Haytowitz, D.B. and Holden, J.M. (2014) USDA database for the flavonoid content of selected foods, release 3.1. US Department of Agriculture, Washington, DC.

Bhatia, D., Thoppil, R.J., Mandal, A., Samtani, K.A., Darvesh, A.S. *et al.* (2013) Pomegranate bioactive constituents suppress cell proliferation and induce apoptosis in an experimental model of hepatocellular carcinoma: role of Wnt/β-catenin signaling pathway. *Evidence-based Complementary and Alternative Medicine* 15, 1–16.

Bhowmik, D., Gopinath, H., Kumar, B.P. and Kumar, K. (2013) Medicinal uses of *Punica granatum* and its health benefits. *Journal of pharmacognosy and phytochemistry* 1(5), 28–35.

Bialonska, D., Kasimsetty, S.G., Khan, S.I. and Ferreira, D. (2009) Urolithins, intestinal microbial metabolites of pomegranate ellagitannins, exhibit potent antioxidant activity in a cell-based assay. *Journal of Agricultural and Food Chemistry* 57, 10181–10186.

Binyamin, O., Larush, L., Frid, K., Keller, G., Friedman-Levi, Y. *et al.* (2015) Treatment of a multiple sclerosis animal model by a novel nanodrop formulation of a natural antioxidant. *International Journal of Nanomedicine* 10, 7165–7174.

Bishayee, A., Bhatia, D., Thoppil, R.J., Darvesh, A.S., Nevo, E. *et al.* (2011a) Pomegranate-mediated chemoprevention of experimental hepatocarcinogenesis involves Nrf_2-regulated antioxidant mechanisms. *Carcinogenesis* 32, 888–896.

Bishayee, A., Mbimba, T., Thoppil, R.J., Háznagy-Radnai, E., Sipos, P. *et al.* (2011b) Anthocyanin-rich black currant (*Ribes nigrum* L.) extract affords chemoprevention against diethylnitrosamine-induced hepatocellular carcinogenesis in rats. *The Journal of Nutritional Biochemistry* 22, 1035–1046.

Boukef, K., Souissi, H. and Balansard, G.J. (1982) Contribution to the study of plants used in traditional medicine in Tunisia. *Plant Medical Phytotherapy* 16, 260–279.

Braga, L., Leite, A.A., Xavier, K.G., Takahashi, J., Bemquerer, M. *et al.* (2005) Synergic interaction between pomegranate extract and antibiotics against *Staphylococcus aureus*. *Canadian Journal of Microbiology* 51, 541–547.

Braidy, N., Subash, S., Essa, M.M., Vaishnav, R., Al-Adawi, S. *et al.* (2014) Neuroprotective effects of a variety of pomegranate juice extracts (PJE) against the excitotoxin quinolinic acid in human primary neurons. *Journal of Preventation of Alzheimers Disease* 1, 84–90.

Braidy, N., Essa, M.M., Poljak, A., Selvaraju, S., Al-Adawi, S. *et al.* (2016) Consumption of pomegranates improves synaptic function in a transgenic mice model of Alzheimer's disease. *Oncotarget* 7, 64589–64604.

Bustamante, A., Hinojosa, A., Robert, P. and Escalona, V. (2017) Extraction and microencapsulation of bioactive compounds from pomegranate (*Punica granatum* var. Wonderful) residues. *International Journal of Food Science & Technology* 52, 1452–1462.

Butterfield, D.A., Castegna, A., Lauderback, C.M. and Drake, J.J. (2002) Evidence that amyloid beta-peptide-induced lipid peroxidation and its sequelae in Alzheimer's disease brain contribute to neuronal death. *Neurobiology of Aging* 23, 655–664.

Cemeroglu, B., Artik, N. and Erbas, S. (1992) Gewinnung von Granatapfelsaft und seine Zusammensetzung. *Flussiges Obst* 59, 335–340.

Chauhan, D. and Chauhan, J. (2001) Flavonoid diglycoside from *Punica granatum*. *Pharmaceutical Biology* 39, 155–157.

Cheng, S.H., Shih, C.C., Lee, I.H., Hou, Y.W., Chen, K.C. *et al.* (2012) A study on the sleep quality of incoming university students. *Psychiatry Research* 197, 270–274.

Chistiakov, D.A., Melnichenko, A.A., Orekhov, A.N. and Bobryshev, Y.V. (2017) Paraoxonase and atherosclerosis-related cardiovascular diseases. *Biochimie* 132, 19–27.

Choi, S.J., Lee, J.H., Heo, H.J., Cho, H.Y., Kim, H.K. *et al.* (2011) *Punica granatum* protects against oxidative stress in PC12 cells and oxidative stress-induced Alzheimer's symptoms in mice. *Journal of Medical Food* 14, 695–701.

Cáceres, A., Girón, L.M., Alvarado, S.R. and Torres, M.F. (1987) Screening of antimicrobial activity of plants popularly used in Guatemala for the treatment of dermatomucosal diseases. *Journal of Ethnopharmacology* 20(3), 223–237.

Dahlawi, H., Jordan-Mahy, N., Clench, M.R. and Le Maitre, C.L. (2012) Bioactive actions of pomegranate fruit extracts on leukemia cell lines in vitro hold promise for new therapeutic agents for leukemia. *Nutrition and Cancer* 64, 100–110.

Dana, N., Javanmard, S.H. and Rafiee, L. (2016) Role of peroxisome proliferator-activated receptor alpha and gamma in antiangiogenic effect of pomegranate peel extract. *Iranian Journal of Basic Medical Sciences* 19, 106.

Das, D. and Maulik, N. (2000) Protection against free radical injury in the heart and cardiac performance. In: Sen, C., Packer, L. and Hänninen, O. (eds) *Handbook of Oxidants and Antioxidants in Exercise*. Elsevier, London.

Dastjerdi, E.V., Abdolazimi, Z., Ghazanfarian, M., Amdjadi, P., Kamalinejad, M. *et al.* (2014) Effect of *Punica granatum* L. flower water extract on five common oral bacteria and bacterial biofilm formation on orthodontic wire. *Iranian Journal of Public Health* 43, 1688.

De Nigris, F., Williams-Ignarro, S., Lerman, L.O., Crimi, E., Botti, C. *et al.* (2005) Beneficial effects of pomegranate juice on oxidation-sensitive genes and endothelial nitric oxide synthase activity at sites of perturbed shear stress. *Proceedings of the National Academy of Sciences* 102, 4896–4901.

De Pascual-Teresa, S., Santos-Buelga, C. and Rivas-Gonzalo, J.C. (2000) Quantitative analysis of flavan-3-ols in Spanish foodstuffs and beverages. *Journal of Agricultural and Food Chemistry* 48, 5331–5337.

Deng, Y., Li, Y., Yang, F., Zeng, A., Yang, S. *et al.* (2017) The extract from *Punica granatum* (pomegranate) peel induces apoptosis and impairs metastasis in prostate cancer cells. *Biomedicine & Pharmacotherapy* 93, 976–984.

Devi, A., Singh, V. and Bhatt, A.J. (2011) Comparative antibacterial study of different extract of pomegranate and its wild variety. *International Journal of Pharmaceutical Science and Research* 2, 2647–2650.

Dikmen, M., Ozturk, N. and Ozturk, Y. (2011) The antioxidant potency of *Punica granatum* L. Fruit peel reduces cell proliferation and induces apoptosis on breast cancer. *Journal of Medicinal Food* 14, 1638–1646.

Dixit, S., Rana, S. and Mittal, A. (2017) Screening of phytochemicals and bioactive compounds in *punica granatum* peel to evaluate its hematological potential. *International Journal of Current Advanced Research* 6, 2524–2529.

Dkhil, M.A. (2013) Anti-coccidial, anthelmintic and antioxidant activities of pomegranate (*Punica granatum*) peel extract. *Parasitology Research* 112, 2639–2646.

Du, C., Wang, P. and Francis, F. (1975) Anthocyanins of pomegranate, *Punica granatum*. *Journal of Food Science* 40, 417–418.

Eikani, M.H., Golmohammad, F. and Homami, S.S. (2012) Extraction of pomegranate (*Punica granatum* L.) seed oil using superheated hexane. *Food and Bioproducts Processing* 90, 32–36.

El-Nemr, S., Ismail, I. and Ragab, M. (1992) The chemical composition of the juice and seeds of pomegranate fruits. *Fluessiges Obst* 59(11), 162–164.

El-Sakka, M.A. (2010) *Phytochemistry Alkaloids*, 3rd edition. Al Azhar University, Cairo, Egypt, pp. 7–22.

El-Toumy, S.A. and Rauwald, H.W. (2002) Two ellagitannins from *Punica granatum* heartwood. *Phytochemistry* 61, 971–974.

Elfalleh, W., Nasri, N., Marzougui, N., Thabti, I., M'rabet, A. *et al.* (2009) Physico-chemical properties and DPPH-ABTS scavenging activity of some local pomegranate (*Punica granatum*) ecotypes. *International Journal of Food Sciences and Nutrition* 60, 197–210.

Elfalleh, W., Tlili, N., Ying, M., Sheng-Hua, H., Ferchichi, A. *et al.* (2011) Organoleptic quality, minerals, proteins and amino acids from two Tunisian commercial pomegranate fruits. *International Journal of Food Engineering* 7(4), 1–13. DOI: 10.2202/1556-3758.2057.

Elfalleh, W., Hannachi, H., Guetat, A., Tlili, N., Guasmi, F. *et al.* (2012a) Storage protein and amino acid contents of Tunisian and Chinese pomegranate (*Punica granatum* L.) cultivars. *Genetic Resources and Crop Evolution* 56, 999–1014.

Elfalleh, W., Hannachi, H., Tlili, N., Yahia, Y., Nasri, N. *et al.* (2012b) Total phenolic contents and antioxidant activities of pomegranate peel, seed, leaf and flower. *Journal of Medicinal Pants Research* 6, 4724–4730.

Endo, E.H., Costa, G.M., Makimori, R.Y., Ueda-Nakamura, T., Nakamura, C.V. *et al.* (2018) Anti-biofilm activity of *Rosmarinus officinalis*, *Punica granatum* and *Tetradenia riparia* against methicillin-resistant *Staphylococcus aureus* (MRSA) and synergic interaction with penicillin. *Journal of Herbal Medicine* 14, 48–54.

Esakkimuthu, S., Darvin, S.S., Mutheeswaran, S., Paulraj, M.G., Pandikumar, P. *et al.* (2018) A study on food-medicine continuum among the non-institutionally trained siddha practitioners of Tiruvallur district, Tamil Nadu, India. *Journal of Ethnobiology and Ethnomedicine* 14, 45.

Esmaillzadeh, A., Tahbaz, F., Gaieni, I., Alavi-Majd, H. and Azadbakht, L. (2006) Cholesterol-lowering effect of concentrated pomegranate juice consumption in type II diabetic patients with hyperlipidemia. *International Journal of Vitamin Nutrition Research* 76, 147–151.

Essa, M., Subash, S., Braidy, N., Al-Adawi, S., Al-Asmi, A. *et al.* (2013) Neuroprotective effects of pomegranate juice extracts on quinolinic acid–induced excitotoxicity in human neurons. *Alzheimer's & Dementia* 9, P800–P801.

Fadavi, A., Barzegar, M. and Azizi, M.H. (2006) Determination of fatty acids and total lipid content in oilseed of 25 pomegranates varieties grown in Iran. *Journal of Food Composition and Analysis* 19, 676–680.

Farkas, D., Oleson, L.E., Zhao, Y., Harmatz, J.S., Zinny, M.A. *et al.* (2007) Pomegranate juice does not impair clearance of oral or intravenous midazolam, a probe for cytochrome P450-3A activity: comparison with grapefruit juice. *The Journal of Clinical Pharmacology* 47, 286–294.

Fawole, O.A., Makunga, N.P. and Opara, U.L. (2012) Antibacterial, antioxidant and tyrosinase-inhibition activities of pomegranate fruit peel methanolic extract. *BMC Complementary and Alternative Medicine* 12, 200.

Ferrara, G., Gianaspro, A., Mazzeo, A., Giove, S.L., Stella Matarrese, A.M. *et al.* (2014) Characterization of pomegranate (*Punica granatum* L.) genotypes collected in Puglia region, southeastern Italy. *Scientia Horticulturae* 178, 70–78.

Fischer, U.A., Carle, R. and Kammerer, D.R. (2011a) Identification and quantification of phenolic compounds from pomegranate (*Punica granatum* L.) peel, mesocarp, aril and differently produced juices by HPLC-DAD–ESI/MSn. *Food Chemistry* 127, 807–821.

Fischer, U.A., Dettmann, J.S., Carle, R. and Kammerer, D.R. (2011b) Impact of processing and storage on the phenolic profiles and contents of pomegranate (*Punica granatum* L) juices. *European Food Research and Technology* 233, 797.

Fischer, U.A., Jaksch, A.V., Carle, R. and Kammerer, D.R. (2011c) Determination of lignans in edible and nonedible parts of pomegranate (*Punica granatum* L.) and products derived therefrom, particularly focusing on the quantitation of isolariciresinol using HPLC-DAD-ESI/MSn. *Journal of Agricultural and Food Chemistry* 60, 283–292.

Frawley, D. and Lad, V. (1986) *The Yoga of Herbs*. Lotus Press, Twin Lakes, Wisconsin.

Fuhrman, B., Volkova, N. and Aviram, M. (2005) Pomegranate juice inhibits oxidized LDL uptake and cholesterol biosynthesis in macrophages. *The Journal of Nutritional Biochemistry* 16, 570–576.

Gabbasova, L.A. and Abdurazakova, S.K. (1969) Chemical composition of pomegranate juice. *Izv.Vyssh. Ucheb. Zaved. Pishch. Tekhnol* 4, 30–31.

George, J., Singh, M., Srivastava, A.K., Bhui, K. and Shukla, Y. (2011) Synergistic growth inhibition of mouse skin tumors by pomegranate fruit extract and diallyl sulfide: evidence for inhibition of activated MAPKs/NF-κB and reduced cell proliferation. *Food and Chemical Toxicology* 49, 1511–1520.

Gil, M.I., García-Viguera, C., Artés, F. and Tomás-Barberán, F.A. (1995) Changes in pomegranate juice pigmentation during ripening. *Journal of the Science of Food and Agriculture* 68, 77–81.

Gil, M.I., Tomás-Barberán, F.A., Hess-Pierce, B., Holcroft, D.M. and Kader, A.A. (2000) Antioxidant activity of pomegranate juice and its relationship with phenolic composition and processing. *Journal of Agricultural and Food Chemistry* 48(10), 4581–4589. DOI: 10.1021/jf000404a.

Ginsberg, Y., Khatib, N., Saadi, N., Ross, M.G., Weiner, Z. *et al.* (2018) Maternal pomegranate juice attenuates maternal inflammation-induced fetal brain injury by inhibition of apoptosis, neuronal nitric oxide synthase, and NF-kappaB in a rat model. *American Journal of Obstetrics and Gynecology* 219, 113.e1–113.e9.

Gongbao, D., Luo, Q., Yu, Y., Wencheng, D., Zeng, S. *et al.* (2018) Analysis on medication regularity and action mechanism of Tibetan medicine in treatment of spleen and stomach diseases based on data mining and integrated pharmacology. *Europe PMC* 43, 3368–3375.

Gundogdu, M. and Yilmaz, H. (2012) Organic acid, phenolic profile and antioxidant capacities of pomegranate (*Punica granatum* L.) cultivars and selected genotypes. *Scientia Horticulturae* 143, 38–42.

Gözlekçi, Ş., Saraçoğlu, O., Onursal, E. and Özgen, M. (2011) Total phenolic distribution of juice, peel, and seed extracts of four pomegranate cultivars. *Pharmacognosy Magazine* 7, 161.

Hagir, G., Elaleem, A., Al Sheikh, A., Albasheer, Khadiga, G. and Elaleem, A. (2016) Phytochemical screening and antibacterial activity of *Punica granatum* fruit rind extracts. *Global Journal of Medicinal Plant Research* 4, 9–15.

Haidari, M., Ali, M., Casscells, S.W. and Madjid, M.J. (2009) Pomegranate (*Punica granatum*) purified polyphenol extract inhibits influenza virus and has a synergistic effect with oseltamivir. *Phytomedicine* 16, 1127–1136.

Hajimahmoodi, M., Shams-Ardakani, M., Saniee, P., Siavoshi, F., Mehrabani, M. *et al.* (2011) In vitro antibacterial activity of some Iranian medicinal plant extracts against *Helicobacter pylori*. *Natural Product Research* 25, 1059–1066.

Hajipour, S., Sarkaki, A., Mohammad, S., Mansouri, T., Pilevarian, A. *et al.* (2014) Motor and cognitive deficits due to permanent cerebral hypoperfusion/ischemia improve by pomegranate seed extract in rats. *Pakistan Journal of Biological Science* 17, 991–998.

Hanley, M., Masse, G., Harmatz, J., Court, M. and Greenblatt, D.J. (2012) Pomegranate juice and pomegranate extract do not impair oral clearance of flurbiprofen in human volunteers: divergence from in vitro results. *Clinical Pharmacology & Therapeutics* 92, 651–657.

Haque, N., Sofi, G., Ali, W., Rashid, M. and Itrat, M. (2015) A comprehensive review of phytochemical and pharmacological profile of anar (*Punica granatum* Linn): a heaven's fruit. *Journal of Ayurvedic and Herbal Medicine* 1, 22–26.

Hassan, N.A., El-Halwagi, A.A. and Sayed, H. (2012) Phytochemicals, antioxidant and chemical properties of 32 pomegranate accessions growing in Egypt. *World Applied Sciences Journal* 16, 1065–1073.

Heftmann, E., Ko, S.T. and Bennett, R.D. (1966) Identification of estrone in pomegranate seeds. *Phytochemistry* 5, 1337–1339.

Hidaka, M., Okumura, M., Fujita, K.I., Ogikubo, T., Yamasaki, K. *et al.* (2005) Effects of pomegranate juice on human cytochrome P450 3A (CYP3A) and carbamazepine pharmacokinetics in rats. *Drug Metabolism and Disposition* 33, 644–648.

Holland, D., Hatib, K. and Bar-Ya'akov, A, I. (2009) Pomegranate: botany, horticulture, breeding. *Horticultural Review* 35, 127–191.

Hong, M.Y., Seeram, N.P. and Heber, D. (2008) Pomegranate polyphenols down-regulate expression of androgen-synthesizing genes in human prostate cancer cells overexpressing the androgen receptor. *The Journal of Nutritional Biochemistry* 19, 848–855.

Hora, J.J., Maydew, E.R., Lansky, E.P. and Dwivedi, C. (2003) Chemopreventive effects of pomegranate seed oil on skin tumor development in CD1 mice. *Journal of Medicinal Food* 6, 157–161.

Hossain, H., Ahmed, T., Howlader, M.S.I., Dey, S.K., Hira, A. *et al.* (2017) In-vitro antioxidant potential from the leaves of *Punica granatum* Linn. grown in Bangladesh. *International Journal of Pharmaceutical and Phytopharmacological Research* 2, 160–166.

Hou, D.X., Fujii, M., Terahara, N. and Yoshimoto, M. (2004) Molecular mechanisms behind the chemopreventive effects of anthocyanidins. *BioMed Research International* 2004, 321–325.

Huang, T.H., Yang, Q., Harada, M., Li, G.Q., Yamahara, J. *et al.* (2005a) Pomegranate flower extract diminishes cardiac fibrosis in Zucker diabetic fatty rats: modulation of cardiac endothelin-1 and nuclear factor-kappaB pathways. *Journal of Cardiovascular Pharmacology* 46, 856–862.

Huang, T.H.W., Peng, G., Kota, B.P., Li, G.Q., Yamahara, J. *et al.* (2005b) Pomegranate flower improves cardiac lipid metabolism in a diabetic rat model: role of lowering circulating lipids. *British Journal of Pharmacology* 145, 767–774.

Hussein, S.A., Barakat, H.H., Merfort, I. and Nawwar, M.A. (1997) Tannins from the leaves of *Punica granatum*. *Phytochemistry* 45, 819–823.

Ignarro, L.J., Byrns, R.E., Sumi, D., De Nigris, F. and Napoli, C. (2006) Pomegranate juice protects nitric oxide against oxidative destruction and enhances the biological actions of nitric oxide. *Nitric Oxide* 15, 93–102.

Ismail, T., Sestili, P. and Akhtar, S.J. (2012) Pomegranate peel and fruit extracts: a review of potential anti-inflammatory and anti-infective effects. *Journal of Ethnopharmacology* 143, 397–405.

Ito, H., Li, P., Koreishi, M., Nagatomo, A., Nishida, N. *et al.* (2014) Ellagitannin oligomers and a neolignan from pomegranate arils and their inhibitory effects on the formation of advanced glycation end products. *Food Chemistry* 152, 323–330.

Izzo, A.A. (2005) Herb–drug interactions: an overview of the clinical evidence. *Pharmacology* 19, 1–16.

Jaeger, P.A., Pickford, F., Sun, C.H., Lucin, K.M., Masliah, E. *et al.* (2010) Regulation of amyloid precursor protein processing by the Beclin 1 complex. *PLoS ONE* 5(6), 11102.

Jaiswal, V., Dermarderosian, A. and Porter, J.R. (2010) Anthocyanins and polyphenol oxidase from dried arils of pomegranate (*Punica granatum* L.). *Food Chemistry* 118, 11–16.

Jarvis, S., Li, C. and Bogle, R.. and (2010) Possible interaction between pomegranate juice and warfarin. *Emergency Medicine Journal* 27, 74–75.

Jayaprakash, A. and Sangeetha, R. (2015) Phytochemical screening of *Punica granatum* Linn. peel extracts. *Journal of Academia and Industrial Research* 4, 160.

Jeune, M.L., Kumi-Diaka, J. and Brown, J. (2005) Anticancer activities of pomegranate extracts and genistein in human breast cancer cells. *Journal of Medicinal Food* 8, 469–475.

Johanningsmeier, S.D. and Harris, G.K. (2011) Pomegranate as a functional food and nutraceutical source. *Annual Food Science Technology* 2, 181–201.

Joseph, M.M., Aravind, S., Varghese, S., Mini, S. and Sreelekha, T. (2012) Evaluation of antioxidant, antitumor and immunomodulatory properties of polysaccharide isolated from fruit rind of *Punica granatum*. *Molecular Medicine Reports* 5, 489–496.

Jurenka, J. (2008) Anti-inflammatory properties of curcumin, a major constituent of *Curcuma longa*: a review of preclinical and clinical research. *Alternative Medicine Review* 14(2), 141–153.

Kaminsky, L.S. and Zhang, Z.Y. (1997) Human P450 metabolism of warfarin. *Therapeutics* 73, 67–74.

Kaplan, M., Hayek, T., Raz, A., Coleman, R., Dornfeld, L. *et al.* (2001) Pomegranate juice supplementation to atherosclerotic mice reduces macrophage lipid peroxidation, cellular cholesterol accumulation and development of atherosclerosis. *The Journal of Nutrition* 131, 2082–2089.

Kawaii, S. and Lansky, E.P. (2004) Differentiation-promoting activity of pomegranate (*Punica granatum*) fruit extracts in HL-60 human promyelocytic leukemia cells. *Journal of Medicinal Food* 7, 13–18.

Khalil, A., Khan, M., Shabbir, M. and Rahman, K. (2017) Comparison of antioxidative potential and punicalagin content of pomegranate peels. *Journal of Animal & Plant Sciences* 27(2), 522–527.

Khan, G.N., Gorin, M.A., Rosenthal, D., Pan, Q., Bao, L.W. *et al.* (2009) Pomegranate fruit extract impairs invasion and motility in human breast cancer. *Integrative Cancer Therapies* 8, 242–253.

Khuu, T., Hickey, A. and Deng, M.C. (2013) Pomegranate-containing products and tacrolimus: a potential interaction. *The Journal of Heart and Lung Transplantation* 32(2), 272–274. DOI: 10.1016/j.healun.2012.10.015.

Khan, M.P.Z. and Ahmad, M. (2015) Traditional preference of wild edible fruits (WEFs) for digestive disorders (DDs) among the indigenous communities of Swat Valley-Pakistan. *Journal of Ethnopharmacology* 174, 339–354. DOI: 10.1016/j.jep.2015.08.024.

Khan, N., Hadi, N., Afaq, F., Syed, D.N., Kweon, M.-H. *et al.* (2007) Pomegranate fruit extract inhibits prosurvival pathways in human A549 lung carcinoma cells and tumor growth in athymic nude mice. *Carcinogenesis* 28(1), 163–173. DOI: 10.1093/carcin/bgl145.

Kim, N.D., Mehta, R., Yu, W., Neeman, I., Livney, T. *et al.* (2002) Chemopreventive and adjuvant therapeutic potential of pomegranate (*Punica granatum*) for human breast cancer. *Breast Cancer Research and Treatment* 71, 203–217.

Kim, Y.E., Hwang, C.J., Lee, H.P., Kim, C.S., Son, D.J. *et al.* (2017) Inhibitory effect of punicalagin on lipopolysaccharide-induced neuroinflammation, oxidative stress and memory impairment via inhibition of nuclear factor-kappaB. *Neuropharmacology* 117, 21–32.

Kohno, H., Suzuki, R., Yasui, Y., Hosokawa, M., Miyashita, K. *et al.* (2004) Pomegranate seed oil rich in conjugated linolenic acid suppresses chemically induced colon carcinogenesis in rats. *Cancer Science* 95, 481–486.

Kokare, C.R., Chakraborty, S., Khopade, A. and Mahadik, K. (2009) Biofilm: importance and applications. *Indian Journal of Biotechnology* 8, 159–168.

Komperda, K.E. (2009) Potential interaction between pomegranate juice and warfarin. *Pharmacotherapy: The Journal of Human Pharmacology and Drug Therapy* 29, 1002–1006.

Langley, P.J. (2000) Why a pomegranate? *British Medical Association Journal* 321, 1153–1154.

Lansky, E.P., Harrison, G., Froom, P. and Jiang, W.G. (2005) Pomegranate (*Punica granatum*) pure chemicals show possible synergistic inhibition of human PC-3 prostate cancer cell invasion across Matrigel™. *Investigational New Drugs* 23, 121–122.

Lansky, E.P. and Newman, R.A. (2007) *Punica granatum* (pomegranate) and its potential for prevention and treatment of inflammation and cancer. *Journal of Ethnopharmacology* 109, 177–206.

Lee, W.J., Ou, H.C., Hsu, W.C., Chou, M.M., Tseng, J.J. *et al.* (2010) Ellagic acid inhibits oxidized LDL-mediated LOX-1 expression, ROS generation, and inflammation in human endothelial cells. *Journal of Vascular Surgery* 52, 1290–1300.

Lee, K.H., Morris-Natschke, S.L., Yang, X., Huang, R., Zhou, T. *et al.* (2012) Recent progress of research on medicinal mushrooms, foods, and other herbal products used in traditional Chinese medicine. *Journal of Traditional and Complementary Medicine* 2, 1–12.

Lee, S.T., Lu, M.H., Chien, L.H., Wu, T.F., Huang, L.C. *et al.* (2013) Suppression of urinary bladder urothelial carcinoma cell by the ethanol extract of pomegranate fruit through cell cycle arrest and apoptosis. *BMC Complementary and Alternative Medicine* 13, 364.

Lei, F., Zhang, X.N., Wang, W., Xing, D.M., Xie, W.D. *et al.* (2007) Evidence of anti-obesity effects of the pomegranate leaf extract in high-fat diet induced obese mice. *International Journal of Obesity* 31(6), 1023–1029. DOI: 10.1038/sj.ijo.0803502.

Les, F., Prieto, J.M., Arbonés-Mainar, J.M., Valero, M.S. and López, V. (2015) Bioactive properties of commercialised pomegranate (*Punica granatum*) juice: antioxidant, antiproliferative and enzyme inhibiting activities. *Food & Function* 6, 2049–2057.

Li, J., Ou-Lee, T.M., Raba, R., Amundson, R.G. and Last, R.L. (1993) Arabidopsis flavonoid mutants are hypersensitive to UV-B irradiation. *The Plant Cell* 5, 171–179.

Li, H., Wang, Z. and Liu, Y. (2003) Review in the studies on tannins activity of cancer prevention and anti-cancer. *Journal of Chinese Medicinal Materials* 26, 444–448.

Li, Y., Guo, C., Yang, J., Wei, J., Xu, J. *et al.* (2006) Evaluation of antioxidant properties of pomegranate peel extract in comparison with pomegranate pulp extract. *Food Chemistry* 96(2), 254–260. DOI: 10.1016/j.foodchem.2005.02.033.

Li, Y., Yang, F., Zheng, W., Hu, M., Wang, J. *et al.* (2016a) *Punica granatum* (pomegranate) leaves extract induces apoptosis through mitochondrial intrinsic pathway and inhibits migration and invasion in non-small cell lung cancer in vitro. *Biomedicine & Pharmacotherapy* 80, 227–235.

Li, Y., Ye, T., Yang, F., Hu, M., Liang, L. *et al.* (2016b) *Punica granatum* (pomegranate) peel extract exerts potent antitumor and anti-metastasis activity in thyroid cancer. *RSC Advances* 6, 84523–84535.

Liu, G., Xu, X., Hao, Q. and Gao, Y. (2009) Supercritical CO2 extraction optimization of pomegranate (Punica granatum L.) seed oil using response surface methodology. *LWT-Food Science and Technology* 42, 1491–1495.

Lo Scalzo, R., Iannoccari, T., Summa, C., Morelli, R. and Rapisarda, P. (2004) Effect of thermal treatments on antioxidant and antiradical activity of blood orange juice. *Food Chemistry* 85(1), 41–47. DOI: 10.1016/j.foodchem.2003.05.005.

Loren, D.J., Seeram, N.P., Schulman, R.N. and Holtzman, D.M. (2005) Maternal dietary supplementation with pomegranate juice is neuroprotective in an animal model of neonatal hypoxic-ischemic brain injury. *Pediatrics Research* 57, 858–864.

Lozoya, J., Aguilar, A. and Camacho, J.R. (1987) Encuesta sobre el uso actual de plantas en la medicina tradicional mexicana. *Archivos Argentinos de Pediatra* 25, 283–291.

Lucci, P., Pacetti, D., Loizzo, M.R. and Frega, N.G. (2015) *Punica granatum* cv. Dente di Cavallo seed ethanolic extract: antioxidant and antiproliferative activities. *Food Chemistry* 167, 475–483.

Lucin, K.M., O'Brien, C.E., Bieri, G., Czirr, E., Mosher, K.I. *et al.* (2013) Microglial beclin 1 regulates retromer trafficking and phagocytosis and is impaired in Alzheimer's disease. *Neuron* 79, 873–886.

Ma, G.Z., Wang, C.M., Li, L., Ding, N. and Gao, X.L. (2015) Effect of pomegranate peel polyphenols on human prostate cancer PC-3 cells in vivo. *Food Science and Biotechnology* 24, 1887–1892.

Mahdihassan, S. (1984) Outline of the beginnings of alchemy and its antecedents. *The American Journal of Chinese Medicine* 12(1-4), 32–42. DOI: 10.1142/S0192415X84000039.

Mahon, K.L. and Escott-Stump, S. (1996) *Krause's Food, Nutrition, and Diet Therapy*, 9th edn. Saunders, Philadelphia, Pennsylvania.

Mansour, E., Ben Khaled, A., Lachiheb, B., Abid, M., Bachar, K. *et al.* (2013) Phenolic compounds, antioxidant, and antibacterial activities of peel extract from Tunisian pomegranate. *Journal of Agricultural Science and Technology* 15, 1393–1403.

Masci, A., Coccia, A., Lendaro, E., Mosca, L., Paolicelli, P. *et al.* (2016) Evaluation of different extraction methods from pomegranate whole fruit or peels and the antioxidant and antiproliferative activity of the polyphenolic fraction. *Food Chemistry* 202, 59–69.

Matthäus, B., Guillaume, D., Gharby, S., Haddad, A., Harhar, H. *et al.* (2010) Effect of processing on the quality of edible argan oil. *Food Chemistry* 120, 426–432.

McMahon, J.B., Currens, M.J., Gulakowski, R.J., Buckheit, R., Lackman-Smith, C. *et al.* (1995) Michellamine B, a novel plant alkaloid, inhibits human immunodeficiency virus-induced cell killing by at least two distinct mechanisms. *Antimicrobial Agents and Chemotherapy* 39, 484–488.

Melgarejo, P., Salazar, D.M. and Artes, F. (2000) Organic acids and sugars composition of harvested pomegranate fruits. *European Food Research and Technology* 211, 185–190.

Melgarejo, P. and Artes, F. (2000) Total lipid content and fatty acid composition of oilseed from lesser known sweet pomegranate clones. *Journal of the Science of Food and Agriculture* 80, 1452–1454.

Mena, P., Calani, L., Dall'asta, C., Galaverna, G., García-Viguera, C. *et al.* (2012) Rapid and comprehensive evaluation of (poly) phenolic compounds in pomegranate (*Punica granatum* L.) juice by UHPLC-MSn. *Molecules* 17, 14821–14840.

Merzouki, A., Ed-Derfoufi, F. and Mesa, J.M. (2000) Contribution to the knowledge of Rifian traditional medicine. II: folk medicine in Ksar Lakbir district (NW Morocco). *Fitoterapia* 71, 278–307.

Miranda, S., Opazo, C., Larrondo, L.F., Muñoz, F.J., Ruiz, F. *et al.* (2000) The role of oxidative stress in the toxicity induced by amyloid β-peptide in Alzheimer's disease. *Progress in Neurobiology* 62, 633–648.

Mirdehghan, S.H. and Rahemi, M. (2007) Seasonal changes of mineral nutrients and phenolics in pomegranate (*Punica granatum* L.) fruit. *Scientia Horticulturae* 111, 120–127.

Misaka, S., Nakamura, R., Uchida, S., Takeuchi, K., Takahashi, N. *et al.* (2011) Effect of 2 weeks' consumption of pomegranate juice on the pharmacokinetics of a single dose of midazolam: an open-label, randomized, single-center, 2-period crossover study in healthy Japanese volunteers. *Clinical Therapeutics* 33, 246–252.

Miyamoto, K.I., Nomura, M., Sasakura, M., Matsui, E., Koshiura, R. *et al.* (1993) Antitumor activity of oenothein B, a unique macrocyclic ellagitannin. *Japanese Journal of Cancer Research* 84, 99–103.

Mizrahi, M., Friedman-Levi, Y., Larush, L., Frid, K., Binyamin, O. *et al.* (2014) Pomegranate seed oil nanoemulsions for the prevention and treatment of neurodegenerative diseases: the case of genetic CJD. *Nanomedicine: Nanotechnology. Biology and Medicine* 10, 1353–1363.

Mo, J., Panichayupakaranant, P., Kaewnopparat, N., Nitiruangjaras, A. and Reanmongkol, W. (2013) Topical anti-inflammatory and analgesic activities of standardized pomegranate rind extract in comparison with its marker compound ellagic acid in vivo. *Journal of Ethnopharmacology* 148, 901–908.

Modaeinama, S., Abasi, M., Abbasi, M.M. and Jahanban-Esfahlan, R. (2015) Anti tumoral properties of *Punica granatum* (pomegranate) peel extract on different human cancer cells. *Asian Pacific Journal of Cancer Preventation* 16, 5697–5701.

Mohajer, S., Taha, R. and Azmi, S.Z. (2016) Phytochemical screening and potential of natural dye colourant from pomegranate (*Punica granatum* L). *Pigment & Resin Technology* 45, 38–44.

Mohan, M., Patankar, P., Ghadi, P. and Kasture, S. (2010) Cardioprotective potential of *Punica granatum* extract in isoproterenol-induced myocardial infarction in Wistar rats. *Journal of Pharmacology & Pharmacotherapeutics* 1, 32.

Mollazadeh, H., Sadeghnia, H.R., Hoseini, A., Farzadnia, M. and Boroushaki, M.T. (2016) Effects of pomegranate seed oil on oxidative stress markers, serum biochemical parameters and pathological findings in kidney and heart of streptozotocin-induced diabetic rats. *Renal Failure* 38, 1256–1266.

Moneam, N., El Sharaky, A. and Badreldin, M. (1988) Oestrogen content of pomegranate seeds. *Journal of Chromatography A* 438, 438–442.

Moorthy, K., Punitha, T., Vinodhini, R., Sureshkumar, B.T., Vijayalakshmi, P. *et al.* (2013) Antimicrobial activity and qualitative phytochemical analysis of *Punica granatum* Linn. *(PERICARP) Journal of Medicinal Plants Research* 7, 474–479.

Morzelle, M.C., Salgado, J.M., Telles, M., Mourelle, D., Bachiega, P. *et al.* (2016) Neuroprotective effects of pomegranate peel extract after chronic infusion with amyloid-β peptide in mice. *PLoS ONE* 11, e0166123.

Mphahlele, R.R., Fawole, O.A., Makunga, N.P. and Opara, U.L. (2017) Functional properties of pomegranate fruit parts: influence of packaging systems and storage time. *Journal of Food Measurement and Characterization* 11, 2233–2246.

Mubaraki, M.A., Hafiz, T.A., Dkhil, M.A. and Al-Quraishy, S.J. (2016) Beneficial effect of *Punica granatum* peel extract on murine malaria-induced spleen injury. *BMC Complementary and Alternative Medicine* 16, 221.

Mushtaq, M., Sultana, B., Anwar, F., Adnan, A. and Rizvi, S.S. (2015) Enzyme-assisted supercritical fluid extraction of phenolic antioxidants from pomegranate peel. *The Journal of Supercritical Fluids* 104, 122–131.

Nagaraju, N. and Rao, K.J. (1990) A survey of plant crude drugs of Rayalaseema, Andhra Pradesh, India. *Journal of Ethnopharmacology* 29, 137–158.

Nagata, M., Hidaka, M., Sekiya, H., Kawano, Y., Yamasaki, K. *et al.* (2006) Effects of pomegranate juice on human cytochrome P450 2C9 (CYP2C9) and tolbutamide pharmacokinetics in rats. *Drug Metabolism and Disposition* 35(2), 302–305.

Nagy, P., Shaw, P.E. and Wordowski, W.F. (1990) *Fruit of Tropical and Subtropical Origin*. Florida Science Source, Florida, USA, pp. 328–347.

Nair, V., Dai, Z., Khan, M. and Ciolino, H.P. (2011) Pomegranate extract induces cell cycle arrest and alters cellular phenotype of human pancreatic cancer cells. *Anticancer Research* 31, 2699–2704.

Namdar, H., Emaratkar, E. and Hadavand, M.B. (2015) Persian traditional medicine and ocular health. *Medical Hypothesis, Discovery and Innovation in Ophthalmology* 4, 162.

Napoli, C. and Ignarro, L.J. (2001) Nitric oxide and atherosclerosis. *Nitric Oxide* 5(2), 88–97. DOI: 10.1006/niox.2001.0337.

Naqvi, S., Khan, M. and Vohora, S.J. (1991) Anti-bacterial, anti-fungal and anthelmintic investigations on Indian medicinal plants. *ScienceOpen* 62, 221–228.

Nawwar, M.A.M., Hussein, S.A.M. and Merfort, I. (1994) Leaf phenolics of *Punica granatum*. *Phytochemistry* 37(4), 1175–1177. DOI: 10.1016/S0031-9422(00)89552-7.

Nemr, S., Ismail, A. and Ragab, M. (2001) Chemical of juice and seeds of pomegranate fruit. Effects of shrink film wrapping and storage temperature on the shelf life and quality of pomegranate fruits. *Postharvest Biology and Technology* 22, 61–69.

Neuhöfer, H., Witte, L., Gorunovic, M. and Czygan, F.C. (1993) Alkaloids in the bark of *Punica granatum* L. (pomegranate) from Yugoslavia. *Pharmazie* 48, 389–391.

Neurath, A.R., Strick, N., Li, Y.-Y. and Debnath, A.K. (2004) *Punica granatum* (pomegranate) juice provides an HIV-1 entry inhibitor and candidate topical microbicide. *BMC Infectious Diseases* 4(1), 41. DOI: 10.1186/1471-2334-4-41.

Newman, R.A. and Lansky, E.P. (2007) *Pomegranate: the Most Medicinal Fruit*. Basic Health Publications, Laguna Beach, CA.

Noda, Y., Kaneyuki, T., Mori, A. and Packer, L. (2002) Antioxidant activities of pomegranate fruit extract and its anthocyanidins: delphinidin, cyanidin, and pelargonidin. *Journal of Agricultural and Food Chemistry* 50(1), 166–171. DOI: 10.1021/jf0108765.

Norwood, D.A., Parke, C.K. and Rappa, L.R. (2015) A comprehensive review of potential warfarin-fruit interactions. *Journal of Pharmacy Practice* 28(6), 561–571. DOI: 10.1177/0897190014544823.

Nuamsetti, T., Dechayuenyong, P. and Tantipaibulvut, S. (2012) Antibacterial activity of pomegranate fruit peels and arils. *ScienceAsia* 38(3), 319–322. DOI: 10.2306/scienceasia1513-1874.2012.38.319.

Núñez-Sánchez, M.A., González-Sarrías, A., Romo-Vaquero, M., García-Villalba, R., Selma, M.V. *et al.* (2015) Dietary phenolics against colorectal cancer – from promising preclinical results to poor translation into clinical trials: pitfalls and future needs. *Molecular Nutrition & Food Research* 59(7), 1274–1291. DOI: 10.1002/mnfr.201400866.

Ono, N.N., Bandaranayake, P.C. and Tian, L. (2012) Establishment of pomegranate (*Punica granatum*) hairy root cultures for genetic interrogation of the hydrolyzable tannin biosynthetic pathway. *Planta Medica* 236, 931–941.

Orak, H.H., Yagar, H. and Isbilir, S.S. (2012) Comparison of antioxidant activities of juice, peel, and seed of pomegranate (*Punica granatum* L.) and inter-relationships with total phenolic, tannin, anthocyanin, and flavonoid contents. *Food Science and Biotechnology* 21(2), 373–387. DOI: 10.1007/s10068-012-0049-6.

Ouelbani, R., Bensari, S., Mouas, T.N. and Khelifi, D. (2016) Ethnobotanical investigations on plants used in folk medicine in the regions of Constantine and Mila (north-east of Algeria). *Journal of Ethnopharmacology* 194, 196–218. DOI: 10.1016/j.jep.2016.08.016.

Ozgen, M., Durgaç, C., Serçe, S. and Kaya, C. (2008) Chemical and antioxidant properties of pomegranate cultivars grown in the Mediterranean region of Turkey. *Food Chemistry* 111(3), 703–706. DOI: 10.1016/j.foodchem.2008.04.043.

Özgül-Yücel, S. (2005) Determination of conjugated linolenic acid content of selected oil seeds grown in Turkey. *Journal of the American Oil Chemists' Society* 82(12), 893–897. DOI: 10.1007/s11746-005-1161-7.

Pagliarulo, C., De Vito, V., Picariello, G., Colicchio, R., Pastore, G. *et al.* (2016) Inhibitory effect of pomegranate (*Punica granatum* L.) polyphenol extracts on the bacterial growth and survival of clinical isolates of pathogenic *Staphylococcus aureus* and *Escherichia coli*. *Food Chemistry* 190, 824–831. DOI: 10.1016/j.foodchem.2015.06.028.

Pande, G. and Akoh, C.C. (2009) Antioxidant capacity and lipid characterization of six Georgia-grown pomegranate cultivars. *Journal of Agricultural and Food Chemistry* 57(20), 9427–9436. DOI: 10.1021/jf901880p.

Panichayupakaranant, P., Tewtrakul, S. and Yuenyongsawad, S. (2010) Antibacterial, anti-inflammatory and anti-allergic activities of standardised pomegranate rind extract. *Food Chemistry* 123(2), 400–403. DOI: 10.1016/j.foodchem.2010.04.054.

Pantuck, A.J., Pettaway, C.A., Dreicer, R., Corman, J., Katz, A. *et al.* (2015) A randomized, double-blind, placebo-controlled study of the effects of pomegranate extract on rising PSA levels in men following primary therapy for prostate cancer. *Prostate Cancer and Prostatic Diseases* 18(3), 242–248. DOI: 10.1038/pcan.2015.32.

Pickford, F., Masliah, E., Britschgi, M., Lucin, K., Narasimhan, R. *et al.* (2008) The autophagy-related protein Beclin 1 shows reduced expression in early Alzheimer disease and regulates amyloid β accumulation in mice. *Journal of Clinical Investigation* 120, 2190–2199. DOI: 10.1172/JCI33585.

Polagruto, J.A., Schramm, D.D., Wang-Polagruto, J.F., Lee, L. and Keen, C.L. (2003) Effects of flavonoid-rich beverages on prostacyclin synthesis in humans and human aortic endothelial cells: association with ex *vivo* platelet function. *Journal of Medicinal Food* 6(4), 301–308. DOI: 10.1089/109662003772519840.

Poyrazoğlu, E., Gökmen, V. and Artιk, N. (2002) Organic acids and phenolic compounds in pomegranates (*Punica granatum* L.) grown in Turkey. *Journal of Food Composition and Analysis* 15(5), 567–575. DOI: 10.1016/S0889-1575(02)91071-9.

Praticò, D. and Delanty, N. (2000) Oxidative injury in diseases of the central nervous system: focus on Alzheimer's disease. *The American Journal of Medicine* 109(7), 577–585. DOI: 10.1016/S0002-9343(00)00547-7.

Qu, W., Breksa III, A.P., Pan, Z. and Ma, H. (2012) Quantitative determination of major polyphenol constituents in pomegranate products. *Food Chemistry* 132(3), 1585–1591. DOI: 10.1016/j.foodchem.2011.11.106.

Rafiq, Z., Narasimhan, S., Vennila, R. and Vaidyanathan, R. (2016) Punigratane, a novel pyrrolidine alkaloid from *Punica granatum* rind with putative efflux inhibition activity. *Natural Product Research* 30(23), 2682–2687. DOI: 10.1080/14786419.2016.1146883.

Recchia, A., Debetto, P., Negro, A., Guidolin, D., Skaper, S.D. *et al.* (2004) α-Synuclein and Parkinson's disease. *Cold Spring Harbor Laboratory Press* 18, 617–626.

Reddy, B.U., Mullick, R., Kumar, A., Sudha, G., Srinivasan, N. *et al.* (2015) Small molecule inhibitors of HCV replication from pomegranate. *Scientific Reports* 4(1), 5411. DOI: 10.1038/srep05411.

Rettig, M.B., Heber, D., An, J., Seeram, N.P., Rao, J.Y. *et al.* (2008) Pomegranate extract inhibits androgen-independent prostate cancer growth through a nuclear factor-κB-dependent mechanism. *Molecular Cancer Therapeutics* 7(9), 2662–2671. DOI: 10.1158/1535-7163.MCT-08-0136.

Robbins, M.P., Bavage, A.D., Strudwicke, C. and Morris, P. (1998) Genetic manipulation of condensed tannins in higher plants: II. Analysis of birdsfoot trefoil plants harboring antisense dihydroflavonol reductase constructs. *Plant Physiology* 116, 1133–1144.

Rosas-Burgos, E.C., Burgos-Hernández, A., Noguera-Artiaga, L., Kačániová, M., Hernández-García, F. *et al.* (2017) Antimicrobial activity of pomegranate peel extracts as affected by cultivar. *Journal of the Science of Food and Agriculture* 97(3), 802–810. DOI: 10.1002/jsfa.7799.

Rosenblat, M., Volkova, N. and Presser, D. (2002) Pomegranate juice flavonoids inhibit low-density lipoprotein oxidation and cardiovascular diseases: studies in atherosclerotic mice and in humans. *Drugs Under Experimental and Clinical Research* 28, 49–62.

Rosenblat, M., Volkova, N. and Aviram, M. (2013) Pomegranate phytosterol (β-sitosterol) and polyphenolic antioxidant (punicalagin) addition to statin, significantly protected against macrophage foam cells formation. *Atherosclerosis* 226(1), 110–117. DOI: 10.1016/j.atherosclerosis.2012.10.054.

Rosenblat, M., Volkova, N., Abassi, Z., Britton, S.L., Koch, L.G. *et al.* (2015) High intrinsic aerobic capacity and pomegranate juice are protective against macrophage atherogenecity: studies in high- vs. low-capacity runner (HCR vs. LCR) rats. *The Journal of Nutritional Biochemistry* 26(10), 1015–1021. DOI: 10.1016/j.jnutbio.2015.04.001.

Rouhi-Boroujeni, H., Rouhi-Boroujeni, H., Heidarian, E., Mohammadizadeh, F. and Rafieian-Kopaei, M.J. (2015) Herbs with anti-lipid effects and their interactions with statins as a chemical anti-hyperlipidemia group drugs: a systematic review. *Arya Atherosclerosis* 11, 244.

Saad, H., Charrier-El Bouhtoury, F., Pizzi, A., Rode, K., Charrier, B. *et al.* (2012) Characterization of pomegranate peels tannin extractives. *Industrial Crops and Products* 40, 239–246. DOI: 10.1016/j.indcrop.2012.02.038.

Samad, L.H. and Barzegar, M. (2006) Investigation of physicochemical properties of ten varieties of Yazd pomegranate seed. *Iranian Journal of Food Science and Technology* 3, 19–26.

Schofield, P., Mbugua, D.M. and Pell, A.N. (2001) Analysis of condensed tannins: a review. *Animal Feed Science and Technology* 91(1–2), 21–40. DOI: 10.1016/S0377-8401(01)00228-0.

Schubert, S.Y., Lansky, E.P. and Neeman, I. (1999) Antioxidant and eicosanoid enzyme inhibition properties of pomegranate seed oil and fermented juice flavonoids. *Journal of Ethnopharmacology* 66(1), 11–17. DOI: 10.1016/S0378-8741(98)00222-0.

Seidi, K., Jahanban-Esfahlan, R., Abasi, M. and Abbasi, M.M. (2016) Anti tumoral properties of *Punica granatum* (pomegranate) seed extract in different human cancer cells. *Asian Pacific Journal of Cancer Prevention* 17(3), 1119–1122. DOI: 10.7314/APJCP.2016.17.3.1119.

Sharma, K., Akansha. and Chauhan, E.S. (2018) Comparative studies of proximate, mineral and phytochemical compositions of pomegranate (*Punica granatum*) in peel, seed and whole fruit powder. *International Journal of Food Science and Nutrition* 3, 192–196.

Sharma, K. and Akansha, C.E. (2018) Comparative studies of proximate, mineral and phytochemical compositions of pomegranate (*Punica granatum*) in peel, seed and whole fruit powder. *Methods* 17, 18.

Sharma, J. and Maity, A. (2010) Pomegranate phytochemicals: nutraceutical and therapeutic values. *Fruit, Vegetable and Cereal Science and Biotechnology* 4, 56–76.

Shattock, R. and Solomon, S. (2004) Microbicides--aids to safer sex. *The Lancet* 363(9414), 1002–1003. DOI: 10.1016/S0140-6736(04)15876-5.

Shema-Didi, L., Kristal, B., Ore, L., Shapiro, G., Geron, R. *et al.* (2013) Pomegranate juice intake attenuates the increase in oxidative stress induced by intravenous iron during hemodialysis. *Nutrition Research* 33(6), 442–446. DOI: 10.1016/j.nutres.2013.04.004.

Shirode, A.B., Bharali, D.J., Nallanthighal, S., Coon, J.K., Mousa, S.A. *et al.* (2015) Nanoencapsulation of pomegranate bioactive compounds for breast cancer chemoprevention. *International Journal of Nanomedicine* 10, 475.

Singh, J.P., Kaur, A., Shevkani, K. and Singh, N. (2016) Composition, bioactive compounds and antioxidant activity of common Indian fruits and vegetables. *Journal of Food Science and Technology* 53(11), 4056–4066. DOI: 10.1007/s13197-016-2412-8.

Singh, D. and Sethi, V. (2003) Screening of pomegranate genotypes for the preparation of quality grade anardana. *Journal of Food Science and Technology* 40, 236–238.

Song, B., Li, J. and Li, J. (2016) Pomegranate peel extract polyphenols induced apoptosis in human hepatoma cells by mitochondrial pathway. *Food and Chemical Toxicology* 93, 158–166. DOI: 10.1016/j.fct.2016.04.020.

Sorokin, A.V., Duncan, B., Panetta, R. and Thompson, P.D. (2006) Rhabdomyolysis associated with pomegranate juice consumption. *The American Journal of Cardiology* 98(5), 705–706. DOI: 10.1016/j.amjcard.2006.03.057.

Spina, E. and de Leon, J. (2007) Metabolic drug interactions with newer antipsychotics: a comparative review. *Basic & Clinical Pharmacology & Toxicology* 100(1), 4–22. DOI: 10.1111/j.1742-7843.2007.00017.x.

Sreedevi, P., Vijayalakshmi, K. and Venkateswari, R. (2017) Phytochemical evaluation of *Punica granatum* L. leaf extract. *International Journal of Current Pharmaceutical Research* 9(4), 14–18. DOI: 10.22159/ijcpr.2017v9i4.1159.

Sreekumar, S., Sithul, H., Muraleedharan, P., Azeez, J.M. and Sreeharshan, S. (2014) Pomegranate fruit as a rich source of biologically active compounds. *BioMed Research International* 2014(25), 1–12. DOI: 10.1155/2014/686921.

Srinivas, N.R. (2013) Is pomegranate juice a potential perpetrator of clinical drug–drug interactions? Review of the in vitro, preclinical and clinical evidence. *European Journal of Drug Metabolism and Pharmacokinetics* 38(4), 223–229. DOI: 10.1007/s13318-013-0137-x.

Srivastava, R., Chauhan, D. and Chauhan, J.S. (2001) A flavonoid diglycoside from *Punica granatum*. *Indian Journal of Chemistry B* 40B, 170–172.

Stojanović, I., Šavikin, K., Đedović, N., Živković, J., Saksida, T. *et al.* (2017) Pomegranate peel extract ameliorates autoimmunity in animal models of multiple sclerosis and type 1 diabetes. *Journal of Functional Foods* 35, 522–530. DOI: 10.1016/j.jff.2017.06.021.

Stone, A. (2002) Microbicides: a new approach to preventing HIV and other sexually transmitted infections. *Nature Reviews Drug Discovery* 1(12), 977–985. DOI: 10.1038/nrd959.

Syed, D.N., Malik, A., Hadi, N., Sarfaraz, S., Afaq, F. *et al.* (2006) Photochemopreventive effect of pomegranate fruit extract on UVA-mediated activation of cellular pathways in normal human epidermal keratinocytes. *Photochemistry and Photobiology* 82(2), 398–405. DOI: 10.1562/2005-06-23-RA-589.

Syed, D., Afaq, F. and Mukhtar, H. (2007) Pomegranate derived products for cancer chemoprevention. *Seminars in Cancer Biology* 17(5), 377–385. DOI: 10.1016/j.semcancer.2007.05.004.

Tanaka, T., Nonaka, G.-I. and Nishioka, I. (1985) Punicafolin, an ellagitannin from the leaves of *Punica granatum*. *Phytochemistry* 24(9), 2075–2078. DOI: 10.1016/S0031-9422(00)83125-8.

Tanaka, T., Nonaka, G.E.N.-I. and Nishioka, I. (1986) Tannins and related compounds. XLI. Isolation and characterization of novel ellagitannins, punicacorteins a, B, C and D, and punigluconin from the bark of Punica granatum L. *Chemical & Pharmaceutical Bulletin* 34(2), 656–663. DOI: 10.1248/cpb.34.656.

Tanaka, Y., Sasaki, N. and Ohmiya, A. (2008) Biosynthesis of plant pigments: anthocyanins, betalains and carotenoids. *The Plant Journal* 54(4), 733–749. DOI: 10.1111/j.1365-313X.2008.03447.x.

Tavakkoli-Kakhki, M., Motavasselian, M., Mosaddegh, M., Esfahani, M.M., Kamalinejad, M. *et al.* (2014) Omega-3 and omega-6 content of medicinal foods for depressed patients: implications from the Iranian traditional medicine. *Avicenna Journal of Phytomedicine* 4, 225.

Tezcan, F., Gültekin-Özgüven, M., Diken, T., Özçelik, B. and Erim, F.B. (2009) Antioxidant activity and total phenolic, organic acid and sugar content in commercial pomegranate juices. *Food Chemistry* 115(3), 873–877. DOI: 10.1016/j.foodchem.2008.12.103.

Tripathi, A.K. and Kohli, S. (2011) Pharmacognostic and phytochemical studies on the flowers of *Punica granatum* L. *International Journal of Pharmaceutical Research & Development* 3, 1–7.

Tzulker, R., Glazer, I., Bar-Ilan, I., Holland, D., Aviram, M. *et al.* (2007) Antioxidant activity, polyphenol content, and related compounds in different fruit juices and homogenates prepared from 29 different pomegranate accessions. *Journal of Agricultural and Food Chemistry* 55(23), 9559–9570. DOI: 10.1021/jf071413n.

Uttara, B., Singh, A., Zamboni, P. and Mahajan, R. (2009) Oxidative stress and neurodegenerative diseases: a review of upstream and downstream antioxidant therapeutic options. *Current Neuropharmacology* 7(1), 65–74. DOI: 10.2174/157015909787602823.

van Elswijk, D.A., Schobel, U.P., Lansky, E.P., Irth, H. and van der Greef, J. (2004) Rapid dereplication of estrogenic compounds in pomegranate (*Punica granatum*) using on-line biochemical detection coupled to mass spectrometry. *Phytochemistry* 65(2), 233–241. DOI: 10.1016/j.phytochem.2003.07.001.

Veres, M. (1976) Study of the mechanical and chemical composition of cultivated pomegranate. *Hrana Ishrana* 17, 426–432.

Vicinanza, R., Zhang, Y., Henning, S.M. and Heber, D. (2013) Pomegranate juice metabolites, ellagic acid and urolithin A, synergistically inhibit androgen-independent prostate cancer cell growth via distinct effects on cell cycle control and apoptosis. *Evidence-Based Complementary and Alternative Medicine* 2013(10), 1–12. DOI: 10.1155/2013/247504.

Viuda-Martos, M., Fernández-López, J. and Pérez-Álvarez, J.A. (2010) Pomegranate and its many functional components as related to human health: a review. In: *Comprehensive Reviews in Food Science and Food Safety*. Vol.9, pp. 635–654. Available at: https://onlinelibrary.wiley.com/doi/full/10.1111/j.1541-4337.2010.00131.x DOI: 10.1111/j.1541-4337.2010.00131.x.

Wafa, B.A., Makni, M., Ammar, S., Khannous, L., Hassana, A.B. *et al.* (2017) Antimicrobial effect of the Tunisian Nana variety *Punica granatum* L. extracts against *Salmonella enterica* (serovars Kentucky and Enteritidis) isolated from chicken meat and phenolic composition of its peel extract. *International Journal of Food Microbiology* 241, 123–131. DOI: 10.1016/j.ijfoodmicro.2016.10.007.

Wang, R., Wang, L., Liu, R., Wang, L, Ding, Y. *et al.* (2006) Constituents of the flowers of *Punica granatum*. *Fitoterapia* 77(7–8), 534–537. DOI: 10.1016/j.fitote.2006.06.011.

West, T., Atzeva, M. and Holtzman, D.M. (2007) Pomegranate polyphenols and resveratrol protect the neonatal brain against hypoxic-ischemic injury. *Developmental Neuroscience* 29(4-5), 363–372. DOI: 10.1159/000105477.

Wilson, A. and Baietto, M. (2009) Applications and advances in electronic-nose technologies. *Sensors* 9(7), 5099–5148. DOI: 10.3390/s90705099.

Wu, S. and Tian, L. (2017) Diverse phytochemicals and bioactivities in the ancient fruit and modern functional food pomegranate (*Punica granatum*). *Molecules* 22(10), 1606. DOI: 10.3390/molecules22101606.

Xie, Y., Morikawa, T., Ninomiya, K., Imura, K., Muraoka, O. *et al.* (2008) Medicinal flowers. XXIII. New taraxastane-type triterpene, punicanolic acid, with tumor necrosis factor-α inhibitory activity from the flowers of Punica granatum. *Chemical and Pharmaceutical Bulletin* 56(11), 1628–1631. DOI: 10.1248/cpb.56.1628.

Youdim, K.A., Mcdonald, J., Kalt, W. and Joseph, J.A. (2002) Potential role of dietary flavonoids in reducing microvascular endothelium vulnerability to oxidative and inflammatory insults. *The Journal of Nutritional Biochemistry* 13, 282–288.

Yuan, T., Wan, C., Ma, H. and Seeram, N. (2013) New phenolics from the flowers of *Punica granatum* and their in vitro α-glucosidase inhibitory activities. *Planta Medica* 79(17), 1674–1679. DOI: 10.1055/s-0033-1350925.

Zhao, X., Yuan, Z., Fang, Y., Yin, Y. and Feng, L. (2013) Characterization and evaluation of major anthocyanins in pomegranate (*Punica granatum* L.) peel of different cultivars and their development phases. *European Food Research and Technology* 236(1), 109–117. DOI: 10.1007/s00217-012-1869-6.

Zhao, X., Yuan, Z., Fang, Y., Yin, Y. and Feng, L. (2014) Flavonols and flavones changes in pomegranate (*Punica granatum* L.) fruit peel during fruit development. *Journal of Agricultural Science and Technology* 16, 1649–1659.

Çam, M. and Hişil, Y. (2010) Pressurised water extraction of polyphenols from pomegranate peels. *Food Chemistry* 123, 878–885.

18 World Pomegranate Market

Z. Shokoohi[1]* and M. Asgari[2]
[1]Department of Agricultural Economics, School of Agriculture, Shiraz University, Shiraz, Iran; [2]Freelance agricultural economist, Tucson, AZ, USA

18.1 Introduction

Pomegranate (*Punica granatum* L., Punicaceae) is an ancient plant that originated in central Asia, with high adaptability to various conditions. The shrub's height reaches typically 5–7 m and grows very well in arid, semi-arid and Mediterranean climates. Commercial orchards of pomegranate have expanded from the Mediterranean basin, Asia and South Africa to the USA and South America. The fruit develops to an almost round shape with a distinguishable crown on top. The external colour of the fruit ranges from yellow, green and pink to deep red, indigo and deep purple, while the colour of the edible part (arils) varies from white to deep red, depending on the cultivar (Holland *et al*., 2009).

As well as the fruit, almost all other parts of the pomegranate plant such as the flowers, fruit skin, root, bark and seeds have long been used in medicine and the dyeing industry. The fruit is consumed as fresh fruit, wine, juice and concentrate, and paste, and can be stored after harvest for a long time. Recent research on the medicinal and health benefits of pomegranate has renewed the interest in and demand for this product across the world. However, most academic publications have focused on the health benefits and biology of the pomegranate tree and less on the economics of production and trade of the products. In this chapter, we will review the economic aspects of pomegranate production in major production countries and the trade flow in the international markets. Analysing market structure and behaviour is important to the industry prior to making any strategic marketing decisions. However, it seems that very limited information has been published regarding the marketing, trade and economics of pomegranates.

18.2 World Pomegranate Production and Trade

Pomegranate is native to a vast region in west Asia from Turkey to Iran, northern India and the Himalayas. Iran, India and Turkey are historically the major producers and exporters of the pomegranate fruit. In recent decades, other countries like Spain and the USA have extensively expanded their commercial orchards to meet the increasing global demand. We review the production and exports of pomegranate fruit in this section for major countries based on the available published and online information.

18.2.1 India

Current statistics show India as the largest producer of pomegranates in the world. In 2017, India produced nearly 2.8 million metric

*Corresponding author: z_shokoohi@shirazu.ac.ir

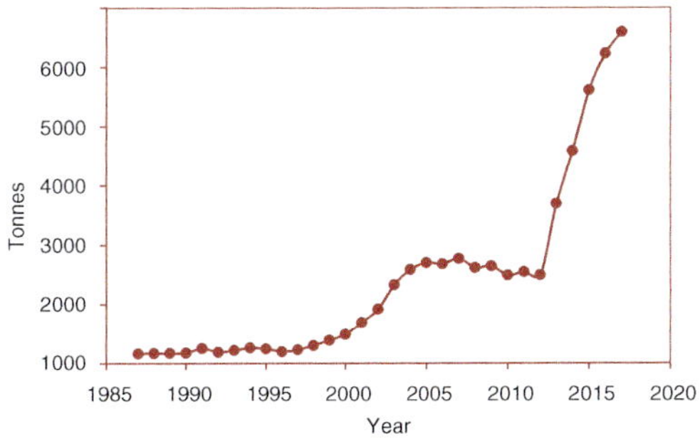

Fig. 18.1. Pomegranate production in India (1987–2016). (From: APEDA.)

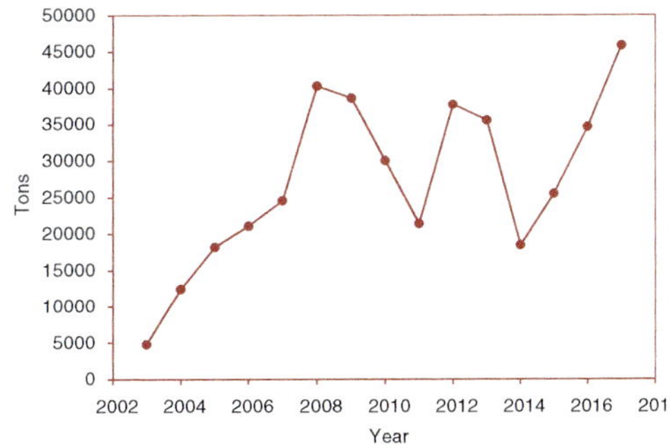

Fig. 18.2. Pomegranate exports from India (2003–2017). (From: APEDA.)

tonnes (t) of pomegranate from 220,000 ha. It is expected that the cultivation area will be expanded to 750,000 ha by 2025 (Agricultural Finance Corporation, 2007). The average yield improved from 6.9 t/ha in 2010 to 12.7 t/ha in 2017. Based on the statistics provided by the Agricultural and Processed Food Products Export Development Authority (APEDA) in India, pomegranate production in India has experienced three different phases since the early 1980s (Fig. 18.1). The first phase, 1980–1998, was fairly stable with the average annual production of 110,000 t followed by an expansion period in the second phase, 1999–2012, that increased the average annual production to 800,000 t. The last phase, which started in 2013, shows a fast growth due to the increased cultivation area and improved yield per hectare. Maharashtra, Karnataka and Gujarat, with 1.5, 0.32 and 0.28 million t, respectively, were the major pomegranate-producing states in 2016, which together accounted for more than 90% of the total pomegranate production of India. The variety 'Ganesh' cultivated in Maharashtra is most suitable for export purposes. Other common cultivars in India are 'Ruby', 'Arakta' and 'Bhagwa'.

Despite its significant production level, India exports a small proportion of its pomegranates to Asia and some European countries. Recent investments in high-quality cultivars and packaging could improve India's export position (Bala and Sudhakar, 2017). The average unit value is reported to be US$1.8/kg for Indian-grown pomegranate. India exported, on average, 30,000 t/year in the past 15 years but, except for the 2008–2014 period, the trend has been increasing and reached 46,000 t in 2017 (Fig. 18.2). The United Arab Emirates, Nepal, Saudi Arabia, Bangladesh, the Netherlands and Sri Lanka are major destinations for India's pomegranates (International Trade Center, 2017).

18.2.2 Iran

Iran is one of the major producers of pomegranate fruit in the world. Its total production in 2016 reached over 1 million t. The production trend has been increasing since the 1980s in Iran, except for the two frostbite incidents in 2002 and 2007 (Fig. 18.3). The total cultivated area in 2016 exceeded 75,000 ha with an average yield of more than 12.5 t/ha. Currently, Fars province produces more than 26% of Iran's total pomegranates, followed by Markazi, Khorasan Razavi, Yazd, Isfahan and Semnan provinces. Common cultivars in Fars are 'Rabab', 'Faroogh' and 'Atabaki' (Iran Ministry of Agriculture-Jahad, 2017).

Iran exported about 1.5% of its pomegranates during 2010–2017 (Fig. 18.4). It seems that drought stress during recent decades and frost damage to the trees in some years has reduced the export of Iran's pomegranate fruit during recent decades (Trade Promotion Organization of Iran, 2018). Pakistan, Turkmenistan, Iraq, Afghanistan and South Korea formed 80% of Iran's export market in 2016, with an average unit value of US$0.87/kg (Iran Customs Administration, 2017). The main exporting pomegranate cultivars of Iran

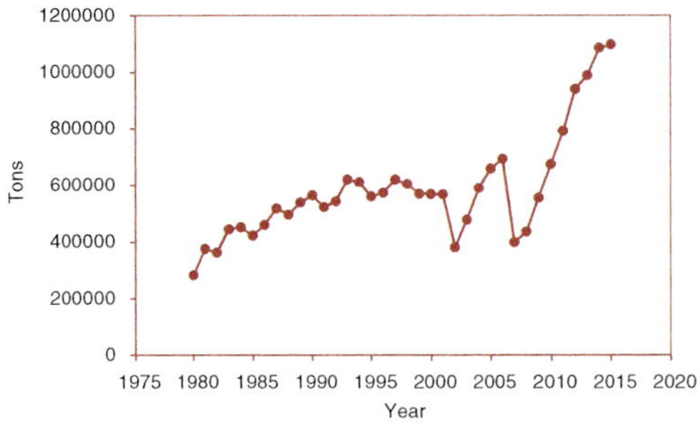

Fig. 18.3. Pomegranate production in Iran (1980–2015). (Source: Iran Ministry of Agriculture-Jahad.)

are 'Rabab-e-Neyriz', 'Sishe-Kap-e-Ferdwos', 'Malas-e-Saveh', 'Malas-e-Yazdi' and 'Naderi-e-Badrud' (Varasteh *et al.*, 2008). Recent financial sanctions and a hike in domestic prices could explain the significant export drop of Iran in 2017.

18.2.3 Turkey

Pomegranate production gradually increased in Turkey from 45,000 t in 1988 to almost 100,000 t in 2006 (Fig. 18.5). However, since 2007 production has experienced significant growth and exceeded 500,000 t in 2017. Due to constant investments during the past decade, Antalya is now the top producer region in Turkey, providing almost 33% of the country's pomegranates. Increasing consumer awareness around the world and easy access to the European market has stimulated pomegranate production in Turkey (Yilmaz *et al.*, 2015).

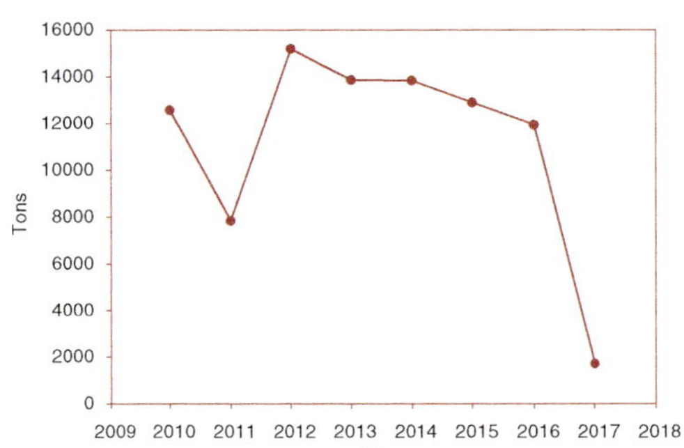

Fig. 18.4. Pomegranate exports from Iran (2010–2017). (From: Iran Customs Administration.)

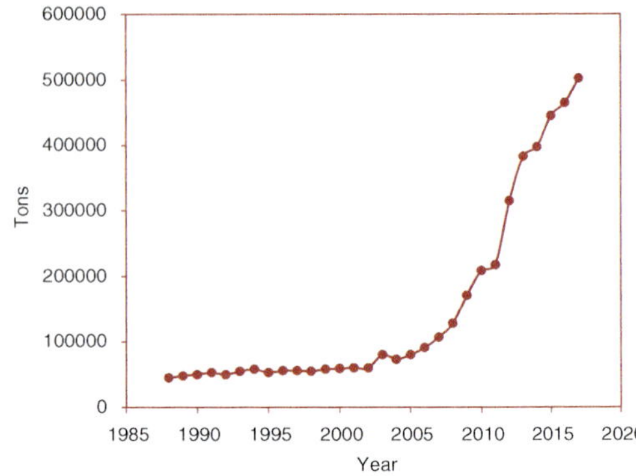

Fig. 18.5. Pomegranate production in Turkey (1988–2017). (From: Turkey Ministry of Food, Agriculture and Livestock.)

The main planted pomegranates in Turkey are 'Hicaznar', 'Cekirdeksiz', 'Lifani', 'Yufka Kabuki' and 'Izmir'.

Expanding commercial orchards for export purposes has been a goal in Turkey in recent years. Although they have lower production compared with India and Iran, Turkey's producers have managed to export higher shares with steady growth rates since 2007 (Fig. 18.6). Total exports in 2017 were 173,000 t, with the average unit price of US$0.6/kg. Germany, Belarus, Iraq, Russia and Ukraine are the major importers of Turkey's pomegranates (Turkish Statistical Institute, 2017).

18.2.4 United States

California is the major producer of pomegranate in the USA. Fresno and Tulare counties account for a significant share of California production. While the total annual production remained below 30,000 t in the period from 1980–2000, it increased dramatically during 2001–2009 and reached its peak at 283,000 t in 2013. Total production seems more volatile in recent years (Fig. 18.7). 'Wonderful' is the dominant pomegranate that is commercially planted in California. The harvested area has been declining in the past decade and dropped from 31,000 acres (almost 12,560 hectares) in 2010 to 15,000 acres (6,070 hectares) in 2017 (USDA, 2019).

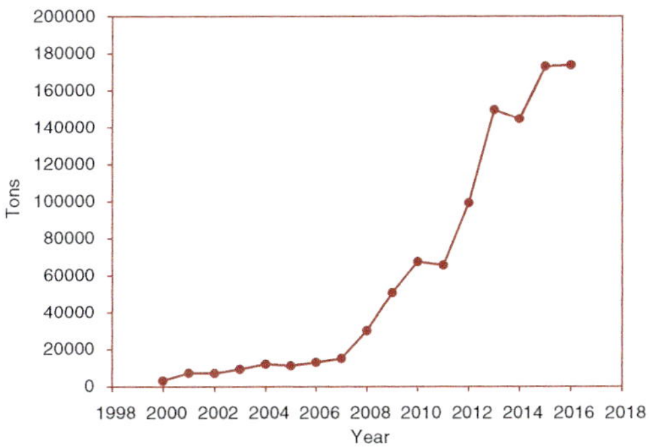

Fig. 18.6. Pomegranate exports from Turkey (2000–2016). (From: Turkey Ministry of Food, Agriculture and Livestock.)

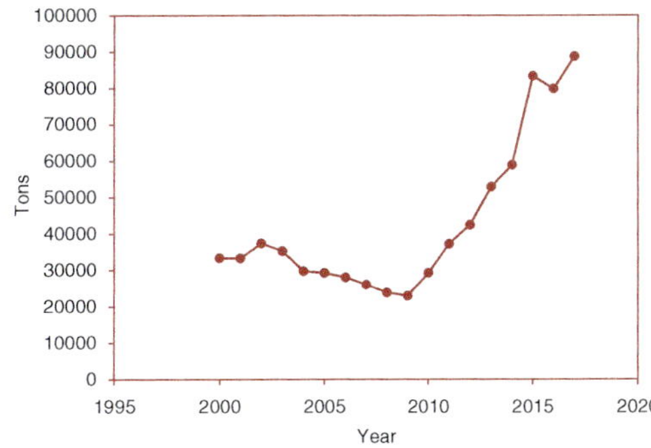

Fig. 18.8. Pomegranate production in Spain (2000–2017). (From: Spain Ministry of Agriculture, Fisheries and Food.)

18.2.5 Spain

From 2000–2009, the annual production rate was decreasing in Spain, and the production dropped from 33,000 t to 23,000 t during this period (Fig. 18.8). Then, total production increased substantially and reached its highest in 2017. Yet, the annual production level remains under 100,000 t. 'Valencia' and 'Murcia' are the major producing regions and 'Mollar de Elche', 'Mollar de Valencia' and 'Wonderful' are the most common varieties in Spain (Spain Ministry of Agriculture, Fisheries and Food, 2017).

The quantity of Spain's pomegranate exports has not been stable in the past 15 years and a small share of the production has been sent to international markets (Fig. 18.9). The highest export volume was in 2010 with more than 5000 t. Russia, the United Arab Emirates, Malaysia, Bahrain, Canada and Belarus are major exporting destinations for Spain. Despite the competition in several destination markets, the unit price for pomegranate from Spain, with the average of US$1.4/kg, has been higher than that of Turkey (Spain Ministry of Industry, Commerce and Tourism, 2017).

18.2.6 South Africa and Chile

Most of the pomegranate producers in the world are located in the northern hemisphere. The harvest time in these countries is from September to January. Although South Africa, and also Chile, are smaller pomegranate producers, their location in the southern hemisphere brings the advantage of supporting the markets

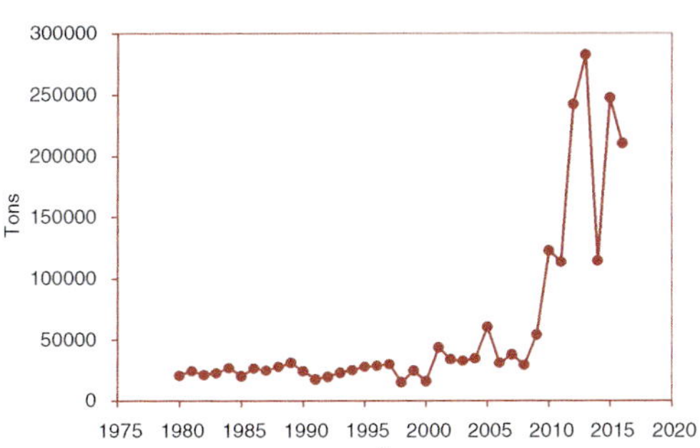

Fig. 18.7. Pomegranate production in California, USA (1980–2016). (From: US Department of Agriculture.)

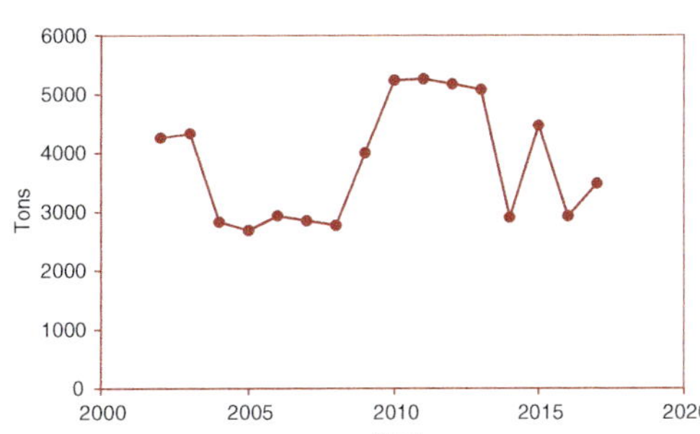

Fig. 18.9. Pomegranate exports from Spain (2002–2017). (From: Spain Ministry of Industry, Commerce and Tourism.)

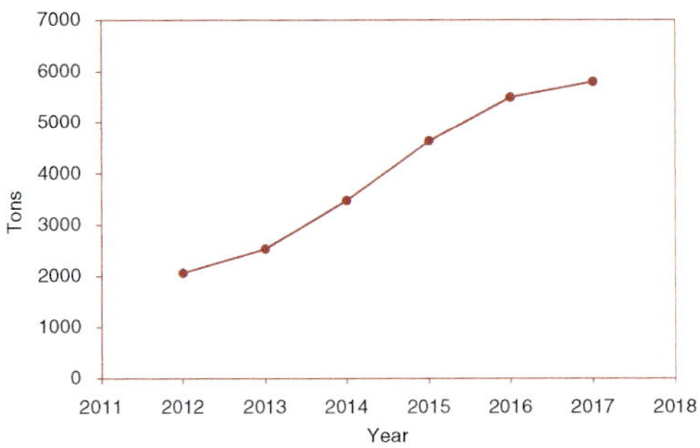

Fig. 18.10. Pomegranate production in South Africa (2012–2017). (From: Pomegranate Association of South Africa.)

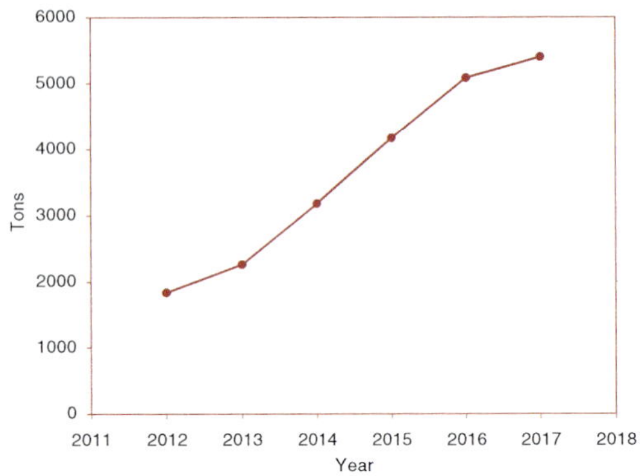

Fig. 18.11. Pomegranate exports from South Africa (2012–2017). (From: Pomegranate Association of South Africa.)

from February to June because of their late harvest time.

The production of pomegranate has experienced a rapid increase since 2012 in South Africa (Fig. 18.10). The area under cultivation quadrupled in 2011, compared with the previous year, increasing from 180 to 780 ha, and continued to grow and reached 826 ha in 2017. The production level has increased from slightly more than 2000 t to 5800 t since 2012 (Pomegranate Association of South Africa, 2017). 'Wonderful' is the primary cultivar in South Africa, forming 65% of the area planted, followed by 'Acco', 'Hershkovitz' and 'Kessari/Baghwa', with about 10% each. The first three cultivars are mostly exported while the latter is consumed domestically due to its inferior quality (Louw, 2018). Comparing the production and export quantities, it implies that South African production is mainly aimed for international markets. South Africa expanded its exports from 2000 t in 2012 to more than 5000 t in 2017 (Fig. 18.11). Similar to South Africa, Chile has exported on average more than 4000 t per year since 2012 (Fig. 18.12) (Chile Ministry of Agriculture, 2018).

18.2.7 European Union

Europe is a major destination for pomegranates. Total imports by the European Union have been increasing since 2002 and exceeded 100,000 t in 2017 (Fig. 18.13). The average value of the European imports from 2002 to 2016 was €109 million per year. An increasing level of research and publications about the benefits of pomegranate consumption have positively affected the demand in Europe. Turkey is the main exporter of pomegranate to the European countries, with an average of more than 48,000 t/year, followed by Peru, Israel, Egypt, India, Chile and Morocco in 2017. The unit price in European retail markets ranges from €1.7–2.2 with an average of €1.9/kg (International Trade Center, 2017).

18.3 Price Discovery

Price information is an essential factor in strategic marketing decision making. However, there is no comprehensive and reliable source of real-time price dynamics available to producers and exporters in the pomegranate markets. Each

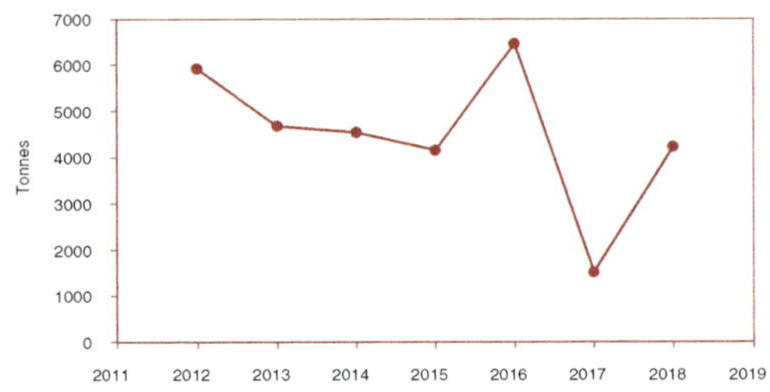

Fig. 18.12. Pomegranate exports from Chile (2012–2018). (From: Chile Ministry of Agriculture.)

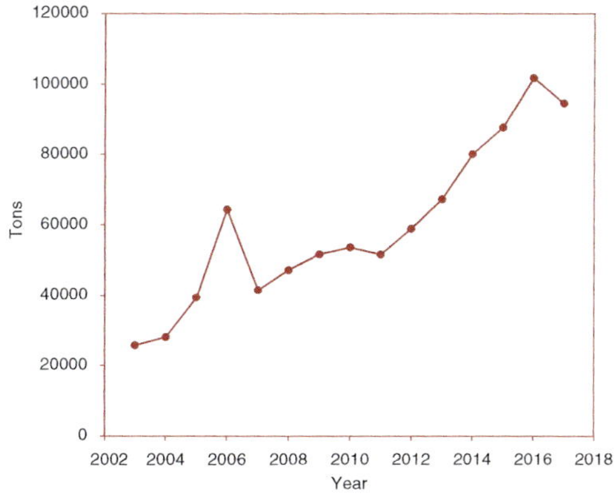

Fig. 18.13. Pomegranate imports to the European Union (2003–2016). (From: International Trade Center, Trade Maps.)

entity has to collect its own data, which can be time-consuming and costly. We used Fresh Fruit Portal to compile some information about the price range and the average unit price of pomegranate in the European market in 2016 (Table 18.1) (Fresh Fruit Portal, 2019). This table shows how the prices vary year-round depending on the origin, the destination market and the cultivar.

18.4 Competitiveness and Market Power in Pomegranate Export Market

There are a few different measures to gauge the competitiveness of a market. For instance, we can measure how concentrated is that market. Market concentration depends on how top suppliers divide the market share. In an export market, if a few exporters control higher market shares compared with the rest of the exporters, the market is concentrated. Higher concentration also indicates less export opportunities available to new suppliers. Comparative advantage, on the other hand, is a measure that weighs the value of a certain product in a country's export portfolio relative to the value of that product in the international market as a whole. It is important to study how the comparative advantage of exporters in a certain market changes over time. If the index is decreasing, the exporter is losing its competitive edge in that market.

To measure how concentrated the export market for a specific commodity is, it is common to use the Concentration Ratio (CR) index, which is the share of the top exporter countries in that market. The export market can be defined either as the total monetary value or the volume of traded goods within a specific geographic location. The CR index can be calculated as $CR = \sum_{j=1}^{n} s_j$ in which s_j is the share of country j in the market and n denotes the number of top exporting countries. It is more likely to observe high concentration if a market has only two major exporters, that is $n = 2$, compared with a market that has multiple players. In general, a CR index of less than 0.4 implies less concentration and higher competition. As the index gets closer to one, the concentration in the market rises, which means each major exporter gains a higher share of the market and thus there is less room for rivals to compete.

One of the common ways to calculate comparative advantage is the Revealed Comparative Advantage (RCA) index (Balassa, 1965). In general, the RCA index for a country (or region) is calculated as a ratio that implies the significance of the export value of a product for that country relative to the significance of the trade of that product in international markets. The index can be calculated as $RCA_{ij} = \frac{X_{ij}/X_j}{\frac{X_{iw}}{X_w}}$ in which X denotes export value; i denotes the specific good or services under the study; j is the country or the region under the study; and w denotes the world. In the case of pomegranate, for example, the numerator indicates the relative share of pomegranate export value to the total export value for country X and the denominator represents the export value of pomegranate in international markets relative to the total world trade. An increasing RCA, over time, indicates that the comparative advantage of a country is improving, current export markets are optimized or better export opportunities are created (in the international or regional markets, depending on the scope of the data).

When RCA >1 the country under the study has a comparative advantage in the international markets and when 0 < RCA < 1, the country does not have a comparative advantage. Therefore, RCA is asymmetric. To address this issue, RCA is modified to the Revealed Symmetric

Table 18.1. Price range in selected European markets in 2016.

Month	Price (Euro/kg) Average	Highest	Lowest	Country of origin	Destination market	Cultivar
January	2.62	3.8	1.44	Israel	Italy	'Wonderful'
				Turkey	UK	-
February	2.35	3.71	1	India	Holland	'Bhagwa'
				Turkey	Holland	'Hicaznar'
March	2.34	3.68	1	Peru	Holland	'Wonderful'
				Turkey	France	'Hicaznar'
April	2.84	3.57	2.11	India	Holland	'Bhagwa'
				South Africa	Holland	'Acco'
May	2.25	2.71	1.8	India	Holland	'Bhagwa'
				South Africa	France	'Acco'
June	2.49	3.68	1.3	Chile	Switzerland	-
				South Africa	UK	-
July	1.91	2.63	1.19	Chile	Holland	'Wonderful'
				Egypt	UK	-
August	2.7	3.57	2	Israel	Holland	'Emek'
				Peru	France	'Wonderful'
September	2.78	3.76	1.8	Israel	Switzerland	-
				Peru	France	'Wonderful'
October	1.64	2.1	1.19	Israel	Holland	-
				Egypt	UK	-
November	2.05	3	1.13	Iran	Holland	'Purple'
				Egypt	Holland	'Wonderful'
December	1.77	2.67	1	Iran	Holland	'Mini Purple'
				Egypt	France	'Wonderful'

This information is obtained from https://www.freshfruitportal.com/. We used (-) when the information was not available. The average price is calculated as the simple average of the lowest and highest price in each month.

Comparative Advantage (RSCA) index and calculated as $RSCA_{ij} = \frac{RCA_{ij}-1}{RCA_{ij}+1}$, $(-1 < RSCA_{ij} < 1)$ where the positive index implies comparative advantage and the negative index implies disadvantage.

We calculated and applied both CR and RSCA indices to evaluate the competitiveness of the four major exporters in the global pomegranate market. We used export data from 2009–2016 for four major exporters: Turkey, Iran, India and Spain (Table 18.2). The first three countries maintained high levels of comparative advantage during these years. Turkey was the most stable one, showing much competition with Iran. While Turkey has gained the highest share in export markets, it has managed to increasingly expand to new markets. Turkey exported more than half of its production in 2017, a considerable increase from 36% in 2009. At the same time, Turkey's rival, Iran, had faced challenges to continue its exports. In Iran although the comparative advantage in export markets has remained relatively high, the total exports has continuously declined since 2012 (Fig. 18.4) such that the share of exports from total production has declined from 11% in 2009 to almost 1% in 2016.

Table 18.2. Revealed symmetric comparative advantage and concentration ratio in pomegranate export market.

	RSCA				
Year	Iran	Turkey	India	Spain	CR
2009	0.98	0.93	0.87	0.25	0.27
2010	0.89	0.95	0.82	0.32	0.33
2011	0.95	0.95	0.72	0.29	0.37
2012	0.95	0.94	0.77	0.25	0.26
2013	0.89	0.95	0.76	0.07	0.35
2014	0.94	0.95	0.81	−0.22	0.40
2015	0.94	0.94	0.85	−0.04	0.41
2016	0.93	0.95	0.87	−0.30	0.41

India is following Turkey and Iran and increasing its comparative advantage steadily. It has doubled its market share in the world, from 17% to 34% in recent years. Spain, on the other hand, has lost its advantage in pomegranate exports since 2014 (negative RSCA measures). Spanish exports to European countries, despite the geographical proximity, have been volatile. The most significant spike happened in 2012 when the share of exports from Spain to the European Union jumped from 0.03% to almost 2%. Although the production level is increasing in Spain, it seems that the exporters have difficulty in creating a stable flow of supply to the international markets.

Although the competition in the international market is higher than before and more countries are producing and exporting pomegranates, the consumer market is also growing rapidly. Therefore, the market is growing as a whole, and major exporters have the opportunity to increase their exports and obtain larger shares. The trend of the concentration ratio for 2009 to 2016 implies that fewer exporters have gained more market share. As the last column in the table shows, the ratio increased from 0.27 in 2009 to 0.41 in 2016, which indicates that the market is moving towards higher concentration and thus less competitiveness (Table 18.2).

18.5 Export Standards

Production and export of pomegranate, like other agricultural products, are subject to sanitary, health and marketing standards. Each producing and importing country may have different sets of standards. There are also frameworks that define acceptable production procedures, chemical uses and residues at the regional and international levels. European countries are normally more restrictive when it comes to health standards. To assure compliance with different standard modules, exporting countries often develop a framework to monitor production and packaging procedures internally. Such internal standards also define sorting and grading criteria that will help exporters to supply the best products to the international market.

In India, the Central Insecticide Board and Registration Committee (CIB&RC) has developed a surveillance procedure that monitors the agrochemicals used in the production of pomegranates for export, and verifies the compliance with importing country standards in terms of residues and other contaminant levels. In the event of any violations from permitted levels or any internal phytosanitary alerts, the CIB&RC would enforce the required corrective actions. The Department of Marketing and Inspection (DMI) provides guidelines for grade classification of pomegranates through granting a Certificate of Agmark Grading (CAG) according to the European Union standards (APEDA, 2018).

The Trade Promotion Organization of Iran provides recommendations to exporters to follow in order to be successful in the international markets. These recommendations include pre- and postharvest procedures to control residue, contamination, waste, storage, sorting, packaging, labelling and branding. They also provide country-specific standards, rules and regulations that are significant for exporters to reach their target markets (Trade Promotion Organization of Iran, 2018).

The European Union and OECD member countries, as the largest group of net importers of pomegranate, have adopted several sets of rules and regulations that define the expectations with regard to the quality (product appearance, cleanness and damage), size (the number of fruits in the packaging and the grading of the fruit based on diameter and weight), packaging (the specifications, allowable material, stickers and arrangements of the packaging), and labelling of fruits and vegetables.

For pomegranates in particular, some of these standards include General Marketing Standards of Regulation No. 543/2011, CODEX STAN 310-2013, and OECD International Standards for pomegranate, in which minimum quality requirements are laid out, the classification system is explained and size grading is described. According to these standrads, 'Extra' class pomegranates are classified as fruits of superior quality free of defects; class I have slight defects in shape and colouring; and class II have defects including skin cracks. Other provisions with regard to packaging, labelling and container identification are explained in detail (CODEX Alimentarius, 2013; OECD, 2014). In the US domestic markets standards are less rigorous. For instance, California requires pomegranates to be mature and free from rot, decay and serious damages including sunburn, cracks, and bruises (Fresno Agricultural Commissioner, 2019). Also, according to Regulation No. 1169/2011, the labelling regulation establishes the principles of food labelling to protect European consumers' right to access useful information concerning the food products labelling (The Centre for the Promotion of Imports, 2019).

18.5.1 Summary

Consumption of pomegranate and its processed products is gaining attention in the international markets. However, lack of comprehensive data and economic analysis challenges the decision making by producers, exporters and policy makers. Based on the currently available data, fewer major exporters are gaining higher market shares, but more countries are exporting high-quality pomegranate fruit, which will eventually impact the average price negatively. Traditional exporters like Iran and India have to accept new rivals like Turkey, Spain and the USA. Further research on production, cultivars, marketing strategies, consumer preferences and trade opportunities will help producers, consumers and policy makers to analyse and understand this rapidly growing pomegranate market. Better knowledge of the market preferences leads to informed decisions, which, hopefully, results in improved welfare for all parties.

Key information about the markets, especially the price dynamics, is critical for decision making by the producers, handlers and exporters of agricultural products. There seems to be a gap in this area in the pomegranate market since the trading volume is not comparable with other major commodities. However, for this evolving market a reliable, up-to-date and comprehensive source to provide information is critical for investors, traders and consumers. Such datasets would also benefit the research community, which ultimately would inform production decisions, marketing strategies and consumer behaviour.

References

Agricultural Finance Corporation (2007) Project report on export promotion of pomegranate from India. Agricultural Finance Corporation, Mumbai, India.

APEDA (2018) Agricultural and processed food products export development authority of India. Available at: http://apeda.in (accessed 12 June 2020).

Bala, D.M.L. and Sudhakar, K. (2017) An overview of export performance of agricultural products in India. *IOSR Journal of Business and Management* 19(02), 01–05. DOI: 10.9790/487X-1902010105.

Balassa, B. (1965) Trade liberalisation and 'revealed' comparative advantage. *The Manchester School* 33(2), 99–123. DOI: 10.1111/j.1467-9957.1965.tb00050.x.

Chile Ministry of Agriculture (2018) Foreign trade statistics. Available at: https://www.odepa.gob.cl/estadisticas-del-sector/comercio-exterior (accessed 12 June 2020).

CODEX Alimentarius (2013) Standard for pomegranate (Codex STAN 310-2013). Available at: www.fao.org/fao-who-codexalimentarius/en/ (accessed 12 June 2020).

Fresh Fruit Portal (2019) European prices. Available at: https://www.freshfruitportal.com/ (accessed 12 June 2020).

Fresno Agricultural Commissioner (2019) Fruit and vegetable quality control. Available at: https://www.co.fresno.ca.us/departments/agricultural-commissioner (accessed 12 June 2020).

Holland, D., Hatib, K. and Ya'akov, B.I. (2009) Pomegranate: botany, horticulture, breeding. *Horticultural Reviews* 35, 127–191.

International Trade Center (2017) Trade maps. Available at: www.trademap.org (accessed 12 June 2020).

Iran Customs Administration (2017) Customs statistical yearbook. Available at: www.irica.gov.ir/index.php?newlang=eng (accessed 12 June 2020).

Iran Ministry of Agriculture-Jahad (2017) Agricultural statistical yearbook. Available at: www.maj.ir/index.aspx?tempname=NewEnMain&lang=2&sub=0 (accessed 12 June 2020).

Louw, M. (2018) Pomegranate production in South Africa. Available at: http://southafrica.co.za (accessed 12 June 2020).

OECD (2014) International standards for fruits and vegetables: pomegranate. Available at: www.oecd.org/agriculture/fruit-vegetables/ (accessed 12 June 2020).

Pomegranate Association of South Africa (2017) Pomegranate industry overview. Available at: https://www.sapomegranate.co.za/ (accessed 12 June 2020).

Spain Ministry of Agriculture, Fisheries and Food (2017) Statistical yearbooks. Available at: https://www.mapa.gob.es/en/ (accessed 12 June 2020).

Spain Ministry of Industry, Commerce and Tourism (2017) Foreign trade statistics. Available at: http://datacomex.comercio.es/ (accessed 12 June 2020).

The Centre for the Promotion of Imports (2019) CBI market intelligence. Available at: www.cbi.eu/market-information (accessed 12 June 2020).

Trade Promotion Organization of Iran (2018) Iran's trade statistics in brief. Available at: http://eng.tpo.ir/ (accessed 12 June 2020).

Turkish Statistical Institute (2017) Homepage. Available at: www.turkstat.gov.tr/Start.do (accessed 4 January 2019).

USDA (2019) County agricultural commissioners' data listing. Available at: https://www.nass.usda.gov/Statistics_by_State/California/Publications/AgComm/index.php (accessed 12 June 2020).

Varasteh, F., Arzani, K., Zamani, Z. and Tabatabaei, S.Z. (2008) Physico-chemical seasonal changes of pomegranate (Punica granatum L.) fruit 'Malas-e-Torsh-e-Saveh' in Iran. *Acta Horticulturae* 769,255–258. DOI: 10.17660/ActaHortic.2008.769.36.

Yilmaz, I., Ozalp, A. and Yilmaz, S. (2015) Marketing structure of pomegranate in Turkey. *Acta Horticulturae* 1089,205–212. DOI: 10.17660/ActaHortic.2015.1089.25.

Index

Note: The locators in bold and italics represents the tables and figures respectively.